INTRODUCTION TO PROBABILITY

INTRODUCTION TO PROBABILITY

Douglas G. Kelly

University of North Carolina at Chapel Hill

Macmillan Publishing Company
New York

Maxwell Macmillan Canada
Toronto

Maxwell Macmillan International
New York Oxford Singapore Sydney

Editor: Robert Pirtle
Production Supervisor: Susan L. Reiland
Production Managers: Aliza Greenblatt and Francesca Drago
Text Designer: Proof Positive/Farrowlyne Associates, Inc.
Cover Designer: Robert Freese
Illustrations: Academy ArtWorks, Inc.

This book was set in Sabon by The Clarinda Company, and was printed and bound by R. R. Donnelley and Sons. The cover was printed by Lehigh Press.

Printed in the United States of America

Macmillan Publishing Company
866 Third Avenue, New York, New York 10022

Macmillan Publishing Company is
part of the Maxwell Communication
Group of Companies.

Maxwell Macmillan Canada, Inc.
1200 Eglinton Avenue East
Suite 200
Don Mills, Ontario M3C 3N1

Library of Congress Cataloging-in-Publication Data

Kelly, Douglas G.
Introduction to probability / Douglas G. Kelly.
p. cm.
Includes bibliographical references and index.
ISBN 0-02-363145-7
1. Probabilities. I. Title.
QA273.K328 1994
519.2—dc20 93-13808
CIP

Printing: 1 2 3 4 5 6 7 Year: 4 5 6 7 8 9 0

To my parents,
Frank A. Kelly and Dorothy M. Kelly,
with love and gratitude

Contents

1 PROBABILITY MODELS: DEFINITIONS AND EXAMPLES 1

1.1 A Preliminary Example 2
1.2 Discrete Probability Spaces 11
1.3 Absolutely Continuous Probability Spaces 30
1.4 Two Other Kinds of Probability Space: Bernoulli Processes and Poisson Processes 57
1.5 The General Definitions of Probability Space and Random Variable 73
Supplementary Exercises 78

2 THE ALGEBRA OF EVENTS AND PROBABILITIES 83

2.1 Basic Formulas and Inequalities for Probabilities 84
2.2 Conditional Probabilities 97
2.3 Conditioning and Bayes's Theorem 113
2.4 Independence 124
2.5 Limits of Probabilities of Nested Events 134
Supplementary Exercises 142

3 PROBABILITY DISTRIBUTIONS 147

3.1 Distributions on $\mathbb{R}$ 148
3.2 Cumulative Distribution Functions 156
3.3 The Distribution of a Random Variable 168
3.4 The Joint Distribution of Two Discrete Random Variables 177
3.5 The Joint Distribution of Two Absolutely Continuous Random Variables 190
3.6 The Joint Distribution of More Than Two Random Variables 209
Supplementary Exercises 214

4 EXPECTED VALUES 217

4.1 Definitions and Examples of Expected Values 218
4.2 The Algebra of Expected Values 243
4.3 Variance and Moments 261
4.4 Covariance and the Variance of a Sum of Random Variables 271
4.5 Laws of Large Numbers 287
Supplementary Exercises 298

5 FUNCTIONS OF RANDOM VARIABLES 303

5.1 Introduction: Finding Distributions 304
5.2 Using Cumulative Distribution Functions 306
5.3 The Change-of-Variable Theorem for Densities 323
5.4 The Multivariate Change-of-Variable Theorem 333
Supplementary Exercises 344

6 NORMAL DISTRIBUTIONS AND THE CENTRAL LIMIT THEOREM 347

6.1 Sums and Averages of i.i.d. Random Variables 348
6.2 Normal Distributions 351

6.3 The Central Limit Theorem 361
6.4 The Poisson and Normal Approximations for Binomial Distributions 372
Supplementary Exercises 384

7 SOME IMPORTANT DISTRIBUTIONS ON THE NONNEGATIVE INTEGERS 387

7.1 Probability Generating Functions 388
7.2 Binomial and Bernoulli Distributions 406
7.3 Geometric and Negative Binomial Distributions 415
7.4 Poisson Distributions 429
7.5 Hypergeometric Distributions 433
7.6 Multinomial Distributions 443
Supplementary Exercises 448

8 SOME IMPORTANT ABSOLUTELY CONTINUOUS DISTRIBUTIONS 453

8.1 Moment Generating Functions 454
8.2 Normal Distributions 468
8.3 Exponential and Erlang Distributions 478
8.4 Gamma Distributions 489
8.5 Uniform Distributions 499
8.6 Beta Distributions 506
Supplementary Exercises 514

9 CONDITIONING AND BAYES'S THEOREM FOR RANDOM VARIABLES 517

9.1 Conditioning on a Discrete Random Variable 518
9.2 Conditioning on an Absolutely Continuous Random Variable 539
Supplementary Exercises 559

Suggested Reading and Bibliography 561

A APPENDIX: REVIEW OF SOME TOPICS FROM CALCULUS AND SET THEORY 565

A.1 Sets of Numbers 566
A.2 Set Operations 567
A.3 Finite, Countably Infinite, and Uncountable Sets 569
A.4 Basic Combinatorics 571
A.5 Sums and Series 575
A.6 L'Hôpital's Rule 579
A.7 Integration by Substitution and Integration by Parts 580
A.8 Improper Integrals 581
A.9 Integrals over Subsets of the Plane 584

B APPENDIX: SUMMARIES OF SOME IMPORTANT DISTRIBUTIONS 591

C APPENDIX: STANDARD NORMAL CUMULATIVE DISTRIBUTION FUNCTION 601

Index 603

Preface

This book is intended for an introductory course in probability theory, for students who have had calculus, but not advanced calculus or measure theory. It can serve as a prerequisite for courses in statistical inference or stochastic modeling. I had four primary goals in writing it:

to emphasize the aspects of probability theory that are needed to use probability models in applications to real situations, particularly in statistics and in stochastic modeling;

to present material from the concrete to the abstract rather than vice versa, motivating definitions and theorems with examples and discussions that show the need for them;

to be honest about the mathematical foundations and details that cannot be covered at this level, while at the same time giving the student an intuitive idea of what the issues are;

to introduce all the important ideas as early in the course as possible, so that every major topic can be rehearsed often, in examples and exercises, before the course ends.

The fourth goal is the one that determined the order of presentation of topics; in my mind it is the main reason for this book. I decided to write it after many semesters of teaching the course, in which there was always at least one important concept that I could not introduce until near the end of the course, and which consequently the students scarcely learned. The things I think of as essential for students to take away from this course are the following:

probability spaces as models for chance experiments; the probability of an event as a quantity representing its *expected long-run relative frequency*

the notion of random variable: a function on an outcome set, used to model a quantity determined by the outcome of a chance experiment

the expected value of a random variable, representing its *expected long-run average value*

the limit theorems (laws of large numbers and the Central Limit Theorem) that make precise the intuitive notions of probability and expected value

the idea of a *distribution* as an assignment of probabilities to intervals, and thence to sets that can be made from intervals by set operations; the distribution of a random variable

the two main kinds of distribution, discrete and absolutely continuous, and the various kinds of calculations involving them

the standard or "brand-name" distributions and models most used in practice, including, if possible, the Poisson process

joint, marginal, and conditional distributions

distribution theory: finding the distribution of a function of one or more random variables; including, if possible, the use of generating functions

conditional probabilities and independence; the use of conditioning and Bayes's Theorem

All these topics are introduced in the first four chapters of this book, and within six chapters everything has been covered except for generating functions and a full treatment of conditioning and Bayes's Theorem. The table of contents makes it appear that the important brand-name distributions are not covered until Chapters 7 and 8, but in fact most of them are given in examples and exercises from the very beginning, and Chapters 7 and 8 are mostly summaries, collecting ideas the students have already worked with.

A full treatment of conditioning on a random variable and Bayes's Theorem for bivariate distributions, both discrete and absolutely continuous, does not appear until Chapter 9. But Chapter 2 contains an introduction to the subject, including conditioning on a finite or countable partition and Bayes's Theorem in that setting. And throughout the text are scattered examples and exercises making use of conditional probabilities as well as the notion of independence.

I do not include "combinatorics" or "finite probability" as a separate topic on the list above. Problems involving finite probability spaces with equally likely outcomes do appear often in examples and exercises, and Appendix A includes a brief exposition of the combinatorics needed for these examples. But I have concentrated on problems that arise in real-world statistics and stochastic modeling, and I have mostly avoided examples that require tricky counting—poker hands, people arranged randomly in circles, and the like. When I have taught such material, it has tended to dominate the early part of the course and to give students the impression that counting is the heart of the subject. Indeed, some students and lay people seem to believe that probability is the study of permutations and combinations, of counting favorable and unfavorable outcomes. I think nothing should get in the way of learning to use integrals, sums, and infinite series to calculate probabilities and expected values. I also think that combinatorial probability, like geometric probability, appears in its proper perspective, and is not nearly so formidable, when it is encountered after one has studied the fundamental concepts of probability.

The third goal mentioned above, being honest about mathematical details while conveying an intuitive idea of what is involved, is reflected most typically in the treatment of the idea of a distribution. This concept is central to the subject, but it is not easy to define it correctly and rigorously at this level. Yet the term is used so commonly that students deserve a definition. I have tried to steer a course between two extremes: at one end, either not defining distributions at all or else simply talking about assignments of probabilities to intervals; and at the other, including coverage of σ-fields and the Borel sets. I have done so by discussing the Borel sets in the optional subsection (1.3.2); this discussion can be omitted or skimmed lightly if one simply remembers two informally stated principles given there. Throughout the book, the reader is reminded from time to time that a distribution is "a consistent assignment of probabilities to the intervals, and thence to the Borel sets," that the Borel sets can be thought of informally as "the sets that can be made from countably many intervals by set operations," but that the full truth, discussed in (1.3.2), requires mathematics beyond the level of this text.

In general, when there is conflict between the goal of mathematical honesty and that of emphasizing the useful features of the theory, I have tried to discuss the mathematics, as fully as possible, in subsections that can be skipped or covered lightly if one is willing to go by certain informal statements and remember that they are only informal.

Organization of material. The first four chapters are the core, and Chapters 1–6 constitute a basic course. Chapters 7 and 8 cover the standard distributions and models systematically, and include treatment of probability generating functions and moment generating functions. However, it is possible to omit the coverage of generating functions in the first section of each of these chapters, and use the remainders of the chapters for review. Chapter 9 is for students who will need to use conditioning arguments and Bayes's Theorem in a serious way after this course.

Chapter 1 introduces most of the ideas of probability, in concrete form, and at the same time introduces, through examples, many of the standard distributions and models. The goal is for students to begin doing calculations immediately that will develop both calculation skills and intuition about real chance experiments. By the end of Chapter 1 they have calculated binomial, geometric, normal, exponential, Poisson, and one- and two-dimensional uniform probabilities, and they can solve problems involving waiting times and numbers of arrivals for Poisson processes, as well as waiting times and numbers of successes in Bernoulli processes. They have worked with the concepts and notations of probability mass functions and probability density functions, and, in an introductory way, with random variables. Thus there is a large supply of practical and useful distributions available for examples in the rest of the course.

Chapter 2 deals with the algebra of probabilities, including conditioning and Bayes's Theorem and the idea of independence. Then in Chapter 3 the ideas introduced in Chapter 1 are reinforced again, in the slightly more formal context of distributions on $\mathbb{R}$—and, most importantly, distributions of random variables. Joint distributions are covered as well, although this material can be skimmed.

Chapter 4 is concerned with expected values, their calculation and their uses, and with laws of large numbers, which make precise our interpretation of expected value as expected long-run average value. In Section 4.5 the Kolmogorov strong law of large numbers is stated and discussed but not proved, and then the Chebyshev weak law is presented from the ground up, in a way intended to motivate the abstract notion of convergence in probability, and to make it relatively easy to absorb. The chapter also covers covariance and correlation.

Chapter 6 covers the normal family and the Central Limit Theorem; Chapter 5, on distribution theory, comes first so that properties of the normal family can be derived. Only the first two sections of Chapter 5 need to be covered for use in Chapter 6, and even these two sections could be omitted; see **Choosing material to fit a term** below.

The Central Limit Theorem is not proved in Chapter 6; it is stated informally and used extensively there, but the proof is deferred to the section on normal distributions in Chapter 8. Along with the Central Limit Theorem in Chapter 6 goes the normal approximation to the binomial distribution, with the Poisson approximation thrown in for good measure. I find that holding the proof of the Central Limit Theorem until Chapter 8, or even omitting it entirely, does not keep the students from understanding its meaning, importance, and use. The proof of the Central Limit Theorem in Section 8.2 uses moment generating functions and not characteristic functions, and consequently does not cover all cases; but apart from this it is essentially rigorous, except for a lemma or two from advanced calculus and the theorem on convergence of generating functions, which cannot be covered in this course.

When we get to Chapters 7 and 8 on the brand-name distributions, discrete in Chapter 7 and absolutely continuous in Chapter 8, all but a few of them have already been seen—most of them several times over—and the students have become familiar with calculations involving them. I find that these two chapters tend to pull the material of the course together in a satisfying way. Each of Chapters 7 and 8 begins with a rather long section on generating functions, probability generating functions in Chapter 7 and moment generating functions in Chapter 8. There are several possibilities for the use of these two chapters, with or without these sections. See **Choosing material to fit a term** below.

Chapter 9 covers conditional distributions rather fully, using as a starting point the "principle of conditioning" that was introduced in the simplest case in Chapter 2: to compute some unconditional quantity, multiply an appropriate conditional quantity by the density or mass function of the conditioning random variable, and then sum or integrate. The definitions of conditional mass function, density, expected value, and so on are motivated by this principle. This seems to me to be the best way to introduce the otherwise disquieting idea of conditioning on an event $(X = x)$ whose probability equals zero. The chapter goes as far as discussing the properties of a conditional expectation $E(Y|X)$ as a random variable that is a function of X, and the remarkable economy of notation and calculation that this notion provides. It also contains some discussion of prior and posterior distributions.

It is possible to cover Chapter 9 ahead of Chapters 7 and 8 if this material is considered more important. Some of the distributions from Chapters 7 and 8 are mentioned by name in Chapter 9, but most of them have appeared in earlier chapters, and one can also rely on Appendix B at this point.

There are three appendices. Appendix C is just a table of the standard normal cumulative distribution function. Appendix A covers a number of topics from calculus and discrete mathematics: integration by parts, improper integrals, integrals over subsets of the plane, summing some familiar series, elementary counting of permutations and combinations, the use of binomial coefficients in counting and in the binomial theorem, and the like. Many students have had no opportunity to use or otherwise reinforce such material—especially the material from calculus—since they first saw it in earlier courses; fortunately, probability at this level gives them a good chance to do so. When I teach the course I do not spend class time on this appendix; instead, I take care to refer to it when some technique is first needed in an example in class, and I take a little extra time to explain the technique at that point.

Appendix B consists of summaries of the important families of distributions. I encourage students to begin using them for reference when we get to Chapter 4. I also make copies of Appendix B and let students use them on the final examination; thus they are relieved from having to memorize a large miscellany of formulas, and the exam can test their understanding of the ideas and their ability to do the important kinds of calculations. Of course, there is nothing about the text that makes such use of the summaries necessary on examinations.

Choosing material to fit a term. Like most texts, this one contains more material than I can cover in a semester. The basic course is in the first six chapters, and one can omit Sections 1.5, 2.5, 3.6, 5.3, 5.4, and 6.4 without losing material needed later. One can even omit all of Chapter 5, by taking a little extra time with the summary of essential facts about the normal family in Section 6.2, if finding the distribution of a function of a random variable is not crucial. Similarly one can go lightly over Sections 3.4 and 3.5, emphasizing only the case of independent random variables, if the course is not required to cover joint distributions. This basic course can be covered at a comfortable pace with three to four weeks to spare in a semester. It should also fit comfortably into a one-quarter course.

The remaining time in a one-semester course can be filled in several ways. If there is plenty of time, one can cover most of Chapters 7 and 8, including the beginning sections on generating functions, by going quickly over the familiar parts of those chapters. (The new material in Chapters 7 and 8 is: negative binomial distributions in 7.3; all of 7.6 on multinomial distributions; part of 7.5 on hypergeometric distributions; Erlang distributions in 8.3; all of 8.4 on gamma distributions with arbitrary real parameter; and 8.6 on beta distributions.)

There are other possibilities. One is to omit Sections 7.1 and 8.1 entirely, and choose material from the rest of these chapters for review. One could add some of the new distributions—perhaps the negative binomial and the Erlang—along with this review. Another option I have used is not to cover these chapters at all, but simply to select exercises for the students to use on their own for review. This would permit coverage of Chapter 9 if it is needed.

For a more rigorous or abstract course, after the first six chapters (again, Sections 5.3 and 5.4 are optional) one can cover the first two sections of Chapter 8, on moment generating functions and the proof of the Central Limit Theorem, in some detail, and continue to Chapter 9. Other combinations are no doubt possible.

It is impossible, also, to spend time in class on all the text material, even in sections that one does not omit. I usually cover a section in class by first telling the students the highlights, pointing out what the various subsections are about and which are most important. (The Instructor's Manual includes a summary of each section.) Then I spend most of the class time on examples. Instructors will likely want to include their own favorite examples; alternatively one could work through some of the text's examples and do some of the exercises.

An example of the kind of material that instructors may want to skip is the material on the Borel sets and σ-fields in Section 1.3. This is intended to describe, as completely as possible at this level, what it means to assign probabilities to all the intervals and to be assured that they are then determined for any event one would encounter. I do not cover this in class; I simply tell the students that they can informally, though not rigorously, take the term "Borel set" to mean "interval or set made from intervals using set operations"; that assigning probabilities in a consistent way to intervals is sufficient for any probability model on the real line; that there are more details in the text; and that the full truth of the subject is covered in courses in measure theory. From then on, talking about the intervals and the Borel sets does not seem to cause the students any difficulty.

There are numerous digressions and side comments, usually in small type; some are historical notes; others present alternative viewpoints or amplifications for the curious, and still others are comments on the philosophical and scientific foundations of the subject. Occasionally an optional proof appears in small type. Instructors will have their own favorite digressions; the ones in the text can be omitted or left to the students.

Exercises. Every section in the text (except 5.1) is followed by a set of exercises, and there is a set of supplementary exercises at the end of each chapter. Following each set are the answers to all exercises that have answers; but solutions are not given, and the answers seldom give hints on how to work the exercises, so that most exercises can be used in graded assignments. The Instructor's Manual contains solutions to nearly all the exercises.

I have talked to a few students about studying the material on their own using this book; they seem to find it sufficiently detailed that they can read the text at reasonable speed, and then concentrate on the exercises. The exercises, as well as the exposition in the text, are intended to be complete enough to make this kind of study practicable. In a course, too, I think it is best for the bulk of both class time and students' study time to be spent doing as many exercises and examples as possible. Certainly students ought to do more exercises than one could hope to collect and grade.

Most of the exercises are intended for practice in the essential kinds of calculations. Occasionally even these routine exercises aim to provide a new viewpoint, invite a conjecture, or otherwise stimulate thought; even so, for the most part they are straightforward. Some exercises appear to introduce important new topics, but these are anticipatory; if the topic is essential, it is covered later in the text.

Other exercises are there primarily because they concern puzzling, curious, or amusing questions, or because of some historical interest. These usually begin with titles or descriptive phrases in boldface type, and they have entries pointing to them in the index. Browsing through the index will give an idea of the types of questions that appear. The main purpose of these exercises is to motivate the subject and its uses by entertaining. They can be time-consuming if covered in detail in class; I have found it effective simply to spend a minute or two telling the students what such an exercise is about, and then leave it to those who are interested to pursue it on their own. (I have not, however, been able to avoid class discussions of Exercise 20 in Section 2.2, on the "Monty Hall paradox," as well as some of the other puzzling exercises at various places.)

Many exercises are referred to later, in other exercises or in the text. I have tried to indicate, as much as possible, when this will happen. Sometimes minor themes run through the exercise sets. An example is the question of the number of items in their "proper places" in a random permutation; this appears in at least eight exercises in six different sections in Chapters 2, 3, 4, and 7. Because of such exercises, students might be well advised to keep notebooks of their solutions to exercises.

Thanks. I am grateful to many people for various things in connection with this book. First are my parents, for giving me the environment, opportunities, and encouragement that made an academic career possible; and similarly my patient and supporting wife and companion Nancy Kelly. My friend and editor Bob Pirtle of Macmillan has made even the difficult parts of the work enjoyable, with his competence, expertise, good sense, and humor. Susan Reiland's abilities and talents in both statistics and publishing, and her forbearance, have been appreciated and invaluable. Walter L. Smith has been a mentor, friend, and inspiration for many years. Ed Carlstein and Karl Petersen have provided valuable conversations, insight, and encouragement. The UNC Department of Statistics, and especially Stamatis Cambanis, its chairman while I was writing the book, allowed me to teach the course often and provided a superb environment for academic work of all kinds. John Pfaltzgraff, who chaired the Department of Mathematics during much of the same period, gave me needed encouragement and support. Warren Wegner, friend and extraordinary student, worked through the entire book—every paragraph, every exercise and example—but more importantly, he has added much to my understanding of teaching and writing. The reviewers who read drafts of this book, Zhidong Bai (Temple University), James R. Bock (California State University), Thomas Boullion (University of Southwestern Louisiana), Steven Chiappari (Santa Clara University), Edgar Enochs (University of Kentucky), Anant P. Godbole (Michigan Technological University), Michael J. Kahn (St. Olaf College), Eric Key (University of Wisconsin-Milwaukee), David Mason (University of Utah), Ian McKeague (Florida State University), Roger Pinkham (Stevens Institute of Technology), and C. Bradley Russell (Clemson University), provided important suggestions and comments. And I owe special thanks to all the students who had drafts of this book as their texts. They helped to shape the book, and I could not have written it without their wisdom about learning and teaching.

D.G.K.

CHAPTER

1

Probability Models: Definitions and Examples

1.1 A Preliminary Example

1.2 Discrete Probability Spaces

1.3 Absolutely Continuous Probability Spaces

1.4 Two Other Kinds of Probability Space: Bernoulli Processes and Poisson Processes

1.5 The General Definitions of Probability Space and Random Variable

Supplementary Exercises

1.1 A PRELIMINARY EXAMPLE

We begin with an example of a probability model for a simple chance experiment, that of tossing a coin three times. This example contains all the main ingredients of probability theory: ***outcome set, events, probabilities of events,*** and ***random variables.*** Because it introduces so much of the theory, we will describe it at some length. If you get bogged down, or if you have seen these ideas before, you should consult the summary near the end of this section to keep track of what the model really is.

(1.1.1) **Example: Three tosses of a fair coin.** Consider the experiment of tossing a fair coin three times, noting the sequence of heads and/or tails that results, and finally observing certain numerical features of the outcome such as the total number of heads, or the number of tails preceding the first head. For example, we might get the sequence TTH, representing two tails followed by a head. The total number of heads would then be 1, and the number of tails preceding the first head would be 2.

A little thought should convince you that there are eight ***possible outcomes:***

HHH, HHT, HTH, HTT, THH, THT, TTH, TTT,

and that if the coin is fair then all eight of them are equally likely. What we mean by *equally likely* is the following: over a large number of trials, we expect each of them to occur about the same number of times. Thus, for example, if we made 8,000 performances of this experiment we would expect each of these eight outcomes to occur not exactly, but approximately, 1,000 times.

For illustration, a computer simulation of 8,000 performances of this experiment produced the following results:

Outcome:	HHH	HHT	HTH	HTT	THH	THT	TTH	TTT
Frequency:	947	1,017	1,054	994	996	1,042	979	971

These observed frequencies are all fairly near 1,000—within 5.4%, in fact. (Whether you think they are close enough depends on your own preconceived notions of probability. Questions like this one are the subject of statistical inference, where tests are developed that help us to decide in some sense how suspicious we should be of these eight deviations from 1,000.)

It is common in studying such situations to divide all the frequencies by the total number of trials—8,000 in this case—to get the ***relative frequencies*** of each of the eight outcomes. We expect each of the relative frequencies to be close to $1{,}000/8{,}000 = .125$. We get (to four decimal places)

Outcome:	HHH	HHT	HTH	HTT	THH	THT	TTH	TTT
Rel. freq.:	.1184	.1271	.1318	.1242	.1245	.1302	.1224	.1214

Of course we have really gained nothing by this calculation, because the relative frequencies are just as far from .125 as the frequencies were from 1,000. The advantage is simply that we do not have to keep the number 8,000 in mind; we would expect these relative frequencies to be close to .125, no matter how many trials we made, as long as the number of trials was large.

So far we have two things: a set of possible outcomes, and the assumption that the outcomes in the set are equally likely—that is, that they have the same ***expected long-run relative frequency.*** We incorporate this in our model by assigning a ***probability*** of $\frac{1}{8} = .125$ to each of the eight outcomes; we are thinking of probability as a measure of expected long-run relative frequency.

The next ingredient in our model consists of the ***events*** whose probability of occurrence we might be interested in. For example, we might want the probability that the first and second tosses agree. This event occurs if, and only if, the outcome is one of the four outcomes HHH, HHT, TTH, and TTT, and so its expected long-run relative frequency is $\frac{1}{8} + \frac{1}{8} + \frac{1}{8} + \frac{1}{8}$, or $\frac{1}{2}$. (In the experiment described above, this event actually occurred $947 + 1{,}017 + 979 + 971 = 3{,}914$ times, so its observed relative frequency was $\frac{3{,}914}{8{,}000} = .48925$. Again we should not be distressed that it is not exactly .5; *observed* relative frequencies are seldom exactly equal to probabilities, which are *expected long-run* relative frequencies.)

In dealing with probability models, it is customary to forget the distinction between an event and the set of outcomes for which the event occurs. That is, we say that events *are* subsets of the set of all possible outcomes of the experiment. We usually use capital letters to denote events and P to denote probabilities. Thus, if we use A to denote the event we mentioned above, "first and second tosses agree," then we would say that $A = \{\text{HHH, HHT, TTH, TTT}\}$ and $P(A) = \frac{1}{2}$.

Some other events, the corresponding sets of outcomes, and their probabilities are as follows:

Event B: Exactly two heads and one tail

Set of outcomes: $B = \{\text{HHT, HTH, THH}\}$

Probability: $P(B) = \frac{1}{8} + \frac{1}{8} + \frac{1}{8} = \frac{3}{8} = .375$

(Observed relative frequency of B in experiment:
$(1{,}017 + 1{,}054 + 996)/8{,}000 = .383375$)

Event C: Tails on the first toss and exactly one other tail

Set of outcomes: $C = \{\text{THT, TTH}\}$

Probability: $P(C) = \frac{1}{8} + \frac{1}{8} = \frac{1}{4} = .25$

(Observed relative frequency of C in experiment:
$(1{,}042 + 979)/8{,}000 = .252625$)

Event D: Three heads

Set of outcomes: $D = \{HHH\}$ (a "singleton" set)

Probability: $P(D) = \frac{1}{8} = .125$

(Observed relative frequency of D in experiment: 947/8,000 = .118375)

We now have most of the ingredients of our full model:

a set, called the ***outcome set*** (***sample space*** is another commonly used term), whose members represent the possible outcomes of the experiment;

events, which correspond to subsets of the sample space; and

some sort of assumption that provides a recipe for computing the ***probabilities of the events.***

This much of a model (outcome set, events, and a probability recipe) is called a ***probability space.*** We will give the general definition in Section 1.5, and in Sections 1.2 and 1.3 we will formally define the two special kinds of probability space that we will study in this text.

What we have not yet put into the model is the result of counting the number of heads. Let us think informally about this first. If X denotes the number of heads, then X has four possible values—namely, 0, 1, 2, and 3—and each of these values corresponds to an event (subset of the outcome set), as follows:

"$X = 0$," or "no heads," is the event {TTT}, whose probability is $\frac{1}{8}$;

"$X = 1$," or "one head," is the event {HTT, THT, TTH}, whose probability is $\frac{3}{8}$;

"$X = 2$" is the event {HHT, HTH, THH}, whose probability is $\frac{3}{8}$;

"$X = 3$" is the event {HHH}, whose probability is $\frac{1}{8}$.

We can say these more concisely as follows:

$$P(X = 0) = P\{\text{TTT}\} = \tfrac{1}{8},$$
$$P(X = 1) = P\{\text{HTT, THT, TTH}\} = \tfrac{3}{8},$$
$$P(X = 2) = P\{\text{HHT, HTH, THH}\} = \tfrac{3}{8},$$
$$P(X = 3) = P\{\text{HHH}\} = \tfrac{1}{8}.$$

In addition, there are other events that can be described in terms of X. For example:

$$P(X \leq 2) = P\{\text{HHT, HTH, HTT, THH, THT, TTH, TTT}\} = \tfrac{7}{8},$$
$$P(X \text{ is even}) = P\{\text{HHT, HTH, THH, TTT}\} = \tfrac{1}{2}.$$

There are also events that *cannot* be described in terms of X. For example, the event A above, whose verbal description is "first and second tosses agree," cannot be described by saying anything about the value of X alone. You should also convince yourself that among the three other events B, C, and D described along with A above, C cannot be described in terms of X, whereas B and D can.

X is called a ***random variable;*** it has a set of possible values, and certain events can be given concise verbal descriptions by making statements about which of its possible values X takes on. For example, "$X \le 1$" is a short way of saying "either three tails or one head and two tails." But what *is* X? What kind of mathematical object is it? We are in an odd position: we know what "$X = 2$" is—it is an event, i.e., a subset of the outcome set—but we may not see what X itself is.

The answer is simple but perhaps surprising: ***a random variable is a function on the outcome set.*** This makes sense when we look at the following table:

Outcome:	HHH	HHT	HTH	HTT	THH	THT	TTH	TTT
Value of X:	3	2	2	1	2	1	1	0

That is, X is the function defined by this table; its domain is the outcome set, and its range is the set $\{0, 1, 2, 3\}$ of possible values, defined by the table above. $X(\text{HHT}) = 2$, $X(\text{THT}) = 1$, $X(\text{TTT}) = 0$, and so forth.

Now we go a step further with out notation. It is common to use an uppercase omega, Ω, to denote an outcome set, and a lowercase omega, ω, to denote a generic outcome (member of Ω). With this notation, we see that, for example,

$$\text{The event "}X = 2\text{" is the set } \{\text{HHT, HTH, THH}\} = \{\omega\text{: } X(\omega) = 2\};$$

$$\text{the event "}X \le 1\text{" is the set } \{\text{HTT, THT, TTH, TTT}\} = \{\omega\text{: } X(\omega) \le 1\}.$$

That is, any statement about X, such as "X is so and so," defines an event that, as a subset of Ω, equals $\{\omega$: $X(\omega)$ is so and so$\}$. We can also use this kind of notation to describe the function X a little more formally: we can say, for example, that if ω is a member of Ω, then ω is a three-letter word from the two-letter alphabet $\{\text{H, T}\}$, and $X(\omega)$ is the number of H's in ω.

There is another thing we can do with this random variable. If we make a table of its possible values and their probabilities, then we can use the table to find quickly the probability of any event that is describable in terms of X. The table is the following:

(1.1.2)

x:	0	1	2	3
$P(X = x)$:	$\frac{1}{8}$	$\frac{3}{8}$	$\frac{3}{8}$	$\frac{1}{8}$

Notice how simple it is to see from this, for example, that $P(X \le 2) = \frac{1}{8} + \frac{3}{8} + \frac{3}{8} = \frac{7}{8}$; we found this probability earlier by listing the seven members of the event "$X \le 2$."

Finally, we consider one more random variable: Y, the number of tails preceding the first head, if there is a head; let us agree to say that $Y = 3$ if no heads at all come up. (That is, Y is simply the number of tails at the start of the sequence.) It is a simple matter then to make the table:

Outcome:	HHH	HHT	HTH	HTT	THH	THT	TTH	TTT
Value of Y:	0	0	0	0	1	1	2	3

From this we make the table of possible values of Y and their probabilities:

(1.1.3)

y:	0	1	2	3
$P(Y = y)$:	$\frac{4}{8}$	$\frac{2}{8}$	$\frac{1}{8}$	$\frac{1}{8}$

And from this we can easily find the probability of any event that can be described in terms of Y. For example:

$$P(Y > 1) = \tfrac{1}{8} + \tfrac{1}{8} = \tfrac{1}{4},$$
$$P(Y \le 2) = \tfrac{4}{8} + \tfrac{2}{8} + \tfrac{1}{8} = \tfrac{7}{8}.$$

We can even find probabilities of events that are described in terms of both X and Y, although it is important to note that they *cannot* be found from the tables (1.1.2) and (1.1.3):

$$P(X \le 2 \text{ and } Y = 0) = P\{\text{HHT, HTH, HTT}\} = \tfrac{3}{8},$$
$$P(X = 1 \text{ and } Y = 1) = P\{\text{THT}\} = \tfrac{1}{8},$$
$$P(X = 2 \text{ and } Y = 2) = P(\varnothing) = 0.$$

Notice that it is impossible for X to be 2 and Y to be 2 at the same time, and accordingly, the event "$X = 2$ and $Y = 2$" is a set with no members—the empty set, denoted by $\varnothing$—which has probability 0.

In Chapter 3 we will consider at greater length events that are describable in terms of two or more random variables. For most of this chapter and the next, we will deal with just one random variable at a time.

Summary of Example (1.1.1): To model the experiment of tossing a fair coin three times, we use the *outcome* set

$$\Omega = \{\text{HHH, HHT, HTH, HTT, THH, THT, TTH, TTT}\}.$$

Events are subsets of this set, and if the coin is fair we are justified in the assumption that these eight outcomes are equally likely. Consequently, for any event A, the *probability* of A is

$$P(A) = (\text{the number of members of } A) \cdot \tfrac{1}{8}.$$

This structure, consisting of the outcome set, the events, and the probabilities of the events, is the *probability space* for the model.

If X is the *random variable* representing the number of heads that appear in the three tosses, then X is the function whose domain is Ω and whose range is $\{0, 1, 2, 3\}$ (the set of possible values of X), defined by the following table:

ω:	HHH	HHT	HTH	HTT	THH	THT	TTH	TTT
$X(\omega)$:	3	2	2	1	2	1	1	0

From this we can make a table of the possible values of X and their probabilities:

x:	0	1	2	3
$P(X = x)$:	$\frac{1}{8}$	$\frac{3}{8}$	$\frac{3}{8}$	$\frac{1}{8}$

And from this table we can find the probability of any event that is describable in terms of X. ∎

The exercises at the end of this section deal with several other models that are like the one in the example above, in that the outcome sets consist of finitely many equally likely outcomes. In such models, probabilities are found by the simple formula

$$P(A) = \frac{\text{number of outcomes in } A}{\text{number of outcomes in } \Omega},$$

because the probability of any single outcome is

$$P(\omega) = \frac{1}{\text{number of outcomes in } \Omega}.$$

In the next two sections we will see examples of the other kinds of probability models we will be studying in the text, starting with finitely many outcomes that are not equally likely, then a countable infinity of outcomes, and finally an uncountable infinity of outcomes.

EXERCISES FOR **SECTION 1.1**

[The answers to most exercises follow the exercise sets. The superscript "a" preceding the number of an exercise, as with Exercises 10, 11c, and 11d of this set, indicates that the answer contains enough detail to provide a hint for the solution.]

1. In the chance experiment modeled in this section, a fair coin is tossed three times, X is the number of heads, and Y is the number of tails preceding the

first head. For each of the following verbal descriptions of events, list the members of the corresponding subset of Ω. Then find the probability of the event. [The important part of this exercise is to make the lists corresponding to the events, not to find the probabilities. Begin by listing the eight members of Ω, along with the values of X and Y for each. You may be able to reason out some of the probabilities without making the lists, and that is good mental gymnastics; but this exercise is about the correspondence between events and sets of outcomes.]

a. At least as many heads as tails
b. $X = Y$
c. Y is even.
d. X is even and Y is odd.
e. The first toss comes up heads or the second toss comes up tails.
f. It is not true that $X = 2$ or $Y = 0$.
g. The statement "There is at least one tail, and the first toss is a head" is false.

2. **Two tosses of a fair coin.** For the experiment of tossing a fair coin two times, the outcome set Ω has only four members.
 a. List them.
 b. For each of the following verbal descriptions of events, list the members of the event and (assuming the four members of Ω are equally likely) find its probability.
  i. At least one head
  ii. More heads than tails
  iii. The first toss results in heads.
 c. Now let the random variable X represent the number of *tails* that show in the two tosses. Describe X as a function on Ω by listing the members of Ω and listing $X(\omega)$ next to each ω in Ω.
 d. Make a table of the possible values of X and their probabilities.

3. Repeat Exercise 2 for the experiment of tossing a fair coin four times. [Ω now has 16 members. It is always a good idea to try to list outcomes in some systematic way.]

4. **Random permutations of a set.** Consider the following experiment. In a hat are three slips of paper, with the numbers 1, 2, and 3 on them. These slips are drawn out of the hat one at a time, at random. The result is a random permutation of the set $\{1, 2, 3\}$.
 a. There are six possible outcomes of this experiment; list them. [Two of them are 132 and 321.]
 b. For each of the following verbal descriptions of events, list the members of the event and (assuming all six outcomes are equally likely) find its probability.
  i. Slip 2 is drawn first.
  ii. Slip 3 is drawn second.

   iii. Slip 2 is drawn first and slip 3 is drawn second.
   iv. Either slip 2 is drawn first or slip 3 is drawn second, or both.
   v. No slip is drawn in its "proper place" (that is, 1 is not first, 2 is not second, and 3 is not third).

c. Let X be the number of slips that are in their "proper places." Give $X(\omega)$ for each ω in Ω.

d. Make a table of the possible values of X and their probabilities.

This is the first of several exercises in this text on random permutations, and especially on the question of the number of objects in their proper places. The subject is not overly important, and you will not miss a major aspect of the theory of probability if you omit them; however, the methods and the results are often surprising.

5. **Two fair dice.** For the experiment of rolling two fair dice and noting the two faces that come up, the model that is most often used has an outcome set of 36 equally likely members:

11	12	13	14	15	16
21	22	23	24	25	26
31	32	33	34	35	36
41	42	43	44	45	46
51	52	53	54	55	56
61	62	63	64	65	66

a. Find the probability of each of the following events:
   i. Both dice show numbers greater than 3.
   ii. The sum of the two numbers is 6.
   iii. The two dice show the same number. (That is, a "double" is rolled.)
   iv. At least one 4 shows.
   v. At least one number greater than 4 shows.

b. Let X be the sum of the two numbers that show. For each possible value of X, say x, identify the members of the event described by "$X = x$." Make a table of the possible values of X and their probabilities.

6. Repeat Exercise 4 but with four slips of paper, numbered 1, 2, 3, and 4, instead of three. [There are now 24 possible outcomes.]

7. **Samples of size 2, without replacement, from a population of size 4.** In a hat are four slips of paper, numbered 1, 2, 3, and 4. The experiment consists of pulling out two slips at random at the same time. There are six possible outcomes, one for each of the six subsets of size 2 in the set {1, 2, 3, 4}.

a. Find the probability of each of the following events:
   i. Slip 2 is drawn.
   ii. Slip 3 is drawn but slip 4 is not.
   iii. The larger of the two numbers is not more than 3.

b. Let X denote the sum of the two numbers drawn. Make a table of the possible values of X and their probabilities.

8. Repeat Exercise 7, but assume that the slips are drawn one at a time. The first slip is not replaced before the second is drawn, so there are now 12 possible outcomes, two for each of the outcomes in Exercise 7.
   c. In addition to parts a and b, find the probability of the following events, which make no sense in the model of Exercise 7. (Do you see why they make no sense?)
      i. The first slip drawn is 3.
      ii. The second slip drawn is 3.

The point of Exercises 7 and 8 is this: The two experiments they model are only slightly different; our intuition is that events like those in parts a and b, which do not involve a "first" or "second" draw, should have the same probability in both—and they do.

There is an additional point to note in Exercise 8c: The two probabilities are the same, despite the fact that the first slip is not replaced before the second is drawn. Some people find this troubling at first sight; they think that the probability on the second draw ought to be either $\frac{1}{3}$, if the first slip is not a 3, or 0, if the first is a 3. But these probabilities are *conditional* probabilities, which we will study in Chapter 2. The fact is that if we do the experiment many times, the second slip will be a 3 about $\frac{1}{4}$ of the time.

9. **Samples of size 2, with replacement, from a population of size 4.** Repeat Exercise 8, but now assume that the first slip is replaced before the second is drawn. Now there are 16 possible outcomes, and the answers are different.

[a]10. So far we have seen several tables of possible values and their probabilities for random variables [two of them are (1.1.2) and (1.1.3), and others are in the exercises above]. In every table the probabilities are nonnegative numbers; what additional property do the numbers in each table have?

11. **A better model for two indistinguishable fair dice?** You may be uncomfortable with the outcome set in Exercise 5 because when dealing with two dice you cannot, or at least you do not, distinguish between them. Thus, for example, the two outcomes 13 and 31 should perhaps be considered as the same outcome. This argument leads to the following 21-member outcome set:

11	12	13	14	15	16
	22	23	24	25	26
		33	34	35	36
			44	45	46
				55	56
					66

   a. Using this outcome set and assuming that all 21 outcomes are equally likely, find the probabilities of the events described in Exercise 5a.
   b. Why do you think the model in Exercise 5 is the one that is successfully used in practice and this one is not? [It must be that it gives more accurate predictions of long-run relative frequencies; why do you think this is true?]
   [a]**c.** How could you convince someone (perhaps yourself) empirically that the model of Exercise 5 is better by using the results of, say, 25,200 rolls of

two fair dice? [We chose 25,200 because it is a large multiple of both 21 and 36.]

[a]**d.** It turns out that the 21-member outcome set is not useless for the roll of two fair dice, however; if we assign the appropriate probabilities to its members (not $\frac{1}{21}$ each), it works perfectly well. What are the appropriate probabilities? [Be certain that your 21 probabilities add to 1.]

ANSWERS

1. a. $\frac{1}{2}$ **b.** $\frac{1}{8}$ **c.** $\frac{5}{8}$ **d.** $\frac{1}{4}$ **e.** $\frac{3}{4}$ **f.** $\frac{3}{8}$ **g.** $\frac{5}{8}$

2. b. $\frac{3}{4}, \frac{1}{4}, \frac{1}{2}$ **d.** Values 0, 1, 2; probs. $\frac{1}{4}, \frac{1}{2}, \frac{1}{4}$

3. b. $\frac{15}{16}, \frac{5}{16}, \frac{1}{2}$ **d.** Values 0, 1, 2, 3, 4; probs. $\frac{1}{16}, \frac{4}{16}, \frac{6}{16}, \frac{4}{16}, \frac{1}{16}$

4. b. $\frac{1}{3}, \frac{1}{3}, \frac{1}{6}, \frac{1}{2}, \frac{1}{3}$ **d.** Values 0, 1, 3; probs. $\frac{1}{3}, \frac{1}{2}, \frac{1}{6}$

5. a. $\frac{1}{4}, \frac{5}{36}, \frac{1}{6}, \frac{11}{36}, \frac{5}{9}$ **b.** Values 2, 3, 4, 5, 6, 7, 8, 9, 10, 11, 12; probs. $\frac{1}{36}, \frac{2}{36}, \frac{3}{36}, \frac{4}{36}, \frac{5}{36}, \frac{6}{36}, \frac{5}{36}, \frac{4}{36}, \frac{3}{36}, \frac{2}{36}, \frac{1}{36}$

6. b. $\frac{1}{4}, \frac{1}{4}, \frac{1}{12}, \frac{5}{12}, \frac{9}{24}$ **d.** Values 0, 1, 2, 4; probs. $\frac{9}{24}, \frac{8}{24}, \frac{6}{24}, \frac{1}{24}$

7. a. $\frac{1}{2}, \frac{1}{3}, \frac{1}{2}$ **b.** Values 3, 4, 5, 6, 7; probs. $\frac{1}{6}$ except for $P(X = 5) = \frac{1}{3}$

8. a, b. Same answers as 7a and b. **c.** $\frac{1}{4}, \frac{1}{4}$

9. a. $\frac{7}{16}, \frac{5}{16}, \frac{9}{16}$ **b.** Values 2, 3, 4, 5, 6, 7, 8; probs. $\frac{1}{16}, \frac{2}{16}, \frac{3}{16}, \frac{4}{16}, \frac{3}{16}, \frac{2}{16}, \frac{1}{16}$ **c.** $\frac{1}{4}, \frac{1}{4}$

10. They add to 1.

11. a. $\frac{2}{7}, \frac{1}{7}, \frac{2}{7}, \frac{2}{7}, \frac{11}{21}$ **c.** Record the frequencies of the 21 outcomes of the model in Exercise 11. If this model is appropriate, they should be nearly equal—all about 25,200/21 = 1,200. If the model of Exercise 5 is appropriate, the doubles should have frequencies near 25,200/36 = 700, and the others near 1,400.
d. $\frac{1}{36}$ to each of the doubles, $\frac{2}{36}$ to each of the rest

1.2 DISCRETE PROBABILITY SPACES

The outcome sets in Section 1.1—the example and all the exercises—have finitely many members, and (except in Exercise 11d) all members of each outcome set are equally likely. But there are many important models in which the outcomes are not all equally likely, and also models in which it is necessary to consider infinitely many possible outcomes. In this section we will consider examples of such models. Remember that a ***probability space*** is described informally as something that consists of

a set, called the ***outcome set,*** whose members represent the possible outcomes of the experiment;

events, which correspond to subsets of the outcome set; and

a recipe for computing the ***probabilities of the events.***

In the previous section, and in the examples of this section, the assumption for computing the probabilities is just an assignment of probabilities to the individual members of the outcome set, taking care that the probabilities of all the outcomes add to 1. Once we have made such an assignment, the recipe for computing the probability of an event is simply to add up the probabilities we have assigned to the members of that event.

In this section, as in the last one, we proceed quite slowly, digressing to introduce informally many of the important ideas of probability theory. We conclude most of the examples with summaries so that you can see quickly what the main points are. And we suggest that you return to this section and the previous one at some time after covering Chapters 2 and 3, to reread some of the points that may have been lost in the welter of comments.

In the following example there are still finitely many outcomes, but they are not equally likely.

(1.2.1) **Example:** Three Bernoulli trials with a success probability other than .5. Suppose we do three performances of some experiment that has two possible results, but that one of the two is more probable than the other. This could happen, for example, if we were tossing a coin that was somehow weighted so that heads came up 60% of the time instead of 50%. It could happen if we were rolling a fair die ("die" is the singular of "dice") three times, but we cared only about the number of 6's we got; we would count a 6 as a success and anything other than a 6 as a failure; then "success" would occur one-sixth of the time, not one-half. It could also happen if we were a polling agency asking three randomly chosen people which of two candidates they intended to vote for. Unless the population were equally divided between the two candidates, the two results would not be equally likely.

When such repeated two-outcome trials can have no physical influence on each other, they are called *Bernoulli trials* after James Bernoulli (1654–1705), the most mathematical member of the famous Bernoulli family of Swiss scientists. James Bernoulli's book *Ars conjectandi* (*The Art of Guessing*), published in 1713, laid much of the foundation for probability theory. In Bernoulli trials, the lack of physical influence makes it possible to work out probabilities as we do below. In technical language, we call the trials *independent;* this term will be defined formally in Section 2.4.

In such experiments, it is customary to call the two possible results "success" and "failure" and to denote them by S and F. The three-trial experiment then has an outcome set of eight members:

$$\Omega = \{\text{SSS, SSF, SFS, SFF, FSS, FSF, FFS, FFF}\}.$$

(If we use 1 and 0 instead of S and F, and list these outcomes in the reverse order, we get the familiar binary sequence 000, 001, 010, 011, 100, 101, 110, 111.)

But these eight outcomes are not equally likely unless S and F are. If, for example, S is more likely than F, then we would expect SSS to occur more often than FFF in the long run. So what probabilities are appropriate?

To make the example specific, let us suppose that in a long series of *single* trials, we expect 65% S's and 35% F's. Then, for example, the probability that we should assign to SFS is $.65 \cdot .35 \cdot .65$.

> Why? The reasoning is as follows. Over an extremely long series of performances of the three-trial experiment, about 65% of them should begin with an S. These 65% still amount to a large number of performances, so on 35% of *them* we expect the second trial to result in F. (The assumption of independence implies that in the long run the relative frequency of outcomes with F on the second trial is the same among those with S on the first trial as among all outcomes.) So we expect $.65 \cdot .35$ of the performances to begin SF. This is still a large number, so we expect 65% of *them* to finish with an S. This is $.65 \cdot .35 \cdot .65$ of the total number of performances.

Using similar reasoning, we can assign probabilities to all eight outcomes, as follows.

(1.2.2)

ω	SSS	SSF	SFS	SFF
Prob.	$.65^3$	$.65^2 \cdot .35$	$.65^2 \cdot .35$	$.65 \cdot .35^2$

ω	FSS	FSF	FFS	FFF
Prob.	$.35 \cdot .65^2$	$.35^2 \cdot .65$	$.35^2 \cdot .65$	$.35^3$

And from these we can compute probabilities of events like the following:

$$
\begin{aligned}
P(\text{at most one F}) &= P\{\text{SSS, SSF, SFS, FSS}\} \\
&= .65^3 + 3 \cdot .65^2 \cdot .35 = .71825; \\
P(\text{S on the second trial}) &= P\{\text{SSS, SSF, FSS, FSF}\} \\
&= .65^3 + 2 \cdot .65^2 \cdot .35 + .65 \cdot .35^2 = .65.
\end{aligned}
$$

A computer simulation of 10,000 performances of this experiment produced the frequencies shown in the third row of the following table. The first two rows are the same as table (1.2.2) except that the probabilities have been multiplied out. Notice that the observed relative frequencies are really quite close to the probabilities given by the model.

ω	SSS	SSF	SFS	SFF	FSS	FSF	FFS	FFF
Prob.	.274625	.147875	.147875	.079625	.147875	.079625	.079625	.042875
Freq.	2,714	1,438	1,532	823	1,473	765	814	441
Rel. freq.	.2714	.1438	.1532	.0823	.1473	.0765	.0814	.0441

We can also find probabilities for random variables connected with this probability space as we did in the previous section. For example, if X is the number of S's, then the table of possible values of X and their probabilities is as follows:

(1.2.3)

x:	0	1	2	3
$P(X = x)$:	$.35^3$	$3 \cdot .35^2 \cdot .65$	$3 \cdot .35 \cdot .65^2$	$.65^3$

You ought to be sure you see where the 3's come from in the middle two probabilities, and why they are not there in the other two. You might also want to check that these four probabilities add to 1. You can do it with a calculator, or you can do it cleverly with the binomial formula $a^3 + 3a^2b + 3ab^2 + b^3 = (a + b)^3$. [See (A.4.7) in Appendix A.] In Example (1.2.13) we will see more about the connection between the binomial theorem and probabilities like these, which are called *binomial probabilities.*

Summary of Example (1.2.1): Ω = {SSS, SSF, SFS, SFF, FSS, FSF, FFS, FFF}; events are subsets of Ω, and the probability of any event is the sum of the probabilities of its members, where the probabilities assigned to the individual members are given in (1.2.2). The random variable X representing the number of successes in three trials is defined by saying that $X(\omega)$ is the number of S's in ω. Its possible values are 0, 1, 2, and 3, and the probabilities of the four events $(X = x)$ are given by table (1.2.3) above.

We can be much more concise than (1.2.2) in describing our assignment of probabilities to the members of Ω, if we make use of our random-variable notation: the probability assigned to ω is equal to

(1.2.4)
$$(.65)^{X(\omega)}(.35)^{3-X(\omega)}.$$

It is not essential, but it may be worth your while to be sure you understand that (1.2.4) says the same thing as (1.2.2). For example, if ω = FSS, then $X(\omega) = 2$ and (1.2.4) gives $(.65)^2(.35)^1$. ∎

In the next example we need a countable infinity of outcomes. You may want to look at Section A.3 of Appendix A on countably infinite sets.

(1.2.5) **Example: Waiting for the second head.** Suppose we return to a fair coin, but now we toss it repeatedly until two heads have come up. The following is a model that works successfully for this experiment. We list the outcomes in rows according to the number of tosses needed to get the second head; all the outcomes on any one row have the same probability.

(1.2.6)

	Outcomes	Probabilities
(Two tosses)	HH	$\frac{1}{4}$
(Three tosses)	HTH, THH	$\frac{1}{8}$ each
(Four tosses)	HTTH, THTH, TTHH	$\frac{1}{16}$ each
(Five tosses)	HTTTH, THTTH, TTHTH, TTTHH	$\frac{1}{32}$ each
(Six tosses)	HTTTTH, THTTTH, TTHTTH, TTTHTH, TTTTHH	$\frac{1}{64}$ each
	etc.	

Notice first that the outcome set, which consists of all the outcomes on all the rows, is infinite; there is a row for any number of tosses, no matter how large. The row for n tosses contains $n - 1$ strings of H's and T's that end with an H and contain just one other H. (For example, the six-toss row has five outcomes on it, the million-toss row has 999,999 outcomes, and so on.) And even though we might do this experiment all our lives and never need more than, say, 75 tosses to get two heads, we cannot ignore the possibility that 76, 77, 175, 500, or even 5,000,000 tosses might be needed. These events are very improbable, but they are possible; and over a really huge number of performances of the experiment, we would expect them to occur with tiny relative frequencies approximately equal to their tiny probabilities.

Next, think about what an infinite outcome set means for this experiment. It does *not* mean that infinitely many tosses might be needed to get two heads. What it says is that the number of tosses will be finite, but that there is no number M large enough to guarantee that there will be fewer than M tosses. There are infinitely many outcomes, but there are no outcomes that represent infinitely many tosses. You may want to reflect awhile on the fact that, in this model, the number of tosses needed to get two heads is sure to be finite, but the number of possible *numbers* of tosses is infinite. (The set of integers is like this; every integer is a finite number, but the number of integers is infinite.)

Another thing we should check is whether the probabilities of all the outcomes in our model add to 1. If they should add to more than 1, then something is wrong with our assignment. If they add to less than 1, perhaps we need to consider other possibilities, such as infinitely many tosses without two heads. We have an infinite series to evaluate:

$$\begin{aligned}\tfrac{1}{4} + \tfrac{2}{8} + \tfrac{3}{16} + \tfrac{4}{32} + \tfrac{5}{64} + \cdots &= \left(\tfrac{1}{2}\right)^2 + 2 \cdot \left(\tfrac{1}{2}\right)^3 + 3 \cdot \left(\tfrac{1}{2}\right)^4 + 4 \cdot \left(\tfrac{1}{2}\right)^5 + 5 \cdot \left(\tfrac{1}{2}\right)^6 + \cdots \\ &= \left(\tfrac{1}{2}\right)^2 \left(1 + 2\left(\tfrac{1}{2}\right) + 3\left(\tfrac{1}{2}\right)^2 + 4\left(\tfrac{1}{2}\right)^3 + 5\left(\tfrac{1}{2}\right)^4 + \cdots\right) \\ &= \left(\tfrac{1}{2}\right)^2 \cdot \frac{1}{\left(1 - \tfrac{1}{2}\right)^2} = 1 \qquad \text{[See (A.5.9) in Appendix A.]}\end{aligned}$$

This is comforting; it serves as a partial check that we have assigned the probabilities in a reasonable way, and it hints to us that there is no "extra room" for outcomes in which we fail to get two heads in infinitely many tosses.

Note, though, that it does not *prove* that we have chosen the right model. No mathematical truth about a model can be used to prove that the model fits a real-world situation. Models do not prove things about the world; they only predict that if the model fits the world, then we might expect the world to behave in some way.

Now we can find the probability of any event as we did in Example (1.2.1), by adding the probabilities of its members. For example:

$$\begin{aligned} P(\text{fewer than five tosses are needed}) &= P\{\text{HH, HTH, THH, HTTH, THTH, TTHH}\} \\ &= \tfrac{1}{4} + \tfrac{1}{8} + \tfrac{1}{8} + \tfrac{1}{16} + \tfrac{1}{16} + \tfrac{1}{16} = \tfrac{11}{16}; \\ P(\text{more than three tosses are needed}) &= 1 - P\{\text{HH, HTH, THH}\} \\ &= 1 - \left(\tfrac{1}{4} + \tfrac{1}{8} + \tfrac{1}{8}\right) = \tfrac{1}{2}. \end{aligned}$$

(You might want to sum the infinite series of probabilities of members *not* in this event, just for practice, to be sure that $\frac{1}{2}$ is correct.)

$$\begin{aligned} &P(\text{an even number of tosses are needed}) \\ &\quad = P(\text{HH, HTTH, THTH, TTHH,} \\ &\qquad\qquad \text{HTTTTH, THTTTH, TTHTTH, TTTHTH, TTTTHH}, \ldots) \\ &\quad = \tfrac{1}{4} + 3 \cdot \tfrac{1}{16} + 5 \cdot \tfrac{1}{64} + \cdots \\ &\quad = \left(\tfrac{1}{2}\right)^2 \cdot (1 + 3x^2 + 5x^4 + 7x^6 + \cdots) \qquad \left(\text{where } x = \tfrac{1}{2}\right) \\ &\quad = \left(\tfrac{1}{2}\right)^2 \cdot (\text{derivative of } x + x^3 + x^5 + x^7 + \cdots) = \left(\tfrac{1}{2}\right)^2 \cdot \left(\text{derivative of } \frac{x}{1 - x^2}\right) \\ &\quad = \left(\tfrac{1}{2}\right)^2 \cdot \frac{1 + x^2}{(1 - x^2)^2} = \tfrac{5}{9}. \end{aligned}$$

(See Section A.5 of Appendix A for a review of calculations like this one.)

Finally, suppose we want to study the number of tails that precede the second head. We are counting all the tails that come up before the first head as well as those immediately preceding the second. So the number of tails is two less than the number of tosses needed to get the second head. Calling the number of tails X, we see that X is the function on Ω defined by

$$\begin{gathered} X(\text{HH}) = 0, \\ X(\text{HTH}) = X(\text{THH}) = 1, \\ X(\text{HTTH}) = X(\text{THTH}) = X(\text{TTHH}) = 2, \\ \text{and for any } \omega,\ X(\omega) \text{ is the number of T's in } \omega. \end{gathered}$$

Using the probabilities given above for the individual outcomes, we see that

$$P(X = 0) = \tfrac{1}{4}, \quad P(X = 1) = 2 \cdot \tfrac{1}{8}, \quad P(X = 2) = 3 \cdot \tfrac{1}{16}, \quad \text{etc.}$$

We can express this in general as follows:

The possible values of X are 0, 1, 2, . . . ; and if x is any of these possible values, then $P(X = x) = (x + 1)\left(\frac{1}{2}\right)^{x+2}$.

It is customary to be even more succinct and say

$$P(X = x) = (x + 1)\left(\tfrac{1}{2}\right)^{x+2}, \quad \text{for } x = 0, 1, 2, \ldots . \tag{1.2.7}$$

But it is very important to remember that ***the formula*** $P(X = x) = (x + 1)\left(\frac{1}{2}\right)^{x+2}$ ***by itself is not enough; we must also give the set of values of x for which this formula is valid*** (in this case, $x = 0, 1, 2, \ldots$). For example, $P(X = 1.5)$ equals 0, not $(2.5)\left(\frac{1}{2}\right)^{3.5}$. This may seem pedantic, but it will come back to haunt you later if you do not get into the habit of including a specification of the domain in your description of a function. That is, ***when you give a formula—any formula—think about, and write down, the set of values of the variables for which the formula is valid.***

The results of another computer simulation follow. This time the experiment was repeated 10,000 times and the value of X was calculated on each repetition. As it happened, the value of X was never greater than 17 in the 10,000 trials. The probabilities, observed frequencies, and observed relative frequencies of the possible values of X, 0, 1, 2, . . . , 17, are shown. The probabilities have been rounded to four decimal places. The observed relative frequencies are strikingly close to the probabilities given by the model.

x	$P(X = x)$	Obs. freq.	Rel. freq.	x	$P(X = x)$	Obs. freq.	Rel. freq.
0	.2500	2,484	.2484	9	.0049	44	.0044
1	.2500	2,559	.2559	10	.0027	27	.0027
2	.1875	1,904	.1904	11	.0015	16	.0016
3	.1250	1,228	.1228	12	.0008	9	.0009
4	.0781	758	.0758	13	.0004	3	.0003
5	.0469	454	.0454	14	.0002	3	.0003
6	.0273	256	.0256	15	.0001	2	.0002
7	.0156	160	.0160	16	.0001	0	.0000
8	.0088	92	.0092	17	.0000	1	.0001

Summary of Example (1.2.5): Ω is the set whose members are listed under "outcomes" in (1.2.6) above. Events are subsets of Ω, and the probability of any event is the sum of the probabilities assigned to the individual members, which are given in the right-hand column of (1.2.6). The random variable X is defined by $X(\omega)$ = the number of T's in ω; its possible values are 0, 1, 2, . . . , and the probabilities of the events $(X = x)$ are given by (1.2.7). ∎

At last it is time for a formal definition.

(1.2.8) **Definition.** A ***discrete probability space*** consists of a finite or countable set Ω called the ***outcome set*** and a function p from Ω to $\mathbb{R}$ such that

(1.2.9)
$$p(\omega) \geq 0 \text{ for all } \omega \in \Omega \quad \text{and} \quad \sum_{\omega \in \Omega} p(\omega) = 1.$$

A function with these two properties is called a ***probability mass function.*** The ***events*** are the subsets of Ω, and the ***probability*** of any event $A \subseteq \Omega$ is defined by

(1.2.10)
$$P(A) = \sum_{\omega \in A} p(\omega).$$

Finally, a ***random variable*** on a discrete probability space is a function from Ω to the set $\mathbb{R}$ of real numbers.

Of course the probability mass function p is just the assignment of probabilities to the outcomes; $p(\omega)$ is nothing but shorthand for the probability assigned to ω. Notice, by the way, that there is a distinction between P and p: P applies to events and the probability of event A is $P(A)$, whereas p applies to single outcomes only. To be precise and a little pedantic about it, $p(\omega)$ is equal to $P(\{\omega\})$, the probability of the singleton event $\{\omega\}$. Writing $p(A)$ when you mean $P(A)$, or $P(\omega)$ when you mean $p(\omega) = P(\{\omega\})$, amounts to a grammatical error, just as if you wrote "the answer are 6" when you meant "the answer is 6." People know what you mean, but they may judge you by the way you say it.

Be sure that you understand how the examples we have seen so far fit Definition (1.2.8). In Example (1.1.1), Ω has eight members and $p(\omega)$ is $\frac{1}{8}$ for each ω. In Example (1.2.1), Ω again has eight members, but the function p is given by table (1.2.2), or, equivalently, by formula (1.2.4). In Example (1.2.5), Ω has a countable infinity of members, and the function p is described in (1.2.6): $p(\text{HH}) = \frac{1}{4}$, etc.

Notice also that the definition says nothing about what the function p should be. *Any* nonnegative function p whose values add to 1 qualifies as a mass function, and it is up to the modeler to pick one that works for the situation at hand.

The rest of this section deals with three important and commonly encountered discrete probability assignments: the Poisson probabilities, given in (1.2.11); the binomial probabilities, in (1.2.13); and the hypergeometric probabilities, in (1.2.17). We will use these, especially the Poisson and binomial, many times in examples throughout the text.

The Poisson probability assignment has a countably infinite outcome set, but the outcomes are integers instead of "words." Furthermore, we give the probabilities without justifying their appropriateness for any real experiment. But they are famous probabilities, and they apply in many different contexts.

(1.2.11) **Example: The Poisson probabilities.** Let Ω be the set $\{0, 1, 2, 3, \ldots\}$ of nonnegative integers. Let λ (lowercase lambda) be a positive number. The value of λ will depend on the situation being modeled; it is called a *parameter* of the model. Define the function p on Ω by

(1.2.12)

$$p(\omega) = \frac{\lambda^{\omega} e^{-\lambda}}{\omega!}, \quad \omega = 0, 1, 2, 3, \ldots$$

Of course these are nonnegative numbers; λ is positive, and exponentials and factorials are always positive numbers. But do they add to 1? That is, is it true that

$$\frac{\lambda^0 e^{-\lambda}}{0!} + \frac{\lambda^1 e^{-\lambda}}{1!} + \frac{\lambda^2 e^{-\lambda}}{2!} + \cdots = 1?$$

Yes, because

$$\frac{\lambda^0}{0!} + \frac{\lambda^1}{1!} + \frac{\lambda^2}{2!} + \cdots = e^{\lambda} \quad \text{for any real number } \lambda.$$

This is the familiar power series expansion of the exponential function; see (A.5.11) in Appendix A.

Now we can use (1.2.10) and (1.2.12) to find the probabilities of events. For example,

$$P\{0, 1, 2\} = \frac{\lambda^0 e^{-\lambda}}{0!} + \frac{\lambda^1 e^{-\lambda}}{1!} + \frac{\lambda^2 e^{-\lambda}}{2!} = e^{-\lambda}\left(1 + \lambda + \frac{\lambda^2}{2}\right),$$

$$P\{2, 3, 4, \ldots\} = \sum_{\omega=2}^{\infty} \frac{\lambda^{\omega} e^{-\lambda}}{\omega!} = 1 - \left(\frac{\lambda^0 e^{-\lambda}}{0!} + \frac{\lambda^1 e^{-\lambda}}{1!}\right) = 1 - e^{-\lambda} - \lambda e^{-\lambda}.$$

In the last line we replaced an infinite sum by 1 minus a finite sum; what we were doing, of course, was using the fact that $P\{2, 3, 4, \ldots\} + P\{0, 1\} = 1$. This is valid because the events $\{0, 1\}$ and $\{2, 3, 4, \ldots\}$ are disjoint and their union is Ω.

Figure 1.2a is a graph showing these probabilities for $\lambda = 5.4$. The probability $p(\omega)$ is shown as the height of the vertical line above the point ω on the x-axis. There appear to be only 15 probabilities, for $\omega = 0, 1, \ldots, 14$; but

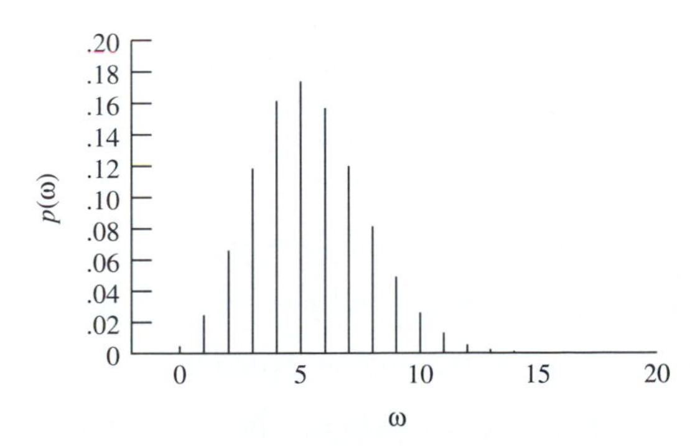

FIGURE 1.2a Poisson mass function with $\lambda = 5.4$

that is only because the others are too small to show on the graph. Regardless of the value of λ, every nonnegative integer ω is assigned a positive probability, $\lambda^{\omega}e^{-\lambda}/\omega!$. Notice that $\omega = 5$ is the most probable integer.

These probabilities were first studied extensively by the French mathematician Siméon Poisson (1781–1840). Poisson and subsequent probability modelers have used them to model the number of occurrences of some "rare phenomenon" when the phenomenon is given a large number of chances to happen. For example, it works well in some situations for the number of automobile accidents in a week in a city. Automobile accidents are rare phenomena, in the sense that only seldom do cars that are near each other actually collide. But in a large city over the course of a week, there are very many situations where cars are near each other, and in some of these situations they do collide. If the average number of automobile accidents per week in a certain city is 5.7, for example, then it is usually not a bad approximation to let $\lambda = 5.7$ and to calculate the probability of, say, four accidents in a given week as $5.7^4e^{-5.7}/4!$, which is approximately .147. (What this means, of course, is that over a long series of weeks, we expect about 14.7% of the weeks to have four accidents in them.) We will encounter Poisson probabilities throughout the text.

You might have noticed that we have not defined a random variable on this probability space. Strictly speaking we do not need one, because the outcome of the experiment itself is a number. In earlier examples, the outcomes were symbolic objects like "HTTTH," and $X(\omega)$ was some numerical feature of ω. If we wish, though, we can define the random variable X as the function on $\Omega = \{0, 1, 2, 3, \ldots\}$ defined by $X(\omega) = \omega$. That is, X is the identity function on Ω. Then we again have the economy of language that random variables provide. For example, the probability P(the number is greater than 7) can be written $P(X > 7)$, which is shorter than $P(\{8, 9, 10, \ldots\})$. ▮

(1.2.13) Example: Bernoulli trials and the binomial probabilities. In this example the important probabilities are not the ones assigned to the outcomes, but the ones that arise as the probabilities of the various possible values of the random variable X.

The experiment consists of performing n Bernoulli trials (trials with two outcomes, S and F, in which the outcome of one trial has no physical influence over that of another) with success probability p on each trial, and then letting X be the number of successes. Special cases of this situation include the following:

Example (1.1.1), where $n = 3$ and $p = \frac{1}{2}$;

Example (1.2.1), where $n = 3$ and $p = .65$;

Exercise 1 at the end of this section, where $n = 3$ and $p = \frac{1}{3}$;

Exercise 2 at the end of this section, where $n = 4$ and $p = \frac{1}{6}$;

and Exercise 3 at the end of this section, where $n = 3$ and p is arbitrary.

(If you want to proceed slowly, it might be a good idea to do Exercises 1, 2, and 3 before continuing with this example.)

First, the outcome set and the probabilities: A typical outcome ω is a string like SSFSFF . . . FSS, made up of n S's and/or F's—that is, an "n-letter word" from the "alphabet" {S, F}. There are 2^n of these [see (A.4.1) in Appendix A]. But they are not equally likely unless $p = \frac{1}{2}$; the probability assigned to any outcome ω is $p^u(1 - p)^v$, where u is the number of S's in ω and v is the number of F's. [See the first few paragraphs in (1.2.1) above for a discussion.] It is convenient, and customary, to let q denote $1 - p$ in this context; thus $p(\omega) = p^u q^v$.

Thus, if n were 7, there would be $2^7 = 128$ possible outcomes, and their probabilities would be, for example, $p(\text{SSFSSSF}) = p^5q^2$ and $p(\text{SFFSFFS}) = p^3q^4$.

Next, the random variable X: For any outcome ω, $X(\omega)$ is the number of S's in ω; thus, for example, $X(\text{SSFSSSF}) = 5$ and $X(\text{SFFSFFS}) = 3$. The possible values of X are $0, 1, 2, 3, \ldots, n$. What is the probability that $X = k$? The event "$X = k$" consists of all the n-letter words of S's and F's that happen to have k S's and $n - k$ F's. All these words have the same probability, p^kq^{n-k}. So the probability of the event "$X = k$" is p^kq^{n-k} times the number of such words, which is $\binom{n}{k}$. [See (A.4.5) in Appendix A.] Putting this together, we get the probabilities for the various possible values of X:

(1.2.14) $$P(X = k) = \binom{n}{k}p^kq^{n-k} \quad \text{for } k = 0, 1, 2, \ldots, n.$$

These are the famous *binomial probabilities*. The name "binomial" comes from the binomial theorem, used in the following calculation, which shows that the probabilities add to 1. The binomial theorem [(A.4.7) in Appendix A] says that for any numbers a and b and any positive integer n,

$$(a + b)^n = \binom{n}{0}a^0b^n + \binom{n}{1}a^1b^{n-1} + \binom{n}{2}a^2b^{n-2} + \cdots + \binom{n}{n}a^nb^0.$$

If we put in p for a and $1 - p = q$ for b, we get just what we want:

$$1^n = \binom{n}{0}p^0q^n + \binom{n}{1}p^1q^{n-1} + \binom{n}{2}p^2q^{n-2} + \cdots + \binom{n}{n}p^nq^0.$$

Figure 1.2b (page 22) is a graph of the binomial probabilities for $n = 20$ and $p = .28$. There are 21 positive probabilities, for $k = 0, 1, 2, \ldots, 20$, although only about a dozen of them appear on the graph. The rest are too small to show.

Here are three examples of the use of (1.2.14).

What is the probability that nine rolls of a fair die produce two 1's and seven non-1's? *Solution:* We have $n = 9$ Bernoulli trials with success probability $p = \frac{1}{6}$, so the probability of two successes is

$$P(X = 2) = \binom{9}{2}\left(\tfrac{1}{6}\right)^2\left(\tfrac{5}{6}\right)^7 = 36 \cdot \frac{5^7}{6^9} \doteq .2791.$$

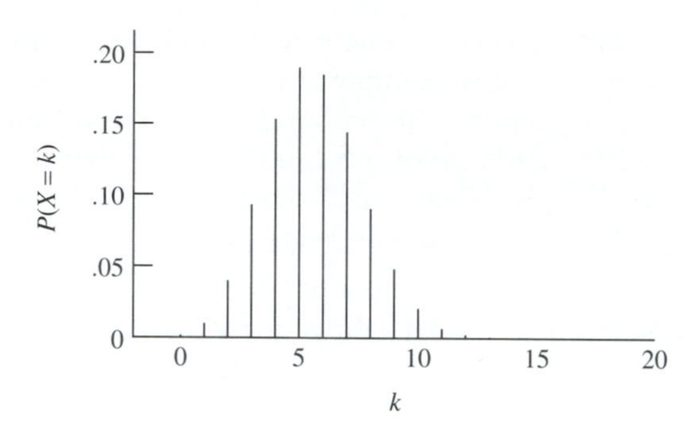

FIGURE 1.2b Binomial mass function with $n = 20$, $p = .28$

What is the probability that 36 rolls of a fair die produce exactly six 1's? *Solution:* We have $n = 36$ Bernoulli trials with success probability $p = \frac{1}{6}$; so the probability of six successes is

$$P(X = 6) = \binom{36}{6}\left(\tfrac{1}{6}\right)^6\left(\tfrac{5}{6}\right)^{30} \doteq .1759.$$

What is the probability that ten tosses of a fair coin produce at least two heads? *Solution:* $n = 10$ and $p = \frac{1}{2}$, and what we want is

$$1 - \Big(P(X = 0) + P(X = 1)\Big) = 1 - \left(\binom{10}{0}\left(\tfrac{1}{2}\right)^0\left(\tfrac{1}{2}\right)^{10} + \binom{10}{1}\left(\tfrac{1}{2}\right)^1\left(\tfrac{1}{2}\right)^9\right)$$

$$= 1 - \left(\left(\tfrac{1}{2}\right)^{10} + 10 \cdot \left(\tfrac{1}{2}\right)^{10}\right) = 1 - \frac{11}{2^{10}} \doteq .9893.$$

(In these calculations, as elsewhere, a dot over an equal sign indicates that the following number is rounded to the number of decimal digits given. In Section 1.4 we will have more to say about the art of getting numerical answers with a desired number of accurate significant digits.) ∎

(1.2.15) **When is the answer $p^k q^{n-k}$ and when is it multiplied by $\binom{n}{k}$?** The binomial probabilities $\binom{n}{k}p^k q^{n-k}$ are appropriate in questions involving the number of successes in a given number of Bernoulli trials; but there are questions that look similar, whose answers do not involve the binomial coefficient, and you must look closely at the event to see which is the correct probability.

Take seven Bernoulli trials, with success probability p, for example, and consider events involving two successes and five failures. The thing to remember is this: p^2q^5 is the probability of any *single* outcome that consists of two successes and five failures. There are $\binom{5}{2}$ such single outcomes, and each one of them has this probability. Thus, for example,

P(two successes followed by five failures) $= p^2q^5$,
because this event consists of the single outcome SSFFFFF;

P(successes on trials 3 and 6, and failures on the other trials) $= p^2q^5$,
because this event consists of the single outcome FFSFFSF;

P(two successes and five failures) $= \binom{5}{2} p^2q^5$,
because this event consists of *all* the single outcomes involving two successes and five failures.

(1.2.16) **Simulating Bernoulli trials with a computer.** Most computer-programming languages have a statement that produces "pseudo-random" numbers between 0 and 1—that is, numbers that can be modeled by the uniform probability model on [0, 1). We will discuss this in the next section. The important feature of such a random-number generator for our purposes is the following: The random number will have probability $\frac{1}{2}$ of being less than $\frac{1}{2}$, probability $\frac{2}{3}$ of being less than $\frac{2}{3}$, and so forth. For any number a between 0 and 1, the probability that the random number is less than a, is a.

Suppose we have in mind a given probability p, say $p = .372$, and we want to do Bernoulli trials with p as our success probability. We want the computer to be able to give us at random one of the two outcomes S and F. Over the long run, S should appear .372 of the time and F should appear .628 of the time. All we have to do is to ask the computer for a random number between 0 and 1. We count the outcome as a success if the number is less than .372 and a failure if not. The probability of a success will then be .372. Exercise 21 in this section asks you to write a program based on this idea.

(1.2.17) **The hypergeometric probabilities.** Here is a setup that looks like Bernoulli trials, but is not because the trials can influence each other. Consider a population of two types of objects; call the types S and F. Suppose that the number of S's in the population is σ (sigma, a Greek s) and the number of F's is ϕ (phi, a Greek f), and that the population size is $\sigma + \phi = \pi$ (pi, a Greek p).

Now suppose we sample n objects from the population at random, one at a time, **without replacement,** and let X equal the number of S's in the sample. The possible values of X are $0, 1, 2, \ldots, n$. What are their probabilities? [Note that if we sampled **with replacement,** and each draw were made at random, then the successive draws would have no influence on one another, and we would have Bernoulli trials. The probability that the sample contained k S's and $n - k$ F's would be given by (1.2.14).]

To find the probabilities for X we consider the set Ω of all outcomes, as we did for Bernoulli trials. Now the outcomes can be regarded as subsets of the population; even though we draw the objects in order, what matters is the set of objects we get, not the order in which we got them. So Ω is the set of all subsets of size n from the population. Since the population has π members, the number of subsets of size n is $\binom{\pi}{n}$. If the sampling is done at random, without bias, then all of these are equally likely.

[To make things concrete, suppose that $\sigma = 2$ and $\phi = 3$, so that $\pi = 5$, and that the sample size n is 2. The five-member population can be thought of as the set $\{S_1, S_2, F_1, F_2, F_3\}$, and then the outcome set Ω consists of all ten subsets of size 2:

$$\Omega = \{\{S_1, S_2\}, \{S_1, F_1\}, \{S_1, F_2\}, \{S_1, F_3\}, \{S_2, F_1\}, \{S_2, F_2\}, \{S_2, F_3\}, \{F_1, F_2\}, \{F_1, F_3\}, \{F_2, F_3\}\}.$$

There are ten possible samples because there are $\binom{5}{2} = 10$ subsets of size 2 in a set of size 5.]

Now what is the event $(X = k)$? It consists of all the subsets of size n that happen to contain k S's and $n - k$ F's. The number of these is just the number of subsets of size k from the S's in the population, *times* the number of subsets of size $n - k$ from the F's in the population (by the so-called "multiplication rule" from elementary counting). This is just $\binom{\sigma}{k}\binom{\phi}{n-k}$. So we see that

(1.2.18)
$$P(X = k) = \frac{\binom{\sigma}{k}\binom{\phi}{n-k}}{\binom{\pi}{n}} \quad \text{for } k = 0, 1, 2, \ldots, n.$$

We will have a little more to say about these probabilities in Section 7.5. They are called "hypergeometric" for technical reasons having to do with their appearance as coefficients in a class of functions bearing that name; as far as probability is concerned, there is little that is geometric in the hypergeometric distribution.

There may appear to be a technical difficulty with the formula (1.2.18) if the number of S's in the population, or the number of F's, is small. For example, if σ is less than n, then there are fewer than n S's in the population, and so the sample cannot contain more than σ S's. The possible values of X will then go only up to σ, not to n. However, the binomial coefficient notation takes care of this nicely for us, if we just agree to the convention that $\binom{\sigma}{k} = 0$ when σ and k are positive integers and k is greater than σ. A similar thing happens if there are fewer than n F's in the population; in this case, X cannot equal k if k is too small, because then $n - k$ will exceed ϕ. But again the notation saves the formula: $\binom{\phi}{n-k} = 0$ by convention if $n - k$ exceeds ϕ.

(1.2.19) **Example.** A hat contains 12 tickets; seven are blue and five are white. Three tickets are drawn at random, one after the other. Let us find the probability that two are blue and one is white, under two different conditions: first, *sampling without replacement,* where each ticket is kept out of the hat after it is drawn and subsequent tickets are drawn from what is left; and second, *sampling with replacement,* where each ticket is replaced after its color is noted, and the tickets in the hat are mixed again, before the next one is drawn. In both cases we consider a blue ticket as an S and a white ticket as an F.

For sampling without replacement we have $\sigma = 7$, $\phi = 5$, and $\pi = 12$, and our sample size is $n = 3$; we want $P(X = 2)$ where X is the number of S's. According to (1.2.18),

$$P(X = 2) = \frac{\binom{7}{2}\binom{5}{1}}{\binom{12}{3}} \doteq .4773 \quad \text{for sampling without replacement.}$$

For sampling with replacement, we have $n = 3$ Bernoulli trials and the success probability is $p = \frac{7}{12}$; so

$$P(X = 2) = \binom{3}{2}\left(\tfrac{7}{12}\right)^2\left(\tfrac{5}{12}\right)^1 \doteq .4253 \quad \text{for sampling with replacement.}$$

You may find it interesting or instrutive to compare the two answers above by looking at them as follows:

$$\binom{3}{2}\left(\tfrac{7}{12}\right)^2\left(\tfrac{5}{12}\right)^1 = \frac{3 \cdot 2}{2 \cdot 1} \cdot \frac{7 \cdot 7}{12 \cdot 12} \cdot \frac{5}{12},$$

and

$$\frac{\binom{7}{2}\binom{5}{1}}{\binom{12}{3}} = \frac{7 \cdot 6}{2 \cdot 1} \cdot \frac{5}{1} \cdot \frac{3 \cdot 2 \cdot 1}{12 \cdot 11 \cdot 10}, \text{ which can be rearranged as } \frac{3 \cdot 2}{2 \cdot 1} \cdot \frac{7 \cdot 6}{12 \cdot 11} \cdot \frac{5}{10}.$$

You may also want to ponder the fact that the event "two blue and one white" is more probable when the sampling is done without replacement. Surely not *all* events can be more probable when the sampling is without replacement; which events are more probable, and which are less? If you are interested, just work out the probabilities of the four events "no blue, three white," "one blue, two white," and so on under both sampling schemes; you will then be able to see what becomes more likely and why. ▮

EXERCISES FOR **SECTION 1.2**

1. Suppose we roll a fair die three times and count each roll a success if a 1 or a 2 comes up and a failure otherwise. Let X be the number of successes in the three rolls.

a. Give the outcome set and the probability mass function.

b. List the possible values of X and their probabilities.

c. Find the probability that X is at least 1.

2. **Four Bernoulli trials.** Repeat Exercise 1 but with four rolls of the die instead of three, and count only a 1 as a success. In part a there are 16 outcomes to list; in part c you are to find the probability of at least one success (one 1) in four rolls.

3. Exercise 1 involves three Bernoulli trials in which the probability of success on each trial is one-third. Do it again for the case where the success probability is some arbitrary number p between 0 and 1.

   Unfortunately, the symbol p has two different uses here: the mass function $p(\omega)$ and the success probability p. For example, $p(\text{SFS}) = p^2q$, where, as always in Bernoulli trials, $q = 1 - p$. Such ambiguities occur frequently in mathematical writing. When you encounter one, the best thing to do is to try to gain some understanding by thinking about the difference between the two uses of the same symbol.

4. For the model of Example (1.2.5), where a fair coin is tossed until the second head comes up, find the probability of each of the following events.
   a. More than two tails come up before the second head. [That is, more than four tosses are needed to get the second head.]
   b. More than three tails come up.
   c. More than four tails come up.
   d. The number of tails that come up is odd. [You can sum the series, or you can use one of the probabilities that were found in Example (1.2.5).]
   e. The first toss comes up heads. [You know the answer intuitively without having to use this model, but it would be nice to identify all the outcomes in this event, sum the series, and get the same answer.]

   [Note that in doing parts a, b, c, and d, it is easier to use (1.2.7) than (1.2.6).]

5. **Waiting for the first head.** Suppose we repeatedly toss a fair coin until the *first* head shows, rather than the second.
   a. List the possible outcomes and assign probabilities to them. Be sure to check that the probabilities you assigned add to 1.
   b. Find the probability that more than two tosses are needed; more than three tosses; more than k tosses (for an arbitrary positive integer k).
   c. Find the probability that the first head shows on an odd-numbered toss.

6. Repeat Exercise 5 for the experiment of waiting for the first success, where success is the roll of a 1 with a fair die.

7. Repeat Exercise 5 for Bernoulli trials with two outcomes, S and F, where the probability of S on each trial is an arbitrary number p between 0 and 1. (Again the symbol p is being used for two different things here; the mass function $p(\omega)$, and the number p.) These probabilities are called the *geometric probabilities;* we will encounter them often in examples.

8. Suppose the number of automobile accidents in a randomly chosen week in a certain city follows the Poisson distribution with $\lambda = 3$; that is, the probability of ω accidents is given by (1.2.12) with 3 in place of λ. For a randomly chosen week:
   a. Find the probability that there are no accidents.
   b. Find the probability that there are fewer than three accidents.

c. Find the probability that there are exactly three accidents.
d. Find the probability that there are six accidents.
e. Find the probability that there are more than six accidents.
f. What is the most likely number of accidents? What is its probability? [*Hint:* Compute the first several probabilities—of no accidents, one accident, and so on. After five or six you will know which number of accidents is most likely.]

9. For five Bernoulli trials with success probability .47, find the probability of each of the following events:
a. Two successes and three failures
b. Two successes, followed by three failures
c. Five successes
d. At least one success

10. Let $\Omega = \{1, 2, 3, 4, 5, 6\}$ and define the probability mass function p by $p(\omega) = \omega^2/91$ for $\omega \in \Omega$. Find the probabilities of the following events:
a. The outcome is an even number.
b. The outcome is not greater than 2.
c. The outcome is a perfect square.

11. Nine slips of paper are put into a hat; four of the slips are marked with a dot and the other five are blank. Slips are drawn, one after the other, and after each slip is drawn it is replaced and the slips are mixed before the next draw. A success is counted each time a slip with a dot is drawn. Ten draws are made. Find the probability of each of the following events:
a. Three successes
b. Three successes, which appear on the last three draws
c. No successes
d. At least two failures

12. **Don't count on equal numbers of heads and tails with a fair coin.** A fair coin is tossed repeatedly. Find the probabilities of the following. (At least, find the first two and look at the answers for the others.)
a. In six tosses, there are three heads and three tails.
b. In ten tosses, there are five heads and five tails.
c. In 20 tosses, there are ten heads and ten tails.
d. In 30 tosses, there are 15 heads and 15 tails.

Note that three heads in six tosses, five heads in ten tosses, ten heads in 20 tosses, etc., are what most people expect, but they are not likely—and they get less likely as the number of tosses increases. The error is in expecting *exactly* the right number of heads. If you were to work out the probability of being *within ten percent* of the expected number of heads—for example, within one of ten heads in 20 tosses, within five of 50 heads in 100 tosses, within 25 of 250 heads in 500 tosses, etc.—you would find the probabilities increasing as the number of tosses increases. Later we will see a way to find these probabilities approximately, without using a computer, and we will learn that they in fact tend to 1 as the number of tosses increases. This fact is a form of what probabilists call the *law of large numbers.*

13. **Samuel Pepys's problem.** A fair die is rolled repeatedly; we are interested in the number of 6's. Which is most likely: at least one 6 in six rolls, at least two 6's in 12 rolls, or at least three 6's in 18 rolls?

Pepys wrote Isaac Newton for the answer to this question; apparently, it seemed to Pepys that the three probabilities should be the same. According to Frederick Mosteller, in his entertaining and instructive book *Fifty Challenging Problems in Probability* (see the bibliography for information), this was Newton's only personal involvement with probability theory. A more famous problem, called the "Chevalier de Méré's paradox," is very similar to this one; see Exercise 8 of Section 7.2.

14. Which of the following functions are probability mass functions? For those that are not, find (if possible) a constant a so that $ap(\omega)$ is a probability mass function.

a. $p(\omega) = \frac{\omega^2}{55}, \quad \omega = 1, 2, 3, 4, 5$
b. $p(\omega) = \left(\frac{1}{3}\right)\left(\frac{2}{3}\right)^{\omega}, \quad \omega = 3, 4, 5, 6, \ldots$
c. $p(\omega) = 1$ for each ω in a nine-member set Ω
d. $p(\omega) = 1$ for each ω in a countably infinite set Ω
e. $p(\omega) = \omega, \quad \omega = 1, 2, 3, 4, \ldots, N$
f. $p(\omega) = \left(\frac{1}{4}\right)^{\omega}, \quad \omega = 0, 1, 2, 3, 4, \ldots$
g. $p(\omega) = \frac{1}{\omega}, \quad \omega = 1, 2, 3, 4, \ldots$
h. $p(\omega) = \frac{1}{3}(\omega - 2), \quad \omega = 0, 1, 2, 3, 4, 5$

15. Consider Bernoulli trials with two outcomes, S and F, with $P(\mathrm{S}) = p$ and $P(\mathrm{F}) = q = 1 - p$ on each trial. Suppose we do repeated trials until the second S occurs. [This is exactly like Example (1.2.5), except that the probabilities are p and q instead of both being $\frac{1}{2}$.] Find the probability that there are n F's preceding the second S. [The n F's are to include all F's preceding the first S as well as those between the first and the second.]

[a]16. Two Bernoulli trials with success probability p (and failure probability $q = 1 - p$) are performed. A is the event that the outcomes of the two trials are the same. Show that $P(A)$ is always at least $\frac{1}{2}$.

17. Find the largest of the Poisson probabilities, and give the corresponding value(s) of ω, for each of the following values of λ. Then make a conjecture as to the most likely ω for an arbitrary λ. [You can do it by simply examining the first few probabilities. In part b, for example, $p(0) \doteq .1003$ and $p(1) \doteq .2306$; and if you compute these along with $p(2)$, $p(3)$, and $p(4)$, it will become clear to you which one is largest. Part d was already done as Exercise 8f above.]

a. $\lambda = 2$ **b.** $\lambda = 2.3$ **c.** $\lambda = 2.7$
d. $\lambda = 3$ **e.** $\lambda = 3.4$

18. A club consists of 35 members, 15 men and 20 women. A committee of ten is selected at random. What is the probability that the committee has four men and six women?

19. In a small town of 1,245 people a small survey is taken; seven people are sampled as to their political preference. As it happens (and unknown to the pollsters), only 20 people in the town favor candidate A and the other 1,225 favor candidate B. What is the probability that the sample contains at most one supporter of A,

a. If the sampling is without replacement?

b. If the sampling is with replacement, so that a person has a chance of being included more than once in the sample?

The answers are very nearly equal, showing what you might guess: If the population is large compared to the sample, there is little difference between sampling with and without replacement.

20. Let probabilities of subsets of $\Omega = \mathbb{N}$ be given by the mass function $p(k) = Cke^{-3k}$ ($k = 1, 2, 3, \ldots$), where C is the constant $e^3(1 - e^{-3})^2$.

a. Find the probability of the event $\{1, 2, 3\}$.

b. Find the probability of the event $\{2, 3, 4, \ldots\}$.

c. Check that the probabilities really add to 1. [*Hint:* Ignore the constant C for the moment and show that the terms ke^{-3k} add to its reciprocal. To do this, replace the 3 by an x and notice that the sum of ke^{-kx} is the derivative of another series that you can evaluate as a geometric series (see (A.5.8) in Appendix A). Differentiate the sum of that series and then put 3 back in for x.]

21. (*Computer exercise*) In (1.2.16) we described how Bernoulli trials with a given success probability, say $p = .372$, can be simulated on a computer. If you can use a programming language that has a random number generator, this exercise will provide a little experience with running and looking at the results of simulation experiments, as well as showing the relation between binomial probabilities and relative frequencies of observed outcomes.

a. Write a program that makes ten Bernoulli trials with success probability $p = .372$, and prints a string such as "SFFSFFFSFF" that shows the results. Run it several times. Notice that you get a different result each time, but that there are usually more F's than S's.

b. Modify it so that the probability of success is .761 instead of .372. Run it again several times and notice that there tend to be more S's.

c. Modify it so that instead of printing "S" or "F" after each trial, it simply keeps track of the number of S's and prints that number at the end. Run it several times and notice the frequencies of 6's, 7's, 8's, and 9's as compared with those of lower numbers.

d. Modify the program in part c so that it repeats the ten-trial experiment 100 times, printing the number of S's after each repetition. (In other words, put your ten-trial loop inside a 100-trial loop.) Look at the 100 numbers you get. What seem to be the most probable numbers?

e. Modify the program in part d so that instead of printing 100 numbers, it keeps track of 11 frequencies: the number of 0's among the 100, the number of 1's, and so on up to the number of 10's. Make it print only these 11 numbers at the end of the 100 experiments.

f. The program in part e prints out 11 frequencies. Modify it to print relative frequencies.

g. Now modify the program in part f so that instead of 100 repetitions of the ten-trial experiment, it does some large number like 5,000 or 10,000.

h. Using a calculator, compute the 11 probabilities

$$p(k) = \binom{10}{k}(.372)^k(.628)^{10-k} \quad \text{for } k = 0, 1, 2, \ldots, 10.$$

Compare them with the 11 relative frequencies obtained in part g.

ANSWERS

1. **b.** Values 3, 2, 1, 0; probs. $\frac{1}{27}, \frac{6}{27}, \frac{12}{27}, \frac{8}{27}$ **c.** $\frac{19}{27}$
2. **b.** Values 4, 3, 2, 1, 0; probs. $\frac{1}{1{,}296}, \frac{20}{1{,}296}, \frac{150}{1{,}296}, \frac{500}{1{,}296}, \frac{625}{1{,}296}$
3. **b.** Values 3, 2, 1, 0; probs. p^3, $3p^2q$, $3pq^2$, q^3 **c.** $1 - q^3$
4. **a.** $\frac{20}{64}$ **b.** $\frac{12}{64}$ **c.** $\frac{7}{64}$ **d.** $\frac{4}{9}$ **e.** $\frac{1}{2}$
5. **b.** $\left(\frac{1}{2}\right)^k$ **c.** $\frac{2}{3}$
6. **b.** $\left(\frac{5}{6}\right)^k$ **c.** $\frac{6}{11}$
7. **b.** q^k **c.** $\dfrac{p}{1 - q^2}$
8. **a.** .049787 **b.** .423190 **c.** .224042 **d.** .050409 **e.** .033509 **f.** $\omega = 2$ and $\omega = 3$; .224042
9. **a.** .3289 **b.** .0329 **c.** .0229 **d.** .9582
10. **a.** $\frac{56}{91}$ **b.** $\frac{5}{91}$ **c.** $\frac{17}{91}$
11. **a.** .1721 **b.** .0014 **c.** .0028 **d.** .9959
12. **a.** .3125 **b.** .2461 **c.** .1762 **d.** .1445
13. At least 1 in 6: .665; at least 2 in 12: .619; at least 3 in 18: .597
14. **a.** Yes **b.** $a = \frac{27}{8}$ **c.** $a = \frac{1}{9}$ **d.** Impossible **e.** $a = \dfrac{2}{N(N + 1)}$ **f.** $a = \frac{3}{4}$ **g.** Impossible **h.** Impossible
15. $(n + 1)p^2q^n$ for $n = 0, 1, 2, \ldots$
16. $p^2 + q^2 \geq 2pq$ because $(p - q)^2 \geq 0$.
17. **a.** $\omega = 1$ and $\omega = 2$ **b.** $\omega = 2$ **c.** $\omega = 2$ **d.** $\omega = 2$ and $\omega = 3$ **e.** $\omega = 3$
18. .2882
19. **a.** .9951 **b.** .9949
20. **a.** .9995 **b.** .0971

1.3 ABSOLUTELY CONTINUOUS PROBABILITY SPACES

In a discrete probability space we assign probabilities to the individual members of the outcome set, and from them we calculate the probabilities of events by adding probabilities of outcomes. This is possible because the outcome sets are finite or countably infinite. In the outcome sets of this section, it is not possible, because there is an uncountable infinity of outcomes. (See Section A.3 of

Appendix A for a brief description of countable and uncountable sets.) Uncountably infinite outcome sets do not fit the definition of discrete probability spaces, and we will define a second type of space, called an *absolutely continuous* probability space, to fit some of them. We will say a word about the origin of the awkward term "absolutely continuous" when we make the definition, at (1.3.7).

(1.3.1) **Example: The uniform distribution on [0, 1).** We want to model the experiment of choosing a number at random from the interval [0, 1), in a "uniform" way, so that no part of the interval is more likely to contain the number than any other part of the same size. For example, the number should have the same probability of falling in the interval [0, .2) as in [.7, .9), because those two intervals have the same length, and if they did not have the same probability then the choice would favor one part of the interval over another.

This kind of "uniform" choice is what a computer's random number generator tries to do when we ask it for a random number. We might also do this experiment by taking the kind of spinner used in board games and marking numbers around its circumference so that 0 is at the right, .25 at the top, .5 on the left, .75 at the bottom, etc. (See Figure 1.3a.) If the spinner is balanced well, it ought to come to rest on a uniformly distributed random point on the circle.

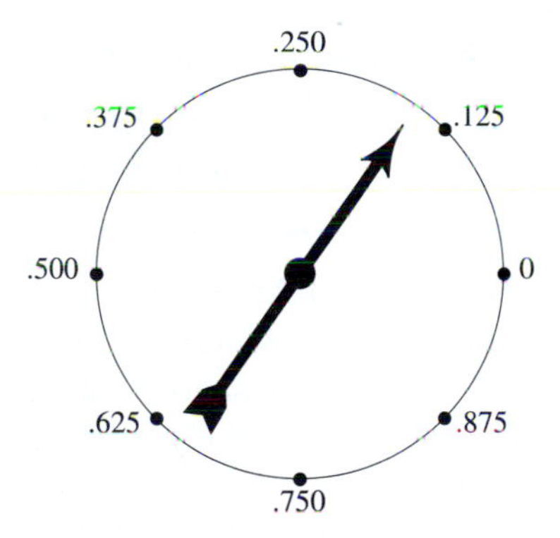

FIGURE 1.3a **A spinner intended to produce a uniform random number from [0, 1)**

It is clear that the outcome set should be $\Omega = [0, 1)$. And as in all probability spaces, events are subsets of Ω, this time in an obvious way:

The event "the number chosen is less than $\frac{1}{3}$" is the subset $\left[0, \frac{1}{3}\right)$;

the event "the number chosen is not more than $\frac{1}{3}$" is the subset $\left[0, \frac{1}{3}\right]$;

the event "the number is not in the open middle third" is the subset $\left[0, \frac{1}{3}\right] \cup \left[\frac{2}{3}, 1\right)$;

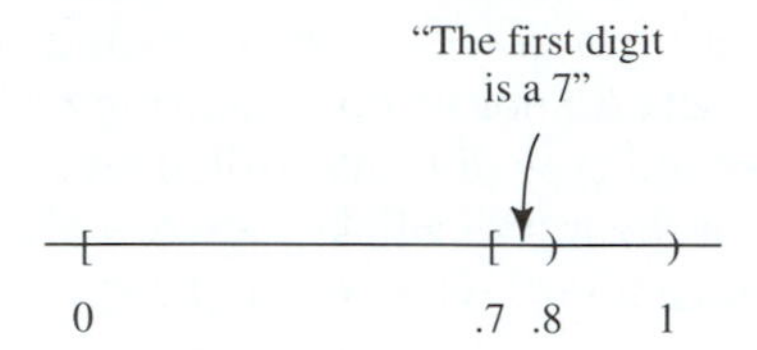

FIGURE 1.3b An event in the outcome set $\Omega = [0, 1)$

the event "the first digit of the number is a 7" is $[.7, .8)$ (see Figure 1.3b);

the event "the second digit is a 7" is $[.07, .08) \cup [.17, .18) \cup [.27, .28) \cup \cdots \cup [.97, .98)$; etc.

As for the probabilities of the events, as we remarked above, it is not possible to assign probabilities to all the individual outcomes and then say that the probability of an event is the sum of the probabilities of its members. The reason is that we have no way of adding a set of numbers unless the size of the set is finite or countably infinite. (It may seem to you that the integral is a way of extending addition to an uncountable set of numbers—and in a sense it is, but not in a sense that works here. Later in the section we will see what integrals *can* do in the way of determining probabilities.)

So how can we translate the notion of "uniformity" into an assumption about the probabilities of events? Think first of the three intervals $\left[0, \frac{1}{3}\right)$, $\left[\frac{1}{3}, \frac{2}{3}\right)$, and $\left[\frac{2}{3}, 1\right)$. Their probabilities must add to 1, and uniformity requires that they all have the same probability; so each of them should have probability $\frac{1}{3}$. Indeed, all intervals of the form $[a, b)$ of length $\frac{1}{3}$ should have probability $\frac{1}{3}$, because any two intervals of the same length should have the same probability. Otherwise the model would not be "uniform;" some parts of $[0, 1)$ would be more probable than other parts of the same size. Similar reasoning leads to a simple assumption about the probabilities of intervals:

In the model of this example, the probability of any interval of the form $[a, b)$ is its length.

Intervals of the form $[a, b)$ are called *half-open intervals,* although "left-closed, right-open" would tell more of the whole truth about them. In any case, the intervals we are dealing with contain their left endpoints but not their right ones. You may wonder why we are restricting our attention to half-open intervals. Presently we will know the probabilities of *all* intervals, but the argument in the previous paragraph worked because a union of adjacent half-open intervals is another half-open interval.

Is the assumption that the probability of any half-open interval is its length enough to enable us to find the probabilities of other events? Certainly if an event is the union of a finite or countably infinite collection of disjoint half-open intervals, then its probability ought to be the sum of the probabilities of those intervals. This should be true of probabilities because it is true of relative frequencies: The frequency of any union of disjoint events is the sum of the frequencies of the events. Thus, for example, we can find the probability that a uniform random number has 7 as its second digit:

$$P(\text{the second digit is a } 7) = \\ P\big([.07, .08) \cup [.17, .18) \cup [.27, .28) \cup \cdots \cup [.97, .98)\big) = 10 \cdot .01 = .1;$$

the second digit has a $\frac{1}{10}$ chance of being a 7, just as the first digit has.

The statement in the previous paragraph is an instance of a fundamental property of probabilities, which they must have because relative frequencies do, namely:

> ***If an event A is the union of finitely or countably many disjoint events A_1, A_2, A_3, . . . , then P(A) must equal***
> $$P(A_1) + P(A_2) + P(A_3) + \cdots.$$

This is part of the definition of a general probability space, which we will see in Section 1.5, and again in Section 2.1. The only other requirements on probabilities in the general definition are that they must be nonnegative, because relative frequencies are; and that the total probability of all outcomes, which is $P(\Omega)$, must be 1, because 1 is always the sum of the relative frequencies of all outcomes.

What about open intervals? We have decided that $P[a, b) = b - a$, but does this enable us to find $P(a, b)$? (In other words, what does the endpoint contribute to the probability?)

By the time we get to Example (2.5.6), we will be able to get the answer quickly. For now we have to use a little cleverness, but the exercise is worthwhile because it gives us a chance to think about the implications of our model. Notice that any open interval can be expressed as the union of a countably infinite set of disjoint half-open intervals. Take the open interval (.5, 1) as an example. It is equal to the following union (see Figure 1.3c):

$$(.5, 1) = \left[\tfrac{3}{4}, 1\right) \cup \left[\tfrac{5}{8}, \tfrac{3}{4}\right) \cup \left[\tfrac{9}{16}, \tfrac{5}{8}\right) \cup \left[\tfrac{17}{32}, \tfrac{9}{16}\right) \cup \cdots.$$

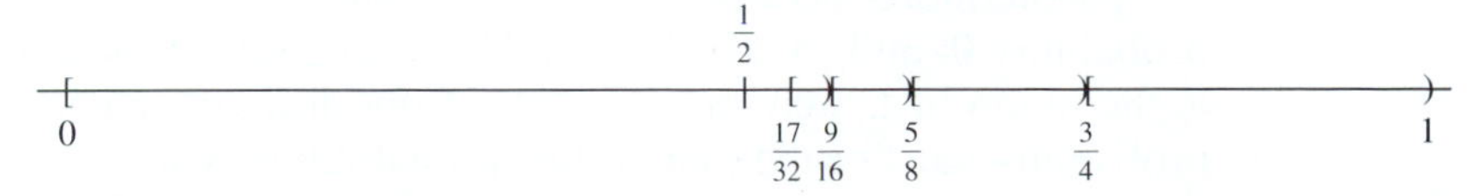

FIGURE 1.3c Interval $\left(\frac{1}{2}, 1\right)$ is a union of disjoint intervals of the form $[a, b)$.

(Think about it: Every number greater than $\frac{1}{2}$ and less than 1 is in one of these intervals; the closer the number is to $\frac{1}{2}$, the farther out in the sequence of intervals it is. But $\frac{1}{2}$ itself is not included in any of them.)

Therefore, $P(.5, 1)$ must be the sum of the probabilities of these intervals, that is,

$$\frac{1}{4} + \frac{1}{8} + \frac{1}{16} + \frac{1}{32} + \cdots = \frac{1}{2}.$$

We see that the interval $(.5, 1)$ has the same probability as the interval $[.5, 1)$. Most people find this disturbing, because it implies that the "singleton" event $\{.5\}$ has probability 0. Similar reasoning shows that *all* singletons have probability 0. That is, whenever you do the experiment, whatever outcome you get has probability 0 according to this model. How can such a model be of any use? Who will take us seriously?

One answer is that we do not interpret "probability 0" as meaning "impossible." Remember that we are interpreting probability as *expected long-run relative frequency,* and that an event with probability 0 is simply one that we expect to occur a very small proportion of the time if we do the experiment a very large number of times. Indeed, if we ask a computer's random number generator for a million random numbers, it is not likely that .5 will be one of them. Many of them will be close to .5, but probably none will be exactly .500000 . . . . And if one *is* exactly equal to .500000 . . . , we will think that there is something wrong with the generator if we see the same number again soon.

Another answer is that in a "uniform" model with infinitely many outcomes, it is impossible for any single outcome to have a probability other than 0. The reason is that in a uniform model all outcomes must have the same probability, and if that probability were positive, then any countably infinite subset of Ω would have infinite probability. But no subset can have probability exceeding 1. So all singletons must have probability 0 in a uniform model with infinitely many outcomes.

Notice also that because all singletons have probability 0, all intervals with the same endpoints have the same probability, regardless of whether they contain the endpoints or not. That is,

In this model, for any a and b in $[0, 1)$ with $a < b$, the four intervals $[a, b]$, $[a, b)$, $(a, b]$, and (a, b) all have probability $b - a$.

Furthermore, because all singletons have probability 0, all finite sets have probability 0, and so do all countably infinite sets. This is so because such a set is the union of finitely or countably many disjoint singletons, and therefore its probability must be the sum of their probabilities, which is a sum of 0's. You may find it even more disturbing that countably infinite sets have probability 0 than that singletons do. After all, the set of rational numbers is countable; so this model tells us that when we choose a number uniformly from $[0, 1)$, the probability that it is rational is 0. This is just another example of the failure of the model to fit reality perfectly. We will discuss this failure a little more at the end of the example.

Where does all this take us? So far we have decided that to model the uniform choice of a random number in $[0, 1)$, we need the following: Ω is the interval $[0, 1)$; the probability of any interval is its length; and the probability of any event that is the union of finitely or countably many disjoint intervals is the sum of the probabilities of those intervals.

But what about subsets that are neither intervals nor unions of finitely or countably many disjoint intervals? (It may not be obvious that there are such subsets, but the set of irrational numbers is one.) Can we find their probabilities from the assumptions that we have made so far? The answer involves both good news and bad news. We will discuss it more fully in (1.3.2) and again in Section 3.1, although we will not be able to prove it in this text; the proof requires measure theory, a subject usually studied at the graduate level.

> *Good news:* If A is any set that can be made starting from intervals by using finitely or countably many set operations of union, intersection, and complement, then the probability of A can be found if the probabilities of the intervals are known.
>
> *Bad news:* Not all sets can be made in this way. Furthermore, it is *impossible* to assign probabilities to all sets in a consistent way and to satisfy the additional requirement that the probabilities of the intervals be their lengths.
>
> *Good* (but strange) *news:* Although there are sets that cannot be given probabilities, no one can exhibit such a set—that is, it is known that there are such sets, but there is no way to construct an example of one. Therefore, there is no point in worrying about the probabilities of these sets, and we simply do not call them events.

Thus, ***all events are subsets of Ω, but in this probability space*** (***and in all the common uncountable probability spaces***) ***not all subsets of Ω are events.*** The sets that we do include as events are called the "Borel sets," and we will describe them more fully in (1.3.2) below.

Notice that the recipe for computing probabilities in this model is very different from those in the previous sections, which had finite or countably infinite outcome sets. There the recipe for the probability of an event was to add the probabilities we had previously assigned to the individual members of the event—that is, to add the values of the probability mass function $p(\omega)$ for all the

ω's in the event. Here, the individual members each have probability 0 and the only sets that qualify as events are those that can be made by set operations starting from intervals. The probabilities of the intervals are given, and the probabilities of other events need to be calculated from them in some way.

In this example, as in Example (1.2.11), we did not define a random variable on the probability space. If we are interested only in studying the value of the number chosen at random, there is no need to do so, because the outcomes in Ω are themselves numbers. As in Example (1.2.11), though, it might be convenient to define $X(\omega) = \omega$ so that we can use language such as "$P\left(X < \frac{1}{2}\right)$," instead of "$P\left(\text{number chosen is less than } \frac{1}{2}\right)$" or "$P\left\{\omega: \omega < \frac{1}{2}\right\}$."

It is worth noting, though, that there will be times when other random variables on this probability space are of interest. For example, the random variables $Y(\omega) = \tan(\pi(\omega - .5))$ and $Z(\omega) = -\ln(\omega)$ have interesting properties, which we will discuss in later chapters.

Summary of Example (1.3.1): $\Omega = [0, 1)$; the events are the Borel subsets of Ω, described in (1.3.2) below; the probability of any interval is its length; and this is enough information to find the probabilities of all events. (We will see in Section 1.4 and in Chapter 2 just how some of these probabilities can be found.)

A final comment of possible interest here concerns the difference between the idealized experiment modeled by the probability space of this example and the real experiment of asking a computer's random number generator for a number. First, the computer cannot give *any* number in [0, 1); it can give only rational numbers whose denominators are 2^{15} or 2^{31} or some other number, depending on the machine and the programming language. Second, it does not really give anything random at all; what it gives is just the next number in a sequence of numbers that is so long and so mixed up that the long-run relative frequencies of most events are close to what the model gives as their probabilities. It may seem surprising, given these differences, that the computer's random number generator and the uniform model can be said to resemble each other at all. Yet the differences, in most cases, are unnoticeable to the computer user, provided that the computer's generator is well designed.

Indeed, it is interesting to consider the question of whether there are *any* experiments that can be done in the real world for which this model is an exact match. It is worth noting, for example, that no one has ever directly measured any quantity in any system of units and found it to be an irrational number of units; any real-world measurement has only a finite number of digits and, thus, is given as a rational number. The set of rational numbers is a countable set, and so no model with uncountably many outcomes can accurately reflect any experiment involving the measurement of some physical quantity.

However, these questions are in the realm of philosophy and physics rather than of mathematics or statistics. The model we have used, even though it may not fit exactly the real situations we use it for, is nevertheless easier to use than models that do, and it fits closely enough to be of use in many real situations. This should not surprise us; after all, Euclidean geometry does not exactly fit measurements of lengths and angles in the real world, both because of measurement errors and because of the curvature of space–time. Yet we happily use it to find areas, circumferences, volumes, distances, and so on. We would be foolish to use anything else—unless, of course, we are measuring objects that are very large, very small, or moving very fast. ∎

(1.3.2) **The Borel sets and σ-fields.** It is very convenient to have a term for the sets that are allowed as events in absolutely continuous probability spaces. We call them the Borel sets, and this subsection is devoted to a brief discussion of what they are. It contains informal descriptions of some abstract ideas that are covered in measure theory; if you get bogged down reading them, just turn to the two facts about the Borel sets listed at the end.

What sets can be allowed as events—that is, what sets can have probabilities? We need to have a probability for any interval, and also for any set that is a union of (finitely many) intervals, and other such sets. But as we said, for technical reasons not every set can be assigned a probability. So what gets included and what does not?

The technical reasons for not being able to assign probabilities to all subsets of an interval consist essentially of two things that go wrong if we try. First, assigning probabilities to the subintervals does not uniquely determine the probabilities of all the other subsets. We would have to describe the assignment in some way for other sets; yet it is impossible even to describe all of these sets, not to mention deciding on their probabilities. Second, even if we could assign probabilities to the rest of the subsets in some way, our assignment would be inconsistent, in that there will be some countable collection of disjoint sets whose union has a probability that is not the sum of the individual sets' probabilities. (We cannot prove here that these things must go wrong, but they will. In the small print below are some comments about the precise statement of what goes wrong.)

Think for a moment about what conditions the family of events in a probability model should satisfy, other than that all the intervals should be included. It might seem reasonable to require that any set constructed as a union, intersection, or complement of other events should also be an event. After all, the union of a collection of events is just the event that one or more of them occurs, and, if we can talk about their probabilities, we ought to be able to talk about the probability that at least one of them occurs. Similarly, the intersection of a collection of events is just the event that they all occur; that too ought to have a probability. Also, the complement of an event is just the event that it does not occur, which certainly deserves a probability if the original event does.

In fact, these requirements are too strong to be reasonable; they are more than we actually need to form a usable family of events, a family about which useful theorems can be proved. What is essential is only that the union and intersection of any *countable* collection of events should themselves be events, and also that the complement of an event should be an event. And then there is a pleasant surprise, proved in courses in measure theory: If we take just enough sets as events to guarantee this essential requirement, then (i) the probabilities of all events are determined once the probabilities of intervals have been specified; and (ii) any set that can be described explicitly is an event.

Accordingly, the definition is the following:

> The ***family of Borel sets*** is the smallest family of sets that contains the intervals and is closed under complementation and under the formation of unions and intersections of countably many sets.
>
> Those Borel sets that are subsets of a given interval (or Borel set) Ω are called the ***Borel subsets*** of Ω.

A family of sets that is closed under these operations—complementation, and countable unions and intersections—is called a ***σ-field.*** (The letter σ is a lowercase Greek sigma; it is commonly used to symbolize the idea of countably many operations.) So we can shorten the definition:

> The ***Borel sets*** are the members of the smallest σ-field that contains all the intervals.

The concept of the "smallest" family of sets with certain properties—in this case, the properties are that the family be a σ-field and contain the intervals—refers to the intersection of all families with these properties. The use of the concept is not justified unless there is a theorem saying that such an intersection itself is a family with these properties. There is such a theorem in this case.

We should ask another question at this point: How do we know that there are any non-Borel sets? Perhaps requiring that the family be closed under countable unions and intersections will force us to include all sets. That this is not so is a theorem in measure theory; see the small print below for an additional comment on it.

However, we will not need to worry about any of these technical details. For us it is enough to remember that the following two facts can be proved about the Borel sets:

1. All commonly encountered sets—intervals of all kinds, unions of intervals, finite sets, countable sets, etc.—are Borel sets. In fact, any set that can be constructed or described explicitly is a Borel set, along with many other sets that cannot be described. It is perhaps useful to remember the Borel sets in this way:

   > ***The Borel sets are those sets that can be made from intervals using countably many set operations of union, intersection, and complementation,*** together with additional sets that must be included to make the family of all events a σ-field.

   The additional sets are troublesome and not to be ignored—there are uncountably many of them, even though not a single example of one can be given. But as a slogan to demystify the Borel sets, the statement in bold italics is useful.

2. If probabilities are assigned to intervals in a consistent way, then (in principle) it is possible to find the probabilities of all Borel sets from those of the intervals. This fact will be stated precisely in Theorem (3.1.2). ("In a consistent way" means that if we break up an interval into the union of countably many disjoint intervals, then its assigned probability must be the sum of their assigned probabilities.)

Émile Borel, by the way, was a French mathematician who lived from 1871 to 1956 and did important early work, especially in the 1890s, in the field of measure theory, which was emerging at that time.

More concise definitions and "lapidary style": We defined the family of Borel sets as the smallest σ-field containing the intervals, but it can be shown that we get the same family if we start with only the half-open intervals, or only the open intervals, or only the closed intervals. The reason is that countably infinite set operations allow us to reconstruct all intervals from any of these special subclasses of intervals. So we could define the family of Borel sets as the smallest σ-field containing the half-open intervals, or as the smallest σ-field containing the open intervals. (The latter is the most common definition in advanced courses in analysis or measure theory.)

If we wanted to be even more succinct, we could shorten the definition of a σ-field by dropping the requirement that it be closed under intersections, because, by De Morgan's Laws, intersections follow automatically if we are allowed to use unions and complements. So, if maximum efficiency is what we want, we will say the following:

> *A* σ-field is a nonempty set family closed under complementation and countable intersection;
>
> and the ***Borel sets*** are the members of the smallest σ-field containing the open intervals.

This kind of reduction of definitions and assumptions to a minimum is not a frivolous exercise. As we progress in mathematics and the concepts pile on one another, such brevity is welcome and necessary. Furthermore, we usually do not understand a theorem fully unless we have thought about the absolute minimum hypothesis needed to reach the conclusion. But such reduction, while important in mathematical *thought,* can lead in mathematical *writing* to what some critics call a "lapidary style," whereby everything is polished, like a stone, until it attains great beauty and elegance, but remembering where it came from is difficult.

It is really a matter of personal taste: You may find it more satisfying, and clearer, to remember definitions, assumptions, and theorems in their most concise forms; or you may prefer more detailed statements that reveal a little more of the substance of the subject. In this text we usually attempt some sort of compromise, looking for informal statements that are easy to keep in mind, but remembering that we cannot ignore the precise details behind them.

Note on the impossibility of assigning probabilities to all subsets. At the beginning of (1.3.2), we asserted that it is impossible to assign probabilities consistently to all subsets of an interval Ω; no matter what assignment we make, there will be some countable collection of disjoint sets, the sum of whose probabilities is not equal to the probability of their union. The theorems that make this assertion precise are a bit technical, and they all require some additional hypotheses.

The best known such theorem says that no consistent assignment is possible if we require that two events A and B have the same probability if they are *translates* of each other—that is, if there is some constant c such that the outcomes in B are just the outcomes in A with c added. (We must imagine Ω as being wrapped on a circle for such translation to make sense; otherwise, adding c to the outcomes in A might produce a B that is not contained in Ω.) This requirement, called "translation invariance," is essentially what we used in Example (1.3.1) to argue that in the uniform model on $[0, 1)$ the probability of any interval must be its length.

The proof of this theorem shows the existence of a remarkable family of sets; there are countably many of them, they are all disjoint, they are all translates of each other, and their union is Ω. Think about what this implies: If these sets have probabilities, then they must all have the same probability; if that probability is 0, then the probability of Ω is 0, but if that probability is positive, then Ω has infinite probability. Since $P(\Omega)$ must equal 1, it follows that none of the sets in this family can be assigned a probability. Therefore they cannot be Borel sets. This settles the question we mentioned earlier of whether there really are any non-Borel sets.

Another theorem with the same conclusion is the Banach–Kuratowski theorem, according to which no consistent assignment of probabilities is possible if we require that singletons have probability 0. This assumption is necessary for the uniform model of Example (1.3.1) also. The proof of the Banach–Kuratowski theorem, however, also requires an assumption called the Continuum Hypothesis, which we will stop short of discussing here.

If you are interested in learning more about the Borel sets and the strange things that happen when we consider *all* subsets of an interval, a course in measure theory usually includes at least a little on the subject. Section 2 in Chapter 1 of Patrick Billingsley's book, *Probability and Measure,* will give you a perspective on the subject and its relation to probability. You could also look at Section 3.4 of H. L. Royden's book *Real Analysis.* If you are curious about the Continuum Hypothesis, a good place to start is in Rudy Rucker's book *Infinity and the Mind.* However, Rucker does not mention what the Continuum Hypothesis has to do with the impossibility of assigning probabilities to all sets. [See the bibliography for information on books cited.]

In the next two examples, the outcome sets Ω are again intervals and the events are the Borel subsets of Ω, as in Example (1.3.1); but the recipes for finding the probability of an event are no longer simply related to the length of the event.

(1.3.3) Example: An exponential distribution on $[0, \infty)$. Let Ω be the interval $[0, \infty)$ of all nonnegative numbers; let the events be the Borel subsets of Ω. Let λ be a positive number; this will be a parameter of the model. Define the probabilities of the intervals by the formula

$$P[a, b) = e^{-\lambda a} - e^{-\lambda b} \quad \text{for real numbers } a \text{ and } b \text{ with } a < b. \tag{1.3.4}$$

As in Example (1.3.1), it turns out that this is enough information to determine the probabilities of all the events. Thus, our probability space has been specified—provided, of course, that the probability of any event is nonnegative and the probability $P(\Omega)$ of the set of all outcomes is 1. It should be clear that (1.3.4) gives a nonnegative probability to any interval, because $e^{-\lambda a}$ is greater than $e^{-\lambda b}$ if a is less than b. (λ is positive.) But how can we check that $P(\Omega)$ is 1?

You may be satisfied by the calculation

$$P[0, \infty) = e^{-\lambda\cdot 0} - e^{-\lambda\cdot\infty} = 1 - 0 = 1,$$

but this is not rigorously justified because ∞ is not a number; the formula (1.3.4) applies only to bounded intervals. A better check is to notice that Ω is the union of the disjoint intervals $[0, 1)$, $[1, 2)$, $[2, 3)$, etc., and so its probability is the sum of their probabilities:

$$\begin{aligned} P[0, \infty) &= P[0, 1) + P[1, 2) + P[2, 3) + \cdots \\ &= (e^{-\lambda\cdot 0} - e^{-\lambda\cdot 1}) + (e^{-\lambda\cdot 1} - e^{-\lambda\cdot 2}) + (e^{-\lambda\cdot 2} - e^{-\lambda\cdot 3}) + \cdots, \end{aligned}$$

a "telescoping series" whose sum is $e^{-\lambda\cdot 0} = 1$. [See (A.5.17) in Appendix A for other examples of telescoping series.]

This probability space is often used to model *waiting times* for the kind of "rare phenomena" mentioned earlier in connection with Poisson probabilities [see Example (1.2.11)]. If phenomena such as automobile accidents are occurring at random time points with an average of λ occurrences per unit time, then the above is a good model for the experiment of starting at some given point in time and measuring the time elapsed before the next occurrence of the phenomenon.

Another "rare phenomenon" to which the model is successfully applied is the arrival of customers at a bank. (Customer arrivals are not rare in the usual sense, but they come close enough to satisfying the necessary assumptions so that the model is useful.) There is a branch of applied probability, called the *theory of queues,* that studies the likely behavior of such quantities as the length of the queue (line) that forms, the average time that customers wait on line, the average time a teller must go without a rest, etc. The particular model in this example applies to the time elapsed between a fixed instant and the arrival of the next customer.

In the next section we will discuss this kind of situation a little more fully. It is modeled by what is called a *Poisson process.* It is customary to refer to the occurrences in this model as "arrivals," even though they may be automobile accidents or other kinds of phenomena.

We illustrate the computation of probabilities of some events in this probability space. To make things concrete and simple, let us suppose $\lambda = 1$. Then,

P(at least 1 but not more than 4 time units until the next arrival)

$$= e^{-1} - e^{-4} \doteq .3496;$$

P(less than 3 time units to next arrival)

$$= P[0, 3) = e^{-0} - e^{-3} \doteq .9502;$$

P(at least 3 time units to next arrival)

$$= P[3, \infty) = 1 - P[0, 3) = 1 - (e^{-0} - e^{-3}) = e^{-3} \doteq .0498;$$

P(at least a time units to next arrival)

$$= e^{-a}.$$

This example is much like Example (1.3.1), with two differences. First, Ω is an unbounded interval, $[0, \infty)$, instead of a bounded one, $[0, 1)$. Second, and more important, probabilities of intervals are not their lengths, but are given instead by the formula (1.3.4). Yet we still have the property that made us uncomfortable in Example (1.3.1)—that all singletons have probability 0 and, consequently, that all finite and countable sets have probability 0. Also, the four intervals $[a, b]$, $[a, b)$, $(a, b]$, and (a, b) have the same probability, although it is not simply $b - a$ as it was in Example (1.3.1). ∎

(1.3.5) **Example: The standard normal distribution.** Let Ω be the interval $(-\infty, \infty)$—that is, the set $\mathbb{R}$ of all real numbers—and again let the events be the Borel sets. As in the previous two examples, to give a recipe for probabilities of events, it is enough to give a recipe for the probabilities of intervals, checking only that all probabilities are nonnegative and that the probability assigned to the whole outcome set is 1. We specify the probabilities of the intervals for this model by

$$P[a, b) = \int_a^b \varphi(x)dx, \qquad (1.3.6)$$

where φ is the function

$$\varphi(x) = \frac{1}{\sqrt{2\pi}} e^{-x^2/2} \quad (-\infty < x < \infty)$$

whose graph is the famous "bell-shaped curve." See Figure 1.3d for a picture of this curve. In Figure 1.3e the shaded area is equal to the probability (1.3.6). [The symbol φ is an alternative form of the lowercase Greek letter phi, and is a standard symbol for this important function. The other phi is written ϕ.]

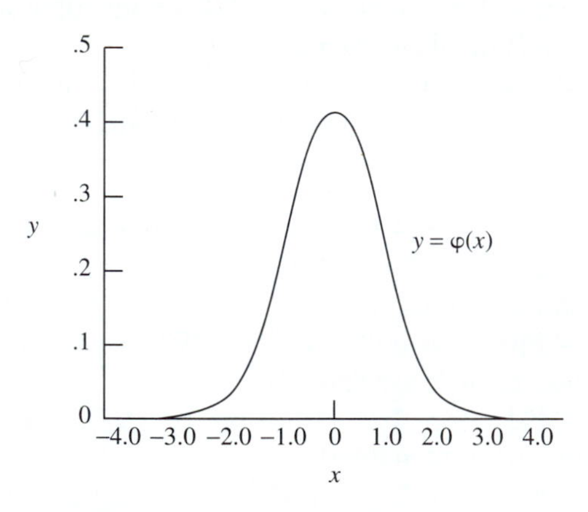

FIGURE 1.3d Standard normal density

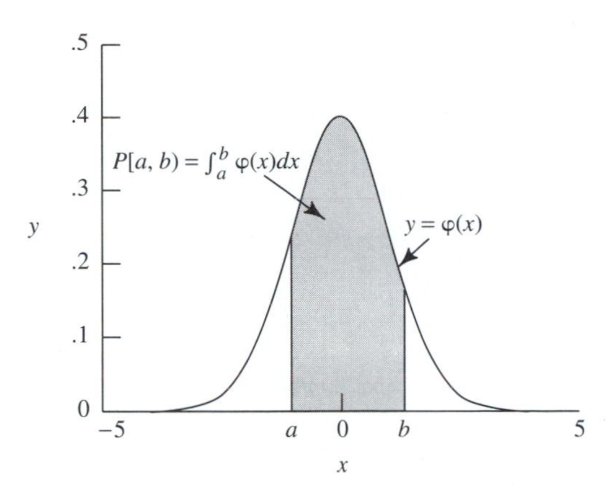

FIGURE 1.3e The probability of an interval in the standard normal model is defined to be the integral of the standard normal density over that interval.

As before, we need to check two things. First, the probability assigned to any interval must be nonnegative. This is true because the function $\varphi(x)$ is nonnegative. Second, the probability assigned to the interval Ω must be 1; in this case that means checking that

$$\int_{-\infty}^{\infty} \frac{1}{\sqrt{2\pi}} e^{-x^2/2} \, dx = 1.$$

This is done in Section A.9 of the Appendix; see (A.9.3), which uses a clever trick that you may remember from a calculus course. Strangely enough, although the trick works to find the integral of $\varphi(x)$ from $-\infty$ to ∞, there is no way to find a formula for (1.3.6) in terms of a and b. That is, there is no formula that gives an antiderivative of φ in terms of standard functions. However, numerical calculations have enabled the construction of tables that we can use to evaluate integrals like (1.3.6), at least to several decimal places. These are called *normal distribution tables,* and one can be found in Appendix C. What it gives is the value of the integral

$$\int_{-\infty}^{z} \frac{1}{\sqrt{2\pi}} e^{-x^2/2} \, dx$$

to four decimal places for various values of z. This integral is the probability of the interval $(-\infty, z]$, which is also the probability of the interval $(-\infty, z)$. It is equal to the shaded area in Figure 1.3f (page 44). As we will learn in Section 3.2, this integral, as a function of z, is denoted by $\Phi(z)$—Φ is an uppercase Greek phi—and is called the cumulative distribution function for the normal probabilities. It gives the cumulative probability of the set of all possible values less than or equal to x.

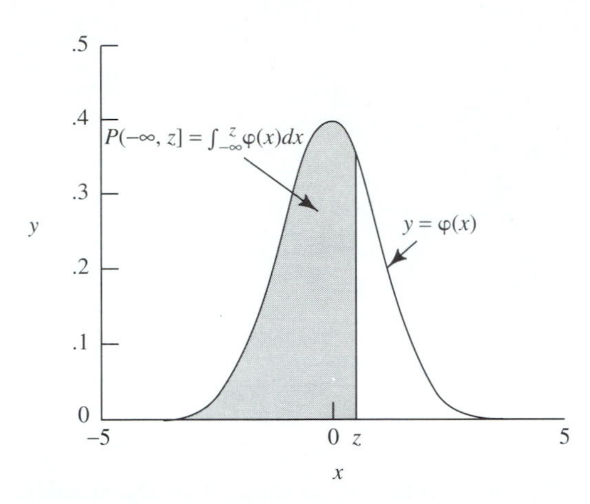

FIGURE 1.3f **The probability given for z in the standard normal distribution table**

Using the table in Appendix C, we can calculate, for example,

$$P(-\infty, 2) = .9772;$$
$$P[1.5, 2.8) = P(-\infty, 2.8) - P(-\infty, 1.5) = .9974 - .9932 = .0642.$$

We will study this probability distribution at length throughout the text. ∎

We are nearly ready for the definition of an absolutely continuous probability space. But first, let us think about what the three examples above have in common.

Each of them has an outcome set that is an interval in $\mathbb{R}$. They all have the same events: not all subsets, but only the Borel subsets, which are described in (1.3.2). In addition—and we are taking this on faith, because the proof requires measure theory—if we just specify the probabilities of the intervals, we can in principle compute the probabilities of all Borel sets. Furthermore, while there are sets that are not Borel sets, none can be described explicitly, and so it does not concern us that they are not events.

But what about the recipes for the probabilities of the intervals in these examples? They do not seem to have much in common: The probability of an interval is,

in Example (1.3.1), equal to its length;

in Example (1.3.3), given by the formula $e^{-\lambda a} - e^{-\lambda b}$;

and, in Example (1.3.5), given by the integral of $\varphi(x)$ over the interval.

But these recipes are really all of the same form: ***In each, the probability of a basic set (an interval) is the integral of some function $f(x)$ over the set.*** To be specific:

The length of the subinterval $[a, b)$ of $[0, 1)$ is $\int_a^b 1\,dx$, so in Example (1.3.1), $f(x) = 1$ for all $x \in [0, 1)$. (See Figure 1.3g.)

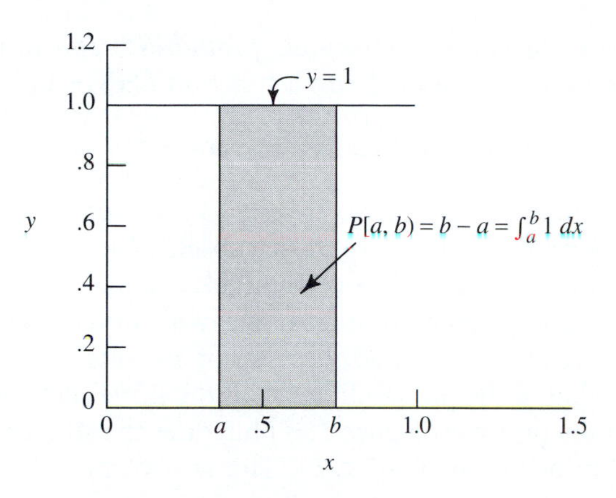

FIGURE 1.3g **In the uniform model on [0, 1), the probability of an interval is its length, which is the integral of the uniform density (the constant 1) over that interval.**

The probability $e^{-\lambda a} - e^{-\lambda b}$ is $\int_a^b \lambda e^{-\lambda x}\,dx$, so in Example (1.3.3), $f(x) = \lambda e^{-\lambda x}$ for all $x > 0$. (See Figure 1.3h.)

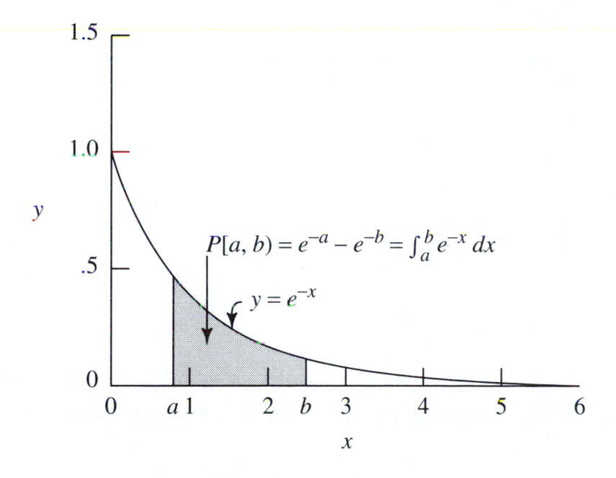

FIGURE 1.3h **In the model of Example (1.3.3), the probability of an interval $[a, b)$ is $e^{-a} - e^{-b}$; this is also the integral of a density function over $[a, b)$.**

Of course, $f(x) = \varphi(x)$ in Example (1.3.5).

The function $f(x)$ is called a *probability density function* and is one of the ingredients in an absolutely continuous probability space. The only restrictions on it are that it be nonnegative (so that intervals are assigned nonnegative probabilities) and that it integrate to 1 over all of Ω (so that the probability of the set of all outcomes is 1).

(1.3.7) **Definition.** *An **absolutely continuous probability space*** in $\mathbb{R}$ consists of an interval Ω called the ***outcome set*** and a function $f(x)$ from Ω to $\mathbb{R}$ such that

$$f(x) \geq 0 \quad \text{for each } x \in \Omega \quad \text{and} \quad \int_{\Omega} f(x)dx = 1.$$

A function with these two properties is called a ***probability density function.*** The ***events*** are the ***Borel sets,*** which are those subsets of $\mathbb{R}$ that can be made starting from intervals by using finitely or countably many set operations of union, intersection, and complementation [together with others that must be added to make a σ-field; see (1.3.2)]. The ***probability*** of any interval is the integral of $f(x)$ over that interval, and the probabilities of all the Borel sets are determined by the probabilities of the intervals. Figure 1.3i illustrates the situation. The shaded area is the probability of the interval $[a, b)$. The area of the whole region under the graph is 1.

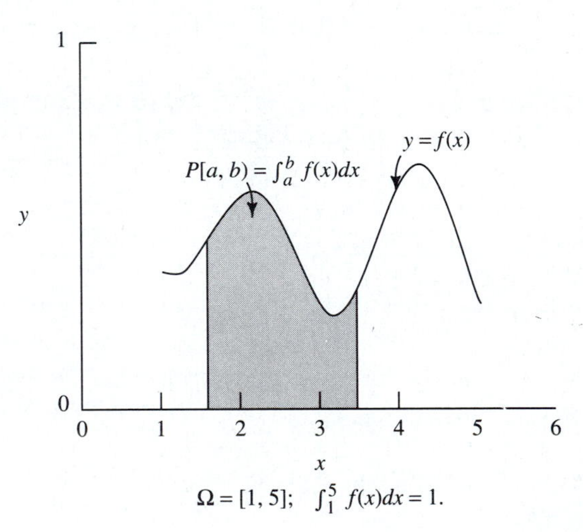

FIGURE 1.3i **In any absolutely continuous model the probability of an interval is the integral of the model's density over that interval.**

All absolutely continuous models have the property that singletons—and, therefore, also finite and countably infinite sets—have probability 0. It seems apparent by now that this must be so, because the integral of a function over a singleton set, $\int_a^a f(x)dx$, is 0. It can be proved more rigorously by arguments like the one we used in Example (1.3.1) to show that (.5, 1) must have the same probability as [.5, 1), and, therefore, {.5} must have probability 0.

The term "absolutely continuous" is unfortunately cumbersome. Many texts shorten it to "continuous"; but more than continuity is operating here, and so we choose to stay with the full expression. The term comes from the mathematical foundations of the subject in real analysis and measure theory. In Chapter 3, after (3.2.6), we will say a little more about why "continuous" is not enough to describe such distributions.

(1.3.8) **Example: A beta density on [0, 1].** It is easy to check that $12x^2(1 - x)$ is nonnegative when x is between 0 and 1, and that $\int_0^1 12x^2(1 - x)dx = 1$. Consequently,

$$f(x) = 12x^2(1 - x), \quad \text{for } 0 \le x \le 1,$$

defines a probability density function. Our model is specified; Ω is [0, 1], probabilities of subintervals of [0, 1] are found by integrating $f(x)$, and probabilities of other Borel sets are found somehow from the probabilities of the intervals. (There will be more in Chapter 2 on how to do this.)

It would be a good idea at this point to do Exercise 8 at the end of this section. There you are asked to graph the density and compute the probabilities assigned by this model to various intervals of the same length, and to notice that they are larger or smaller depending on whether the density is large or small on them.

Incidentally, the term "beta density" refers to a family of densities that includes this one; the family is named after the beta function, which figures in the constants used to make the densities integrate to 1. We will study this family in Section 8.6; this example reappears in Exercise 5 of that section. ▮

(1.3.9) **What does the value of $f(x)$ signify?** The role of the density function in an absolutely continuous model takes some getting used to. ***The value of the density function $f(x)$ at a point x in Ω is not the probability of the outcome x; in fact, the probability of the outcome x is 0, and $f(x)$ is not to be interpreted as a probability at all.*** (See Exercise 8b at the end of this section.) The only probabilities we can find from a density function are probabilities of intervals and we find them by integrating the function. Thus, the values of $f(x)$ tell us nothing, except that probabilities are high in regions where $f(x)$ is high and low where it is low, because the probability of $[a, b]$ is the integral of $f(x)$ over $[a, b]$.

But there is an *approximate* sense in which densities are related to probabilities. If I is a very small interval, say $I = [c, c + h]$ where h is a small number, and if the density $f(x)$ is continuous at $x = c$, then the integral of $f(x)$ over I is approximately equal to $f(c) \cdot h$, which is the area of the rectangle of height $f(c)$ over I. Figure 1.3j (page 48) illustrates this. It shows that $f(c) \cdot h$ differs from the integral of $\int_c^{c+h} f(x)dx$ only by the area of the small triangular region at the top of the rectangle. But $f(x)$ is *not* continuous at $x = d$, and the difference between $\int_d^{d+h} f(x)dx$ and $f(d) \cdot h$ is considerable.

Stated a little more generally, the principle is this:

If the density $f(x)$ is continuous at $x = c$, then the probability of a small interval of width h containing c is approximately $f(c) \cdot h$.

This conclusion is sometimes expressed informally by saying "$f(x)dx$ is the probability of a vanishingly small interval of width dx containing x."

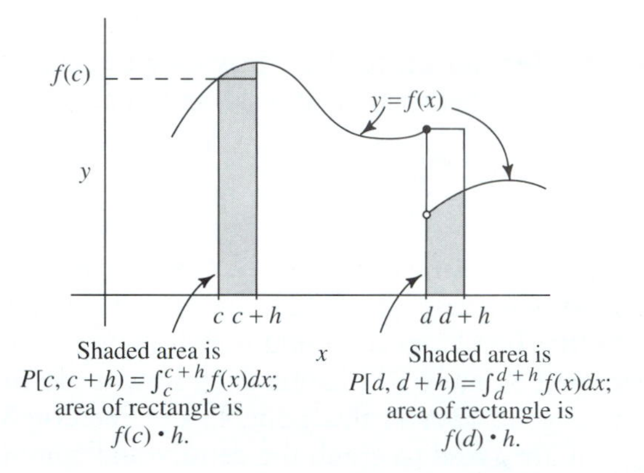

FIGURE 1.3j A density does not give probabilities, but if it is continuous at x, then $f(x) \cdot h$ is the approximate probability of a small interval of width h containing x.

(1.3.10) Random variables on absolutely continuous probability spaces. Remember that for technical reasons it is not possible in some outcome sets to allow every subset to be an event. If we did so in $\Omega = [0, 1)$, for example, there would be no way to define probabilities for all subsets so that the probability of any interval was its length. That was the bad news; the good news was that the sets we are forced to exclude are sets no one can describe.

Because of this, not every function on an absolutely continuous probability space can qualify as a random variable. The reason is simple. We want to be able to work with probabilities like $P(3 \le X \le 5)$, for example. This refers to the probability of the subset $\{\omega: 3 \le X(\omega) \le 5\}$ of Ω, and so this subset must be an event in order that we can discuss its probability. That is, if not all subsets of Ω are events, then X cannot qualify as a random variable if it is so complicated a function that subsets like the one above are not events in Ω.

We will formalize this requirement in the next section. For now we will simply mention that the good news about events is true about random variables as well: any function that can be described is a random variable. Thus, for example, if Ω is an interval in $\mathbb{R}$, then all the functions of calculus—polynomials, rational functions (ratios of polynomials), algebraic functions (involving roots and other fractional powers), and the standard transcendental functions (logarithms, exponential functions, and trigonometric functions, for example)—are random variables. So are any of the "strange" functions you may have seen defined in calculus texts or elsewhere: the function $X(\omega)$ that is 1 if ω is rational and 0 if ω is irrational, for example. There are many functions that are not random variables, but it is impossible to exhibit an example of one.

Next we give two examples of probability calculations for random variables on absolutely continuous probability spaces.

(1.3.11) Example: The square of a uniformly distributed random number. As in Example (1.3.1), let Ω be the unit interval [0, 1), let events be Borel sets, and let the probability of an interval be its length. (As we have seen, this is the same as saying: Let probabilities be given by the density $f(x) = 1$ for $0 \le x < 1$.) Now let $X(\omega) = \omega^2$. What are the probabilities of events describable in terms of X?

First, we ought to think about the set of possible values of X. If you take all the numbers in [0, 1) and square them, what set of numbers do you get? The same set, of course. This does not happen for all functions, but in this case the square of 0 is 0, the square of 1 is 1, and the square of a number between 0 and 1 is also between 0 and 1.

Next, what are the probabilities that X is in various subsets of its set of possible values? For example, what is $P(X \le .5)$? This is not difficult if we approach it this way:

$$\begin{aligned} P(X \le .5) &= P\{\omega: X(\omega) \le .5\} \\ &= P\{\omega: \omega^2 \le .5\} \\ &= P\{\omega: \omega \le \sqrt{.5}\} \qquad ☡ \\ &= P[0, \sqrt{.5}] = \sqrt{.5} \doteq .7071. \end{aligned}$$

The key step is the one marked with the sign ☡; there we translated the statement about $X = \omega^2$ (namely, "$\omega^2 \le .5$") into a statement about an outcome ω in the original probability space, whose probability we could find from the original recipe for probabilities in that space. This kind of step will be used often; we will have many occasions to take an event described in terms of one random variable and translate its description into terms involving another random variable, or involving the outcomes in the original space.

But the sign ☡ means something else: It is the European sign for a dangerous stretch of road and was first used in mathematical writing by the "mathematician" Nicolas Bourbaki to indicate something about which a warning is necessary. Mathematicians call it the "hazard sign." Here the warning is this: It is true that "$\omega^2 \le .5$" is the same statement as "$\omega \le \sqrt{.5}$," but only because our outcome set contains no negative numbers. The true solution to the inequality "$\omega^2 \le .5$" is the double inequality "$-\sqrt{.5} \le \omega \le \sqrt{.5}$." If Ω had happened to contain any of the numbers between $-\sqrt{.5}$ and 0, we would have had to include them in the translated statement. This situation arises in the next example. ∎

Nicolas Bourbaki was not a real mathematician because he was not a real person. The name was adopted by a group of French mathematicians who, in the 1930s, set about trying to put all mathematics on a rigorous and unified foundation of set theory and logic. Bourbaki's monumental work, *Elements of Mathematics,* comprises 33 volumes—the first one was published in 1939 and the last in 1967. It has had a huge influence, for better or worse, on taste and style in mathematics.

(1.3.12) Example: The square of a number having the standard normal distribution. Let $\Omega = \mathbb{R}$ and let probabilities be given by the standard normal density, as in Example (1.3.5):

$$P[a, b) = \int_a^b \varphi(x)dx, \quad \text{where } \varphi(x) = \frac{1}{\sqrt{2\pi}} e^{-x^2/2} \quad (-\infty < x < \infty).$$

Let Y be the random variable on Ω defined by $Y(\omega) = \omega^2$. What is $P(Y \leq 6)$?

We proceed as in the previous example, except that the translation step is different because Ω contains the negative numbers:

$$\begin{aligned} P(Y \leq 6) &= P\{\omega\colon Y(\omega) \leq 6\} \\ &= P\{\omega\colon \omega^2 \leq 6\} \\ &= P\{\omega\colon -\sqrt{6} \leq \omega \leq \sqrt{6}\} \\ &\doteq P\{\omega\colon -2.45 \leq \omega \leq 2.45\} \doteq .9858. \end{aligned}$$

The rest of this section deals with absolutely continuous probability spaces in two, three, and more dimensions. (Yes, there is a need for probability spaces of dimension greater than three. If you measure the temperature, barometric pressure, humidity, and wind vector at noon on a random September day, your outcome is a point in a five-dimensional space.) Our example is in two dimensions.

(1.3.13) Example: The uniform distribution on a disk. Consider the experiment of picking a point at random from inside the circle of radius 1 centered at the origin in the plane. You might think of doing this by throwing a dart randomly at a circular target, although it is difficult to throw randomly and still hit the circle. But this is only a model, and models always depict idealized situations that may not be exactly realizable. Let us study the model without worrying for now about what it fits.

The outcome set is, of course, the set of points within the circle itself: $\Omega = \{(x, y)\colon x^2 + y^2 < 1\}$. In proper mathematical terminology, this set is the *open unit disk*. (The unit circle is $\{(x, y)\colon x^2 + y^2 = 1\}$, which is the boundary of the disk.)

The events turn out to be similar to those in the other models of this section, except that we are dealing with sets in two dimensions instead of one. And the probabilities are similar to those in Example (1.3.1), where probabilities are lengths. Here probabilities are areas, except that we have to scale them because the area of the whole outcome set Ω is not 1, as it is in (1.3.1).

The assumption of uniformity—that no part of the circle is favored over any other part of the same size—translates to the requirement that any two sets having the same area should have the same probability, just as any two intervals having the same length did in Example (1.3.1). But this probability cannot simply be the area, because the area of Ω is π and not 1, and we need $P(\Omega) = 1$ in any probability model. We need to scale it, and define

$$P(A) = \frac{\text{area of } A}{\pi} \quad \text{for all events } A.$$

Then sets with the same area have the same probabilities, and $P(\Omega) = 1$.

But what sets are events? As in the one-dimensional case, there are certain basic sets whose areas we can find easily. Any set that is describable at all can be made, starting from basic sets, by using finitely or countably many set operations of union, intersection, and complementation. In one dimension we use the intervals as our basic sets, although we could use the half-open intervals, or some other family of intervals, and get the same family of events. Here also there are several choices, but a common choice is the set of *half-open rectangles* that are subsets of the disk: sets of the form (see Figure 1.3k)

$$A = \{(x, y): a \le x < b \text{ and } c \le y < d\} = [a, b) \times [c, d).$$

The area of such a set is obviously $(b - a) \cdot (d - c)$, so its probability in this example—assuming that it is wholly contained in the disk Ω—is $(b - a) \cdot (d - c)/\pi$.

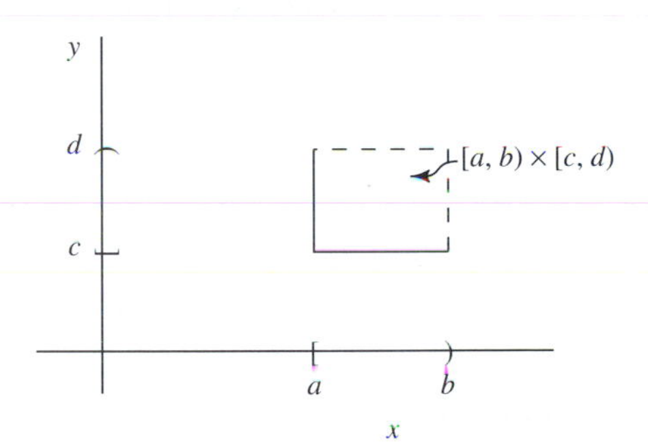

FIGURE 1.3k A half-open rectangle

Again we use the term ***Borel sets,*** this time to refer to the subsets of the plane that can be made, starting from half-open rectangles, by using finitely or countably many set operations of union, intersection, and complementation [again, together with others that must be added to make a σ-field; see (1.3.2)].

There is no need to go into the details of how to express events in terms of half-open rectangles. The reason is that, in practice, the events whose probabilities we need are nearly always simple geometric objects like smaller circles, triangles, sectors of circles, etc., whose areas we can find using formulas from Euclidean geometry. For example, the probability of the event "the distance of the point from the origin is less than .5" is the area of the circle of radius .5, divided by π; this is $\pi(.5)^2/\pi = .25$. (You might have thought that a random point in a disk of radius 1 had a 50% chance of being within .5 of the center, but it has only a 25% chance.)

However, we must observe that, in this model as in the previous example, singletons and therefore also finite and countably infinite sets have probability 0. But even more: line segments have probability 0. To see this, consider the two rectangles $[a, b) \times [c, d)$ and $(a, b) \times [c, d)$. We can show that these two rectangles have the same probability by using an argument like the one we used in Example (1.3.1) to show that $[.5, 1)$ and $(.5, 1)$ have the same probability. Therefore, the line segment along the left edge, which is in the first rectangle and not the second, must have probability 0. Again this is disturbing at first, but it becomes less so when we remember that "probability 0" does not mean "impossible"; it simply means "having vanishingly small expected long-run relative frequency."

Let us stop for a moment and summarize what we have so far. It is much like Example (1.3.1):

> ***Ω is the unit disk $\{(x, y): x^2 + y^2 < 1\}$; the events are the Borel subsets of Ω; and the probability of any event is its area divided by π.***

What function will serve as a density for this probability assignment? Remember that a density is a function whose integral over a set equals the probability assigned to that set. So here we want a function $f(x, y)$ of two variables with the property that, for any Borel subset of the unit disk,

$$\iint_A f(x, y)dx\,dy = \frac{\text{area of } A}{\pi}.$$

It is obvious that the function to integrate is the constant function $\frac{1}{\pi}$. So instead of saying that probabilities are areas divided by π, we could say in this example that the probability of any event is the integral of the function $f(x, y) = \frac{1}{\pi}$ over the event.

The density in this example is not the best way to specify probabilities; it is simpler just to think of areas divided by π. But that is only because the model is uniform and the density is therefore constant. In most models the density is a nonconstant function, and then it is often the most efficient way to specify the probability recipe.

For an example of a random variable on this probability space, let R denote the distance of the point chosen from the center of the circle. That is, suppose ω is a point in Ω with coordinates x and y; then $R(\omega) = \sqrt{x^2 + y^2}$. Then as before, statements about the value of R translate to events: for example, "$R \leq .5$" is the event that the point is within .5 of the center—that is, the disk of radius .5 concentric with Ω. For another example, "$.3 \leq R < .8$" is a ring (*annulus* is the technical term).

The possible values of R are, of course, the members of the set $[0, 1)$. But it will not do to try to list the possible values and their probabilities; the set of possible values is uncountable, and we run into the same problem as we did in one dimension. Still, we can find probabilities for various events; for example,

$$P(R \le .5) = \frac{1}{\pi} \cdot (\text{area of the concentric disk of radius } .5) = .25;$$

$$P(.3 \le R < .8) = \frac{1}{\pi} \cdot \Big((\text{area of disk of radius } .8) - (\text{area of disk of radius } .3) \Big)$$

$$= \frac{1}{\pi} \cdot \Big(\pi(.8)^2 - \pi(.3)^2 \Big) = .55. \qquad \blacksquare$$

(1.3.14) **Absolutely continuous probability spaces, and random variables on them, in more than one dimension.** It is simple to modify Definition (1.3.7) for two-, three-, and higher-dimensional absolutely continuous probability spaces. Instead of an interval in $\mathbb{R}$, Ω can be any Borel set in $\mathbb{R}^2$, $\mathbb{R}^3$, or $\mathbb{R}^n$. In most models, it is some simple geometric figure—a circle, a rectangle, a box in $\mathbb{R}^3$, or some such. And the density function f, instead of being $f(x)$, is $f(x_1, x_2, \ldots, x_n)$, but it must still be nonnegative and integrate to 1 over all of Ω. The Borel sets in $\mathbb{R}^n$ are defined as they are in (1.3.2) for $\mathbb{R}$, except that the basic sets are not just half-open intervals, but Cartesian products of half-open intervals, like $[a_1, b_1) \times [a_2, b_2) \times \cdots \times [a_n, b_n)$. These basic sets are of course rectangles in $\mathbb{R}^2$, boxes (rectangular parallelepipeds) in $\mathbb{R}^3$, and their counterparts in $\mathbb{R}^n$.

As for random variables on such spaces, all we need to say here is that they are real-valued functions on the outcome set Ω, that not all functions qualify as random variables, but that any function that can be described does qualify.

EXERCISES FOR **SECTION 1.3**

1. Suppose that the time, in minutes, from the moment you arrive at a bus stop until the arrival of the next bus is modeled by the exponential model of Example (1.3.3), with $\lambda = .37$. Find the probabilities of the following events:
   a. The next bus arrives between 2 and 4 minutes after you arrive.
   b. You wait more than 2 minutes for the first bus.
   c. The first bus comes within the first 90 seconds.
   d. There is no bus in the first 5 minutes after you arrive.
   e. There is at least one bus in the first minute.

2. Let Ω be the interval $(0, 1)$ and let probabilities of subintervals of Ω be given by integrating the density function $f(x) = 4x^3$. Find the probabilities of the following subintervals:
   a. $(.2, .3)$ **b.** $(.7, .8)$ **c.** $[.5, .8]$ **d.** $(0, .5]$

3. Let Ω be the interval $[0, 1)$, with the uniform distribution as in Example (1.3.1). Find the probabilities of the following events:
   a. The outcome is between .3 and .55.

b. The square of the outcome is less than .5.
c. The natural logarithm of the outcome is less than -4.

4. Using the normal distribution table in Appendix C, which gives $P(-\infty, z]$ for various values of z, find the probabilities that the standard normal distribution assigns to the following intervals:
a. $(-\infty, 2]$ **b.** $(-\infty, 3]$ **c.** $(2, 3]$ **d.** $(3, \infty)$ **e.** $(0, 1)$

5. The standard normal distribution is *symmetric about* 0, in that $P(-b, -a)$ is equal to $P(a, b)$—that is, the probability of any subinterval of $(-\infty, 0]$ equals the probability of its mirror image in $[0, \infty)$. Using this symmetry, find the probabilities of the following intervals under the standard normal distribution:
a. $(-\infty, -3)$ **b.** $(-2, -1)$ **c.** $(-2, \infty)$ **d.** $(-1, 1)$ **e.** $(-2, 2)$ **f.** $(-3, 3)$

6. In Example (1.3.3), on the exponential distribution, do the calculation showing that the formula (1.3.4) for $P[a, b)$ is the same as that given by integrating the density $f(x) = \lambda e^{-\lambda x}$ $(x > 0)$ over $[a, b)$.

7. Which of the following are probability density functions? For those that are not, find a number a (if possible) so that $af(x)$ is a density.
a. $f(x) = x^4$ for $0 < x < 1$
b. $f(x) = x^4$ for $0 < x < 2$
c. $f(x) = x^{-2}$ for $1 < x$
d. $f(x) = \sin x$ for $0 < x < \pi$
e. $f(x) = \cos x$ for $0 < x < \pi$
f. $f(x) = \frac{1}{x}$ for $0 < x$
g. $f(x) = xe^{-x^2/2}$ for $x > 0$

8. Graph the beta density of Example (1.3.8) and notice that its maximum is at $x = \frac{2}{3}$.
a. Find the following probabilities. Notice that the intervals all have the same length, but that the probability is higher where the density is higher.
i. $P\left[0, \frac{1}{3}\right]$ **ii.** $P\left(\frac{1}{6}, \frac{1}{2}\right]$ **iii.** $P\left(\frac{1}{3}, \frac{2}{3}\right]$ **iv.** $P\left(\frac{1}{2}, \frac{5}{6}\right]$ **v.** $P\left(\frac{2}{3}, 1\right]$
b. Find $f\left(\frac{2}{3}\right)$ and $P\left(\left\{\frac{2}{3}\right\}\right)$. Notice that they are not the same and, in fact, that $f\left(\frac{2}{3}\right)$ cannot be the probability of anything. How do we know this?
c. Find $P[0, x]$ as a function of x. Now notice that because $P(a, b] = P[0, b] - P[0, a]$, you could have found the probabilities in part a above without computing five separate integrals.

9. In the model of Example (1.3.1), the uniform distribution on $[0, 1)$, let A be the event "the third digit is a 7."
a. A is the union of 100 disjoint intervals of the form $[a, b)$; list several of them (enough to be sure you understand what they all are and why there are 100).
b. What is $P(A)$? [Compare with the probability that the second digit is a 7, found in Example (1.3.1).]

[a]**10.** In the model of Example (1.3.1), express the interval (0, 1) as the union of countably many disjoint intervals of the form $[a, b)$ and show that their probabilities add to 1. (Thus, [0, 1) and (0, 1) must have the same probability, so the singleton {0} must have probability 0.) [*Hint:* Let the first of your intervals be $\left[\frac{1}{2}, 1\right)$.]

11. Let $\Omega = [0, 1)$, with the uniform distribution as in Example (1.3.1). Define the random variable Z by $Z(\omega) = -\ln(\omega)$.

a. Find $P(Z \leq 2)$.
b. Find $P(Z > 3)$.
c. Find $P(3 \leq Z \leq 4)$.
d. Find $P(a \leq Z \leq b)$ (assuming $0 \leq a < b$).
e. What distribution is it that gives probabilities for Z?

12. Using the standard normal probability space as in Example (1.3.5), let $W(\omega) = \sqrt{|\omega|}$. Find the following probabilities. [This is like Example (1.3.12) except that we are taking the square root instead of the square.]

a. $P(W \leq 1.3)$
b. $P(W \geq .7)$
c. $P(W \leq -1)$
d. $P(.5 \leq W \leq 1.5)$

13. In the absolutely continuous model in which Ω is the interval [0, 3] and probabilities are determined by the density function $f(x) = cx^2$, what is the value of c?

14. Find the probability that, if a point is chosen at random from the uniform distribution on a disk of radius 1 [Example (1.3.13)], the distance of the point from the center is

a. Less than $\frac{1}{3}$.
b. Greater than $\frac{1}{3}$ but not greater than $\frac{2}{3}$.
c. Greater than $\frac{2}{3}$.
d. Less than or equal to r, where r is an arbitrary number in (0, 1).

15. **The uniform distribution on a triangle.** Let Ω be the triangle whose vertices are at (0, 0), (0, 1), and (1, 0) in the plane. Suppose we want to model the choice of a random point from this triangle in a "uniform" manner, so that the probability of any basic set is proportional to its area, as it was in Example (1.3.13).

a. What is the constant of proportionality? That is, what is the number c for which $P(A) = c \cdot (\text{area of } A)$? [$c$ was $\dfrac{1}{\pi}$ in Example (1.3.13).]
b. Find the probabilities of the following events in this model:
  i. The x-coordinate of the point is greater than $\frac{1}{2}$.
  ii. At least one coordinate is greater than $\frac{3}{4}$.
  iii. Both coordinates are less than $\frac{3}{4}$.

iv. Both coordinates are greater than $\frac{3}{4}$.

v. The sum of the two coordinates is less than $\frac{1}{2}$.

vi. The distance of the point from the origin is less than $\frac{1}{2}$.

16. Consider the probability space of Example (1.3.3) with $\lambda = 1$, in which probabilities are determined by the exponential density function $f(x) = e^{-x}$ $(0 < x < \infty)$.

a. Let the random variable Z be defined by $Z(\omega) = \lfloor \omega \rfloor$, the greatest integer less than or equal to the outcome ω. [Thus "$Z = 0$" is the event $[0, 1)$, "$Z = 2$" is $[2, 3)$, etc.] Find $P(Z = k)$ for $k = 0, 1, 2, \ldots$.

b. Do the same for $W(\omega) = \lfloor \omega/a \rfloor$, where a is a positive constant.

c. Suppose $a = -\ln\left(\frac{5}{6}\right)$. What is $P(W = k)$ for $k = 0, 1, 2, \ldots$?

d. Compare the result of part c with that of Exercise 6 in Section 1.2. Describe a way to simulate the experiment of waiting for a 1 with a fair die using a computer that can generate random numbers having the exponential distribution of Example (1.3.3).

17. (*Computer exercise*) This is an exercise in simulating a number of observations from a uniform distribution.

a. Write a program that generates 100 numbers from the uniform distribution on $[0, 1)$. Do not print out the 100 random observations; instead, have the program keep track of the number of observations that fall in each of the ten intervals $[0, .1), [.1, .2), \ldots, [.9, 1)$. Run the program several times and see how close to being equal the ten frequencies are. (On some runs they may be nearly equal, but often they will be far from equal.)

b. Modify your program so that it prints relative frequencies (frequencies divided by 100) instead of frequencies.

c. Modify the program in part b so that it generates some large number of observations—say 1,000, 5,000, or 10,000—instead of 100. The relative frequencies should be more nearly equal for larger samples.

[a]**18.** (*Computer exercise*) Repeat Exercise 17, but generate the squares of the uniformly distributed numbers. Now you should no longer expect the ten frequencies to be nearly equal. Can you work out the numbers that they should be close to? [See Example (1.3.11).]

ANSWERS

1. a. .2495 **b.** .4771 **c.** .4259 **d.** .1572 **e.** .3093

2. a. .0065 **b.** .1695 **c.** .3471 **d.** .0625

3. a. .25 **b.** .7071 **c.** .0183

4. a. .9772 **b.** .9987 **c.** .0215 **d.** .0013 **e.** .3413

5. a. .0013 **b.** .1359 **c.** .9772 **d.** .6826 **e.** .9544 **f.** .9974

7. a. $a = 5$ **b.** $a = \frac{5}{32}$ **c.** Yes **d.** $a = \frac{1}{2}$ **e.** Impossible **f.** Impossible **g.** Yes

8. a. i. .1111 **ii.** .2963 **iii.** .4815

iv. .5556 **v.** .4074 **b.** $\frac{16}{9}$; 0
c. $4x^3 - 3x^4$, $0 \le x \le 1$

9. a. One of them is [.637, .638). **b.** .1

10. $(0,1) = \left[\frac{1}{2}, 1\right) \cup \left[\frac{1}{4}, \frac{1}{2}\right) \cup \left[\frac{1}{8}, \frac{1}{4}\right) \cup \cdots$;
$\frac{1}{2} + \frac{1}{4} + \frac{1}{8} + \frac{1}{16} + \cdots = 1$

11. a. .8647 **b.** .04979 **c.** .03147
d. $e^{-a} - e^{-b}$ **e.** Exponential with $\lambda = 1$

12. a. .9090 **b.** .6242 **c.** 0 **d.** .7782

13. $\frac{1}{9}$

14. a. $\frac{1}{9}$ **b.** $\frac{1}{3}$ **c.** $\frac{5}{9}$ **d.** r^2

15. a. 2 **b.** $\frac{1}{4}, \frac{1}{8}, \frac{7}{8}, 0, \frac{1}{4}, \frac{\pi}{8}$

16. a. $e^{-k}(1 - e^{-1})$ **b.** $e^{-ak}(1 - e^{-a})$
c. $\left(\frac{5}{6}\right)^k \cdot \frac{1}{6}$

18. They should be close to the probabilities, $\sqrt{.1} \doteq .316$, $\sqrt{.2} - \sqrt{.1} \doteq .131$, $\sqrt{.3} - \sqrt{.2} \doteq .101$, etc.

1.4 TWO OTHER KINDS OF PROBABILITY SPACE: BERNOULLI PROCESSES AND POISSON PROCESSES

The previous two sections have been concerned with the two most common types of probability space:

> discrete spaces, in which the outcome set is finite or countable, all subsets are events, and probabilities are found by summing a mass function; and
>
> absolutely continuous spaces, in which the outcome set is an interval in $\mathbb{R}$ (or some suitable set in $\mathbb{R}^2$ or $\mathbb{R}^n$); the events are the Borel sets, whose probabilities are all known once we have specified the probabilities of certain basic sets; and probabilities of basic sets are found by integrating a density function.

Here we introduce two very important probability spaces that are not either of these types. One, a Bernoulli process, generalizes the Bernoulli-trials model of Example (1.2.13), allowing for a potentially infinite sequence of trials. The other, a Poisson process, is a model for the times of occurrence of "rare phenomena," like those we discussed in Examples (1.2.11) and (1.3.3).

BERNOULLI PROCESSES

We have already seen Bernoulli trials several times. They are repeated trials, under identical conditions, of an experiment that has two outcomes called S and F, in which on each trial the probability of S (the "success probability") is p and the probability of F is $q = 1 - p$. We looked at two kinds of experiment: doing a fixed number of Bernoulli trials and counting the number of S's, and doing repeated trials until a desired number of S's have appeared. In the first kind of experiment, the outcome set is finite; if we do n trials, there are 2^n outcomes. In the second, there is countably infinite number of outcomes, but each outcome involves only a finite number of trials. For example, if we are waiting for the first S, we use the outcome set $\Omega = \{S, FS, FFS, FFFS, \ldots\}$, which contains infinitely many finite strings. The model does not recognize the possibility of an infinite sequence of trials without an S. (There is no need to do so, because as we saw in Exercise 5a of Section 1.2, the probabilities add to 1 without such an outcome.)

Here we want to model the experiment of making an infinite sequence of Bernoulli trials and recording the result (S or F) of each one. A single outcome of such an experiment is an infinite sequence of S's and F's. There is an uncountable infinity of such sequences. We recognize that such an experiment cannot be performed in real life, but there are three important advantages of considering it anyway. One is that it gives us a single probability model that covers both kinds of experiment described above. (To study three Bernoulli trials in a model of infinitely many trials, just look at the first three letters in the sequence; there are eight possibilities, resulting in eight subsets for study.)

A second advantage is that it enables us to consider questions that cannot be asked in models of the simpler types. For example, we might be interested in the long-run proportion of S's, expressed as the limit as $n \to \infty$ of the relative frequency of S's in the first n trials. This limit, although it is purely theoretical, is of great interest, because we have defined probability in an attempt to represent it. It is impossible to study this limit adequately using only models involving finitely many trials.

The third advantage is a practical one: Some experiments that *can* be done are modeled more simply with this full model than with a smaller one. For example, suppose we toss a fair coin until we get either three successive heads or three successive tails. We will make only a finite number of tosses and so we might think of an outcome set in which the outcomes are the finite strings of H's and T's that end with three H's or three T's. But a simple listing of these outcomes along with their probabilities is not easy to make, and we gain nothing by doing so. It is more convenient to start with a Bernoulli process and ask for the set of sequences whose first appearance of SSS or FFF is at the nth position in the sequence.

So the outcome set Ω of a Bernoulli process is the (uncountably infinite) set of all infinite sequences of S's and F's. We will represent a typical outcome by writing something like $\omega = \text{SFSSSFFSSFS} \ldots$; listing only the first few letters is the best we can do, unless the outcome consists of some recognizable pattern of S's and F's.

What are the events, and how do we specify their probabilities? The answer is similar to that for absolutely continuous probability spaces: There are certain basic events and a rule for their probabilities; all the other events can be made from countably many basic events by unions, intersections, and complements; and we can find (in principle) the probability of any event from the probabilities of the basic events it is made of. Here are some basic events and their probability assignments:

$$P(\text{the first four trials result in SSFS}) = p^3q,$$
$$P(\text{the fifth, sixth, and seventh trials result in FSF}) = pq^2,$$

$$P(\text{the 231st trial results in S}) = p,$$
$$P(\text{the first 4,000 trials all result in S's}) = p^{4,000}.$$

Notice what these events really are, as subsets of Ω. For example, the event "the fifth, sixth, and seventh trials result in FSF" is really the set of all infinite sequences of S's and F's that happen to have FSF in positions 5, 6, and 7. Its probability is the probability that three Bernoulli trials produce FSF, namely, $q \cdot p \cdot q$, just as in Example (1.2.13).

The definition of a Bernoulli process formalizes the preceding discussion.

(1.4.1) **Definition.** The ***Bernoulli process*** with success probability p is a probability space whose outcome set is the set of all infinite sequences of S's and F's; a typical basic event is the set of all infinite sequences that have a specified string in a specified set of positions; and its probability is $p^u q^v$, where u and v are the numbers of S's and F's in the specified string. (As usual, q denotes $1 - p$.) The other events are those that can be made from countably many basic events by unions, intersections, and complements.

For example, let A be the basic event "SFFSF on trials 9 through 13." Then $P(A) = p^2q^3$. As a set of outcomes, A is the set of infinite sequences of S's and F's that have SFFSF in positions 9, 10, 11, 12, and 13.

Notice that we have not specified the probabilities of the individual outcomes of a Bernoulli process; as we will see later, they are all 0. This happens also with absolutely continuous probability spaces, in which we specify the probabilities of the intervals and the singletons have probability 0. [Actually, there is a surprising connection between Bernoulli processes and a certain absolutely continuous model. See (1.4.12) below.]

Using Definition (1.4.1), it is possible to derive rules for calculating the probabilities of most kinds of events encountered in dealing with Bernoulli processes. We will not derive these rules, but will simply list some of them here.

B1. For any sequence of n consecutive trials (say, trials $a, a + 1, a + 2, \ldots, a + n - 1$), the probability that there are k S's and $n - k$ F's on these trials is the binomial probability $\binom{n}{k}p^kq^{n-k}$.

B2. For any trial, say, trial a, the probability that the first S after trial a comes after k F's (that is, that it occurs on trial $a + k + 1$) is the geometric probability q^kp, for $k = 0, 1, 2, \ldots$.

B3. The probability that there are k F's *between* any two successive S's (that is, the probability that the waiting time from a given S to the next is $k + 1$ trials) is also q^kp.

B4. Given any two nonoverlapping sequences of trials, one consisting of n trials and the other of m trials, the probability that there are k S's in the first sequence and j in the second is the product $\binom{n}{k}p^kq^{n-k} \cdot \binom{m}{j}p^jq^{m-j}$.

The thing to remember when finding probabilities for Bernoulli processes is that, although the outcomes may represent infinite sequences of trials, the basic probability calculations involve only a finite set of trials, and thus they are essentially like those in Section 1.2.

Following are some examples of the use of these rules.

(1.4.2) Example. In repeated rolls of a fair die, what is the probability that the first 6 occurs on or before the fourth roll?

Solution 1. We want the probability that there is at least one S in the first four trials—that is, 1 minus the probability that there are no successes in the first four trials. We can rule B1 to find this probability. The success probability is $p = \frac{1}{6}$, so the answer is

$$1 - \binom{4}{0}\left(\frac{1}{6}\right)^0\left(\frac{5}{6}\right)^4 = 1 - \left(\frac{5}{6}\right)^4 \doteq 1 - .482253 = .517747,$$

which rounds to .5177. (Remember that the dot over the equal sign indicates that what follows has been rounded to the number of significant digits shown.)

Solution 2. For any j, the probability that the first S occurs on the jth trial is, by rule B2, $\left(\frac{5}{6}\right)^{j-1}\left(\frac{1}{6}\right)$. So the answer is found by summing the values of this expression for $j = 1, 2, 3$, and 4:

$$\left(\frac{5}{6}\right)^0\left(\frac{1}{6}\right) + \left(\frac{5}{6}\right)^1\left(\frac{1}{6}\right) + \left(\frac{5}{6}\right)^2\left(\frac{1}{6}\right) + \left(\frac{5}{6}\right)^3\left(\frac{1}{6}\right)$$
$$\doteq .166667 + .138889 + .115741 + .096451 = .517748,$$

which again rounds to .5177.

We have used the following rule of thumb in getting a numerical answer: ***Throughout a calculation, carry at least two more significant digits than you will report in the answer.*** Notice that if we had carried only five digits in the previous calculation, the first four digits would not be correct; we would get $1 - .48225 = .51775$, which rounds to .5178.

Even the "two more than needed" rule of thumb is inadequate for certain types of calculations. If you multiply together several numbers that are accurate to six digits, the answer may not be accurate to four digits. Or if you subtract two numbers that are nearly equal, you lose the first few significant digits in the answer. For example, $1 - .999914 = .000086$, and although this has six decimal places, it has only two significant digits.

The question of accuracy in the use of a hand calculator or a computer is not a simple one; unless you have studied such matters in a course in numerical analysis, it may be difficult to know how much accuracy you are losing in a calculation. Some people advocate a stronger rule of thumb: *When using a calculator, do not do any rounding until you get the final answer.* This is often tedious, especially if you do not use your calculator's memory and have to write down intermediate results; but it has the advantage of safety. ∎

(1.4.3) Example. If the success probability is $p = .45$, what is the probability that there are eight S's in the first 25 trials, but that they all come after the fourth trial?

Solution. The trick is to express the event in terms of nonoverlapping sequences of trials and use rule B4. We want the probability that there are no S's in the first four trials and eight in the next 21 trials. By rule B4, this is

$$\binom{4}{0}.45^0.55^4 \cdot \binom{21}{8}.45^8.55^{13}.$$

To get a numerical answer, don't compute the powers separately, but combine them as much as possible to minimize roundoff error. The first two factors, $\binom{4}{0}.45^0$, just give 1, so the answer is

$$\binom{21}{8}.45^8.55^{17} \doteq 203{,}490 \cdot .00168151 \cdot .0000385625 \doteq .0131949,$$

which we round to .01320.

Notice again that "six significant digits" does not mean "six digits after the decimal point"; it means "six digits beginning with the first nonzero digit." ∎

(1.4.4) **Example.** Items coming off an assembly line are selected randomly and inspected, and defective items are discarded. Suppose that 3% of the items are actually defective. What is the probability that the third defective item is detected in the first 25 inspections?

Solution. This is the probability that there are three ***or more*** defectives in 25 inspections. We find this as 1 minus the probability of two or fewer; that is,

$$\begin{aligned} &1 - \left[\binom{25}{0}.03^0.97^{25} + \binom{25}{1}.03^1.97^{24} + \binom{25}{2}.03^2.97^{23}\right] \\ &= 1 - .97^{23} \cdot (.97^2 + 25 \cdot .03 \cdot .97 + 300 \cdot .03^2) \\ &\doteq 1 - .496306 \cdot (.9409 + .7275 + .27) = 1 - .962040 \doteq .037960. \end{aligned}$$

Do not fall into the trap of finding the probability of *exactly* three defectives in 25; notice that even if there were seven or eight, or 21, or even 25 defectives in the first 25, we would still have encountered the third defective.

Notice that we carried six significant digits through the calculation (the numbers with fewer than six digits are exact); but because the final calculation subtracted two numbers that are close to each other, we ended up with only five significant digits. If the last digit of .037960 had been 4 or 5 instead of 0, then rounding to four significant digits would have been unreliable. This shows why we should carry six to have any confidence in four, and also why even that might not be enough. ∎

(1.4.5) **Example.** In Bernoulli trials with success probability $p = .26$, what is the probability that the fourth success does not appear until some time after the 20th trial?

Solution. We want the probability that there are three or fewer successes in 20 trials. This is

$$\binom{20}{0}.26^0.74^{20} + \binom{20}{1}.26^1.74^{19} + \binom{20}{2}.26^2.74^{18} + \binom{20}{3}.26^3.74^{17}.$$

As an exercise, find this number to four significant digits. The answer is .1962. ∎

POISSON PROCESSES

A Poisson process is a model for the times of occurrence of "rare phenomena" of the kind described in connection with Examples (1.2.11) (the Poisson distribution) and (1.3.3) (the exponential distribution). As we did previously, we call the rare phenomena *arrivals*. It is useful to think of them in concrete terms, such as the times of arrival of customers at a bank or the times of occurrence of automobile accidents in a city.

In Example (1.2.11) we used the Poisson distribution to model the number of arrivals in a given time period; recall that if λ is the average number of arrivals per week, then the probability of k arrivals in a given week is the Poisson probability $\lambda^k e^{-\lambda}/k!$.

And in Example (1.3.3) we used the exponential distribution to model the waiting time before an arrival. Recall that if λ is the average number of arrivals per unit time, then the waiting time for the first arrival has probability $\int_a^b \lambda e^{-\lambda x}\,dx = e^{-\lambda a} - e^{-\lambda b}$ of being in the time interval $[a, b)$.

The Poisson process combines these two kinds of model by considering the set of all arrival times in the time interval $[0, \infty)$ as a single outcome of the experiment. As with Bernoulli processes (where we model all the results of an infinite sequence of trials), it turns out to be more convenient to model all the infinitely many arrivals in the time interval $[0, \infty)$ than to model the times of the first 100 arrivals, say, or the arrivals in the finite time period $[0, 24)$ representing one day.

We will not give a complete description of this probability space; such a description is given in a course in stochastic processes, which usually follows a course from a text like this one. Instead we will think carefully about what the outcomes are, and then describe the probabilities of certain important types of events, without going into detail about exactly what subsets of the outcome set qualify as events.

What is a typical outcome of the experiment of marking the times of all the automobile accidents that occur in a given city from now on? It is a countably infinite set $\{a_1, a_2, a_3, \ldots\}$ of arrival times (accident times), with $a_1 < a_2 < a_3 < \cdots$ (see Figure 1.4a); and it has the additional property that, in any bounded interval $[t_1, t_2]$ of times, there are only finitely many a's. Such a set, $\omega = \{a_1, a_2, a_3, \ldots\}$, is a *member*, not a subset, of Ω; the outcome set Ω is the set of all such infinite sets of numbers.

To avoid a lot of words, let us agree for the moment to call a set of real numbers "Poissonian" if it is countably infinite but if each finite interval has only

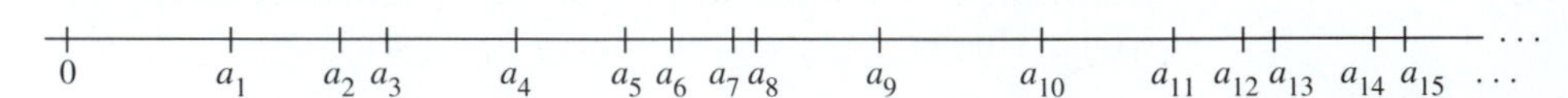

FIGURE 1.4a **Typical outcome of a Poisson process: a countably infinite set of real numbers $a_1 < a_2 < a_3 < \cdots$, with only finitely many a_j's in any bounded interval**

finitely many members of the set in it. For example, the set of integers is Poissonian, but the set of rational numbers is not, because any interval (except for singletons) contains infinitely many rational numbers. The set of rational numbers with denominators less than 100 is Poissonian, but the set of rational numbers with odd denominators is not. The set of reciprocals of integers is not Poissonian, because an interval $[0, a]$ contains infinitely many of them (although any closed interval not containing 0 contains only finitely many of them).

So the outcome set Ω of a Poisson process is the set of all Poissonian subsets of $[0, \infty)$. And events are subsets of Ω—that is, sets of Poissonian subsets. As with absolutely continuous models and Bernoulli processes, there are some basic events and the other events are made from them by set operations. But there is no need to get involved with this in order to make use of Poisson processes. Certain types of events have natural verbal description, and it will suffice to specify the probabilities of such events. Some examples of the kinds of events we will deal with are:

"more than six arrivals in the first minute,"

"at most 1 hour between the fifth and sixth arrivals,"

"less than 10 seconds between time $t = 4$ and the next arrival after $t = 4$."

It may be worthwhile to consider what such events actually are, as subsets of Ω. The first one, for example, "more than six arrivals in the first minute," is the set of all Poissonian subsets that contain at least six points in the interval $[0, 1]$. The third, "less than 10 seconds between time $t = 4$ and the next arrival," is the set of Poissonian subsets that contain at least one point in the interval $\left[4, 4\frac{1}{6}\right]$.

But thinking about the outcomes and events in such terms is seldom necessary when one works with a Poisson process as a model. The rules stated below for finding probabilities are given in terms of the verbal descriptions of the events, not in terms of the events as sets of Poissonian subsets. (Remember also that we used the word "Poissonian" only to make the preceding discussion a little shorter; it is not a standard term in modeling or theory.)

Before specifying the probabilities, we need to decide on the unit of time we are using. Does $t = 3$ represent 3 seconds, 3 minutes, 3 hours, or 3 days after the start of observations? And we also need to agree on the value of the parameter λ, the average number of arrivals per unit time. λ is called the *arrival rate* of the process. Different choices of these two quantities produce different Poisson processes. We might say, for example, that we are working with a Poisson process whose arrival rate is $\lambda = 15$ arrivals per hour. If we wished to convert the time unit to minutes, we would rescale, and say that the arrival rate is .25 arrival per minute; the process and all its probabilities would be the same. (A process with $\lambda = 15$ arrivals per hour is of course very different from one with $\lambda = 15$ arrivals per minute.)

Finally, then, here are the essential facts about the probabilities for a Poisson process with an arrival rate of λ arrivals per unit time.

P1. If $I = [t, t + s]$ is any time interval of length s, then for any integer $k \geq 0$ the probability that there are k arrivals in I is the Poisson probability $(\lambda s)^k e^{-\lambda s}/k!$ for $k = 0, 1, 2, \ldots$. (This is the same as (1.2.12), but with λs in place of λ.)

P2. For any time point t, the waiting time until the next arrival after t has an exponential distribution, given by the density $f(x) = \lambda e^{-\lambda x}$ for $x > 0$. (This is the same as saying that the probability that the waiting time is between a and b is $e^{-\lambda a} - e^{-\lambda b}$, which is the same as (1.3.4).)

P3. The time *between* any two consecutive arrivals also has the exponential distribution, given by the same density as in rule P2.

P4. Given any two nonoverlapping time intervals, say I and J, of lengths r and s, the probability that there are k arrivals in I and l arrivals in J (for nonnegative integers k and l) is the product

$$\frac{(\lambda r)^k e^{-\lambda r}}{k!} \cdot \frac{(\lambda s)^l e^{-\lambda s}}{l!}.$$

These four rules are analogous to the rules B1–B4 for Bernoulli processes. They too are derivable from the formal definition of a Poisson process, which we are not worrying about here.

Following are a few examples of the kinds of probability calculations we can make with these rules.

(1.4.6) Example. In a Poisson process with arrival rate $\lambda = 2.6$ arrivals per hour, what is the probability that there are exactly three arrivals between 1:00 and 2:30?

Solution. The length of the time period is $s = 1.5$ hours, so $\lambda s = 2.6 \cdot 1.5 = 3.9$ and the probability of three arrivals is, by rule P1,

$$\frac{(\lambda s)^k e^{-\lambda s}}{k!} = \frac{(3.9)^3 e^{-3.9}}{3!}.$$

Suppose we want a numerical answer to three significant digits. Then we proceed with the previous equation as follows, carrying five or more significant digits in each rounded number:

$$\doteq \frac{59.319 \cdot .020242}{6} \doteq .20012,$$

which we report as .200, accurate to three significant digits.

Remember that we use a dot over an equal sign to indicate that the number to follow has been rounded to the number of decimal places given. Notice that we used .020242 instead of .02024, because the latter has only four significant digits, not five. ∎

(1.4.7) Example. If automobile accidents occur in a certain town at an average rate of 4.2 accidents per week, and if we assume they occur according to a Poisson process, what is the probability that there will be more than two accidents tomorrow?

Solution. The length of the time period is $s = \frac{1}{7}$ week, so λs is .6, and we want 1 minus the probability of 0, 1, or 2 accidents. By rule P1, this is

$$1 - \left(\frac{(.6)^0 e^{-.6}}{0!} + \frac{(.6)^1 e^{-.6}}{1!} + \frac{(.6)^2 e^{-.6}}{2!}\right).$$

The best way to get an accurate numerical answer is to combine as much as possible. If we need four significant digits in our answer we carry six:

$$= 1 - e^{-.6}\left(1 + .6 + \frac{.6^2}{2}\right) = 1 - e^{-.6}(1.78) \doteq 1 - .548812 \cdot 1.78$$
$$\doteq 1 - .976885 = .023115,$$

which we report as .02312. (Reporting .0231 would be giving only three significant digits.) ∎

(1.4.8) Example. If customers arrive at a bank according to a Poisson process with a rate of $\lambda = 3$ customers per minute, what is the probability that the first customer arrives within 10 seconds after the bank opens?

Solution 1. We want the probability that the waiting time for the first arrival is less than $\frac{1}{6}$ minute—that is, in the interval $\left[0, \frac{1}{6}\right)$. By rule P2, this is

$$e^{-3\cdot 0} - e^{-3\cdot(1/6)} = 1 - e^{-.5} \doteq .3935.$$

Solution 2. This is also the probability that there are one or more arrivals in the first $s = \frac{1}{6}$ minute, which is 1 minus the probability of no arrivals in that interval. We can find this using rule P1, with $\lambda s = .5$; the probability is

$$1 - \frac{(.5)^0 e^{-.5}}{0!} \doteq .3935.$$ ∎

(1.4.9) Example. In modeling the behavior of neurons (nerve cells) it is often assumed, as a first approximation, that a neuron "fires" (that is, releases a voltage burst, called an action potential, or a "spike," to other neurons) at random times in accordance with a Poisson process. Assuming this model is valid, and that a neuron is firing at the rate of 300 spikes per second, what is the probability that there will be six spikes in the next 10 milliseconds, and then no spikes in the next 5 milliseconds after that?

Solution. We have two nonoverlapping time intervals of lengths $r = .010$ and $s = .005$ second. We want the probability that there are six spikes in the first interval and none in the second. We have $\lambda r = 3$ and $\lambda s = 1.5$, so the probability is, by rule P4,

$$\frac{(3)^6 e^{-3}}{6!} \cdot \frac{(1.5)^0 e^{-1.5}}{0!} = \tfrac{729}{720} \cdot e^{-4.5} \doteq .0112. \quad \blacksquare$$

(1.4.10) Example. A certain piece of equipment is subject to breakdowns, which occur at the average rate of five per year. Assuming, for a first approximation, that the time needed to repair a breakdown is negligible, and that the times of breakdown follow a Poisson process, what is the probability that there will be at least 3 months between the next two breakdowns?

Solution. The arrival rate (breakdown rate) is $\lambda = 5$ per year, and so the time between breakdowns has the exponential distribution with $\lambda = 5$. We want the probability that this time is in the interval $[.25, \infty)$. By rule P3, this is

$$\int_{.25}^{\infty} 5e^{-5x}\, dx = e^{-5 \cdot .25} \doteq .2865. \quad \blacksquare$$

A few comments about the rules for probabilities in a Poisson process are in order here. First, they show the intimate connection between the Poisson and exponential distributions. We see this connection in the two solutions to Example (1.4.8) above; they produce the same answer by looking at the question in two apparently different ways.

Second, rules P2 and P3 may seem to contradict each other: How can it be that the waiting time *between* two arrivals has the same probability properties as the waiting time from a *fixed* time point to the next arrival? Suppose customers have been arriving at a bank all day at $\lambda = 3$ per minute, but that you start observing them at 2:00. By rule P2, the time until you see the first customer has the exponential distribution with $\lambda = 3$. But by rule P3, the time between the previous arrival, before 2:00, and the next arrival, after 2:00, seems also to have the same distribution. In fact, the resolution of this apparent contradiction appears at first stranger than the contradiction itself: They do not have the same distribution. The time between two arbitrary arrivals, say the 20th and the 21st, has the exponential distribution, but the time between the two arrivals on either side of 2:00 has a different distribution and is likely to be longer. This indeed seems strange; but it is a property of Poisson processes, and Poisson processes have passed many tests as successful models of real situations. This "paradox" is discussed further in Exercise 11 of Section 9.1. It is related to several other properties that we will discover later in the text, properties that are concerned with what is called the "lack of memory" of a Poisson process. The amount of time since the last arrival before 2:00 has no influence on the time between 2:00 and the first arrival after 2:00.

We should also remark that the four rules, P1–P4, are not the whole truth about Poisson processes; there are many other facts that can be used to compute various kinds of probabilities for them. But all the facts can be proved from those four rules; indeed, they can be proved from a smaller set of assumptions than those four. Such a set of assumptions would be a formal, abstract definition of a Poisson process, which we will not cover in this text.

(1.4.11) The connection between Bernoulli processes and Poisson processes. This connection is important for several reasons, one of which is that it allows us to use a Bernoulli process to get an approximate simulation of a Poisson process, as follows.

Suppose, for example, that we want to simulate a 1-hour time interval of a Poisson process with a rate of λ arrivals per hour. Choose a large integer n, and divide the hour up into n equal subintervals. Do n Bernoulli trials, each with success probability $p = \lambda/n$, one for each subinterval. For each trial, if the result of the trial is S, put an arrival at the end of the corresponding subinterval. If the result is F, there is no arrival in that subinterval.

If n is small, this simulation will not look much like a Poisson process. If $n = 60$, for example, there will be only 60 possible arrival times; all the arrivals that occur will be at the ends of whole minutes. But if n is 3,600, then the end of every second is a possible arrival time, and if $n = 36{,}000$, then the arrival times will be measured in tenths of seconds.

It seems clear that we can make the simulated process look very much like a set of random arrival times. But why should this Bernoulli simulation have the probabilities of a true Poisson process? For example, why should the probability of four arrivals in the simulated 1-hour process equal the Poisson probability $\lambda^4 e^{-\lambda}/4!$?

Well, it doesn't, at least not exactly. The real probability that the simulation produces four arrivals in an hour is

$$P(\text{four arrivals in simulated process}) = \binom{n}{4}\left(\frac{\lambda}{n}\right)^4\left(1-\frac{\lambda}{n}\right)^{n-4},$$

because we have n trials with success probability λ/n, and this is the probability of four successes. But watch what a little algebra does to this probability; it equals

$$\begin{aligned}
&\frac{n(n-1)(n-2)(n-3)}{4!}\cdot\frac{\lambda^4}{n^4}\cdot\left(1-\frac{\lambda}{n}\right)^n\cdot\left(1-\frac{\lambda}{n}\right)^{-4}\\
&=\left[\frac{\lambda^4}{4!}\cdot\left(1-\frac{\lambda}{n}\right)^n\right]\cdot\left[\frac{n}{n}\cdot\frac{n-1}{n}\cdot\frac{n-2}{n}\cdot\frac{n-3}{n}\right]\cdot\left[\left(1-\frac{\lambda}{n}\right)^{-4}\right].
\end{aligned}$$

Now we need only to recall some limits from calculus. The middle bracketed quantity is the product of four factors that approach 1 as $n \to \infty$, so for large n

it is near 1. The third bracketed quantity is also near 1 for large n, because it is the fourth power of a quantity that approaches 1 as $n \to \infty$. And finally,

$$\lim_{n\to\infty}\left(1-\frac{\lambda}{n}\right)^n = e^{-\lambda}$$

(see Appendix A.5). As a result,

$$\lim_{n\to\infty}\binom{n}{4}\left(\frac{\lambda}{n}\right)^4\left(1-\frac{\lambda}{n}\right)^{n-4} = \frac{\lambda^4 e^{-\lambda}}{4!},$$

and so if n is large, then the probability of four arrivals in the simulated Poisson process is close to the desired probability for a real Poisson process. Needless to say, this calculation works just as well for any number of arrivals other than four.

The general simulation method is as follows. Exercise 17 at the end of this section asks you to do some experimenting with it.

> To simulate a Poisson process with a rate of λ arrivals per unit time, over the time interval $[0, T]$, let the possible arrival times be the n equally spaced time points $\frac{T}{n}, \frac{2T}{n}, \frac{3T}{n}$, and so on. Do n Bernoulli trials, with success probability $p = \lambda T/n$. If the jth trial results in S, put an arrival at time point jT/n.

(1.4.12) The connection between Bernoulli processes and the uniform model on [0, 1]. There is a surprising connection if the success probability in the Bernoulli process is $\frac{1}{2}$. Remember that an outcome of a Bernoulli process is an infinite sequence of S's and F's. Suppose we write 1 for S and 0 for F, and put a decimal point—really a binary point—in front of the sequence. Then we have the binary expansion of a number in [0, 1]. So there is a correspondence between outcomes of a Bernoulli process and numbers in [0, 1]. It is not quite a one-to-one correspondence, because some numbers have two different binary expansions. But this technical detail can be overcome, because the set of these troublesome numbers has probability 0.

Now here is the connection: If the probabilities of S and F are both $\frac{1}{2}$, then the number that results from the string of trials can be modeled by the uniform probability model on [0, 1]. That is, we can simulate an infinite sequence of Bernoulli trials $\left(\text{with } p = \frac{1}{2}\right)$ by choosing a number from [0, 1] according to the uniform model and looking at its binary expansion. Similarly, we can simulate a uniform random number from [0, 1] by tossing a fair coin repeatedly and writing down a 1 for each head and a 0 for each tail; the result is the binary expansion of the random number.

This is not too difficult to prove, but we omit the proof here. Something really interesting happens if the success probability is not $\frac{1}{2}$: The S–F sequence still converts to a number in [0, 1], but it is no longer uniformly distributed; in fact, its distribution is one of those that are neither discrete nor absolutely continuous. For more on the subject, see the first few pages of Billingsley's book *Probability and Measure* (see the bibliography for information).

EXERCISES FOR **SECTION 1.4**

1. A fair die is rolled repeatedly and each roll counts as a success if a 6 appears. Find the probabilities of the following events, correct to four significant digits.
   a. Exactly three 6's in the first 12 rolls
   b. Exactly three 6's in rolls 51 through 62
   c. The third 6 comes within the first 12 rolls.
   d. The fourth 6 does not appear until after the twelfth roll.

   Warning: If you use .16 or .17 for $\frac{1}{6}$, your answers will not be reliable to more than one significant digit. To get four digits of accuracy, carry at least six. If possible, compute $\frac{1}{6}$ and store it in the calculator's memory to get all the digits the calculator can give for $\frac{1}{6}$. Better yet, simplify all your answers to fractions and do not divide until the end of the calculation.

2. A certain space agency makes a sequence of attempts to launch shuttles. Suppose that the probability of a successful launch is .85, and that the sequence of attempts can be modeled by a Bernoulli process. Find the probability that:
   a. There are at least eight successes in the first ten attempts.
   b. The first successful launch does not occur within the next three attempts.
   c. There are two or more failures between the third and fourth successful launches.
   d. There are six successes in the first ten attempts, but the first two of the ten are failures.

3. For a Poisson process with an arrival rate of $\lambda = 3.5$ arrivals per minute, find the probabilities of the following events.
   a. Exactly three arrivals in the first 2 minutes
   b. No arrivals in the first 45 seconds
   c. The fifth arrival comes in the interval between 40 seconds and 50 seconds after the fourth.

4. In a large factory building, where the fluorescent lights are kept on day and night, the lights burn out at the rate of four per day. Assume for simplicity that they are replaced as soon as they burn out, and that the times of burnout follow a Poisson process. Find the probabilities of:
   a. More than two burnouts between noon and 1 P.M. tomorrow.
   b. Exactly four burnouts tomorrow.
   c. Exactly 28 burnouts next week.
   d. No burnouts tomorrow and then three burnouts between midnight and noon the next day.

5. Tests of a certain brand of car show that its breakdown rate is one breakdown in 48 months. A salesperson tells you that, therefore, you can expect to go 4 years without a breakdown.
   a. What is the actual probability that you will go 48 months without a breakdown? Assume that successive months represent Bernoulli trials, with a "success" being a breakdown in that month, so the success probability is $\frac{1}{48}$.
   b. What is the probability that the first breakdown comes in the first year?
   c. What is the probability that you will have no breakdowns in the first year, but one or more breakdowns in the second year?

6. Repeat Exercise 5, but use a Poisson process as a model; you can use the year as a time unit, with $\lambda = \frac{1}{4}$, or the month, with $\lambda = \frac{1}{48}$. [Notice how close the answers are to those of Exercise 5. See (1.4.11) for an explanation of why this should not be surprising.]

7. Suppose that during the tornado season, tornadoes occur in a given section of Kansas according to a Poisson process with a rate of .9 tornado per week. Find the probability that:
   a. The waiting time for the first tornado after the start of the season is longer than 2 weeks.
   b. Less than 2 days elapse between the first and second tornadoes.
   c. There are three or more tornadoes in the last 2 weeks of the season.

8. During the years from 1851 to 1890, coal-mine disasters (defined as explosions resulting in more than ten deaths) occurred in Great Britain according to a Poisson process (approximately) with a rate of 3.1418 disasters per year. Assuming this model was still valid in the 1910s and 1920s, find in two ways the probability that there were no disasters in the years 1919, 1920, and 1921. [*Hint:* See the two solutions for Example (1.4.8). The difference is that there we found the probability that the next arrival occurs *before* some time, whereas here we want the probability that it occurs *after* the end of the 3-year period.]

   There were in fact no disasters in that 3-year period. So if the model was still valid by 1919, then an event of very small probability occurred. This leads us to believe that the model was *not* still valid by 1919. The simplest explanation is that λ was much less than 3.1418 by then, because safety methods had improved.

   The data for this problem were taken from *Data* by D. F. Andrews and A. M. Herzberg, a useful and instructive collection of 71 data sets from a wide range of sources. See the bibliography for information on books cited in this text.

9. A radio disk jockey announces at precisely 10:06 that the fourth person to call the station will win a prize. Telephone calls then come in to the station in accordance with a Poisson process, at the rate of 9.3 calls per minute. Find the probabilities of the following events:
   a. The first call comes in between 10:06:05 and 10:06:15.

b. The second call comes in within the first 10 seconds after 10:06.
c. The fourth call comes in after the first 30 seconds.

10. In a certain small town, robberies are reported to the police at random times that follow a Poisson process with $\lambda = .6$ per day. What is the probability that, next week, four robberies are reported, but none are reported on Monday or Tuesday?

11. In a Bernoulli process with success probability $p = \frac{1}{6}$, what is the probability that the first success occurs after the fourth trial, but on an odd-numbered trial?

12. A certain experiment has numerical outcomes that can be modeled by the standard normal model discussed in Example (1.3.5). The outcome is considered a success if it is between -1 and 1. If the experiment is repeated independently 20 times, what is the probability that there are either 13 or 14 successes?

13. We observe ten successive, nonoverlapping 1-minute intervals of a Poisson process with an arrival rate of $\lambda = 4$ arrivals per minute. Each minute counts as a success if there are two or more arrivals. What is the probability that:
a. There are exactly 8 successful minutes?
b. All 10 minutes are successes?
c. There are at least 8 successful minutes?

14. **The "problem of points" and the origin of probability theory.** Players A and B have been playing a game in which the first player to get 21 points wins. Successive points are Bernoulli trials, and each player has probability $\frac{1}{2}$ of winning a point. But the game is interrupted at a time when A has 19 points and B has 17. What is the probability that A would have won had they been able to continue? [*Hint:* Translate the statement "the second S comes before the fourth F" into a statement about the number of S's in the next five trials.]

This is part of the problem that (according to some people) brought probability theory into being. The original "problem of points" goes like this: Suppose A and B have agreed that the winner (the first one to reach n points) will collect $1 from the loser. If the game is interrupted at a point when A needs a more points to win and B needs b points, how much should B pay A to settle the game fairly? Versions of this problem are known to have been in print as early as 1380. But it is famous because, after Pascal heard of it from the Chevalier de Méré, he wrote about it to Fermat in 1654, and their subsequent exchange of letters led to Pascal's book *Traité du Triangle Arithmetique*. This book dealt with the binomial coefficients (Pascal's triangle) and the binomial distribution—and, in the minds of some, gave probability theory its start.

Of course, people had thought about probabilities earlier, and it is impossible to say when the theory had its beginning. But the Pascal–Fermat correspondence is considered a great milestone; perhaps more than any other event it deserves to be called the birth of the modern subject.

The connection between the original problem and this exercise is that if p is the probability that A will win, then B should pay $p - (1 - p)$ dollars to A. This is true because, as we will see in Chapter 4, this is the expected value of A's gain—that is, the average amount A could expect to win from B if they were to finish the game repeatedly starting at 19–17. See Exercise 13 in Section 4.1.

15. **The Martingale gambling system: Double your bet after each loss and you will eventually be ahead.** Suppose you want to win money playing a casino game in which, on each play, you win the amount you bet with probability p and lose it with probability q, where $p + q = 1$ and p is a little less than $\frac{1}{2}$. The scheme is this. Bet a dollar on the first play. If you win, stop; you are a dollar ahead. If you lose, bet two dollars on the second play. If you win, stop; you are a dollar ahead. If you lose, bet four dollars on the third play. If you win, stop; you are a dollar ahead. Otherwise bet eight dollars on the fourth play. And so on.

As we have suggested, and as we will see rigorously in Section 2.5, as long as p is not 0, you are sure to win eventually; therefore you are guaranteed to win a dollar with this scheme. Do it a million times and you will be a millionaire. Or will you? Surely there must be a catch.

The difficulty is that you may not have enough money to keep doubling your bets if you suffer enough losses before your first win. Suppose you start with \$127; find the probability that the scheme will fail and you will lose all your money before winning the dollar.

This number is quite small. If, for example, $p = .45$ (which is typical of your chance of winning in casino gambling games), then the probability that the scheme fails when you start with \$127 is only about .015. So you have a better than 98% chance of making a dollar. However, if you pocket that dollar and try to win another, then another, and so on, you will eventually fail; then your \$127 will be gone and you will have only the dollars you have managed to win.

Perhaps of more interest here is the probability of eventual failure if you do not simply pocket your one-dollar winnings, but add them to your original \$127 to increase the chance of winning future dollars. That situation is more complicated, but it turns out that your probability of eventual failure is 1 here also. And in this case you are not even left with the dollars you managed to win.

16. In a Poisson process with $\lambda = 3.4$ arrivals per hour, what is the probability that there are five arrivals in the first hour? If you simulate the first hour of this process using $n = 400$ subintervals, what is the probability that the simulation produces five arrivals? Obtain both answers to five significant digits.

17. (*Computer exercise*) Using the approximate simulation method described in (1.4.11), write a computer program that simulates an hour of a Poisson process with arrival rate $\lambda = 25$, using $n = 360$ time points.
    a. Run it several times to see what observed Poisson processes look like.
    b. Modify it so that the arrival rate is 3.2 instead of 25. Run this a few times.
    c. Modify the program in part b so that (i) it only prints out the number of arrivals in the hour, rather than graphing the process, and (ii) it uses $n =$ 3,600 instead of 360.

d. Now put the program in part c into a loop that makes 100 simulations of the hour. Print out the 100 arrival numbers and find the relative frequencies of no arrivals, one arrival, and so on.
e. Using a calculator, find the Poisson probabilities for $\lambda = 3.2$. [The first three are approximately $p(0) = .0408$, $p(1) = .130$, $p(2) = .209$.] Compare them with the observed relative frequencies found in part d.

ANSWERS

1. **a.** .1974 **b.** .1974 **c.** .3226 **d.** .8748
2. **a.** .82020 **b.** .003375 **c.** .0225 **d.** .005346
3. **a.** .05213 **b.** .07244 **c.** .04286
4. **a.** .0006813 **b.** .1954 **c.** .07517 **d.** .003305
5. **a.** .3640 **b.** .2233 **c.** .1734
6. **a.** .3679 **b.** .2212 **c.** .1723
7. **a.** .1653 **b.** .2267 **c.** .2694
8. .000080649
9. **a.** .3629 **b.** .4588 **c.** .3176
10. .05061
11. .2630
12. .3647
13. **a.** .1750 **b.** .3827 **c.** .9436
14. $\frac{26}{32}$
15. q^7
16. True: .12636; simulated: .12675

1.5 THE GENERAL DEFINITIONS OF PROBABILITY SPACE AND RANDOM VARIABLE

All the probability spaces that we have considered have three ingredients:

An *outcome set* Ω, which is
a finite or countable set in the discrete case,
an interval in the absolutely continuous case,
the set of sequences of S's and F's for a Bernoulli process,
a certain set of countable subsets of $[0, \infty)$ for a Poisson process.

A collection of *events*, which are
the subsets of Ω (all of them) in the discrete case;
in the other cases, sets made from some kind of basic sets by union, intersection, and complementation.

A recipe for computing *probabilities* of events, which is determined by
a *probability mass function* $p(\omega)$—that is, an assignment of probabilities to the singletons, in the discrete case;

a *probability density function* $f(x)$, for which $P(I) = \int_I f(x)dx$ for intervals I, in the absolutely continuous case;
a prescription for probabilities of certain kinds of basic sets in the other cases.

In this section we will give the general definitions of a probability space and of a random variable, the ones that are used in probability theory at all levels from the most elementary to the most advanced. Understanding these definitions fully is not essential to your understanding of the rest of the text, because we will be dealing almost exclusively with the special types of probability space that we have already defined. But it is good to know what our spaces have in common and also to know that even the most advanced branches of probability are based on nothing more complicated than this.

It ought to be clear that there need to be some restrictions on what qualifies as a probability space—that not any set, collection of subsets, and recipe for probabilities will do. For example, a collection of subsets would be useless as events if the complement of some event, or the union of two events, were not also events. Likewise, an assignment of probabilities would be useless if some event were assigned a negative probability, or if there were some disjoint events A and B for which the probability assigned to $A \cup B$ were not the sum of the probabilities assigned to A and to B.

The following definition gives exactly the restrictions that have been found to be necessary for probability spaces to model real chance experiments and, at the same time, sufficient for the proof of useful theorems.

(1.5.1) Definition. A ***probability space*** consists of

a set Ω called an ***outcome set;***

a nonempty collection $\mathcal{F}$ of subsets of Ω called ***events,*** having the property that the complement of any event, as well as the union and intersection of any finite or countable collection of events, are also events (and that $\mathcal{F}$ is nonempty, so that there is at least one event);

a function P from $\mathcal{F}$ to $\mathbb{R}$ (an assignment of ***probabilities*** to events), satisfying:

1. $P(A) \geq 0$ for all events A.
2. $P(\Omega) = 1$.
3. If $A_1, A_2, \ldots$ are finitely or countably many disjoint events, then

$$P(A_1 \cup A_2 \cup \cdots) = P(A_1) + P(A_2) + \cdots.$$

A collection of $\mathcal{F}$ of sets with the properties described is called a ***σ-field*** and a function P as described is called a ***probability measure.*** (The letter σ is a lowercase sigma, commonly used to reflect the idea of countable operations.)

Two things are worth thinking about here. First, discrete and absolutely continuous probability spaces do fit this definition. (So do Bernoulli and Poisson processes, of course.) To check this we would need to confirm that, in both cases, the collection of events and the assignment of probabilities are guaranteed to fill the requirements above. For example, in both cases the family of events is a σ-field: in the discrete case because *all* subsets are events and the family of all subsets is a σ-field, and in the absolutely continuous case because, as we mentioned in (1.3.2), the family of Borel sets is defined as the smallest σ-field containing the intervals. Similarly, in both cases the probability of the union of countably many disjoint events is the sum of their probabilities, essentially because that is the way sums (of mass function values) and integrals (of density functions) behave.

Second—and this is important to keep in mind—***the definition says nothing about how the probabilities should be specified, as long as they satisfy conditions 1, 2, and 3.*** This can be done with a mass function on a countable set or with a density function on an interval, but it can also be done in any way one chooses. Anything satisfying the definition is a probability space, and it is up to the modeler to find one that fits the situation.

In this connection it might be illuminating to consider one rather trivial example of a probability space, just to see that there are other kinds of spaces than those we have seen. Let Ω be the set of real numbers, but let $\mathcal{F}$ consist of just eight events: the empty set $\varnothing$, the whole space $\Omega = \mathbb{R}$, the two sets $A = [-1, 0)$ and $B = [0, 1]$ and their union $[-1, 1]$, and the complements of these three sets. Define P on $\mathcal{F}$ by the following table:

Event E:	$\varnothing$	A	B	$A \cup B$	A^c	B^c	$(A \cup B)^c$	$\mathbb{R}$
$P(E)$:	0	.2	.4	.6	.8	.6	.4	1

Now what we must check is the following: (*i*) the collection $\mathcal{F}$ of these eight sets is a σ-field, and (*ii*) P is a probability measure. That is, (*i*) unions, intersections, and complements of any of these eight sets are also among the eight, and (*ii*) the assigned probabilities are nonnegative with $P(\mathbb{R}) = 1$ and the probability of a disjoint union is the sum of the probabilities. Both of these are easy to confirm.

Could this peculiar probability space model any real chance experiment? Perhaps. Imagine an experiment with real-number outcomes, in which we do not care, or cannot tell, anything about the outcome except which, if either, of the two intervals $[-1, 0)$ and $[0, 1]$ it is in. Then the only sets that need probabilities assigned to them are the eight sets in $\mathcal{F}$.

But we do not have to find a use for every model; the point is that in trying to define models meaningfully we are forced to admit some models that may not be good for anything. This is no different from what happens in other branches of mathematics. For example, there is no denying the usefulness of the concepts of calculus, but the theory of calculus forces us to admit the existence of functions so strange that it is difficult to see that they represent anything. To exclude such "useless" objects from the theory would make many useful theorems invalid, or at least more difficult.

Random variables on general probability spaces. Finally, we take up the question of which functions on a probability space can qualify as random variables. First, it will be worth the time to think again of two ideas we considered earlier.

As we discussed in Section 1.3, for technical reasons it is not possible in some outcome sets to allow every subset to be an event. If we did so in $\Omega = [0, 1)$, for example, there would be no way to define probabilities for all subsets so that the probability of every interval was its length. That was the bad news; the good news was that the sets we are forced to exclude are sets no one can describe.

The second thing to remember is that when we have an outcome set Ω and a random variable X, which is a function from Ω to $\mathbb{R}$, statements about X, such as "$3 \le X \le 5$," translate to subsets of Ω, in this case $\{\omega: 3 \le X(\omega) \le 5\}$. Thus, when we write $P(3 \le X \le 5)$, we mean the probability of this event in Ω.

So what we must require is that all such subsets be events, so that they will have probabilities assigned to them. That is, if Ω is an outcome set and $\mathcal{F}$ is the set of events, then a function X on Ω will not do as a random variable unless, for any interval I, the statement "X is in I" translates to a subset of Ω that is in $\mathcal{F}$. This requirement is formalized as follows.

(1.5.2) Definition. If Ω is an outcome set and $\mathcal{F}$ is the σ-field of subsets of Ω that are the events for a probability space, then a ***random variable*** is a function X from Ω to $\mathbb{R}$ with the property that, for every interval I in $\mathbb{R}$, the subset $\{\omega: X(\omega) \in I\} = (X \in I)$ is a member of $\mathcal{F}$.

Fortunately, this abstract definition need not concern us as we proceed to work with discrete and absolutely continuous probability spaces, or Bernoulli or Poisson processes, just as the question of which subsets are events need not concern us. For just as the sets that are not allowed to be events are sets so complicated that they cannot be described explicitly, so the functions that are disallowed as random variables are not describable. Any function you can specify explicitly on $\mathbb{R}$ or higher-dimensional spaces $\mathbb{R}^n$ satisfies the condition in the definition (as long as the events are the Borel sets, which they are in absolutely continuous models).

Thus, the concepts of σ-fields of events, and of random variables as functions that satisfy the definition above, are necessary in the general theory; but in the kinds of spaces we work with they boil down to the following:

In a discrete probability space:
- Ω is finite or countable;
- every subset of Ω is an event; and
- every function from Ω to $\mathbb{R}$ is a random variable.

In an absolutely continuous probability space, in a Bernoulli process, or in a Poisson process:
- Ω is an interval, or the set of sequences of S's and F's, or the set of "Poissonian" (see Section 1.4) subsets of $[0, \infty)$;
- the events are the sets that can be made from basic events by set operations,

and subsets that are *not* events are not describable; and
the random variables are the functions X satisfying the definition above, and functions that are *not* random variables are not describable.

EXERCISES FOR **SECTION 1.5**

1. Let $\Omega = [0, \infty)$ and define X on Ω by $X(\omega) = 1/(1 + \omega^2)$. For each of the following intervals I, find the event $(X \in I) = \{\omega: X(\omega) \in I\}$. [*Hint:* Graph the function $1/(1 + \omega^2)$ for $0 \le \omega$.]
a. $I = \left[\frac{1}{5}, \frac{1}{2}\right]$ **b.** $I = \left(0, \frac{1}{10}\right)$ **c.** $I = (2, 7)$ **d.** $I = [0, 1]$ **e.** $I = [-6, 1]$

2. Let $\Omega = \{0, 1, 2, 3, 4, \ldots\}$ and define X on Ω by $X(\omega) = (-1)^{\omega}$. For each of the following intervals I, find the event $(X \in I) = \{\omega: X(\omega) \in I\}$.
a. $I = [0, 1.5)$ **b.** $I = (0. 1.5)$ **c.** $I = (-3, 3)$ **d.** $I = [-1, 1]$
e. $I = (-1, 1)$

3. Let Ω be the three-element set $\{a, b, c\}$ and let $\mathscr{F}$ be the σ-field of all eight subsets of Ω. Define probabilities on these eight events by means of the following mass function:

$$p(a) = \tfrac{1}{7}, \quad p(b) = \tfrac{4}{7}, \quad p(c) = \tfrac{2}{7}.$$

Let X be the random variable from Ω to $\mathbb{R}$ defined by

$$X(a) = 1, \quad X(b) = 5, \quad X(c) = 0.$$

Find the following.
a. The probability $P\{a, b\}$
b. The event $\{\omega: X(\omega) \in I\}$ where I is the interval $[3, 6]$
c. The probability of the event in part b—that is, $P(3 \le X \le 6)$
d. The probability of the event $\{\omega: X(\omega) = n\}$, that is, $P(X = n)$, for each $n = 0, 1, 2, 3, \ldots$.

4. Let $\Omega = \mathbb{R}$ and let A be the subset $[1, 3]$ of Ω.
a. Show that the four sets $\varnothing, A, A^c, \Omega$ constitute a σ-field.
b. Show that the following assignment of probabilities to the four events satisfies the requirements for a probability measure.

Event E:	$\varnothing$	A	A^c	Ω
$P(E)$:	0	.32	.68	1

c. Define the function $X(\omega)$ on Ω by $X(\omega) = \omega^2$. Show that X is *not* a random variable if these four sets are the only events. [*Hint:* Pick an interval, say $I = [0, 2]$, and show that $[\omega: X(\omega) \in I\,]$ is not an event.]
d. Define $Y(\omega)$ by $Y(\omega) = 2$ if ω is in A and $Y(\omega) = 3$ if ω is not in A. Show that Y is a random variable. [*Hint:* No matter what interval I you pick, it

either contains both 2 and 3, or 2 but not 3, or 3 but not 2, or neither 2 nor 3. In each of these cases, find the set $\{\omega: X(\omega) \in I\}$ and show that it is one of the four events.]

5. Referring to Exercise 4, let Ω be any set and A any subset, and replace the probability values .32 and .68 by p and $1 - p$, where p is any number between 0 and 1. Show that parts a and b of Exercise 4 are still valid—that is, that we still have a probability space.

ANSWERS

1. a. $[1, 2]$ **b.** $(3, \infty)$ **c.** $\varnothing$ **d.** Ω **e.** Ω

2. a. $\{0, 2, 4, 6, \ldots\}$
b. $\{0, 2, 4, 6, \ldots\}$ **c.** Ω **d.** Ω **e.** $\varnothing$

3. a. $\frac{5}{7}$ **b.** $\{b\}$ **c.** $\frac{4}{7}$
d. $\frac{2}{7}$ for $n = 0$, $\frac{1}{7}$ for $n = 1$, $\frac{4}{7}$ for $n = 5$, 0 for all other n

SUPPLEMENTARY EXERCISES FOR CHAPTER 1

1. Let probabilities be defined on the Borel sets in $\mathbb{R}$ by the standard normal density, as in Example (1.3.5). Using the normal distribution table in Appendix C, find the probabilities of the following events:
   a. $(-1.64, 1.64)$ **b.** $(-1.65, 1.65)$
   c. $(-2, 2)$ **d.** $(-1.96, 1.96)$
   e. $(2.58, \infty)$ **f.** $(2.71, \infty)$
   g. $(-\infty, -2)$ **h.** $(-2, 0) \cup (2, \infty)$

2. Using the standard normal probability space, as in Example (1.3.5), find the following probabilities:
   a. $P(X \leq 28)$ where X is the random variable defined by $X(\omega) = 20 + 5\omega$
   b. $P(Y > 10)$ where $Y(\omega) = 6.2 + 1.4\omega$ (Compare with Exercise 1f above.)
   c. $P(Z$ is within 8 of 12) where $Z(\omega) = 12 + 4\omega$ (Compare with Exercise 1c above.)
   d. $P(W$ is within 2σ of $\mu)$ where $W(\omega) = \mu + \sigma\omega$

   If you have worked with normal probabilities before and these seem familiar, look ahead to Section 6.2. The Greek letters μ (mu) and σ (sigma) stand for arbitrary numbers here. As we will see later, these two letters have special meaning in connection with normal probabilities, and indeed with random variables of all kinds.

3. Let probabilities be given on the Borel subsets of $(0, \infty)$ by the density $f(x) = xe^{-x^2/2}$ $(0 < x)$. Find the following. You can do each of them by integrating the density, but you can save time by first finding a formula for the probability of $(0, z]$ as a function of $z > 0$.
   a. $P(0, 1]$ **b.** $P(.5, 2]$
   c. $P[.5, 2)$ **d.** $P[4, \infty)$
   e. The number a for which $P(a, \infty) = \frac{1}{2}$

4. For the experiment of choosing a point at random from the unit square $\Omega = [0, 1] \times [0, 1]$ in the plane, the model to use is the one that assigns to any Borel subset of Ω its area. (No scaling is necessary as it was in Example (1.3.13) because the area of Ω is 1 in this case.) Find the probabilities of the following events:
   a. $\left(U \le \frac{3}{4}\right)$, where $U(\omega)$ is the sum of the two coordinates of the point ω
   b. $(U \le z)$, where z is a number between 0 and 1
   c. $(U \le z)$, where z is a number between 1 and 2
   d. $\left(V \le \frac{3}{4}\right)$, where $V(\omega)$ is the *larger* of the two coordinates of ω [*Hint:* The larger of the two coordinates is less than or equal to $\frac{3}{4}$ if and only if both coordinates are.]
   e. $(V \le z)$, where z is a number between 0 and 1

5. You have a coin that has been somehow weighted so that a head comes up only 10% of the time. You toss it until the first head comes up and count the number of tails that precede the first head. Find the probability that the number of tails is:
   a. Equal to 2.
   b. Equal to 9.
   c. Less than 3.
   d. Greater than or equal to 2.
   e. Greater than or equal to 3.
   f. Greater than or equal to k, for $k = 0, 1, 2, 3, \ldots$.
   g. Even.
   h. In the interval $[1, \infty)$.
   i. In the interval $(1, \infty)$.

6. Let probabilities be given on the Borel subsets of $[0, 1]$ by the density function $f(x) = 4(1 - x)^3$ $(0 \le x \le 1)$. Find the following. You can find each of them as an integral of the density, but you can save time by first finding $P[0, z]$ as a function of z between 0 and 1.
   a. $P[0, .5]$
   b. $P(.25, 1]$
   c. $P[0, .1]$
   d. $P(.9, 1]$

7. If $f(x) = Cx^2(1 - x)^2$ $(0 < x < 1)$ is a density function, find each of the following:
   a. The value of C
   b. $P(0, z]$ as a function of z for $0 < z < 1$
   c. $P(.5, 1)$

8. For a Poisson process with an arrival rate of $\lambda = 4.5$ per minute, find the probability that:
   a. There are nine arrivals in the next 2 minutes.
   b. There is at least one arrival in the next 30 seconds.

c. There are at most three arrivals in the next 90 seconds.
d. There are no arrivals in a given 20-second period.
e. There is at least 1 minute between the fifth and sixth arrivals.
f. The second arrival comes after the first minute is over.
g. The waiting time for the first arrival is less than 10 seconds.

9. A certain experiment has probability .3 of success and .7 of failure. It is performed repeatedly in such a way that the outcome of one trial cannot influence that of another. What is the probability that:
a. there are two successes in five trials?
b. there are three successes in ten trials?
c. there is at least one success in ten trials?
d. there are more than three successes in five trials?
e. the third success comes within the first eight trials?
f. the first 18 trials produce five successes, all of which come before the 16th trial?

10. Two fair dice are rolled and a success is counted if the sum is 7 or 11.
a. What is the probability of success for a single roll of the two dice?
b. What is the probability of three successes in ten rolls?
c. What is the probability that there are no successes in eight rolls?
d. What is the probability that there are five successes in 20 rolls?

11. You use a computer's random number generator to get a sequence of numbers; for some reason you decide to count a number as "good" if its first digit is a 1 or its second digit is a 2 or both. Find the following.
a. The probability that none of the first three numbers is good
b. The probability that the third number is the first good one
c. The probability that there are four good numbers in the first ten

12. In Bernoulli trials with success probability $\frac{1}{4}$, the probability that the first success comes on the kth trial is $\frac{1}{4} \cdot \left(\frac{3}{4}\right)^{k-1}$ for $k = 1, 2, 3, \ldots$. Find the probability that the first success comes on a trial whose number is divisible by 3—that is, on the third, sixth, ninth, . . . trial.

13. In Bernoulli trials with success probability p and failure probability $q = 1 - p$, find the probability that the number of failures before the first success is divisible by 5.

14. In a Poisson process with an arrival rate of 2.7 per hour, find the probability that there are three arrivals in the next hour, but that they all come within the first 40 minutes.

15. If $p(x) = c \cdot (.3)^x$ for $x = 1, 2, 3, \ldots$ (and 0 otherwise) defines a probability mass function, what is the value of c?

16. A 911 emergency line in a large city receives calls for help that come in according to a Poisson process with an arrival rate of 1.7 per minute. What is the probability that there are at least two calls in the first 30 seconds after noon tomorrow?

17. In a Poisson process with an arrival rate of 3.2 per hour, what is the probability that the third arrival occurs within the first half hour?

18. a. Consider a Bernoulli process with success probability p. If the probability of getting at least one success in ten trials is exactly $\frac{1}{2}$, what is p?
    b. Consider a Poisson process with arrival rate λ per minute. If the probability of at least one arrival in 10 minutes is exactly $\frac{1}{2}$, what is λ?

ANSWERS

1. a. .8990 b. .9010 c. .9544 d. .9500 e. .0049 f. .0034 g. .0228 h. .5000
2. a. .9452 b. .0034 c. .9544 d. .9544
3. a. .3935 b. .7472 c. .7472 d. .0003 e. 1.1774
4. a. $\frac{9}{32}$ b. $\frac{z^2}{2}$ c. $1 - \frac{1}{2}(2 - z)^2$ d. $\frac{9}{16}$ e. z^2
5. a. .081 b. .0387 c. .2710 d. .81 e. .729 f. $(.9)^k$ g. .5263 h. .9 i. .81
6. a. .9375 b. .3164 c. .3439 d. .0001
7. a. 30 b. $30\left(\frac{z^3}{3} - \frac{z^4}{2} + \frac{z^5}{5}\right)$ c. .5
8. a. .1318 b. .8946 c. .0958 d. .2231 e. .0111 f. .0611 g. .5276
9. a. .3087 b. .2668 c. .9718 d. .0308 e. .44823 f. .070703
10. a. $\frac{8}{36}$ b. .2267 c. .1339 d. .1937
11. a. .5314 b. .1247 c. .07729
12. $\frac{9}{37}$
13. $p/(1 - q^5)$
14. .0653
15. $\frac{7}{3}$
16. .2093
17. .2166
18. a. .0670 b. .0693

CHAPTER

2

The Algebra of Events and Probabilities

2.1 Basic Formulas and Inequalities for Probabilities

2.2 Conditional Probabilities

2.3 Conditioning and Bayes's Theorem

2.4 Independence

2.5 Limits of Probabilities of Nested Events

Supplementary Exercises

2.1 BASIC FORMULAS AND INEQUALITIES FOR PROBABILITIES

This section contains a few of the many identities and inequalities that hold in any probability space because they follow from the general definition (1.5.1). (If you omitted Section 1.5, no harm was done, because we repeat the definitions below, and that is all you need.) Some of them we have already used in Chapter 1, almost without thinking about the need for their justification. We include their proofs, which are either quite simple or just a little tricky. In all cases, though, it is easy to understand what the formulas say. You might want to refer to Section A.2 in Appendix A for a brief description of the basic set operations.

Really there are only two basic formulas in this section, (2.1.1) and (2.1.2)—although the latter has generalizations, like (2.1.3), that are often useful. But it would be a serious mistake to think that, because those two formulas are clear and simple, you understand the material of the section. What is important is the use of the formulas in various ways to answer a wide range of questions. *As in most sections of this text, the examples and exercises are the main things to study.*

First, we give the definition [a repetition of (1.5.1)] of a probability space. Especially important for our purposes here are the three requirements for P.

Definition. A ***probability space*** consists of

a set Ω called an ***outcome set***;

a nonempty collection $\mathcal{F}$ of subsets of Ω called ***events***, having the property that the complement of any event, and the union and intersection of any finite or countable collection of events, are also events

a function P from $\mathcal{F}$ to $\mathbb{R}$ (an assignment of ***probabilities*** to events), satisfying:

1. $P(A) \geq 0$ for all events A;
2. $P(\Omega) = 1$; and
3. if $A_1, A_2, \ldots$ are finitely or countably many disjoint events, then

$$P(A_1 \cup A_2 \cup \cdots) = P(A_1) + P(A_2) + \cdots.$$

Notice that $\mathcal{F}$ is assumed to be nonempty; that is, there is at least one event. A probability space without events would be useless, and would also have to be treated as an exceptional case in many definitions and theorems.

The fact that there is at least one event implies that ***both Ω and $\varnothing$ are events.*** Why? Because if A is an event, then so is A^c and, therefore, so are $A \cup A^c$, which is Ω, and $A \cap A^c$, which is $\varnothing$.

In Chapter 1 we described two special kinds of probability space, discrete and absolutely continuous. In discrete spaces Ω is finite or countable and $\mathcal{F}$ consists of all subsets of Ω; in absolutely continuous spaces Ω is an interval in $\mathbb{R}$ or some such set in $\mathbb{R}^n$ and $\mathcal{F}$ is the family of Borel sets. You may remember from (1.3.2) that a family $\mathcal{F}$ with the properties described above is called a ***σ-field*** and that the family of Borel sets is the smallest σ-field containing the intervals. We also considered two other probability spaces that belong to neither of these types: Bernoulli processes and Poisson processes. Here we are considering all probability spaces at once; what we derive is true not only for the spaces of Chapter 1, but also for any other probability spaces we might ever encounter.

Our first formula relates the probability of the complement of a set to the probability of the set itself. Notice that if A is any event, then one verbal description of the event A^c is "A does not occur."

(2.1.1) $$P(A^c) = 1 - P(A) \quad \textbf{for any event } A.$$

This is of course true because A and A^c are disjoint and their union is Ω, and therefore, $1 = P(\Omega) = P(A) + P(A^c)$. We used (2.1.1) several times in Chapter 1 without mentioning it. See, for example, the calculations of the probability that more than three tosses are needed in Example (1.2.5), or the probability of the event $\{2, 3, 4, \ldots\}$ in Example (1.2.11).

Formula (2.1.1) is often useful in finding the probability of an event whose verbal description contains the words "at least": for example, in tossing a fair coin a certain number of times, $P(\text{at least one head}) = 1 - P(\text{no heads})$. We will give some examples after presenting the next identity.

An obvious corollary of (2.1.1) is that the empty set must have probability 0 in any probability space, because $P(\varnothing) = 1 - P(\varnothing^c) = 1 - P(\Omega) = 1 - 1 = 0$.

The next identity gives the probability of the union of two events. If A and B are any events, their union $A \cup B$ has the verbal description "A occurs or B occurs," where, as always in mathematics, "or" is interpreted to mean "and/or." Another verbal description of $A \cup B$ is "at least one of A and B occurs." This latter description works as well for unions of more than two events: the union of any collection of events is the event that at least one of them occurs.

(2.1.2) $$P(A \cup B) = P(A) + P(B) - P(A \cap B) \quad \textbf{for any events } A \textbf{ and } B.$$

Notice that if A and B are disjoint, then we already have a formula for $P(A \cup B)$, the one given by the third property of probabilities in Definition (1.5.1): $P(A \cup B) = P(A) + P(B)$. But (2.1.2) does not conflict with this, because if A and B are disjoint, then $A \cap B$ is empty, so its probability is 0.

Proof of (2.1.2): This proof is easier to follow if you refer to a Venn diagram (see Figure 2.1a, page 86). We observe the following facts:

The three events $(A \cap B^c)$, $(B \cap A^c)$, and $(A \cap B)$ are disjoint;

$$A \cup B = (A \cap B^c) \cup (A \cap B) \cup (B \cap A^c);$$

$$A = (A \cap B^c) \cup (A \cap B); \quad \text{and}$$

$$B = (B \cap A^c) \cup (A \cap B).$$

Therefore,

$$P(A) + P(B) = P(A \cap B^c) + P(A \cap B) + P(B \cap A^c) + P(A \cap B);$$

that is,

$$P(A) + P(B) = P(A \cup B) + P(A \cap B),$$

which is the same as (2.1.2). ▯

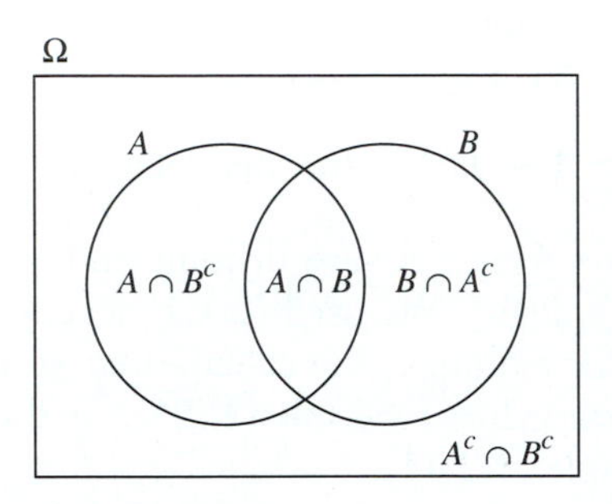

FIGURE 2.1a **Venn diagram illustrating the proof of** (2.1.2)

There is a version of (2.1.2) for three events, and indeed one for n events, for any integer $n \geq 2$. The three-event version looks formidable but has a certain order about it, which suggests the n-event version:

$$\begin{aligned} P(A \cup B \cup C) = P(A) &+ P(B) + P(C) \\ &- P(A \cap B) - P(A \cap C) - P(B \cap C) \\ &+ P(A \cap B \cap C). \end{aligned} \tag{2.1.3}$$

In general, the probability of the union of n events is

the sum of the probabilities of the n events,

minus the sum of the probabilities of all the intersections of two events,

plus the sum of the probabilities of all the intersections of three events,

minus the sum of the probabilities of all the intersections of four events,

etc., and finally,

plus or minus the probability of the intersection of all n events.

This identity is called the ***principle of inclusion and exclusion***, for obvious reasons. Notice that there are $n = \binom{n}{1}$ single events, $\binom{n}{2}$ intersections of two events, $\binom{n}{3}$ intersections of three events, and so forth, up to $\binom{n}{n} = 1$ intersection of all n events. [See (A.4.5) in Appendix A.]

You may want to check the three-event version (2.1.3) by looking at the Venn diagram in Figure 2.1b. For simplicity we have labeled the eight regions with letters, so that

$$P(A) = p + q + s + t, \quad P(B) = q + r + t + u, \quad P(C) = s + t + u + v;$$

$$P(A \cap B) = q + t, \quad P(A \cap C) = s + t, \quad P(B \cap C) = t + u;$$

$$\text{and } P(A \cap B \cap C) = t.$$

Now write down the right side of (2.1.3) using these. The result will be $p + q + r + s + t + u + v$, which is precisely $P(A \cup B \cup C)$.

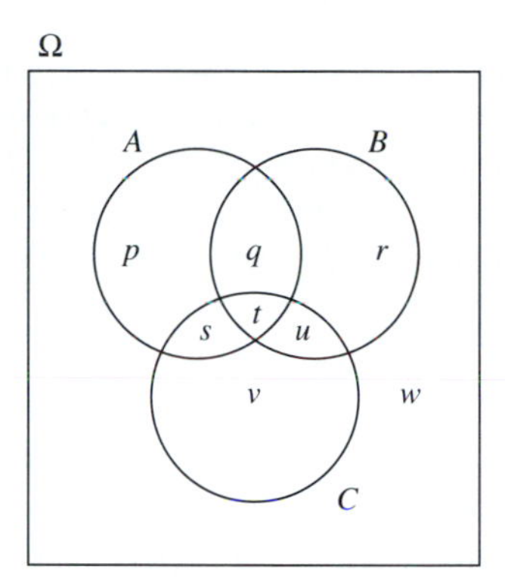

FIGURE 2.1b Venn diagram illustrating the proof of (2.1.3)

Following are a few examples of the uses of (2.1.1) and (2.1.2). In some cases, as in the first example, either formula can be used, although you may prefer one over the other. In other examples, one is decidedly more complicated than the other. Do not expect to learn immediately the best way to approach such problems or to decide which formula to use; such things are a matter of experience.

(2.1.4) **Example.** In a (very small) lottery there are ten tickets, numbered 1, 2, 3, . . . , 10. Two numbers are drawn for prizes. You hold the tickets numbered 1 and 2. What is the probability that you win at least one prize?

Solution. The event—call it B—that you win at least one prize can be thought of in two ways: as the union of the two events, A_1 = "ticket 1 is drawn" and A_2 = "ticket 2 is drawn," or as the complement of the event that neither ticket is drawn. We give both solutions. In each we use as our outcome set the set Ω of all the $10 \cdot 9 = 90$ "two-letter words" with no repeated letter from the "ten-letter alphabet" {1, 2, 3, . . . , 10} and we assume that these 90 outcomes are equally likely. [See Section A.4 in Appendix A, and especially Theorem (A.4.3), if this is unfamiliar to you.]

1. $P(B) = P(A_1 \cup A_2)$, so we find $P(A_1)$, $P(A_2)$, and $P(A_1 \cap A_2)$ and then apply (2.1.2). A_1 has 18 members (nine in which 1 is the first letter in the word, and nine in which it is the second). Similarly, A_2 has 18 members. And $A_1 \cap A_2$ has just two members, 12 and 21. So

$$P(A_1 \cup A_2) = \tfrac{18}{90} + \tfrac{18}{90} - \tfrac{2}{90} = \tfrac{34}{90}.$$

2. $P(B) = 1 - P(B^c)$, and B^c is the event that neither 1 nor 2 is drawn. It has $8 \cdot 7 = 56$ members; these are the two-letter words from the eight-letter alphabet $\{3, 4, 5, 6, 7, 8, 9, 10\}$. So its probability is $\frac{56}{90}$, and by (2.1.1),

$$P(B) = 1 - \tfrac{56}{90} = \tfrac{34}{90}.$$ ∎

In problems like the previous example, people often reason incorrectly as follows: "I have two tickets, so if there were just one draw for a prize, my chance of winning would be 2 in 10, or $\frac{1}{5}$. Since there are two draws, my chance is $\frac{2}{5}$, or 40%." The actual chance, $\frac{34}{90}$, is about 37.8%. The error in the naive reasoning is that it counts twice the probability of winning both prizes. That probability is $\frac{2}{90}$, which is exactly the difference between $\frac{2}{5}$ and $\frac{34}{90}$.

(2.1.5) Example. If the numbers 1, 2, 3, . . . , 10 are permuted randomly, what is the probability that at least one of the numbers 1 and 2 is in its proper place—that is, that 1 is first or 2 is second or both? You can get a random permutation of the numbers 1, 2, 3, . . . , 10 by putting ten slips of paper in a hat (with the numbers on them, of course) and drawing them out one at a time without replacement. [See Exercise 4 in Section 1.1. and (A.4.4) in Appendix A.] The obvious probability model for this experiment has an outcome set consisting of the 10!= 3,628,800 permutations, all equally likely.

Solution. Let B be the event described, that 1 is first or 2 is second (or both) in the permutation. In this example it turns out to be more difficult to use the complement formula (2.1.1.); B^c is the event that neither 1 nor 2 is in its proper place, and it is a little tricky to find its probability correctly. But let A_1 be the event that 1 is first and A_2 the event that 2 is second; then $B = A_1 \cup A_2$. A_1 consists of the 9! permutations that begin with 1, so $P(A_1) = \frac{9!}{10!} = \frac{1}{10}$. Similarly, A_2, the event that 2 is second, contains 9! permutations, one for each way to rearrange the rest of the numbers around a 2 in the second place. So, $P(A_2) = \frac{1}{10}$ also. And $A_1 \cap A_2$, the event that the permutation begins with 1, then 2, has 8! members, one for each permutation of the last eight numbers. So $P(A_1 \cap A_2) = \frac{8!}{10!} = \frac{1}{10 \cdot 9}$. Therefore, by (2.1.2),

$$P(B) = P(A_1 \cup A_2) = \tfrac{1}{10} + \tfrac{1}{10} - \tfrac{1}{90} = \tfrac{17}{90}.$$

It is a good general rule of thumb that to find the probability of an event whose description involves the words "at least," you should look for the probability of the complement of the event and subtract from 1. But this example is an exception. You might want to try it this way and see if your answer agrees with the solution above. If you do, it is best to break up the event that neither 1 nor 2 is in its proper place into four events: 2 is first and 1 is second; 2 is first and something else is second; 1 is second and something else is first; neither 1 nor 2 is in either of the first two places. ▮

Questions of permutations with items in or not in their proper positions are generally tricky, but tractable, using the principle of inclusion and exclusion, which generalizes (2.1.2) and (2.1.3). For example, a permutation with *no* numbers in their proper places is called a *derangement*. For permutations of 1, 2, 3, . . . , 10, the event that the permutation is a derangement is the complement of the event $A_1 \cup A_2 \cup \cdots \cup A_{10}$. It turns out not to be too difficult to find the probability of this event using the inclusion–exclusion principle. See Exercise 18 at the end of this section.

As we mentioned after Exercise 4 in Section 1.1, there are several exercises and examples concerning random permutations, and the number of items in their proper places, at various places in this text.

(2.1.6) Example. A standard deck of playing cards contains 52 cards, among which are 12 "face cards": four kings, four queens, and four jacks. Suppose you are given three draws (without replacement), and you win a prize if you can draw a face card. What is the probability that you win? (You get only one prize even if you draw more than one face card.)

Some people would approach this problem intuitively by saying that there is a $\frac{12}{52}$ chance of getting a face card on one draw, so there ought to be three times this chance, or $\frac{36}{52} \doteq .6923$, of getting a face card in three draws. This is wrong—just as the naive reasoning mentioned after Example (2.1.4) was wrong—because it tries to find the probability of the union of three events by adding their probabilities, but neglects the fact that the events are not disjoint. It is possible to get a face card on both the first and second draws, for example.

Solution. This problem is a bit complicated if we try to do it using (2.1.3), involving the union of three events. But it is quite simple if we look at the complement of the event in question. Say A is the event that you win; then A^c is the event that you get no face card in three draws. The outcome set Ω consists of $52 \cdot 51 \cdot 50$ members, which are the "three-letter words without repeated letters" from the "alphabet" of the 52 cards. The event A^c has $40 \cdot 39 \cdot 38$ members; they are words from the reduced 40-letter alphabet that has the face cards taken out. So,

$$P(A) = 1 - \frac{40 \cdot 39 \cdot 38}{52 \cdot 51 \cdot 50} \doteq .5529.$$

It is indeed more likely than not that you will succeed in getting a face card in three draws; but it is not nearly so likely as the .6923 given by the incorrect solution would indicate. ▮

Try to guess whether your chances of winning would be greater or less than .5529 if you replaced each card before drawing the next. Then, as an exercise (Exercise 13 at the end of this section), find the probability in this case. The only changes from the solution above are that Ω and A consist of words with repeated letters allowed.

(2.1.7) Example. A small bank has just one teller, who is rather lazy and likes to keep track of half-hour periods in which no customers arrive. We suppose that customers arrive according to a Poisson process with an arrival rate of $\lambda = 4$ per hour. Let X be the number of arrivals in the half hour from 1:00 to 1:30 P.M. tomorrow, and let Y be the number of arrivals in the half hour from 1:20 to 1:50. What is the probability that either $X = 0$ or $Y = 0$ (or both), so that at least one of those two overlapping half-hour periods will be a success for our teller?

Solution. Let A be the event "$X = 0$" and B the event "$Y = 0$"; we want the probability of $C = A \cup B$. Here the complement formula (2.1.1) is difficult to use, because the complement of C is the event $(X > 0) \cap (Y > 0)$, and this can happen in many different ways. (For example, an arrival between 1:20 and 1:30 will do it, but so will an arrival before 1:20 and an arrival after 1:30. And there are other combinations of these that will do it as well: several arrivals before 1:20 and some between 1:20 and 1:30, etc.)

But it is easy to find the probabilities of A, B, and $A \cap B$, and then use (2.1.2). Remember (see the rules for a Poisson process in Section 1.4) that for any time interval of length s, the probability of k arrivals in that interval is $(\lambda s)^k e^{-\lambda s}/k!$. So for our process, with $\lambda = 4$, the probability of no arrivals in a time interval of length s hours is $(4s)^0 e^{-4s}/0! = e^{-4s}$. Therefore,

$$\begin{aligned} P(A) &= P(\text{no arrivals between 1:00 and 1:30}) = e^{-2}; \\ P(B) &= P(\text{no arrivals between 1:20 and 1:50}) = e^{-2}; \\ P(A \cap B) &= P(\text{no arrivals between 1:00 and 1:50}) = e^{-(5/6)\cdot 4}. \end{aligned}$$

[See the following note if $P(A \cap B)$ does not seem right to you.] So by (2.1.2)

$$P(C) = e^{-2} + e^{-2} - e^{-(5/6)\cdot 4} \doteq .2350. \qquad \blacksquare$$

Some people are troubled by the probability of $P(A \cap B)$ given in the previous example. The fact is that if A = "no arrivals in interval I" and B = "no arrivals in interval J," then $A \cap B$ is the event "no arrivals in $I \cup J$," and not "no arrivals in $I \cap J$."

To see this, you need to think about what the verbal description "no arrivals in I *and* no arrivals in J" really says. If you learn from one person that there were no arrivals in I, and then hear from another that there were no arrivals in J, you have not learned simply that there were no arrivals in the interval common to both; you have learned that there were no arrivals at any time point that is in I or J or both.

Thinking that $A \cap B$ should be "no arrivals in $I \cap J$" just because an intersection should go with an intersection, amounts to wishfully making up an algebraic rule based on form and not on substance. It is no more valid than saying that "no blue crayons and no yellow crayons" is the same as "no green crayons," just because blue and yellow make green.

Notice, by the way, that if the events were other than they are, then $A \cap B$ might involve an intersection. For example, if A were "exactly one arrival in I" and B were "exactly one arrival in J," and if the intervals I and J were overlapping, then $A \cap B$ would be "exactly one arrival in $I \cap J$." But if they did not overlap, then $A \cap B$ would require two arrivals, one in each.

The lesson is that the algebraic rules that are valid in a given situation depend on the situation, and that real trouble lies ahead if you try to avoid thinking about the substance of the problem by looking only at its form.

Our next identity is another obvious one, but one that will provide the basis for two important formulas in Section 2.3. It involves the notion of a ***partition*** of the outcome set Ω, which is a collection of disjoint events whose union is Ω. Think for a moment about what it means for events $A_1, A_2, A_3, \ldots$ to form a partition of Ω. They are disjoint; so any two of them have an empty intersection. That is, no outcome is in more than one of the A_i's. But their union is all of Ω, so every outcome is in one of the A_i's. That is, every member of Ω is in one and only one of the events of the partition. When the experiment is done, one and only one of the A_i's occurs.

(2.1.8) **If a finite or countably infinite collection of events $A_1, A_2, A_3, \ldots$ form a partition of Ω, then for any event B,**

$$P(B) = P(B \cap A_1) + P(B \cap A_2) + P(B \cap A_3) + \cdots.$$

The reason, of course, is that if $A_1, A_2, A_3, \ldots$ partition Ω, then $B \cap A_1$, $B \cap A_2$, $B \cap A_3, \ldots$ partition B. See Figure 2.1c.

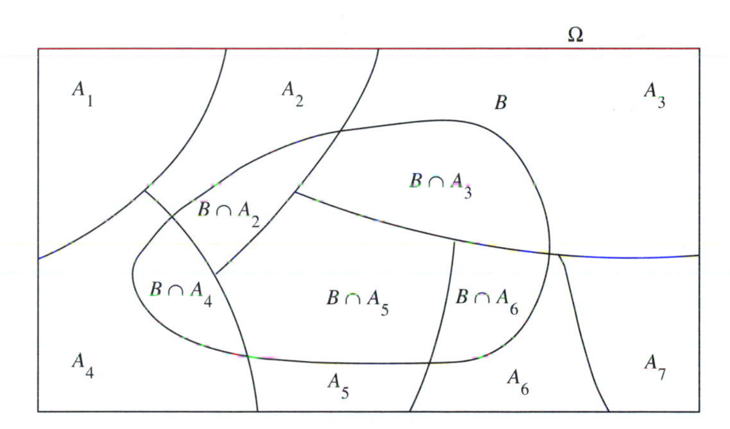

FIGURE 2.1c **If $A_1, A_2, \ldots$ partition Ω, then $B \cap A_1$, $B \cap A_2, \ldots$ partition B.**

The following example involves a chance experiment that arises frequently in probability textbooks but seldom in real life—drawing balls from urns. We imagine that one or more urns are filled with balls that are identical except for their colors, and that we choose an urn and draw some balls at random from it; we are interested in the colors of the balls drawn.

The reason such an idealized experiment is worth considering, of course, is that it is a model for a great many real experiments. For example, if we poll a population of voters to find out which candidate is preferred, we can model the experiment by thinking of the voters as balls in an urn, the colors corresponding to the candidates. Taking a poll involves drawing balls at random. This next example could be thought of as sampling from a population made up of equal numbers of three types of voters—say Democrats, Republicans, and independents—each type having its own mix of candidate preferences. To keep the calculations simple, though, we make the numbers small.

(2.1.9) Example. Suppose we have three urns, containing balls that are identical except that some are white and some are blue. Urn 1 contains one blue ball and three white balls; urn 2 contains three blue balls and seven white balls; and urn 3 contains 80 blue balls and 20 white balls. We choose one of the urns at random, in such a way that all three urns are equally likely; then we draw a ball at random from it. What is the probability that the ball is blue?

You might think this could be found simply, as follows: The three urns contain among them 114 balls—84 blues and 30 whites—so the probability of a blue ball is $\frac{84}{114} \doteq .7368$. But this solution assumes an outcome set with 114 equally likely outcomes, corresponding to the 114 balls, and a moment's thought ought to convince you that the 114 balls in this experiment are not equally likely. If the experiment is repeated many times, a given ball in urn 1 will be drawn roughly $\frac{1}{12}$ of the time, whereas a given ball in urn 3 will be drawn only about $\frac{1}{300}$ of the time.

The simplest analysis of this experiment focuses not on the whole outcome set as we did in Sections 1.1 and 1.2, but rather on certain events to which we can assign probabilities appropriate for the experiment. Specifically, let A_1 be the event "urn 1 is chosen," and similarly A_2 and A_3; and let B be the event "a blue ball is drawn." Then A_1, A_2, and A_3 form a partition—together, they include all outcomes, but only one of them can occur at a time. Consequently, according to (2.1.8),

$$P(B) = P(B \cap A_1) + P(B \cap A_2) + P(B \cap A_3).$$

Now what are the three numbers $P(B \cap A_i)$? Take $B \cap A_2$, for example; it is the event that urn 2 is chosen and the ball that is drawn from it is blue. Over a long series of trials, this will happen about $\frac{1}{3} \cdot \frac{3}{10}$ of the time, because urn 2 will be chosen on about $\frac{1}{3}$ of the trials and on those trials the ball will be blue about $\frac{3}{10}$ of the time. [We will think more carefully about this kind of multiplication of probabilities in Section 2.2. There we will see that the $\frac{1}{3}$ is $P(A_2)$ and the $\frac{3}{10}$ is the *conditional probability* of B, given A_2. See (2.2.2) and (2.2.8).]

Doing the same thing for $B \cap A_1$ and $B \cap A_3$ gives us the answer:

$$P(B) = \tfrac{1}{3} \cdot \tfrac{1}{4} + \tfrac{1}{3} \cdot \tfrac{3}{10} + \tfrac{1}{3} \cdot \tfrac{80}{100} = \tfrac{135}{300} = .45,$$

which is not even close to .7368. In fact, the naive solution even makes blue more probable than white, when in fact white is more probable.

Why is the naive solution so far off? The reason is that the large number of balls in urn 3, which is the only urn to favor blue, gives it undue weight if we think all 114 balls are equally likely. In fact, the probability $P(B \cap A_3)$ that urn 3 is chosen and a blue ball is drawn is the same, $\frac{1}{3} \cdot .8$, whether urn 3 contains eight blues out of ten, 80 blues out of 100, or 8 million blues out of 10 million. ∎

We conclude this section with an obvious inequality that is nevertheless worthwhile to notice: if A is a subset of B, then its probability cannot exceed that of B.

(2.1.10) **If $A \subseteq B$, then $P(A) \leq P(B)$.**

Proof. If $A \subseteq B$, then B is the union of the two disjoint sets A and $B \cap A^c$, so $P(B)$ is the sum $P(A) + P(B \cap A^c)$, which cannot be less than $P(A)$. □

Notice what it means for event A to be a subset of event B: every outcome in A is in B. That is, if A occurs, then B occurs. That is, A implies B.

EXERCISES FOR SECTION 2.1

1. Suppose you do three Bernoulli trials with success probability p on each trial. [See Example (1.2.1).]
   a. What is the probability of getting at least one success in the three trials?
   b. For which value of p is this probability exactly $\frac{1}{2}$? That is, for what success probability is there an even chance of getting at least one success in three trials?

2. On a certain Monday morning, a class of 50 students includes ten who failed to prepare. The instructor chooses five people at random to work problems at the chalkboard. What is the probability that at least one of the ten unprepared students is chosen?

3. If two cards are drawn at random from a standard deck, without replacement, what is the probability that they are both aces? (Four of the 52 cards in the deck are aces.) Do the problem two ways to see whether it makes a difference. In each case, give the size of the outcome set, the size of the event, and the probability.
   a. Suppose the two cards are drawn one after the other.
   b. Suppose the two cards are drawn together; that is, a random subset of size 2 is taken. [See also Exercise 12 below.]

4. If five cards are drawn from a standard deck, without replacement, what is the probability that at least one of them is a face card or an ace? (Sixteen of the 52 cards in the deck are face cards or aces.)

5. A garlic press factory has four machines turning out garlic presses. Machine 1 makes 35% of the total output, and 5% of the presses turned out by machine 1 are defective. (Thus, the probability that a randomly chosen press happens to be a defective press made by machine 1 is $.05 \cdot .35 = .0175$.) Similarly, machines 2, 3, and 4 make 30%, 20%, and 15% of the total output, and their

percentages of defective presses are 4%, 2%, and 2%, respectively. Find the probability that a randomly chosen garlic press from the output of these four machines is

a. Defective and made by machine i ($i = 2, 3, 4$).
b. Defective.
c. Not defective.
d. Made by machine 2 and not defective.

[*Hint:* Look at Example (2.1.9).]

6. Suppose that 28% of the Republicans, 75% of the Democrats, and 42% of the independents in the country favor candidate A. Suppose further that 40% of the voters are Republicans, 43% are Democrats, and 17% are independents. What proportion of the voters favor candidate A?

7. In a certain population, half the people are male and half female, and 10% are left-handed. If 6% of the population are left-handed males, what proportion are right-handed females? [You can probably puzzle this out without using any formulas. But also be sure you see how (2.1.2) applies.]

8. In a certain probability space are two events, A and B, with the following properties: $P(A) = .4$, $P(A \cup B) = .7$, and $P(B^c) = .55$. Find the probability of $A \cap B$.

9. Given that $P(A) = .4$, $P(A \cap B) = .1$, and $P((A \cup B)^c) = .2$, find $P(B)$.

10. Consider the standard normal probability space given in Example (1.3.5). Let A be the event "outcome is less than 2," and B the event "outcome is greater than 1."

    a. Using the normal distribution table in Appendix C, find $P(A)$, $P(B)$, and $P(A \cap B)$.
    b. Using the results of part a, find $P(A \cup B)$.
    c. Of course you are not surprised by the result of part b. Why?

11. You and a friend, along with an eccentric rich probabilist, are observing a Poisson process whose arrival rate is $\lambda = .5$ per hour. The probabilist offers to pay you $100 if there is at least one arrival between noon and 2 P.M., and also offers to pay your friend $100 if there is at least one arrival between 1 P.M. and 3 P.M.

    a. What is the probability that either you or your friend, or both, gets $100?
    b. What is the probability that one of you wins $100, but not both?

12. In Example (2.1.6) we supposed the three cards were drawn from the deck one at a time, so Ω contained $52 \cdot 51 \cdot 50$ outcomes. Repeat the problem, assuming that the three cards are drawn all at once: now Ω is the set of all subsets of size 3 from the 52-card deck. Intuition tells us that this will not change the probability; be sure you see why it does not. [Note that the answer is the same whether you think of the draws as being made one at a time or all together. This is generally true in sampling without replacement from a finite population.]

13. Find the probability asked for in small print at the end of Example (2.1.6).

14. You and a friend are arguing over the political preferences of the 100 members of a certain group on campus. You select three of them at random; all three favor candidate A over candidate B. You say this proves that a majority of the group favor A; your friend replies that you probably selected the only three members of the group who favor A.

a. If in fact only three members of the group favor A and 97 favor B, find the probability that the three chosen at random are those three.

b. What is the probability that, if three favor A and 97 favor B, there is at least one in the sample who favors A?

15. Eight college graduates are being interviewed by a corporation that has four job openings. Call them $G_1, G_2, \ldots, G_8$, and suppose their grade-point averages are in the order given; that is, G_1 has the highest and G_8 has the lowest average. Unfortunately, the corporation, after all the interviews, chooses four of the eight candidates at random.

a. Find the probability that neither G_1 nor G_2 is chosen for the job.

b. Find the probability that G_3 has the highest average among the four who are chosen.

16. A fair die is rolled three times.

a. Find the probability that the three outcomes are all different.

b. Find the probability that the three outcomes are all the same.

17. **The birthday problem.** In a group of n people, what is the probability that there are at least two with the same birthdate (but not necessarily the same year of birth)? For simplicity, assume that there are 365 possible birthdays and that all possible assignments of n birthdays to the n people are equally likely. [*Hint:* The probability of the complement is much easier to find.]

This famous problem is not at all difficult, but it is surprising to see how large the probability of a match is. It would seem that n must be quite large for a match to be likely; but, in fact, the probability of a match is greater than $\frac{1}{2}$ when n is as small as 23. The following table gives the probability of a match for some values of n:

n	10	20	22	23	30	40	50	60
P(match)	.1170	.4114	.4757	.5073	.7063	.8912	.9704	.9941

You have a very good chance of winning if you bet that there is at least one birthday match in a group of 40 or more people.

18. **The probability of a derangement.** The probability that a random permutation of the n integers $1, 2, \ldots, n$ is *not* a derangement—that is, that at least one of the integers is in its proper place [see the comment at the end of Example (2.1.5)]—is equal to

(2.1.11)
$$P(\text{not a derangement}) = \frac{1}{1!} - \frac{1}{2!} + \frac{1}{3!} - \frac{1}{4!} + - \cdots \pm \frac{1}{n!}.$$

a. Verify this for $n = 5$ using the principle of inclusion and exclusion, by first noting that the event in question is $A_1 \cup A_2 \cup A_3 \cup A_4 \cup A_5$, where A_i is the event "i is in the ith place," and then checking the following:

There are $5 = \binom{5}{1}$ individual A_i's, each with probability $\frac{4!}{5!}$;

there are $\binom{5}{2}$ two-event intersections $A_i \cap A_j$, each with probability $\frac{3!}{5!}$;

there are $\binom{5}{3}$ three-event intersections $A_i \cap A_j \cap A_k$, each with probability $\frac{2!}{5!}$;

etc.

So the principle of inclusion and exclusion gives

$$P(A_1 \cup A_2 \cup A_3 \cup A_4 \cup A_5) = \binom{5}{1} \cdot \frac{4!}{5!} - \binom{5}{2} \cdot \frac{3!}{5!} + \binom{5}{3} \cdot \frac{2!}{5!} - \binom{5}{4} \cdot \frac{1!}{5!} + \binom{5}{5} \cdot \frac{0!}{5!},$$

which equals the sum given in (2.1.11) for the case $n = 5$.

b. Use a calculator to compute the probability of a derangement, which is 1 minus the sum in (2.1.11), for some small values of n, say $n = 1, 2, 3, \ldots, 9$. Notice how quickly they seem to be approaching a limit. Can you tell what this limit is? [Look at the reciprocals of the numbers and see if they seem to be approaching anything familiar. You can find the limit of (2.1.11) for large n, by using an appropriately chosen value of x in the series (A.5.11) in Appendix A.]

In courses in probability or combinatorial mathematics, the derangement problem often appears in many guises that serve to emphasize how surprising the result is. For example, if n letters are typed, along with envelopes, but then the letters are put into the envelopes at random, then the above is the probability that no letter goes into the correct envelope. This probability is within .0001 of .3679 for all n greater than 6; it is essentially the same for seven letters as it is for 7,000 or 7 trillion.

See also Exercise 19 in Section 4.2, in which we find the (surprising) expected number of items in their proper places in a random permutation. Above we found the probability that that number is 0.

ANSWERS

1. **a.** $1 - (1 - p)^3$ **b.** $p \doteq .2063$
2. .6894
3. **a.** $\frac{1}{221}$ **b.** $\frac{1}{221}$
4. .85495
5. **a.** .0120, .0040, .0030 **b.** .0365 **c.** .9635 **d.** .2880
6. .5059
7. .46
8. .15
9. .5
10. **a.** .9772, .1587, .1359 **b.** 1 **c.** $A \cup B = \mathbb{R}$

11. a. .7769 **b.** .2895

13. .5448

14. a. .000006184 **b.** .08819

15. a. $\frac{15}{70}$ **b.** $\frac{10}{70}$

16. a. $\frac{120}{216}$ **b.** $\frac{6}{216}$

18. b. $\frac{1}{e}$

2.2 CONDITIONAL PROBABILITIES

Suppose that a chance experiment has been performed, but we still do not know the outcome ω that has occurred. (Imagine that someone has tossed a coin three times, but that you have not seen the result.) In this case the probabilities of the events are still appropriate numbers to know; the probability of an event B is just the probability that the (as yet unknown) outcome ω is in B.

But it may happen that, without learning which ω has occurred, we learn some partial information about ω. Then the probability of B may not be appropriate; we may need to revise the probability to take into account the new information. One option is to change the model; but it turns out to be more convenient to define a new set of probabilities in the same model. The following example shows what we mean.

(2.2.1) Example. Suppose we are tossing a fair coin three times and we are interested in the event B that there are two or more heads. Remember that the probability space for this experiment has eight equally likely outcomes:

$$\Omega = \{\text{HHH, HHT, HTH, HTT, THH, THT, TTH, TTT}\}.$$

The probability of B is $\frac{1}{2}$, because B (which equals {HHH, HHT, HTH, THH}) contains four of the eight outcomes, each having probability $\frac{1}{8}$. But now suppose we learn that the first toss came up heads. What should we now think about the probability of B?

Your reaction to this may well be as follows: given that the first toss came up heads, there are now four equally likely outcomes (the first four listed in Ω), and on three of them B occurs. So the revised probability of B, given that the first toss came up heads, is $\frac{3}{4}$. If this is your reaction, good; you are correct and your intuition agrees with the way most probabilists think about this question. [But do not trust this kind of intuition too far; see Example (2.2.5) below.]

Another way to think about the same question is in terms of expected long-run relative frequencies, which is what probabilities represent. Suppose a great many trials of this experiment are made. Then we expect B—two or more heads—to occur on about half the trials. But, *among those trials on which the first toss came up heads,* what proportion do we expect to have two or more heads? To answer this, imagine looking at the list of outcomes of all the trials. Cross out all those trials on which the first toss did not come up heads. What proportion of the rest will have two or more heads? Well, we are left with approximately equal numbers of the four outcomes HHH, HHT, HTH, HTT, so about $\frac{3}{4}$ of them should have two or more heads. Again we get $\frac{3}{4}$ as our revised probability.

The previous two paragraphs represent two common ways of thinking about conditional probabilities. The first paragraph might be called the "revised outcome set" view; the information about the first toss caused us to throw away some members of the outcome set as irrelevant. The second might be called the "revised record of outcomes" view; we imagine a long series of trial outcomes and cross out those outcomes not in the given event. You can see that these two views are really the same.

To formalize this a little, let A be the event that the first toss comes up heads. We have

$$A = \{\text{HHH, HHT, HTH, HTT}\},$$

and, as noted above,

$$B = \{\text{HHH, HHT, HTH, THH}\}.$$

The set of outcomes in A that are also in B is then the set

$$A \cap B = \{\text{HHH, HHT, HTH}\}.$$

To get to the conditional probability of B given A, $\frac{3}{4}$, we divided $\frac{3}{8}$ by $\frac{4}{8}$; that is, we ***divided the probability of the outcomes in A that are also in B by the probability of A itself.*** You might be thinking of it as dividing the *number* of outcomes in both A and B, namely 3, by the *number* of outcomes in A, namely 4, but it is better to think of it as $\frac{3}{8}$ divided by $\frac{4}{8}$. The quotient is the expected relative frequency of B, among those trials on which A occurred. (It would not do to define the conditional probability as the *number* of outcomes in one event divided by the *number* of outcomes in another; first, that is valid only if all outcomes are equally likely, and second, in many probability spaces, the events of interest have infinitely many outcomes.) ∎

The formal definition is just what the previous paragraph suggests.

(2.2.2) Definition. Let B be any event in a probability space, and let A be any event with $P(A) > 0$. The ***conditional probability of B given A*** is the quantity denoted by $P(B|A)$ and defined by

$$P(B|A) = \frac{P(B \cap A)}{P(A)}.$$

(2.2.3) Example. Suppose we have a Poisson process with arrival rate $\lambda = 3$ per minute. Given that there was at least one arrival in the previous 30 seconds, what is the conditional probability that there were fewer than three arrivals?

Solution. First, remember that the probability of k arrivals in a given half-minute for this process is the Poisson probability (see rule P1 in Section 1.4) with $\lambda s = 3 \cdot \frac{1}{2} = 1.5$:

(2.2.4)
$$p(k) = \frac{1.5^k e^{-1.5}}{k!}, \quad k = 0, 1, 2, \ldots .$$

That is, probabilities are given by this mass function $p(k)$ on $\Omega = \{0, 1, 2, \ldots\}$.

Now the given event A is the event that there was at least one arrival in the half-minute; that is, $A = \{1, 2, 3, \ldots\}$. And we want the conditional probability of the event B that there were fewer than three arrivals; that is, $B = \{0, 1, 2\}$. The definition says that

$$P(B|A) = \frac{P(B \cap A)}{P(A)} = \frac{P\{1, 2\}}{P\{1, 2, 3, \ldots\}}.$$

The numerator here is

$$P\{1, 2\} = \frac{1.5^1 e^{-1.5}}{1!} + \frac{1.5^2 e^{-1.5}}{2!} = 2.625e^{-1.5},$$

and the denominator is

$$1 - P(\{0\}) = 1 - \frac{1.5^0 e^{-1.5}}{0!} = 1 - e^{-1.5};$$

so

$$P(B|A) = \frac{2.625e^{-1.5}}{1 - e^{-1.5}} \doteq .7539.$$

Can we make sense of this in terms of the two views of conditional probability given in the previous example?

The "revised outcome set" view would say that, given A, we throw away the outcome 0, leaving a new outcome set $\{1, 2, 3, \ldots\}$. What probabilities should we assign in the new outcome set? We cannot use the old ones, because they do not add to 1. But if we divide them by what they *do* add to, then they will add to 1, and they will stay proportional to what they were before. What they add to is $P(A)$, which we computed as $1 - e^{-1.5}$. So the conditional probability of B should be the sum of the revised probabilities of those members of B that are in the revised outcome set. That is exactly the same as summing the original probabilities of the members of $B \cap A$, divided by $P(A)$.

And the "revised record of outcomes" view also leads to the same result. Imagine a very long record of trial outcomes; we expect it to contain the outcomes in Ω with relative frequencies close to the probabilities given by (2.2.4). Cross out all the 0's on this list. Then what remains are the outcomes in A, with relative frequencies close to the original relative frequencies divided by $1 - e^{-1.5}$, because that was the relative frequency of the outcomes other than 0 on the full record. So the revised relative frequency of B will be close to the original probability of $B \cap A$ divided by $1 - e^{-1.5}$—the answer is the same.

Notice that nowhere in this example was it necessary to calculate $P(B)$. But it is often of interest to compare $P(B)$ and $P(B|A)$, to see how our view of the probability of the occurrence of B is changed by the information that A has occurred. In the context of conditional probabilities, an ordinary probability like this is called the *unconditional probability* of the event in question.

In this example, the unconditional probability of B is

$$P(B) = P(\{0, 1, 2\}) = \frac{1.5^0 e^{-1.5}}{0!} + \frac{1.5^1 e^{-1.5}}{1!} + \frac{1.5^2 e^{-1.5}}{2!} \doteq .8088.$$

Comparing this with $P(B|A) \doteq .7539$, we see that the information that there was at least one arrival diminishes the probability that there are fewer than three. This is no surprise; that information removes one way for there to be fewer than three arrivals. ▮

(2.2.5) Example: The "older child paradox." Suppose we pick a family at random from among all families with two children, assuming that all four gender distributions—FF, FM, MF, and MM—are equally likely. We are using "FM," for example, to refer to the outcome that the older child is a girl and the younger one is a boy. (This experiment is essentially like tossing a fair coin twice; we might call the two experiments "probabilistically equivalent." But the seeming paradox we will encounter is more interesting in this language than in the coin-tossing language.)

Let B be the event that both children in the family are girls. Then of course $B = \{\text{FF}\}$, and $P(B) = \frac{1}{4}$.

Question 1: What is the conditional probability of B, given the event A that there is at least one girl?

Question 2: What is the conditional probability of B, given the event C that the *older* child is a girl?

Many people think that the answers to these two questions should be the same, because the information given seems to be similar in both cases. After all, if you are talking to the parents and you learn that one of their two children is a girl, you may think you have simply learned that either the older child is a girl or the younger one is; the probability that the other child is also a girl does not seem to depend on whether the girl you know about is the older or the younger child. But read on.

Solution. A is the event {FF, FM, MF}, $B \cap A = \{\text{FF}\}$, and so

$$P(B|A) = \frac{P(B \cap A)}{P(A)} = \frac{1/4}{3/4} = \frac{1}{3}.$$

C is the event {FF, FM}, $B \cap C = \{\text{FF}\}$, and so

$$P(B|C) = \frac{P(B \cap C)}{P(C)} = \frac{1/4}{2/4} = \frac{1}{2}.$$

Even though it seems irrelevant to know whether the daughter mentioned is the older one or the younger one, it nevertheless represents information that changes the expected relative frequencies. Among all two-child families with a female older child, half have a female younger child. But among all two-child families with at least one girl, only one-third are FF families, one-third are FM, and one-third are MF. ∎

The next example shows that in some cases the additional information that A has occurred does not change the probability of B.

(2.2.6) **Example.** Suppose we roll a fair die twice (which is probabilistically equivalent to rolling two fair dice). Given that the number that shows on the first die is 3 or less, what is the conditional probability that the sum of the two numbers is odd? That is, let A be the event "number on first die is 3 or less" and B the event "sum is odd." What is $P(B|A)$?

Before doing this, let us note that the unconditional probability of B is $\frac{1}{2}$: if you look at the 36 equally likely possible outcomes (see Exercise 5 in Section 1.1), you will see that half of them have an odd sum.

The conditional probability is easily found, again just by counting outcomes: $A \cap B$ is the event that the first number is 1, 2, or 3 and the sum is odd; this has nine members, so $P(A \cap B) = \frac{9}{36}$. The event A itself has 18 members, so $P(A) = \frac{18}{36}$. Thus,

$$P(B|A) = \frac{9/36}{18/36} = \frac{1}{2},$$

which is the same as $P(B)$.

This is a very special kind of connection between events, which we will study in Section 2.4. Knowing that A occurred restricts the ways in which B can occur, but it does not change the probability that B occurs. We call A and B *independent* in this case, although it is important to remember that the word "independent" is being used in a special, technical sense. ∎

(2.2.7) **Example.** In a Poisson process with arrival rate λ per minute, suppose we know that there was just one arrival in the first minute. What is the conditional probability that the arrival came in the first 20 seconds?

Solution. The event A, that there is exactly one arrival in the first minute, has probability $(\lambda \cdot 1)e^{-\lambda \cdot 1}/1! = \lambda e^{-\lambda}$. We want $P(B|A)$, where B is the event that the first arrival is in the first one-third minute. So we find $P(B \cap A)$ and divide by $\lambda e^{-\lambda}$.

$B \cap A$ is the event that the first arrival is in the first 20 seconds and that there is just one arrival in the first minute; that is, there is one arrival in the first one-third minute and no arrival in the next two-thirds. We have expressed $B \cap A$ in terms of two nonoverlapping intervals; so rule P4 for Poisson processes, in Section 1.4, applies. We get

$$P(B \cap A) = \frac{(\lambda/3)^1 e^{-\lambda/3}}{1!} \cdot \frac{(2\lambda/3)^0 e^{-2\lambda/3}}{0!} = \frac{\lambda}{3} e^{-\lambda}.$$

Dividing by $P(A) = \lambda e^{-\lambda}$ gives $P(B|A) = \frac{1}{3}$.

It is interesting that the answer does not depend on λ. No matter what the arrival rate, fast or slow, if there happens to be one arrival in the first minute, the chance that it is in the first third of a minute is one-third. And it is easy to run through this calculation again with some other number in place of $\frac{1}{3}$, say .6345, and get the answer .6345. In general, given that there is one arrival in a minute, the conditional probability that it is in the interval $[0, a]$ is a. That is, *no matter what the arrival rate, given that there is one arrival in a minute, then the time of that arrival is conditionally the same as a uniformly distributed random number from* $[0, 1]$. This is a surprising and important property of Poisson processes, and it is shared by Bernoulli processes. See Exercise 17 at the end of this section for the Bernoulli-process version of the result; and see Exercise 16 for another consequence of the same property for Poisson processes. ∎

The chain rule for probabilities. In all of the preceding examples we used Definition (2.2.2) to calculate a conditional probability. But strangely enough, the most common use of that formula is in a different form, which is sometimes called the ***chain rule*** for probabilities:

(2.2.8) $P(A \cap B) = P(A)P(B|A)$ for any events A and B [as long as $P(B|A)$ is defined].

The reason this formula is useful is that often we know the two probabilities on the right side of (2.2.8) and can then find $P(A \cap B)$ from them.

For example, recall Example (2.1.9), in which we have three urns with various combinations of blue and white balls. We pick an urn at random—each urn has probability $\frac{1}{3}$ of being chosen—and then we pick a ball from the chosen urn. B is the event that the ball is blue. A_i ($i = 1, 2, 3$) is the event that urn i is chosen. Urn 2, which is chosen one-third of the time, contains three blue balls and seven white balls. In Example (2.1.9) we argued that the probability of the event $B \cap A_2$, that urn 2 was picked and the ball chosen was blue, was $\frac{1}{3} \cdot \frac{3}{10}$. The argument was that, over a long series of trials, urn 2 will have been chosen on about one-third of the trials, and on those we will get a blue ball about three-tenths of the time. Now we can see what we were really doing there:

$$P(A_2) = P(\text{urn 2 chosen}) = \frac{1}{3}$$

and

$$P(B|A_2) = P(\text{blue ball}|\text{urn 2 chosen}) = \frac{3}{10},$$

and so

$$P(B \cap A_2) = P(A_2)P(B|A_2) = \tfrac{1}{3} \cdot \tfrac{3}{10}.$$

That is, $\frac{1}{3}$ is the probability that urn 2 is chosen and $\frac{3}{10}$ is the *conditional* probability of a blue ball, given that the ball was drawn from urn 2. We did not compute the conditional probability from Definition (2.2.2); rather, we could take $P(B|A_2) = \frac{3}{10}$ as an *assumption* about our model. We will see much more of this in the next section.

The chain rule is often useful in finding probabilities for sampling without replacement, as in the following examples.

(2.2.9) **Example.** An urn contains five blue balls and ten white balls. Two are drawn, one after the other, without replacement. What is the probabiltity that both are blue?

Solution. You could probably get the correct answer, $\frac{5}{15} \cdot \frac{4}{14}$, without knowing anything about conditional probabilities. You would reason as follows, although you would leave out the italicized words. The probability that the first ball is blue is $\frac{5}{15}$, obviously. And given that the first ball is blue, there are now 14 balls left in the urn, of which 4 are blue; so the *conditional* probability that the second ball is blue *given that the first ball is blue* is $\frac{4}{14}$.

To put the argument into symbols: let B_1 be the event that the first ball is blue and B_2 the event that the second is blue. Then

$$P(B_1) = \tfrac{5}{15} \quad \text{and} \quad P(B_2|B_1) = \tfrac{4}{14},$$

so

$$P(B_1 \cap B_2) = P(B_1)P(B_2|B_1) = \tfrac{5}{15} \cdot \tfrac{4}{14}.$$

We could have done this calculation another way avoiding conditional probabilities altogether, using an argument like the ones in Section 1.2. It goes as follows. The set Ω of all possible outcomes of the two-draw experiment has $15 \cdot 14$ members; these members are "two-letter words without repeated letters" from the "alphabet" consisting of the 15 balls in the urn. The event that both balls are blue consists of those "words" in which both "letters" are blue balls; there are $5 \cdot 4$ of these. So the probability is $\frac{5 \cdot 4}{15 \cdot 14}$. ∎

The chain rule (2.2.8) extends to three and more events in an obvious and easy-to-use way. The three-event version is

$$P(A \cap B \cap C) = P(A)P(B|A)P(C|A \cap B), \tag{2.2.10}$$

as long as the two conditional probabilities are defined. You ought to check that this identity is valid. You will find it simple and fun if you start with the right-hand side, using Definition (2.2.2) for both conditional probabilities.

The n-event version is

$$P(A_1 \cap A_2 \cap A_3 \cap \cdots \cap A_n) = P(A_1)P(A_2|A_1)P(A_3|A_1 \cap A_2)P(A_4|A_1 \cap A_2 \cap A_3) \cdots P(A_n|A_1 \cap \cdots \cap A_{n-1}).$$

(2.2.11) Example. You draw three cards in a row from an ordinary deck, without replacement. What is the probability that you get two kings and a queen, in that order?

Solution. Let K_1 be the event "first card drawn is a king," K_2 "second card is a king," and Q_3 "third card is a queen." Then

$$P(K_1) = \tfrac{4}{52}, \quad P(K_2|K_1) = \tfrac{3}{51}, \quad \text{and} \quad P(Q_3|K_1 \cap K_2) = \tfrac{4}{50}$$

and so

$$P(K_1 \cap K_2 \cap Q_3) = P(K_1)P(K_2|K_1)P(Q_3|K_1 \cap K_2) = \tfrac{4}{52} \cdot \tfrac{3}{51} \cdot \tfrac{4}{50}.$$ ∎

(2.2.12) Example: Pólya's urn scheme. An urn initially contains some combination of blue and white balls. We draw a ball and note its color; then we replace it and add one more ball of the same color. Thus each ball drawn slightly increases the chance that the next ball will be of the color just drawn. This turns out to be a good first-approximation model for the spread of contagious diseases, at least during the expanding phase of an epidemic.

Suppose for this example that the urn originally contains three blue balls and seven white balls. What is the probability that the first two balls drawn are blue and the third is white?

Solution. As in previous examples let B_1, B_2, W_3 be the events "first ball blue," "second ball blue," and "third ball white." Then it is not hard to see that

$$P(B_1) = \tfrac{3}{10}, \quad P(B_2|B_1) = \tfrac{4}{11}, \quad \text{and} \quad P(W_3|B_1 \cap B_2) = \tfrac{7}{12}.$$

(*Explanation:* Given B_1, the blue ball drawn was replaced along with another blue ball, so the urn then contained four blues and seven whites. Given the first and second balls were blue, the urn then contained five blues and seven whites.) Therefore,

$$P(B_1 \cap B_2 \cap W_3) = P(B_1)P(B_2|B_1)P(W_3|B_1 \cap B_2) = \tfrac{3}{10} \cdot \tfrac{4}{11} \cdot \tfrac{7}{12}.$$

You might want to guess whether the probability of two blues followed by a white is the same as the probability of blue, then white, then blue—that is, of $B_1 \cap W_2 \cap B_3$. The white ball in the middle changes the conditional probability that the third ball is blue; but what happens to the final answer? You are asked to do this in Exercise 14. ∎

George Pólya was born in Hungary in 1887 and died in the United States in 1985. He was a professor of mathematics in Zurich until 1940, and then in the United States at Brown University, Smith College, and Stanford University. His book *How to Solve It* has been read by students in science and mathematics for decades. He made important contributions to many branches of mathematics, from analysis to combinatorics.

The next observation, (2.2.13), says essentially that conditional probabilities are probabilities, in that they satisfy the general definition of probabilities found at the beginning of this chapter. It is important because it allows us to apply all the formulas of Section 2.1 to conditional probabilities as well as to unconditional probabilities. For example, (2.1.1) and (2.2.13) together enable us to use

$$P(B^c|A) = 1 - P(B|A);$$

similarly, (2.1.2) and (2.2.13) provide us with

$$P(B_1 \cup B_2|A) = P(B_1|A) + P(B_2|A) - P(B_1 \cap B_2|A),$$

and so forth.

The fact that conditional probabilities obey the rules for probabilities is one that, like (2.1.1), you would probably use without thinking about. Mathematics is full of "obvious" facts that really need to be thought about and checked, even if you are a user and not a creator, for two reasons. First, there are quite a number of false propositions that seem true. Second, to use a mathematical fact properly, it is important to know whether it is true because it follows from other facts or because we are simply assuming it.

(2.2.13) **Conditional probabilities are probabilities.** For any fixed event A with $P(A) > 0$, the conditional probabilities $P(B|A)$, for all events B, satisfy the axioms for probabilities given in Definition (1.5.1) (and repeated at the beginning of this chapter). That is,

$$P(B|A) \geq 0 \text{ for all events } B;$$

$$P(\Omega|A) = 1; \text{ and}$$

$$P(B_1 \cup B_2 \cup B_3 \cup \cdots|A) = P(B_1|A) + P(B_2|A) + P(B_3|A) + \cdots$$

for finitely or countably many disjoint events $B_1, B_2, B_3, \cdots$.

Proof. The first two of these are very easy to check: $P(B|A)$ is nonnegative because by definition it is a ratio of two probabilities, and all probabilities are nonnegative. And $P(\Omega|A)$ is the ratio of two probabilities that are equal.

To check the third, notice that by the definition of conditional probability

$$P(B_1 \cup B_2 \cup B_3 \cup \cdots|A) = \frac{P((B_1 \cup B_2 \cup B_3 \cup \cdots) \cap A)}{P(A)}.$$

By the distributive law (see the end of Section A.2 in the Appendix), the numerator of this fraction equals

$$P((B_1 \cap A) \cup (B_2 \cap A) \cup (B_3 \cap A) \cup \cdots).$$

Because the B_j's are disjoint, so are the sets $B_j \cap A$. Therefore, the probability of their union equals the sum of their probabilities, $P(B_1 \cap A) + P(B_2 \cap A) + P(B_3 \cap A) + \cdots$. When we divide by $P(A)$, we get $P(B_1|A) + P(B_2|A) + P(B_3|A) + \cdots$. ∎

We close this section with a somewhat obscure point concerning the requirement that $P(A)$ be positive in order for $P(B|A)$ to be defined. There is often a way to make sense out of $P(B|A)$ even when $P(A)$ is 0, if we can see a way to assume its value without having to compute it from Definition (2.2.2).

For example, recall that if we do three Bernoulli trials with success probability .35, and let B be the event "two successes and one failure," then $P(B) = 3(.35)^2(.65)$. [See Examples (1.2.1) and (1.2.13) and Exercise 3 of Section 1.2.] But now suppose that we choose p at random according to the uniform distribution on the interval $(0, 1)$ [see Example (1.3.1)], and *then* do three Bernoulli trials with success probability p. Then $3(.35)^2(.65)$ is the *conditional* probability of two successes, given the event A that $p = .35$, and this is an event of probability 0. The point is that we did not find $P(B|A)$ by dividing $P(A \cap B)$ by $P(A)$; rather, we *assumed* its value as part of our model.

Thus, we can sometimes use the chain rule, or generalizations of it, even when $P(A)$ is 0. We can never use Definition (2.2.2) in this case. Admittedly, (2.2.8) is somewhat trivial when $P(A)$ is 0; both sides are 0. But the generalizations we will encounter in Chapter 9, when we study absolutely continuous joint distributions, will be more interesting.

EXERCISES FOR **SECTION 2.2**

1. A fair coin is tossed three times. Find the following:
   - **a.** The conditional probability that the first toss comes up heads, given that there was exactly one head [Guess the answer first, then use the definition of conditional probability.]
   - **b.** The conditional probability that there was exactly one head, given heads on the first toss
   - **c.** The conditional probability that there are three heads, given that there are at least two heads
   - **d.** The conditional probability that there are three heads, given that the first two tosses come up heads

2. For two rolls of a fair die, guess whether any of the following three conditional probabilities are equal. Then find them.
   - **a.** The conditional probability that the sum is 7, given that at least one 3 is rolled
   - **b.** The conditional probability that the sum is 7, given that exactly one 3 is rolled
   - **c.** The conditional probability that the sum is 7, given that the first roll is a 3

Now guess whether the following two probabilities will be equal, or whether either of them will be equal to the answer in Example (2.2.6). Then find them.

d. The conditional probability that the sum is odd, given that there is at least one 3

e. The conditional probability that the sum is odd, given that the first roll is a 3

3. Two of the integers from 1 to 100 are chosen at random, one at a time, without replacement. Find the following:

a. The probability that 1 is chosen

b. The conditional probability that 1 is chosen, given that 2 is not chosen

c. The conditional probability that 1 is the second number chosen, given that neither 1 nor 2 is the first

d. The conditional probability that 1 is the second number chosen, given that 2 is not the first

4. A fair die is rolled ten times, and the number of 5's is counted.

a. Find the conditional probability that there are at least two 5's, given that there is at least one 5. Compare with the unconditional probability that there are at least two 5's.

b. Find the conditional probabilities of zero, one, and two 5's, given that there were two or fewer 5's. Compare with the unconditional probabilities of zero, one, and two 5's, if only two rolls are made. [Guess in advance whether they will be the same.]

5. Think about what the following conditional probabilities mean and decide what their values ought to be. Then find them using Definition (2.2.2).

a. $P(B|B)$

b. $P(B|B^c)$

c. $P(B|\Omega)$

d. $P(B|\varnothing)$

e. $P(B|A)$ when A is a subset of B

f. $P(B|A)$ when B is a subset of A

6. In a Poisson process with arrival rate $\lambda = 3$ per hour, let X be the number of arrivals in the first hour. Find the following probabilities:

a. $P(X < 3|X \leq 5)$

b. $P(X \geq 2|X \geq 1)$

c. $P(X = k|X \leq 3)$ (Do this for each $k = 0, 1, 2,$ and 3.)

7. Suppose $P(B|A) = P(B)$. Must $P(A|B)$ equal $P(A)$? Why? [Assume that both $P(A)$ and $P(B)$ are greater than 0.]

8. For a random permutation of the four digits 1, 2, 3, and 4, the usual probability space has $4! = 24$ equally likely outcomes. It will probably help to write them down before finding the following probabilities. [It is always a good idea to list outcomes systematically. Here there are six permutations

beginning with 1, six beginning with 2, etc. Of the six beginning with 1, two begin with 12, two with 13, and two with 14. And so forth.]

a. The conditional probability that 2 is second, given that 1 is first

b. The conditional probability that no digit is in its proper place, given that 1 is not first

c. The conditional probability that at least one digit is out of place, given that 4 is last

9. **The lack of memory of a Poisson process.** Recall that in a Poisson process whose arrival rate is λ arrivals per hour, the waiting time for the next arrival has the exponential distribution with parameter λ. Consequently, the probability that the waiting time for the next arrival is more than a hours is $P(a, \infty] = e^{-\lambda a}$. Find the following:

a. The unconditional probability that the waiting time is more than 2 hours

b. The conditional probability that the waiting time is more than 5 hours, given that it is at least 3 hours. Compare with part a.

c. The unconditional probability that the waiting time is more than s hours

d. The conditional probability that the waiting time is more than $t + s$ hours, given that it is at least t hours. Compare with part c.

10. In the standard normal probability model find the following conditional probabilities:

a. P(outcome is less than 1 | outcome is less than 2)

b. P(outcome is less than 1 | outcome is positive)

c. P(outcome is positive | outcome is less than 1)

11. Prove the three-event chain rule (2.2.10).

12. Three cards are drawn from an ordinary deck, one at a time and without replacement. Find the probability that:

a. They are all red. [26 of the 52 cards are red and the other 26 are black.]

b. The first two are black and the third is red.

c. The first and third are black and the second is red.

d. The second and third are black and the first is red.

e. Exactly one is red.

f. Exactly one is black. [No calculations are necessary if you know the answer to part e.]

g. The first is an ace and the second is an ace. [There are four aces in the deck.]

h. The first is not an ace and the second is an ace.

i. The second is an ace.

13. If $P(A) = .68$, $P(B) = .53$, and $P(A \cup B) = .75$, find $P(B|A)$.

14. For the Pólya urn scheme of Example (2.2.12), suppose we start with three blues and seven whites. As in the example, let B_i be the event "blue on ith draw" and W_i "white on ith draw."

a. Find $P(B_1 \cap W_2 \cap B_3)$. Compare with the answer in Example (2.2.12).
b. Find $P(W_1 \cap B_2 \cap B_3)$. Compare again.
c. Find $P(B_2)$. Compare with $P(B_1)$. [*Hint for* $P(B_2)$: B_2 is the union of the two disjoint events $B_1 \cap B_2$ and $W_1 \cap B_2$. (B_1 and W_1 form a partition of Ω. See (2.1.8).)]

15. Consider a Poisson process with arrival rate λ per minute. [See Example (2.2.7).]
a. Given that there were two arrivals in the first minute, find the conditional probability that they were both in the first one-third minute.
b. Repeat part a with an arbitrary positive number a in place of one-third.
c. Repeat part b with an arbitrary positive integer n in place of two.

16. Consider a Poisson process with arrival rate λ per minute. Given that there were three arrivals in the first 2 minutes, find the probability that there were k arrivals in the first minute; do this for $k = 0, 1, 2$, and 3.

The result is perhaps surprising; the probabilities are the same as the probabilities of zero, one, two, or three heads in three tosses of a fair coin. Thus, given that there were three arrivals, they can be regarded as independent trials in which each has probability $\frac{1}{2}$ of occurring in the first minute.

17. In a Bernoulli process with success probability p, given that there was just one success in the first ten trials, find the probability that it came in the first four trials.

18. *Die Gleichförmigkeit in der Welt (Uniformity in the World)* is the title of a book written in 1916 by the German philosopher K. Marbe, who believed, as many people do, that in repeated tosses of a fair coin, a run of heads makes a tail more likely on the next toss. (They think that this is implied by the "law of averages.")
a. For three tosses of a fair coin [Example (1.1.1)], find the conditional probability that the third toss comes up tails, given that the first two tosses come up heads.
b. What about a run of ten heads in a row? Does it make tails more likely? Find the conditional probability that the 11th toss comes up tails, given that the first ten tosses come up heads. [To model 11 tosses of a fair coin, the outcome set has $2^{11} = 2{,}048$ equally likely members. There are two outcomes in which the first ten tosses come up heads. On one of these two, the 11th comes up heads, and on the other it comes up tails.]

It is important to realize that these calculations do not disprove the belief that a tail is more likely after a run of heads. They simply show that *if a Bernoulli process is an accurate model for real coin tossing,* then the belief is false.

An approach that might be more satisfactory from the philosopher's viewpoint is to imagine a very long series of performances of the experiment of tossing a coin 11 times. Imagine, say, 2,048,000 trials of the 11-toss experiment. There are 2,048 possible outcomes, so each outcome ought to come up about 1,000 times. In particular, the two outcomes HHHHHHHHHHT and HHHHHHHHHHH should come up about 1,000 times each. We see that ten heads are followed by a head about as often as by a tail.

But even this would not convince K. Marbe, and perhaps not your roommate, or even you. One has to accept some assumptions about expected relative frequencies to be persuaded by this argument. Still, most probabilists and philosophers disagree with Marbe and agree that, if the coin is really fair and the tosses are really independent, then even after a very long run of heads, the conditional probability of tails on the next trial is still $\frac{1}{2}$.

19. **The "prisoner paradox."** A jailer is in charge of three prisoners, called A, B, and C. They learn that two of the prisoners have been chosen at random to be executed the following morning; the jailer knows which two they are, but will not tell the prisoners. Prisoner A realizes that his chance of survival is $\frac{1}{3}$. But he says to the jailer, "Tell me the name of one of the other two who will be executed; this will not give me any information, because I know that at least one of them must die, and I know neither of them." The jailer agrees and tells A that B is going to be executed.

    But now A has apparently raised his chance of survival from $\frac{1}{3}$ to $\frac{1}{2}$, because it will be either A or C who is the other one to be executed. Notice that if the jailer had said C instead of B, A's chances would still apparently have gone up to $\frac{1}{2}$. How can this be? [*Hint:* This famous and baffling puzzle has a simple solution when one looks at the proper sample space and uses conditional probabilities. The experiment to model is not simply the choice of the two prisoners to be executed, but it is that choice followed by the jailer's choice (if he has one) of which prisoner to name. There are four outcomes: AB–B, meaning "A and B are to be executed, and jailer names B," AC–C, BC–B, and BC–C. The first two have probability $\frac{1}{3}$ each and the last two have probability $\frac{1}{6}$ each.]

20. **The "Monty Hall paradox," as popularized in "Ask Marilyn."** Suppose you are a contestant on a game show. At one point you are shown three doors; behind one of them is a valuable prize, and you will win the prize if you choose the right door. Suppose you choose door 1. You know that your probability of winning is $\frac{1}{3}$. But now the host opens door 3, showing you that the valuable prize is not behind that door. You are interested in the conditional probability that the prize is behind door 1. You might think that it is now $\frac{1}{2}$, but show that it is still only $\frac{1}{3}$. Consequently, if you are offered the chance to switch to door 2, you should do so, because the conditional probability that the prize is there is $\frac{2}{3}$.

    Assume that the host knows where the prize is and would never open that door. Assume also that if the prize is behind door 1, the host will choose one of the other two at random. [*Hint:* Use an outcome space with four outcomes, as follows: P1H2, meaning "prize behind door 1, host opens door 2," P1H3, P2H3, and P3H2. The probabilities of P1H2 and P1H3 are $\frac{1}{6}$ each, and the other two outcomes each have probability $\frac{1}{3}$. This is the "prisoner paradox" of Exercise 19 in disguise.]

This is sometimes called the "Let's Make a Deal" puzzle after the television show on which a similar game was played. It is also called the "Monty Hall paradox" after the host of that show. It caused a stir in the popular press in 1990 and 1991, because of its appearance in the "Ask Marilyn" column in *Parade* magazine, 9/9/90, 12/2/90, 2/17/91, and 7/7/91. Marilyn vos Savant, the columnist, correctly solved the problem described in this exercise. A large number of people—including many mathematicians and statisticians—wrote to tell her that she was wrong, but they were wrong.

A possible explanation for their error is that they may have interpreted the problem differently—they might have assumed that the host opens a door at random and does not care whether it is the door with the prize. [See Exercise 21.] However, the problem, as originally posed in the column, mentioned that the host "knows what's behind the doors"; and according to vos Savant, the letters she received show that "the vast majority" of writers understood the conditions to be those described in this exercise. Probably a better explanation is simply that this is a very counterintuitive result.

It is interesting to notice that the "prisoner paradox" appears as an exercise in many textbooks and is familiar to most probabilists and statisticians; and yet the "Monty Hall paradox," which is equivalent to it, fooled so many. Moreover, this problem, in the "Monty Hall" form, was discussed as early as 1975, in letters to the journal *American Statistician* (vol. 29: 67 and 134). The writer, Steve Selvin, gave a slightly different solution from the one suggested here, but came to the same correct answer. Selvin also received several letters from people who thought he was wrong.

The solution suggested in the hint above, as well as Selvin's, provide rigorous arguments for the correctness of vos Savant's answer, but they may not convince people who are not familiar with probability theory and especially with conditional probabilities. There are at least two good ways to convince someone informally of the correctness of this answer. The first is to play the game a large number of times; as vos Savant suggested, you can do it with three cups and a coin. Put the coin under one of the cups; have a friend choose a cup; turn over one of the other cups without the coin; and then have your friend switch to the cup not turned over. That cup will have the coin on about two-thirds of the trials. Alternatively, you could program a computer to simulate the experiment.

If you do this experiment a few times, you may come to the following line of reasoning, which also seems to work well in convincing people. If you played the game many times without the host opening any doors, of course you would find the prize on one-third of the trials. Now on every trial, the host can open a door, other than the one you chose, that does not contain the prize; the fact that this can be done cannot change the fact that you choose the door with the prize only one-third of the time. The other two-thirds of the time the prize is behind one of the other two doors and the host is showing you which one.

We should point out that it is not clear whether the assumptions of this exercise actually held every time the game was played on the television show "Let's Make a Deal." In particular, it is impossible to know whether Monty Hall was committed to open a door every time, or whether he would always open one without a prize (although it seems this is what happened whenever the game was played). It is also impossible to know whether he actually made a random choice of door when the contestant chose the right one, or whether he tended to favor one. See the next two exercises.

It is also true that, on "Let's Make a Deal," Monty Hall never offered to allow the contestant to switch; instead he offered a cash prize in place of what was behind the chosen door. In this case your decision is more subjective, but some people would make the choice that offers the greater expected gain. We will study this concept in Chapter 4, where we will see that the expected gain from staying with the door is one-third the value of the prize behind the door. Consequently, if the prize behind the door is worth more than three times the cash prize, you have a greater expected gain if you stay with the door. See Exercise 4 of Section 4.1.

21. **The "Monty Hall paradox," continued.** Repeat Exercise 20, assuming this time that the host will open one of doors 2 and 3 at random, without worrying about where the prize is. Again find the conditional probability that the prize is behind door 1, given that door 3 was opened and the prize was not there. [*Hint:* Now there are six outcomes: P1H2, P1H3, P2H2, P2H3, P3H2, and P3H3, all equally likely.]

22. **The "Monty Hall paradox," continued.** Now suppose that the host knows where the prize is and will not open that door, but also has the option of not opening a door at all. Again suppose you chose door 1 and the host opened door 3. What is the conditional probability that the prize is behind door 1, given that the host did choose to open a door and it was door 3?
    This cannot be answered without a value for the probability that the host chooses to open a door. Suppose that the probability is p_1 if the prize is behind door 1 and p_2 if it is behind door 2 or door 3. Show that the desired conditional probability is $p_1/(p_1 + 2p_2)$ and, thus, you should switch if $p_1 < 2p_2$. [*Hint:* Now there are seven outcomes: P1H2, P1H3, P1HN (meaning "prize behind door 1, host opens no door"), P2H3, P2HN, P3H2, P3HN.]

23. **The number of items in their proper places in a random permutation.** Let p_{nk} denote the probability that, in a random permutation of n objects (say the integers $1, 2, \ldots, n$), exactly k of them are in their proper places (for $k = 0, 1, 2, \ldots, n$). [So p_{n0} is the probability of a derangement, which we found in Exercise 18 of Section 2.1. That result will be unnecessary here.]
    a. Show that $p_{nk} = p_{n-k,0}/k!$. [*Hint:* The event "k integers are in their proper places" is the union of $\binom{n}{k}$ disjoint events, one for each choice of the k integers. Use the chain rule to show that each of these events has probability $p_{n-k,0}/[n(n-1)\cdots(n-k+1)]$.]
    b. Conclude that, if $n > 1$ and $k > 0$, then $p_{nk} = p_{n-1,k-1}/k$.
    c. Now it is easy to compute (or to write a computer program to compute) all the probabilities p_{nk}, by computing the following triangle row by row. Do so at least to the fifth row.

$$\begin{array}{llll} p_{10} & p_{11} & & \\ p_{20} & p_{21} & p_{22} & \\ p_{30} & p_{31} & p_{32} & p_{33} \\ \cdot & & & \cdot \\ \cdot & & & \quad\cdot \\ \cdot & & & \qquad\cdot \end{array}$$

[*Hint:* The first row is easy: $p_{10} = 0$ and $p_{11} = 1$. Use the result of part b to compute p_{21} and p_{22}; then find p_{20} using the fact that the three entries on the second row must sum to 1. Continue to the next row, using part b, to get all entries but the first, then subtracting from 1 to get the first entry. As a check on your calculations, $p_{50} = .366666. . . .$]

As mentioned, in Exercise 18 of Section 2.1 we found a formula for p_{n0} for any n; here we do not get (directly) a formula, but we can get the values not only of p_{n0}, but also of all the p_{nk}.

If you compute several rows of this triangle (it is not difficult to write a computer program to do so), you will get a pleasant surprise if you compare your last row with the Poisson probabilities for $\lambda = 1$. It can be proved, though not easily, that for any fixed k, the limit of p_{nk} as $n \to \infty$ is $1/k!e$, the Poisson probability. The convergence is quite rapid; for n as small as 10 and $k \le n$, p_{nk} and $1/k!e$ agree to four significant digits.

ANSWERS

1. **a.** $\frac{1}{3}$ **b.** $\frac{1}{4}$ **c.** $\frac{1}{4}$ **d.** $\frac{1}{2}$

2. **a.** $\frac{2}{11}$ **b.** $\frac{2}{10}$ **c.** $\frac{1}{6}$ **d.** $\frac{6}{11}$ **e.** $\frac{1}{2}$

3. **a.** $\frac{2}{100}$ **b.** $\frac{2}{99}$ **c.** $\frac{1}{99}$ **d.** $\frac{98}{9{,}801}$

4. **a.** .6148 (*vs.* .5155)
b. .2083, .4167, .3750 (*vs.* .6944, .2778, .0278)

5. **a.** 1 **b.** 0 **c.** $P(B)$ **d.** Meaningless **e.** 1 **f.** $P(B)/P(A)$

6. **a.** .4620 **b.** .8428 **c.** .0769, .2308, .3462, .3462

7. Yes

8. **a.** $\frac{1}{3}$ **b.** $\frac{9}{18}$ **c.** $\frac{5}{6}$

9. a and **b.** $e^{-2\lambda}$ c and **d.** $e^{-s\lambda}$

10. **a.** .8609 **b.** .6826 **c.** .4057

12. **a.** .1176 **b.** .1275 **c.** .1275 **d.** .1275 **e.** .3824 **f.** .3824 **g.** .0045 **h.** .0724 **i.** .0769

13. .6765

14. **a.** .0636 **b.** .0636 **c.** $\frac{3}{10}$, the same as $P(B_1)$

15. **a.** $\frac{1}{9}$ **b.** a^2 **c.** a^n

17. $\frac{4}{10}$

18. **a.** $\frac{1}{2}$ **b.** $\frac{1}{2}$

19. The conditional probability of surviving is still $\frac{1}{3}$.

21. Now the conditional probabilities are $\frac{1}{2}$ each.

2.3 CONDITIONING AND BAYES'S THEOREM

The two formulas of this section, the law of total probability and Bayes's theorem, apply when the outcome set Ω is partitioned into events A_1, A_2, A_3, . . . , a finite or countable collection of disjoint events whose union is Ω. You may want to look back at the partition formula (2.1.8), at Example (2.1.9), and at Exercise 5 of Section 2.1.

The Law of Total Probability

Using this law is often called ***conditioning;*** we will see in the examples why and how this term is used. The law is as follows.

(2.3.1) If a finite or countable collection of events $A_1, A_2, A_3, \ldots$ form a partition of Ω, and if $P(A_i) > 0$ for each A_i, then for any event B,

$$P(B) = P(B|A_1)P(A_1) + P(B|A_2)P(A_2) + P(B|A_3)P(A_3) + \cdots.$$

Proof. This formula is obtained simply by putting together the partition formula (2.1.8), which says

$$P(B) = P(B \cap A_1) + P(B \cap A_2) + P(B \cap A_3 + \cdots,$$

and the two-event chain rule (2.2.8), according to which, for each i,

$$P(B \cap A_i) = P(B|A_i)P(A_i). \qquad \square$$

Our first example of conditioning has been done twice already, in Example (2.1.9) and immediately following the chain rule (2.2.8). We do it again here for the sake of clarity.

(2.3.2) Example. Three urns contain both blue and white balls. Urn 1 contains one blue and three whites; urn 2 contains three blues and seven whites; and urn 3 contains 80 blues and 20 whites. An urn is chosen at random (all three are equally likely) and a ball is chosen at random from it (all balls in any one urn are equally likely). What is the probability that the ball is blue?

Solution. Let A_i be the event "urn i is chosen" ($i = 1, 2, 3$) and let B be the event "the ball is blue." Then we have

$$\begin{aligned} P(A_1) &= \tfrac{1}{3} \quad \text{and} \quad P(B|A_1) = \tfrac{1}{4}; \\ P(A_2) &= \tfrac{1}{3} \quad \text{and} \quad P(B|A_2) = \tfrac{3}{10}; \\ P(A_3) &= \tfrac{1}{3} \quad \text{and} \quad P(B|A_3) = \tfrac{80}{100}. \end{aligned}$$

The events A_1, A_2, A_3 partition Ω (they are disjoint and their union includes all possible outcomes), so

$$\begin{aligned} P(B) &= P(B|A_1)P(A_1) + P(B|A_2)P(A_2) + P(B|A_3)P(A_3) \\ &= \tfrac{1}{4} \cdot \tfrac{1}{3} + \tfrac{3}{10} \cdot \tfrac{1}{3} + \tfrac{80}{100} \cdot \tfrac{1}{3} = \tfrac{135}{300}. \end{aligned} \qquad \blacksquare$$

We say we have done this by "conditioning on the urn chosen."

(2.3.3) Example. In a certain population 5% of the females and 8% of the males are left-handed; 48% of the population are males. What proportion of the population are left-handed? That is, what is the probability that a randomly chosen member of the population is left-handed?

Solution. We condition on the gender of the person chosen; that means we partition the outcome set into events A_i representing the genders. There are just two: A_1, the event that the person chosen is female, and A_2 that the person is male. Let B be the event "left-handed." Then what we are given is

$$P(A_1) = .52 \quad \text{and} \quad P(B|A_1) = .05;$$
$$P(A_2) = .48 \quad \text{and} \quad P(B|A_2) = .08;$$

and so

$$P(B) = P(B|A_1)P(A_1) + P(B|A_2)P(A_2) = .05 \cdot .52 + .08 \cdot .48 = .0644.$$

Notice that this answer is not quite the average of .05 and .08, which equals .065. It is a "weighted average," where the weights are the probabilities of the two genders. If the population were exactly half female and half male, the weights would have been .50 and .50 instead of .52 and .48, and we would have had the ordinary average of .05 and .08. ▮

(2.3.4) Example. Here is a hypothetical gambling game in which you either win \$1 or lose \$1 on each play. Start with an empty hat (or a box) and a number of identical slips of paper. Now begin tossing a fair coin. For each tail that comes up, put a blank slip into the hat. When the first head comes up, put a slip of paper with an **X** on it into the hat and stop tossing. Now mix the slips in the hat and draw one. If you draw a blank slip, you win \$1; if you get the **X**, you lose \$1. Would you like to play? Try to guess before reading the solution.

Solution. Of interest here is the probability that you win. If it is greater than .5, then playing is to your advantage; otherwise not. It is a little simpler to find the probability that you lose, that is, that you draw the **X**. So let B be this event.

We condition on the number of tosses needed to get the first head. Let A_k be the event that the first head comes on the kth toss ($k = 1, 2, 3, \ldots$). We worked out the probabilities of the A_k's in Exercise 5 of Section 1.2; there we found that $P(A_k) = \left(\frac{1}{2}\right)^k$ for $k = 1, 2, 3, \ldots$.

Now what is $P(B|A_k)$? If A_k occurs, then there are k slips in the hat—the one with the **X** and the $k - 1$ blanks that were added when the tails were tossed. So the conditional probability that you get the **X**, given A_k, is $\frac{1}{k}$. (This is when $k = 1$, as it should be.) Therefore, the probability that you lose this game is

$$\frac{1}{1} \cdot \left(\frac{1}{2}\right)^1 + \frac{1}{2} \cdot \left(\frac{1}{2}\right)^2 + \frac{1}{3} \cdot \left(\frac{1}{2}\right)^3 + \cdots = \sum_{k=1}^{\infty} \frac{1}{k} \cdot \left(\frac{1}{2}\right)^k.$$

Now the series $\sum_{k=1}^{\infty} \frac{x^k}{k}$ is the Maclaurin series for $-\ln(1 - x)$; it converges for $-1 \leq x < 1$. [See (A.5.12) in Appendix A.] So the probability that you lose is $-\ln\left(1 - \frac{1}{2}\right) = \ln(2) \doteq .693$.

You would be foolish to play this game very many times. Imagine the result of a large number of plays, say N plays. You can expect to lose \$1 on about $.693 \cdot N$ of the plays and win \$1 on about $.307 \cdot N$ of the plays, so your net loss will be about $.693 \cdot N - .307$

$\cdot N = .386 \cdot N$ dollars. This is almost 39 cents per play; the more you play, the more you can expect to lose. If you played 10,000 times, you would lose about \$3,860. (This assumes that both you and your opponent start with enough money that you do not go broke before 10,000 plays have been made.) Gambling houses make their owners rich by offering to play games with people and giving them about a 48% chance to win. This game gives you about a 31% chance.

On the other hand, you might be able to get people to play it with you; you will pay them \$1 if they can draw a blank slip and they must pay you \$1 if they get the **X**. Then you will be ahead about \$3,860 after 10,000 plays. (Don't forget that the hat must be emptied and refilled by tossing the coin on each new play.)

Of course, someone who thinks a bit, or plays the game a few times, will realize that on half the plays the first toss will come up heads, and then there will be nothing in the hat but the losing slip. They may refuse to play, knowing that they are guaranteed to lose on at least half the plays. In that case, offer them 2-to-1 odds; that is, offer to pay them \$2 if they draw a blank slip, while they pay you just \$1 if they get the **X**. (Offer this on every play, before the coin is tossed.) Then your net gain after N plays will be about $.693 \cdot N - 2 \cdot .307 \cdot N = .079 \cdot N$. You have to play more to make the same amount—you will be only about \$790 ahead after 10,000 plays—but you will still make money.

These calculations, which estimate the result of a large number of plays, are closely related to the expected value calculations that we will make in Chapter 4. In particular, Exercise 12 of Section 4.1 refers to this example. ∎

Bayes's Theorem

(2.3.5) Example. Let us return once more to Example (2.1.9), but ask for a different conditional probability. Recall that urn 1 has one blue ball and three white balls, urn 2 has three blues and seven whites, and urn 3 has 80 blues and 20 whites. We pick an urn at random and draw a ball from it; B is the event that the ball is blue. The partition consists of events A_1, A_2, and A_3, where A_i is the event "urn i is chosen." We took the following information as given:

$$P(A_1) = \tfrac{1}{3} \quad \text{and} \quad P(B|A_1) = \tfrac{1}{4};$$
$$P(A_2) = \tfrac{1}{3} \quad \text{and} \quad P(B|A_2) = \tfrac{3}{10};$$
$$P(A_3) = \tfrac{1}{3} \quad \text{and} \quad P(B|A_3) = \tfrac{80}{100}.$$

Now suppose someone has done this experiment and tells us that a blue ball was drawn, but says nothing about the urn it came from. What is the probability that it came from urn 1?

Here we are asking for $P(A_1|B)$, which is not one of the given numbers. Just knowing $P(B|A_1)$ does not tell us what it might be. What can we do? When in doubt about a conditional probability, try the definition:

$$P(A_1|B) = \frac{P(A_1 \cap B)}{P(B)}. \tag{2.3.6}$$

Does this help? Well, we know how to find the denominator. In fact, we have already done so; that is what the law of total probability (2.3.1) is for. As for the numerator, what can we use to find the probability of an intersection? The chain rule is all we have, and it works perfectly:

(2.3.7)
$$P(A_1 \cap B) = P(B|A_1)P(A_1),$$

both of which are numbers we know. The answer, which you ought to check, is $P(A_1|B) = \frac{25}{135}$.

If we put (2.3.7) into (2.3.6), we get the formula

$$P(A_1|B) = \frac{P(B|A_1)P(A_1)}{P(B)}.$$

If we continue and put (2.3.1) in for the denominator, we get

$$P(A_1|B) = \frac{P(B|A_1)P(A_1)}{P(B|A_1)P(A_1) + P(B|A_2)P(A_2) + P(B|A_3)P(A_3)},$$

which is the three-event version of Bayes's theorem. The answer is

$$P(A_1|B) = \frac{\frac{1}{4} \cdot \frac{1}{3}}{\frac{1}{4} \cdot \frac{1}{3} + \frac{3}{10} \cdot \frac{1}{3} + \frac{80}{100} \cdot \frac{1}{3}} \doteq .185185.$$ ∎

The general version of Bayes's theorem, for an arbitrary finite or countable partition of Ω, is as follows.

(2.3.8) Bayes's theorem. If a finite or countable collection of events $A_1, A_2, A_3, \ldots$ form a partition of Ω, and if $P(A_i) > 0$ for each A_i, then for any event B and any A_i in the partition,

$$P(A_i|B) = \frac{P(B|A_i)P(A_i)}{P(B|A_1)P(A_1) + P(B|A_2)P(A_2) + P(B|A_3)P(A_3) + \cdots}.$$

This is not so hard to remember as it looks. The desired probability is $P(A_i|B)$; the numerator is the "reverse" of this conditional probability, $P(B|A_i)$, times the probability of the condition, $P(A_i)$. And the denominator is the sum of all possible terms like the numerator.

(2.3.9) **Example.** Let X have possible values 0, 1, 2, and 3, with probabilities .4, .3, .2, and .1, respectively, and let Y be the number of heads in X tosses of a fair coin. (If $X = 0$, then the coin is not tossed, and Y is of course 0.) The experiment is done and the number of heads is 1. We are not told the number of tails. What is the conditional probability that X was equal to 1?

Solution. We condition on the value of X; let A_i be the event "$X = i$," for $i = 0, 1, 2, 3$, and let B be the event "$Y = 1$." We want $P(A_1|B)$. Bayes's formula gives

$$P(A_1|B) = \frac{P(B|A_1)P(A_1)}{P(B|A_0)P(A_0) + P(B|A_1)P(A_1) + P(B|A_2)P(A_2) + P(B|A_3)P(A_3)}.$$

The unconditional probabilities of the A_i's have been given to us, and the conditional probabilities of B, given the A_i's are easy to work out:

$$\begin{aligned} P(B|A_0) &= P(\text{one head in no tosses}) = 0; \\ P(B|A_1) &= P(\text{one head in one toss}) = \tfrac{1}{2}; \\ P(B|A_2) &= P(\text{exactly one head in two tosses}) = \tfrac{1}{2}; \\ P(B|A_3) &= P(\text{exactly one head in three tosses}) = \tfrac{3}{8}. \end{aligned}$$

Putting it all together,

$$P(A_1|B) = \frac{\frac{1}{2} \cdot .3}{0 \cdot .4 + \frac{1}{2} \cdot .3 + \frac{1}{2} \cdot .2 + \frac{3}{8} \cdot .1} = \frac{.15}{.2875} \doteq .5217.$$ ▮

(2.3.10) Example. We have two coins, a fair one and a two-headed one. Someone chooses one of these coins at random, tosses it three times, and reports to us that heads came up all three times. What is the conditional probability that the coin was the two-headed one?

Solution. We condition on the coin that was chosen. The partition consists of only two events: A_1 = "fair coin" and A_2 = "two-headed coin." Let B = "heads on all three tosses." We want $P(A_2|B)$. Bayes's formula says

$$P(A_2|B) = \frac{P(B|A_2)P(A_2)}{P(B|A_1)P(A_1) + P(B|A_2)P(A_2)}.$$

It is easy to check that $P(A_1) = P(A_2) = \frac{1}{2}$, $P(B|A_1) = \frac{1}{8}$, and $P(B|A_2) = 1$. When we put these into the preceding formula, we get $P(A_2|B) = \frac{8}{9}$. ▮

Bayes's theorem is sometimes called the "theorem on the probability of causes," and sometimes the "theorem on inverse probabilities." We can think of the A_i's as different possible causes, each with its own probability of producing the effect B. When the effect B is observed, Bayes's formula gives the conditional probability that it was caused by one or another of the A_i's. The law of total probability simply gives the unconditional probability of the effect B, which is the denominator of Bayes's formula.

The important thing to remember is that, in either case, it is not enough to know just the conditional probabilities of B given the various causes; we must also know the unconditional probabilities of the causes themselves.

If the formulas of this section seem difficult to remember or apply, you might want to keep in mind the following facts about the law of total probability and Bayes's theorem:

1. They both apply in situations involving an event B and a partition of the outcome set into events A_1, A_2, A_3, etc., in which we know $P(A_i)$ and $P(B|A_i)$ for each i.
2. The law of total probability gives $P(B)$, which is the denominator of Bayes's formula.
3. Bayes's formula gives $P(A_i|B)$ for any of the A_i's.

Remember also that Bayes's formula is not hard to memorize, as we mentioned in (2.3.8). But memorizing formulas, while sometimes necessary, is always a poor substitute for understanding them to the extent that you can reconstruct them. If you want to be able to reconstruct Bayes's formula, begin with Definition (2.2.2) of conditional probability, which is fundamental, and the law of total probability (2.3.1). Bayes's theorem follows naturally if you do what we did in Example (2.3.5). That is, use the definition of conditional probability first; then use the chain rule for the numerator and the law of total probability for the denominator.

Thomas Bayes (1702–1761) was a Presbyterian minister in England who was interested in mathematics and probability. In 1731 he wrote a paper called "Divine Benevolence: An Attempt to Prove that the Principal End of the Divine Providence and Government Is the Happiness of His Creatures." In 1736 he wrote a mathematico–theological paper called "Introduction to the Doctrine of Fluxions," which was a response to Bishop Berkeley's attack on Newton's calculus (which had been discovered in 1665 but which Leibniz had first published a few years later).

He discovered the theorem that bears his name toward the end of his life. It was not published until 1763, in his "Essay Towards Solving a Problem in the Doctrine of Chances." In that paper Bayes set himself (and solved) what he called the "converse problem": "*Given* the number of times in which an unknown event has happened and failed: *Required* the chance that the probability of its happening on a single trial lies somewhere between any two degrees of probability that can be named."

In modern language, this problem involves Bernoulli trials with unknown success probability p, which is considered as a random variable. Bayes's "converse problem" is to find the conditional probability that p is between a and b, say, given that there have been k successes in n trials. (The version of Bayes's theorem in this section does not apply to this problem; it needs a more general version that we will study in Chapter 9. See (9.2.20) for a solution to Bayes's problem.)

EXERCISES FOR **SECTION 2.3**

1. In a certain population of voters, 80% of the Democrats, 45% of the Republicans, and 55% of the independents say the president is doing a good job. The voters are 35% Democrat, 40% Republican, and 25% independent.
   a. What proportion of the voters think the president is doing a good job?
   b. A randomly chosen voter thinks the president is doing a good job. What is the probability that the voter is a Democrat?

2. A fair die is rolled. An urn is then loaded with one blue ball and as many white balls as there were spots showing on the die. The balls are mixed and one is drawn. What is the probability that it is blue?

3. You like to climb stairs by taking either one or two stairs at a time, and you decide at random before each step. If you have a choice, you take two stairs with probability p and one stair with probability $q = 1 - p$. So if you have a flight of only three stairs to climb, you will do it in either two or three steps. What is the probability that you do it in two? [Condition on the first choice you make.]

4. Someone has done three Bernoulli trials after first choosing the success probability p at random. With probability .4, p was chosen to be $\frac{1}{2}$, with probability .3 it was $\frac{1}{4}$, and with probability .3 it was $\frac{3}{4}$. Given that all three trials resulted in successes, what is the conditional probability that p was $\frac{3}{4}$?

5. You roll a fair die and let λ be the number of spots that show. Then you turn to a Poisson process whose arrival rate is λ per minute, and you watch it for a minute. What is the probability that you see no arrivals?

6. An urn contains one red ball and two green balls. You draw one at random. If it is red, you put it back along with three more red balls; if it is green, you put it back along with five more green balls. Then you mix the balls and draw one at random. What is the probability that it is green?

7. **The paradox of false positives.** One form of EIA (enzyme immunoassay) test for HIV (the AIDS virus) gives either a positive result, indicating likely infection with the virus, or a negative result, indicating no infection. This test is accurate to the extent that it gives a positive with probability 98% in infected subjects and it gives a negative with probability 99.5% in noninfected subjects.
   a. In a certain population, the proportion of people who are infected is .0003. Find the conditional probability that you are infected, given that you tested positive.
   b. Find the same conditional probability assuming that 10% of the population have the virus.

Notice how crucial is the proportion of the population who are infected. When it is small, even a very reliable test gives a high proportion of false positives. This is a common phenomenon in medical testing. It has two implications worth thinking about. For the individual who tests positive, the news is not so surely bad as one might think. For society as a whole, any large-scale testing program is going to involve far more false positives than true ones. In practice, all positives are retested, usually with a different, more expensive, test. The cost of retesting so many healthy people, along with the emotional cost to those who are false positives, must be borne in mind.

The 98% probability of a "true positive" is called the *sensitivity* of the test, and the 99.5% probability of a "true negative" is called the *specificity*. These numbers, along with the estimate .0003 of the prevalence of HIV, were used as assumptions in a 1990 study of

AIDS testing (Schwartz et al., *Journal of the American Medical Association,* vol. 264: 1704–1710). They are typical values, but should not be taken as the true values for a given test or for a particular population.

8. In Pólya's urn scheme as in Example (2.2.12), an urn contains balls of two colors. A ball is drawn, then replaced along with another ball of the same color. Let us suppose that initially the urn has five blue balls and five white balls. Given that the second ball is blue, what is the conditional probability that the first was blue?

Compare the answer with the "reverse" conditional probability—that the second is blue, given that the first was. Some people find this surprising.

9. Recall Example (1.2.5), in which X is the number of fair-coin tosses needed to get the second head. The event $(X = 2)$ has one outcome, HH, with probability $\frac{1}{4}$; $(X = 3)$ has two outcomes, HTH and THH, with probability $\frac{1}{8}$ each; $(X = 4)$ has three outcomes with probability $\frac{1}{16}$ each, and so on. Recall from (1.2.7) that $P(X = k) = (k - 1)\left(\frac{1}{2}\right)^k$ for $k = 2, 3, 4, \ldots$. Now let Y be the number of the toss on which the *first* head appeared.

a. What is the conditional probability that $Y = 1$, given that $X = 5$?
b. What is $P(Y = j \mid X = 5)$? Give the answer for $j = 1, 2, 3, 4$.
c. What is $P(Y = j \mid X = n)$? Answer for all positive integers j and n with $1 \leq j < n$.
d. What is $P(Y = 1)$? [Use the law of total probability, conditioning on the value of X. After you get the answer, think of a simpler way to get it.]

10. We have two urns. Urn 1 contains seven red balls and three green balls; urn 2 contains one red and one green. A ball is drawn at random from urn 1 and put into urn 2. The balls in urn 2 are then mixed, and one is drawn at random.

a. What is the probability that the ball drawn from urn 2 is red? Green? [Condition on the color of the ball moved.]
b. Given that the ball drawn from urn 2 is red, what is the probability that the ball moved from urn 1 was red?
c. What is the probability that the ball moved and the ball drawn are the same color? [*Hint:* Ignore parts a and b; it is much easier to use the formulas of the previous section.]

11. In a box are nine fair coins and one two-headed coin. One coin is chosen at random and tossed twice. Given that heads show both times, what is the conditional probability that the coin is the two-headed one? What if it were tossed three times and heads came up on all three tosses? Four times? n times?

12. Here is a variation on the Pólya urn scheme. An urn contains four blue balls and six white balls. A ball is drawn. If it is blue, it is replaced along with another blue ball. If it is white, it is not replaced and nothing is added. Then the balls are mixed and a second ball is drawn.

a. What is the probability that both balls drawn are blue?
b. What is the probability that the second ball drawn is blue?

13. Suppose that the number of children in a randomly chosen family is modeled by the Poisson distribution with $\lambda = 1.8$. (That is, the probability that there are k children is $\lambda^k e^{-\lambda}/k!$.) Suppose that within a family, the genders of the children are determined according to a Bernoulli process with $p = q = \frac{1}{2}$. Find the probability that a randomly chosen family has two female children. [*Hint:* Condition on the number of children in the family.]

14. We have three cards; each card has a dot on each side that is purple or green. Card 1 has a purple dot on both sides, card 2 has a green dot on both sides, and card 3 is purple on one side and green on the other. One of these cards is drawn at random (all equally likely) and you are shown one side. It is green. We are interested in the conditional probability that the other side is green—that is, that you are looking at card 2. Which of the following solutions gives the correct answer?

Solution 1: Since you see a green dot, you are looking at card 2 or card 3. Since they were both equally likely, the probability is $\frac{1}{2}$ that it is card 2.

Solution 2: There are six dots in all. The green dot you see is one of three green dots. Two of the three green dots are on card 2, so the probability is $\frac{2}{3}$ that you see card 2.

[*Hint:* Do it correctly using Bayes's theorem, conditioning on the card that was drawn.]

15. **The gambler's ruin problem.** Conditioning can sometimes be used to find a *recursion,* which can then be used to find a sequence of probabilities. The solution to this classical problem shows the method. Suppose you have k dollars. You start playing repeatedly a game in which you win a dollar with probability p and lose a dollar with probability q. You set yourself a goal of n dollars; you will stop when you either are ruined (lose all your money) or get to n dollars. We want to find the probability that you will be ruined rather than reach your goal.

The way to solve this is to let r_k denote the probability of ruin if you start with k dollars, and to find all of the numbers $r_0, r_1, r_2, \ldots, r_n$ at once. Of course, $r_0 = 1$ and $r_n = 0$.

a. By conditioning on the outcome of the first play, find r_k (for $k = 1, 2, 3, \ldots, n-1$) in terms of r_{k-1} and r_{k+1}. [This is the only part of this exercise that relates to the material of this section.]
b. Check that $r_k = (q^n - q^k p^{n-k})/(q^n - p^n)$ satisfies both the relation found in part a and the two boundary conditions $r_0 = 1$ and $r_n = 0$. Therefore, this must be the probability of ruin, starting with k dollars. [This formula is

usually written $(a^n - a^k)/(a^n - 1)$, where a is the ratio q/p. In courses in combinatorial mathematics, one usually learns how to find such solutions, rather than merely checking a given solution.]

c. Notice that the formula in part b makes no sense if $p = q = \frac{1}{2}$. But check that in this case $r_k = 1 - k/n$ satisfies both the recurrence and the boundary conditions.

d. Suppose that $p = .4737$ and $q = .5263$. [These are approximately your chances of winning and losing if you bet a dollar on "even" at roulette in an American casino.] Use the result of part b to check that, if you start with \$20, your chance of being ruined before reaching \$21 is only about .1122.

Is this a way to beat the odds at a casino? If you have \$20 and your goal is only to gain a dollar, you have a much better chance of succeeding than of failing, and you could do this time after time. Is there a catch? Yes, as we will see in Exercise 21 of Section 4.1.

16. **Bold play or cautious play?** Suppose you have \$10 and badly need another \$10. Your only source of money at the time happens to be a gambling game in which you can bet any amount; you win that amount with probability $p = .47$ and you lose the same amount with probability $q = .53$. Should you boldly make a single bet of \$10, or should you cautiously bet a dollar at a time? [That is, which gives you the higher probability of reaching \$20 before being ruined? Use the result of Exercise 15b.]

17. Repeat Exercises 19 and 20 in the previous section (the "prisoner paradox" and the "Monty Hall paradox") using Bayes's theorem. Some people find this method clearer.

ANSWERS

1. **a.** .5975 **b.** .4686
2. .2655
3. $p(1 + q)$
4. .6983
5. .09676
6. .6944
7. **a.** .0556 **b.** .9561
8. $\frac{6}{11}$, the same as $P(B_2|B_1)$
9. **a.** .25 **b.** .25 for each j **c.** $1/(n - 1)$ for each j **d.** $\frac{1}{2}$
10. **a.** $\frac{17}{30}, \frac{13}{30}$ **b.** $\frac{14}{17}$ **c.** $\frac{2}{3}$
11. $\frac{4}{13}, \frac{8}{17}, \frac{16}{25}$, and, in general, $2^n/(2^n + 9)$
12. **a.** $\frac{2}{11}$ **b.** .4485
13. $(.9)^2e^{-.9}/2!$
14. $\frac{2}{3}$ is correct.
15. **a.** $r_k = pr_{k+1} + qr_{k-1}$
16. With a bold bet of \$10, prob. of success is exactly .47. With cautious bets of \$1 at a time it is about .2312.

2.4 INDEPENDENCE

This section has three parts, which cover different aspects of the same concept. First, we define and give examples for the independence of two events; then, we cover the independence of more than two events; and, finally, we discuss the independence of random variables, which is really just independence of the events describable in terms of the random variables.

We can think about independence in different ways. One is through the equation $P(B|A) = P(B)$; if *the conditional probability equals the unconditional probability,* then the known occurrence of A does nothing to change the probability of B. Essentially equivalent to this is the equation $P(A \cap B) = P(A)P(B)$; the *multiplication rule for the probability of an intersection* holds if and only if the events are independent. [Otherwise, the chain rule is used to find $P(A \cap B)$.] Finally, independence has to do with the *lack of physical influence of the events on each other;* if there is no such influence in the situation being modeled, then we assume independence in the model. (But not conversely; as we will see in Exercise 1 at the end of this section, it is possible for A to affect the occurrence of B without affecting its probability.)

We have already encountered independence on three occasions. One was in Example (2.2.6), where we saw that for two rolls of a fair die, the events A = "3 or less on the first roll" and B = "sum is odd" satisfy $P(B|A) = P(B)$.

Another was in Example (1.2.1), three Bernoulli trials with success probability .65. There we decided on a mass function that had, for example, $p(\text{SFS}) = .65 \cdot .35 \cdot .65$. We gave a justification in terms of the expected relative frequency of SFS over a long series of performances of the three-trial experiment. What we were really justifying was the equation $P(S_1 \cap F_2 \cap S_3) = P(S_1)P(F_2)P(S_3)$, where S_1 is the event "success on first trial," F_2 is "failure on second trial," and S_3 is "success on third trial." That is, we were justifying the assumption that the multiplication rule holds for these three events, using the fact that the three trials can have no physical influence on each other.

The same idea appeared again in Example (1.2.13), where we introduced the binomial distribution for the number of successes in n Bernoulli trials; and, of course, also when we covered Bernoulli processes in Section 1.4. The essence of these probability assignments is the fact that the probability of an outcome like SFFSFF is the product of the probabilities for the individual trials, p^2q^4.

Independence of Two Events

If it happens that $P(B|A) = P(B)$, then the information that A has occurred does nothing to change our view of the probability that B has also occurred. Intuitively, this implies that for the real events being modeled, the occurrence of A does not physically influence the chance of B's occurrence. The events are called *independent* in that case, but we do not use $P(B|A) = P(B)$ as the formal definition. Instead, we rewrite it a little. Using the definition of conditional probability for $P(B|A)$ and then multiplying by $P(A)$, we get the following.

(2.4.1) **Definition.** Events A and B are called ***independent*** if $P(A \cap B) = P(A)P(B)$.

This definition has advantages over $P(B|A) = P(B)$. For one, it is symmetric and does not assign A and B unequal roles. More importantly, $P(B|A) = P(B)$ makes no sense unless $P(A)$ is nonzero, whereas $P(A \cap B) = P(A)P(B)$ makes sense for any events whatever. It may be true or it may be false, depending on the events, but it can be checked, whether the probabilities are 0 or not. It is, of course, equivalent to $P(B|A) = P(B)$ if $P(A)$ is nonzero. In addition, it is equivalent to $P(A|B) = P(A)$, as long as $P(B)$ is nonzero. (See Exercise 7 in Section 2.2.)

Notice that the term *independent* is being used here in a sense that is precisely defined; it may not entirely agree with any previous notions you may have had about what the "independence" of two sets might mean. In particular, "independent" is *not* the same as "disjoint." If A and B are disjoint, then $A \cap B$ is the empty set, whose probability is 0, whereas, if they are independent, then $A \cap B$ has a probability equal to the product of the two probabilities $P(A)$ and $P(B)$. See Exercise 6 at the end of the section for a little more on this.

Many authors like to use the term *probabilistic independence,* or *stochastic independence,* instead of simply "independence," to emphasize the difference between the technical and everyday uses of the word. We do not do so, but you are cautioned to remember that, like many mathematical terms, the word has a different meaning here from that in everyday usage.

(2.4.2) **Example.** If the digits 1, 2, 3, and 4 are permuted randomly, are the events A_1 = "1 is first" and A_2 = "2 is second" independent? (Try to decide intuitively first. The unconditional probability that 2 is second is $\frac{1}{4}$. If you know 1 is first, does the conditional probability that 2 is second stay the same, or does it change?)

Solution. $P(A_1)$ and $P(A_2)$ are both $\frac{1}{4}$ (check this if you need to by listing all 24 equally likely outcomes and noting that 1 is first on six of them and 2 is second on six of them). So they are independent if $P(A_1 \cap A_2) = \frac{1}{4} \cdot \frac{1}{4} = \frac{1}{16}$. But the event $A_1 \cap A_2$ has two outcomes, 1234 and 1243, so its probability is $\frac{1}{12}$ and not $\frac{1}{16}$. The two events are not independent.

[The conditional probability of A_2 given A_1 is $\frac{1}{3}$; you may have decided that even before reading the solution, or you can check it now as the ratio of $P(A_1 \cap A_2) = \frac{1}{12}$ and $P(A_1) = \frac{1}{4}$. Knowing that 1 is first gives 2 a $\frac{1}{3}$ chance of being second; it is as though we were just permuting the remaining three digits, and each has a $\frac{1}{3}$ chance of being in its proper place.] ∎

(2.4.3) **Example.** An urn contains six blue balls and four white balls. Two are drawn, one after the other. Are the events B_1 = "first ball is blue" and W_2 = "second ball is white" independent?

Solution. Of course, the answer depends on whether the sampling is done with or without replacement. You can probably guess the correct answer in both cases. The simplest way to do it rigorously is to compare $P(W_2|B_1)$ with $P(W_2)$.

Without replacement. $P(W_2|B_1)$ is simple. Given that the first ball was blue, the urn then contained five blues and four whites, so the conditional probability of a white on the second draw is $\frac{4}{9}$.

What is the unconditional probability of W_2? We use the law of total probability, conditioning on the color of the first ball. The partition has two events, B_1 and W_1; thus

$$P(W_2) = P(B_1)P(W_2|B_1) + P(W_1)P(W_2|W_1) = \frac{6}{10} \cdot \frac{4}{9} + \frac{4}{10} \cdot \frac{3}{9} = \frac{36}{90} = \frac{4}{10}.$$

Since $\frac{4}{9} \neq \frac{4}{10}$, B_1 and W_2 are not independent. (Notice that the unconditional probability of a white ball on the second draw, $\frac{4}{10}$, is the same as the probability of a white ball on the first draw. If this puzzles you, you might want to go back to Exercise 12i in Section 2.2.)

With replacement. The calculations follow the same pattern as in the without-replacement solution, except that the numbers are a little different because the first ball is replaced. $P(W_2|B_1) = \frac{4}{10}$, because after a blue ball has been drawn and replaced there are four whites and six blues in the urn.

As for the unconditional probability $P(W_2)$, you may not think that any calculation is necessary, because before the second draw, the urn is just as it was before the first, so $P(W_2)$ should be $\frac{4}{10}$. But just to be safe, let us do it with the law of total probability. Only the numbers are different:

$$P(W_2) = P(B_1)P(W_2|B_1) + P(W_1)P(W_2|W_1) = \frac{6}{10} \cdot \frac{4}{10} + \frac{4}{10} \cdot \frac{4}{10} = \frac{4}{10}.$$

So, in this case, B_1 and W_2 are independent.

Notice that if we repeatedly sample *with* replacement from an urn containing balls of two colors, we are performing Bernoulli trials. The success probability is just the proportion of balls that are of the color we count as a success.

(2.4.4) Example: $\varnothing$ and Ω are independent of all events. Let B be any event. The events $\varnothing$ and B are independent, because $\varnothing \cap B = \varnothing$ and, therefore,

$$P(\varnothing \cap B) = P(\varnothing) = 0 \quad \text{and} \quad P(\varnothing)P(B) = 0 \cdot P(B) = 0.$$

Similarly, Ω and B are independent, because $\Omega \cap B = B$ and, therefore,

$$P(\Omega \cap B) = P(B) \quad \text{and} \quad P(\Omega)P(B) = 1 \cdot P(B) = P(B).$$

This seems rather trivial, but some people find it disturbing that an event can be independent of *every* event, including itself. Yes, the event $\varnothing$ is independent of itself, and so is Ω. Can there be any other events that are independent of themselves? Well, if A were independent of itself, then $P(A \cap A)$ would have to equal $P(A)P(A)$. Since $A \cap A = A$, this says that the number $P(A)$ has to equal its own square. The only numbers that equal their own squares are 0 and 1. So it follows that the only events that are independent of themselves are events whose probabilities are 0 and 1.

Recall that there *are* such events, other than $\varnothing$ and Ω, in some probability spaces. In an absolutely continuous space, for example, all singletons and also all finite and countably infinite sets have probability 0, and their complements have probability 1. ∎

We make one more observation before proceeding to the independence of three or more events.

(2.4.5) If events A and B are independent, then so are A^c and B.

Proof. B is the union of the two disjoint events $A \cap B$ and $A^c \cap B$, so

$$P(A^c \cap B) = P(B) - P(A \cap B) = P(B) - P(A)P(B) = \big(1 - P(A)\big)P(B) = P(A^c)P(B).$$ □

Of course, it also follows that A and B^c are independent, and so are A^c and B^c. Thus, in Example (2.4.3), since we found that B_1 and W_2 are independent when sampling with replacement, it follows that B_1 and B_2 are also independent, since B_2 equals W_2^c. Similarly, W_1 and W_2 are independent, as are W_1 and B_2.

Independence of More Than Two Events

(2.4.6) Definition. Three events A, B, and C are ***independent*** if the following hold:

$$P(A \cap B) = P(A)P(B), \quad P(A \cap C) = P(A)P(C), \quad P(B \cap C) = P(B)P(C),$$
$$\text{and} \quad P(A \cap B \cap C) = P(A)P(B)P(C).$$

n events $A_1, A_2, A_3, \ldots, A_n$ are ***independent*** if the following hold:

$P(A_i \cap A_j) = P(A_i)P(A_j)$ for all pairs A_i and A_j (with $i \neq j$),

$P(A_i \cap A_j \cap A_k) = P(A_i)P(A_j)P(A_k)$ for all triples A_i, A_j, A_k (with i, j, and k all different),

$P(A_i \cap A_j \cap A_k \cap A_l) = P(A_i)P(A_j)P(A_k)P(A_l)$ for all quadruples A_i, A_j, A_k, A_l, etc.; and, finally,

$P(A_1 \cap A_2 \cap A_3 \cap \cdots \cap A_n) = P(A_1)P(A_2)P(A_3) \cdots P(A_n)$.

The above is cumbersome, but it can be stated more economically, as follows: A collection of n events (for any $n \geq 2$) is ***independent*** if for every subcollection of two or more of the events, the following product rule holds:

$$P(A_1 \cap A_2 \cap A_3 \cap \cdots \cap A_r) = P(A_1)P(A_2)P(A_3) \cdots P(A_r).$$

In fact, we will see in Definition (2.4.12) below that this single statement serves to define independence even for infinite collections of events. But this does not lessen the complexity of the definition; it merely states several equations in one line. The independence of three events entails four equations; the independence of four events involves eleven equations; and in general $2^n - n - 1$ equations are needed to express the independence of n events. (Can you see why?) This is a lot of equations: $2^{10} - 10 - 1$ equals 1,013, for example. You may wonder why some smaller set of equations would not do.

For example, why would not $P(A \cap B \cap C) = P(A)P(B)P(C)$ by itself be sufficient for the independence of three events? The following example shows why. It gives three events that satisfy this condition, but obviously do not deserve to be called independent.

(2.4.7) Example. Let A and B be any two events that are *not* independent in some probability space and let C be the empty set $\varnothing$. Then $A \cap B \cap C = A \cap B \cap \varnothing = \varnothing$, so $P(A \cap B \cap C) = 0$, and also $P(A)P(B)P(C) = P(A)P(B)P(\varnothing) = 0$. ∎

So $P(A \cap B \cap C) = P(A)P(B)P(C)$ by itself is not enough. What about the three two-event equations $P(A \cap B) = P(A)P(B)$, $P(A \cap C) = P(A)P(C)$, and $P(B \cap C) = P(B)P(C)$? You might think that they would imply the three-event equation. But the following example shows three events that satisfy all three of the two-event equations, but should clearly not be called independent.

(2.4.8) Example. Let the experiment consist of two rolls of a fair die and let A be the event "first die shows an odd number," B the event "second die shows an odd number," and C "the sum is an odd number." Then you can check that

$$P(A) = P(B) = P(C) = \tfrac{1}{2} \quad \text{and} \quad P(A \cap B) = P(A \cap C) = P(B \cap C) = \tfrac{1}{4},$$

and so the two-event equations are all satisfied. But the three-event equation is not satisfied, because $A \cap B \cap C = \varnothing$ and so $P(A \cap B \cap C) = 0$, which is not the product of the probabilities of the three events.

Moreover, we would think something was wrong if the definition allowed these sets to be independent, because if we know that A and B have happened, our conditional probability of C is 0, whereas its unconditional probability is $\tfrac{1}{2}$. A and B together clearly influence the chance of C occurring, even though A alone does not, nor does B. ∎

No, there is no help for it. It can be shown that not one of the equations in the definition of independence is redundant. If you leave out any one of them from the definition, it is possible to make up an example in which all the others hold but the one you left out is violated.

But the good news is this: seldom is anyone asked to *verify* that a collection of events are independent by checking all those equations. Instead, in most situations involving independence, we *assume* that a collection of events are independent because we know from the situation being modeled that they can have no physical influence on each other. When we *assume* events are independent, then we have all those equations at our disposal; we can use them as we wish.

(2.4.9) **Example.** A common example of such a situation, where we need not verify the equations but instead can use them because we assume independence, is in Bernoulli trials. These, you recall, are trials with two outcomes, S and F, in which the probability of S is the same on all trials and the trials have no physical influence on each other. So we assume independence. That is how we justify our decision, in the Bernoulli-trial Examples (1.2.1) and (1.2.13), to assign an outcome like SSFSF, say, the probability p^3q^2. The singleton {SSFSF} is really the intersection of five events that we assume to be independent, $S_1 \cap S_2 \cap F_3 \cap S_4 \cap F_5$, where as usual S_1 is the event "S on trial 1," and so forth. So the probability of that singleton is

$$P(S_1)P(S_2)P(F_3)P(S_4)P(F_5) = p \cdot p \cdot q \cdot p \cdot q = p^3q^2.$$

The assumption of independence, stated formally, says that events described in terms of nonoverlapping sets of trials are independent. This is just a restatement, and generalization, of rule B4 for Bernoulli processes in Section 1.4. For example, if A is the event "three S's in the first five trials" and B is "no S's in the seventh through the tenth trials," then we assume A and B are independent. ∎

(2.4.10) **Example.** Three people independently play some game against a common opponent, once each. Because of their different levels of skill, A has probability $\frac{1}{2}$ of winning, B's probability is $\frac{1}{3}$, and C's is $\frac{1}{4}$. What is the probability that they all win? What is the probability that at least one of them wins? What is the probability that exactly one of them wins?

Solution. The probability that they all win is $P(A \text{ wins} \cap B \text{ wins} \cap C \text{ wins})$. Because we are given that their plays are independent, we can use the three-set product formula and get the answer: $\frac{1}{2} \cdot \frac{1}{3} \cdot \frac{1}{4} = \frac{1}{24}$.

The event that at least one of them wins is the complement of the event that they all lose. So the probability is $1 - P(A \text{ loses} \cap B \text{ loses} \cap C \text{ loses})$. Again we can assume that these three events are independent (they are the complements of independent events, for one thing), so the probability is $1 - \frac{1}{2} \cdot \frac{2}{3} \cdot \frac{3}{4} = \frac{3}{4}$.

To find the probability that exactly one person wins, we break the event into the union of three disjoint events and find their probabilities by independence:

$$\begin{aligned} P(\text{exactly one wins}) &= P(A \text{ wins} \cap B \text{ loses} \cap C \text{ loses}) + P(A \text{ loses} \cap B \text{ wins} \cap C \text{ loses}) \\ &\quad + P(A \text{ loses} \cap B \text{ loses} \cap C \text{ wins}) \\ &= \frac{1}{2} \cdot \frac{2}{3} \cdot \frac{3}{4} + \frac{1}{2} \cdot \frac{1}{3} \cdot \frac{3}{4} + \frac{1}{2} \cdot \frac{2}{3} \cdot \frac{1}{4} = \frac{11}{24}. \end{aligned}$$

∎

(2.4.11) **Example: Independence of nonoverlapping intervals in Poisson processes.** In Section 1.4 we gave some rules for finding probabilities related to Poisson processes. Rule P4, like rule B4 for Bernoulli processes, was a little clumsy because we did not have the language of independence; it said that for two nonoverlapping time intervals I and J of lengths r and s, the probability of k arrivals in I and l arrivals in J is the product of Poisson probabilities

$$\frac{(\lambda r)^k e^{-\lambda r}}{k!} \cdot \frac{(\lambda s)^l e^{-\lambda s}}{l!}.$$

Now we can shorten this rule by saying that for any nonoverlapping time intervals I and J and any nonnegative integers k and l, the events "k arrivals in I" and "l arrivals in J" are independent. Since rule P1 tells what the probabilities of those intervals are, this is all that needs to be said.

Actually, more could be said. For any finite collection of nonoverlapping time intervals $I_1, I_2, \ldots, I_n$, and any nonnegative integers $k_1, k_2, \ldots, k_n$, the n events "k_1 arrivals in I_1," "k_2 arrivals in I_2," etc. are independent. ∎

Before we turn to the independence of random variables, we will say a word about the independence of infinitely many events. It turns out that something nice happens here: the definition of independence that works best requires only that all *finite* collections of the events satisfy the multiplication rule for the probability of the intersection. Actually, this definition serves to define independence for finitely many events as well, so it supersedes the previous two definitions, (2.4.1) and (2.4.6).

(2.4.12) **Definition.** An arbitrary collection of events is ***independent*** if, for every finite subcollection of two or more of the events, say $A_1, A_2, \ldots, A_r$,

$$P(A_1 \cap A_2 \cap \cdots \cap A_r) = P(A_1)P(A_2) \cdots P(A_r).$$

Independence of Random Variables

Recall that a random variable X is a function from an outcome set Ω to $\mathbb{R}$. The "events describable in terms of X" are events such as

$$(X = 2) = \{\omega: \ X(\omega) = 2\},$$
$$(X \geq 1.4) = \{\omega: \ X(\omega) \geq 1.4\},$$

and, in general,

$$(X \in A) = \{\omega: \ X(\omega) \in A\}$$

for any interval (or other Borel set) A.

(2.4.13) **Definition.** Random variables X and Y are ***independent*** if, for all possible choices of two Borel subsets A and B of $\mathbb{R}$, the events $(X \in A)$ and $(Y \in B)$ are independent (that is, if any event describable in terms of X is independent of any event describable in terms of Y).

More generally, any collection $X_1, X_2, \ldots$ of random variables are ***independent*** if, for all choices of Borel subsets $A_1, A_2, \ldots,$ the events $(X_1 \in A_1), (X_2 \in A_2), \ldots$ are independent.

This is another definition that looks formidable. In most cases, there are infinitely many events describable in terms of a random variable X, and so it appears that we have a tremendous infinity of equations to deal with if we want to think about the independence of random variables. But here there are two pieces of good news.

First, as before, we seldom have to *verify* that random variables are independent; rather, we *assume* they are independent, and then we get to use any of those equations we wish.

But second, it turns out that there are some theorems (which we will see later) that enable us to check that random variables are independent without checking independence of events. We simply have to check a product formula for any of a number of different functions (densities, mass functions, cumulative distribution functions, generating functions) that arise in connection with the random variables.

Following are examples of some calculations made possible by assuming the independence of random variables.

(2.4.14) Example. In a Poisson process with arrival rate $\lambda = 2.6$ per hour, let X be the number of arrivals in the first 2 hours and Y the number in the next 3 hours. What is the probability that X is greater than 1 but Y is less than 3?

Solution. Probabilities for X are Poisson probabilities with $\lambda \cdot 2 = 5.2$ in place of λ, and probabilities for Y are Poisson with $\lambda \cdot 3 = 7.8$. So

$$\begin{aligned} &P((X > 1) \cap (Y < 3)) \\ &\quad = P(X > 1) \cdot P(Y < 3) \qquad \text{(because of independence)} \\ &\quad = \left(1 - \frac{(5.2)^0 e^{-5.2}}{0!} - \frac{(5.2)^1 e^{-5.2}}{1!}\right) \cdot \left(\frac{(7.8)^0 e^{-7.8}}{0!} + \frac{(7.8)^1 e^{-7.8}}{1!} + \frac{(7.8)^2 e^{-7.8}}{2!}\right) \\ &\quad \doteq .9658 \cdot .0161 \doteq .0155. \end{aligned}$$

∎

(2.4.15) Example. Let X have the possible values 0, 1, 2, and 3, with probabilities .1, .3, .25, and .35, respectively. Let Y have possible values 0, 1, and 2, with probabilities .3, .5, and .2, respectively. Assuming X and Y are independent, we will find the probabilities of some events describable in terms of X and Y. Note that we would have no way of finding these probabilities without the assumption of independence, unless we had some other information about what are called the "joint probabilities" for X and Y.

a. What is the probability that $X = 3$ and $Y = 0$?

Solution.

$$P(X = 3 \cap Y = 0) = P(X = 3)P(Y = 0) = .35 \cdot .3 = .105.$$

b. What is the probability that $X + Y = 3$?

Solution. It equals

$$P(X = 3 \cap Y = 0) + P(X = 2 \cap Y = 1) + P(X = 1 \cap Y = 2),$$

and because of independence each of these three is a product:

$$.35 \cdot .3 + .25 \cdot .5 + .3 \cdot .2 = .29.$$

c. What is the probability that $X = Y$?

Solution.

$$P(X = 0 \cap Y = 0) + P(X = 1 \cap Y = 1) + P(X = 2 \cap Y = 2)$$
$$= .1 \cdot .3 + .3 \cdot .5 + .25 \cdot .2 = .23.$$

■

EXERCISES FOR **SECTION 2.4**

1. Consider three tosses of a fair coin. Which of the following pairs of events are independent?
   a. A = "heads on the first toss" and B = "an even number of heads"
   b. A = "no heads in the first two tosses" and B = "no heads in the second and third tosses"
   c. A = "even number of heads in the first two tosses" and B = "even number of heads in the second and third tosses"

   Note that although events with no influence on each other are independent, independent events may have some influence on each other. In the independent pairs in this exercise, for example, the occurrence of A will influence *how* B can occur, but cannot influence the *probability* that B occurs.

2. Let X and Y be independent random variables; suppose X has the Poisson distribution with $\lambda = 2$ and Y has the Poisson distribution with $\lambda = 3$. Find the following probabilities:
   a. $P(X = 2 \text{ and } Y = 1)$
   b. $P(X = 2 \text{ and } Y < X)$
   c. $P(X \geq 2 \text{ and } Y \leq 1)$
   d. $P(X + Y = 3)$

3. Let X and Y be independent random variables. Each has possible values 0, 1, and 2, with probabilities .2, .3, and .5, respectively. Find the probabilities of the following events:
   a. $X = 2$ and $Y = 2$
   b. $X = Y$

c. $X + Y = 1$
d. $X > Y$ [You may see how to get this from the answer to part b with very little calculation.]
e. $XY = 0$
f. $|X - Y| = 1$

4. Let A, B, and C be independent events; suppose $P(A) = \frac{2}{3}$, $P(B) = \frac{3}{4}$, and $P(C) = \frac{1}{5}$. Find the probabilities of the following events:
a. $A \cap B \cap C$
b. $A \cup B \cup C$
c. $A \cap B^c \cap C^c$
d. Exactly one of the three events occurs.
e. Exactly k of the three events occur (for $k = 0, 1, 2, 3$). [*Hint:* You already have the probabilities for $k = 3$ and $k = 1$, and you can get the one for $k = 0$ from part b. Then you can get the one for $k = 2$ by subtraction, but it would be prudent to do it separately as a check.]

5. a. Suppose your probability of winning a certain game is $\frac{1}{10}$. You play it 10 times. If successive plays are independent, what is the chance that you win at least once?
b. Repeat part a, but with n plays and $\frac{1}{n}$ as the probability of winning on each play.
c. What is the limit of the answer to part b as n increases to ∞?

6. It is a common error to confuse "independent" with "disjoint." They do not mean the same thing; disjoint events have an empty intersection, whereas independent events have an intersection whose probability is exactly the product of the probabilities of the events. The question here is: can two disjoint events *ever* be independent? If not, explain why. If it is possible, give an example of two disjoint independent events.

7. Is it possible for A and B to be independent if A is a subset of B? If it is impossible, explain why; if it is possible, give an example.

8. For random permutations of the integers $1, 2, 3, \ldots, n$, let A be the event "1 is first" and B the event "2 is second." Find $P(A)$, $P(B)$, and $P(A \cap B)$.

9. Suppose Ω is a finite outcome space in which all outcomes are equally likely, and A and B are events in Ω.
a. Suppose A has five members, B has eight, and $A \cap B$ has two. If A and B are independent, how many members has Ω?
b. Suppose A has seven members, B has nine, and $A \cap B$ has four. Show that A and B cannot possibly be independent.

[a]**10.** Let $X_1, X_2, \ldots, X_{10}$ be independent random variables. Suppose each of them has just two possible values, 1 and 0, and for each X_i, $P(X_i = 1) = p$ and $P(X_i = 0) = 1 - p$. Let $Y = X_1 + X_2 + \cdots + X_{10}$. What are the possible values of Y, and what are their probabilities?

11. Let a be a number between 0 and 1.

a. If n points are chosen independently and at random with the uniform distribution on the interval (0, 1), find the probability that they are all in the subinterval $(0, a)$.

b. In a Poisson process with arrival rate λ, given that there are exactly n arrivals in the interval (0, 1), find the conditional probability that they are all in the subinterval $(0, a)$.

These two probabilities are the same. This is a special case of a more general fact about Poisson processes: given that there were n arrivals in an interval I, probabilities for them are the same as for n points chosen independently and at random from the uniform distribution on I. (There is a slight technical difference, in that the arrival times are automatically ordered, while n uniformly distributed points will not be chosen in order from least to greatest.) The proof of the full result is usually given in a textbook on stochastic processes. The result is often expressed informally by saying that *given the number of arrivals in a time interval of a Poisson process, the arrival times are conditionally uniformly distributed.* This will be discussed again in Exercise 13 of Section 8.6.

12. If events A and B are not independent, then either $P(B|A) > P(B)$, in which case we say that A is *favorable for* B, or else $P(B|A) < P(B)$, and we say that A is *unfavorable for* B.

a. Show that A is favorable for B if and only if $P(A \cap B) > P(A)P(B)$.

b. Show that if A is favorable for B, then B is favorable for A.

c. Show that if A is unfavorable for B, then A is favorable for B^c.

[Of course you can see why these terms are used: $P(B|A) > P(B)$ says that knowledge of A's occurrence increases B's chance of occurring.]

ANSWERS

1. **a.** Yes **b.** No **c.** Yes
2. **a.** .0404 **b.** .0539 **c.** .1183 **d.** .1404
3. **a.** .25 **b.** .38 **c.** .12 **d.** .31 **e.** .36 **f.** .42
4. **a.** $\frac{6}{60}$ **b.** $\frac{56}{60}$ **c.** $\frac{8}{60}$ **d.** $\frac{21}{60}$ **e.** $\frac{4}{60}, \frac{21}{60}, \frac{29}{60}, \frac{6}{60}$
5. **a.** .6513 **b.** $1 - \left(1 - \frac{1}{n}\right)^n$ **c.** $1 - \frac{1}{e}$
6. It is possible.
7. It is possible.
8. $1/n$, $1/n$, $1/n(n-1)$
9. **a.** 20
10. Values 0, 1, 2, . . . , 10; probabilities are binomial with $n = 10$ and the given p.
11. a^n in both parts

2.5 LIMITS OF PROBABILITIES OF NESTED EVENTS

The purpose of this section is to present two formulas, given in (2.5.3), that we will use directly only once or twice in later chapters, but that need to be included because of their usefulness in the theory, as well as in some applications. Together, these two formulas are sometimes called the "nested sets theorem," because they deal with countably infinite collections of events $A_1, A_2, A_3, \ldots$ in which either $A_1 \subset A_2 \subset A_3 \subset \cdots$ (a nested increasing sequence) or $A_1 \supset A_2 \supset A_3 \supset \cdots$ (decreasing).

At first glance you might not think that there could be a nested decreasing sequence that does not eventually reach the empty set; but when the sets are infinite it can happen easily. Similarly, there are plenty of examples of nested increasing sequences that do not eventually reach Ω, if Ω is infinite.

(2.5.1) Example: Decreasing nested sets

a. Let $\Omega = \mathbb{R}$, and let $A_1 = [0, 1]$, $A_2 = \left[0, \frac{1}{2}\right]$, $A_3 = \left[0, \frac{1}{3}\right]$, and in general $A_n = \left[0, \frac{1}{n}\right]$. Then the A_n's are nested and decreasing in that $A_1 \supset A_2 \supset A_3 \supset \cdots$. In this case we might well be interested in the intersection of all the A_n's, which is in some sense their limit. That intersection is the singleton set $\{0\}$.
b. Again with $\Omega = \mathbb{R}$, let $A_1 = (-\infty, -1]$, $A_2 = (-\infty, -2]$, and in general $A_n = (-\infty, -n]$. Then again $A_1 \supset A_2 \supset \cdots$. In this case, the intersection is the empty set, even though each A_n is an infinitely long interval.
c. With $\Omega = \mathbb{R}$, let x be some number, and let $A_1 = (-\infty, x + 1]$, $A_2 = \left(-\infty, x + \frac{1}{2}\right]$, and, in general, $A_n = \left(-\infty, x + \frac{1}{n}\right]$. Again each A_n is a subset of the previous one. Now the intersection is $(-\infty, x]$.
d. Change the previous example so that the A_n's are *open* intervals [that is, $A_n = \left(-\infty, x + \frac{1}{n}\right)$ for each n]. Perhaps surprisingly, the intersection is still $(-\infty, x]$, not $(-\infty, x)$. Be sure you see why the intersection contains its right endpoint even though the A_n's do not contain theirs.
e. This time let Ω be the outcome set for a Bernoulli process—that is, the set of all infinite sequences of S's and F's. Let A_n be the event that there are no S's in the first n trials; that is, A_n is the set of all sequences whose first n letters are F's. Then each A_n is a subset of the previous one, because any sequence whose first n letters are F's is also a sequence whose first $n - 1$ letters are F's. In this case, the intersection of all the A_n's is the singleton event whose only member is the sequence containing nothing but F's. See (2.5.4) for more on this. ∎

(2.5.2) Example: Increasing nested sets

a. Let $\Omega = \mathbb{R}$, and let $A_1 = [1, \infty)$, $A_2 = \left[\frac{1}{2}, \infty\right)$, $A_3 = \left[\frac{1}{3}, \infty\right)$, and, in general, $A_n = \left[\frac{1}{n}, \infty\right)$. Then the A_n's are nested and increasing in that $A_1 \subset A_2 \subset A_3 \subset \cdots$. In this case, their union is considered as their limit. That union is $(0, \infty)$. (Notice that the A_n's all contain their left endpoints, but the union does not contain its left endpoint.)
b. Let $\Omega = \mathbb{R}$, and let $A_1 = (-\infty, 1]$, $A_2 = (-\infty, 2]$, and, in general, $A_n = (-\infty, n]$. Again $A_1 \subset A_2 \subset \cdots$ and, in this case, their union is $\mathbb{R}$.
c. Let $\Omega = \mathbb{R}$, let x be some number, and let $A_1 = (-\infty, x - 1]$, $A_2 = \left(-\infty, x - \frac{1}{2}\right]$, and, in general, $A_n = \left(-\infty, x - \frac{1}{n}\right]$. Each A_n is again a subset of

the next one, and their union is $(-\infty, x)$. [If the A_n's were open, the union would be the same. Compare this with parts c and d in Example (2.5.1).]

d. Let Ω be the set of all infinite sequences of S's and F's (the outcome space for a Bernoulli process) and let A_n be the event "there are at least two S's in the first n trials"; that is, A_n is the set of all sequences whose first n letters include at least two S's. (A_1 is empty, but the rest of the A_n's all contain infinitely many sequences.) Now you can see that $A_1 \subset A_2 \subset A_3 \subset \cdots$, because if a sequence has two or more S's in its first n letters, then it certainly has two or more in its first $n + 1$. The union of all the A_n's is the set of all sequences that contain two or more S's. ▮

It is good to compare Examples (2.5.1c) and (2.5.2c). In both cases the sets A_n contain their right endpoints. When those endpoints decrease to x, as in Example (2.5.1c), the limit (intersection) contains x. But when they increase to x, as in Example (2.5.2c), the limit (union) does not contain x.

You will find two more examples like these at the end of Section A.3 in Appendix A.

(2.5.3) Nested sets theorem. Let $A_1, A_2, A_3, \ldots$ be events in any probability space.

a. If $A_1 \subset A_2 \subset A_3 \subset \cdots$, then $P(A_1 \cup A_2 \cup A_3 \cup \cdots) = \lim_{n\to\infty} P(A_n)$.

b. If $A_1 \supset A_2 \supset A_3 \supset \cdots$, then $P(A_1 \cap A_2 \cap A_3 \cap \cdots) = \lim_{n\to\infty} P(A_n)$.

The answer is the same in both cases. The probability of the "limiting set" is the limit of the probabilities as long as the events are nested. But the limiting set is the union if the events are increasing, and the intersection if they are decreasing.

A proof of this theorem is at the end of the section. First we illustrate its use.

(2.5.4) **Example: The probability of no successes, ever, in a Bernoulli process, is 0.** Suppose the success probability in a Bernoulli process is extremely small, though not 0. Is there any guarantee that we will eventually see a success? That is, if A is the event that we get an infinite string of F's without an S, what is the probability of A? If it is 0, then an eventual success is certain if the Bernoulli process model is valid. Here we prove that it is 0.

Proof. We use the nested sets theorem. The event A, "an infinite string of F's with no S"—that is, the singleton {FFFFF. . .}—is the intersection of a nested decreasing sequence of events. In particular, as we noted in Example (2.5.1e),

> if A_n is the event "no S's in the first n trials," then $A_1 \supset A_2 \supset A_3 \supset \cdots$, and $A_1 \cap A_2 \cap A_3 \cap \cdots$ is the singleton $A =$ {FFFFF. . .}.

And the probabilities of the A_n's are easy to find: A_n is just the event that the first n trials result in failure, so its probability is q^n. Therefore, by the nested sets theorem, $P(A) = \lim_{n\to\infty} q^n = 0$. □

Thus, in a Bernoulli process model, the probability of the singleton event {FFFFF. . .} is 0. Therefore, if this model accurately represents repeated trials, then no matter how unlikely a success is, there will eventually be a success. Informally, if something *can* happen, then it *will* happen (with probability 1) if we do enough trials.

You may recall that we have previously seen something that led us to this same conclusion. In Exercise 7 of Section 1.2 we modeled the experiment of waiting for the first success; the probability that it comes after n failures is $q^n p$ (for $n = 0, 1, 2, \ldots$). These probabilities also add to 1, without the need for an additional outcome representing no successes at all, because their sum is

$$q^0 p + q^1 p + q^2 p + \cdots = p \cdot (1 + q + q^2 + q^3 + \cdots) = \frac{p}{1-q} = \frac{p}{p} = 1.$$

Even in that early example we saw that there was no need for the model to include the possibility of no S's in infinitely many trials. Now we have seen that the more general Bernoulli process model also assigns probability 0 to that outcome.

(2.5.5) Remark on the certainty of an eventual success. People are sometimes disturbed by the implications of the fact that "if something has positive probability of happening, then it *will* eventually happen, no matter how small the probability." For example, if the standard physical model of gases is applicable, then within a given day there is a positive probability that all the air molecules in your room will at some time find themselves in one half of the room. The nested sets theorem implies that, with probability 1, there will come a day on which that happens. Should you move your bed to the middle, or try in some other way to plan for this possibility? No, because although it is certain to happen over an *infinite* sequence of days, the probability that it will happen in the *finite* sequence of days in your life, or even in the life of the universe, is very small.

To see how small, let us take another example of a rare event whose probability we can calculate—a "perfect deal" in the game of bridge. Suppose we deal a random bridge hand; that is, we distribute 52 cards at random to four people so that each person gets 13 cards. Then the probability of a perfect deal, in which someone gets all 13 hearts, someone gets all 13 spades, and so forth, turns out to be

$$\frac{4!}{\binom{52}{13} \cdot \binom{39}{13} \cdot \binom{26}{13} \cdot \binom{13}{13}} \doteq 4.474 \cdot 10^{-28}$$

$$= .0000000000000000000000000004474.$$

[The numerator, 4!, is the number of perfect deals, and the denominator is the number of possible deals—that is, the numer of ways to assign 13 cards to one player, 13 from the remaining 39 to a second player, and so on.]

Now an infinite sequence of bridge deals is certain to contain at least one perfect deal, but what is the probability that there is a perfect deal in the first billion billion (10^{18}) deals? The answer is 1 minus the probability that there are no perfect deals in the first billion billion, and

$$P(\textit{no}\text{ perfect deal in the first billion billion}) = q^N = (1 - p)^N,$$

where p is $4.474 \cdot 10^{-28}$ and $N = 10^{18}$.

How can we get an idea of the size of such a number? We can approximate it by remembering that if N is large compared to x, then $(1 - x/N)^N$ is close to e^{-x}. So

$$\begin{aligned} P(\textit{no}\text{ perfect deal in } N \text{ deals}) &= (1 - p)^N \\ &= \left(1 - \frac{Np}{N}\right)^N \approx e^{-Np} = e^{-(4.474 \cdot 10^{-10})}, \end{aligned}$$

which is e to such a small negative power that it is essentially 1. (It is .99999999955, to 11 decimal places.) Thus, although a perfect deal is certain to appear eventually, there is almost no chance that it will appear in the first billion billion deals.

Now a billion billion is somewhere near the number of seconds in the current age of the universe. So even if someone had been dealing bridge hands at the rate of one per second since the universe began, there would be almost no chance of a perfect deal up to now. Another way of looking at it is this: if 400 million bridge players, at 100 million tables, each play 100 hands every evening, it will take almost 274,000 years to complete a billion billion deals.

It is interesting in this connection to notice that reports of perfect bridge deals do appear from time to time. Between 1952 and 1963 there were at least *eight* reports of perfect deals in newspapers in the United States and England. This was discussed by N. T. Gridgeman in *The American Statistician* (vol. 18, February and April 1964). (There was actually a ninth report, by a bridge writer who claimed to have seen five or six perfect deals.) What are we to make of this? There are two possible explanations. Either something occurred that is so unlikely as to be considered miraculous (that is, eight perfect deals within 11 years), or else the model of random distribution of cards was not a valid one for the situations that produced the perfect deals. A basic principle of statistical inference is that the "model doesn't fit" explanation is preferred to the "rare occurrence" explanation, when the probability assigned by the model to the rare occurrence is very small. Here the probability assigned to the occurrence "eight perfect deals in 11 years" by the model of independent random deals is well under .000000001, which is certainly very small.

So most people would decide that the model of independent random deals is not applicable. What do we conclude? Either that the reports were fakes, or else that they were true reports but that the cards were not being dealt independently and randomly. Actually, it is quite difficult to shuffle a deck of cards effectively, expecially in a game like bridge were the suits tend to be together at the end of a hand. So we might choose the charitable interpretation and say that the perfect deals might really have happened, but if so the shuffling left something to be desired.

On the other hand, Gridgeman does a bit of analysis in his *American Statistician* paper showing that the reported number of perfect deals is very unlikely even if we allow for poor shuffling. And he makes a further point. A "semiperfect deal," in which two players each get all the cards of one suit but the other two do not, is 30 million times more likely than a perfect deal; and yet there are apparently no reports of semiperfect deals.

(2.5.6) **Example.** In Example (1.3.1), the model for a uniform random number from $[0, 1)$, we knew at one point that $P[a, b)$ was $b - a$, but we did not yet know that $P(a, b)$ was also $b - a$. We showed that $P(.5, 1)$ was $\frac{1}{2}$, the same as $[.5, 1)$, by expressing $(.5, 1)$ as a union of disjoint intervals of the form $[a, b)$ and showing that their probabilities add to $\frac{1}{2}$.

We had to be clever in this way because we did not have the nested sets theorem at that point. Now we can do it more simply. Suppose we know that $P[a, b) = b - a$ for all subintervals $[a, b)$ of $[0, 1)$, and we want to show that $P(a, b)$ is also $b - a$. We just notice that (a, b) is the union of the increasing nested events $\left[a + \frac{1}{n}, b\right)$; so its probability is the limit of their probabilities, which is the limit of $b - a - \frac{1}{n}$, which is $b - a$. ∎

Example (2.5.6) shows how to find the probability of an interval that does not contain its right endpoint, in terms of the probabilities of intervals that do contain their right endpoints. In fact, this is our main use for the nested sets theorem—finding probabilities of intervals of various kinds when all we know is the probabilities of intervals of the form $(-\infty, x]$. If we knew this for every x but needed the probability of $(-\infty, a)$, for example, we could use the fact that $(-\infty, a)$ is the union of the increasing nested events $A_1 = (-\infty, a - 1]$, $A_2 = \left(-\infty, a - \frac{1}{2}\right]$, . . . , $A_n = \left(-\infty, a - \frac{1}{n}\right]$, ... The probability of $(-\infty, a)$ would be the limit of their probabilities, which we could find.

The probability of $(-\infty, x]$ as a function of x is so useful an idea that we give it a name, the *cumulative distribution function.*

(2.5.7) Cumulative distribution functions. Let us suppose that we have some assignment of probabilities to the Borel subsets of $\mathbb{R}$. All this means is that the intervals have been assigned probabilities in a consistent way, because from such an assignment we can in principle find the probabilities of sets that can be made from intervals. This was discussed in Section 1.3, especially in (1.3.2). Such probability assignments are called probability distributions; they are the subject of Chapter 3.

Suppose that such an assignment has been made and we are given $P(-\infty, x]$ for each real number x, as a function of x, but that we are not given any other information about the probabilities. What can we say about this function of x? Let us give it a name, $F(x) = P(-\infty, x]$. It is called the *cumulative distribution function* of the probability assignment. (Some authors call it simply the *distribution function,* but the word "cumulative" helps to remind us that it gives the probability of all values up to and including x.)

Cumulative distribution functions have several important properties and uses, which we will see in the next chapter. Here we mention three properties that follow from the nested sets theorem.

a. $F(x) = P(-\infty, x]$, but this does not necessarily equal $P(-\infty, x)$. The difference is the probability of the singleton $\{x\}$, which may or may not be positive.

However, we can find $P(-\infty, x)$ in terms of the values of $F(x)$. It equals the left-hand limit of F at x. This limit can be found by taking any sequence $a_1, a_2, a_3, \ldots$ of positive numbers that decrease to 0 $\left(a_n = \frac{1}{n} \text{ will do}\right)$ and evaluating $\lim_{n\to\infty} F(x - a_n)$. This works because the intervals $(-\infty, x - a_n]$ are increasing nested events whose union is $(-\infty, x)$.

Since the left-hand limit of F at x need not equal $F(x)$, we say that $F(x)$ is not necessarily *continuous from the left.*

b. But $F(x)$ is automatically *continuous from the right;* that is, if $a_1, a_2, a_3, \ldots$ is a sequence of positive numbers that decrease to 0, then for any x, $\lim_{n\to\infty} F(x + a_n) = F(x)$.

Proof. The intervals $(-\infty, x + a_n]$ are decreasing nested events whose intersection is $(-\infty, x]$. □

c. $\lim_{n\to\infty} F(x) = 1$ and $\lim_{x\to-\infty} F(x) = 0$. This is the same as saying that if $c_1, c_2, c_3, \ldots$ is a sequence of positive numbers that increase to ∞, then $\lim_{n\to\infty} F(c_n) = 1$ and $\lim_{n\to\infty} F(-c_n) = 0$.

Proof. The intervals $(-\infty, c_n]$ are increasing nested events whose union is $\mathbb{R}$, and the intervals $(-\infty, -c_n]$ are decreasing nested events whose intersection is empty. □

We close the section with a proof of (2.5.3), just for the record.

(2.5.8) **Proof of the nested sets theorem.** First of all, we need to prove only part a, because we can prove part b from part a by applying part a to the complements of the A_i's. Specifically, if $A_1 \supset A_2 \supset \cdots$, then $A_1^c \subset A_2^c \subset \cdots$, and so by part a,

$$P(A_1^c \cup A_2^c \cup \cdots) = \lim_{n\to\infty} P(A_n^c).$$

But by De Morgan's law (given for two sets in Section A.2 of Appendix A, but valid for any collection of sets), the left side of this is the probability of the complement of $A_1 \cap A_2 \cap A_3 \cap \cdots$, which is $1 - P(A_1 \cap A_2 \cap A_3 \cap \cdots)$. The right side is $\lim_{n\to\infty} (1 - P(A_n)) = 1 - \lim_{n\to\infty} P(A_n)$. Part b follows.

So we need to prove part a. The trick is to consider the disjoint sets $B_1, B_2, B_3, \ldots$ shown in Figure 2.5a; $B_1 = A_1$, $B_2 = A_2 \cap A_1^c$, $B_3 = A_3 \cap A_2^c$, etc. These sets have the following properties:

(2.5.9) $$B_1 \cup B_2 \cup B_3 \cup \cdots = A_1 \cup A_2 \cup A_3 \cup \cdots, \quad \text{and}$$

(2.5.10) $$\text{for each } n, \quad B_1 \cup B_2 \cup B_3 \cup \cdots \cup B_n = A_n.$$

Therefore,

$$\begin{aligned}
P(A_1 \cup A_2 \cup A_3 \cup \cdots) \\
&= P(B_1 \cup B_2 \cup B_3 \cup \cdots) && \text{(by (2.5.9))} \\
&= \sum_{n=1}^{\infty} P(B_n) && \text{(because the } B_n\text{'s are disjoint)} \\
&= \lim_{n\to\infty} (P(B_1) + P(B_2) + \cdots + P(B_n)) && \text{(definition of sum of a series)} \\
&= \lim_{n\to\infty} P(B_1 \cup B_2 \cup \cdots \cup B_n) && \text{(because the } B_k\text{'s are disjoint)} \\
&= \lim_{n\to\infty} P(A_n) && \text{(by (2.5.10)).} \quad \square
\end{aligned}$$

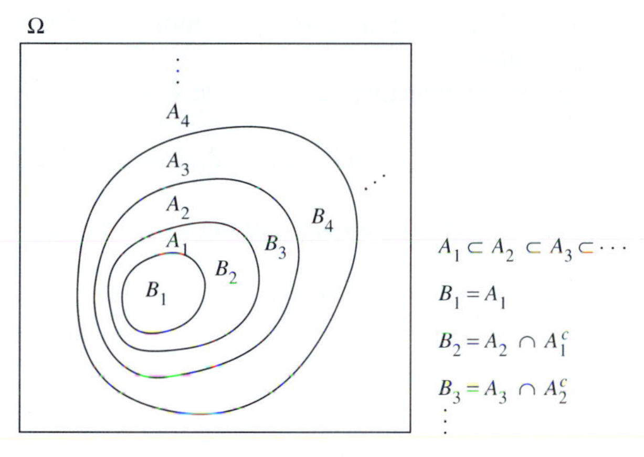

FIGURE 2.5a Nested sets $A_1, A_2, \ldots$ and disjoint sets $B_1, B_2, \ldots$ constructed from them, used in the proof of the nested sets theorem

EXERCISES FOR SECTION 2.5

1. Let $\Omega = \mathbb{R}$. Each of the following defines a nested sequence $A_1, A_2, \ldots$ of subsets of $\mathbb{R}$; find its union if it is increasing and its intersection if it is decreasing.
   a. $A_n = \left(0, 1 - \frac{1}{n}\right]$ **b.** $A_n = \left[\frac{1}{n}, 1 - \frac{1}{n}\right)$ **c.** $A_n = \left(-\frac{1}{n}, \frac{1}{n}\right)$
   d. $A_n = \left[\frac{1}{n}, n\right]$

2. Suppose that for a certain probability assignment on $\Omega = [0, \infty)$, it happens that $P[0, x] = x^2/2$ when $0 \le x < 1$, and that the probability of the singleton $\{1\}$ is $\frac{1}{4}$.
   a. What is $P[0, 1)$? **b.** What is $P[0, 1]$?

3. **An eventual arrival is certain in a Poisson process.** To see that this is true no matter how low the arrival rate λ is, let A_n be the event "no arrival in the first n time units."

a. Are the events A_n increasing or decreasing?
b. What is the probability of A_n?
[a]c. What does the nested sets theorem imply about their union or intersection?
[a]d. What is this event? Give a verbal description.

[a]4. **No matter what the success probability is, there will be infinitely many successes in a Bernoulli process.** Consider a Bernoulli process with success probability p. Assume p is positive. In Example (2.5.4) we showed that with probability 1 there will eventually be a success. In fact, there will be infinitely many, which we can show as follows.
a. What is the probability that there is at most one S and all the rest of the trials result in F? [*Hint:* Let A_n be the event "at most one success in the first n trials."]
b. What is the probability that there are at most two S's in the infinite sequence of trials?
c. Let B_k be the event that there are at most k successes in the infinite sequence of trials. What is the probability of B_k?
d. $B_1, B_2, B_3, \ldots$ form a nested sequence of events. Are they increasing or decreasing?
e. What does the nested sets theorem imply about the union or intersection of the B_k's?
f. Give a verbal description of the event whose probability is found in part e.

ANSWERS

1. **a.** $(0, 1)$ **b.** $(0, 1)$ **c.** $\{0\}$ **d.** $(0, \infty)$
2. **a.** $\frac{1}{2}$ **b.** $\frac{3}{4}$
3. **b.** $e^{-\lambda n}$ **c.** The probability of their intersection is 0. **d.** No arrivals in $[0, \infty)$
4. **a.** 0 **b.** 0 **c.** 0, regardless of k **d.** Increasing **e.** The probability of their union is 0. **f.** There are finitely many successes in the infinite sequence of trials.

SUPPLEMENTARY EXERCISES FOR **CHAPTER 2**

1. For a Poisson process with arrival rate $\lambda = .5$ per minute, let A be the event "no arrivals in $[0, 2]$" and B the event "no arrivals in $[1, 3]$." Find $P(A \cap B)$ and $P(A \cup B)$.

2. What is the probability that you get at least one ace in four draws without replacement from a standard deck?

3. For a random permutation of the numbers 1, 2, 3, 4, 5, and 6, what is the probability that at least one of the numbers 1, 2, and 3 is in its proper place?

4. Consider a lottery in which 1,000 tickets are sold.
a. Twenty numbers will be drawn from the numbers $1, 2, \ldots, 1{,}000$ (without replacement), and prizes will be given to the holders of the 20

tickets. If you have bought one ticket, what is the probability that you win a prize?

b. Suppose only ten winners will be drawn instead of 20, but you have bought two tickets. Are your chances of winning at least one prize better or worse than in part a? (Guess first.)

c. Are your chances better or worse than in part b if the lottery has 10,000 tickets with 100 winners and you have two tickets? Guess first.

d. What about a 100-ticket lottery with only one winner, if you have two tickets?

[The point of this exercise is that the situation looks similar in all four parts; your number of tickets times the proportion of winning tickets is .02. Your chance of winning is close to this number in all cases, but it is exactly equal only if you have just one ticket or if there is just one winner.]

5. Show that if $P(A) = .8$ and $P(B) = .7$, then $P(A \cap B)$ cannot be less than .5.

6. In ten rolls of a fair die, find the conditional probability that there are no 1's, given that there are fewer than three 1's.

7. For a standard normal probability model, find the conditional probability that the outcome is positive and greater than 3, given that its absolute value exceeds 2.

8. A point is chosen at random from the unit square [see Supplementary Exercise 4 in Chapter 1].

a. What is the conditional probability that its x-coordinate is less than $\frac{1}{2}$, given that the sum of its two coordinates is less than 1?

b. Are the events "x-coordinate is less than $\frac{1}{2}$" and "x-coordinate is less than y-coordinate" independent?

9. Suppose 8% of the population have a certain symptom—call it symptom A. Of those with symptom A, 15% have disease B, and of those without symptom A, 4% have disease B.

a. What proportion of the population have disease B?

b. What proportion of people with disease B also have symptom A?

10. A company manufactures calculators that are all identical in appearance but contain chips of four different types. Among its calculators, 40% contain type 1 chips, 25% contain type 2, 20% type 3, and 15% type 4; 3% of the type 1 calculators are defective, 5% of the type 2's, 6% of the type 3's, and 10% of the type 4's. You purchase a calculator and find that it is defective. What is the probability that it contains a type 4 chip?

11. Someone draws three cards, without replacement, from an ordinary deck (which contains 52 cards, 26 of which are red). Find the conditional probability that:

a. The third card is red, given that the first two are red.

b. The first card is red, given that the second and third are red.

c. The second card is red, given that the first and third are red.

12. You and a friend each toss a fair coin three times. Let X be the number of heads you get and Y the number your friend gets. Assume X and Y are independent.

a. What is the probability that $X = Y = 2$?
b. What is the probability that $X = Y$?
c. What is the probability that $X > Y$?
d. Given that $X = 2$, what is the probability that $X > Y$?
e. What is the probability that together you get three heads? [Break the event into four disjoint events, $(X = 0 \cap Y = 3)$, $(X = 1 \cap Y = 2)$, etc., and add their probabilities.]
f. Compare the answer to part e to the probability that six tosses of a fair coin produce three heads.

13. You and your friend each do some experiment in which the standard normal probability model is appropriate. X is your result, and Y is your friend's. Assume X and Y are independent.

a. What is the probability that X and Y are both less than 1?
b. What is the probability that one or both of X and Y are less than 1?
c. What is the probability that exactly one of X and Y is less than 1?
d. What is the conditional probability that X is less than 1, given that Y is?

14. Consider a Bernoulli process with success probability p. Find the conditional probability that:

a. The first trial results in S, given that there is at least one S in the first ten trials.
b. The first trial results in S, given that there is exactly one S in the first ten trials.

Find these first by using the definition of conditional probability; then do them again with Bayes's theorem, conditioning on the outcome of the first trial.

15. In a Bernoulli process with success probability p, find the conditional probability that:

a. The first S comes on trial 13, given that there is no S in the first ten trials.
b. The first S comes on trial 103, given that there is no S in the first 100 trials.
c. The second S comes on trial 13, given that the first S comes on trial 10.
d. The second S comes on trial 103, given that the first S comes on trial 100.

[This shows the "lack of memory" of a Bernoulli process. If you look at the answers you will find them easy to believe; but the calculation is different in each case, and you should be sure you see how to get all of them.]

16. A survey shows that at a certain university, 35% of the freshmen, 25% of the sophomores, 20% of the juniors, and 15% of the seniors suffer from "math anxiety." At this university, 30% of the students are freshmen, 25% are sophomores, 25% are juniors, and 20% are seniors. Given that a randomly chosen student suffers from math anxiety, what is the probability that the student is a freshman?

17. At another university, 56% of the students are female, 12% of all students are left-handed, and 15% of males are left-handed. What proportion of left-handed students are male?

18. If A and B are events in an outcome space Ω and we know that $P(A) = .43$, $P(B|A) = .65$, and $P(A \cup B) = .52$, what is $P(B)$?

19. Consider the following experiment. First, you roll a fair die. If a 6 comes up, let $X = 0$; otherwise, toss a fair coin three times and let X equal the number of heads that show. Now suppose someone did that experiment and the resulting value of X was 0. What is the probability that a 6 came up on the die?

20. In a Bernoulli process with success probability $p = .1$, given that there was at least one success in the first ten trials, find the conditional probability that there were no successes in the first five trials.

21. The marketing division of the King soft-drink company has determined that there are three types of soft-drink purchasers. Among all purchasers, 25% are type I—these people choose the company's product, Old King Cola, with probability .75; 55% are type II—they choose Old King with probability .45; the remaining 20% are type III—they choose Old King with probability .15.
   a. What is the probability that a randomly chosen purchaser will choose Old King?
   b. Of all purchasers who choose Old King, what proportion are type I purchasers?

ANSWERS

1. $P(A \cap B) = .2231$; $P(A \cup B) = .5126$
2. .2813
3. .4083
4. **a.** .020000 **b.** A little worse: .019910 **c.** Still worse: .019901 **d.** .020000
6. .2083
7. .0285
8. **a.** $\frac{3}{4}$ **b.** No
9. **a.** .0488 **b.** .2459
10. .2913
11. All are .48
12. **a.** .140625 **b.** .3125 **c.** .34375 $\left(= \frac{1}{2} \cdot (1 - \text{ans. to b})\right)$ **d.** .5 **e.** .3125 **f.** The same
13. **a.** .7078 **b.** .9748 **c.** .2670 **d.** .8413
14. **a.** $p/(1 - q^{10})$ **b.** $\frac{1}{10}$
15. q^2p in all four parts
16. .4242
17. 55%
18. .3695
19. $\frac{8}{13}$
20. .3712
21. **a.** .4650 **b.** .4032

CHAPTER 3

Probability Distributions

3.1 Distributions on $\mathbb{R}$

3.2 Cumulative Distribution Functions

3.3 The Distribution of a Random Variable

3.4 The Joint Distribution of Two Discrete Random Variables

3.5 The Joint Distribution of Two Absolutely Continuous Random Variables

3.6 The Joint Distribution of More Than Two Random Variables

Supplementary Exercises

3.1 DISTRIBUTIONS ON $\mathbb{R}$

We have used the term *probability distribution,* or *distribution* for short, a number of times in Chapters 1 and 2, without saying much about what it means. It refers to an assignment of probabilities to the intervals of the real line $\mathbb{R}$, and hence also to all the Borel subsets of $\mathbb{R}$. Recall that the Borel sets are those subsets of $\mathbb{R}$ that can be made from intervals by using finitely or countably many set operations of union, intersection, and complementation [along with other sets that must be included to make a σ-field; see (1.3.2)].

Another way of thinking about a distribution is as a recipe for computing the probabilities of all events in a particular model, or as the "probability information" that is sufficient to determine the probabilities of all the events. As we said, the recipe need not specify explicitly the probabilities of all the Borel sets. It is enough if it specifies the probabilities of all the intervals, or even just the probabilities of the intervals of the form $(a, b]$, because that information will make it possible to determine the probabilities of all the other Borel sets. (How? Mostly by means of the rules for the algebra of probabilities that we discussed in Sections 2.1 and 2.5.)

For example, the term "standard normal distribution" refers to the probability assignment we get from the standard normal density $\varphi(x) = (1/\sqrt{2\pi})e^{-x^2/2}$ by decreeing that the probability of any interval is the integral of φ over that interval.

What about the exponential distribution? It specifies the probabilities of subintervals of $(0, \infty)$ by the formula $P(a, b] = e^{-\lambda a} - e^{-\lambda b}$, which is the same as the integral of the density $f(x) = \lambda e^{-\lambda x}$ from a to b. You might not think this qualifies, because it assigns probabilities only to the Borel subsets of $(0, \infty)$ and not to all the Borel sets in $\mathbb{R}$. But that is no problem; we simply assign a probability of 0 to any subinterval of $(-\infty, 0]$. For an interval that contains some negative numbers and some positive numbers, say $(-2, 3]$, we just give it the probability of the interval $(0, 3]$, which is its intersection with $(0, \infty)$. We get the same result if we extend the definition of the density function $f(x)$ to include all of $\mathbb{R}$:

$$f(x) = \begin{cases} \lambda e^{-\lambda x} & \text{if } x > 0, \\ 0 & \text{if } x \le 0. \end{cases}$$

The integral of this function over any interval in $\mathbb{R}$ gives the correct probability.

What about the Poisson distribution of Example (1.2.11)? It assigns probability $p(k) = \lambda^k e^{-\lambda}/k!$ to k if k is a nonnegative integer, but it does not appear to assign probabilities to intervals. Again, there is a way around this. We simply say that the probability assigned to an interval I is the sum of the values of $p(k)$ for all the nonnegative integers k that happen to be in I. One way to do this formally is to define the mass function p as a function on *all* real numbers,

$$p(x) = \begin{cases} \dfrac{\lambda^x e^{-\lambda}}{x!} & \text{if } x \text{ is a nonnegative integer}, \\ 0 & \text{otherwise}, \end{cases}$$

and then to define $P(I)$ for any interval I as the sum of the values of $p(x)$ for all $x \in I$: $P(I) = \sum_{x \in I} p(x)$. This sum looks like an invalid expression, because there are uncountably many x's in an interval I, but it makes sense because $p(x)$ is 0 for all but a countable number of them.

Finally, consider the binomial distribution, which arises when we count the successes in a fixed number of Bernoulli trials. We covered it in Example (1.2.13). The members ω of the outcome set Ω are n-letter words from the alphabet {S, F}; there are 2^n of them. The probability assigned to one of these words is $p^k q^{n-k}$, where k is the number of S's in the word (and p and q are the success and failure probabilities for a single trial). There is no distribution so far, because we have assigned probabilities to subsets of Ω, not subsets of $\mathbb{R}$.

But now think of the random variable X, defined by saying that $X(\omega)$ equals the number of S's in the word ω. Then we get probabilities assigned to sets of numbers; the possible values of X are $0, 1, 2, \ldots, n$, and the probability of any of these, say k, is the binomial probability $\binom{n}{k} p^k q^{n-k}$. So to get a full-fledged distribution—that is, an assignment of probabilities to intervals, and hence to all Borel sets—we define $p(x)$ for all real x by

$$p(x) = \begin{cases} \binom{n}{x} p^x q^{n-x} & \text{if } x = 0, 1, 2, \ldots, n, \\ 0 & \text{otherwise,} \end{cases}$$

and then define the probability of any interval to be the sum of the values of $p(x)$ for all x in the interval.

Following are the formal definition of a probability distribution on $\mathbb{R}$ and the basic facts about how distributions can be specified—in particular, how we specify discrete and absolutely continuous distributions. We emphasize that there are no new concepts in this section; we are simply giving a name to something we have been dealing with all along.

(3.1.1) Definition. A ***probability distribution*** (or simply a ***distribution***) is an assignment of probabilities to the Borel subsets of $\mathbb{R}$—that is, a function P from the collection of Borel sets to $\mathbb{R}$ satisfying the probability axioms:

1. $P(B) \geq 0$ for every Borel set B,
2. $P(\mathbb{R}) = 1$, and
3. If $B_1, B_2, \ldots$ are finitely or countably many disjoint Borel sets, then

$$P(B_1 \cup B_2 \cup \cdots) = P(B_1) + P(B_2) + \cdots.$$

Notice that these are the same requirements given in the definition of a probability space at the beginning of Chapter 2. The only difference is that here we are talking only about the Borel sets in $\mathbb{R}$ instead of the events in an arbitrary outcome space.

The following theorem, which we mentioned in (1.3.2), says that we can specify a distribution completely just by giving the probabilities of the intervals, as long as we give them probabilities that are consistent with each other. The third requirement of this theorem is the consistency requirement.

We do not prove this theorem here; its proof requires measure theory. Its importance is that because of it, ***we can think of a distribution as an assignment of probabilities to intervals.***

(3.1.2) **Theorem.** The probabilities of all the Borel sets are uniquely determined if probabilities are assigned to the intervals in such a way that

1. $P(I) \geq 0$ for every interval I,
2. $P(-\infty, \infty) = 1$, and
3. If an interval I is partitioned into finitely or countably many disjoint intervals $I_1, I_2, \ldots$, then $P(I) = P(I_1) + P(I_2) + \cdots$.

Saying that "the probabilities of all the Borel sets are uniquely determined" means that, given such a consistent assignment of probabilities to intervals, there is one and only one assignment of probabilities to all the Borel sets that agrees with it.)

(3.1.3) Note. As we have said, something stronger than this theorem is true: it is not necessary to specify the probabilities of *all* intervals. It can be shown that if we merely specify the probabilities of the intervals of the form $(a, b]$—the "left-open, right-closed" intervals—in such a way that the three requirements of the theorem are satisfied, then the probabilities of all the other intervals and all the Borel sets are uniquely determined. Having this, we see that it is really necessary to specify only the probabilities of the "left-infinite, right-closed" intervals, the ones of the form $(-\infty, b]$, because once we have $P(-\infty, b]$ for every b, then we can get $P(a, b]$ as $P(-\infty, b] - P(-\infty, a]$ for any a and b.

Notice that intervals of the form $(a, b]$ can be partitioned nicely into intervals of the same form. If $a < c_1 < c_2 < c_3 < b$, for example, then $(a, b]$ is the union of the disjoint intervals $(a, c_1]$, $(c_1, c_2]$, $(c_2, c_3]$, and $(c_3, b]$. See Figure 3.1a.

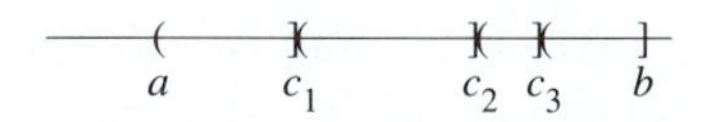

FIGURE 3.1a Intervals of the form $(a, b]$ can be partitioned into intervals of the same form.

The next two theorems say how a distribution in either of the two important classes—discrete and absolutely continuous—can be specified.

(3.1.4) **Theorem (a mass function specifies a discrete distribution).** Any function $p(x)$ on $\mathbb{R}$ having the properties

1. $p(x) = 0$ except for a finite or countable set of values $x_1, x_2, x_3, \ldots$,
2. $p(x_k) \geq 0$ for $k = 1, 2, 3, \ldots$, and
3. $\sum_k p(x_k) = 1$,

determines a distribution, in which the probabilities of intervals are defined by

$$P(I) = \sum_{x \in I} p(x).$$

Such a distribution is called a *discrete distribution* and the function $p(x)$ that specifies it is called the *probability mass function* (or simply the *mass function*) of the distribution. The numbers x_k for which $p(x) > 0$ are called the *possible values* of the distribution.

(3.1.5) **Example.** If we roll a fair die and count the number of spots that show, we get the values 1, 2, 3, 4, 5, and 6, all equally likely. We can express this by saying that this distribution assigns probability $\frac{1}{6}$ to each of the possible values 1, 2, 3, 4, 5, and 6. Or we can say that this is the distribution with mass function

$$p(x) = \begin{cases} \frac{1}{6} & \text{if } x = 1, 2, 3, 4, 5, 6, \\ 0 & \text{otherwise.} \end{cases}$$

This distribution assigns a probability to every interval, as does any distribution, but the probability that it assigns is just $\frac{1}{6}$ times the number of possible values that are in the interval. ∎

(3.1.6) **Theorem (a density function specifies an absolutely continuous distribution).** Any function $f(x)$ on $\mathbb{R}$ having the properties

1. $f(x) \geq 0$ for all $x \in \mathbb{R}$, and
2. $\int_{-\infty}^{\infty} f(x)dx = 1$,

determines a distribution, in which the probabilities of intervals are defined by

$$P(I) = \int_I f(x)dx.$$

Such a distribution is called an *absolutely continuous distribution*, and the function $f(x)$ is called a *probability density function* (or simply a *density function*) for the distribution. The set of values of x for which $f(x)$ is nonzero is called the set of *possible values,* or sometimes the *range,* of the distribution.

The term "possible values" may seem a little strange in this context, because the probability of any one of the possible values is 0. But remember that "probability 0" does not mean "impossible," it simply means "having an expected relative frequency that approaches 0 as the number of trials increases."

(3.1.7) **Example.** For the uniform choice of a random number from [0, 1], we can specify the distribution by simply giving the density function, which in this case is

$$f(x) = \begin{cases} 1 & \text{if } 0 \leq x \leq 1, \\ 0 & \text{otherwise.} \end{cases}$$

We could be more succinct and say simply that this is the distribution with density $f(x) = 1$ on the set $[0, 1]$. ▮

Theorems (3.1.4) and (3.1.6) are not difficult to prove; one simply checks that either of the two ways of specifying probabilities of intervals satisfies the three requirements of Theorem (3.1.2).

Recall the property of absolutely continuous distributions that we discussed in Section 1.3—***in any absolutely continuous distribution, any singleton has probability*** **0.** Consequently, all finite and countable sets have probability 0. Also, the probabilities of intervals are the same whether they contain their endpoints or not: the four intervals (a, b), $(a, b]$, $[a, b)$, and $[a, b]$ all have the same probability.

By contrast, ***in a discrete distribution all the probability is concentrated in singletons.*** Consequently, the endpoints of an interval must be considered carefully in finding its probability, because all the probability is concentrated in a countable set of points, and those endpoints might be among them.

(3.1.8) Example: Comparison of an absolutely continuous distribution with a discrete distribution. The two distributions shown in Figure 3.1b look somewhat similar; also, the density of one and the mass function of the other appear to be almost the same function. Their formulas are the following.

The density function: $f(x) = .2e^{-.2x}$ for all x in the interval $(0, \infty)$;

The mass function: $p(x) = .1813e^{-.2x}$ for $x = 0, 1, 2, \ldots.$

[Only the constant multiple is different; they cannot be the same, because the density must integrate to 1 over the interval $(0, \infty)$, while the mass function must sum to 1 over the nonnegative integers.]

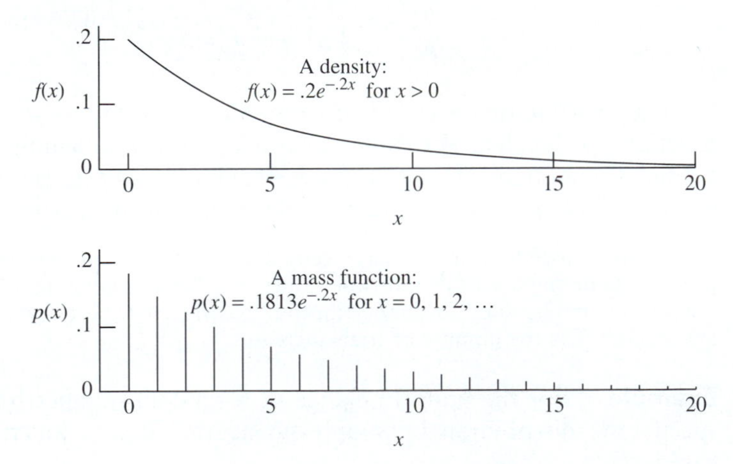

FIGURE 3.1b A density and a mass function, both proportional to $e^{-.2x}$, but producing very different distributions

But the distributions are quite different. Consider the following sets, for example.

a. The interval [1, 2]. The density gives it a probability of $\int_1^2 .2e^{-.2x}dx \doteq .1484$; the mass function gives it a probability of $p(1) + p(2) = .2700$.
b. The interval (1, 2). The density gives it a probability of .1484, the same as [1, 2]; but the mass function gives it a probability of 0 (because it contains no integers).
c. The set $\{1, 2\}$ consisting of just two integers. The density gives it a probability of 0; but the mass function gives it a probability of .2700, the same as the interval [1, 2] (because they both contain the same two integers).
d. The interval (.999, 1.001). The density function gives it a probability of $\int_{.999}^{1.001} .2e^{-.2x}dx \doteq .00033$; but the mass function gives it a probability of $p(1) \doteq .1484$. If the interval were made smaller, the density would give it a still smaller probability, but the mass function would continue to give it a probability of $p(1)$.

There is one similarity in the probabilities assigned by these two distributions. You can check that, for an integer k, the probability that the density assigns to the interval $[k, k + 1)$ is the same as the probability that the mass function assigns to k itself. But the differences between them, as shown by the sets above, are more important than this similarity. ▮

It is crucial, when first encountering a distribution, to notice whether it is absolutely continuous or discrete, and then to fix in your mind the way probabilities are assigned. To repeat:

> ***For an absolutely continuous distribution, probabilities will be integrals, and singletons (in particular, the endpoints of intervals) will have probability 0.***
>
> ***For a discrete distribution, probabilities will be found by adding the probabilities of possible values that are in the set in question, and all the probability will be concentrated in singletons.***

We should note that there do exist distributions that are neither discrete nor absolutely continuous. Example (3.2.18) shows how such a thing might arise. But we will not have much to do with them here; nearly all the distributions that arise in practice are either discrete or absolutely continuous.

Summary. For practical purposes we can summarize this section as follows.

1. Any probability mass function amounts to the complete description of a discrete distribution, if we use the rule $P(I) = \sum_{x \in I} p(x)$.
2. Any probability density function amounts to the complete description of an absolutely continuous distribution, if we use the rule $P(I) = \int_I f(x)dx$.
3. More generally, any recipe that specifies the probabilities of all the intervals, or even all the intervals of the form $(a, b]$, or even just the intervals of the

form $(-\infty, b]$, amounts to the complete description of a distribution, as long as the probabilities satisfy the three requirements of Theorem (3.1.2).

You may sense that you have seen all of this before, and you are essentially right; it was in Sections 1.2 and 1.3. The differences between the probability spaces in Chapter 1 and the distributions in this chapter are very slight; perhaps it will be helpful to compare them.

In Section 1.3 we defined an absolutely continuous (one-dimensional) probability space as having an interval as its outcome set, with probabilities of the Borel subsets of that interval being given as the integrals of a density function. That is almost exactly what we defined as an absolutely continuous distribution, except that for a distribution we always take the whole of $\mathbb{R}$ as the outcome space, and we define the density to be 0 on those parts of $\mathbb{R}$ where the probability is to be 0. Essentially, in an absolutely continuous probability space, the assignment of probabilities is an absolutely continuous distribution.

There is a little more difference between discrete probability spaces (Section 1.2) and discrete distributions on $\mathbb{R}$. First of all, a discrete probability space may have an outcome set Ω that is not a set of numbers, but a set of words or permutations or cards or other objects. In this case we still have a probability mass function on Ω, but the probability assignment does not qualify as a distribution because the events are sets of words, or permutations, or something else, but not subsets of $\mathbb{R}$.

Second, in any discrete probability space the outcome set is finite or countable, and all subsets are events. By contrast, a discrete distribution is like a probability space that has $\mathbb{R}$ as its outcome set and Borel sets as events; but all the probability is concentrated on some finite or countable set of numbers, and there is a mass function on that set (and zero elsewhere) that determines the probabilities. This seems a needless technicality; we are complicating a nice countable outcome set by throwing in all the rest of $\mathbb{R}$, but with probability 0. It is reminiscent of Lewis Carroll's plan in *Through the Looking-Glass* to "dye one's whiskers green/and always wear so large a fan/that they could not be seen." But it turns out to be convenient to have *all* distributions defined on the same set of events, and so we settle for the complication.

EXERCISES FOR **SECTION 3.1**

1. The function p defined by $p(x) = x/10$ for $x = 1, 2, 3, 4$ (and 0 otherwise) is a probability mass function and therefore determines probabilities for all intervals and thence for all Borel sets. Find the following probabilities:
   a. $P(1, 3)$ b. $P[1, 3)$ c. $P(1.5, 3)$
   d. $P[1.5, 3)$ e. $P[1.5, 1.99]$ f. $P(-\infty, 1]$
   g. $P(-\infty, 1)$ h. $P([0, 2] \mid [1, 3])$ i. $P\{2, 3\}$
   j. $P\{1, 2, 3, 4\}$ k. $P(\mathbb{N})$ ($\mathbb{N}$ is the set of all positive integers.)
   l. $P(\mathbb{Q})$ ($\mathbb{Q}$ is the set of all rational numbers.)

2. The function f defined by $f(x) = \frac{4}{3}x^{-2}$ for $1 \le x \le 4$ (and 0 otherwise) is a probability density function and therefore determines probabilities for all intervals. For this distribution, find the probabilities listed in Exercise 1. [*Hint:* You can save yourself time finding the probabilities of intervals if you first find a formula for $P[1, z]$, which in this case is the same as $P(-\infty, z]$, in terms of z.]

3. The Poisson distribution with parameter λ has the mass function defined by $p(x) = \lambda^x e^{-\lambda}/x!$ if x is a nonnegative integer (and 0 otherwise). Find the probability it assigns to each of the following sets:

a. $[0, 2)$ **b.** $(-\infty, 1]$ **c.** $(-\infty, 1.5]$
d. $(-\infty, 2)$ **e.** $(-\infty, 2]$ **f.** $(.5, \infty)$
g. $\{0, 1, 2\}$

4. The exponential distribution with parameter λ has the density function $f(x) = \lambda e^{-\lambda x}$ for $x \geq 0$ (and 0 otherwise). Find the probability it assigns to each of the sets in Exercise 3.

5. Find the value of a that makes the following function a probability density function: $f(x) = ax^2(2 - x)$ for $0 < x < 2$ (and 0 otherwise). [*Hint:* The integral of a density must be 1.]

6. Find the value of a that makes the following function a probability mass function:

$$p(x) = \frac{a}{x + 1} \quad \text{for } x = 0, 1, 2, 3 \text{ (and 0 otherwise).}$$

[*Hint:* The four probabilities must add to 1.]

7. The possible values of a certain discrete distribution are 1, 2, 3, 4, and 5, and the probability assigned to a possible value x is proportional to x^2.
a. What is the mass function of this distribution?
b. What probability does this distribution assign to the interval $(1, 3]$?

8. A certain absolutely continuous distribution has range $[1, 5]$, and has a density proportional to x^2.
a. What is this density function?
b. What probability does this distribution assign to the interval $(1, 3]$?

[a]9. For what experiment is the distribution given by the following mass function appropriate?

$$p(x) = \begin{cases} \dfrac{x - 1}{36} & \text{if } x = 2, 3, 4, 5, 6, 7, \\ \dfrac{13 - x}{36} & \text{if } x = 8, 9, 10, 11, 12, \\ 0 & \text{otherwise.} \end{cases}$$

10. A puzzle: Consider the distribution that assigns probabilities to the intervals according to the following rules:

If neither 1 nor 2 is in the interval I, then $P(I) = 0$.

If I contains 1 but not 2, then $P(I) = .25$.

If I contains 2 but not 1, then $P(I) = .75$.

If both 1 and 2 are in I, then $P(I) = 1$.

What is this distribution? Give its set of possible values. If it is absolutely continuous, give its density; if it is discrete, give its mass function.

ANSWERS

1. **a.** .2 **b.** .3 **c.** .2 **d.** .2 **e.** 0 **f.** .1 **g.** 0 **h.** .5 **i.** .5 **j.** 1 **k.** 1 **l.** 1
2. **a.** $\frac{8}{9}$ **b.** $\frac{8}{9}$ **c.** $\frac{4}{9}$ **d.** $\frac{4}{9}$ **e.** .21887 **f.** 0 **g.** 0 **h.** $\frac{3}{4}$ **i.** 0 **j.** 0 **k.** 0 **l.** 0
3. **a–d.** $(1 + \lambda)e^{-\lambda}$ **e.** $\left(1 + \lambda + \frac{\lambda^2}{2}\right)e^{-\lambda}$ **f.** $1 - e^{-\lambda}$ **g.** $\left(1 + \lambda + \frac{\lambda^2}{2}\right)e^{-\lambda}$
4. **a.** $1 - e^{-2\lambda}$ **b.** $1 - e^{-\lambda}$ **c.** $1 - e^{-1.5\lambda}$ **d, e.** $1 - e^{-2\lambda}$ **f.** $e^{-.5\lambda}$ **g.** 0
5. $\frac{3}{4}$
6. .48
7. **b.** $\frac{13}{55}$
8. **b.** $\frac{26}{124}$
9. Rolling two fair dice (or one fair die twice) and counting the spots

3.2 CUMULATIVE DISTRIBUTION FUNCTIONS

(3.2.1) Definition. The *cumulative distribution function* (CDF) of a probability distribution is the function $F(x)$ on $\mathbb{R}$ defined by

$$F(x) = P(-\infty, x].$$

Many authors call this function simply the *distribution function* of the distribution; we keep the word "cumulative" because it reminds us that $F(x)$ is the accumulated probability of all numbers less than or equal to x. As x increases, the probability continues to accumulate.

The importance of this function is that, as we pointed out in (3.1.3), if we have a way to find the probabilities of the intervals of the form $(-\infty, x]$, then we can find the probabilities of all intervals. And from the probabilities of the intervals we can find (using rules like those in Sections 2.1 and 2.5) the probabilities of all Borel sets. So knowing $F(x)$ for all x is in some sense a minimal amount of information needed to know everything about the distribution.

We saw this in Exercise 2 of the previous section, as well as in Exercise 8 in Section 1.3 and Supplementary Exercise 6 for Chapter 1. In those exercises we pointed out that if you find $P(-\infty, x]$ as a function of x first, you can save doing the same integral several times with different limits of integration.

There is one important CDF that we have been using since Example (1.3.5): the CDF of the standard normal distribution. The table in Appendix C, which we use for nearly all probabilities given by this distribution, gives for various z the probability of the interval $(-\infty, z]$—that is, the CDF. This particular CDF is denoted by Φ (uppercase Greek phi) instead of F, to emphasize that it is the integral of the standard normal density φ. As we will see in (3.2.5), the CDF of any absolutely continuous distribution is the integral of its density in this way.

We first discussed CDFs formally in Section 2.5, where we derived some of their important limiting properties using the nested sets theorem. But there is no need to go back at this point if you omitted Section 2.5; we will discuss those properties and others in (3.2.11) below.

First, we will look at the CDFs of two familiar distributions, one discrete and one absolutely continuous. Following that, we will think about the relationship between the CDF and the mass function of a discrete distribution, and between the CDF and the density function of an absolutely continuous distribution. Finally, we will consider what kinds of functions CDFs are, in general, and we will review briefly how the probabilities of all intervals can be found if the CDF is known.

(3.2.2) Example. The binomial distribution with $n = 3$ and $p = \frac{1}{2}$ has the probability mass function

$$p(x) = \begin{cases} \binom{3}{x}\left(\frac{1}{2}\right)^x\left(\frac{1}{2}\right)^{3-x} & \text{if } x = 0, 1, 2, 3, \\ 0 & \text{otherwise.} \end{cases}$$

Of course this distribution is an old friend; we have known it since Example (1.1.1), and this is one of the more obscure descriptions of it. It is much simpler to say that the distribution has possible values 0, 1, 2, and 3, with probabilities $\frac{1}{8}, \frac{3}{8}, \frac{3}{8}$, and $\frac{1}{8}$, respectively. A graph of this mass function—that is, of these four probabilities—is shown at the top of Figure 3.2a on page 158.

What is the CDF? It is not difficult to see that

$$F(0) = P(-\infty, 0] = P\{0\} = \tfrac{1}{8},$$

$$F(1) = P(-\infty, 1] = P\{0, 1\} = \tfrac{1}{8} + \tfrac{3}{8} = \tfrac{1}{2},$$

$$F(2) = P(-\infty, 2] = P\{0, 1, 2\} = \tfrac{7}{8}, \text{ and}$$

$$F(3) = P(-\infty, 3] = P\{0, 1, 2, 3\} = 1.$$

But CDFs are to be defined for *all* x, and we need to think about x's other than the four possible values. For example, what is $F(2.36)$? This is simple: $F(2.36) = P(-\infty, 2.36]$, and the only possible values in this interval are 0, 1, and 2; so $F(2.36) = P\{0, 1, 2\} = \frac{7}{8}$. A similar argument shows that $F(x) = \frac{7}{8}$ for *any* x between 2 and 3, but not for $x = 3$. In this way we can see that

for any $x < 0$, $(-\infty, x]$ contains no possible values, so $F(x) = 0$;

if $0 \le x < 1$, then $(-\infty, x]$ contains just the possible value 0, so $F(x) = P\{0\} = \frac{1}{8}$;

if $1 \le x < 2$, then $(-\infty, x]$ contains the possible values 0 and 1, so $F(x) = P\{0, 1\} = \frac{1}{2}$;

if $2 \le x < 3$, then $(-\infty, x]$ contains the possible values 0, 1, and 2, so $F(x) = P\{0, 1, 2\} = \frac{7}{8}$;

and if $x \ge 3$, then $(-\infty, x]$ contains all four possible values, so $F(x) = P\{0, 1, 2, 3\} = 1$.

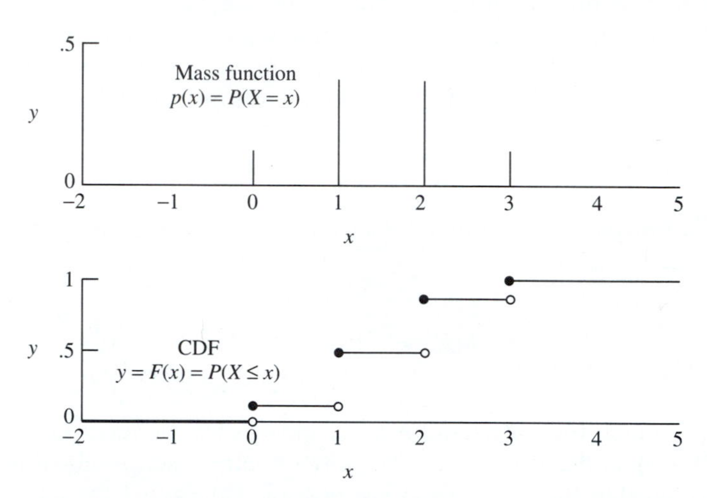

FIGURE 3.2a The mass function and the CDF of the binomial distribution with $n = 3$ and $p = \frac{1}{2}$

A graph of F is shown at the bottom of Figure 3.2a. Notice that it is a *step function;* that is, it is constant on any of the intervals between possible values, and it has a jump at each possible value, of height equal to the probability of that value.

Unfortunately, there is not a convenient formula for this CDF; this is typical of the CDFs of discrete distributions. The best we can do to describe $F(x)$ is something like the following:

$$F(x) = \begin{cases} 0 & \text{if } x < 0, \\ \frac{1}{8} & \text{if } 0 \le x < 1, \\ \frac{1}{2} & \text{if } 1 \le x < 2, \\ \frac{7}{8} & \text{if } 2 \le x < 3, \\ 1 & \text{if } x \ge 3. \end{cases}$$

∎

(3.2.3) Example. The exponential distribution with $\lambda = .8$ has the density function

$$f(x) = \begin{cases} .8e^{-.8x} & \text{if } x > 0, \\ 0 & \text{otherwise.} \end{cases}$$

A graph of this function is shown at the top of Figure 3.2b. As we know, saying that a distribution has this density is the same as saying that the set of possible values is the interval $(0, \infty)$, and that $P(I) = e^{-.8a} - e^{-.8b}$ if I is any subinterval of $(0, \infty)$ whose left and right endpoints are a and b. What is the CDF?

For any $x > 0$, $F(x) = P(-\infty, x] = P(0, x] = e^{-.8 \cdot 0} - e^{-.8x} = 1 - e^{-.8x}$; and for any $x \leq 0$, $F(x) = P(-\infty, x] = 0$.

That is, $F(x)$ is the function pictured at the bottom of Figure 3.2b. It is described formally by

$$F(x) = \begin{cases} 0 & \text{if } x \leq 0, \\ 1 - e^{-.8x} & \text{if } x > 0. \end{cases}$$

The connection between the two functions f and F, as we will see in (3.2.5) below, is that $F(x)$ is the integral of f from 0 to x, and consequently that f is the derivative of F (except at $x = 0$, where F has no derivative). Notice that F is increasing most rapidly where f is largest.

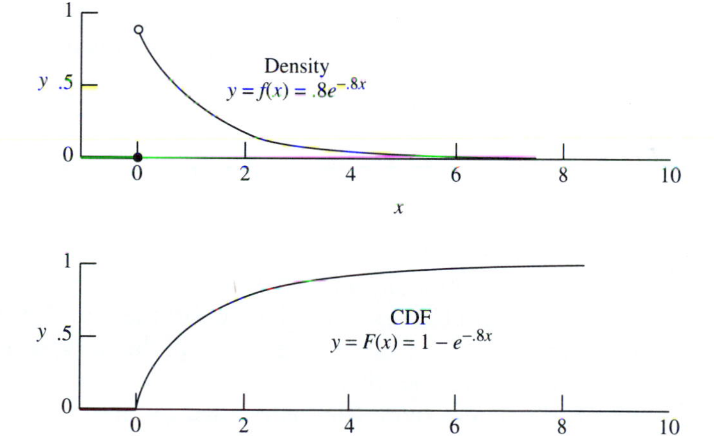

FIGURE 3.2b The density and the CDF of the exponential distribution with $\lambda = .8$ ▮

(3.2.4) **CDFs of discrete distributions.** Example (3.2.2) is enough to show us in general what the connection is between the CDF $F(x)$ and the probability mass function $p(x)$ of a discrete distribution: the CDF is a step function, constant on the intervals between possible values, and having a jump of height $p(x_k)$ at each possible value x_k.

Only one other thing is worth noting (see Figure 3.2a): at each jump point x_k, the value of F is the higher of the two values on either side. For example, the height of the graph jumps from $\frac{1}{2}$ to $\frac{7}{8}$ at $x = 2$, and $F(2)$ is $\frac{7}{8}$, the higher of the values. That is, $F(x)$ is continuous from the right, but not from the left, at the jump points. (Of course, it is continuous from both sides at other points.)

To be honest, we must say that CDFs are not much used in connection with discrete distributions; it is with the absolutely continuous ones that their existence is justified in practical ways. The CDF of a discrete distribution contains the same information as the mass function, but in less usable form; seldom is it helpful to use CDFs to solve a problem involving discrete distributions. The only reason we think about them at all is that CDFs are a unifying device. Some distributions have density functions, some have mass functions, and some are specified in other ways; but all distributions have CDFs. Consequently, it is worth the trouble to find out what they are like for the distributions we work with.

(3.2.5) **CDFs of absolutely continuous distributions.** Let $f(x)$ be a probability density function—that is, a nonnegative function that integrates to 1. It determines a distribution, the one that assigns probability $\int_a^b f(x)dx$ to any interval with endpoints a and b. So what is the CDF $F(x)$ and what is its relation to the density $f(x)$? It is simply that

(3.2.6) $$F(x) = P(-\infty, x] = \int_{-\infty}^{x} f(t)dt.$$

[Notice that we cannot use $f(x)dx$ in the integral, as the variable x is already being used as one of the limits of integration.]

Now (3.2.6) reminds us immediately of the Fundamental Theorem of Calculus: when F is defined as the integral of f in this way, then f is the derivative of F. That is, (3.2.6) implies that $f(x) = F'(x)$.

Actually, this is not quite what is implied by the Fundamental Theorem as presented in most calculus texts, but it is nevertheless true, with one slight reservation. First of all, the integral in (3.2.6) has an infinite left endpoint, whereas the version of the Fundamental Theorem that is given in most calculus texts applies to a bounded interval $[a, b]$. But it can be extended to improper integrals like (3.2.6) as long as the integral of f over all of $\mathbb{R}$ is finite, so that is no problem.

What does cause a slight problem is that the Fundamental Theorem is usually presented only for *continuous* integrands f, but densities are not always continuous. Look back at Figure 3.2b, for example. The density $f(x)$ is not continuous at $x = 0$; it jumps there from 0 to .8. And the function $F(x)$ is not differentiable at $x = 0$; its graph has a "corner" there.

However, that is the worst that can happen. $F'(x)$ may fail to exist at points where $f(x)$ is discontinuous, but we can still say the following.

(3.2.7) If $F(x)$ is the CDF of the absolutely continuous distribution whose density is $f(x)$, then $F'(x) = f(x)$ at all points x where f is continuous.

Incidentally, we see here one of the reasons for using the term "absolutely continuous" instead of the simpler "continuous" for the distributions given by a density. The CDF of such a distribution is indeed a continuous function; that is the same as saying that the probability of any singleton is 0. But the CDF of such a distribution is more than continuous; it is differentiable, except on a small set of points, and it is the integral of its derivative as in (3.2.6). Such functions are called absolutely continuous in real analysis, where results like (3.2.7) are proved.

(3.2.8) **Example.** Consider the absolutely continuous distribution whose density function is

$$f(x) = \begin{cases} \frac{2}{9}(3 - x) & \text{if } 0 < x < 3, \\ 0 & \text{otherwise.} \end{cases}$$

What is the CDF?

Solution. The interval (0, 3) is the set of possible values of the distribution, so the CDF $F(x)$ will be 0 when $x \leq 0$ and 1 when $x \geq 3$. All we need to do is to find $F(x)$ for x between 0 and 3. For such x,

$$F(x) = P(-\infty, x] = P(0, x] = \int_0^x \tfrac{2}{9}(3 - t)dt = \left[-\tfrac{1}{9}(3 - t)^2\right]_0^x = 1 - \tfrac{1}{9}(3 - x)^2.$$

Putting it together, we get the full description of $F(x)$:

$$F(x) = \begin{cases} 0 & \text{if } x < 0, \\ 1 - \frac{1}{9}(3 - x)^2 & \text{if } 0 \leq x < 3, \\ 1 & \text{if } x \geq 3. \end{cases}$$

∎

(3.2.9) **Example.** Suppose we are given that the CDF of a certain distribution is

$$F(x) = \begin{cases} 0 & \text{if } x < 0, \\ x^2 & \text{if } 0 \leq x < 1, \\ 1 & \text{if } x \geq 1. \end{cases}$$

What is the density of the distribution?

Solution. The density is the derivative of the CDF, at least at those points where the density is continuous. But if we don't know the density yet, how can we know where it is continuous? The way out of this apparent dilemma is to notice that F can be differentiated easily everywhere except at 0 and 1. (Actually it has a derivative at 0, but this doesn't matter, and in many similar examples it will not have a derivative at such points.) The derivative (except at 0 and 1) is obviously

$$F'(x) = \begin{cases} 0 & \text{if } x < 0, \\ 2x & \text{if } 0 < x < 1, \\ 0 & \text{if } x > 1. \end{cases}$$

Now here is an important fact about densities: ***changing a density at a finite set of points does not change the distribution,*** because it does not change the integral of the density over any intervals. Consequently, it does not matter whether we get the values of the density "right" at 0 and 1; any values we choose for $f(0)$ and $f(1)$ will give a function that serves equally well as a density for this distribution. Perhaps the easiest thing to do is to define them both to be 0; then we get the function

$$f(x) = \begin{cases} 2x & \text{if } 0 < x < 1, \\ 0 & \text{otherwise.} \end{cases}$$

And now, as an exercise, you ought to find the CDF associated with this density, as we did in the previous example, and check that it is indeed the function $F(x)$ given above. ∎

Note. Because the density can be changed at a finite, or even countable, set of points without changing the probabilities of any interval, it is, strictly speaking, incorrect to speak of *the* density of a distribution; rather, we should refer to *a* density. Thus, in the previous example, the function $f(x)$ given at the end is one density for the distribution; another is

$$f_2(x) = \begin{cases} 2x & \text{if } 0 \le x \le 1, \\ 0 & \text{otherwise.} \end{cases}$$

If we wanted to be annoyingly obscure, we could use the following as a density; we have whimsically changed it at three points ($x = \frac{1}{7}$, 1, and -6), but it has exactly the same integral over any interval as the previous two densities.

$$f_3(x) = \begin{cases} 2x & \text{if } 0 < x < 1 \text{ but } x \ne \frac{1}{7}, \\ 3 & \text{if } x = \frac{1}{7}, \\ \pi & \text{if } x = 1, \\ 17.3 & \text{if } x = -6, \\ 0 & \text{otherwise.} \end{cases}$$

But remember that any distribution has just one CDF.

(3.2.10) Example. Consider the distribution whose CDF is

$$F(x) = \begin{cases} 0 & \text{if } x < 0, \\ 1 - e^{-2x} - 2xe^{-2x} & \text{if } x \ge 0. \end{cases}$$

What is a density for this distribution?

Solution. We can differentiate this function immediately everywhere except perhaps at $x = 0$. Since this gives us only a finite set of points—in this case, the single point $x = 0$—where we might not know the derivative, we can define our density to be anything we like at that point. We choose to define $f(0)$ to be 0. For the rest of the x's, $F'(x)$ is of course 0 when $x < 0$, and $F'(x) = 4xe^{-2x}$ when $x > 0$. Thus, one density for this distribution is

$$f(x) = \begin{cases} 4xe^{-2x} & \text{if } x > 0, \\ 0 & \text{otherwise.} \end{cases}$$

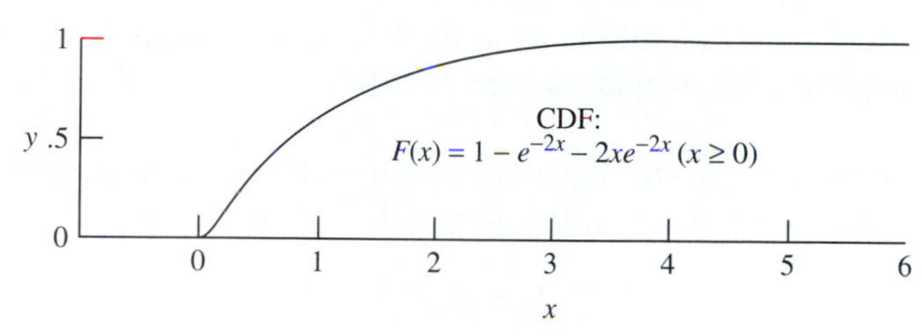

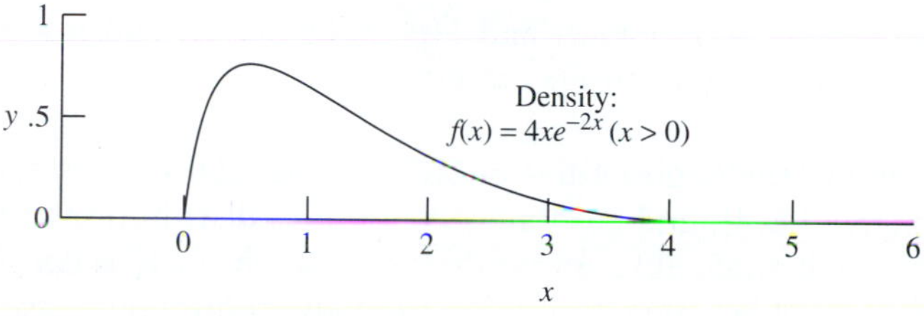

FIGURE 3.2c A CDF and its associated density function

Graphs of the two functions $F(x)$ and $f(x)$ are shown in Figure 3.2c. Notice that F increases most steeply where f is largest.

[As it happens, the given function $F(x)$ *does* have a derivative at $x = 0$, and $F'(0) = 0$. But the point is that it need not concern us.] ∎

(3.2.11) **Limits and monotonicity of CDFs.** The rest of this section is concerned with two things—giving the characteristic properties that all CDFs have in common, and showing how to find the probabilities of arbitrary intervals, not just $(-\infty, x]$ or $(a, b]$, from a CDF. It turns out that both of these are easy to do when we have examined the following four limits:

$$\lim_{x\to\infty} F(x), \text{ denoted by } F(\infty), \qquad \lim_{x\to-\infty} F(x), \text{ denoted by } F(-\infty),$$

$$\lim_{x\downarrow a} F(x), \text{ denoted by } F(a+), \qquad \text{and } \lim_{x\uparrow a} F(x), \text{ denoted by } F(a-).$$

We found these in (2.5.7), as an example of the use of the nested sets theorem. The results are as follows. If $F(x)$ is any CDF, then

(3.2.12)
$$F(\infty) = 1 \quad \text{and} \quad F(-\infty) = 0;$$

and for any number a,

(3.2.13)
$$F(a+) = P(-\infty, a] \quad \text{[which equals } F(a)\text{]}$$

and

(3.2.14)
$$F(a-) = P(-\infty, a) \quad \text{[which equals } F(a) \text{ only if } F \text{ has no jump at } x = a\text{]}.$$

Notice that (3.2.13) says that $F(a+)= F(a)$—that is, that F is right-continuous. Remember, though, that CDFs are not necessarily left-continuous. The right-hand limit (3.2.13) differs from the left-hand limit (3.2.14) by the probability of the singleton $\{a\}$, which may be positive.

Monotonicity. CDFs have one additional important property—they are *monotone increasing*. That is, for any CDF $F(x)$, we have:

(3.2.15)
$$\text{If } x < y, \text{ then } F(x) \leq F(y).$$

Proof. $F(x) = P(-\infty, x]$ and $F(y) = P(-\infty, y]$, and if $x < y$, then $(-\infty, x] \subset (-\infty, y]$; so by (2.1.10) $F(x) \leq F(y)$. ∎

(3.2.16) **The characteristic properties of CDFs.** The CDFs of discrete distributions, as in Example (3.2.2), and absolutely continuous distributions, as in Example (3.2.3), do not look much alike. In the discrete case, the CDF is constant except for jumps at the possible values; in the absolutely continuous case, the CDF increases smoothly through the interval of possible values. What do they have in common?

It turns out that all CDFs share three crucial properties, and these three properties are enough to guarantee that a function is a CDF:

(3.2.17) **Theorem.** If $F(x)$ is the CDF of a distribution, then F has the following three properties. Conversely, if $F(x)$ is a function with these three properties, then $F(x)$ is the CDF of some distribution on the Borel sets of $\mathbb{R}$.

1. F is monotone increasing: if $x < y$, then $F(x) \leq F(y)$.
2. F is right-continuous: $\lim_{x \downarrow a} F(x) = F(a)$ for any $a \in \mathbb{R}$. [That is, $F(a+) = F(a)$.]
3. $\lim_{x \to \infty} F(x) = 1$ and $\lim_{x \to -\infty} F(x) = 0$. [That is, $F(+\infty) = 1$ and $F(-\infty) = 0$.]

We have proved in various places that any CDF must have these three properties. Property 1 is (3.2.15) and the others are (3.2.12) and (3.2.13), which were proved in (2.5.7). The proof of the converse—that any function with these three properties is the CDF of some distribution—involves some measure theory, and we therefore omit it.

Note that although any CDF determines a distribution, the distribution need not be discrete or absolutely continuous. If, for example, F were increasing and differentiable on some intervals but had upward jumps at the endpoints of those intervals, then the distribution would be a "mixture," with some possible values that had positive probabilities, but with probabilities determined by a density function on the rest of the possible values. To see how such a thing could arise, look at the next example.

(3.2.18) **Example.** Consider the following experiment. Start by tossing a fair coin. If it comes up heads, toss it again and take the result of the experiment to be a 1 if you get heads and a 0 if tails. But if the original toss comes up tails, then choose a number from the uniform distribution on (0, 1) and let *that* be the result of the experiment.

Now the outcome set for this experiment is [0, 1]. With probability $\frac{1}{2}$, the outcome is one of the two numbers 0 and 1; they have probability $\frac{1}{4}$ each. But the rest of the time the result is in (0, 1), and its distribution (really its conditional distribution) has a density. The CDF of the distribution of the outcome of this experiment is the following. Figure 3.2d is a graph of this function.

$$F(x) = \begin{cases} 0 & \text{if } x < 0, \\ \dfrac{2x + 1}{4} & \text{if } 0 \leq x < 1, \\ 1 & \text{if } x \geq 1. \end{cases}$$

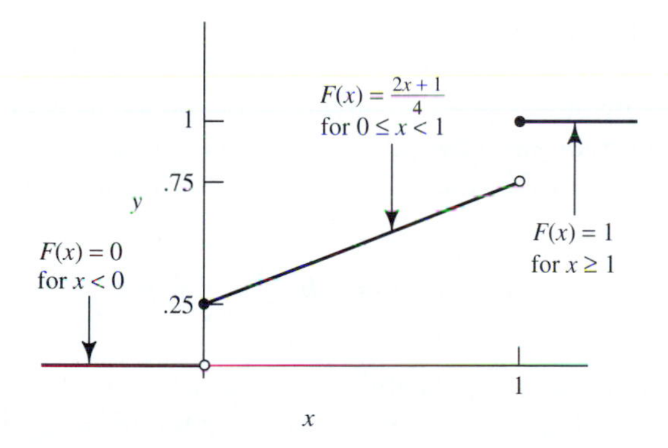

FIGURE 3.2d The CDF of a distribution that is neither discrete nor absolutely continuous

The situation seems rather complicated, with two different kinds of distribution mixed up in one, but notice how easy it is to find the probabilities of intervals from the CDF. For example,

$$P(.5, .8] = F(.8) - F(.5) = .65 - .5 = .15;$$
$$P(.9, 2] = F(2) - F(.9) = 1 - .7 = .3.$$

What about $P(.5, 1)$? It is not the same as $P(.5, 1]$, because the latter interval contains the point 1, which has probability $\frac{1}{4}$, and the former interval does not. It seems pretty obvious that

$$P(.5, 1) = P(.5, 1] - P\{1\} = F(1) - F(.5) - \tfrac{1}{4} = 1 - .5 - .25 = .25.$$

But we knew in advance that $P\{1\}$ was $\frac{1}{4}$; if we had been given only $F(x)$, could we have found $P\{1\}$? This probability equals the height of the jump that F takes at $x = 1$, and to find this from the formula for F we have to evaluate a limit:

$$P\{1\} = F(1) - \lim_{x \uparrow 1} F(x) = 1 - \lim_{x \uparrow 1} \frac{2x + 1}{4} = 1 - \frac{3}{4} = .25.$$

$F(1)$ is the height of the function on the right side of the jump, but there is nothing we can plug into F to get the height on the left; to find it we have to evaluate $F(1-) = \lim_{x \uparrow 1} F(x)$. ∎

(3.2.19) Finding the probability of any interval from the CDF. There are nine different kinds of intervals in $\mathbb{R}$:

$$(a, b), \quad (a, b], \quad [a, b), \quad [a, b],$$
$$(-\infty, b), \quad (-\infty, b], \quad (a, \infty), \quad [a, \infty),$$
$$\text{and } (-\infty, \infty).$$

The CDF of a distribution gives directly only the probability of one kind, $(-\infty, b]$. But the probability of any interval can be found from the CDF. All we need are the definition of the CDF, namely $P(-\infty, b] = F(b)$, and the formula (3.2.14), which says that

$$P(-\infty, b) = F(b-) = \lim_{x \uparrow b} F(x).$$

And once we know the probabilities of the intervals $(-\infty, b]$ and $(-\infty, b)$ for all real b, the probability of any interval can be found using the rules of Section 2.1, as follows:

$$\begin{aligned}
P(a, b] &= P(-\infty, b] - P(-\infty, a]; \\
P(a, b) &= P(-\infty, b) - P(-\infty, a]; \\
P[a, b] &= P(-\infty, b] - P(-\infty, a); \\
P[a, b) &= P(-\infty, b) - P(-\infty, a); \\
P(a, \infty) &= 1 - P(-\infty, a]; \\
P[a, \infty) &= 1 - P(-\infty, a).
\end{aligned}$$

Of course, $P(-\infty, \infty) = 1$; so we can find the probabilities of intervals of all kinds.

What about a singleton, as in the previous example? A singleton is a closed interval, $\{a\} = [a, a]$, and so the third formula above gives the same result as before:

$$\begin{aligned} P\{a\} &= P[a, a] = P(-\infty, a] - P(-\infty, a) \\ &= F(a) - F(a-) = F(a) - \lim_{x \uparrow a} F(x). \end{aligned}$$

This is of course the height of the jump at a if there is a jump, and 0 if not.

EXERCISES FOR **SECTION 3.2**

1. Write down the CDF for the binomial distribution with $n = 2$ and $p = \frac{1}{6}$, in the same form that was used at the end of Example (3.2.2). Also sketch the graph.

2. Find the CDF of the uniform distribution on (0, 1).

3. If $F(x)$ is the CDF of the Poisson distribution with parameter $\lambda = 3.2$, find the following:
   a. $F(1)$ **b.** $F(.99)$ **c.** $F(3.2)$
   d. $F(-4.5)$ **e.** $\lim_{x \uparrow 2} F(x)$

4. Give the CDF of the distribution whose density is defined by $f(x) = x^{-2}$ for $x > 1$ (and 0 otherwise).

5. Find a density for the distribution whose CDF is

$$F(x) = \begin{cases} 0 & \text{if } x < 0, \\ 3x^3 - 2x^4 & \text{if } 0 \le x < 1, \\ 1 & \text{if } x \ge 1. \end{cases}$$

6. Find a density for the distribution whose CDF is

$$F(x) = \begin{cases} 0 & \text{if } x < 0, \\ 1 - (1 + 3x)e^{-3x} & \text{if } x \ge 0. \end{cases}$$

7. For any positive number a, there is a density on (0, 1) proportional to x^a.
   a. Find this density. **b.** Find the associated CDF.

[a]8. Describe in some other way the distribution whose CDF is

$$F(x) = \begin{cases} 0 & \text{if } x < 1, \\ .2 & \text{if } 1 \le x < 3, \\ .35 & \text{if } 3 \le x < 4, \\ .6 & \text{if } 4 \le x < 4.5, \\ 1 & \text{if } x \ge 4.5. \end{cases}$$

9. Sketch the graph of the following CDF:

$$F(x) = \begin{cases} 0 & \text{if } x < 0, \\ \frac{1}{2}(1 + x^2) & \text{if } 0 \le x < 1, \\ 1 & \text{if } x \ge 1. \end{cases}$$

Then find the probability assigned by the distribution to the following sets:
a. $\{0\}$ b. $\{1\}$ c. $(0, .5]$
d. $[0, .5]$ e. $(.5, 1]$ f. $(.5, 1)$

10. If a density for X is $f(x) = \frac{3}{26}(x + 1)^2$ for $0 < x < 2$, find the cumulative distribution function of X.

ANSWERS

1. $F(x) = 0$ for $x < 0$, $\frac{25}{36}$ for $0 \le x < 1$, $\frac{35}{36}$ for $1 \le x < 2$, and 1 for $x \ge 2$.
2. $F(x) = 0$ for $x < 0$, x for $0 \le x < 1$, and 1 for $x \ge 1$.
3. **a.** .1712 **b.** .0408 **c.** .6025 **d.** 0 **e.** .1712
4. $F(x) = 0$ for $x < 1$ and $1 - \frac{1}{x}$ for $x \ge 1$.
5. $f(x) = x^2(9 - 8x)$ for $0 < x < 1$ and 0 otherwise.
6. $f(x) = 9xe^{-3x}$ for $x > 0$ and 0 otherwise.
7. **a.** $f(x) = (a + 1)x^a$ for $0 < x < 1$ and 0 otherwise.
b. $F(x) = 0$ for $x \le 0$, x^{a+1} for $0 < x < 1$, and 1 for $x > 1$.
8. Discrete distribution with possible values 1, 3, 4, 4.5 having probabilities .2, .15, .25, .4, respectively
9. **a.** $\frac{1}{2}$ **b.** 0 **c.** $\frac{1}{8}$ **d.** $\frac{5}{8}$ **e.** $\frac{3}{8}$ **f.** $\frac{3}{8}$
10. 0 for $x < 0$, $\frac{1}{26}((x + 1)^3 - 1)$ for $0 \le x < 1$, and 1 for $x > 1$

3.3 THE DISTRIBUTION OF A RANDOM VARIABLE

Recall from Chapter 1 that a random variable is meant to represent a numerical quantity associated with the outcome of a chance experiment, like the number of heads that come up when a fair coin is tossed three times, or the waiting time between the third and fourth arrivals in a Poisson process. We model the experiment with an outcome set Ω that has probabilities $P(A)$ assigned to its subsets in some way, and the random variable is a function X on Ω: $X(\omega)$ is the value of the quantity when ω is the outcome.

Then statements about X translate to subsets of Ω. For example, $(X \le 3)$ is the subset of Ω defined by $\{\omega\colon X(\omega) \le 3\}$, and in general $(X \in B)$ is the subset $\{\omega\colon X(\omega) \in B\}$. Indeed, the only requirement for a function X to qualify as a random variable is that this subset be an event for every Borel set B.

But this means that a probability has been assigned to each Borel set. Borel set B gets probability $P(X \in B)$, which means $P\{\omega: X(\omega) \in B\}$. If we denote this probability by $P_X(B)$, then P_X is a probability distribution. We call it the *probability distribution of X.*

Notice that we have two different, though related, probability assignments, P and P_X. P is the original assignment of probabilities to events in Ω, and P_X is an assignment of probabilities to Borel sets in $\mathbb{R}$. They are related to each other, as we said in the previous paragraph, by

$$P_X(B) = P(A), \quad \text{where } A = \{\omega: X(\omega) \in B\}.$$

We will give the formal definition in (3.3.5), after we look at two examples.

(3.3.1) Example: The distribution of the number of heads in three tosses of a fair coin. We return to Example (1.1.1).

The probability space: Ω has the familiar eight members (HHH, HHT, etc.); all subsets of Ω are events; and P is the assignment of probabilities defined by

$$P(A) = \tfrac{1}{8} \cdot (\text{number of outcomes in } A).$$

The random variable X and its distribution: $X(\omega)$ is the number of H's in ω. X has four possible values, 0, 1, 2, and 3, with probabilities $\frac{1}{8}$, $\frac{3}{8}$, $\frac{3}{8}$, and $\frac{1}{8}$, respectively. That is, for any Borel set B, $P_X(B) = P(X \in B)$ is the sum of whichever of those four probabilities correspond to possible values that happen to be in B.

So the distribution of X is the discrete distribution whose probability mass function is $p(0) = \frac{1}{8}$, $p(1) = \frac{3}{8}$, etc. That is, the distribution of X is the binomial distribution with $n = 3$ and $p = \frac{1}{2}$. ∎

(3.3.2) Example: The distribution of the square of a uniformly distributed random number. Now we return to Example (1.3.11). There Ω is $[0, 1)$ and $P(I)$ is the length of I for subintervals I of Ω, so that the probability space models a uniformly distributed random number; and the random variable Y is defined by $Y(\omega) = \omega^2$. In Example (1.3.11) we worked out $P(Y \leq .5)$, which we can now call $P_Y[0, .5]$ (or $P_Y(-\infty, .5]$, since the probability that Y is in $(-\infty, 0)$ is 0).

But what is the distribution of Y? That is, how do we find $P_Y(B) = P(Y \in B)$ for an arbitrary Borel set B? As with any distribution, we will know it completely if we know $P_Y(-\infty, y]$ for any $y \in \mathbb{R}$—that is, if we know the CDF of the distribution. That CDF is denoted by F_Y:

(3.3.3)
$$F_Y(y) = P_Y(-\infty, y] = P(Y \leq y).$$

It is not difficult to see what this function is. First of all, the set of possible values of Y is the interval $[0, 1)$, and so $F_Y(y)$ is 0 if $y < 0$ and 1 if $y \geq 1$. And if $0 \leq y < 1$, then

(3.3.4)
$$F_Y(y) = P\{\omega\colon \omega^2 \leq y\} = P\{\omega\colon \omega \leq \sqrt{y}\} = P[0, \sqrt{y}] = \sqrt{y}.$$

It is worth the time to compare and contrast P and P_Y in (3.3.3) and (3.3.4). P always refers to the probability assignment in the original Ω, and P_Y refers to the probability that Y is in some Borel set or other. Confusion is possible in a situation like this one, because the original Ω is a subset of $\mathbb{R}$.

Putting it together, the CDF of Y is

$$F_Y(y) = \begin{cases} 0 & \text{if } y < 0, \\ \sqrt{y} & \text{if } 0 \leq y < 1, \\ 1 & \text{if } y \geq 1. \end{cases}$$

If we want to understand the distribution further, we can find a density for it [denoted by $f_Y(y)$, naturally] by differentiating as we did in the previous section. The derivative exists except perhaps at 0 and 1, and we can define the density to be 0 at those two points and F'_Y elsewhere:

$$f_Y(y) = \begin{cases} \dfrac{1}{2\sqrt{y}} & \text{if } 0 < y < 1, \\ 0 & \text{otherwise.} \end{cases}$$

A sketch of the graph of this density function will show you where Y is likely to be (near 0) and where it is not so likely (near 1). ∎

(3.3.5) Definition. Given any random variable X, the ***distribution of X*** is the assignment of probabilities to Borel sets denoted by P_X and defined by

$$P_X(B) = P(X \in B).$$

The cumulative distribution function of this distribution is called the ***cumulative distribution function (CDF) of X*** and is denoted by F_X:

$$F_X(x) = P_X(-\infty, x] = P(X \leq x).$$

If this distribution is discrete, then its probability mass function is called the ***probability mass function of X*** and denoted by $p_X(x)$. In this case X is called a ***discrete random variable.***

If it is absolutely continuous, then a density for it is called a ***density for X*** and denoted by $f_X(x)$. In this case X is called an ***absolutely continuous random variable.***

A note on notation: The subscript X is used to make it clear that the function—density, mass function, or CDF—describes the distribution of X. Such subscripting is especially useful when two or more random variables are under discussion. But we will often omit it if there is no possibility of confusion.

(3.3.6) **Examples.** Here we summarize some earlier examples, but using the language of random variables and their distributions.

a. **Bernoulli trials and the binomial distribution** [Example (1.2.13)]. Consider n Bernoulli trials with outcomes S and F and success probability p on each trial (and, of course, failure probability $q = 1 - p$). If X is the number of successes, then the result of Example (1.2.13) says simply that X is a discrete random variable whose mass function is

$$p_X(x) = \begin{cases} \binom{n}{x} p^x q^{n-x} & \text{if } x = 0, 1, 2 \ldots, n, \\ 0 & \text{otherwise.} \end{cases}$$

Notice what this means: when we want to compute probabilities for numbers of successes in Bernoulli trials, this formula is all we need; we can forget about the outcome set Ω with its 2^n members that are "n-letter words" from the "alphabet" {S, F}, the probabilities of various words like SSFSFF, and so forth. The distribution of X, however we specify it (in this case via its mass function), contains all the "probability information" we need to do calculations concerning X. ∎

b. **Waiting for the second head with a fair coin** [Example (1.2.5)]. If we toss a fair coin repeatedly until heads have come up twice, and let X denote the number of tails we get, then X is a discrete random variable whose mass function is [see (1.2.7)]

$$p_X(x) = \begin{cases} (x+1)\left(\frac{1}{2}\right)^{x+2} & \text{if } x = 0, 1, 2, \ldots, \\ 0 & \text{otherwise.} \end{cases}$$

Once again, if we want to find probabilities of events describable in terms of X, we can ignore the probability space Ω and its members (which are "words" like TTHTTTH); the distribution of X, specified by means of its mass function, is all we need. ∎

c. **The square of a uniformly distributed random number** [Examples (1.3.11) and (3.3.2)]. If we choose a number from the uniform distribution on [0, 1) and let Y be its square, then Y is an absolutely continuous random variable with density

$$f_Y(y) = \begin{cases} \dfrac{1}{2\sqrt{y}} & \text{if } 0 < y < 1, \\ 0 & \text{otherwise.} \end{cases}$$

Again, all we need is the distribution of Y, specified here by a density function; we can forget about the original Ω and probabilities as lengths.

In this case, though, the density may not be the most convenient way to specify the distribution. As with many absolutely continuous random variables, the CDF is a simpler means of finding probabilities. Here the CDF, as we found in Example (3.3.2), is

$$F_Y(y) = \begin{cases} 0 & \text{if } y < 0, \\ \sqrt{y} & \text{if } 0 \le y < 1, \\ 1 & \text{if } y \ge 1. \end{cases}$$

∎

Before you decide that densities are useless, however, let us point out that they will be very much used for finding expected values in the next chapter. And at the least, a sketch of the density shows where the probability is high and where it is low.

d. The distance from the center to a random point in the unit disk [Example (1.3.13) and Exercise 14 of Section 1.3]. Suppose we choose a point at random from the uniform distribution on the unit disk $\Omega = \{(x, y): x^2 + y^2 < 1\}$, and let X be its distance from the origin. In Exercise 14d of Section 1.3 we found that for $0 < r < 1$, $P(X \le r) = r^2$. That was essentially the CDF of X; its full description is

$$F_X(x) = \begin{cases} 0 & \text{if } x < 0, \\ x^2 & \text{if } 0 \le x < 1, \\ 1 & \text{if } x \ge 1. \end{cases}$$

Again, this contains all the "probability information" about X; as long as we just want probabilities for events concerning the distance X from the center, we can forget about the disk Ω, the uniform distribution on it, the distance formula, etc.

Of course, if it suits us we can notice that $F_X(x)$ is differentiable except perhaps at 0 and 1, and therefore X is an absolutely continuous random variable. A density for it is

$$f_X(x) = \begin{cases} 2x & \text{if } 0 < x < 1, \\ 0 & \text{otherwise.} \end{cases}$$

This shows that the distance is more likely to be near 1 than near 0; the probability that it is in a small interval containing x increases in proportion to x (approximately) as x increases. ∎

e. A uniformly distributed random variable [Example (1.3.1)]. In our original model for the random choice of a number from $\Omega = [0, 1)$, Borel sets are events and the probability of an interval is its length. We introduced the random variable $X(\omega) = \omega$ at the end of Example (1.3.1) and said that it was not really needed, except that it gave us economy of language. We can say "$P(X \le .5)$," for example, instead of "P(the number chosen is less than .5)."

Now the economy really pays off. We can specify the situation fully by saying simply that the distribution of X is the uniform distribution on $[0, 1)$. The original probability space is still there, but our discussion need not include it; we just say "let X have the uniform distribution on $[0, 1)$," and then immediately we have probabilities like

$$P(.2 < X \leq .74) = .74 - .2 = .54.$$

Incidentally, we already know a density for X; we first saw it in Section 1.3, a few paragraphs before (1.3.7). It is the uniform density

$$f(x) = \begin{cases} 1 & \text{if } 0 < x < 1, \\ 0 & \text{otherwise.} \end{cases}$$

We also found the CDF, in Exercise 2 of Section 3.2; it is

$$F_X(x) = \begin{cases} 0 & \text{if } x < 0, \\ x & \text{if } 0 \leq x < 1, \\ 1 & \text{if } x \geq 1. \end{cases}$$

■

Note. Sometimes, in specifying density functions and probability mass functions, we omit the line that says "0 otherwise" and take it as understood. Thus in part d of Example (3.3.6), we could simply say that the density of X is $2x$ for $0 < x < 1$; and in part b, we could describe the mass function as $(x + 1)\left(\frac{1}{2}\right)^{x+2}$ for $x = 0, 1, 2, \ldots$. Similar comments apply to all five parts of Example (3.3.6). We will often do this in what follows.

But there are two things to be remembered. First, we do not usually omit anything of this sort when specifying a CDF. Second, and most important, although it is permissible to omit "0 otherwise," ***we can never omit the set of values of x for which the expression is valid.*** In part d, for example, it would never do to say that the density of X is $f(x) = 2x$ without saying that it is $2x$ only when x is between 0 and 1.

(3.3.7) **Random variables associated with a Poisson process.** In Section 1.4 we introduced Poisson processes and gave four rules for computing the probabilities of various events. Now we can restate those rules more simply by defining certain random variables and phrasing the rules in terms of the distributions of the random variables, as follows.

Consider a Poisson process with arrival rate λ per unit time. (Remember that this is only an average arrival rate; the actual number of arrivals in any time interval of unit length is a random variable.) Then the four rules given in Section 1.4 are these:

P1. For any time interval $I = [t, t + s]$, the number of arrivals in I is a discrete random variable whose probability mass function is $p(x) = (\lambda s)^x e^{-\lambda s}/x!$ for

$x = 0, 1, 2, \ldots$. (The distribution of this random variable is the *Poisson distribution with parameter* λs.)

P2. For any time point t, the waiting time from t to the next arrival is an absolutely continuous random variable whose density function is $f(t) = \lambda e^{-\lambda t}$ for $t > 0$. (This distribution is the *exponential distribution with parameter* λ.)

P3. The waiting time from any one arrival to the next is also a random variable having the exponential distribution with parameter λ.

P4. Given two nonoverlapping time intervals I and J, if X and Y are the numbers of arrivals in I and J, respectively, then X and Y are independent random variables.

There is also a fifth rule that is sometimes useful, which we did not mention earlier:

P5. Nonoverlapping waiting times are independent random variables.

For example, the time between the fifth and sixth arrivals is independent of the time between the sixth and the seventh, or between the fourth and the fifth. In addition, for any fixed time point t, any waiting time involving arrivals prior to t is independent of any waiting time involving arrivals after t.

For examples and practice in using these rules, you can now go back to previous sections and repeat any of the exercises involving Poisson processes, looking at these versions of P1–P4 instead of the ones given in Section 1.4. Such exercises can be found in Sections 1.4 and 2.2, and in the Supplementary Exercises for both Chapters 1 and 2. There are one or two at the end of this section as well.

(3.3.8) **Random variables without explicitly described probability spaces.** One of the lessons of Examples (3.3.6) is that once we know the distribution of a random variable X we can forget where it came from; that is, we do not need to think about the outcome space Ω, the probabilities on events in Ω, or the definition of the function $X(\omega)$ on Ω.

We can carry this a step further. It is possible to talk about a random variable without having any outcome space in mind at all. For example, we might begin a discussion by saying, "Let X have the density $f(x) = 3x^2$ for $0 < x < 1$." We can then proceed to find probabilities for events describable in terms of X without ever considering what the experiment or the outcome set Ω might be.

Or recall that in Example (1.3.5), we let Ω be $\mathbb{R}$ and defined probabilities on Ω by the standard normal density $\varphi(x) = (1/\sqrt{2\pi})e^{-x^2/2}$ for $x \in \mathbb{R}$. We then found probabilities of various intervals using the standard normal distribution table in Appendix C; for example, the probability that the outcome is less than or equal to 1 is $P(-\infty, 1] = .8413$. Now we can talk about the same distribution without mentioning an Ω; all we need to say is "let X have the density $(1/\sqrt{2\pi})e^{-x^2/2}$ for $x \in \mathbb{R}$," or more simply "let X have the standard normal density." Then the probability that the outcome is less than or equal to 1 is simply $P(X \leq 1)$.

There is a little more on this point in the comment after Exercise 11 at the end of this section.

EXERCISES FOR **SECTION 3.3**

1. Let X have the density $f_X(x) = 3x^2$ for $0 < x < 1$.
   a. Find the CDF of X, and then use it to find the following probabilities.
   b. $P(X \leq .5)$
   c. $P(X > .25)$

2. Suppose the CDF of X is $F(x) = 1 - e^{-x^2}$ for $x > 0$ (and 0 for $x \leq 0$). Find a density for X.

3. Customers are arriving at a bank according to a Poisson process; the arrival rate is $\lambda = 4.7$ per hour. Let X be the number of customers arriving in the first half hour.
   a. What is the distribution of X? [See rule P1 as stated in (3.3.7).]
   b. What is $p_X(2)$?
   c. What is the probability that X is positive?

4. In Exercise 3 let Y be the time of arrival of the first customer. Find the following:
   a. The distribution of Y [See rule P2 as stated in (3.3.7).]
   b. $P(Y \leq .5)$ [Compare with the answer to Exercise 3c and explain.]
   c. $P(Y > a)$ where a is some positive number
   d. $P(X < 3 \text{ and } Y \geq .25)$

5. Suppose the probability mass function of X is $p_X(x) = x^2/14$ for $x = 1, 2, 3$ (and 0 otherwise). Find the following probabilities:
   a. $P(X = 3)$
   b. $P(X \leq 2.7614)$
   c. $P(X = 2.7614)$
   d. $P(X \leq 2)$
   e. $P(X < 2)$

[a]6. Let X be the number of 6's in three rolls of a fair die. Give the mass function of X, both with a formula and with a table.

7. Let X have the exponential density $f_X(x) = 2e^{-2x}$ for $x > 0$. Find the following probabilities. [Write down the CDF first to save work.]
   a. $P(2X + 1 > 3)$
   b. $P(X^2 \leq 7)$
   c. $P(X \leq 2 | X > .5)$
   d. $P(X > .5 | X \leq 2)$

8. In a random permutation of the numbers 1, 2, 3, 4, let S be the number of numbers that are in their proper places. Find the probability mass function of S. [*Hint:* If you did Exercise 8 of Section 2.2 you wrote down the 24-member

outcome space. Alternatively, if you did Exercise 23 of the same section, you found a way to compute the probabilities without thinking about the outcome space.]

9. Let X be the number of arrivals in the first hour of a Poisson process, and let Y be the number of arrivals in the first 2 hours. Suppose the arrival rate is $\lambda = 1.6$ per hour. Find

a. $P(Y \geq 2)$ **b.** $P(Y \geq 2 | X = 0)$
c. $P(Y = X)$ **d.** The mass function of $Y - X$

10. If the mass function of the random variable X is $p(k) = ck(.42)^{k-1}$ ($k = 1, 2, 3, \ldots$), what is the value of c?

11. The lifetimes of items like lightbulbs that are subject to sudden failure are often assumed to have exponential distributions. Suppose X, the lifetime in hundreds of hours of a randomly chosen bulb made in a certain factory, has the exponential density with parameter $\frac{1}{10}$, which is $f(x) = \frac{1}{10}e^{-x/10}$ for $x > 0$. Find the following probabilities:

a. $P(X \geq 4)$ [Remember that time is measured in hundreds of hours.]
b. $P(X \geq 14 | X \geq 10)$

This is another example, like those mentioned in (3.3.8), in which we can work with a random variable without thinking about what outcome set it comes from. In fact, it is difficult to think of any good outcome set for the experiment of choosing a lightbulb at random from a given factory's output. Is it the set of all lightbulbs ever made in this factory? The set of all bulbs made on the day the random bulb is chosen? No, because those are finite sets, and so any random variable on them will be discrete. How about the set of all hypothetical bulbs that *could* be made in this factory? This seems a bit metaphysical. An artificial solution is to let the outcome set be the set of all possible *lifelengths* that a lightbulb could have. But that is really no different from saying, "Suppose X has the exponential distribution. . . ," as we did in the exercise. It is convenient to be able to talk about a random variable without specifying the Ω it comes from.

If it is absolutely necessary to have an outcome space for this problem, the proper thing to do is to imagine a finite outcome space consisting of all the actual bulbs that are of concern. Then the random variable X really is a discrete random variable, but the exponential distribution serves well as an approximation to its distribution.

12. A point is chosen at random from the triangle whose vertices are at (0, 0), (0, 2), and (2, 0); X is the sum of the two coordinates of the point.

a. Find the CDF of X.
b. Find a density for X.
c. Find $P(X \leq 1)$. [You can do part c in three ways, and it is good to see why they are really the same. $P(X \leq 1)$ can be found from the CDF as $F(1)$; it can be found by integrating the density from 0 to 1; or it can be found by dividing the area of the set of points whose coordinates sum to 1 or less by the area of the triangle Ω.]

13. Let X and Y be independent random variables with the following mass functions:

x:	0	1	2
$p_X(x)$:	.3	.5	.2

y:	0	1	2
$p_Y(y)$:	.6	.1	.3

a. Make a table showing $P(X = x \text{ and } Y = y)$ for each pair (x, y) of possible values.
b. Find $P(X = Y)$.
c. Find $P(X + Y = 2)$.
d. Find the probability mass function of the random variable $Z = X + Y$.

14. A certain random variable X has possible values 1, 2, 3, . . . , and $P(X > k) = \left(\frac{1}{2}\right)^k$ for $k = 1, 2, 3, \ldots$. What is the mass function p_X? [*Hint:* $P(X = k) = P(X > k - 1) - P(X > k)$ if k is a positive integer.]

ANSWERS

1. a. 0 for $x < 0$, x^3 for $0 \le x < 1$, and 1 for $x \ge 1$ **b.** $\frac{1}{8}$ **c.** $\frac{63}{64}$

2. $2xe^{-x^2}$ for $x > 0$ (and 0 otherwise)

3. b. .2633 **c.** .9046

4. b. .9046 **c.** $e^{-4.7 \cdot a}$ **d.** .2733

5. a. $\frac{9}{14}$ **b.** $\frac{5}{14}$ **c.** 0 **d.** $\frac{5}{14}$ **e.** $\frac{1}{14}$

6. Function: $p(k) = P(X = k) = \binom{3}{k}\left(\frac{1}{6}\right)^k\left(\frac{5}{6}\right)^{3-k}$ for $k = 0, 1, 2, 3$;

Table:

k:	0	1	2	3
$p(k)$:	$\frac{125}{216}$	$\frac{75}{216}$	$\frac{15}{216}$	$\frac{1}{216}$

7. a. .1353 **b.** .9950 **c.** .9502 **d.** .3561

8. $p(0) = \frac{9}{24}$, $p(1) = \frac{8}{24}$, $p(2) = \frac{6}{24}$, $p(4) = \frac{1}{24}$

9. a. .8288 **b.** .4751 **c.** .2019 **d.** Poisson with $\lambda = 1.6$

10. .3364

11. a. .6703 **b.** .6703

12. a. $F_X(x) = 0$ for $x < 0$, $x^2/4$ for $0 \le x < 2$, and 1 for $x \ge 2$ **b.** $f_X(x) = x/2$ for $0 < x < 2$ **c.** $\frac{1}{4}$

13. b. .29 **c.** .26 **d.** Values 0, 1, 2, 3, 4; probabilities .18, .33, .26, .17, .06

14. $p_X(x) = \left(\frac{1}{2}\right)^x$ for $x = 1, 2, 3, \ldots$

3.4 THE JOINT DISTRIBUTION OF TWO DISCRETE RANDOM VARIABLES

The main point to keep in mind for the remainder of this chapter is that, when we are dealing with two or more random variables, ***knowing the distributions of the individual random variables is not enough information*** for finding probabilities involving more than one of them. What is needed is the information contained in what we call the ***joint distribution*** of the random variables.

In this section we will introduce the joint distribution of two discrete random variables. We will show how their individual distributions can be obtained from the joint distribution as what are called ***marginal distributions,*** but that it cannot be found from them. Finally, we will discuss the independence of two discrete random variables and will see that it is equivalent to the product formula for the *joint mass function*. The definitions are in (3.4.9); first we look at an example.

(3.4.1) **Example.** Recall the two random variables X and Y that we introduced toward the end of Example (1.1.1). The experiment consists of three tosses of a fair coin, so Ω is the familiar eight-member outcome set {HHH, HHT, etc.}. The random variable X is the number of heads in the three tosses, and Y is the number of tails that precede the first head. (If no heads come up, we let $Y = 3$.) We saw in Example (1.1.1) that the mass functions of X and Y are the following:

(3.4.2)

x:	0	1	2	3
$P(X = x)$:	$\frac{1}{8}$	$\frac{3}{8}$	$\frac{3}{8}$	$\frac{1}{8}$

y:	0	1	2	3
$P(Y = y)$:	$\frac{4}{8}$	$\frac{2}{8}$	$\frac{1}{8}$	$\frac{1}{8}$

And at the very end of Example (1.1.1), right before the summary, we found a few probabilities of events involving both X and Y. For example,

$$P(X = 1 \text{ and } Y = 1) = P(\text{THT}) = \tfrac{1}{8},$$

because the event "$X = 1$ and $Y = 1$" occurs only if the outcome of the experiment is THT. But notice that this answer, $\frac{1}{8}$, cannot be found from the tables (3.4.2) alone. For another example,

$$P(X = 2 \text{ and } Y = 2) = P(\varnothing) = 0,$$

because this event cannot occur; there are no outcomes ω for which $X(\omega) = 2$ and $Y(\omega) = 2$. Again, there is no way to see from tables (3.4.2) that this probability is 0.

The tables (3.4.2) are fine for computing probabilities for X alone or Y alone. For example, we can see that $P(X \geq 2)$ is $\frac{3}{8} + \frac{1}{8}$ from table (3.4.2) without going back to the outcome set Ω. But if those tables were all we knew about X and Y, there would be no way to know that $P(X = 2 \text{ and } Y = 2) = 0$. What *do* we need? It seems clear that in this example, the probabilities in the following table will give us enough information:

(3.4.3)

	$p(x, y)$	y = 0	1	2	3
x	0	0	0	0	$\frac{1}{8}$
	1	$\frac{1}{8}$	$\frac{1}{8}$	$\frac{1}{8}$	0
	2	$\frac{2}{8}$	$\frac{1}{8}$	0	0
	3	$\frac{1}{8}$	0	0	0

The notation "$p(x, y)$" refers to $P(X = x \text{ and } Y = y)$; thus, for example, $p(2, 0) = P(X = 2 \text{ and } Y = 0) = \frac{2}{8}$. Notice that the 16 probabilities in the table add to 1. Be sure you see how they are found from the original outcome set. For example, $P(X = 2 \text{ and } Y = 0) = \frac{2}{8}$ because the event "$X = 2$ and $Y = 0$" is {HTH, HHT}; these are the two ways that we can get two heads with no tails before the first head.

The entries in table (3.4.3) amount to the specification of a function of x and y, $p(x, y) = P(X = x \text{ and } Y = y)$. It is called the *joint probability mass function,* or just the *joint mass function,* of X and Y. The formal definition is given at (3.4.10).

The next thing to notice about table (3.4.3) is that from it we can recover the two tables (3.4.2)—the individual mass functions of X and Y—by adding across the rows and down the columns:

(3.4.4)

$p(x, y)$	$y = 0$	1	2	3	$p_X(x) = P(X = x)$
$x = 0$	0	0	0	$\frac{1}{8}$	$\frac{1}{8}$
1	$\frac{1}{8}$	$\frac{1}{8}$	$\frac{1}{8}$	0	$\frac{3}{8}$
2	$\frac{2}{8}$	$\frac{1}{8}$	0	0	$\frac{3}{8}$
3	$\frac{1}{8}$	0	0	0	$\frac{1}{8}$
$p_Y(y) = P(Y = y)$	$\frac{4}{8}$	$\frac{2}{8}$	$\frac{1}{8}$	$\frac{1}{8}$	

For this reason the individual mass functions are called *marginal mass functions* of X and Y when the context involves a joint mass function. The word *marginal* adds no meaning here—a marginal mass function is just a mass function. The word merely distinguishes the mass functions of the individual random variables from their joint mass function. Notice also the importance of the subscripts on the names p_X and p_Y of the marginal mass functions.

Furthermore, we can find the probabilities of all events describable in terms of X and Y from the joint mass function. For example, $P(X \geq 2 \text{ and } Y \leq 1)$ is the sum of the appropriate entries in the table: $\frac{2}{8} + \frac{1}{8} + \frac{1}{8} + 0 = \frac{4}{8}$.

Finally, to drive home the point that the marginal mass functions of X and Y are not enough to determine the joint mass function, here are two other joint mass functions, each of which has the same pair of marginal mass functions as table (3.4.4).

(3.4.5)

$p(z, w)$	$w = 0$	1	2	3	$p_Z(z)$
$z = 0$	$\frac{1}{8}$	0	0	0	$\frac{1}{8}$
1	$\frac{3}{8}$	0	0	0	$\frac{3}{8}$
2	0	$\frac{2}{8}$	$\frac{1}{8}$	0	$\frac{3}{8}$
3	0	0	0	$\frac{1}{8}$	$\frac{1}{8}$
$p_W(w)$	$\frac{4}{8}$	$\frac{2}{8}$	$\frac{1}{8}$	$\frac{1}{8}$	

(3.4.6)

$p(u, v)$	$v = 0$	1	2	3	$p_U(u)$
$u = 0$	$\frac{4}{64}$	$\frac{2}{64}$	$\frac{1}{64}$	$\frac{1}{64}$	$\frac{1}{8}$
1	$\frac{12}{64}$	$\frac{6}{64}$	$\frac{3}{64}$	$\frac{3}{64}$	$\frac{3}{8}$
2	$\frac{12}{64}$	$\frac{6}{64}$	$\frac{3}{64}$	$\frac{3}{64}$	$\frac{3}{8}$
3	$\frac{4}{64}$	$\frac{2}{64}$	$\frac{1}{64}$	$\frac{1}{64}$	$\frac{1}{8}$
$p_V(v)$	$\frac{4}{8}$	$\frac{2}{8}$	$\frac{1}{8}$	$\frac{1}{8}$	

∎

The joint mass function (3.4.6) is a special one; each entry was obtained as the product of the two corresponding marginal entries. So (3.4.6) has the property—not shared by (3.4.4) and (3.4.5)—that

$$p(u, v) = p_U(u) \cdot p_V(v) \quad \text{for all } u \text{ and } v. \tag{3.4.7}$$

That is, $P(U = u \text{ and } V = v) = P(U = u) \cdot P(V = v)$ for all possible values u of U and v of V. This reminds us of the definition of independence in Section 2.4:

$$U \text{ and } V \text{ are independent if}$$

$$P(U \text{ is in } A \text{ and } V \text{ is in } B) = P(U \text{ is in } A) \cdot P(V \text{ is in } B) \tag{3.4.8}$$

$$\text{for all Borel sets } A \text{ and } B.$$

Indeed, it can be proved [Theorem (3.4.13) below] that for two discrete random variables, (3.4.7) is enough to imply (3.4.8). This is what you would expect, knowing that the joint mass function is sufficient to determine the probabilities of all events describable in terms of U and V.

But it is important to remember that (3.4.7) does not hold in general for two random variables. Only when two random variables are independent can we find their joint mass function from the marginal mass functions, by multiplying. If they are not independent, we will first have to get the joint mass function somehow, and then get the marginal mass functions from it by adding across the rows and down the columns.

The general definition of the joint mass function of two discrete random variables contains nothing that was not in the previous example.

(3.4.9) **Definition.** The ***joint probability mass function*** of two discrete random variables X and Y is the function $p(x, y)$, defined for all pairs of real numbers x and y by

(3.4.10)
$$p(x, y) = P(X = x \text{ and } Y = y).$$

(This is of course 0 if x and y are not possible values of X and Y, respectively.) In this context the probability mass functions of X and Y are called ***marginal probability mass functions*** and denoted by $p_X(x)$ and $p_Y(y)$.

As in Example (3.4.1), the marginal mass functions can be obtained from the joint mass function:

$$P(X = x) = \sum_y P(X = x \text{ and } Y = y).$$

That is,

(3.4.11)
$$p_X(x) = \sum_y p(x, y).$$

Of course this is because the event $(X = x)$ is the union of the disjoint events $(X = x \text{ and } Y = y)$. The counterpart of (3.4.11) for the other marginal mass function is

(3.4.12)
$$p_Y(y) = \sum_x p(x, y).$$

And as we mentioned earlier, we can use the joint and marginal mass functions to decide whether discrete random variables are independent, as follows.

(3.4.13) **Theorem.** Two discrete random variables X and Y, with joint mass function $p(x, y)$ and marginal mass functions $p_X(x)$ and $p_Y(y)$, are independent if and only if

(3.4.14)
$$p(x, y) = p_X(x) \cdot p_Y(y) \quad \text{for all } x \text{ and } y.$$

Proof. Remember that X and Y are independent, by definition, if and only if

(3.4.15)
$$P(X \text{ is in } A \text{ and } Y \text{ is in } B) = P(X \text{ is in } A) \cdot P(Y \text{ is in } B)$$

for all intervals (and hence all Borel sets) A and B. It is clear that (3.4.15) implies (3.4.14), because (3.4.14) is just the special case in which A and B are singletons. To see that (3.4.14) is enough to imply (3.4.15), notice that the left side of (3.4.15) is

$$\sum_{\substack{x \in A \\ y \in B}} p(x, y) = \sum_{x \in A} \sum_{y \in B} p_X(x) \cdot p_Y(y) = \sum_{x \in A} p_X(x) \cdot \sum_{y \in B} p_Y(y),$$

and this last expression is the right side of (3.4.15). □

(3.4.16) **Example: The first and second successes in a Bernoulli process.** In a Bernoulli process with success probability p and failure probability $q = 1 - p$, let X be the number of failures preceding the first S, and let Y be the number of failures preceding the second, including those preceding the first. We have already found the distributions of X and Y in Section 1.2; we know that

$$P(X = x) = pq^x \quad \text{for integers } x \geq 0$$

and

$$P(Y = y) = (y + 1)p^2q^y \quad \text{for integers } y \geq 0.$$

But here let us find the joint mass function of X and Y first. Then, as an exercise in finding marginal mass functions from a joint mass function, we will use (3.4.11) and (3.4.12) to rederive these results.

First, we want $p(x, y) = P(X = x \text{ and } Y = y)$ for all possible values x of X and y of Y. For which pairs (x, y) will this be nonzero? Answer: x and y must be nonnegative integers and x must be less than or equal to y, because Y counts failures that include failures counted by X.

So suppose x and y are nonnegative integers and $x \leq y$; then $p(x, y)$ is the probability that the first S comes after x F's and the second comes after $y - x$ more F's that follow the first S. As an example, $p(2, 5)$ is the probability that we get two F's before the first S, then three more F's, then the second S. That is, $p(2, 5)$ is the probability that the first seven trials result in FFSFFFS; this is p^2q^5.

In general, for $0 \leq x \leq y$, $p(x, y)$ is the probability that the first $y + 2$ trials result in FF. . .FSFF. . .FS, with x F's before the first S and $y - x$ F's between the first and the second S's. The probability of this event is p^2q^y. Therefore,

(3.4.17)
$$p(x, y) = \begin{cases} p^2q^y & \text{if } x \text{ and } y \text{ are integers and } 0 \leq x \leq y, \\ 0 & \text{otherwise.} \end{cases}$$

It may be disconcerting that the formula p^2q^y does not involve x. But $p(x, y)$ is indeed a function of x as well as of y; remember that a function is not just a formula, but a formula together with the set of values to which the formula applies. Changing x will not change the value of p^2q^y, but it can change the value of $p(x, y)$. [For example, although $p(0, 5) = p(2, 5) = \cdots = p(5, 5) = p^2q^5$, $p(6, 5) = 0$.]

Now for the marginal mass functions. Let us start with $p_X(x) = P(X = x)$. This is nonzero for all nonnegative integers x, and we find it by adding up $p(x, y)$ for all y. For a given nonnegative integer x, $p(x, y)$ is nonzero only for values of y that are greater than or equal to x; so we get

$$\begin{aligned} p_X(x) &= \sum_y p(x, y) = \sum_{y=x}^{\infty} p^2 q^y \\ &= p^2(q^x + q^{x+1} + q^{x+2} + \cdots) \\ &= p^2 q^x (1 + q + q^2 + \cdots) \\ &= p^2 q^x \frac{1}{1-q} = pq^x \quad \text{for } x = 0, 1, 2, 3, \ldots. \end{aligned}$$

Now we find $p_Y(y) = P(Y = y)$. This is nonzero for any nonnegative integer y, and we find it by adding $p(x, y)$ over all x. For a given y, $p(x, y)$ is nonzero only when x is less than or equal to y; so

$$\begin{aligned} p_Y(y) &= \sum_x p(x, y) = \sum_{x=0}^{y} p^2 q^y \\ &= (y + 1)p^2 q^y \quad \text{for } y = 0, 1, 2, 3, \ldots. \end{aligned}$$

This last step may be confusing; we are summing an expression that does not involve x, p^2q^y, for all x from 0 to y. The result is just $y + 1$ times the expression. ∎

(3.4.18) **Example.** Suppose that the machines in a large factory are subject to breakdowns at times that follow a Poisson process with a rate of $\lambda = 5$ breakdowns per week. Let X be the number of breakdowns in the first week of observations, and Y the number in the first 3 weeks (including those in the first week). What is the joint mass function of X and Y? That is, for given numbers x and y, what is the probability $p(x, y)$ of the event $(X = x \text{ and } Y = y)$? Are X and Y independent?

We have done several calculations like this previously. For example,

$$\begin{aligned} P(X = 4 \text{ and } Y = 11) &= P(\text{4 breakdowns in the first week and 7 in the second and third weeks}) \\ &= \frac{5^4 e^{-5}}{4!} \cdot \frac{10^7 e^{-10}}{7!}. \end{aligned}$$

As you may remember, the trick with such a probability for a Poisson process is to express the event in terms of events involving nonoverlapping intervals. In general, we see that the joint mass function is

(3.4.19) $$p(x, y) = \frac{5^x e^{-5}}{x!} \cdot \frac{10^{y-x} e^{-10}}{(y-x)!} \quad \text{if } x \text{ and } y \text{ are nonnegative integers and } x \le y,$$

and $p(x, y) = 0$ otherwise. This can be simplified to $(1/2)^x 10^y e^{-15}/x!(y - x)!$, although (3.4.19) reveals the situation a little more clearly.

To decide whether X and Y are independent, we must get their marginal mass functions and check whether (3.4.14) holds. We could get the marginal mass functions using (3.4.11) and (3.4.12) as we did in the previous example, but this is unnecessary; we already know them, because we know [see (3.3.7), rule P1] that X has the Poisson distribution with $\lambda = 5$ and Y has the Poisson distribution with $\lambda = 15$. So we have

(3.4.20) $$p_X(x) = \frac{5^x e^{-5}}{x!} \quad (x = 0, 1, 2, \ldots) \quad \text{and} \quad p_Y(y) = \frac{15^y e^{-15}}{y!} \quad (y = 0, 1, 2, \ldots)$$

Now we can check whether X and Y are independent. The answer is "yes" if and only if the function in (3.4.19) is the product of the two functions in (3.4.20). But you can see that it is not, because, when $x = 2$ and $y = 6$, for example,

$$\frac{5^2 e^{-5}}{2!} \cdot \frac{15^6 e^{-15}}{6!} \quad \text{is not equal to} \quad \frac{5^2 e^{-5}}{2!} \cdot \frac{10^{6-2} e^{-10}}{(6-2)!}.$$

But it is even easier to see in this example by noticing that the function in (3.4.19) is 0 if $y < x$, but the product of the two functions in (3.4.20) is nonzero for any nonnegative integers x and y. For example, the events $(X = 5)$ and $(Y = 4)$ have positive probabilities, but the event $(X = 5 \text{ and } Y = 4)$ has probability 0. The point is this:

> X and Y cannot be independent if there are values x and y for which $P(X = x)$ and $P(Y = y)$ are nonzero but $P(X = x \text{ and } Y = y) = 0$.

Even though it is unnecessary in this example, let us now go through the exercise of recovering the two mass functions in (3.4.20) from the joint mass function (3.4.19), as we did in Example (3.4.16). Exercise 1 at the end of this section asks you to do this for some other joint mass functions.

To find $p_X(x)$ from (3.4.19) we sum over all $y \geq x$:

$$\begin{aligned} p_X(x) &= \sum_{y=x}^{\infty} \frac{5^x e^{-5}}{x!} \cdot \frac{10^{y-x} e^{-10}}{(y-x)!} = \frac{5^x e^{-5}}{x!} \cdot \sum_{y=x}^{\infty} \frac{10^{y-x} e^{-10}}{(y-x)!} \\ &= \frac{5^x e^{-5}}{x!} \cdot \left(\frac{10^0 e^{-10}}{0!} + \frac{10^1 e^{-10}}{1!} + \frac{10^2 e^{-10}}{2!} + \cdots \right) = \frac{5^x e^{-5}}{x!}. \end{aligned}$$

The quantity in parentheses is 1 because it is the sum of all the Poisson probabilities for $\lambda = 10$.

To find $p_Y(y)$ from (3.4.19) we sum over all $x \leq y$:

$$p_Y(y) = \sum_{x=0}^{y} \frac{5^x e^{-5}}{x!} \cdot \frac{10^{y-x} e^{-10}}{(y-x)!} = e^{-15} \sum_{x=0}^{y} \frac{1}{x!} \cdot \frac{1}{(y-x)!} 5^x 10^{y-x}.$$

Now this last sum would be a binomial expansion if only we had the binomial coefficient $\binom{y}{x}$ instead of the two factorials in the denominator. So we multiply and divide by $y!$, and get

$$\frac{e^{-15}}{y!} \sum_{x=0}^{y} \frac{y!}{x!(y-x)!} 5^x 10^{y-x} = \frac{e^{-15}}{y!} (5 + 10)^y = \frac{15^y e^{-15}}{y!}. \quad \blacksquare$$

(3.4.21) **Conditional probabilities in a joint discrete distribution.** Given the joint mass function of X and Y, we can find conditional probabilities like $P(X = x \mid Y = y)$ and $P(Y = y \mid X = x)$, using the definition of conditional probability. For instance, in Example (3.4.1), where probabilities are given by table (3.4.4), suppose we learn that the experiment was done and $X = 1$. What are the conditional probabilities for the various values of Y, given $X = 1$? They are easy to see:

$$P(Y = 0 \mid X = 1) = \frac{P(Y = 0 \text{ and } X = 1)}{P(X = 1)} = \frac{1/8}{3/8} = \frac{1}{3},$$
$$P(Y = 1 \mid X = 1) = \frac{P(Y = 1 \text{ and } X = 1)}{P(X = 1)} = \frac{1/8}{3/8} = \frac{1}{3},$$

and, similarly, we see that $P(Y = 2 \mid X = 1) = \frac{1}{3}$, but that $P(Y = 3 \mid X = 1) = 0$.

As an exercise at this point, find the conditional probabilities of the four possible values of X, given that $Y = 0$. The answers are: $P(X = 0 \mid Y = 0) = 0$; $P(X = 1 \mid Y = 0) = \frac{1}{4}$; $P(X = 2 \mid Y = 0) = \frac{1}{2}$; and $P(X = 3 \mid Y = 0) = \frac{1}{4}$.

It is customary to use notation such as $p_X(x \mid Y = y)$, or $p_{X|Y}(x \mid y)$, or simply $p(x \mid y)$, for the conditional probability $P(X = x \mid Y = y)$. If we look at this as a function of x for some given fixed value of y, it turns out to be a probability mass function; this is a consequence of (2.2.13), which says that conditional probabilities given a fixed event satisfy all the axioms of unconditional probabilities. The formal definition is as follows.

(3.4.22) **Definition.** Let discrete random variables X and Y have the joint mass function $p(x, y)$ and marginal mass functions $p_X(x)$ and $p_Y(y)$.

If y is a number such that $p_Y(y)$ is positive, then the function of x defined by

(3.4.23)
$$p_X(x \mid Y = y) = \frac{p(x, y)}{p_Y(y)}$$

is a probability mass function, called the ***conditional mass function of X, given Y = y.*** Similarly, for a fixed x such that $p_X(x)$ is positive, the following function of y,

$$p_Y(y \mid X = x) = \frac{p(x, y)}{p_X(x)},$$

is the ***conditional mass function of Y, given X = x.***

The notation for conditional mass functions is cumbersome; whenever we can, we will omit the subscripts. For example, instead of $p_X(x \mid Y = y)$ we will write simply $p(x \mid y)$ if no confusion is possible.

(3.4.24) **Example.** In Example (3.4.16) we considered X and Y, the numbers of failures preceding the first and second successes in a Bernoulli process. The joint mass function of X and Y is given by (3.4.17), which says

$$p(x, y) = \begin{cases} p^2q^y & \text{if } x \text{ and } y \text{ are integers and } 0 \leq x \leq y, \\ 0 & \text{otherwise.} \end{cases}$$

Let us find the conditional mass function of X, given that $Y = 5$. We want

$$p(x|5) = P(X = x|Y = 5) = \frac{p(x, 5)}{p_Y(5)}$$

for all the possible values x of X. Now the numerator of this is p^2q^5 if x is one of the integers 0, 1, 2, 3, 4, 5. And we can get the denominator from the formula that was given at the beginning of Example (3.4.16) (actually we first saw it in Section 1.2):

$$P(Y = y) = (y + 1)p^2q^y \quad \text{for integers } y \geq 0.$$

For $y = 5$, this is $6p^2q^5$. Therefore, the conditional mass function is

$$p(x|5) = \begin{cases} \frac{1}{6} & \text{if } x = 0, 1, 2, 3, 4, 5, \\ 0 & \text{otherwise.} \end{cases}$$

That is, given that the second success came after six trials (five failures and one success in some order), the first success is conditionally equally likely to have occurred on any of the first six trials. ∎

(3.4.25) **Example.** Suppose we start with a fair die and a fair coin. We roll the die; whatever number shows, we toss the coin that many times. What is the distribution of the number of heads that appear? One way to find it is with the law of total probability (conditioning) from Section 2.3. To find the probability of, say, four heads, we can condition on the number that comes up on the die:

(3.4.26)
$$\begin{aligned} P(4 \text{ heads}) &= \sum_k P(4 \text{ heads}|\text{die shows } k) \cdot P(\text{die shows } k) \\ &= \sum_{k=4}^{6} \binom{k}{4}\left(\frac{1}{2}\right)^k \cdot \frac{1}{6} = \binom{4}{4} \cdot \left(\frac{1}{2}\right)^4 \cdot \frac{1}{6} + \binom{5}{4} \cdot \left(\frac{1}{2}\right)^5 \cdot \frac{1}{6} + \binom{6}{4} \cdot \left(\frac{1}{2}\right)^6 \cdot \frac{1}{6} \\ &= \frac{1}{6} \cdot \left(\frac{1}{16} + \frac{5}{32} + \frac{15}{64}\right) = \frac{1}{6} \cdot \frac{29}{64} = \frac{29}{384}. \end{aligned}$$

We can now see this calculation in the language of conditional and joint mass functions. Let Y be the number that shows on the die, and let X be the number of heads; we are looking for the mass function of X. What we know is the mass function of Y—possible values 1, 2, 3, 4, 5, 6, probabilities $\frac{1}{6}$ each—and the ***conditional*** mass function of X, given values of Y. In particular, given that

$Y = y$, X conditionally has the binomial distribution with $n = y$ and $p = \frac{1}{2}$. That is,

$$p_Y(y) = P(Y = y) = \tfrac{1}{6} \quad (y = 1, 2, 3, 4, 5, 6)$$

and

$$p(x|y) = P(X = x|Y = y) = \binom{y}{x}\left(\frac{1}{2}\right)^y \quad (x = 0, 1, \ldots, y).$$

Now we can use (3.4.23) to find the joint mass function:

$$\begin{aligned} p(x, y) &= P(X = x \text{ and } Y = y) \\ &= p_Y(y) \cdot p(x|y) \\ &= \tfrac{1}{6} \cdot \binom{y}{x}\left(\tfrac{1}{2}\right)^y \quad \text{for } y = 1, 2, \ldots, 6 \text{ and } x = 0, 1, \ldots, y. \end{aligned}$$

Finally, to find $p_X(x)$ for any given x, we add the values of $p(x, y)$ over all y from x up to 6:

(3.4.27)
$$p_X(x) = P(X = x) \sum_{y=x}^{6} \tfrac{1}{6} \cdot \binom{y}{x}\left(\tfrac{1}{2}\right)^y \quad (x = 0, 1, 2, \ldots, 6).$$

This is exactly what the conditioning argument in (3.4.26) did for $x = 4$.

Just in case you are curious to know what values of X are most likely, here is the mass function, calculated from (3.4.27):

x	0	1	2	3	4	5	6
p_X	$\frac{63}{384}$	$\frac{120}{384}$	$\frac{99}{384}$	$\frac{64}{384}$	$\frac{29}{384}$	$\frac{8}{384}$	$\frac{1}{384}$

The point of this example is that (3.4.23) simply repeats the definition of conditional probability. Like that definition, it can be used two ways: to find a conditional probability, or to find the probability of an intersection if a conditional probability is known. In the language of mass functions, we can use (3.4.23) to find a conditional mass function, or to find a joint mass function if a conditional mass function is known. ∎

EXERCISES FOR **SECTION 3.4**

1. For each of the following joint probability mass functions, make a table, find the two marginal mass functions, and decide whether X and Y are independent. (In every case the mass function is defined to be 0 at values of x and y not specified.)

   a. $p(x, y) = \frac{1}{36}$ if both x and y are in the set $\{1, 2, 3, 4, 5, 6\}$

   b. $p(x, y) = \dfrac{x^2 y}{84}$ for $x = 1, 2, 3$ and $y = 1, 2, 3$

   c. $p(x, y) = \frac{1}{14}$ if x is an integer between 0 and 3 (inclusive) and y is an integer between x and 4 (inclusive)

2. A fair coin is tossed four times; X is the number of heads that come up on the first three tosses, and Y is the number of heads that come up on tosses 2, 3, and 4.
   a. What is the distribution of X? Of Y?
   b. Make a table of the joint mass function of X and Y. [It has 16 entries.]
   c. Find the marginal mass functions and check that they agree with your answer to part a.
   d. Are X and Y independent?
   e. Find the conditional probability that $X = x$, given that $Y = 2$, for $x = 0, 1, 2$, and 3.

3. A hat contains four slips of paper, numbered 1, 2, 3, and 4. Two slips are drawn at random, one after the other, without replacement. X is the number on the first slip drawn and Y is the sum of the two numbers drawn. Make a table of the joint mass function of X and Y, find the marginal mass functions, and check whether X and Y are independent.

4. Suppose X and Y are independent, each with the following probability mass function:

x:	0	1	2	3
$p(x)$:	$\frac{1}{10}$	$\frac{2}{10}$	$\frac{3}{10}$	$\frac{4}{10}$

   Make a table of the joint mass function of X and Y and use it to find $P(X + Y = a)$ for $a = 0, 1, 2, 3, \ldots$.

5. Suppose X and Y are independent Poisson random variables, with parameters 3 and 4, respectively. Find the probability that $X + Y = 5$. [*Hint:* This event is the union of six disjoint events: ($X = 0$ and $Y = 5$), ($X = 1$ and $Y = 4$), etc. Use the binomial formula to find the sum of the six probabilities.]

6. In Exercise 3 above find the conditional mass function of Y, given $X = 2$.

7. The joint mass function of X and Y is $p(x, y) = (\lambda/2)^y e^{-\lambda}/x!(y - x)!$ when x and y are integers and $0 \le x \le y$.
   a. Find the marginal mass function of X. [*Hint:* Write $p(x, y)$ as

$$\frac{(\lambda/2)^x}{x!} \cdot e^{-\lambda} \cdot \frac{(\lambda/2)^{y-x}}{(y - x)!},$$

   and write out the sum from $y = x$ to ∞.]
   b. Find the marginal mass function of Y. [*Hint:* Multiply and divide $p(x, y)$ by $y!$, producing a binomial coefficient, and then sum from $x = 0$ to y. When the parts not involving x are factored out of the sum, the rest is a binomial sum.]
   c. Are X and Y independent?

8. Let N be a positive integer. Let X and Y be discrete random variables whose joint mass function is

$$p(x, y) = \frac{x^y e^{-x}}{N \cdot y!} \quad \text{for } x = 1, 2, 3, \ldots, N \text{ and } y = 0, 1, 2, 3, \ldots.$$

Find the marginal mass function of X.

In Supplementary Exercise 8 for Chapter 9 we will describe the true situation here: X is a random choice from the set $\{1, 2, 3, \ldots, N\}$ and, given $X = x$, the *conditional* distribution of Y is the Poisson distribution with $\lambda = x$.

9. You roll a fair die twice and let X be the number of 6's that appear. Then you turn to a Poisson process whose arrival rate is $\lambda = 3$ per minute, and watch it for $X + 1$ minutes, counting the arrivals. Y is the number of arrivals.
   a. What is the distribution of X? Give its name and also make a table of the mass function.
   b. What is the conditional distribution of Y, given $X = k$? Do this for $k = 0$, 1, 2; give the name of the distribution and the mass function.
   c. What is the unconditional mass function of Y? That is, find $P(Y = j)$ for $j = 0, 1, 2, \ldots$. [Condition on the value of X.]
   d. What is the joint mass function of X and Y? [You were finding this function while doing part c.]

10. Breakdowns occur in a factory according to a Poisson process, with an average of λ breakdowns per week. The breakdowns are classified as damaging or nondamaging, depending on whether they hold up production for more than 5 minutes. The fraction of breakdowns that are damaging is p; that is, p is the probability that a given breakdown is damaging. Let X be the number of breakdowns in a given week, and let Y be the number of damaging breakdowns in that same week. We know the distribution of X; it is Poisson with parameter λ; we want the distribution of Y.
    a. Given that $X = x$, assume that the x breakdowns represent independent trials; the conditional probability that $Y = y$ given $X = x$ is the probability that y of the x breakdowns are damaging. Write down this probability.
    b. Find the joint mass function $p(x, y) = P(X = x \text{ and } Y = y)$.
    c. For a fixed y, sum the mass function over all x ($= y, y + 1, y + 2, \ldots$) to get the unconditional mass function of Y.

11. Let X and Y be independent, each with the mass function $p(k) = pq^k$ ($k = 0$, 1, 2, . . .). (This is the geometric mass function with parameter p; as usual, $0 < p < 1$ and $q = 1 - p$.) Let Z be the smaller of X and Y. We want to find the mass function of Z.
    a. For $n = 0, 1, 2, \ldots$, find $P(X \geq n)$.
    b. For $n = 0, 1, 2, \ldots$, find $P(Z \geq n)$. [*Hint:* $Z \geq n$ if and only if both X and Y are greater than n.]
    c. Using the result of part b, find $P(Z = k)$.

ANSWERS

1. **a.** $p_X(x) = \frac{1}{6}$ for $x = 1, 2, 3, 4, 5, 6$; they are independent.
b. X and Y each have possible values 1, 2, 3; probabilities for X are $\frac{1}{14}, \frac{4}{14}, \frac{9}{14}$; they are independent.
c. Y has possible values 0, 1, 2, 3, 4 with probabilities $\frac{1}{14}, \frac{2}{14}, \frac{3}{14}, \frac{4}{14}, \frac{4}{14}$. They are not independent.
2. **a.** Each has the binomial distribution with $n = 3$ and $p = \frac{1}{2}$. **d.** No **e.** $0, \frac{1}{3}, \frac{1}{2}, \frac{1}{6}$
3. Possible values of X: 1, 2, 3, 4; marginal probabilities for X: $\frac{1}{4}$ each; possible values of Y: 3, 4, 5, 6, 7; marginal probabilities for Y: $\frac{1}{6}$ for 3, 4, 6, 7 and $\frac{2}{6}$ for 5; X and Y are not independent.
4. Possible values of $X + Y$: 0, 1, 2, 3, 4, 5, 6; probabilities: .01, .04, .10, .20, .25, .24, .16
5. $\dfrac{7^5 e^{-7}}{5!}$
6. Possible values 3, 5, 6; probabilities $\frac{1}{3}$ each
7. **a.** X is Poisson with parameter $\lambda/2$.
b. Y is Poisson with parameter λ. **c.** No
8. $p(x) = 1/N$ for $x = 1, 2, 3, \ldots, N$
9. **a.** X has the binomial distribution with $n = 2$ and $p = \frac{1}{6}$. The mass function is $P(X = k) = \binom{2}{k}\left(\frac{1}{6}\right)^k\left(\frac{5}{6}\right)^{2-k}$ for $k = 0, 1, 2$; that is, $p(0) = \frac{25}{36}$, $p(1) = \frac{10}{36}$, and $p(2) = \frac{1}{36}$.
b. Given $X = k$, Y has conditionally the Poisson distribution with parameter $\lambda(k + 1)$. The mass function is
$$p_Y(j|X = k) = \frac{(\lambda(k+1))^j e^{-\lambda(k+1)}}{j!}$$
for $k = 0, 1, 2$ and $j = 0, 1, 2, 3, \ldots$.
c. $P(Y = j) = \dfrac{1}{36j!}(25\lambda^j e^{-\lambda} + 10(2\lambda)^j e^{-2\lambda} + (3\lambda)^j e^{-3\lambda})$
d. $P(X = k \text{ and } Y = j) =$
$$\frac{(\lambda(k+1))^j e^{-\lambda(k+1)}}{j!} \cdot \binom{2}{k}\left(\frac{1}{6}\right)^k\left(\frac{5}{6}\right)^{2-k}$$
for $k = 0, 1, 2$ and $j = 0, 1, 2, \ldots$.
10. **a.** $\binom{x}{y} p^y q^{x-y}$ $(y = 0, 1, \ldots, x)$
b. $\dfrac{p^y}{y!} \cdot \lambda^x e^{-\lambda} \cdot \dfrac{q^{x-y}}{(x-y)!}$ $(x = 0, 1, 2, \ldots;$ $y = 0, 1, 2, \ldots, x)$. **c.** Poisson with parameter λp
11. **a.** q^n **b.** q^{2n} **c.** $q^{2n}(1 - q^2)$

3.5 THE JOINT DISTRIBUTION OF TWO ABSOLUTELY CONTINUOUS RANDOM VARIABLES

Now we turn to pairs of absolutely continuous random variables. Remember that the probability of any single possible value is 0 for an absolutely continuous random variable X. To find the probability that X is in a given interval, we integrate the density function of X over that interval.

Here the situation involves more theory and technical details, much of which we will omit, because the central results are very similar to those for discrete random variables. They are as follows. The individual density functions of X and Y by themselves are not sufficient to find probabilities for events involving both

X and Y; what is needed is a *joint density function* $f(x, y)$. We find probabilities from this function by integrating it over appropriate sets in the xy-plane. The individual density functions of X and Y, called *marginal density functions* in this context, can be found by integrating the joint density. And finally, X and Y are independent if and only if the joint density is the product of the marginal densities.

As with single random variables, the main feature of the absolutely continuous case is that a density function gives us probabilities not directly, but only when we integrate it. This contrasts with the discrete case, in which the joint mass function $p(x, y)$ is itself the probability that $X = x$ and $Y = y$.

In the case of a single absolutely continuous random variable X, $\int_I f(x)dx$ is the probability that X is in the interval I, a subset of the real line. In the case of a pair of random variables, X and Y, subsets of the line are replaced by subsets of the plane, because the set of possible values of a pair (X, Y) of random variables is a set of pairs (x, y) of numbers—that is, a set of points in the plane. We saw something like this at the end of Section 1.3, where we looked at examples of two-dimensional absolutely continuous probability models determined by density functions over regions of the plane.

The following example illustrates all the concepts of this section.

(3.5.1) **Example.** Consider this function:

(3.5.2)
$$f(x, y) = \begin{cases} 6xy^2 & \text{if } 0 \le x \le 1 \text{ and } 0 \le y \le 1, \\ 0 & \text{otherwise.} \end{cases}$$

Before going any further, stop and recall how to integrate such functions. Section A.9 of Appendix A is a review of such calculations; if the double integrations in this example are not familiar, it will be well worth the time you spend now reviewing this material from third-semester calculus. The first integral we choose to look at is the one over the whole square ($0 \le x \le 1$ and $0 \le y \le 1$) on which $f(x, y)$ is nonzero. It must equal 1 if $f(x, y)$ is to be a density, and it does:

$$\int_0^1 \int_0^1 6xy^2\,dx\,dy = 6\int_0^1 y^2 \int_0^1 x\,dx\,dy = 6\int_0^1 y^2 \cdot \tfrac{1}{2}\,dy = 3\int_0^1 y^2\,dy = 1.$$

(Of course, it must also be nonnegative to be a density; you can see that it is.)

Now suppose we are told that this function gives probabilities for X and Y in the way that densities do. What is the probability that X and Y are both less than $\frac{1}{2}$? This is the probability of the region shaded in Figure 3.5a on page 192, so it is the integral of the density over that region:

$$P\left(X \le \tfrac{1}{2} \text{ and } Y \le \tfrac{1}{2}\right) = \int_0^{1/2}\int_0^{1/2} 6xy^2\,dx\,dy = 6\int_0^{1/2} y^2 \int_0^{1/2} x\,dx\,dy$$
$$= 6\int_0^{1/2} y^2 \cdot \tfrac{1}{8}\,dy = \tfrac{3}{4}\int_0^{1/2} y^2\,dy = \tfrac{3}{4} \cdot \tfrac{1}{24} = \tfrac{1}{32}.$$

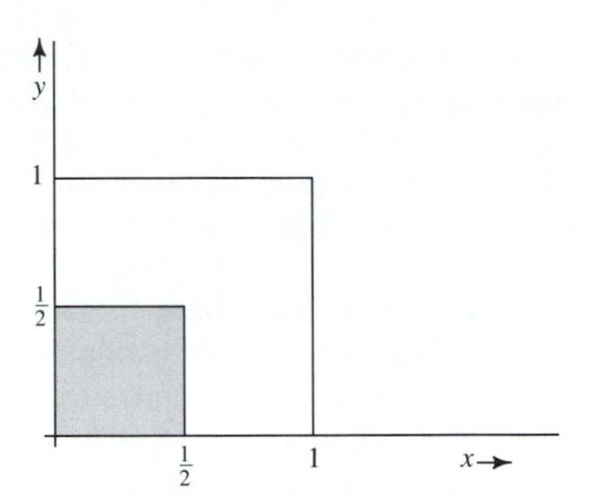

FIGURE 3.5a The set of (x, y) in $[0, 1] \times [0, 1]$ for which x and y are both less than $\frac{1}{2}$

Now what about the probability that X is less than Y? This is the integral of $f(x, y)$ over the set of points whose x-coordinates are less than their y-coordinates, which is the triangle shaded in Figure 3.5b.

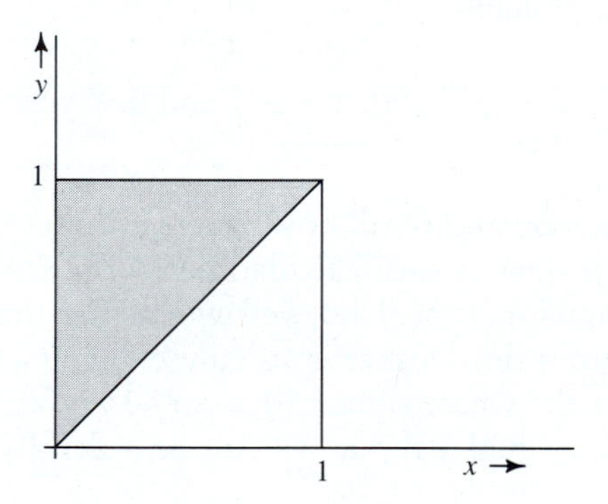

FIGURE 3.5b The set of (x, y) in $[0, 1] \times [0, 1]$ for which x is less than y

The integral is

$$P(X < Y) = \int_0^1 \int_0^y 6xy^2\, dx\, dy = 6\int_0^1 y^2 \int_0^y x\, dx\, dy$$
$$= 6\int_0^1 y^2 \cdot \tfrac{1}{2} y^2\, dy = 3\int_0^1 y^4\, dy = \tfrac{3}{5}.$$

As with any such double integral, we could get the same answer via a different route by reversing the order of integration:

$$P(X < Y) = \int_0^1 \int_x^1 6xy^2\,dy\,dx = 6\int_0^1 x \int_x^1 y^2\,dy\,dx$$
$$= 6\int_0^1 x \cdot \tfrac{1}{3}(1 - x^3)dx = 2\int_0^1 x\,dx - 2\int_0^1 x^4\,dx = 1 - \tfrac{2}{5} = \tfrac{3}{5}.$$

Finally, what about a probability for one of the random variables alone, say X? Let us find $P(X \leq a)$ for a given number a between 0 and 1:

$$P(X \leq a) = P(-\infty < X \leq a \text{ and } -\infty < Y < \infty) = P(0 \leq X \leq a \text{ and } 0 \leq Y \leq 1)$$
$$= \int_0^1 \int_0^a 6xy^2\,dx\,dy = 6\int_0^1 y^2 \int_0^a x\,dx\,dy$$
$$= 6\int_0^1 y^2 \cdot \tfrac{1}{2}a^2\,dy = 3a^2\int_0^1 y^2\,dy = a^2.$$

Notice that what we have found is the CDF of X:

$$F_X(x) = P(X \leq x) = \begin{cases} 0 & \text{if } x < 0, \\ x^2 & \text{if } 0 \leq x \leq 1, \\ 1 & \text{if } x > 1. \end{cases}$$

We can differentiate the CDF to get a density for X (except perhaps at $x = 0$ and $x = 1$; we can define the density as we please at those two points):

(3.5.3) $$f_X(x) = 2x \quad \text{for } 0 \leq x \leq 1 \quad \text{(and 0 otherwise).}$$

In a similar way, you can check that the CDF of Y is

$$F_Y(y) = P(Y \leq y) = \begin{cases} 0 & \text{if } y < 0, \\ y^3 & \text{if } 0 \leq y \leq 1, \\ 1 & \text{if } y > 1, \end{cases}$$

and then differentiate to get a density for Y (except perhaps at $y = 0$ and $y = 1$, but we are free to define a density as we like at a finite set of points):

(3.5.4) $$f_Y(y) = 3y^2 \quad \text{for } 0 \leq y \leq 1 \quad \text{(and 0 otherwise).}$$ ▮

Notice two things about the previous example before we go on to the next. First, it happens that the joint density $f(x, y)$ is the product of the individual (marginal) densities $f_X(x)$ and $f_Y(y)$. As we will see shortly, this is enough to guarantee that X and Y are independent random variables. It does not happen in the next example.

Second, the density we got for X, $2x$ for $0 \leq x \leq 1$, is the same function that we would have got if we had integrated the joint density over all values of y. For any x between 0 and 1,

$$\int_{-\infty}^{\infty} f(x, y)dy = \int_0^1 6xy^2\, dy = 6x\int_0^1 y^2\, dy = 6x \cdot \tfrac{1}{3} = 2x.$$

This happens also if we integrate $f(x, y)$ with respect to x—the resulting function of y is a density for Y. We will see later that this is no coincidence; densities for X and Y can always be found by integrating a joint density.

(3.5.5) **Example.** Suppose we know that probabilities for X and Y can be found by integrating the joint density function

$$f(x, y) = \begin{cases} xe^{-(x+xy)} & \text{if } x \geq 0 \text{ and } y \geq 0, \\ 0 & \text{otherwise.} \end{cases}$$

This is nonnegative for all x and y and it integrates to 1:

$$\int_0^{\infty}\int_0^{\infty} xe^{-(x+xy)}\, dy\, dx = \int_0^{\infty} xe^{-x}\int_0^{\infty} e^{-xy}\, dy\, dx = \int_0^{\infty} xe^{-x} \cdot \tfrac{1}{x}\, dx = \int_0^{\infty} e^{-x}\, dx = 1.$$

(Reversing the order of integration gives a different calculation but of course the same answer. Try it for practice; you will have to use integration by parts in the inner integral.)

The set of possible values of the pair (X, Y) is the set of (x, y) for which $x \geq 0$ and $y \geq 0$—that is, the positive quadrant in the plane. So the set of possible values of X is $(0, \infty)$, and so is the set of possible values of Y.

Let us find the CDF of X and then differentiate to find a density. We will find $F_X(a) = P(X \leq a)$ for a positive number a. (Because the integral involves an x, it is convenient to use some other letter for the variable in the CDF.)

$$\begin{aligned} F_X(a) &= P(X \leq a) = P(0 \leq X \leq a \text{ and } 0 \leq Y < \infty) = \int_0^a\int_0^{\infty} xe^{-(x+xy)}\, dy\, dx \\ &= \int_0^a xe^{-x}\int_0^{\infty} e^{-xy}\, dy\, dx = \int_0^a xe^{-x} \cdot \tfrac{1}{x}\, dx = \int_0^a e^{-x}\, dx = 1 - e^{-a}. \end{aligned}$$

Now we can switch back to x and say that

$$F_X(x) = \begin{cases} 0 & \text{if } x < 0, \\ 1 - e^{-x} & \text{if } x \geq 0. \end{cases}$$

You probably recognize this as the CDF of the exponential distribution with $\lambda = 1$; the corresponding density is the derivative of the CDF (except possibly at $x = 0$):

$$f_X(x) = e^{-x} \quad \text{for } x \geq 0 \quad \text{(and 0 otherwise).}$$

Again, we can check that we would get the same function of x by integrating $f(x, y)$ over all y. For any $x > 0$,

$$\int_{-\infty}^{\infty} f(x, y)dy = \int_{0}^{\infty} xe^{-(x+xy)}\,dy = xe^{-x}\int_{0}^{\infty} e^{-xy}\,dy = e^{-x}.$$

Now let us do the same thing for Y. First, we will find Y's CDF as a double integral and differentiate it to get a density; then we will check that we get the same result by integrating $f(x, y)$ with respect to x over all values.

If a is any possible value of Y—that is, if $a \geq 0$, then

$$\begin{aligned}F_Y(a) &= P(Y \leq a) = P(0 \leq X < \infty \text{ and } 0 \leq Y \leq a) = \int_0^{\infty}\int_0^{a} xe^{-(x+xy)}\,dy\,dx \\ &= \int_0^{\infty} xe^{-x}\int_0^{a} e^{-xy}\,dy\,dx = \int_0^{\infty} xe^{-x} \cdot \tfrac{1}{x}(1 - e^{-xa})\,dx \\ &= \int_0^{\infty} e^{-x}\,dx - \int_0^{\infty} e^{-x(1+a)}\,dx = 1 - \frac{1}{1+a}.\end{aligned}$$

So the CDF of Y is

$$F_Y(y) = \begin{cases} 0 & \text{if } y < 0, \\ 1 - \dfrac{1}{1+y} & \text{if } y \geq 0, \end{cases}$$

and the derivative gives us a density (except at $y = 0$, where we can define it as we please):

$$f_Y(y) = \frac{1}{(1+y)^2} \quad \text{for } y \geq 0.$$

Now we can get the same result by integrating the joint density over all values of x:

$$\int_{-\infty}^{\infty} f(x, y)dx = \int_0^{\infty} xe^{-(x+xy)}\,dx = \int_0^{\infty} xe^{-(1+y)x}\,dx;$$

integrating by parts with $u = x$ and $dv = e^{-(1+y)x}\,dx$ gives

$$\left[-\frac{x}{1+y}e^{-(1+y)x}\right]_{x=0}^{x=\infty} + \frac{1}{1+y}\int_0^{\infty} e^{-(1+y)x}\,dx = 0 + \frac{1}{(1+y)^2}.$$

As we mentioned before, this is no coincidence, but is another instance of Theorem (3.5.7). ∎

Notice that $f(x, y)$ does *not* equal $f(x) \cdot f(y)$ in the preceding example; the joint density of X and Y is not the product of the individual (marginal) densities. This happens, as we will state formally in Theorem (3.5.8), only in the special case in which X and Y are independent random variables.

The following definition summarizes the ideas of joint and marginal densities. The two theorems after it state the facts mentioned in the examples above: that densities for X and for Y can always be found by integrating the joint density, and that X and Y are independent if and only if the product of densities for X and Y is a joint density. We will discuss the proofs of the two theorems at the end of the section.

(3.5.6) Definition. Random variables X and Y are ***(jointly) absolutely continuous*** if there is a function $f(x, y)$, defined and nonnegative for all pairs (x, y) of real numbers, such that for any Borel set B in the plane,

$$P((X, Y) \in B) = \iint_B f(x, y)dx\,dy$$

Such a function is called a ***joint (probability) density function*** for the pair (X, Y). Its integral over the plane, $\int_{-\infty}^{\infty}\int_{-\infty}^{\infty} f(x, y)\,dx\,dy$, equals 1.

(3.5.7) Theorem. If X and Y are jointly absolutely continuous random variables, with joint density $f(x, y)$, then X and Y themselves are absolutely continuous. Densities for them, denoted by $f_X(x)$ and $f_Y(y)$, are called ***marginal density functions,*** and can be found by integrating $f(x, y)$ as follows:

$$f_X(x) = \int_{-\infty}^{\infty} f(x, y)dy \quad \text{and} \quad f_Y(y) = \int_{-\infty}^{\infty} f(x, y)dx.$$

The proof is at (3.5.22) at the end of the section.

(3.5.8) Theorem. Suppose X and Y are jointly absolutely continuous random variables. Then X and Y are independent if and only if, given any two densities for X and Y, their product is a joint density for the pair. Equivalent to this is the statement that any joint density $f(x, y)$ for the pair can be written as a product of two functions that are densities for X and for Y (except possibly on a set of points whose probability is 0).

We will give a partial proof of this theorem in (3.5.24). The theorem says that in some sense

(3.5.9) X and Y are independent if and only if $f(x, y) = f_X(x) \cdot f_Y(y)$.

But this is a little slipshod, because (3.5.9) sounds as though X and Y are independent if and only if *the* joint density is the product of *the* marginal densities. As we know, there is no one density for a given random variable—it can be changed at single points without changing the distribution. If the three densities in (3.5.9) are wantonly changed at some points, then the product formula might not hold at those points.

Look back at (3.5.3) and (3.5.4), for example. We could just as easily have come up with the following marginal densities, which differ at $x = 1$ from the ones we wrote down:

$$f_X(x) = 2x \quad \text{for } 0 < x < 1 \quad \text{and} \quad f_Y(y) = 3y^2 \quad \text{for } 0 < y < 1.$$

The product of these two functions is not equal to the joint density (3.5.2), but it differs only at the point $(x, y) = (1, 1)$, and the probability $P(X = 1 \text{ and } Y = 1)$ is 0.

The truth of the matter is expressed in the following statement: Given any function $f(x, y)$ that is a joint density for the pair (X, Y), and any functions $f_1(x)$ and $f_2(y)$ that are densities for X and for Y, the random variables are independent if and only if $f(x, y) = f_1(x) \cdot f_2(y)$ for all pairs (x, y) except those in a set whose probability is 0.

But these technicalities need not concern us much; the product rule (3.5.9), though true only in the sense described in the preceding paragraph, is a good way to think informally about the independence of absolutely continuous random variables.

Another comment should be made here. Although it is usually better to speak of a function as being *a* density rather than *the* density for a random variable, when we refer to the density $f_X(x)$ that we get by integrating some given joint density $f(x, y)$, we will usually call it *the* marginal density that comes from that particular joint density. There are still many other densities for X, just as there are many joint densities for X and Y; but this one is *the* one that comes from the given $f(x, y)$.

(3.5.10) **Example.** Let the following function be a joint density for X and Y:

$$f(x, y) = \begin{cases} 4(x + y^2) & \text{if } x \text{ and } y \text{ are positive and } x + y < 1, \\ 0 & \text{otherwise.} \end{cases}$$

We will find the marginal densities for X and Y and determine whether they are independent.

The first thing to notice is that the set of possible values of the pair (X, Y) is a triangle; $f(x, y)$ is positive if and only if the pair (x, y) is in the positive quadrant but below the line $x + y = 1$. The set of possible values is the triangle shaded in Figure 3.5c. Thus, for any given possible value x of X, the possible values of Y range from 0 to $1 - x$. When we want to find $f_X(x)$ by integrating over all y, the limits of integration depend on x.

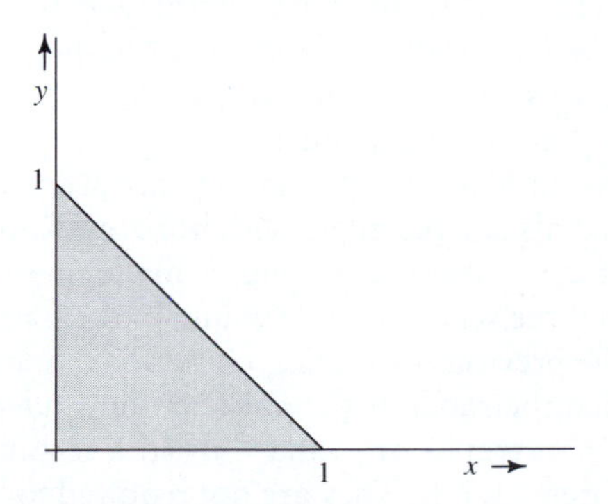

FIGURE 3.5c The set of (x, y) for which x and y are positive and $x + y < 1$

For $0 < x < 1$,

$$f_X(x) = \int_{-\infty}^{\infty} f(x, y)dy = \int_0^{1-x} 4(x + y^2)dy = 4x\int_0^{1-x} dy + 4\int_0^{1-x} y^2\, dy$$
$$= 4x(1 - x) + \tfrac{4}{3}(1 - x)^3 = \tfrac{4}{3}(1 - x^3).$$

So a density for X is

$$f_X(x) = \tfrac{4}{3}(1 - x^3) \quad \text{for } 0 < x < 1 \quad \text{(and 0 otherwise).}$$

Similarly, to find a density for Y we integrate over all x, and the limits depend on y. For $0 < y < 1$,

$$f_Y(y) = \int_{-\infty}^{\infty} f(x, y)dx = \int_0^{1-y} 4(x + y^2)dx = 4\int_0^{1-y} x\, dx + 4y^2\int_0^{1-y} dx$$
$$= 2(1 - y)^2 + 4y^2(1 - y) = 2(1 - 2y + 3y^2 - 2y^3);$$

so a density for Y is

$$f_Y(y) = 2(1 - 2y + 3y^2 - 2y^3) \quad \text{for } 0 < y < 1 \quad \text{(and 0 otherwise).}$$

We see that X and Y are not independent, because the product $f_X(x) \cdot f_Y(y)$ differs from the joint density $f(x, y)$ at all but perhaps a few points. ∎

(3.5.11) **Independence requires a Cartesian product.** In the previous example, we could have seen that X and Y are not independent even without finding the marginal densities, as follows. Look at the set of possible values of (X, Y) in Figure 3.5c, and consider a point (x, y) that is in the square, but above the triangle of possible values—say $(x, y) = \left(\frac{3}{4}, \frac{3}{4}\right)$. You can see that $x = \frac{3}{4}$ is a possible value of X, and $y = \frac{3}{4}$ is a possible value of Y, but $(x, y) = \left(\frac{3}{4}, \frac{3}{4}\right)$ is not a possible value of the pair. In other words, we can see, even without computing the marginal densities, that $f_X\left(\frac{3}{4}\right)$ and $f_Y\left(\frac{3}{4}\right)$ are both positive, but $f\left(\frac{3}{4}, \frac{3}{4}\right)$ is 0.

In general, if $f(x, y) = f_X(x) \cdot f_Y(y)$, then $f(x, y)$ must be nonzero whenever $f_X(x)$ and $f_Y(y)$ are nonzero; that is, ***the set of points where the joint density is positive must be the Cartesian product of the sets of points where the marginal densities are positive.*** If the set of (x, y) for which $f(x, y) > 0$ is not a Cartesian product, then X and Y cannot be independent.

And it is usually easy to identify sets that are not Cartesian products. The Cartesian product of intervals is a rectangle with horizontal and vertical sides; in general, the boundary of any Cartesian product is made up of horizontal and/or vertical line segments. So if the set of possible values of the pair (X, Y) is bounded by a sloping line (as in the previous example), or a curved line (perhaps the set is a circle or an ellipse), then the random variables are not independent.

(The preceding two paragraphs are valid only in a technical sense like that described following Theorem (3.5.8). They are not required to be true for all pairs

(x, y), but only for those outside some set whose probability is 0. Again, we will not trouble ourselves with these matters.)

As we remarked, the observation that independence requires a Cartesian product often makes it easy to see that two random variables are *not* independent. But do not fall into the trap of thinking that X and Y *are* independent just because their set of possible values is a Cartesian product. Example (3.5.5) above shows that they need not be independent. If you like slogans, you might want to remember that

> Independence requires a Cartesian product, but a Cartesian product does not guarantee independence.

(3.5.12) Example. Let X and Y be independent random variables, each having the exponential distribution with parameter λ. Recall that this is the distribution whose density is $\lambda e^{-\lambda x}$ for $x > 0$. Let us find the probability that $X + Y \leq a$ for some positive number a; this will give us the CDF of the random variable $Z = X + Y$. We can then differentiate that to get a density for Z.

First, we write down the joint density. Because we are assuming that X and Y are independent, it is the product of the marginal densities. They are given by $f_X(x) = \lambda e^{-\lambda x}$ and $f_Y(y) = \lambda e^{-\lambda y}$ when x and y are positive, so

$$f(x, y) = f_X(x) \cdot f_Y(y) = \begin{cases} \lambda^2 e^{-\lambda(x+y)} & \text{if } x > 0 \text{ and } y > 0, \\ 0 & \text{otherwise.} \end{cases}$$

The set of possible values of the pair (X, Y) is the positive quadrant in the plane. Now to find $P(X + Y \leq a)$ we need to integrate the density over the set of points (x, y) in the positive quadrant for which $x + y \leq a$; this is the set of points on or below the line $x + y = a$. If a is a positive number, it is the shaded triangle in Figure 3.5d. (If a is negative, then there are no points in the positive quadrant with $x + y \leq a$.)

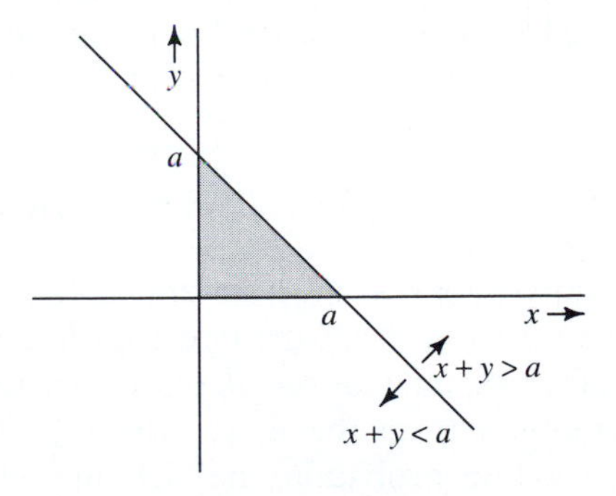

FIGURE 3.5d The line $x + y = a$ divides the plane into two regions; below the line $x + y < a$, while above it $x + y > a$. If a is positive then the positive quadrant contains points in the region where $x + y < a$.

The integration goes as follows:

$$P(X + Y \le a) = \int_0^a \int_0^{a-y} \lambda^2 e^{-\lambda(x+y)}\, dx\, dy = \int_0^a \lambda e^{-\lambda y} \int_0^{a-y} \lambda e^{-\lambda x}\, dx\, dy$$

$$= \int_0^a \lambda e^{-\lambda y} \cdot \left(1 - e^{-\lambda(a-y)}\right) dy = \int_0^a \lambda e^{-\lambda y}\, dy - \lambda e^{-\lambda a} \int_0^a dy$$

$$= 1 - e^{-\lambda a} - \lambda a e^{-\lambda a}.$$

That is, the CDF of $Z = X + Y$ is

$$F_Z(z) = \begin{cases} 0 & \text{if } z \le 0, \\ 1 - e^{-\lambda z} - \lambda z e^{-\lambda z} & \text{if } z > 0. \end{cases}$$

Differentiating this gives us a density for Z:

$$f_Z(z) = \lambda^2 z e^{-\lambda z} \quad \text{for } z > 0 \quad \text{(and 0 otherwise).}$$ ∎

The main point of the preceding example is to illustrate how to find probabilities by integrating a joint density; but the result is also an important one in the theory of Poisson processes. Remember that in a Poisson process with arrival rate λ, the waiting times between successive arrivals are independent random variables having the exponential distribution with parameter λ. Thus the random variable $Z = X + Y$ in the preceding example represents the time of the *second* arrival in the process. The distribution of Z is one of the Erlang, or gamma, distributions that we will encounter in Chapter 8.

(3.5.13) Example: The uniform distribution on a disk, revisited. We return to Example (1.3.13), which modeled the choice of a point at random from the unit disk—the set of points (x, y) in the plane for which $x^2 + y^2 \le 1$. In that example, the disk is the outcome set Ω, events are Borel subsets of Ω, and the probability of an event is its area divided by π. This probability is the same as the integral of the function $f(x, y) = 1/\pi$ over the event, so in that example probabilities are found by integrating the density

$$f(x, y) = \tfrac{1}{\pi} \quad \text{if } x^2 + y^2 \le 1 \quad \text{(and 0 otherwise).}$$

Now we are in a position to look at this function as the joint density of two random variables, X and Y, that represent the two coordinates of the randomly chosen point. We can ask, for example, for the marginal density of X. (Guess first. If a point is chosen at random from the disk, how will its x-coordinate be distributed? That is, where will its probability be high and where will it be low? Your intuition is probably correct here.) We can also ask for the marginal density of Y, and about the independence of X and Y.

Notice first that the set of possible values of X and Y is not bounded by horizontal and vertical straight lines, so we know X and Y are not independent even without finding their densities. This is what we talked about in (3.5.11).

To find the marginal density of X, we identify the set of possible values of X, pick a typical possible value x, identify the set of y's at which $f(x, y)$ is positive (for the given x), and integrate $f(x, y)$ over that set of y's. This is easier than it sounds.

> The set of possible values of X is $[-1, 1]$.
>
> For any x in this set, the y's for which $f(x, y) > 0$ are the y's between $-\sqrt{1 - x^2}$ and $\sqrt{1 - x^2}$.
>
> The integral of $f(x, y) = 1/\pi$ from $y = -\sqrt{1 - x^2}$ to $\sqrt{1 - x^2}$ is just $\frac{2}{\pi} \cdot \sqrt{1 - x^2}$.

So

$$f_X(x) = \frac{2}{\pi} \cdot \sqrt{1 - x^2} \quad \text{for } -1 \le x \le 1 \quad \text{(and 0 otherwise).}$$

You can see that this density is highest at $x = 0$ and lowest at $\pm\ 1$; so the x-coordinate of a random point in the unit disk is more likely to be near 0 than near 1 or -1. Of course, Y has exactly the same density as X; this model has x–y symmetry. ∎

(3.5.14) **Example: Jointly normal random variables.** If X and Y are independent standard normal random variables, their joint density is easy to write down; it is just the product of the marginal densities:

$$f(x, y) = \frac{1}{2\pi} e^{-(x^2+y^2)/2} \quad \text{for } x \in \mathbb{R} \text{ and } y \in \mathbb{R}.$$

But there are other joint densities for which the marginals are both the standard normal density. Here is a class of them, one for every number ρ (Greek rho) in the interval $(-1, 1)$:

$$f(x, y) = \frac{1}{2\pi\sqrt{1 - \rho^2}} e^{-(x^2+y^2-2\rho xy)/2(1-\rho^2)} \quad \text{for } x \in \mathbb{R} \text{ and } y \in \mathbb{R}.$$

You can see that if $\rho = 0$ we get the previous density, for which X and Y are independent. To see that the marginal densities are standard normal even when ρ is not 0, let us look at the marginal density of Y. To find it we need to integrate with respect to x from $-\infty$ to ∞. A little judicious rewriting of the exponent in the joint density, followed by the right change of variables, does the trick. The exponent equals

$$-\frac{1}{2(1-\rho^2)}(x^2 - 2\rho xy + \rho^2 y^2 + (1-\rho^2)y^2)$$
$$= -\frac{1}{2(1-\rho^2)}(x^2 - 2\rho xy + \rho^2 y^2) - \frac{y^2}{2}$$
$$= -\frac{(x-\rho y)^2}{2(1-\rho^2)} - \frac{y^2}{2}.$$

Using this, we see that

$$f_Y(y) = \int_{-\infty}^{\infty} f(x, y)dx = \int_{-\infty}^{\infty} \frac{1}{2\pi\sqrt{1-\rho^2}} e^{-(x-\rho y)^2/2(1-\rho^2)} \cdot e^{-y^2/2}\, dx$$
$$= \frac{1}{\sqrt{2\pi}} e^{-y^2/2} \cdot \int_{-\infty}^{\infty} \frac{1}{\sqrt{2\pi}\sqrt{1-\rho^2}} e^{-(x-\rho y)^2/2(1-\rho^2)}\, dx.$$

Now to compute this integral we make the substitution $z = (x - \rho y)/\sqrt{1-\rho^2}$ (remember that y is a constant in the integral). The integrand becomes the standard normal density, the integral is 1, and the result is that Y has the standard normal density.

We will return to this important joint density, and the more general class of *bivariate normal densities,* in Example (5.4.10). ∎

(3.5.15) **Conditional densities for jointly absolutely continuous random variables.** In the previous section on discrete joint distributions, given a joint mass function $p(x, y) = P(X = x \text{ and } Y = y)$ and a value y for which $p_Y(y) > 0$, we defined the conditional mass function of X, given $Y = y$, by $p(x|y) = p(x, y)/p_Y(y)$. We could do this because the denominator is nonzero.

The analogous thing to do with density functions is to define, for any value y at which the density $f_Y(y) > 0$, the conditional density of X, given $Y = y$, by

(3.5.16)
$$f(x|y) = \frac{f(x, y)}{f_Y(y)}.$$

This is indeed what we do; the difference is that it is not a conditional probability at all. Moreover, there are no conditional probabilities of anything, given $Y = y$, because the event "$Y = y$" has probability 0 for any y when Y is absolutely continuous.

Nevertheless, the function defined in (3.5.16) turns out to be useful, and it can be used as though it were something conditional. There are versions of the law of total probability and of Bayes's theorem (from Section 2.3) that involve this function. However, we will need to learn about expected values before we can understand them fully. The subject will be covered in Chapter 9.

(3.5.17) Independence and CDFs. In this and the previous section we have seen that two discrete random variables are independent if and only if their joint mass function is the product of their marginal mass functions; and we have seen the same fact for absolutely continuous random variables, with densities in place of mass functions. There is also a factorization theorem like this for CDFs, which works for all random variables—discrete, absolutely continuous, or anything.

The ***joint cumulative distribution function*** (joint CDF) of two random variables X and Y is the function F of two variables defined by

(3.5.18)
$$F(x, y) = P(X \leq x \text{ and } Y \leq y).$$

This function is sometimes useful for finding probabilities, but not quite as useful for that purpose as in the case of a single random variable. We bring it up here only because of its role as a unifying device; we can state a condition for independence that is the same for all random variables, as follows.

(3.5.19) **Theorem.** Let X and Y be random variables; let $F(x, y)$ be their joint CDF and let $F_X(x)$ and $F_Y(y)$ be their marginal CDFs. Then X and Y are independent if and only if

(3.5.20)
$$F(x, y) = F_X(x) \cdot F_Y(y) \quad \text{for all } x \text{ and } y.$$

For a hint of the use of this theorem, do Exercise 1 at the end of the section without thinking about CDFs; then find the joint CDF of X and Y and go through the problem again, making use of (3.5.20).

Partial proof. As with Theorem (3.5.8) on independence and densities, one half of the proof of this is easy and the other half is easy to believe but needs measure theory to make it rigorous. The easy half is that if X and Y are independent, then (3.5.20) holds. To see this, notice that if X and Y are independent, then by definition

(3.5.21)
$$P(X \in A \text{ and } Y \in B) = P(X \in A) \cdot P(Y \in B) \quad \text{for any Borel sets } A \text{ and } B.$$

In particular, (3.5.21) holds for the Borel sets $A = (-\infty, x]$ and $B = (-\infty, y]$. That is,

$$P(X \leq x \text{ and } Y \leq y) = P(X \leq x) \cdot P(Y \leq y) \quad \text{for any numbers } x \text{ and } y.$$

This is exactly what (3.5.20) says.

As for the other half, the proof has to show that if (3.5.20) holds, then so does (3.5.21). As we noted above, (3.5.20) is just the special case of (3.5.21) where the Borel sets A and B are the intervals $(-\infty, x]$ and $(-\infty, y]$. It is not hard to check that this implies the case where A and B are any intervals at all. To show that it follows for any Borel sets is the step that requires a measure-theoretic argument, which we have to omit here. ▯

We close the section with the proofs of the two theorems on joint and marginal densities. As with the previous theorem, we have to omit the measure-theoretic part of the proof of Theorem (3.5.8) on independence.

(3.5.22) **Proof of Theorem (3.5.7) on marginal densities.** We need to show that the following function of x,

$$g(x) = \int_{-\infty}^{\infty} f(x, y)dy,$$

is a density for X. It would be enough to show that integrating $g(x)$ will give the CDF of X; that is, that for any number a,

$$P(X \leq a) = \int_{-\infty}^{a} g(x)dx, \tag{3.5.23}$$

because then it would follow that we could find the probability that X is in any given integral by integrating $g(x)$ over that interval. But by the definition of $g(x)$, the right side of (3.5.23) is just

$$\int_{-\infty}^{a}\int_{-\infty}^{\infty} f(x, y)dy\,dx,$$

which is just the integral of the joint density over the set of points whose x-coordinates are less than or equal to a. This is indeed $P(X \leq a)$, so (3.5.23) is proved. □

(3.5.24) **Partial proof of Theorem (3.5.8) on independence.** This theorem says two things. First, if the joint density of X and Y factors as the product of the marginal densities, then X and Y are independent. Second, if X and Y are independent, then the joint density factors in that way. The first part is simple to prove. Because of the technical difficulties we talked about earlier, we would need measure theory to prove the second part rigorously; even so, we will be able to see why it should be true.

First, suppose the joint density factors. To show X and Y are independent, we need to show that $P(X \in I \text{ and } Y \in J) = P(X \in I) \cdot P(Y \in J)$ for any two intervals I and J. But the factoring of the density is just what makes this work. For suppose $I = [a, b]$ and $J = [c, d]$. Then

$$\begin{aligned}
P(X \in I \text{ and } Y \in J) &= \int_a^b\int_c^d f(x, y)dy\,dx = \int_a^b\int_c^d f_X(x) \cdot f_Y(y)dy\,dx \\
&= \int_a^b f_X(x) \int_c^d f_Y(y)dy\,dx = \int_a^b f_X(x) \cdot P(Y \in J)dx \\
&= P(Y \in J) \int_a^b f_X(x)dx \\
&= P(Y \in J) \cdot P(X \in I).
\end{aligned}$$

To prove the second part, we suppose X and Y are independent and show that $f_X(x) \cdot f_Y(y)$ is a function of x and y that works as a joint density—that is, a function we could integrate to find probabilities of events involving both X and Y. A measure-theoretic argument shows that it is sufficient to do this for "rectangular" events—that is, events of the form ($X \in I$ and $Y \in J$) for intervals I and J. That is, if $I = [a, b]$ and $J = [c, d]$, it is enough to show that

$$P(X \in I \text{ and } Y \in J) = \int_a^b \int_c^d f_X(x) \cdot f_Y(y)dy\, dx. \tag{3.5.25}$$

Notice that in the first part of the proof we were assuming something about the right side of (3.5.25)—namely, that the integrand is the joint density. In this part we are trying to *prove* that; what we are assuming is that X and Y are independent, so the left side factors:

$$P(X \in I \text{ and } Y \in J) = P(X \in I) \cdot P(Y \in J) = \int_a^b f_X(x)dx \cdot \int_c^d f_Y(y)dy.$$

And this product of integrals is the same as the double integral on the right side of (3.5.25). ▯

EXERCISES FOR **SECTION 3.5**

1. Let X and Y have the joint density

$$f(x, y) = \tfrac{3}{32}x^2y^3 \quad \text{for } 0 < x < 2 \text{ and } 0 < y < 2 \quad \text{(and 0 otherwise).}$$

a. Find the probability that X and Y are both less than 1.
b. Find the probability that X is greater than $\frac{1}{2}$.
c. Find $P(Y \le a)$ for a given number a between 0 and 2.
d. Find a density for Y in two ways:
  i. By differentiating the CDF of Y (which you found in part c)
  ii. By integrating the joint density over all values of x
e. Find a density for X.
f. Are X and Y independent random variables?

2. Let X and Y have the joint density

$$f(x, y) = cx^2y \quad \text{for } 0 < x < 2 \text{ and } 0 < y < 2 - x \quad \text{(and 0 otherwise),}$$

where c is the appropriate constant.
a. Draw a picture of the set of possible values of the pair (X, Y).
b. Show that c must equal $\frac{15}{8}$ if $f(x, y)$ is to integrate to 1 over the set of possible values.
c. Show that X and Y are not independent before finding their marginal densities.
d. Find a density for X by integrating $f(x, y)$. [For any x between 0 and 2, you

must integrate over all values of y from 0 to $2 - x$, because those are the y's for which $f(x, y)$ is positive.]

e. For a given possible value y of Y, what is the set of x for which the joint density is positive?

f. Find a density for Y by integrating $f(x, y)$ over the values of x found in part e.

3. Let X and Y be the coordinates of a point that has the uniform distribution on the triangle of Exercise 2. That is, the joint density of X and Y is a constant: $f(x, y) = c$ for $0 < x < 2$ and $0 < y < 2 - x$ (and 0 otherwise).
   a. What is the value of c?
   b. What is the probability that X and Y are both less than 1?
   c. Find marginal densities for X and Y.

4. Let X and Y be independent, with $f_X(x) = 2x$ if $0 < x < 1$ (and 0 otherwise) and $f_Y(y) = 2/y^3$ if $y > 1$ (and 0 otherwise).
   a. Draw a picture of the set of possible values of the pair (X, Y).
   b. Write down the joint density.
   c. Find $P(X + Y \leq 2)$. [First identify the set of points on the picture, then integrate the joint density over that set.]

5. Let X and Y jointly have the uniform density on the quarter-disk of points (x, y) in the positive quadrant for which $x^2 + y^2 < 1$.
   a. Draw a picture of the set of possible values of the pair (X, Y). Are X and Y independent?
   b. The joint density of X and Y is a constant c on the quarter-disk. What is the value of c?
   c. Find a marginal density for X.
   d. Find the probability that X is less than Y. [You should be able to do it without calculations, by identifying the appropriate region and using the fact that probabilities are areas times c.]

6. Let X and Y be as in Example (3.5.12), independent, each having the exponential distribution with parameter λ, so that a joint density is $f(x, y) = \lambda^2 e^{-\lambda(x+y)}$ when $x > 0$ and $y > 0$ (and 0 otherwise). In Example (3.5.12) we found the CDF of $Z = X + Y$ and then differentiated to find a density for Z. Do the same for the following two random variables:
   a. $U = \max(X, Y)$, the larger of X and Y [*Hint:* For any positive a, $U \leq a$ if and only if both $X \leq a$ and $Y \leq a$.]
   b. $V = \min(X, Y)$, the smaller of X and Y [*Hint:* For a given positive a, find $P(V > a)$ and from it get $P(V \leq a)$.]

7. Let X and Y be independent random variables, each with the standard normal density

$$\varphi(x) = \frac{1}{\sqrt{2\pi}} e^{-x^2/2} \quad (x \in \mathbb{R}).$$

a. Write down the joint density of X and Y.
b. For a given positive number a, find the integral of this density over the disk (the interior of the circle) circle of radius a centered at the origin. [What you are finding is $P(R \le a)$ where R is the random variable $\sqrt{X^2 + Y^2}$. To do it you change to polar coordinates; see (A.9.2) in Appendix A.]
c. Differentiate the CDF of R to find a density.

8. Let X and Y be independent, each with the uniform distribution on the interval $[0, 1]$. In the following calculations you do not need to do any integrals, because the set of possible values is a square in the plane and the joint density equals 1 on that square, and so integrals are just areas.
a. Write down their joint density.
b. Find $P(2X \le Y)$.
c. For a number a between 0 and 1, find the probability that the larger of X and Y is less than or equal to a. This is the CDF of $M = \max(X, Y)$. Differentiate it to get a density for M.
d. Find $P(X + Y \le a)$ for a number a between 0 and 1. [This and the next part were done in Supplementary Exercise 4 of Chapter 1.]
e. Find $P(X + Y \le a)$ for a number a between 1 and 2.
f. Combine the results of parts d and e to get the CDF of $Z = X + Y$. Differentiate to get a density for Z.

9. Suppose the joint density function of X and Y is $f(x, y) = 27y^2e^{-3x}/x^2$ for $0 < x < \infty$ and $0 < y < x$.
a. Find the marginal density for X.
b. Are X and Y independent? Why or why not?

10. Let X and Y be independent random variables, each with the uniform distribution on $[0, 1]$. Consider the random polynomial defined by $p(z) = z^2 - 2Xz + Y^2$.
a. What is the probability that $p(z)$ has two different real roots?
b. What is the probability that $p(z)$ has exactly one real root?
c. Now let X and Y be independent with any absolutely continuous distribution on any subinterval of $[0, \infty)$. Show that the answers to parts a and b do not change.

11. Let X and Y be independent random variables with the same distribution; that distribution is not assumed to be absolutely continuous or discrete. Show that $P(X < Y) \le \frac{1}{2} \le P(X \le Y)$. [*Hint:* Consider the three quantities $P(X < Y)$, $P(X = Y)$, and $P(X > Y)$.] This result has a surprising corollary, given in Exercise 18 of Section 9.2.

12. **Why do I choose the wrong line so often?** You and a friend are shopping at a grocery store, each with your own basket. You finish at the same time and you choose two checkout counters; there is one person ahead of each of you at these counters. We want to find the probability that you will have to wait three times as long as your friend, or more, before you can begin checking out. To do this let X be the time you wait before being served and Y the time your friend waits. Assume that X and Y are independent random variables having the exponential distribution with parameter λ, as in Exercise 6, and find $P(X \geq 3Y)$.

The answer is $\frac{1}{4}$. Notice that the probability that your friend has to wait three times as long as you, or more, is also $\frac{1}{4}$. So there is an even chance that one of you will have a waiting time that is at least three times as long as the other's. If you do the problem again with "twice as long" instead of "three times," you will see that there is a $\frac{2}{3}$ chance that one of you has to wait twice as long as the other. This may help to make it plausible that it really is fairly easy to "choose the wrong line."

Banks in recent years have been alleviating this problem by having only one line; the customer at the head of the line goes to the first available teller. Grocery stores apparently do not have the space to use this scheme. See Exercise 13 in Section 8.3 for a comparison of the two schemes.

It should also be noted that in modeling real problems of this type, involving waiting lines for service (the subject is called *queueing theory*), the exponential distribution is often not adequate for service times. One sometimes gets a better model by using gamma distributions, which are discussed in Chapter 8, or other distributions.

This problem, and some others relating to the common perception that one's luck is consistently bad, are discussed by William Feller in Volume 2 of *An Introduction to Probability Theory and Its Applications,* Section I.5.

ANSWERS

1. **a.** $\frac{1}{128}$ **b.** $\frac{63}{64}$ **c.** $a^4/16$ **d.** $f_Y(y) = y^3/4$ for $0 < y < 2$ **f.** Yes

2. **a.** The triangle with vertices at (0, 0), (0, 2), and (2, 0) **d.** $\frac{15}{16}x^2(2 - x)^2$ for $0 < x < 2$ **e.** The interval $(0, 2 - y)$

3. **a.** $\frac{1}{2}$ **b.** $\frac{1}{2}$ **c.** $\frac{1}{2}(2 - x)$ for $0 < x < 2$; $\frac{1}{2}(2 - y)$ for $0 < y < 2$

4. **a.** The infinite rectangle extending upward from (0, 1) and (1, 1) **b.** $4x/y^3$ for $0 < x < 1$ and $y > 1$ **c.** $2 \ln 2 - 1$

5. **a.** No; the quarter-disk is not a Cartesian product. **b.** $\frac{4}{\pi}$ **c.** $\frac{4}{\pi}\sqrt{1 - x^2}$ for $0 < x < 1$ **d.** $\frac{1}{2}$

6. **a.** $F_U(u) = 0$ for $u < 0$ and $(1 - e^{-\lambda u})^2$ for $u \geq 0$; $f_U(u) = 2\lambda e^{-\lambda u}(1 - e^{-\lambda u})$ for $u > 0$ **b.** $F_V(v) = 0$ for $v < 0$ and $1 - e^{-2\lambda v}$ for $v \geq 0$; $f_V(v) = 2\lambda e^{-2\lambda v}$ for $v > 0$

7. **a.** $\dfrac{1}{2\pi}e^{-(x^2+y^2)/2}$ for any $x \in \mathbb{R}$ and $y \in \mathbb{R}$ **b.** $1 - e^{-a^2/2}$ **c.** $f_R(r) = re^{-r^2/2}$ for $r > 0$

8. **a.** $f(x, y) = 1$ for $0 < x < 1$ and $0 < y < 1$ **b.** $\frac{1}{4}$ **c.** a^2; $f_M(x) = 2x$ for $0 < x < 1$ **d.** $a^2/2$ **e.** $1 - (2 - a)^2/2$ **f.** $f_Z(z) = z$ if $0 < z < 1$, $2 - z$ if $1 \leq z < 2$, 0 otherwise.

9. **a.** $9xe^{-3x}$ for $x > 0$ **b.** No; the set of possible values of (X, Y) is not a Cartesian product.

10. **a.** $\frac{1}{2}$ **b.** 0

3.6 THE JOINT DISTRIBUTION OF MORE THAN TWO RANDOM VARIABLES

This section extends what we learned in the previous two sections to the cases of three, four, and more random variables. We begin with a discussion of the general situation, but the only calculations we will be concerned with are described in (3.6.1), which follows the discussion.

We have seen in Sections 3.4 and 3.5 that when we are dealing with two random variables, it is not enough to know their mass functions or their density functions. Instead:

> We must consider their joint mass function $p(x, y)$ if they are discrete, or their joint density function $f(x, y)$ if they are absolutely continuous.
>
> We can find their individual mass functions or densities (the "marginal" mass functions or densities) from the joint mass function or density by adding or integrating.
>
> We can find probabilities of events involving both of them, such as $P(X + Y \leq a)$, by adding appropriate values of the joint mass function or integrating the joint density over the appropriate region of the plane.
>
> We can check whether they are independent by seeing whether the joint mass function or density factors as the product of the marginal mass functions or densities.
>
> Only if we are able to *assume* that X and Y are independent can we get all the information we need from the individual mass functions or densities, because then the joint mass function or density is their product.

For more than two random variables—say, for n random variables $X_1, X_2, \ldots, X_n$—the situation is virtually identical to this, except that the joint mass function or density is now a function of n variables instead of two, $p(x_1, x_2, \ldots, x_n)$ or $f(x_1, x_2, \ldots, x_n)$. To find marginal mass functions or densities we integrate over all the variables other than the ones we want.

For example, suppose that X, Y, and Z are absolutely continuous and their joint density function is $f(x, y, z)$. Then the marginal density of X is

$$f_X(x) = \int_{-\infty}^{\infty}\int_{-\infty}^{\infty} f(x, y, z)dy\, dz,$$

the joint density of X and Z is

$$f_{X,Z}(x, z) = \int_{-\infty}^{\infty} f(x, y, z)dy,$$

and so on.

Or suppose we have four discrete random variables, X, Y, Z, and W, whose joint mass function is

$$p(x, y, z, w) = P(X = x, Y = y, Z = z, \text{ and } W = w).$$

Then the joint mass function of Y and Z is found by summing over all x and w:

$$p_{Y,Z}(y, z) = P(Y = y \text{ and } Z = z) = \sum_x \sum_w p(x, y, z, w).$$

This seems complicated, and it can indeed get quite intricate. For example, if we had three absolutely continuous random variables as above and we needed the probability that X was the largest and Y was the smallest, we would find it by integrating $f(x, y, z)$ over the region in three-dimensional space consisting of the points (x, y, z), for which $x > z > y$. This is not as difficult as it seems at first; but in this text we will not deal much with such situations.

Our dealings with more than two random variables will be simplified in two ways. First, we will deal almost exclusively with *independent* random variables that all have the same distribution. In this case, the joint mass function or density is the product of the marginals and is easy to write down. Second, most of our calculations will involve only the sum of the random variables in question, or sometimes the product of just two of them at a time; and we will develop methods for handling such calculations easily.

In this section we will confine ourselves to a few examples illustrating one simple idea: that of writing down the joint mass function or the joint density of a set of independent, identically distributed random variables. By the way, the phrase "independent and identically distributed" occurs often enough in probability and statistics that there is a common abbreviation for it: "i.i.d."

You may wonder why working with a large collection of i.i.d. random variables is so important. The reason is that this is a fundamental setup in statistical inference. Suppose we want to know something about a large population—say the blood cholesterol levels of teenagers in the United States. The usual first step would be to take a random sample of teenagers, which we hope is typical of the population as a whole, and measure their cholesterol levels. Experience and statistical theory show that for most purposes the best thing to do is to take as large a sample of *independent* measurements as possible—"independent" implying, for example, that we do not include two siblings in the sample, or any teenagers whose cholesterol levels can be expected to be correlated in some way.

If we take a sample of 300 independent cholesterol level measurements, then what we get are 300 numbers $x_1, x_2, \ldots, x_{300}$. These numbers are observed values of 300 independent random variables $X_1, X_2, \ldots, X_{300}$, because the subjects were chosen independently and at random from the population; each X_j is the result of the experiment of choosing a teenager at random and measuring her or his cholesterol level. Since they all are chosen from the same population, they all have the same distribution.

Now why should we be interested in probabilities for $X_1, X_2, \ldots, X_{300}$ when we actually know the numbers $x_1, x_2, \ldots, x_{300}$? To see why, remember that we know nothing about the population except what we learn from the sample. For example, we are probably interested in the average cholesterol level of all teenagers; call it μ. We can never know this number. But we can measure the average of our data, $\bar{x} = (x_1 + x_2 + \cdots + x_{300})/300$, and hope it is close to the unknown average of the population, μ. The number $\bar{x}$ is an observed value of the random variable $\bar{X} = (X_1 + X_2 + \cdots + X_{300})/300$, which has some distribution that depends on our unknown number μ, and it is natural to ask for the probability that $\bar{X}$ is within, say, 5% of the true value of μ. To find probabilities like this for $\bar{X}$, we start with the joint distribution of $X_1, X_2, \ldots, X_{300}$.

We will return to this discussion in Section 4.5.

Because of the statistical setting described in the preceding discussion, a collection $X_1, X_2, \ldots, X_n$ of i.i.d. random variables—independent random variables with the same distribution—is often called a ***sample*** from that distribution. Thus, we might say "let $(X_1, X_2, \ldots, X_n)$ be a sample from the exponential distribution with parameter λ" to mean that $X_1, X_2, \ldots, X_n$ are independent random variables and the density of each of them is $f(x) = \lambda e^{-\lambda x}$ for $x > 0$.

We repeat that we will not do much calculation in this section, except to write down the joint mass functions or densities. The only facts we use are that (i) the joint mass function of independent discrete random variables is the product of their individual mass functions, and (ii) a joint density for independent absolutely continuous random variables is the product of densities for the individual random variables. That is:

If $X_1, X_2, \ldots, X_n$ are independent discrete random variables, then their joint mass function is

(3.6.1)

$$p(x_1, x_2, \ldots, x_n) = p_{X_1}(x_1)p_{X_2}(x_2) \cdots p_{X_n}(x_n).$$

If $X_1, X_2, \ldots, X_n$ are independent absolutely continuous random variables, then a joint density is

$$f(x_1, x_2, \ldots, x_n) = f_{X_1}(x_1)f_{X_2}(x_2) \cdots f_{X_n}(x_n).$$

Notice that if the random variables are i.i.d.—that is, not only independent but also identically distributed—then their individual densities or mass functions are all the same function. This is what happens in the following examples. You will notice that they, and the exercises in this section, are really nothing but practice in working algebraically with a certain kind of function of n variables.

(3.6.2) **Example: A sample from a Poisson distribution.** If $X_1, X_2, \ldots, X_n$ are i.i.d., each having the Poisson distribution with parameter λ, then the mass function of one of them is

$$p_{X_i}(x) = P(X_i = x) = \frac{\lambda^x e^{-\lambda}}{x!} \quad \text{for } x = 0, 1, 2, 3, \ldots.$$

So their joint mass function is a function of n variables $x_1, x_2, \ldots, x_n$ that is the product of the individual mass functions:

$$\begin{aligned} p(x_1, x_2, \ldots, x_n) &= p_{X_1}(x_1)p_{X_2}(x_2) \cdots p_{X_n}(x_n) \\ &= \frac{\lambda^{x_1}e^{-\lambda}}{x_1!}\frac{\lambda^{x_2}e^{-\lambda}}{x_2!} \cdots \frac{\lambda^{x_n}e^{-\lambda}}{x_n!} \\ &= \frac{\lambda^{(x_1+x_2+\cdots+x_n)} \cdot e^{-n\lambda}}{x_1!x_2!\cdots x_n!} \quad \text{if each } x_i \text{ is a nonnegative integer.} \end{aligned}$$

(3.6.3) **Example: A sample from the standard normal distribution.** The density of each of the X_i's is

$$f_{X_i}(x) = \varphi(x) = \frac{1}{\sqrt{2\pi}} e^{-x^2/2} \quad \text{for any real } x.$$

So the joint density is

$$\begin{aligned} f(x_1, x_2, \ldots, x_n) &= f_{X_1}(x_1)f_{X_2}(x_2) \cdots f_{X_n}(x_n) \\ &= \frac{1}{\sqrt{2\pi}} e^{-x_1^2/2} \frac{1}{\sqrt{2\pi}} e^{-x_2^2/2} \cdots \frac{1}{\sqrt{2\pi}} e^{-x_n^2/2} \\ &= \frac{1}{(2\pi)^{n/2}} e^{-(x_1^2+x_2^2+\cdots+x_n^2)/2} \quad \text{for any real numbers } x_1, x_2, \ldots, x_n. \end{aligned}$$

(3.6.4) **Example: A sample from a uniform distribution.** Suppose first that the distribution is the uniform distribution on [0, 1]. Then the density of an individual X_i is

$$f_{X_i}(x) = 1 \quad \text{if } 0 \leq x \leq 1 \quad \text{(and 0 otherwise).}$$

So the joint density is

$$\begin{aligned} f(x_1, x_2, \ldots, x_n) &= f_{X_1}(x_1) f_{X_2}(x_2) \cdots f_{X_n}(x_n) \\ &= 1 \quad \text{if each } x_i \text{ is in } [0, 1] \quad \text{(and 0 otherwise).} \end{aligned}$$

The set of possible values of the sample $(X_1, X_2, \ldots, X_n)$ is an n-dimensional box, and the joint density function $f(x_1, x_2, \ldots, x_n)$ is the constant 1 as long as $(x_1, x_2, \ldots, x_n)$ is in that box. So integrals of the density are just n-dimensional volumes and we do not need to integrate, unless there is no other way to find a volume we need. Compare this with Exercise 8 in Section 3.5, where we had the case $n = 2$ and the box was a square.

What if we had some other uniform distribution, such as the uniform distribution on $[-2, 2]$, or in general on $[a, b]$? The set of possible values of the sample would then be a different box in n-space: the set of all $(x_1, x_2, \ldots, x_n)$ for which each x_i is in the interval $[a, b]$. And the joint density is still a constant; but it is no longer 1, because the density of an individual X_i is

$$f_{X_i}(x) = \frac{1}{b-a} \quad \text{if } a \le x \le b \quad \text{(and 0 otherwise),}$$

so the joint density is the product

$$\begin{aligned} f(x_1, x_2, \ldots, x_n) &= f_{X_1}(x_1)f_{X_2}(x_2) \cdots f_{X_n}(x_n) \\ &= \frac{1}{(b-a)^n} \quad \text{if each } x_i \text{ is in } [a, b] \quad \text{(and 0 otherwise).} \end{aligned}$$

Probabilities are not just volumes, but volumes divided by $(b-a)^n$, which is the volume of the box of possible values. ∎

EXERCISES FOR **SECTION 3.6**

[a]**1.** Suppose that $X_1, X_2, \ldots, X_n$ are independent and each has the exponential distribution with parameter λ; the density is $\lambda e^{-\lambda x}$ for $x > 0$. Write down their joint density.

[a]**2.** Let $X_1, X_2, \ldots, X_n$ be independent discrete random variables, each having the geometric distribution with parameter p; the mass function is $p(x) = pq^x$ for $x = 0, 1, 2, 3, \ldots$. Write down their joint mass function.

Note (not necessary to do the exercise): Remember where the geometric distribution arises. In a Bernoulli process, it is the distribution of the number of failures preceding the first success, and also of the number of failures after the first success but before the second, and so on. So the joint mass function $p(x_1, x_2, \ldots, x_n)$ gives the probability that the first success comes after x_1 failures, the second comes after x_2 more failures, and so on. As always with a Bernoulli process, q denotes $1 - p$.

3. If we take two independent random variables X and Y with the standard normal distribution and let $R = \sqrt{X^2 + Y^2}$, then a density for R is $f(r) = re^{-r^2/2}$ for $r > 0$. [We found this in Exercise 7 in the previous section.] Suppose the experiment of choosing X and Y is repeated independently 25 times, and $R_1, R_2, \ldots, R_{25}$ are the values of R that result. What is the joint density of these 25 random variables?

4. What is the joint mass function of 25 independent random variables, each having the binomial distribution with parameters n and p?

ANSWERS

1. $f(x_1, x_2, \ldots, x_n) = \lambda^n e^{-\lambda(x_1 + \cdots + x_n)}$ if all x_i are positive

2. $p(x_1, x_2, \ldots, x_n) = p^n q^{(x_1 + \cdots + x_n)}$ if all x_i are nonnegative integers

SUPPLEMENTARY EXERCISES FOR **CHAPTER 3**

1. For any positive integer n, the following function is a CDF:

$$F(x) = \begin{cases} 0 & \text{if } x < 0, \\ x^n & \text{if } 0 \le x < 1, \\ 1 & \text{if } x \ge 1. \end{cases}$$

   a. Sketch this function.
   b. Is the distribution discrete, absolutely continuous, or neither? If discrete, give the mass function. If absolutely continuous, give a density.
   c. Find (in terms of n) the probability that this distribution assigns to the interval $\left(0, \frac{1}{2}\right)$.

2. Let X be an absolutely continuous random variable whose density is $f(x) = \lambda^2 x e^{-\lambda x}$ for $x > 0$ (and 0 for $x \le 0$), where λ is a positive constant. Find the cumulative distribution function of X.

[a]3. Write down the CDF of a random variable that has possible values 2 and 5, with probabilities .7 and .3, respectively.

4. Let X have the following CDF:

$$F(x) = \begin{cases} 0 & \text{if } x < 0, \\ \dfrac{x+1}{4} & \text{if } 0 \le x < 2, \\ 1 & \text{if } x \ge 2. \end{cases}$$

   a. Sketch this function.
   b. Is X discrete, absolutely continuous, or neither? If discrete, give the mass function; if absolutely continuous, give a density.
   c. Find $P(0 \le X \le 1)$.
   d. Find $P(0 < X \le 1)$.
   e. Find $P(0 < X < 2)$.

5. Let X have the uniform density $f_X(x) = 1$ for $0 < x < 1$, and let Y have the density $f_Y(y) = 3y^2$ for $0 < y < 1$. We toss a fair coin. If heads comes up we observe X, and if tails comes up we observe Y.
   a. What is the probability that the outcome is less than $\frac{1}{2}$? [Condition on the outcome of the coin toss.]
   b. Given that the outcome is less than $\frac{1}{2}$, what is the conditional probability that we observed X? That we observed Y?

6. Let X have the standard normal distribution. Find the following probabilities:
   a. $P(10 + 4X > 19)$
   b. $P(X \ge 2 | X \le 3)$
   c. $P(| X | > 1.96)$

7. Let X be the number of successes in eight Bernoulli trials, in which the success probability on each trial is .3. Find the following probabilities:
a. $P(X = 2)$
b. $P(X \geq 2)$
c. $P(X = 2 | X \geq 2)$

8. Let X have the Poisson distribution with parameter 1.3; that is, $p_X(x) = (1.3)^x e^{-1.3}/x!$ for $x = 0, 1, 2, \ldots$. Find the following probabilities:
a. $P(X = 1.3)$
b. $P(X = 1)$
c. $P(X \geq 1)$
d. $P(X = 1 | X \geq 1)$
e. $P(X \leq 5)$

9. Let $X_1, X_2, X_3, \ldots, X_{10}$ be independent random variables, each with the standard normal distribution. Let Y be the number of X_j's that are less than 1. What is the distribution of Y? [*Hint:* Think of "$X_j < 1$" as a success on the jth trial.]

10. Let $X_1, X_2, \ldots, X_{10}$ be independent random variables, all having the Poisson distribution with parameter λ.
a. What is the probability that $X_1 + X_2 + \cdots + X_{10} = 0$?
b. What is the probability that $X_1 + X_2 + \cdots + X_{10} = 1$? [*Hint:* This event is the union of ten disjoint events that all have the same probability.]
c. What is the probability that $X_1 + X_2 + \cdots + X_{10} = 2$? [*Hint:* This is more difficult. The event is the union of $\binom{10}{2} + \binom{10}{1}$ disjoint events of two types.]
d. From the answers to parts a, b, and c, guess the distribution of $X_1 + X_2 + \cdots + X_{10}$. [We will see how to find it in (7.4.7).]

[a]11. In Exercise 10 what is the joint mass function of $X_1, X_2, \ldots, X_{10}$?

12. Let X and Y have the joint density $f(x, y) = \lambda^3 x e^{-\lambda(x+y)}$ for $x > 0$ and $y > 0$.
a. Find the marginal densities and show that X and Y are independent.
b. For given positive numbers a and b find $P(X \leq a$ and $Y \leq b)$.
c. Find $P(X \leq a)$.
d. Find the probability that $X + Y \leq a$ for a given positive number a.
e. Find a density for $Z = X + Y$.

13. A hat contains six slips of paper, with numbers on them; there are one 1, two 2's, and three 3's. One slip is drawn at random, and Y is the number on it. Now a fair coin is tossed Y times, and X is the number of heads. Find the mass function of X. [It is probably simplest to do this as four separate probability calculations: $P(X = 0)$, $P(X = 1)$, $P(X = 2)$, and $P(X = 3)$.]

14. Consider the following probability mass function: $p(k) = ck(.3)^k$ for $k = 1, 2, 3, \ldots$.

a. In terms of c, what is the probability of the event $A = \{3, 4, 5, 6, \ldots\}$?
b. What is the value of c?

15. Let probabilities be assigned to subintervals of $[1, \infty)$ according to the density function $f(x) = 2/3x^2 + 2/3x^3$. What probability is assigned to the interval $(3, \infty)$?

16. Let X be obtained as follows: a number is chosen from the uniform distribution on (0, 1), and X is its cube root. Find the CDF and then a density for X.

17. Let X and Y be independent, with densities $f_X(x) = 2(1 - x)$ (for $0 < x < 1$) and $f_Y(y) = 2y$ (for $0 < y < 1$). Find the probability that X is greater than Y.

18. The possible values of the pair (X, Y) are the points inside the triangle whose vertices are at (0, 0), (0, 1), and (1, 0). A joint density for X and Y is $f(x, y) = 60x^2y$ on that triangle.
a. Find the marginal density for Y.
b. Are X and Y independent? Why?
c. Find the CDF of $Z = X + Y$.

19. Let X and Y have the joint density $f(x, y) = \frac{2}{3}(x + 2y)$ on the unit square. Find $P(X > Y)$.

ANSWERS

1. b. Absolutely continuous; density is $f(x) = nx^{n-1}$ for $0 < x < 1$ **c.** 2^{-n}

2. $F(x) = 0$ for $x \le 0$ and $1 - (1 + \lambda x)e^{-\lambda x}$ for $x > 0$

3. $F(x) = 0$ for $x < 2$, .7 for $2 \le x < 5$, and 1 for $x \ge 5$

4. b. Neither **c.** .5 **d.** .25 **e.** .5

5. a. $\frac{5}{16}$ **b.** $\frac{4}{5}, \frac{1}{5}$

6. a. .0122 **b.** .0215 **c.** .0500

7. a. .2965 **b.** .7447 **c.** .3981

8. a. 0 **b.** .3543 **c.** .7275 **d.** .4870 **e.** .9978

9. Binomial with $n = 10$ and $p = .8413$

10. a. $e^{-10\lambda}$ **b.** $10\lambda e^{-10\lambda}$ **c.** $\dfrac{(10\lambda)^2 e^{-10\lambda}}{2!}$

11. $p(x_1, x_2, \ldots, x_{10}) = \dfrac{\lambda^{(x_1+x_2+\cdots+x_{10})}e^{-10\lambda}}{x_1!x_2!\cdots x_{10}!}$

12. b. $(1 - (1 + \lambda a)e^{-\lambda a}) \cdot (1 - e^{-\lambda b})$
c. $1 - (1 + \lambda a)e^{-\lambda a}$
d. $1 - \left(1 + \lambda a + \dfrac{\lambda^2 a^2}{2}\right)e^{-\lambda a}$
e. $\dfrac{\lambda^3 z^2}{2} e^{-\lambda z}$ $(z > 0)$

13. X has possible values 0, 1, 2, and 3, with probabilities $\frac{11}{48}, \frac{21}{48}, \frac{13}{48}$, and $\frac{3}{48}$.

14. a. $1 - .48 \cdot c$ **b.** 1.6333 . . .

15. $\frac{7}{27}$

16. Density is $f(x) = 3x^2$ for $0 < x < 1$ and 0 otherwise.

17. $\frac{1}{6}$

18. a. $f_Y(y) = 20y(1 - y)^3$ for $0 < y < 1$
b. No **c.** 0 for $z < 0$, z^5 for $0 \le z < 1$, and 1 for $z > 1$

19. $\frac{4}{9}$

CHAPTER 4

Expected Values

4.1 Definitions and Examples of Expected Values

4.2 The Algebra of Expected Values

4.3 Variance and Moments

4.4 Covariance and the Variance of a Sum of Random Variables

4.5 Laws of Large Numbers

Supplementary Exercises

4.1 DEFINITIONS AND EXAMPLES OF EXPECTED VALUES

Suppose we have a chance experiment that produces a random variable X. If we repeated the experiment many times, getting a value for X each time, and then averaged the values, what would we expect to get? The expected value of X is a number, computed from the probability mass function or the density function of X, that represents this ***expected long-run average observed value*** of X.

We will begin with the two definitions—(4.1.1) for a discrete random variable and (4.1.4) for an absolutely continuous one—and an example of how the expected value is computed in each case, without explaining *why* these definitions are the right ones to represent the expected long-run average value. Then we will give two examples that explain why. We will conclude with a few more examples of calculating expected values.

Because the expected value of an absolutely continuous random variable is defined as an integral, we will be doing a lot of computing of integrals. We will need the techniques of substitution and of integration by parts, and we will also need to pay attention to what happens when an integral is an improper integral (that is, one in which the interval of integration is unbounded or in which the integrand is undefined at one or both endpoints). You will probably want to review Sections A.7 and A.8 in the Appendix before finishing this section.

Furthermore, the expected value of a discrete random variable is either a finite sum or an infinite series, depending on whether the random variable has finitely or infinitely many possible values. With that in mind, it might be wise also to look at Section A.5 of Appendix A.

(4.1.1) **Definition:** Expected value of a discrete random variable. Let X be a discrete random variable whose probability mass function is $p(x)$. The ***expected value*** of X is the number, denoted by EX or $E(X)$, defined by

$$EX = \sum_x xp(x),$$

provided that, if this sum is an infinite series, then it converges absolutely.

Notice that although EX appears to be defined as a sum over all real x, which makes no sense, it is really a perfectly legal finite sum or infinite series, because the mass function $p(x)$ is nonzero only on a finite or countable set of x's.

The requirement of absolute convergence means not only that the series converges, but that it would converge even if we replaced every term $xp(x)$ by its absolute value, which equals $|x|p(x)$ (because the probabilities $p(x)$ are never negative).

If X has only finitely many possible values, then of course there is no question about convergence; the expected value is simply a sum of finitely many numbers. If X has a countable infinity of possible values but they are all nonnegative, then convergence and absolute convergence are the same; the only restriction is that the series converge. We will say more later about what to do if X has infinitely many possible values including some positive and some negative ones.

(4.1.2) **Example.** Let X have possible values 0, 1, 2, and 3, with probabilities .1, .3, .2, and .4, respectively. Then the expected value of X is

$$EX = 0 \cdot .1 + 1 \cdot .3 + 2 \cdot .2 + 3 \cdot .4 = 1.9. \quad \blacksquare$$

(4.1.3) **Example: The expected value of a geometrically distributed random variable is q/p.** Let X be the number of failures before the first success in a Bernoulli process with success probability p. Then the possible values of X are the nonnegative integers 0, 1, 2, 3, . . . , and the probabilities are

$$p(k) = P(X = k) = pq^k \quad (k = 0, 1, 2, 3, \ldots),$$

where q denotes $1 - p$. The distribution of X is called the *geometric* distribution with parameter p; the word "geometric" comes from the fact that the probabilities are the terms of a geometric series. A graph of the probabilities, for the particular values $p = .16$ and $q = .84$, is shown in Figure 4.1a. Notice that $k = 0$ is the most probable value. You might want to guess the expected value before looking at the calculation; it is not easy to do so from a picture of this distribution. Note also that the x-axis goes only to $x = 40$. Every nonnegative integer has a positive probability in this distribution, but the probabilities of integers greater than 40 are too small to appear on the graph.

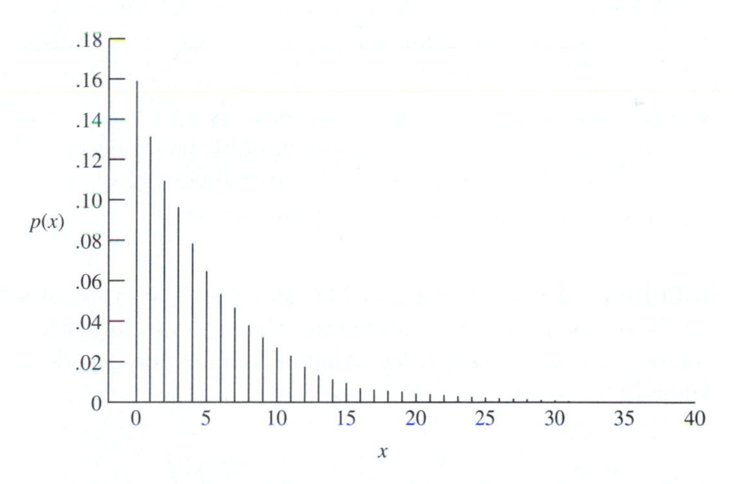

FIGURE 4.1a The mass function of the geometric distribution with $p = .16$

The expected value of X is

$$EX = \sum_{k=0}^{\infty} k \cdot pq^k.$$

All the terms in this series are nonnegative, so there is no worry about absolute convergence. We need only to sum the series (if it converges). We did such sums in Section 1.2; they are also covered at (A.5.9) in Appendix A. It goes like this:

$$\begin{aligned}EX &= 0 \cdot pq^0 + 1 \cdot pq^1 + 2 \cdot pq^2 + 3 \cdot pq^3 + \cdots \\ &= pq \cdot (1 + 2q + 3q^2 + 4q^3 + \cdots) \\ &= pq \cdot \frac{d}{dq}(1 + q + q^2 + q^3 + \cdots) \\ &= pq \cdot \frac{d}{dq}\left(\frac{1}{1-q}\right) = pq \cdot \frac{1}{(1-q)^2}.\end{aligned}$$

But because $1 - q$ is just p, we get $EX = q/p$.

For the particular geometric distribution shown in Figure 4.1a, $q/p = .84/.16 = 5.25$. Notice where this lies on the x-axis; it does not look like a typical value at all, but if you do the experiment many times and average the resulting values of X, the average will likely be close to 5.25. ▮

The answer q/p in Example 4.1.3 is actually quite intuitive. If the chance of success is $\frac{1}{6}$, for example, then $EX = \frac{5}{6}\Big/\frac{1}{6} = 5$; on the average, over many performances of this experiment, the first success will come after five failures—that is, on the sixth trial.

But notice that there is no reason to expect, on a *single* performance, that the first success will come on the sixth trial. The probability that it will is $p(5) = \frac{1}{6} \cdot \left(\frac{5}{6}\right)^5 \doteq .06698$. By contrast, the probability that the first success comes on the first trial (that is, after no failures) is $p(0) = \frac{1}{6} \doteq .1667$.

The *expected* value of a random variable is not necessarily its *most probable* value. "Expected" means "expected in the average over many performances of the experiment," *not* "expected on the next trial."

(4.1.4) Definition: Expected value of an absolutely continuous random variable. Let X be an absolutely continuous random variable whose probability mass function is $f(x)$. The ***expected value*** of X is the number, denoted by EX or $E(X)$, defined by

$$EX = \int_{-\infty}^{\infty} x f(x)dx,$$

provided that, if this is an improper integral, then it converges absolutely.

Notice that although this integral is defined over the interval $(-\infty, \infty)$, the density $f(x)$ will usually be positive only on some subinterval, and then the expected value will be an integral over that subinterval only, as in the examples that follow.

The requirement of absolute convergence means not only that the integral converges, but that it would converge even if we replaced the integrand $xf(x)$ by its absolute value, which equals $|x|f(x)$ (because the density $f(x)$ is never negative).

If the set of possible values of X (that is, the set of x for which $f(x)$ is positive) is a finite interval $[a, b]$, and if $xf(x)$ is bounded, then of course there is no question about convergence; the expected value is simply an integral from a to b. If the set of possible values is an unbounded interval like $[1, \infty)$ in which all possible values are positive, then convergence and absolute convergence are the same, and the only restriction is that the integral converge. We will say more later on what to do if the set of possible values of X is an unbounded interval that contains some positive and some negative values, or when $xf(x)$ is unbounded on a finite interval.

(4.1.5) Example. Let X have the density function

$$f(x) = \begin{cases} 12x^2(1-x) & \text{if } 0 < x < 1, \\ 0 & \text{otherwise.} \end{cases}$$

This density is graphed in Figure 4.1b. You might want to guess, by looking at the graph, where the expected value will be.

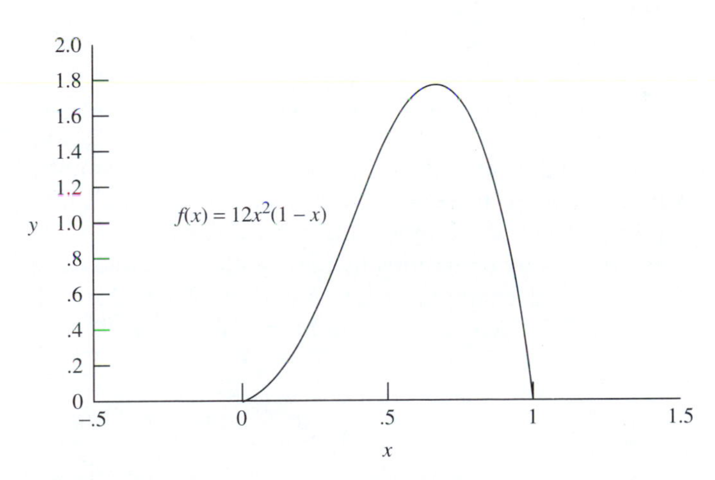

FIGURE 4.1b The density of Example (4.1.5)

The expected value of X is

$$EX = \int_0^1 x \cdot 12x^2(1-x)dx,$$

which we can evaluate easily:

$$EX = 12 \int_0^1 x^3 dx - 12 \int_0^1 x^4 dx = 12 \cdot \tfrac{1}{4} - 12 \cdot \tfrac{1}{5} = .6.$$

It is worth noting that $x = .6$ is not the point where the density is largest; as you can check, that occurs at $x = \frac{2}{3}$. Nor is it the value of x that has half the probability to the left and half to the right; that point happens to be at about .6145. [See Exercises 10 and 11 for other examples where EX does not look like a "typical" or "common" value of X.] ▮

(4.1.6) Example: The expected value of an exponential random variable is $1/\lambda$. Let X be the time between two consecutive arrivals in a Poisson process with an arrival rate of λ. What is EX? We will see that it is $1/\lambda$. This is the natural answer to expect: If arrivals are coming at an average of λ per minute, then there should be an average of $1/\lambda$ minutes between arrivals.

We know that X has the exponential distribution with parameter λ; that is, X is an absolutely continuous random variable with the density

$$f(x) = \begin{cases} \lambda e^{-\lambda x} & \text{if } x > 0, \\ 0 & \text{otherwise.} \end{cases}$$

So Definition (4.1.4) says that

$$EX = \int_0^\infty x \cdot \lambda e^{-\lambda x} dx.$$

This is an improper integral (it has an infinite endpoint), but the integrand is always positive, so there is no concern about absolute convergence. We need only to find the value of the integral if it converges.

We do this by integrating by parts with $u = x$ and $dv = \lambda e^{-\lambda x} dx$. See (A.7.3) and the second example under (A.8.1) in Appendix A for the details of this calculation. We have $du = dx$ and $v = -e^{-\lambda x}$, so

$$\begin{aligned} EX &= \left[-xe^{-\lambda x} \right]_0^\infty + \int_0^\infty e^{-\lambda x} dx \\ &= [-0 + 0] + \left[-\frac{1}{\lambda} e^{-\lambda x} \right]_0^\infty = \frac{1}{\lambda}. \end{aligned}$$

Notice that we did the calculation informally, "plugging in ∞" as an upper endpoint. The correct thing to do is to note that

$$\int_0^\infty x \cdot \lambda e^{-\lambda x} dx = \lim_{A \to \infty} \int_0^A x \cdot \lambda e^{-\lambda x} dx,$$

then use integration by parts on this proper integral, and finally take the limit. The integration by parts gives us

$$\int_0^A x \cdot \lambda e^{-\lambda x} dx = [-0 + Ae^{-\lambda A}] + \left[-\frac{1}{\lambda}e^{-\lambda x}\right]_0^A$$
$$= Ae^{-\lambda A} + \frac{1}{\lambda} - \frac{1}{\lambda}e^{-\lambda A}.$$

When we take the limit of this as $A \to \infty$, we get $0 + \frac{1}{\lambda} - 0 = \frac{1}{\lambda}$. When we "plugged in ∞" in the informal calculation, we were really taking the limit as $A \to \infty$. ∎

(4.1.7) **Why Definition (4.1.1)?** Consider again the discrete random variable X in Example (4.1.2), which has possible values 0, 1, 2, and 3 with corresponding probabilities .1, .3, .2, and .4. Why should the sum

$$EX = 0 \cdot .1 + 1 \cdot .3 + 2 \cdot .2 + 3 \cdot .4 = 1.9$$

represent the expected long-run average value of X?

Imagine a million trials of the experiment, producing a million values of X, numbers which are all 0's, 1's, 2's, and 3's. What would we expect these million numbers to be, and what would be their average? We would expect approximately (though not exactly) 100,000 0's, 300,000 1's, 200,000 2's, and 400,000 3's; the average of these would be

$$\frac{0 \cdot 100{,}000 + 1 \cdot 300{,}000 + 2 \cdot 200{,}000 + 3 \cdot 400{,}000}{1{,}000{,}000},$$

which is exactly $0 \cdot .1 + 1 \cdot .3 + 2 \cdot .2 + 3 \cdot .4$.

In general, if the mass function is $p(x)$ and we make N trials, where N is large, then for any possible value, say x, we expect the N observed values of X to include the value x about $p(x) \cdot N$ times. [This is true because $p(x) = P(X = x)$ is the expected long-run *relative frequency* of occurrence of the event $(X = x)$, and so $p(x) \cdot N$ is the expected *frequency* of that event.] Thus, the sum of the N observed values of X is expected to be about $\sum_x x \cdot p(x) \cdot N$; dividing by N gives us the expected long-run average.

So the interpretation of the probability $p(x) = P(X = x)$ as an ***expected long-run relative frequency*** leads to Definition (4.1.1) for the expected value, which we can interpret as an ***expected long-run average observed value.***

Notice that just as we do not expect the relative frequency of an event to be exactly equal to its probability, so we do not expect the average observed value of X to be exactly equal to EX. The number EX merely represents our prediction of the average, based on our expectation that the relative frequencies will be close to the probabilities.

To illustrate this, here are the results of 10,000 performances of a simulated experiment that produces the random variable X of Examples (4.1.2) and (4.1.7)—986 0's, 3,021 1's, 2,032 2's, and 3,961 3's. The average of these numbers is

$$\frac{986 \cdot 0 + 3{,}021 \cdot 1 + 2{,}032 \cdot 2 + 3{,}961 \cdot 3}{10{,}000} = 1.8968.$$

This is quite close to

$$\frac{1{,}000 \cdot 0 + 3{,}000 \cdot 1 + 2{,}000 \cdot 2 + 4{,}000 \cdot 3}{10{,}000} = 1.9.$$

(4.1.8) Why Definition (4.1.4)? The definition $EX = \int_{-\infty}^{\infty} xf(x)dx$ for absolutely continuous random variables has essentially the same motivation as for discrete ones, except that there is an additional step of approximation between the expected long-run average and the integral EX. Instead of writing down the actual expected average, we write down something that should approximate it closely, and then notice that what we have written is a Riemann sum for the integral $\int_{-\infty}^{\infty} xf(x)dx$.

To simplify the discussion a little, let us suppose that the set of possible values of X is the interval $(0, 1)$, so that the density $f(x)$ of X is 0 except when $0 < x < 1$. What values of X would we expect to observe in a million trials? As an approximation, consider dividing the interval $(0, 1)$ into ten subintervals $(0, .1)$, $(.1, .2)$, . . . , $(.9, 1)$. Looking at Figure 4.1c, we see (at least if the density is continuous) that

the probability that X is in $(0, .1)$ is approximately $f(.05) \cdot .1$,

the probability that X is in $(.1, .2)$ is approximately $f(.15) \cdot .1$,

the probability that X is in $(.2, .3)$ is approximately $f(.25) \cdot .1$,

.
.
.

the probability that X is in $(.9, 1)$ is approximately $f(.95) \cdot .1$.

Thus, if we recorded the value of X on each of a million trials, we would expect to see

about $f(.05) \cdot .1 \cdot 1{,}000{,}000$ numbers that are close to .05,

about $f(.15) \cdot .1 \cdot 1{,}000{,}000$ numbers that are close to .15,

about $f(.25) \cdot .1 \cdot 1{,}000{,}000$ numbers that are close to .25,

and so forth, up to

about $f(.95) \cdot .1 \cdot 1{,}000{,}000$ numbers that are close to .95.

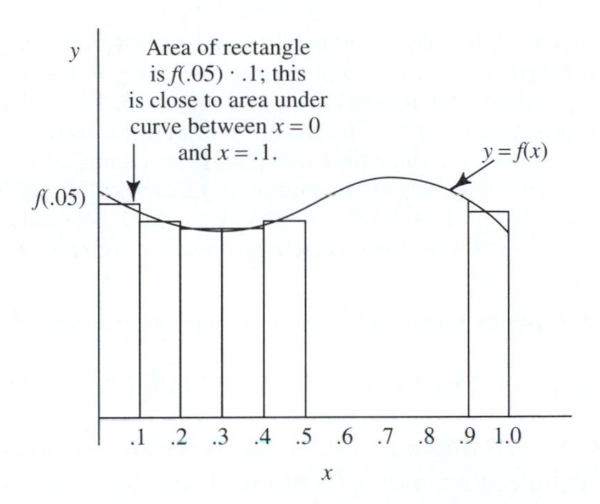

FIGURE 4.1c **The probability that X is in an interval of width .1 is approximately $f(m) \cdot .1$, where m is the interval's midpoint.**

Adding these numbers and dividing by 1,000,000 gives us the expected average value

$$.05 \cdot f(.05) \cdot .1 + .15 \cdot f(.15) \cdot .1 + .25 \cdot f(.25) \cdot .1 + \cdots + .95 \cdot f(.95) \cdot .1.$$

This is an approximating sum, or Riemann sum, based on $n = 10$ subintervals, for the integral $\int_0^1 xf(x)dx$. If we increased the number of subintervals from ten to larger and larger values of n, the sums would approach the integral; but they would also get closer to what we would expect to observe. Hence, we use the integral as the definition of the expected value of X.

Before discussing the problem of absolute convergence, we will look at a few more simple examples of expected value calculations.

(4.1.9) **Example.** In a random permutation of the numbers 1, 2, 3, and 4, let S be the number of numbers that are in their proper places. What is ES?

The random variable S is discrete; in Exercise 6d of Section 1.1 we found that its possible values are 0, 1, 2, and 4, with probabilities $\frac{9}{24}, \frac{8}{24}, \frac{6}{24}$, and $\frac{1}{24}$. So the expected value of S is

$$ES = 0 \cdot \tfrac{9}{24} + 1 \cdot \tfrac{8}{24} + 2 \cdot \tfrac{6}{24} + 4 \cdot \tfrac{1}{24} = 1.$$

Is this a coincidence or is the expected number of numbers in their proper places equal to 1 no matter how many numbers are permuted? If you are curious, you might first find ES for the random permutations of three numbers. (You could also try it for two numbers; it is pretty trivial there, because there are only two permutations. It is even more trivial for just one number.) If it still seems to be true, you would need to try it next for five numbers, then for six. For five numbers there are $5! = 120$ permutations, and the possible values of

X are 0, 1, 2, 3, and 5, with corresponding probabilities 44, 45, 20, 10, and 1, all divided by 120. For six numbers, there are 720 permutations; the possible values are 0, 1, 2, 3, 4, and 6, with corresponding probabilities 265, 264, 135, 40, 15, and 1, all divided by 720. If you want to continue, you could compute the probabilities using the method of Exercise 23 of Section 2.2. But it seems clear that we would need some other method to study the situation in general. Two methods are available—the method of indicators, which we will see in Examples (4.2.17) and (4.2.18), enables us to find ES easily; and the method of Exercise 26 at the end of this section, which produces more results. ▮

(4.1.10) Example. Let X be an absolutely continuous random variable whose density is

$$f(x) = 2x \quad \text{for } 0 < x < 1 \quad \text{(and 0 otherwise)}.$$

You may remember [Example (3.3.6d)] that if we choose a random point having the uniform distribution on a disk of radius 1 and let X be its distance from the center, then this is the density of X. Notice that the density is large when x is near 1, and so X is more likely to be near 1 than near 0. Accordingly, the expected value of X ought to be closer to 1 than 0. And it is:

$$EX = \int_0^1 x \cdot 2x dx = \tfrac{2}{3}.$$ ▮

Before continuing, guess the expected value of a random variable X that has the uniform distribution on [0, 1]; then find it using the definition. (This is Exercise 18a at the end of the section.) The density is of course

$$f(x) = \begin{cases} 1 & \text{if } 0 \le x \le 1, \\ 0 & \text{otherwise.} \end{cases}$$

(4.1.11) Example: The expected value of a Poisson random variable is λ. Let X have the Poisson distribution with parameter λ; that is, let X be a discrete random variable with the mass function

$$p(x) = \begin{cases} \dfrac{\lambda^x e^{-\lambda}}{x!} & \text{for } x = 0, 1, 2, \ldots, \\ 0 & \text{otherwise.} \end{cases}$$

What is EX? According to Definition (4.1.1), it is the following sum, provided the sum converges absolutely:

$$EX = 0 \cdot \frac{\lambda^0 e^{-\lambda}}{0!} + 1 \cdot \frac{\lambda^1 e^{-\lambda}}{1!} + 2 \cdot \frac{\lambda^2 e^{-\lambda}}{2!} + 3 \cdot \frac{\lambda^3 e^{-\lambda}}{3!} + \cdots.$$

Again, absolute convergence is the same as ordinary convergence because all the terms are already nonnegative. But what is the sum? The first term is 0, and there is a nice cancellation in all the other terms; because $k/k! = 1/(k-1)!$, we get

$$EX = 0 + \frac{\lambda^1 e^{-\lambda}}{0!} + \frac{\lambda^2 e^{-\lambda}}{1!} + \frac{\lambda^3 e^{-\lambda}}{2!} + \cdots,$$

and if we factor out λ, what remains is just the sum of all the Poisson probabilities:

$$EX = \lambda \cdot \left(\frac{\lambda^0 e^{-\lambda}}{0!} + \frac{\lambda^1 e^{-\lambda}}{1!} + \frac{\lambda^2 e^{-\lambda}}{2!} + \cdots \right) = \lambda \cdot 1 = \lambda. \quad \blacksquare$$

(4.1.12) Note on the Poisson process. The previous example shows why the parameter λ of a Poisson process should be chosen as the average number of arrivals per unit time. Recall that for a Poisson process with arrival rate λ per minute, the number of arrivals in a given minute has the Poisson distribution with parameter λ. The preceding example shows that the expected number of arrivals in a given minute is indeed λ. This says that over a large number of minutes, the average number of arrivals per minute is expected to be close to λ, which is exactly why we called λ the arrival rate.

Example (4.1.6) gives the companion result: the expected time between arrivals is $1/\lambda$. That is, the arrival rate, which is the expected number of arrivals per unit time, is the reciprocal of the expected time between arrivals.

(4.1.13) **Example.** Let X be absolutely continuous, with the density

$$f(x) = \begin{cases} x^{-2} & \text{if } x > 1, \\ 0 & \text{otherwise.} \end{cases}$$

Then EX is the following integral, provided the integral converges absolutely:

$$EX = \int_1^\infty x \cdot x^{-2} dx = \int_1^\infty \frac{1}{x} dx.$$

Once again absolute convergence is the same as ordinary convergence; the integrand is positive, so if it converges at all, then it converges absolutely. But it does not converge:

$$\int_1^\infty \frac{1}{x} dx = \lim_{A \to \infty} \int_1^A \frac{1}{x} dx = \lim_{A \to \infty} (\ln(A) - \ln(1)) = \infty.$$

So this random variable does not have an expected value. We could say that the expected value is $+\infty$, or, more properly, that it diverges to $+\infty$. The main point to remember is that this random variable has no number that we can call its expected value. $\blacksquare$

The fact that some random variables simply do not have expected values may cause you to wonder if something is wrong with our theory. Surely, if we make many repetitions of an experiment that gives the X of the previous example, and average the results, we will get *some* number; yet the definition refuses to give us a number to represent what we expect.

The answer is that of course there will be an average after any number of trials, but there is no one number that we can expect the averages to be close to when the number of trials is large. In fact, there are some theorems, called *laws of large numbers,* which predict that if we perform a sequence of trials and recompute the average observed value of X after each trial, then that sequence of averages will tend to EX if the expected value exists, and will not have a limit if it does not. We will study such theorems in Section 4.5.

In the previous example, because the integral diverges to ∞, it can be shown that the sequence of observed averages will diverge to ∞.

Here is an example of a discrete random variable that has no expected value.

(4.1.14) Example: My luck is bad and getting worse. Let X have the possible values 1, 2, 3, . . . , and the probability mass function

$$p(n) = \frac{1}{n(n+1)} \quad \text{for } n = 1, 2, 3, \ldots .$$

After doing the calculation we will see what this has to do with my bad luck (and yours).

The definition says that

$$EX = \sum_{n=1}^{\infty} n \cdot \frac{1}{n(n+1)},$$

provided this series converges absolutely. All the terms are positive, so the only question is whether the series converges at all. It does not, of course. It equals

$$\sum_{n=1}^{\infty} \frac{1}{(n+1)} = \frac{1}{2} + \frac{1}{3} + \frac{1}{4} + \cdots,$$

which is the famous harmonic series with the first term missing; it diverges to $+\infty$. So EX does not exist, or diverges to $+\infty$. According to the strong law of large numbers (a theorem that we will see at the end of this chapter), if we do repeated performances of an experiment that produces X and if, after each new X comes in, we average the X's obtained so far, then our sequence of averages will grow without bound.

What does this result have to do with bad luck? The following discussion is a bit impractical, but the result is amusing. The example was noted by William Feller in Volume 2 of *An Introduction to Probability Theory and Its Applications,* Section I.5.

Suppose I and a number of my acquaintances are all exposed to some situation in which something unfortunate happens to us—perhaps we all take the same course from some instructor who is less than perfectly competent. Suppose further that we are all able to measure our own degree of bad luck—in the case of the miserable instructor, perhaps we each end with a certain average score on tests which in no way reflects our ability. (To take another example, suppose we all invest money in some attractive scheme, and we all lose a certain amount. Or say we all are assigned priority numbers at random for basketball tickets.)

Now the experiment is as follows. I know how badly I did and I begin asking my friends how badly they did; X is the number of friends I have to ask before finding one who did worse than I did. We will see below that X has the distribution given in this example, and consequently that EX diverges to ∞.

What does it mean that EX diverges to ∞? Suppose I do this experiment over and over throughout my life, as I encounter various unfortunate situations. Each time I do I record X, the number of people I have to ask before finding someone unluckier than I am, and I also calculate the average of all the X's I have recorded to that point. Then my average value of X will grow without bound; if I lived infinitely long, my average X would tend to ∞. Hence the title of this example.

We close the example by showing why the X described actually has the distribution given. It is fairly simple: For any n, the probability that X is *greater than* n is just the probability that the first n people were luckier than I was—that is, that I was the unluckiest among $n + 1$ people. Assuming that the bad luck is distributed fairly, this probability is $1/(n + 1)$. So

$$P(X > n) = \frac{1}{n + 1},$$

and, therefore, for $n = 1, 2, 3, \ldots,$

$$\begin{aligned} p(n) = P(X = n) &= P(X > n - 1) - P(X > n) \\ &= \frac{1}{n} - \frac{1}{n + 1} = \frac{1}{n(n + 1)}. \end{aligned}$$

∎

(4.1.15) Checking absolute convergence for absolutely continuous random variables. Now we consider random variables that have both positive and negative possible values, where absolute convergence is not identical to ordinary convergence. The typical situation in the absolutely continuous case is one in which the density is positive for all real x, so that

$$EX = \int_{-\infty}^{\infty} xf(x)dx,$$

provided this converges absolutely—that is, provided that the improper integral $\int_{-\infty}^{\infty} |x|f(x)dx$ converges. One way to check whether it does is to consider the two integrals

(4.1.16)
$$EX^+ = \int_0^{\infty} xf(x)dx \qquad \text{and} \qquad EX^- = \int_{-\infty}^{0} xf(x)dx.$$

The integrand in EX^+ is always positive, and the one in EX^- is always negative. So for both of these integrals, absolute convergence is the same as ordinary convergence; the integral converges absolutely as long as they both converge to finite numbers. And it is easy to see that

$$\int_{-\infty}^{\infty} xf(x)dx = EX^+ + EX^-,$$

and also that

$$\int_{-\infty}^{\infty} |x| f(x)dx = EX^+ - EX^-,$$

which implies that the integral converges absolutely if and only if both EX^+ and EX^- are finite. Summarizing:

(4.1.17) If X is an absolutely continuous random variable with density $f(x)$, then EX exists if and only if both EX^+ and EX^- are finite; and if they are, then $EX = EX^+ + EX^-$.

(4.1.18) **Example: The standard Cauchy distribution has no expected value.** Let X be an absolutely continuous random variable with the following density:

$$f(x) = \frac{1}{\pi(1 + x^2)} \quad \text{for } x \in \mathbb{R}.$$

This is called the standard Cauchy density; its graph is shown in Figure 4.1d. Its expected value is defined as the integral

$$EX = \int_{-\infty}^{\infty} \frac{x}{\pi(1 + x^2)} dx,$$

provided this integral converges absolutely; our job is to check whether it does by computing EX^+ and EX^-; if they are finite, then EX is their sum. We have

$$EX^+ = \int_0^{\infty} \frac{x}{\pi(1 + x^2)} dx \quad \text{and} \quad EX^- = \int_{-\infty}^{0} \frac{x}{\pi(1 + x^2)} dx.$$

EX is the sum of these two numbers, if they are both finite; if either or both of them are infinite, then EX does not exist. In this case, they are *not* both finite; in fact, they are both infinite:

$$\begin{aligned} EX^+ &= \int_0^{\infty} \frac{x}{\pi(1 + x^2)} dx = \lim_{A \to \infty} \int_0^{A} \frac{x}{\pi(1 + x^2)} dx \\ &= \lim_{A \to \infty} \frac{1}{2\pi} \Big[\ln(1 + x^2) \Big]_0^A = \lim_{A \to \infty} \frac{1}{2\pi} \ln(1 + A^2), \end{aligned}$$

which diverges to ∞, and

$$EX^- = \int_{-\infty}^{0} \frac{x}{\pi(1 + x^2)} dx,$$

which is just the same as

$$\int_0^\infty \frac{-x}{\pi(1 + x^2)} dx = -EX^+.$$

So EX does not exist. We cannot even say that it diverges to ∞ as in Examples (4.1.13) and (4.1.14), because in this case there is divergence to both $+\infty$ and $-\infty$. If we observed a sequence of values of X, and kept track of the current average after each observation, then the law of large numbers (which we will study later) would predict a sequence of averages which not only does not converge, but fails to diverge to either $+\infty$ or $-\infty$; it would swing between increasingly large positive values and increasingly large negative values.

The standard Cauchy distribution is one of the most famous examples of a distribution that has no expected value. The graph of the standard Cauchy density is not very different from that of the standard normal density (see Figure 4.1d), yet the standard normal distribution *does* have an expected value, as we see in the next example.

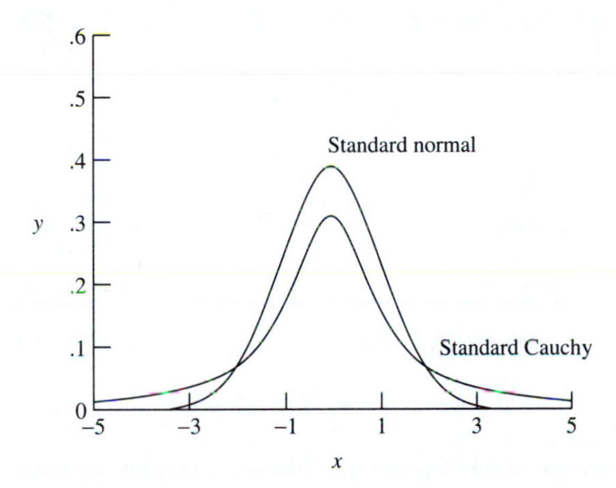

FIGURE 4.1d Normal and Cauchy densities

By the way, you ought to be able to show that the Cauchy density actually integrates to 1—that is, that

$$\int_{-\infty}^\infty \frac{1}{\pi(1 + x^2)} dx = 1.$$

The way to do it, if you cannot remember the indefinite integral of $1/(1 + x^2)$, is to notice that this integral is just twice the integral from 0 to ∞, and then to evaluate that integral using the substitution $x = \tan(u)$. [The indefinite integral of $1/(1 + x^2)$ is $\arctan(x) + C$.] ∎

(4.1.19) **Example: The expected value of the standard normal distribution is 0.** Let X have the standard normal density

$$\varphi(x) = \frac{1}{\sqrt{2\pi}} e^{-x^2/2} \quad \text{for } x \in \mathbb{R}.$$

We want to find EX, if it exists:

$$EX = \int_{-\infty}^{\infty} \frac{x}{\sqrt{2\pi}} e^{-x^2/2} dx,$$

provided this integral converges absolutely. As in the previous example, we consider

$$EX^+ = \int_0^{\infty} \frac{x}{\sqrt{2\pi}} e^{-x^2/2} dx \qquad \text{and} \qquad EX^- = \int_{-\infty}^{0} \frac{x}{\sqrt{2\pi}} e^{-x^2/2} dx;$$

if they are both finite, then EX exists and equals $EX^+ + EX^-$. Now

$$\begin{aligned} EX^+ &= \int_0^{\infty} \frac{x}{\sqrt{2\pi}} e^{-x^2/2} dx = \lim_{A\to\infty} \int_0^{A} \frac{x}{\sqrt{2\pi}} e^{-x^2/2} dx \\ &= \lim_{A\to\infty} \left[-\frac{1}{\sqrt{2\pi}} e^{-x^2/2} \right]_0^A = \lim_{A\to\infty} \left[\frac{1}{\sqrt{2\pi}} - \frac{1}{\sqrt{2\pi}} e^{-A^2/2} \right] = \frac{1}{\sqrt{2\pi}}, \end{aligned}$$

a finite number. It is easy to see that EX^- equals $-EX^+$. So

$$EX = EX^+ + EX^- = EX^+ + (-EX^+) = 0.$$

Thus, if X is a standard normal random variable, then EX is 0. ∎

(4.1.20) Checking absolute convergence for discrete random variables. This situation arises less often in practice, but it does sometimes happen that one needs to deal with a discrete random variable that has infinitely many positive values and infinitely many negative values. In this case, according to Definition (4.1.1),

$$EX = \sum_x xp(x),$$

and this sum includes infinitely many terms in which x is positive and infinitely many in which x is negative. (In most situations like this that are encountered in practice, the set of possible values will be the set of all positive and negative integers.) Then EX exists, provided the sum $\sum_x |x|p(x)$ converges.

As we did in the absolutely continuous case, we consider the two sums

$$EX^+ = \sum_{x>0} xp(x) \qquad \text{and} \qquad EX^- = \sum_{x<0} xp(x).$$

All the terms in EX^+ are positive and all those in EX^- are negative; so in both of them, absolute convergence is the same as ordinary convergence. In addition,

$$\sum_x xp(x) = EX^+ + EX^-$$

and

$$\sum_x |x|p(x) = EX^+ - EX^-.$$

So the prescription (4.1.17) applies to the discrete case as well.

(4.1.21) **Example.** Suppose every integer, positive and negative, is a possible value of X, and that the mass function is

$$p(n) = \begin{cases} \left(\frac{1}{3}\right)^{n+1} & \text{if } n = 0, 1, 2, \ldots, \\ \left(\frac{1}{2}\right)^{|n|+1} & \text{if } n = -1, -2, -3, \ldots. \end{cases}$$

So, for example, $P(X = 6) = \left(\frac{1}{3}\right)^7$ and $P(X = -6) = \left(\frac{1}{2}\right)^7$. (This distribution is contrived just to illustrate the principles.) We consider the two sums

$$EX^+ = \sum_{n=1}^{\infty} n \cdot \left(\frac{1}{3}\right)^{n+1} \quad \text{and} \quad EX^- = \sum_{n=-1}^{-\infty} n \cdot \left(\frac{1}{2}\right)^{|n|+1}.$$

Notice that the term for $n = 0$ appears in neither of these sums; that term is $0 \cdot \left(\frac{1}{3}\right)^1$, which equals 0. You can check the following calculations:

$$EX^+ = \left(\frac{1}{3}\right)^2 \cdot \left(1 + 2 \cdot \frac{1}{3} + 3 \cdot \left(\frac{1}{3}\right)^2 + \cdots\right) = \left(\frac{1}{3}\right)^2 \cdot \frac{1}{\left(1 - \frac{1}{3}\right)^2} = \frac{1}{4}$$

and

$$EX^- = -\left(\frac{1}{2}\right)^2\left(1 + 2 \cdot \frac{1}{2} + 3 \cdot \left(\frac{1}{2}\right)^2 + \cdots\right) = -\left(\frac{1}{2}\right)^2 \cdot \frac{1}{\left(1 - \frac{1}{2}\right)^2} = -1.$$

Both are finite, so EX is their sum:

$$EX = \frac{1}{4} - 1 = -\frac{3}{4}.$$

∎

For readers who will go on to a course in measure theory, we should point out that EX^- will probably be defined there so as to be the negative of our EX^-. This will make X^- a nonnegative random variable, like X^+, and that is more convenient in defining the kinds of integrals needed in measure theory.

Our next example does not involve convergence problems, but it is extremely important.

(4.1.22) **Example: The expected value of a binomial random variable is *np*.** Let X have the binomial distribution with parameters n and p. That is, consider n Bernoulli trials with success probability p (and failure probability $q = 1 - p$) on each trial, and let X be the number of successes. Then to find EX we need to evaluate the following sum:

$$EX = \sum_{k=0}^{n} k \cdot \binom{n}{k} p^k q^{n-k}.$$

There is no concern about absolute convergence because there are only finitely many terms.

Before we find EX, notice why you should not be surprised that np is the answer: If you do $n = 100$ trials with success probability $p = .3$, you should expect to average 30 successes per 100 trials, because the probability .3 is supposed to represent the expected relative frequency of successes over a large number of trials. That is, if you were to do the 100-trial experiment many times and record the number of successes (out of a possible 100) each time, you would expect an average of 30 successes per 100-trial experiment.

In general, if p is the expected relative frequency of successes, then we expect np successes in n trials. Let us do the calculation first for a very simple case, and then in general, to see that Definition (4.1.1) indeed gives us np.

First, suppose $n = 3$ and $p = \frac{1}{6}$, so that $np = .5$. Is EX equal to .5? The possible values of X are 0, 1, 2, and 3, and their probabilities are

$$P(X = 0) = \binom{3}{0}\left(\tfrac{1}{6}\right)^0\left(\tfrac{5}{6}\right)^3 = \tfrac{125}{216}; \qquad P(X = 1) = \binom{3}{1}\left(\tfrac{1}{6}\right)^1\left(\tfrac{5}{6}\right)^2 = \tfrac{75}{216};$$

$$P(X = 2) = \binom{3}{2}\left(\tfrac{1}{6}\right)^2\left(\tfrac{5}{6}\right)^1 = \tfrac{15}{216}; \qquad P(X = 3) = \binom{3}{3}\left(\tfrac{1}{6}\right)^3\left(\tfrac{5}{6}\right)^0 = \tfrac{1}{216}.$$

So

$$EX = 0 \cdot \tfrac{125}{216} + 1 \cdot \tfrac{75}{216} + 2 \cdot \tfrac{15}{216} + 3 \cdot \tfrac{1}{216} = \tfrac{108}{216} = .5.$$

So far, so good; what about general n and p? We will give two different calculations here to show that the answer is np. Later, using the "method of indicators," we will see a third way to show the same thing. We do three different calculations of the same result for two reasons. First, the result is very important. Second, the different calculations illustrate various useful methods of finding sums like this one.

Calculation 1. This is the "brute force" calculation, using a minimum of tricks (although it still takes a little cleverness). The idea is to notice that each term in EX, except the first one, which is 0, has a factor of np. When we divide np out, what is left are the binomial probabilities for parameters $n - 1$ and p, and the binomial probabilities for any pair of parameters must add to 1.

$$EX = 0 \cdot \binom{n}{0}p^0q^n + 1 \cdot \binom{n}{1}p^1q^{n-1} + 2 \cdot \binom{n}{2}p^2q^{n-2} + 3 \cdot \binom{n}{3}p^3q^{n-3} + \cdots + k \cdot \binom{n}{k}p^kq^{n-k} + \cdots + n \cdot \binom{n}{n}p^nq^0.$$

Now the first term is 0, and we proceed to look at the others:

$$1 \cdot \binom{n}{1}p^1q^{n-1} = np \cdot q^{n-1};$$

$$2 \cdot \binom{n}{2}p^2q^{n-2} = n(n-1)p^2q^{n-2} = np \cdot (n-1)pq^{n-2};$$

$$3 \cdot \binom{n}{3}p^3q^{n-3} = n \cdot \frac{(n-1)(n-2)}{2}p^3q^{n-3} = np \cdot \binom{n-1}{2}p^2q^{n-3};$$

in general,

$$\begin{aligned} k \cdot \binom{n}{k}p^kq^{n-k} &= k \cdot \frac{n(n-1)(n-2)\cdots(n-k+1)}{k!} \cdot p^kq^{n-k} \\ &= n \cdot \frac{(n-1)(n-2)\cdots(n-k+1)}{(k-1)!} \cdot p^kq^{n-k} \\ &= np \cdot \binom{n-1}{k-1} \cdot p^{k-1}q^{n-k}; \end{aligned}$$

and finally

$$n \cdot \binom{n}{n}p^nq^0 = np \cdot p^{n-1}.$$

We have isolated a factor of np in every term. If we divide it out and look at what is left of each term, we see

$$\begin{aligned} q^{n-1}, &\quad \text{which equals } \binom{n-1}{0}p^0q^{n-1}; \\ (n-1)pq^{n-2}, &\quad \text{which equals } \binom{n-1}{1}p^1q^{n-1-1}; \\ \binom{n-1}{2}p^2q^{n-3}, &\quad \text{which equals } \binom{n-1}{2}p^2q^{n-1-2}; \end{aligned}$$

in general,

$$\binom{n-1}{k-1} \cdot p^{k-1}q^{n-k}, \quad \text{which equals } \binom{n-1}{k-1}p^{k-1}q^{n-1-(k-1)};$$

and finally

$$p^{n-1}, \quad \text{which equals } \binom{n-1}{n-1}p^{n-1}q^0.$$

These are exactly the binomial probabilities for parameters $n - 1$ and p in place of n and p. So these numbers add to 1, and therefore the terms in EX add to np. ∎

Actually, Calculation 1 can be made shorter, though perhaps a little more opaque, if we use summation notation after first checking the identity

$$k \cdot \binom{n}{k} = n \cdot \binom{n-1}{k-1} \quad \text{for } k = 1, 2, 3, \ldots, n.$$

This is checked by writing out the two binomial coefficients using the definition $\binom{n}{k} = n!/k!(n-k)!$. Using it, we get

$$\begin{aligned}
EX &= \sum_{k=0}^{n} k \cdot \binom{n}{k} p^k q^{n-k} \\
&= \sum_{k=1}^{n} k \cdot \binom{n}{k} p^k q^{n-k} \qquad \text{(because the term for } k = 0 \text{ equals 0)} \\
&= \sum_{k=1}^{n} n \cdot \binom{n-1}{k-1} p^k q^{n-k} \\
&= np \cdot \sum_{k=1}^{n} \binom{n-1}{k-1} p^{k-1} q^{(n-1)-(k-1)} \\
&= np \cdot \sum_{j=0}^{n-1} \binom{n-1}{j} p^j q^{(n-1)-j} \qquad (j \text{ replaces } k - 1) \\
&= np \cdot 1 = np \qquad \text{(the sum of binomial probabilities is 1).}
\end{aligned}$$ ∎

Calculation 2. This calculation uses a trick involving the differentiation of a known sum, like the one used to get (A.5.7) in Appendix A. Recall that the binomial theorem gives us an expansion for $(a + b)^n$. If we write it out with pt in place of a and $q = 1 - p$ in place of b, we get

(4.1.23)
$$(pt + q)^n = \sum_{k=0}^{n} \binom{n}{k} (pt)^k q^{n-k} \quad \text{for any positive real number } t.$$

If we differentiate this with respect to t, we get

$$np(pt + q)^{n-1} = \sum_{k=0}^{n} \binom{n}{k} p^k \cdot k t^{k-1} q^{n-k},$$

and since this is true for all t it is true for $t = 1$. But if we take $t = 1$, we get (because $p + q = 1$)

$$np = \sum_{k=0}^{n} k \cdot \binom{n}{k} p^k q^{n-k},$$

and this sum is just EX. ∎

This beautiful calculation shows one of the uses of what are called *probability generating functions,* which we will study in Chapter 7. The sum in (4.1.23) is actually the expected value of a certain function of X, namely $h(X) = t^X$, as we will see in the next section. ▮

(4.1.24) **The expected value of a constant.** This may seem silly at first, but it is often useful to consider a number a as a random variable that equals a with probability 1. For example, the following statements are both true, and we will see them in the next section: (i) If X and Y are random variables and $Z = X + Y$, then $EZ = EX + EY$. (ii) If X is a random variable and $Z = X + a$ for some number a, then $EZ = EX + a$. Now if the constant a can be viewed as a random variable that always equals a and whose expected value is a, then (ii) can be omitted because (i) already says it.

A more important reason is that we sometimes have to deal with a quantity that is defined as a random variable—that is, as a numerical function of the outcome of an experiment—but that turns out to have only one possible value, and to take that value with probability 1. A good example of this is a theorem that we will see in Section 4.5, called the *law of large numbers.* Informally, it says the following:

> Suppose a certain chance experiment produces a random variable X whose expected value is a, and suppose we do the experiment repeatedly under identical conditions. Let X_n be the observed value of X on the nth repetition, let Y_n be the average of the first n X's (that is, $Y_n = (X_1 + X_2 + \cdots + X_n)/n$), and let $Z = \lim_{n\to\infty} Y_n$. Then $Z = a$ with probability 1.

This theorem justifies our attempt to define EX as the expected long-run average value of X, because it says that the average observed values of X are certain to approach its expected value. It is clear that Z is defined as a random variable, not as a number; but the theorem says that it is a constant. Not to allow constants to be random variables would make it difficult to express this theorem.

What kind of random variable is the constant a? The answer is easy. If the random variable X is equal to the number a with probability 1, then X is a discrete random variable with the mass function

$$p(x) = \begin{cases} 1 & \text{if } x = a, \\ 0 & \text{otherwise.} \end{cases}$$

The expected value is of course $1 \cdot a = a$, by Definition (4.1.1).

(4.1.25) **Expected values of random variables that are neither discrete nor absolutely continuous.** Every random variable we study in this text is either discrete, with a mass function $p(x)$ determining its distribution, or absolutely continuous, with a density function $f(x)$ determining its distribution. If a random variable is of neither of these two types, then of course neither of the two definitions above tells us how to find its expected value. When such random variables are encountered, it is usually possible to find their expected values using "conditioning" arguments like those presented in Chapter 9 of this text.

There is a definition of expected value, however, that works for any random variable. In fact, there are two equivalent definitions, both involving types of integrals that are more general than the Riemann integrals of calculus.

One definition involves what is called a *Riemann–Stieltjes integral:* $EX = \int_{-\infty}^{\infty} x dF(x)$, which involves $F(x)$, the CDF of X. Reimann–Stieltjes integrals are generally studied in upper-level courses in real analysis. Since all random variables have CDFs, this definition is perfectly general.

The other definition involves the measure-theoretic definition of the integral of a function X from a probability space Ω to $\mathbb{R}$, denoted by $\int_\Omega X(\omega)P(d\omega)$. It turns out that random variables are precisely the kind of functions for which such integrals can be defined, and EX is defined to be the value of the integral (as always, provided the integral converges absolutely).

However, we will not consider such general definitions further.

EXERCISES FOR **SECTION 4.1**

1. Let X be the number of spots that show when a fair die is rolled. Find EX.

2. We roll a fair die until an odd number comes up; let X be the number of even numbers we get before the first odd number. Find EX.

3. Find EX if X has the given density.
   a. $f(x) = 6x(1 - x)$ for $0 < x < 1$ (and 0 otherwise)
   b. $f(x) = 3/x^4$ for $x > 1$ (and 0 otherwise)

4. **Accepting the cash prize from Monty Hall.** You are playing a television game show involving three doors; behind one of the three is a valuable prize. You picked door 1 and the host then opened door 3, showing that the valuable prize was not behind it. You are then offered a cash prize instead of the chance to win whatever is behind door 1. As we saw in Exercise 20 of Section 2.2, the (conditional) probability that the prize is behind door 1 is now $\frac{1}{3}$. Show that if the prize has a value of V, then your expected gain in staying with door 1 is $\frac{1}{3}V$.

   Of course, if you take the cash, your gain is a constant—say C dollars is the amount of cash you are offered. Consequently, if you base your decision on maximizing your expected gain, you should stay with door 1 if V is greater than $3C$ and take the cash otherwise.

   However, not all people will base their decisions on the expected gain. Some people do not like risks and will take the cash, if it is a reasonable amount, rather than risk getting nothing. Others like a gamble and will refuse the cash, especially if the hidden prize is valuable. For more on this, see Exercise 10 of Section 6.3.

5. A hat contains five slips of paper, numbered 1, 2, 3, 4, and 5. Two are chosen at random, without replacement, and X is the sum of the two numbers. Write down the possible values of X and their probabilities; from them find EX.

6. Recall that $f(x) = \frac{1}{2}\sin x$ $(0 < x < \pi)$ defines a density function. Find the associated expected value.

7. The mass function of X is given by $p(k) = (k - 1)\left(\frac{1}{2}\right)^k$ for $k = 2, 3, 4, \ldots$. Find EX.

8. In roulette (as played in most American gambling houses) there are 18 squares with (positive) even numbers on them, 18 squares with odd numbers, and two other squares, marked 0 and 00. A wheel is used to choose one of the 38 squares at random. If you bet a dollar on "even," you will get back your dollar plus another dollar if one of the positive even squares is chosen; in this case your gain is $+1$. If an odd square or one of the zero squares is chosen, you lose your dollar; your gain is -1.
   a. If X is your gain on one play, write down the possible values of X and their probabilities.
   b. Find your expected gain on one play.
   c. What is your expected total net gain on 100 plays? On 1,000 plays? 10,000? (Remember the interpretation of expected value as expected long-run average value per play.)

Looking at the result of part c from the gambling house's point of view shows why it is not much of a gamble to own one. Many more than 10,000 bets are made in a typical evening in a large casino; many of them are less favorable to the players than the one described, and most of them are for considerably more than one dollar.

9. **Chuck-a-luck.** In the popular carnival game of chuck-a-luck, three fair dice are rolled. You bet on one of the numbers; say you bet a dollar on number 6. If no 6's show, you lose your dollar. If any 6's show, you get your dollar back, plus an extra dollar for each 6 that shows. Let X be your net gain. So X will be -1 if no 6's show; otherwise, it will be $+1$, $+2$, or $+3$ depending on the number of 6's that show.
   a. Find the probability mass function of X.
   b. Find EX.
   c. What do you expect to happen if you bet a dollar on 6 a thousand times?

Notice that the distribution of X is the same whether you bet on 6 or on some other number. A little thought will convince you that it is the same even if you change your number between bets. So you can expect the same thing to happen in 1,000 plays, no matter what numbers you bet on.

10. **EX does not always represent a "typical" value of X.** Suppose the possible values of X are 1, 2, and 1,000, with probabilities $\frac{1}{4}$, $\frac{1}{2}$, and $\frac{1}{4}$, respectively. Find EX. [Notice that EX is not the most probable value of X; it is not even close to any of the possible values of X. All EX has to recommend it is that if we averaged many observations of X, the average would likely be close to EX.]

11. **EX is not always a "middle" value of the distribution.** Let X have the exponential distribution with parameter λ.
    a. What is the probability that X is greater than EX? [It is not $\frac{1}{2}$.]
    b. Find the number m for which $P(X > m) = \frac{1}{2}$ [it is called the *median* of X] and show that regardless of λ, it is less than EX.

12. In Example (2.3.4) we considered a gambling game in which you lose a dollar with probability .693 (approximately) and win a dollar with probability .307.

[This is the game in which you toss a coin, putting blank slips of paper into a hat after each tail, and putting in a marked slip after the first head; then you draw one slip, and you win if you do not draw the marked one.]

a. If X is your net gain on one play, find EX.

b. Find your expected total gain in 1,000 plays.

c. Now suppose that you lose a dollar with probability .693, as before, but win *two* dollars with probability .307. If Y is your net gain on one play, find EY.

d. Now what is your expected net gain after 1,000 plays?

13. Players A and B are interrupted in the middle of a game; the winner of the game was to have collected a dollar from the loser. Suppose that at the time they are interrupted, the probability that A will win is p. If p is greater than $\frac{1}{2}$, argue that $2p - 1$ dollars is a fair amount for B to pay A to settle the game. [See Exercise 14 of Section 1.4 on the "problem of points" and the origin of probability theory.]

14. The *double exponential* density with parameter λ is $f(x) = \lambda e^{-\lambda|x|}/2$ for all real x. If X has this density, find EX^+ and EX^-. [See (4.1.16).] Does EX exist? If so, what is it?

15. If X has the density $f(x) = xe^{-x^2/2}$ for $x > 0$ (and 0 otherwise), find EX.

16. If X is the number of failures before the second success in a Bernoulli process, then as we have seen earlier, the mass function of X is $p(n) = (n + 1)p^2q^n$ for $n = 0, 1, 2, \ldots$. Find EX.

17. A hat contains 15 slips of paper with numbers on them; there is one 1, two 2's, and so on. If X is the number on a randomly drawn slip from this hat, find EX.

18. **a.** Let X have the uniform distribution on the interval $[0, 1]$. Find EX. [Guess the answer first, then confirm it by integrating x times the density.]

b. Repeat part a for the interval $[a, b]$.

19. Let the set of possible values of X be $[0, 1]$.

a. Suppose the density of X is proportional to x^2. Find EX.

b. Suppose the density of X is proportional to x^k. Find EX.

20. Let X be a nonnegative random variable.

a. Suppose the possible values of X are integers. Show that $\sum_{n=0}^{\infty} P(X > n) = EX$.

b. Suppose X is absolutely continuous. Show that $\int_0^\infty P(X > x)dx = EX$. [*Hint:* Write $P(X > x)$ as an integral from x to ∞, then reverse the order of integration.]

These remarkable formulas often provide useful ways to find expected values. The formula in part b is actually valid for all nonnegative random variables, whether they are absolutely continuous, discrete, or neither.

21. **Does the gambler's ruin problem point to a way to beat the odds?** In Exercise 8 above, we saw that in roulette at an American gambling house, your expected net gain per play is negative; accordingly, if you play many times, you can expect to lose many dollars. But in Exercise 15 of Section 2.3 we saw that if you start with \$20 and only want to gain \$1, your chance of success is about .8878, because the probability of being ruined before you get to \$21 is only .1122. Apparently you could do this repeatedly, starting with \$20 and aiming for \$21, and succeed almost eight times as often as you fail.

    To see why you will still lose money in the long run, let X be your gain on one trial of this experiment (that is, the experiment of starting with \$20 and quitting when you reach either \$21 or \$0). X has only two possible values, $+1$ and -20. Find EX. What will happen over a long series of trials?

22. **Can the Martingale gambling system beat the odds?** In this gambling scheme, which we first saw in Exercise 15 of Section 1.4, instead of betting a dollar at a time, you double your bet after each loss. We saw that if your win probability on a single play is $p = .45$ and you start with \$127, then you will gain a dollar before being ruined with probability .985 (approximately). Repeat Exercise 21 for the experiment of starting with \$127 and quitting when you reach \$128 or \$0. (The two possible values of X are now $+1$ and -127.)

23. **The St. Petersburg paradox.** Players A and B play a game in which A pays B a fixed amount ("entry fee") at the start; then a fair coin is tossed repeatedly. If heads show on the first toss, B pays A \$1. If the first head shows on the second toss, B pays A \$2; if on the third toss, \$4; the fourth, \$8; and so on. To decide on a fair entry fee, they agree to use EX, where X is the amount B pays A on a single play. What is wrong?

This was considered a paradox in the early years of the subject, because there was not a clear distinction between the two concepts of expected value and fair entry fee. In fact, expectations and even probabilities were first thought of in terms of payments that made games fair; see Exercise 13 above and Exercise 14 in Section 1.4. Probabilists and statisticians now agree that EX should be the number defined as in this section; the problem of a fair entry fee, especially when EX does not exist, is not settled and depends on other concepts like the value ("utility") of money, the price paid for enjoyment of the game, and so on. Utility theory has become an important area in the behavioral sciences, economics, and business decision making.

The question is called the "St. Petersburg paradox" because Daniel Bernoulli's paper on the subject, *Specimen Theoriae Novae de Mensura Sortis (Exposition of a New Theory on the Measurement of Risk)*, was published in the journal of the St. Petersburg Academy in 1738. The problem had first been studied by Daniel Bernoulli's older cousin Nicholas Bernoulli, who had proposed it in a letter to the probabilist Montmort in 1713. Montmort realized that EX diverges to infinity, but apparently was not troubled by the nonexistence of a fair entry fee. A third Bernoulli, Nicholas's uncle James, was also indirectly involved via his student Gabriel Cramer, who proposed some resolutions of the paradox based on the idea of utility; see Exercise 24 of Section 4.2.

An English translation of Daniel Bernoulli's paper appears in the journal *Econometrica* (1954), vol. 22: 23–48. Further reading on this and other historical matters can be found in the *Encyclopedia of the Statistical Sciences,* edited by Samuel Kotz and Norman L. Johnson.

24. (*Computer exercise*) This exercise involves simulating many observations of a random variable, computing their average, and comparing the observed average with the expected value.

a. Write a program that simulates and prints (on the screen) 100 observations of a random variable that has the exponential distribution with parameter $\lambda = 3.2$; in addition, accumulate their sum and at the end compute the average and print it. [A random variable X having the exponential distribution with parameter λ is generated by letting U have the uniform distribution on $(0, 1)$ and taking $X = -(1/\lambda) \cdot \ln(U)$. This is discussed in Example (5.2.12).]

b. Run it two or three times and see if you can get a feeling for the kinds of numbers one gets. You will likely see mostly small numbers, with a few larger ones; some may be quite large.

c. Modify it so that it generates 1,000 numbers instead of 100, but does not print them all. Print only the average at the end. Run this program several times to see whether the averages seem to be near the expected value $\frac{1}{3.2} = .3125$.

d. Try it again, but generate 5,000 or 10,000 numbers instead of only 1,000; run it a few times and see whether the averages are closer to the expected value.

25. (*Computer exercise*) If you did Exercise 21 in Section 1.2, you wrote a program that generated a number of observations of a random variable having the binomial distribution with $n = 10$ and $p = .372$. We now know that the expected value of such a random variable is $10 \cdot .372 = 3.72$. Modify your program to generate several thousand observations of this random variable and print their average. Run this program several times to see how close the averages are to 3.72.

26. **The expected number of items in their proper places in a random permutation.** Let S_n denote the number of items in their proper places in a random permutation of n items. As in Exercise 23 of Section 2.2, denote $P(S_n = k)$ by p_{nk}. In that exercise we showed that $p_{nk} = (1/k)p_{n-1,\,k-1}$ for $n > 1$ and $k > 0$, and we used that formula to find the probabilities p_{nk}. Here, without computing any of the probabilities (but using the same formula), show that $ES_n = 1$ for all n.

It may seem surprising that no matter how large n is, the expected number of items in their proper places equals 1. This problem is often stated in a variety of concrete forms; for example, n people check their hats at a restaurant, but the tickets are lost and the hats are returned to the people at random. Whether n is 2, 2,000, or 2,000,000, the number of people who happen to get their own hats averages 1 over a long series of trials of this experiment.

We will get this same result using another method, the method of indicators, in Exercise 19 of the next section. The method used here is not so well known, and indicators have wide application apart from permutation problems. Still, the method of this exercise produces more results for random permutations than indicators do, as we will see in Exercise 19 of Section 7.1.

ANSWERS

1. 3.5

2. 1

3. **a.** $\frac{1}{2}$ **b.** $\frac{3}{2}$

5. 6

6. $\frac{\pi}{2}$

7. 4

8. **a.** $p(-1) = \frac{20}{38}$; $p(1) = \frac{18}{38}$ **b.** $-\frac{1}{19}$ dollar, or about a 5.2632¢ loss **c.** $-\$5.26$; $-\$52.63$; $-\$526.32$

9. **a.** Possible values $-1, 1, 2, 3$; probabilities $\frac{125}{216}, \frac{75}{216}, \frac{15}{216}, \frac{1}{216}$ **b.** $-\frac{17}{216}$ dollar, or about a 7.87¢ loss **c.** Expect to lose about \$78.70

10. 251.25

11. **a.** $1/e \doteq .3679$ **b.** $(\ln 2)/\lambda$ (which is less than $1/\lambda$)

12. **a.** $-\$.386$ (a 38.6¢ loss) **b.** $-\$386$ **c.** $-\$.079$ (a 7.9¢ loss) **d.** $-\$79$

14. It exists and equals 0.

15. $\frac{1}{2}\sqrt{2\pi}$

16. $2q/p$

17. $3\frac{2}{3}$

18. **a.** $\frac{1}{2}$ **b.** $(a + b)/2$

19. **a.** $\frac{3}{4}$ **b.** $(k + 1)/(k + 2)$

21. $EX = -1.3562$. Over a long series of trials you can expect to be behind by about \$1.36 per trial.

22. $EX = -.92$. Over a long series of trials you can expect to be behind by about 92¢ per trial.

23. EX does not exist (diverges to ∞).

4.2 THE ALGEBRA OF EXPECTED VALUES

This section contains the facts and formulas that are essential for working with expected values. There are really only three basic ideas that we need to know, and we will state them informally at the start. At the end of the section are a few more useful facts, but these three are by far the most important ones.

(4.2.2) If X is a discrete or absolutely continuous random variable and Y is a function of X, say $Y = h(X)$, then EY can be found without finding the distribution of Y, from the formula

$$EY = Eh(X) = \int_0^\infty h(x) f(x) dx \quad \text{or} \quad \sum_x h(x)p(x).$$

More generally, if Y is a function $h(X_1, X_2, \ldots, X_n)$ of several random variables, then EY can be found as a similar integral or sum using the joint density or mass function of the X's.

(4.2.13) If X and Y are random variables and a and b are numbers, then:

$$\text{If } Z = aX + bY, \text{ then } EZ = a \cdot EX + b \cdot EY.$$

More generally, if $X_1, X_2, \ldots, X_n$ are random variables and $a_1, a_2, \ldots, a_n$ are numbers, then:

If $Z = a_1X_1 + a_2X_2 + \cdots + a_nX_n$,

$$\text{then } EZ = a_1 \cdot EX_1 + a_2 \cdot EX_2 + \cdots + a_n \cdot EX_n.$$

(4.2.19) If X and Y are independent random variables and $Z = XY$, then $EZ = EX \cdot EY$.

(4.2.1) Expected value of a function of a random variable. Suppose you have been working with a random variable X, and you have calculated the number EX, but you are also interested in $Y = X^2$ and need its expected value, $EY = E(X^2)$. The first thing that comes to mind is

WRONG! → $$E(X^2) \overset{?}{=} (EX)^2.$$ **Z**

We will see what is right in Theorem (4.2.2) below. For now we want to emphasize what is *not* right—except in very special situations, the expected value of a function of X is not that function of EX.

To convince yourself of this, recall Example (4.1.2), in which X has possible values 0, 1, 2, and 3, with probabilities .1, .3, .2, and .4, and, therefore,

$$EX = 0 \cdot .1 + 1 \cdot .3 + 2 \cdot .2 + 3 \cdot .4 = 1.9.$$

Now if $Y = X^2$, then is $EY = 3.61$, which is the square of 1.9? No; Y has possible values 0, 1, 4, and 9 with probabilities .1, .3, .2, and .4, so

$$EY = 0 \cdot .1 + 1 \cdot .3 + 4 \cdot .2 + 9 \cdot .4 = 4.7.$$

This trap is so easy to fall into that we emphasize it here by giving several instances of it. Except in isolated examples,

$E(X^2)$ is not equal to $(EX)^2$.

$E\left(\dfrac{1}{X}\right)$ is not equal to $\dfrac{1}{EX}$.

$E(e^{tX})$ is not equal to $e^{t(EX)}$.

$E(\sqrt{X})$ is not equal to $\sqrt{EX}$.

$E(\ln X)$ is not equal to $\ln(EX)$.

Actually, there are examples of specific random variables and specific functions h for which $Eh(X) = h(EX)$; also, it is true for any function h if X is a constant random variable. In Exercise 18 at the end of the section we see another way it can happen. But the following is what needs to be remembered:

The only functions $h(x)$ for which $Eh(X) = h(EX)$, in general, are functions of the form $h(x) = ax + b$.

So what can we do to find EY if Y is a function of X? We are presumably given the distribution of X, but to use Definition (4.1.1) or (4.1.4) we would need to find the mass function or a density for Y. In later chapters we will see some techniques for finding the distribution of $Y = h(X)$ if we know the distribution of X, but the next theorem gives a much simpler way to find the expected value of Y without finding its distribution.

(4.2.2) **Theorem.** Let X be a random variable and let $Y = h(X)$.

a. If X is discrete with mass function $p(x)$, then $EY = \sum_x h(x)p(x)$, provided that this sum converges absolutely.
b. If X is absolutely continuous with density $f(x)$, then $EY = \int_{-\infty}^{\infty} h(x)f(x)dx$, provided that this integral converges absolutely.

We will not prove this theorem; the important thing to realize is that it is a theorem and not an automatic fact about expected values. The discrete part of it is not at all difficult to prove; the absolutely continuous part is related to theorems on integration by substitution that are usually proved in advanced calculus courses.

Instead, we will give some examples that might help to convince you of the truth of the theorem, as well as to show how it is used. The first one involves a simple discrete random variable.

(4.2.3) **Example.** Let X have possible values 1, 2, and 3, with probabilities .2, .5, and .3, respectively. Let $Y = 5X - X^2$. Then Theorem (4.2.2a) tells us that EY can be found as

$$(5 \cdot 1 - 1^2) \cdot .2 + (5 \cdot 2 - 2^2) \cdot .5 + (5 \cdot 3 - 3^2) \cdot .3 = 4 \cdot .2 + 6 \cdot .5 + 6 \cdot .3 = 5.6.$$

What could we do to convince ourselves that this is actually EY? We could write down the possible values of Y and their probabilities, and then compute EY from Definition (4.1.1). To do that we make the following table of the values of X, their probabilities, and the corresponding values of Y:

x:	1	2	3
$P(X = x)$:	.2	.5	.3
$y = 5x - x^2$:	4	6	6

Using the last two rows of this table, we see that the possible values of Y are 4 and 6, with probabilities .2 and .8, respectively. So according to the definition, $EY = 4 \cdot .2 + 6 \cdot .8 = 5.6$.

Comparing these two calculations should convince you that they will always lead to the same sum. There is nothing difficult about the proof; it is just a little messy notationally because of the possibility that, as in this example, some possible values of Y correspond to more than one possible value of X. ∎

(4.2.4) Example. Let X have the uniform distribution on (0, 1) and let $Y = h(X) = X^2$. Then according to Theorem (4.2.2b), we can find EY by

(4.2.5)
$$EY = E(X^2) = \int_0^1 x^2 \cdot 1 dx = \tfrac{1}{3}.$$

(The 1 in the integrand is the density of X.)

Can we do anything to convince ourselves that this is correct? If we knew the density of Y, $f_Y(y)$, we could compute EY from Definition (4.1.4) as $\int_0^1 y f_Y(y) dy$. But we do know that density; we found it in Example (3.3.2):

$$f_Y(y) = \begin{cases} \dfrac{1}{2\sqrt{y}} & \text{if } 0 < y < 1, \\ 0 & \text{otherwise.} \end{cases}$$

When we find EY from this, we get the same result:

(4.2.6)
$$\int_0^1 y \cdot \frac{1}{2\sqrt{y}} dy = \tfrac{1}{2} \int_0^1 y^{1/2} dy = \left[\tfrac{1}{2} \cdot \tfrac{2}{3} y^{3/2}\right]_0^1 = \tfrac{1}{3}.$$

To get an idea of how the proof goes in general, evaluate the integral in (4.2.5) by making the change of variable $y = x^2$; you will get the integral in (4.2.6). ∎

(4.2.7) Example: The expected square of an exponential random variable. If X has the exponential distribution with parameter λ, then we know from Example (4.1.6) that $EX = 1/\lambda$; but what is the expected value of $Y = X^2$? We use Theorem (4.2.2b):

$$EY = E(X^2) = \int_0^\infty x^2 \cdot \lambda e^{-\lambda x} dx;$$

to evaluate this we integrate by parts with $u = x^2$ and $dv = \lambda e^{-\lambda x} dx$; we get

$$EY = \left[-x^2 e^{-\lambda x}\right]_0^\infty + 2 \int_0^\infty x e^{-\lambda x} dx.$$

The first part is 0; to find the second part we use another integration by parts with $u = x$. The result is

$$EY = \frac{2}{\lambda^2}.$$

(Or we can do the second part by remembering that $\int_0^\infty x \cdot \lambda e^{-\lambda x} dx = EX = 1/\lambda$.)

∎

(4.2.8) Note on notation. It is customary to drop the parentheses in $E(X^2)$ and refer to it as EX^2. This notation is ambiguous; EX^2 could mean either $E(X^2)$ or $(EX)^2$, and as we pointed out earlier, these are not equal. But it is generally agreed that EX^2 refers to $E(X^2)$, the expected square of X. When we need to refer to the square of EX, we will use parentheses. The same convention holds for any power of X:

$$EX^k \text{ refers to } E(X^k), \text{ which is not the same as } (EX)^k.$$

(4.2.9) **Example: The expected square of a standard normal random variable is equal to 1.** If X is a standard normal random variable, then we know from Example (4.1.19) that $EX = 0$. What is EX^2? Again we use Theorem (4.2.2b):

$$EX^2 = \int_{-\infty}^{\infty} x^2 \cdot \frac{1}{\sqrt{2\pi}} e^{-x^2/2} dx.$$

This is an improper integral, but the integrand is always positive even though x ranges through negative values as well as positive. So absolute convergence is the same as convergence. Furthermore, we can simplify it so that it is improper at only one end by noting that the integrand is an *even* function; its value at $-x$ is the same as at x. So

$$EX^2 = 2 \int_0^{\infty} x^2 \cdot \frac{1}{\sqrt{2\pi}} e^{-x^2/2} dx.$$

The way to evaluate the integral is to integrate by parts with $u = x$ and $dv = xe^{-x^2/2}dx$; we get

(4.2.10)
$$EX^2 = \frac{2}{\sqrt{2\pi}} \cdot \left[-xe^{-x^2/2} \right]_0^{\infty} + \frac{2}{\sqrt{2\pi}} \int_0^{\infty} e^{-x^2/2} dx.$$

The first part equals 0. And the second part is just twice the integral of the standard normal density from 0 to ∞; so the second part is twice $\frac{1}{2}$, or 1. To repeat:

If X has the standard normal distribution, then $EX^2 = 1$. ∎

(4.2.11) Expected value of a function of several random variables. Theorem (4.2.2) has a natural extension to functions of two or more random variables. Suppose, for example, that we know the joint distribution of X and Y, and we would like to know something about $Z = X^2 + Y^2$, or $W = X + Y$, or $U = X \cdot \sin(2\pi Y)$, or some other function of two variables applied to X and Y. The theorem is just like Theorem (4.2.2), except that we have a multiple sum or integral involving a joint mass function or density:

(4.2.12) **Theorem.** Let $X_1, X_2, \ldots, X_n$ be random variables and let $Y = h(X_1, X_2, \ldots, X_n)$.

a. If $X_1, X_2, \ldots, X_n$ are discrete with joint mass function $p(x_1, x_2, \ldots, x_n)$, then

$$EY = \sum_{x_1}\sum_{x_2}\cdots\sum_{x_n} h(x_1, x_2, \ldots, x_n)p(x_1, x_2, \ldots, x_n),$$

provided that this sum converges absolutely.

b. If $X_1, X_2, \ldots, X_n$ are absolutely continuous with joint density function $f(x_1, x_2, \ldots, x_n)$, then

$$EY = \int_{-\infty}^{\infty}\int_{-\infty}^{\infty}\cdots\int_{-\infty}^{\infty} h(x_1, x_2, \ldots, x_n)f(x_1, x_2, \ldots, x_n)dx_1dx_2\ldots dx_n,$$

provided that this integral converges absolutely.

These look unpleasant, but we will not use them except to indicate why the next theorem, the second of our basic facts about expected values, should be true.

(4.2.13) **Theorem: Linearity of expected value.** For any random variables X and Y and any numbers a and b, if EX and EY exist, then so does $E(aX + bY)$, and

$$E(aX + bY) = a \cdot EX + b \cdot EY.$$

A partial proof is shown in (4.2.16). But first we make two comments and give an example.

Note 1. If in this formula we take Y to be a constant random variable that always equals 1, we get $E(aX + b) = a \cdot EX + b$. This says that

$$\text{If } h(x) = ax + b, \text{ then } Eh(X) = h(EX).$$

As we remarked earlier, except for some isolated examples this is the only kind of function for which such a formula holds:

$Eh(X)$ is not in general equal to $h(EX)$ unless $h(X)$ is of the form $aX + b$.

Note 2. Using mathematical induction we can prove from Theorem (4.2.13) that expected value "respects" all linear combinations, in the sense that if $X_1, X_2, \ldots, X_n$ are random variables and $a_1, a_2, \ldots, a_n$ are numbers, then

$$E(a_1X_1 + a_2X_2 + \cdots + a_nX_n) = a_1 \cdot EX_1 + a_2 \cdot EX_2 + \cdots + a_n \cdot EX_n. \quad (4.2.14)$$

If you have had a course in linear algebra, this will look familiar; it says that the expected value operator E is a linear function. The abstract setup is this: the set of all random variables defined on a given outcome set is a vector space, and the random variables whose expected values exist form a subspace. And the expected value operator E is a *linear functional* on this subspace—that is, a linear function from this subspace into the set of real numbers.

We remark that (4.2.14) holds for *finite* linear combinations; its extension to a countable infinity of random variables does *not* hold generally without some additional hypotheses. Expected values do not behave as nicely for infinite series of random variables as probabilities do for infinite sequences of disjoint events. There are some important theorems, called *convergence theorems,* that give conditions under which the counterpart of (4.2.14) holds for an infinite series of random variables. They are usually covered early in a graduate course in probability.

(4.2.15) **Example.** Suppose X is a random variable and r is any number, and we want to know $E(X - r)^2$. Using the linearity theorem, we can say

$$E(X - r)^2 = E(X^2 - 2rX + r^2) = EX^2 - 2rEX + r^2.$$

But notice that we cannot simplify any further and express EX^2 in terms of EX; we have to find them separately to evaluate this expression. ▮

(4.2.16) Partial proof of Theorem (4.2.13). We can prove the theorem for the case when X and Y have a joint distribution that is discrete or absolutely continuous, but we cannot prove it in its full generality without using some measure-theoretic ideas. For example, the theorem holds if X is absolutely continuous and Y is discrete, or even if they have distributions that are of neither type. But we will prove it only in the two cases mentioned.

Suppose X and Y are absolutely continuous, with joint density $f(x, y)$. To show that $E(aX + bY) = a \cdot EX + b \cdot EY$ we apply Theorem (4.2.12b) using the function $h(x, y) = ax + by$:

$$\begin{aligned} E(aX + bY) &= \int_{-\infty}^{\infty}\int_{-\infty}^{\infty} (ax + by) \cdot f(x, y)dx\,dy \\ &= a\int_{-\infty}^{\infty}\int_{-\infty}^{\infty} x \cdot f(x, y)dx\,dy + b\int_{-\infty}^{\infty}\int_{-\infty}^{\infty} y \cdot f(x, y)dx\,dy \\ &= a\int_{-\infty}^{\infty} x\left[\int_{-\infty}^{\infty} f(x, y)dy\right]dx + b\int_{-\infty}^{\infty} y\left[\int_{-\infty}^{\infty} f(x, y)dx\right]dy. \end{aligned}$$

Now the two integrals in brackets are just marginal densities, by Theorem (3.5.7), and so

$$E(aX + bY) = a\int_{-\infty}^{\infty} x \cdot f_X(x)dx + b\int_{-\infty}^{\infty} y \cdot f_Y(y)dy;$$

by the definition of expected value, this equals $aEX + bEY$.

The proof for the case in which X and Y are discrete is essentially identical; sums replace the integrals:

$$\begin{aligned} E(aX + bY) &= \sum_x \sum_y (ax + by)p(x, y) \\ &= a\sum_x x\sum_y p(x, y) + b\sum_y y \sum_x p(x, y), \end{aligned}$$

and, according to (3.4.11) and (3.4.12), the inner sums are the marginal mass functions. We get

$$E(aX + bY) = a\sum_{x} x p_X(x) + b\sum_{y} y p_Y(y).$$

By the definition of expected value, this equals $aEX + bEY$. □

Notice that at one point in the proof for the absolutely continuous case we reversed the order of integration, from $\int\int dx\, dy$ to $\int\int dy\, dx$. (It was in the first integral on the third line of integrals.) And in the proof for the discrete case we reversed the order of summing, from $\sum_x \sum_y$ to $\sum_y \sum_x$ (in the second sum on the second line of sums). This kind of reversal cannot be done in general without some additional hypotheses. For that matter, evaluating a double integral as an iterated integral, or a double sum as an iterated sum, also is not justified without some assumption. But in our case there is no problem, because the absolute convergence of the double integral or the double sum is such an assumption. And it can be shown that the assumed existence of EX and EY is enough to guarantee this absolute convergence.

(4.2.17) Example: A third calculation of the expected value of a binomial random variable, using indicators. Let X be the number of successes in n Bernoulli trials, where the success probability on each trial is p and the failure probability is $q = 1 - p$. In Example (4.1.22) we showed two different calculations of

$$EX = \sum_{j=0}^{n} j \cdot \binom{n}{j} p^j q^{n-j} = np.$$

One calculation was rather complicated, and the other used a clever trick. But (4.2.14) gives us a really simple and elegant way to see that $EX = np$ without even considering the binomial probabilities $\binom{n}{j} p^j (1 - p)^{n-j}$.

Let X_1 equal 1 or 0, according to whether the first trial results in success or not. That is, let X_1 be the number of successes on the first trial. Similarly, let X_2, $X_3, \ldots, X_n$ each be 1 or 0 according to the results of trials $2, 3, \ldots, n$. Then X, the total number of successes in the n trials, is just $X_1 + X_2 + \cdots + X_n$, and so

$$EX = EX_1 + EX_2 + \cdots + EX_n.$$

But each of these n expected values is easy to find. X_j has possible values 1 and 0 with probabilities p and q, and so for each j,

$$EX_j = 1 \cdot p + 0 \cdot q = p.$$

Therefore, EX is the sum of n p's, which is np. ∎

If this situation seems familiar, you may be remembering Exercise 10 of Section 2.4. The trick is actually a general method, as we now see.

(4.2.18) Indicator random variables and the Bernoulli distribution. The previous example uses what is called the *method of indicators*. The random variable X_j, which is 1 if there is a success on trial j and 0 otherwise, is called the *indicator random variable* of the event "success on trial j."

In general, let Ω be any outcome set and let A be any event in Ω. The random variable defined by

$$X(\omega) = \begin{cases} 1 & \text{if } \omega \in A, \\ 0 & \text{if } \omega \in A^c \end{cases}$$

is called the *indicator random variable,* or the *indicator function,* or simply the *indicator,* of the event A. It is sometimes denoted by I_A: $I_A(\omega) = 1$ if ω is a member of A and 0 if not. Notice that I_A has only two possible values, 1 and 0, and their probabilities are

$$P(I_A = 1) = P\{\omega\colon I_A(\omega) = 1\} = P(A),$$

and

$$P(I_A = 0) = P\{\omega\colon I_A(\omega) = 0\} = P(A^c).$$

That is, the probability mass function of I_A is

$$p(x) = \begin{cases} P(A) & \text{if } x = 1, \\ P(A^c) & \text{if } x = 0. \end{cases}$$

Its expected value is equally simple:

$$EI_A = 1 \cdot P(A) + 0 \cdot P(A^c) = P(A).$$

This makes it easy to find the expected value of a random variable that can be expressed as a sum of indicator random variables, as we did in the previous example.

The distribution of an indicator random variable seems so trivial as not to deserve a name—especially the distinguished name Bernoulli. But we call it that, because it is so fundamental, and because it has a close connection with Bernoulli processes and the binomial distribution. The ***Bernoulli distribution*** with parameter p is the one with possible values 0 and 1 and mass function

$$p(x) = \begin{cases} p & \text{if } x = 1, \\ q & \text{if } x = 0. \end{cases}$$

So the indicator random variable of an event A has the Bernoulli distribution with parameter $p = P(A)$. The expected value of this distribution is of course p.

The connection with Bernoulli trials is fairly obvious—this is the distribution of the number of successes on one Bernoulli trial. That is, a Bernoulli distribution is a binomial distribution with $n = 1$. Notice that

$$\binom{1}{0}p^0q^{1-0} = q \quad \text{and} \quad \binom{1}{1}p^1q^{1-1} = p.$$

And as we saw earlier, any random variable having the binomial distribution with parameters n and p can be thought of as the sum of n independent random variables having the Bernoulli distribution with parameter p.

(4.2.19) **Theorem: The expected product of independent random variables.** This is the last of our three important facts about expected values: the expected value of the product is the product of the expected values, *if* the random variables are independent.

If X and Y are independent and EX and EY exist, then so does $E(XY)$, and $E(XY) = EX \cdot EY$.

Partial proof. Like Theorem (4.2.13) on linearity, this theorem is easy to prove if X and Y are jointly discrete or jointly absolutely continuous, but in this text we cannot prove it in more general cases.

If they are jointly discrete, then their joint mass function is the product of their marginal mass functions: $p(x, y) = p_X(x) \cdot p_Y(y)$. To find their expected product we use Theorem (4.2.12a) with the function $h(x, y) = xy$:

$$\begin{aligned} E(XY) &= \sum_x \sum_y xyp(x, y) = \sum_x \sum_y xyp_X(x)p_Y(y) \\ &= \sum_x xp_X(x)\left[\sum_y yp_Y(y)\right]. \end{aligned}$$

The inner sum in brackets is just EY. When we take it out of the sum over x, that sum is seen to be EX, and the result follows.

If X and Y are jointly absolutely continuous, then we can use the product of their densities as a joint density: $f(x, y) = f_X(x) \cdot f_Y(y)$. So by Theorem (4.2.12b) with $h(x, y) = xy$, we get

$$\begin{aligned} E(XY) &= \int_{-\infty}^{\infty}\int_{-\infty}^{\infty} xyf(x, y)dx\,dy = \int_{-\infty}^{\infty}\int_{-\infty}^{\infty} xyf_X(x)f_Y(y)dx\,dy \\ &= \int_{-\infty}^{\infty} xf_X(x)\left[\int_{-\infty}^{\infty} yf_Y(y)dx\right]dy. \end{aligned}$$

The integral in brackets is EY. When we take it out, we see that the remaining integral is EX. ☐

(4.2.20) **Example: Showing that the converse of (4.2.19) is false.** There do exist pairs of random variables that satisfy $E(XY) = EX \cdot EY$ even though they are not independent. Thus, it is not possible to conclude that two random variables are independent just because their expected values multiply.

The example that is usually given is a pair of discrete random variables with the following joint mass function:

	$p(x, y)$	$y = -1$	$y = 0$	$y = 1$
x	-1	0	$\frac{1}{4}$	0
	0	$\frac{1}{4}$	0	$\frac{1}{4}$
	1	0	$\frac{1}{4}$	0

You can check that X and Y have the same marginal mass function: possible values -1, 0, and 1, with corresponding probabilities $\frac{1}{4}$, $\frac{1}{2}$, and $\frac{1}{4}$. Thus, EX and EY are both equal to $(-1) \cdot \frac{1}{4} + 0 \cdot \frac{1}{2} + 1 \cdot \frac{1}{4} = 0$.

But $E(XY)$ is also 0, because by Theorem (4.2.12a) it equals the following sum of nine terms, all of which are 0:

$$(-1) \cdot (-1) \cdot 0 + (-1) \cdot 0 \cdot \tfrac{1}{4} + (-1) \cdot 1 \cdot 0 + 0 \cdot (-1) \cdot \tfrac{1}{4} + 0 \cdot 0 \cdot 0 + 0 \cdot 1 \cdot \tfrac{1}{4} + 1 \cdot (-1) \cdot 0 + 1 \cdot 0 \cdot \tfrac{1}{4} + 1 \cdot 1 \cdot 0.$$

Another way to see that $E(XY) = 0$ is to notice that the random variable XY is actually a constant; it equals 0 with probability 1.

So $E(XY)$ and $EX \cdot EY$ are equal. But X and Y are clearly not independent, because their joint mass function is not the product of the marginal mass functions. ∎

(4.2.21) **Example: The expected square of a sum.** What do we need to know to find $E(X + Y)^2$? Remember that by convention this means $E[(X + Y)^2]$, not $[E(X + Y)]^2$.

First, we expand $(X + Y)^2$ before taking expected values:

$$E(X + Y)^2 = E(X^2 + 2XY + Y^2).$$

Now we use Theorem (4.2.13) on linearity to get

$$EX^2 + 2E(XY) + EY^2.$$

But here we are stuck, unless we know that X and Y are independent. If they are, then we can go one step further using (4.2.19):

$$E(X + Y)^2 = EX^2 + 2 \cdot EX \cdot EY + EY^2 \quad \text{if } X \text{ and } Y \text{ are independent.}$$

But that is all the simplification we can do; as we have repeated many times, EX^2 is not $(EX)^2$ and EY^2 is not $(EY)^2$ unless X and Y are constants.

This does not look like a very satisfactory conclusion, but as we will see in Section 4.4, it will give us an important simplification for the variance of a sum of independent random variables. ∎

(4.2.22) **Some inequalities for expected values.** The following four properties may seem obvious, but they have important implications and are used often.

(4.2.23) If X is a nonnegative random variable [that is, if $P(X \geq 0) = 1$], then $EX \geq 0$ (assuming EX exists).

It is easy to see why. If X is discrete, then in the sum $\sum_x xp(x)$ all terms are nonnegative. If X is absolutely continuous, then the integral $\int_{-\infty}^{\infty} xf(x)dx$ is just $\int_0^{\infty} xf(x)dx$ [because X has a density that is 0 on $(-\infty, 0)$], and this integral has a nonnegative integrand. The next two inequalities are simple corollaries of (4.2.23).

(4.2.24) If X and Y are random variables and $X \leq Y$, then $EX \leq EY$.

This follows because the random variable $Y - X$ is nonnegative, and so $E(Y - X) \geq 0$; but of course $E(Y - X) = EY - EX$.

Note what "$X \leq Y$" means: that $X(\omega) \leq Y(\omega)$ for all ω. That is, every time the experiment is done, the value of X is less than or equal to that of Y.

(4.2.25) If $X \leq b$, then $EX \leq b$. If $X \geq a$, then $EX \geq a$.

This follows from (4.2.23) because $b - X$ and $X - a$ are nonnegative random variables.

Remembering (4.2.25) sometimes helps in detecting errors in calculation of expected values. Suppose, for example, that X has a density defined on an interval (a, b); if you calculate EX and find it is not between a and b, then you have made an error in the calculation.

(4.2.26) If X is a nonnegative random variable and $EX = 0$, then X is the constant 0; that is, $P(X = 0) = 1$.

This is easy to see if X is already known to be discrete. If all possible values of X are nonnegative but the sum $\sum_x xp(x)$ is 0, then there can be no positive x's in the sum, so there can be no positive x's with positive probability. If X is not assumed to be discrete, the proof is somewhat technical (notice that X *is* discrete if $X \geq 0$ and $EX = 0$, but if we don't assume it we have to prove it). We omit it here.

We conclude the section with the idea of a *symmetric* distribution. When the distribution of X has this important property, the calculation of EX, among other things, is greatly simplified.

(4.2.27) **The expected value of a symmetric random variable.** The density function $\varphi(x) = e^{-x^2/2}/\sqrt{2\pi}$ of the standard normal distribution is ***symmetric about x = 0*** in the sense that $\varphi(-x) = \varphi(x)$ for any x. And it comes as no surprise that $EX = 0$, as we saw in (4.1.19). This is nearly, but not quite, automatically true for any such density. For example, the Cauchy density given in Example (4.1.18) is $f(x) = 1/\pi(1 + x^2)$. This is also symmetric about 0, but EX does not exist.

What is true is that if X has a density $f(x)$ satisfying $f(x) = f(-x)$ for all x, then the two integrals

$$EX^+ = \int_0^\infty xf(x)dx \quad \text{and} \quad EX^- = \int_{-\infty}^0 xf(x)dx$$

either both diverge or else both converge and are negatives of each other. Since [see (4.1.15) and (4.1.20)] $EX = EX^+ + EX^-$ if both converge, the result is that EX is 0 if it exists.

Similarly, suppose X is a discrete random variable whose mass function is symmetric about 0; that is, $P(X = -x) = P(X = x)$ for all x. Then the two sums

$$EX^+ = \sum_{x>0} xp(x) \quad \text{and} \quad EX^- = \sum_{x<0} xp(x)$$

either both diverge or both converge and are negatives of each other. Since $EX = EX^+ + EX^-$ if both converge, EX equals 0 if it exists.

Symmetry about nonzero numbers. Consider the binomial distribution with $n = 3$ and $p = \frac{1}{2}$; the possible values are 0, 1, 2, and 3, and the probabilities are $\frac{1}{8}, \frac{3}{8}, \frac{3}{8}$, and $\frac{1}{8}$. This distribution is symmetric about 1.5 in the sense that $P(X = 1.5 + x) = P(X = 1.5 - x)$ for any x. And sure enough, $EX = 1.5$.

Or consider the following density on (0, 1):

$$f(x) = \begin{cases} 6x(1 - x) & \text{if } 0 < x < 1, \\ 0 & \text{otherwise.} \end{cases}$$

If you graph this function you will see clearly that it is symmetric about $\frac{1}{2}$. The symmetry is expressed algebraically by saying that $f\left(\frac{1}{2} + x\right) = f\left(\frac{1}{2} - x\right)$ for all real x [which is the same as $f(z) = f(1 - z)$]. And you can confirm, by working out the integral, that $EX = \frac{1}{2}$.

It is not difficult to see why EX must equal a if X is symmetric about a, provided that EX exists. If X is symmetric about a, then $X - a$ is symmetric about 0, and so $E(X - a) = 0$; but of course $E(X - a) = EX - a$.

To state all of this a little more formally:

An absolutely continuous random variable is ***symmetric about a*** if it has a density satisfying $f(a - x) = f(a + x)$ for all x. A discrete random variable is ***symmetric about a*** if its probability mass function satisfies $p(a - x) = p(a + x)$ for all x. [In general, a random variable X is ***symmetric about a*** if

$P(X \in A) = P(X \in B)$ whenever A and B are Borel sets such that $a - x$ is in A if and only if $a + x$ is in B.]

If X is any random variable that is symmetric about a, then EX equals a if it exists.

EXERCISES FOR SECTION 4.2

1. Let X have the density $f(x) = \frac{3}{8}x^2$ for $0 < x < 2$ (and 0 otherwise). Find the following.
   a. EX b. EX^2 c. EX^k for an arbitrary positive integer k
2. Repeat Exercise 1 for the case where X has the uniform distribution on $[0, 1]$.
3. If $EX = 4$ and $EX^2 = 21$, find
   a. $E(X^2 - 3X + 2)$ b. $E(X + 1)^2$ c. $E(X - EX)^2$ d. $EX^2 - (EX)^2$
4. Suppose that $EX = \mu$.
   a. Show that $E(X - \mu)^2 = EX^2 - \mu^2$.
   b. Show that for any number a, $E(X - a)^2 = E(X - \mu)^2 + (a - \mu)^2$. [*Hint:* Write $(X - a)^2$ as $(X - \mu + \mu - a)^2$.]
   c. Conclude that the value of a for which $E(X - a)^2$ is minimized is $a = \mu$.

   The number $E(X - EX)^2$ is an important one, called the *variance* of X. We will study it in the next section. Part a shows that it is no coincidence that parts c and d of the previous exercise have the same answer.
5. Let X have the uniform distribution on $(0, \pi)$. [The density is $f(x) = 1/\pi$ for $0 < x < \pi$.]
   a. Find $E(\sin X)$.
   b. Find $E(\sin^2 X)$. [Use $\sin^2 x = \frac{1}{2}(1 - \cos 2x)$.]
   c. Find $E(\cos^2 X)$. [You do not need to do another integral if you use another, more familiar, trigonometric identity.]
6. a. Suppose X is a random variable and $EX^2 = 0$. What then? [*Hint:* Use (4.2.26).]
   b. What if $E(X - a)^2 = 0$?
   c. What if $EX^2 < 0$?
7. a. Let X have the binomial distribution with parameters n and p, and let t be a real number. Find $E(t^X)$.

   This expected value, viewed as a function of t, is called the *probability generating function* of X. It is an important tool for dealing with discrete distributions whose possible values are nonnegative integers. We first encountered it in (4.1.23), where we saw that differentiating it and then setting $t = 1$ gives EX. We will study probability generating functions in Section 7.1.

   b. Let X have the exponential density with parameter λ, defined by $f(x) = \lambda e^{-\lambda x}$ for $x > 0$ (and 0 otherwise). Let s be a real number with $s < \lambda$. Find $E(e^{sX})$. What is $s \geq \lambda$?

This expected value, viewed as a function of s, is called the *moment generating function* of X. It is an important tool for working with absolutely continuous distributions, especially those whose possible values are positive. If you are curious about what it can do, think about differentiating $E(e^{sX})$ with respect to s and then setting s equal to 0. We will study moment generating functions in Section 8.1.

8. In Example (4.1.11) we found that if X has the Poisson distribution with parameter λ, then $EX = \lambda$.

a. Show that $E(X(X - 1)) = \lambda^2$. [*Hint:* Write down the series $\sum k(k - 1) \cdot p(k)$ and notice that the terms for $k = 0$ and $k = 1$ are 0. In all the other terms, the factor $k(k - 1)$ cancels with the denominator to give $(k - 2)!$. Factoring out λ^2 leaves a series that adds to 1.]

b. From part a and EX, find EX^2.

c. Use the result of Exercise 4a with EX and EX^2 to find $E(X - \lambda)^2$.

9. Let X be the number of failures before the first success in Bernoulli trials with success probability p. That is, X has the geometric distribution with parameter p; the mass function is $p(k) = pq^k$ *for* $k = 0, 1, 2, \ldots$. In Example (4.1.3) we found that $EX = q/p$.

a. Show that $E(X(X - 1)) = 2q^2/p^2$.

b. From part a and EX, find EX^2.

c. Use the result of Exercise 4 above with EX and EX^2 to find $E(X - q/p)^2$.

The point of Exercises 8 and 9 is that in some cases it is difficult to evaluate EX^2, but $E(X(X - 1))$ is simpler. As we pointed out after Exercise 4, we will study the number $E(X - EX)^2$, called the *variance* of X, in the next section.

10. Suppose the density of X is $f(x) = \frac{2}{9}x(3 - x)$ for $0 < x < 3$ (and 0 otherwise).

a. Find EX without evaluating the integral by discovering the point about which the density is symmetric.

b. Confirm the result by evaluating $\int_0^3 xf(x)dx$.

11. Find EX if the CDF of X is

$$F(x) = \begin{cases} 0 & \text{if } x \le 0, \\ \frac{1}{8}x^{3/2} & \text{if } 0 < x \le 4, \\ 1 & \text{otherwise.} \end{cases}$$

12. Let the mass function of X be $p(k) = kp^2q^{k-1}$ for $k = 1, 2, 3, \ldots$. Find the expected value of $Y = 1/X$.

13. Two fair dice are rolled and X is the sum of the two numbers that appear. Earlier we found that the possible values of X are 2, 3, 4, 5, 6, 7, 8, 9, 10, 11, 12, with corresponding probabilities 1, 2, 3, 4, 5, 6, 5, 4, 3, 2, 1, all divided by 36. Find EX in three different ways:

a. Directly from the definition, using the mass function.

b. Using the symmetry of the mass function.

c. Using the fact that $X = X_1 + X_2$, where X_1 is the score on the first die and X_2 the score on the second. X_1 has possible values 1, 2, 3, 4, 5, 6, with probabilities $\frac{1}{6}$ each, as does X_2, so their expected values are easy to find.

14. Let S be the number of numbers in their proper places in a random permutation of the numbers 1, 2, 3, and 4. As we saw in Example (4.1.9), the possible values of S are 0, 1, 2, and 4, and the corresponding probabilities are 9, 8, 6, and 1, all divided by 24. We also saw in Example (4.1.9) that $ES = 1$. Find:

a. ES^2 **b.** $E(S - ES)^2$ **c.** $E(S(S - 1))$ **d.** $E(S(S - 1)(S - 2))$

To see which of these answers are coincidences, valid only for $n = 4$, and which might be true for all n, look at the distributions of S for $n = 5$ and 6 given at the end of Example (4.1.9) in the previous section. Or see Exercise 23 of Section 2.2, where we saw how to find the distribution of S_n for any n. Also see Exercise 26 in the previous section and Exercise 19 below. The question will finally be answered in Exercise 19 of Section 7.1.

15. Let X have the exponential density $f(x) = \lambda e^{-\lambda x}$ for $x > 0$ (and 0 otherwise). Recall that in Example (4.1.6) we found that $EX = 1/\lambda$, and in Example (4.2.7) we found that $EX^2 = 2/\lambda^2$.

a. Define $\mu_k = EX^k$ for $k = 1, 2, 3, \ldots$. Show that $\mu_k = (k/\lambda)\mu_{k-1}$ for $k = 2, 3, 4, \ldots$. [*Hint:* Write down the integral for EX^k and integrate by parts, using $u = x^k$ and $dv = \lambda e^{-\lambda x}dx$.]

b. Using the result of part a and the values you know for μ_1 and μ_2, find μ_k in terms of k.

16. Let X and Y be independent random variables. Suppose that $EX = 2$, $EX^2 = 6$, $EY = 3$, and $EY^2 = 13$. Find the following.

a. $E(X + Y)$

b. $E(2XY)$

c. $E(3X - Y)^2$

d. $[E(3X - Y)]^2$

17. Let X be the number of spots that show when a fair die is rolled once. Find EX^2.

18. **A case where $Eh(X) = h(EX)$ even though $h(x)$ is not of the form $ax + b$.** Let X have the uniform distribution on $[-1, 1]$. Show that $E(X^3) = (EX)^3$. More generally, show that if the distribution of X is symmetric about 0 and $h(x)$ satisfies $h(-x) = -h(x)$ (such functions are called *odd* functions), then $E(h(X)) = h(EX)$.

19. **The expected number of items in their proper places in a random permutation.** Consider random permutations of the set $\{1, 2, 3, \ldots, n\}$ and let S be the number of numbers that are in their proper places. In Exercise 26 of the previous section we saw that $ES = 1$, regardless of the value of n. Here we get the same result using the method of indicators, as follows. Let X_1 equal 1 if 1 is first in the permutation, and 0 is not. Let X_2 equal 1 if 2 is second, and 0 if not. And so on for $X_3, X_4, \ldots, X_n$. Then $S = X_1 + X_2 + \cdots + X_n$.

a. Find the probabilities of the two possible values of X_1.

b. Find EX_1.

c. Do parts a and b for $X_2, X_3, \ldots, X_n$.

d. Conclude that $ES = 1$.

As we mentioned after Exercise 26 of the previous section, the method of indicators is not as powerful in dealing with random permutations as the method used in that exercise. Still, the method of indicators is useful in a wide range of problems, and it is worthwhile to see how it works here.

20. **The expected number of children in a family: Lotka's model.** Suppose the probability that a randomly chosen family has k children is

$$p(k) = \begin{cases} \beta p q^k & \text{for } k = 1, 2, 3, \ldots, \\ 1 - \beta q & \text{for } k = 0. \end{cases}$$

Here p and q are positive and add to 1, and β is also between 0 and 1. Show that the expected number of children in a family equals $\beta q/p$.

A simpler model for family sizes would be geometric: $p(k) = pq^k$ for $k = 0, 1, 2, \ldots$; the expected family size would then be q/p, as we saw in Example (4.1.3). However, such a model gives too small a probability to $k = 0$. In Lotka's model, β represents the proportion of families that *can* have children; among those, it is assumed that after each birth q is the probability that the family will continue having children, and p is the probability that they will not. Notice that $p(0) = (1 - \beta) + \beta p$; $1 - \beta$ is the proportion who cannot have children, and βp is the proportion who can but do not.

According to Feller's textbook *An Introduction to Probability Theory and Its Applications,* Volume 1, Lotka, in his book *Théorie analytique des associations biologiques,* gives the values $\beta = .8785$ and $p = .2642$ for American family sizes in the 1930s. This leads to an expected family size of about 2.45 children.

For readers familiar with Feller's book: Our notation here differs from his—we have reversed p and q, and our βp is Feller's α—but the probabilities agree. The mass function is $p(k) = .2321(.7358)^k$ for $k \geq 1$, with $p(0) = .3536$. The value of β has been recovered by combining two different distributions, one for family sizes (p. 141) and one for numbers of male offspring (p. 294); as a by-product we infer that the probability is .5152 that a given child is a male.

21. **What proportion of people are firstborns?** You might think it would be the reciprocal of the expected size of a family; with Lotka's numbers from the previous exercise, this would be 1/2.45, or about 41%. But a clearer way to think about it is to imagine a very large number N of families, and to divide the approximate number of firstborn children in these families by the approximate total number of children. Suppose family sizes are distributed as in the previous exercise. Then the expected number of children in N families is $N \cdot EX$, where X is the number of children in a random family. And the expected number of firstborns is just the expected number of families with one or more children. Find this latter expected value, and show that the ratio of the two equals p. [Thus, according to Lotka's model and Lotka's data, 26.42% of all American children were firstborns in the 1930s.]

We will be able to see more clearly the difference between these two answers, and why the latter one is right, in Exercise 12 of Section 9.1. The correct answer is the reciprocal of the (conditional) expected size of a family chosen at random from families with one or more children.

[a]22. **How to win at roulette: Find a friend who wants to reform you.** If you bet a dollar on a single number at roulette, you will win \$35 with probability $\frac{1}{38}$ and

lose \$1 with probability $\frac{37}{38}$. But suppose a friend bets you \$20 that you will be behind at the end of 36 plays. (That is, you will win \$20 if you are even or ahead and lose \$20 if behind.) Find your expected gain after 36 plays.

Frederick Mosteller gave this exercise, in his entertaining and instructive book *Fifty Challenging Problems in Probability,* as an example of how a well-meaning friend, thinking to teach a gambler a lesson, actually provides a way to make money.

23. **An application of probability to calculus: Monte Carlo approximation of an integral.** Suppose we are interested in the value of $\int_0^1 h(x)dx$, where $h(x)$ is some given function. (Assume the integral converges absolutely.) Let X be a random variable with the uniform distribution on (0, 1). The Monte Carlo method approximates the integral by making a large number of observations of X and averaging the values of $h(X)$. Explain why we could expect the average to be close to the desired integral.

In Exercise 14 of Section 4.5 we will see how the theorems called laws of large numbers guarantee that this method works in some sense. The method, with various improvements, is in much use in modern computational mathematics.

24. **Cramer's proposed solutions to the St. Petersburg paradox.** In Exercise 23 of section 4.1 we considered the St. Petersburg game, in which a fair coin is tossed until a head comes up and Y is the number of tails. Player B must then pay $X = 2^Y$ dollars to player A. The "paradox" is that, because EX does not exist, there does not appear to be a fair entry fee that A should pay B for the privilege of playing. In 1728 Gabriel Cramer suggested that the problem might be resolved by considering that a dollar is worth less to someone who has a great deal of money than to someone who does not.
    a. One proposal was that the real value of a very large sum of money is not the actual amount, but some function of the amount that grows more slowly than linearly—say, the square root of the amount. In this case the real value to A of the amount B must pay A is $Z = (\sqrt{2})^Y$, and a fair entry fee is $(EZ)^2$. Find this.
    b. Another proposal was that, after a person has amassed a huge fortune, further money is of no value at all. Say that any amount over 2^{50} (about a million billion) dollars is simply worth 2^{50} dollars. In this case the real value to A of the amount B must pay A is

$$W = \begin{cases} 2^Y & \text{if } Y \le 50, \\ 2^{50} & \text{if } Y > 50, \end{cases}$$

and a fair entry fee is EW. Find it.

ANSWERS

1. a. $\frac{3}{2}$ **b.** $\frac{12}{5}$ **c.** $3 \cdot 2^k/(k+3)$

2. a. $\frac{1}{2}$ **b.** $\frac{1}{3}$ **c.** $1/(k+1)$

3. a. 11 **b.** 30 **c.** 5 **d.** 5

5. a. $2/\pi$ **b.** $\frac{1}{2}$ **c.** $\frac{1}{2}$

6. **a.** X is the constant 0. **b.** X is the constant a. **c.** $EX^2 < 0$ is impossible.

7. **a.** $(pt + q)^n$ **b.** $\lambda/(\lambda - s)$ If $s > \lambda$ the expected value does not exist.

8. **b.** $\lambda + \lambda^2$ **c.** λ

9. **b.** $(2q^2/p^2) + (q/p)$ **c.** q/p^2

10. **a.** 1.5

11. $\frac{12}{5}$

12. p

13. 7

14. **a.** 2 **b.** 1 **c.** 1 **d.** 1

15. **b.** $k!/\lambda^k$

16. **a.** 5 **b.** 12 **c.** 31 **d.** 9

17. $\frac{91}{6}$

18. Both are 0.

19. **c.** $EX_j = 1/n$ for all j

22. Expected gain from casino in 36 plays: -1.89; expected gain from friend: $+4.69$; total expected gain: $+2.80$

24. **a.** $(2 - \sqrt{2})^{-2} \doteq 2.91$ dollars **b.** \$26

4.3 VARIANCE AND MOMENTS

Suppose X is a random variable; let μ denote the number EX. (This is a lowercase Greek mu, traditionally used for the expected value of a random variable, because of its correspondence with our m for "mean.") Consider the following function of X: $Y = (X - \mu)^2$. This is a nonnegative random variable that is large when X is far from EX and small when it is close. So the expected value of $Y = (X - \mu)^2$ will be relatively small if X has a high probability of being close to μ, and relatively large if not. This expected value, EY, is what we will define shortly as the variance of X.

(4.3.1) **Example.** Here are mass functions for two discrete random variables, X_1 and X_2, with the same set of possible values, $\{1, 2, 3, 4, 5\}$, and the same expected value, $\mu = 3$. Notice that X_2 is more likely to be far from 3 than X_1:

x:	1	2	3	4	5
$P(X_1 = x)$:	.1	.2	.4	.2	.1
$P(X_2 = x)$:	.3	.1	.2	.1	.3

Let $Y_1 = (X_1 - 3)^2$ and $Y_2 = (X_2 - 3)^2$. Then the mass functions of Y_1 and Y_2 are as follows:

y:	0	1	4
$P(Y_1 = y)$:	.4	.4	.2
$P(Y_2 = y)$:	.2	.2	.6

We see that $EY_1 = 1.2$ and $EY_2 = 2.6$. The magnitudes of these two numbers individually do not mean much, but the fact that EY_2 is greater than EY_1 tells us that X_2 has a greater tendency to be far from its expected value than does X_1.

We will see later that even the individual numbers $E(X - \mu)^2$ have some interpretation—or at least their square roots do—but for now we will be content to think of $E(X - \mu)^2$ as a *relative* indicator of the tendency of X to be far from its expected value μ. ▮

(4.3.2) Definition. Let X be a random variable and suppose EX exists and equals μ. The ***variance*** of X is the number denoted by $\text{Var}(X)$, or $V(X)$, or simply VX, and defined by

$$VX = E(X - \mu)^2 = E(X - EX)^2,$$

provided this expected value exists.

But we rarely use the definition to compute a variance; instead we use (4.3.4) below.

(4.3.3) Computing VX. In Example (4.3.1) we computed VX (for both X_1 and X_2) by writing down the possible values of the random variable $Y = (X - \mu)^2$ and their probabilities, and then finding the expected values of Y from them. It would be simpler to use Theorem (4.2.2), according to which

$$VX = \sum_x (x - \mu)^2 p(x) \quad \text{or} \quad \int_{-\infty}^{\infty} (x - \mu)^2 f(x)dx$$

depending on whether X is discrete or absolutely continuous.

However, it is even simpler in nearly every case to use the following formula:

$$VX = EX^2 - \mu^2 = EX^2 - (EX)^2. \tag{4.3.4}$$

That is, the variance is the expected square, minus the squared expected value—or, as some people like to remember it, "mean square minus squared mean."

Exercise 4a in the previous section asked you to prove (4.3.4); here is the proof:

$$VX = E(X - \mu)^2 = E(X^2 - 2\mu X + \mu^2) = EX^2 - 2\mu EX + \mu^2;$$

but because $EX = \mu$, this is $EX^2 - 2\mu^2 + \mu^2 = EX^2 - \mu^2$. □

(4.3.5) Example. For X_1 in Example (4.3.1),

$$EX = 1 \cdot .1 + 2 \cdot .2 + 3 \cdot .4 + 4 \cdot .2 + 5 \cdot .1 = 3$$

and

$$EX^2 = 1^2 \cdot .1 + 2^2 \cdot .2 + 3^2 \cdot .4 + 4^2 \cdot .2 + 5^2 \cdot .1 = 10.2,$$

and so

$$VX = 10.2 - 3^2 = 1.2. \qquad ▮$$

(4.3.6) **Example: The variance of an exponential random variable.** Let X have the exponential distribution with parameter λ. What is the variance of X?

We found EX earlier, in Example (4.1.6), using integration by parts; also see the examples under (A.8.1) in Appendix A. We got

$$EX = \int_0^\infty x \cdot \lambda e^{-\lambda x} dx = \frac{1}{\lambda}.$$

In Example (4.2.7) we found EX^2 using two integrations by parts; it also appears under (A.8.1) in Appendix A:

$$EX^2 = \int_0^\infty x^2 \cdot \lambda e^{-\lambda x} dx = \frac{2}{\lambda^2}.$$

Consequently,

$$VX = EX^2 - (EX)^2 = \frac{2}{\lambda^2} - \left(\frac{1}{\lambda}\right)^2 = \frac{1}{\lambda^2}. \qquad ▮$$

The details of the two integrations in the preceding example are important; if you do not remember them it would be well to look back and review them now. This kind of calculation will appear often in the rest of the text.

(4.3.7) **Example.** Figure 4.3a (page 264) shows the graphs of two density functions; they are

$$f(x) = 12x^2(1 - x) \quad (0 < x < 1) \qquad \text{and} \qquad g(x) = 72x^7(1 - x) \quad (0 < x < 1).$$

Looking at the figure, we might first guess that the expected values of the two distributions are different, but we can also see that the distribution given by $g(x)$ is more concentrated about its expected value, whereas that of $f(x)$ is more spread out or "dispersed." So we will not be surprised to find that the distribution given by f has a larger variance than that of g. And indeed it does:

If X has density f, then

$$EX = \int_0^1 x \cdot 12x^2(1 - x)dx = .6 \quad \text{and} \quad EX^2 = \int_0^1 x^2 \cdot 12x^2(1 - x)dx = .4,$$

and so

$$VX = .4 - (.6)^2 = .04.$$

If Y has density g, then

$$EY = \int_0^1 x \cdot 72x^7(1 - x)dx = .8 \quad \text{and} \quad EY^2 = \int_0^1 x^2 \cdot 72x^7(1 - x)dx \doteq .6545,$$

and so

$$VY \doteq .6545 - (.8)^2 = .0145. \qquad \blacksquare$$

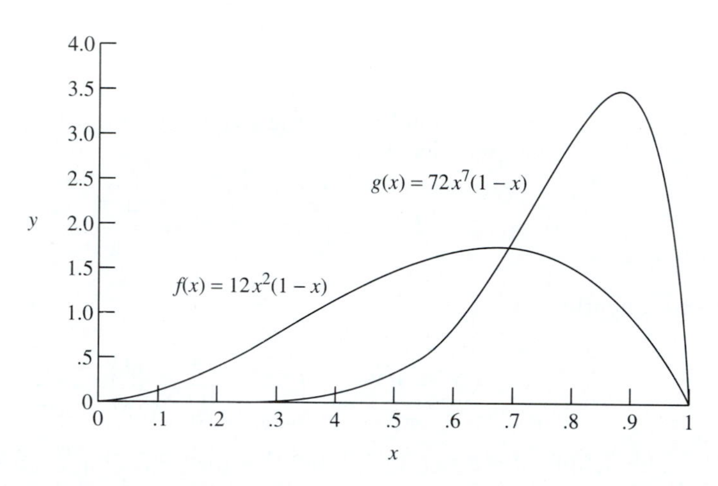

FIGURE 4.3a Two densities; $f(x)$ is more "dispersed" than $g(x)$, and has larger variance.

(4.3.8) **Example.** Let X have the uniform distribution on (0, 1). What are EX and VX? Again the calculations are very simple:

$$EX = \int_0^1 x dx = \tfrac{1}{2}, \quad EX^2 = \int_0^1 x^2 dx = \tfrac{1}{3}, \quad \text{and so} \quad VX = \tfrac{1}{3} - \left(\tfrac{1}{2}\right)^2 = \tfrac{1}{12}. \qquad \blacksquare$$

(4.3.9) **Example: The variance of a standard normal random variable is 1.** Let Z be a standard normal random variable. In Example (4.1.19) we saw that $EZ = 0$, and in Example (4.2.9) we saw that $EZ^2 = 1$. Consequently, $VZ = 1 - 0^2 = 1$. $\blacksquare$

(4.3.10) The variance of a linear function of X. As with expectations, we cannot hope that the variance of $Y = h(X)$ will be related to the variance of X in any simple way; certainly $V(h(X))$ is not $h(VX)$ in general. In fact, variance is not even a linear operator: $V(aX + b)$ is not $aVX + b$ in general. What *is* true for linear functions is fairly simple, however: if VX exists, then for any numbers a and b,

(4.3.11)
$$V(aX + b) = a^2 VX.$$

This may be a little surprising—the variance of $aX + b$ does not depend on b at all. But it is not surprising when we consider that the variance is meant to reflect the tendency for X to deviate from its expected value. If we just add a constant to a random variable, we will not change the extent of that tendency, because the expected value increases by the same constant.

Notice the following special cases of (4.3.11):

$$V(X + b) = VX \text{ for any number } b.$$
$$V(-X) = VX.$$

Proof. By the definition, the variance of $aX + b$ is the expected value of the square of $(aX + b) - E(aX + b)$, and the basic formula (4.2.13) on the algebra of expected values shows that

$$\begin{aligned}(aX + b) - E(aX + b) &= aX - aEX \\ &= a(X - EX).\end{aligned}$$

When we square this and take the expected value, again using (4.2.13), we get

$$V(aX + b) = E[a^2(X - EX)^2] = a^2E[(X - EX)^2] = a^2VX. \qquad \square$$

Recall that in the previous section, along with $E(aX + b)$, we discussed $E(X + Y)$. In both cases the results were simple and just what one would expect. We have just seen that $V(aX + b)$ is not quite so simple; similarly, it turns out that $V(X + Y)$ is not simply $VX + VY$. An additional term, involving what is called their *covariance*, is needed. We will cover this in the next section.

As a summary and reminder:

$$\mathbf{E(aX + b) = aEX + b \quad \text{and} \quad V(aX + b) = a^2VX;}$$

and

If $h(X)$ is any function that is *not* of the form $aX + b$, then we cannot say anything in general about $Eh(X)$ and $Vh(X)$ in terms of EX and VX.

[We have to compute $Eh(X)$ and $E[h(X)^2]$ using Theorem (4.2.2), and $Vh(X)$ using (4.3.4).]

(4.3.12) **Variances are nonnegative.** This is another important fact about variances that you may have already noticed. The reason VX must be nonnegative is that VX is the expected value of $(X - EX)^2$, which is a nonnegative random variable. By (4.2.23), the expected value of a nonnegative random variable is nonnegative.

Could a variance ever be 0? Recall (4.2.26), which says that if X is nonnegative and $EX = 0$, then X must be the constant 0. Now if $VX = 0$, then $(X - \mu)^2$ is a nonnegative random variable whose expected value is 0, where, as usual, $\mu = EX$. So $(X - \mu)^2$ is the constant 0, which implies that X is the constant μ. To summarize:

$$VX \geq 0, \text{ and if } VX = 0, \text{ then } X \text{ is a constant random variable.}$$

Notice also, by the way, that because $VX = EX^2 - (EX)^2$ and $VX \geq 0$, it must always be the case that $EX^2 \geq (EX)^2$.

(4.3.13) **Moments of a random variable.** The expected values of the various powers of a random variable X, namely, EX, EX^2, EX^3, etc., are important enough to be given special names; they are called the *moments* of X. More specifically, for any nonnegative integer k, the ***k*th *moment*** of X is the number denoted by μ_k and defined by

$$\mu_k = EX^k \quad \text{(provided this expected value exists).}$$

Notice that the zero-th moment μ_0 is always 1: $\mu_0 = EX^0 = E1 = 1$. The first moment is just EX. And the variance of X can be found from its first two moments: $VX = EX^2 - (EX)^2 = \mu_2 - \mu_1^2$.

At the risk of running such warnings into the ground, we emphasize once again that you cannot find one moment from another: μ_2 is EX^2, but this is $E(X^2)$ and not $(EX)^2$, which equals μ_1^2. Similarly, $\mu_3 = EX^3$ cannot be found in terms of μ_1 and μ_2, and so forth.

(4.3.14) **Central moments.** In addition to the moments of X, the moments of the random variable $X - EX$ are also of interest. They are called the ***central moments*** of X, because EX is sometimes regarded as a sort of "center" for the distribution of X. The term "moments of X about EX" is probably more expressive of what central moments are. We will denote them by $c_0, c_1, c_2, \ldots$:

$$c_k = E(X - \mu)^k \quad \text{where } \mu = EX.$$

Notice that the zero-th and first central moments are always 1 and 0, respectively:

$$c_0 = E(X - \mu)^0 = E1 = 1,$$

and

$$c_1 = E(X - \mu) = EX - \mu = \mu - \mu = 0.$$

It is only after the first that the central moments get interesting. The second central moment is the variance:

$$c_2 = E(X - \mu)^2 = VX.$$

The third and fourth central moments are related to quantities ("skewness" and "kurtosis") that are sometimes used in statistics to describe the "shape" of a distribution. We will not study them further in this text.

(4.3.15) Relations between moments and central moments. It is a matter of simple algebraic manipulation to derive formulas expressing the central moments c_k in terms of the moments μ_k. For example,

$$\begin{aligned} c_3 &= E(X-\mu_1)^3 = E(X^3 - 3\mu_1 X^2 + 3\mu_1^2 X - \mu_1^3) \\ &= EX^3 - 3\mu_1 EX^2 + 3\mu_1^2 EX - \mu_1^3 = \mu_3 - 3\mu_1\mu_2 + 3\mu_1^3 - \mu_1^3 \\ &= \mu_3 - 3\mu_1\mu_2 + 2\mu_1^3. \end{aligned}$$

It is also possible to go in the other direction and express $\mu_2, \mu_3, \ldots$ in terms of μ_1 and the c_k's, if we proceed inductively: $c_2 = \mu_2 - \mu_1^2$, so

$$\mu_2 = c_2 + \mu_1^2.$$

As for μ_3, we saw above that $c_3 = \mu_3 - 3\mu_1\mu_2 + 2\mu_1^3$, and so

$$\mu_3 = c_3 + 3\mu_1\mu_2 - 2\mu_1^2 = c_3 + 3\mu_1(c_2 + \mu_1^2) - 2\mu_1^3 = c_3 + 3\mu_1 c_2 + \mu_1^3.$$

To continue, we would first find c_4 in terms of μ_1 through μ_4, then solve for μ_4, and then use the two expressions above for μ_2 and μ_3.

There are general formulas that include all the expressions above, involving some coefficients called Stirling numbers that are studied in combinatorial analysis; we will not go into that here. In practice, one seldom encounters moments other than

$$\mu_1 = EX, \quad \mu_2 = EX^2, \quad c_2 = VX,$$

and, more rarely,

$$\mu_3 = EX^3, \quad c_3 = E(X - EX)^3, \quad \mu_4 = EX^4, \quad \text{and } c_4 = E(X - EX)^4.$$

(4.3.16) Existence of moments. You will recall that EX exists only if both EX^+ and EX^- are finite, in which case $EX = EX^+ + EX^-$. [See (4.1.15) and (4.1.20) for the definitions of these quantities.] What about existence of the higher moments EX^2, EX^3, and so on, and the higher central moments $E(X - EX)^2$ (= VX), $E(X - EX)^3$, and so on?

Notice, for one thing, that the existence of EX^2 is a less complicated matter than that of EX, because X^2 has no negative values and, consequently, the quantity that would be called $E(X^2)^-$ is 0. Therefore, EX^2 itself is always either some finite nonnegative number or $+\infty$. Thus, there is only one thing to check; it exists if and only if it is finite. The same is true for all moments EX^k for *even* positive integers k, because even powers are nonnegative. For all the odd moments, of course, we still have two integrals to check, one positive and one negative, and the moment exists if and only if both are finite.

The same holds for central moments: for any even k, $E(X - EX)^k$ is either a finite nonnegative number or $+\infty$, and so it exists if and only if it is finite. For odd k, the central moment $E(X - EX)^k$ exists if and only if both its positive and negative parts are finite.

But there is another fact that simplifies all this considerably, and that is the following.

> ***If for any positive integer k, either the kth moment EX^k or the kth central moment $E(X - EX)^k$ exists, then all moments EX^j and all central moments $E(X - EX)^j$ exist for every $j = 1, 2, 3, \ldots, k$.***

This says that if we are given, or can check, the existence (which implies the finiteness) of any moment or any central moment, then all moments and all central moments of that order and all lower orders exist. It implies, among other things, the following important special case:

> ***If either EX^2 or VX is finite, then so is the other, and so is EX.***

We will not go into the proof of this fact in this text.

(4.3.17) Moments about numbers other than 0 and *EX*. If $EX = \mu$, then the kth central moment of X is, of course, $E(X - \mu)^k$. The kth moment itself can be written as $E(X - 0)^k$. It is natural to think of defining, for any real number a, the quantity $E(X - a)^k$, which is called the ***kth moment of X about a.***

Look back now at Exercise 4b in the previous section, in which you are asked to verify that if $EX = \mu$, then for any a

$$E(X - a)^2 = E(X - \mu)^2 + (\mu - a)^2.$$

That is, the second moment of X about any number a is the variance of X—the second *central* moment—plus the square of $EX - a$. This fact will not be of use to us in this text, but the algebra that produces it is useful in many calculations in statistics. In addition, it implies an interesting fact:

> For any a, the second moment of X about a is at least as large as the variance of X.

Stated another way:

> The number a that minimizes $E(X - a)^2$ is $a = EX$.

EXERCISES FOR **SECTION 4.3**

1. Find EX and VX if X has the uniform distribution on the interval (a, b). The density is $f(x) = 1/(b - a)$ for $a < x < b$ (and 0 otherwise).
2. Let X be a discrete random variable with possible values $1, 2, 3, \ldots, n$, all equally likely. Find EX and VX.

3. Find EX, EX^2, and VX if X has the density $f(x) = \lambda^2 x e^{-\lambda x}$ for $x > 0$ (and 0 otherwise).

4. Let X have the density $f(x) = 2x$ for $0 < x < 1$ (and 0 otherwise).
   a. Find EX and VX.
   b. Find the kth moment μ_k of X, for $k = 0, 1, 2, \ldots$.

5. Let X have the binomial distribution with $n = 4$ and $p = \frac{1}{2}$. Find VX. [We know $EX = np = 2$. Later we will learn that $VX = npq$ for a binomially distributed random variable. But here you are asked to find EX^2 directly; just write down the five possible values of X and their probabilities, and compute EX^2 from them. Then find VX.]

6. Let X have the Poisson distribution with parameter λ. All of the following have been done already, in Example (4.1.11) and in Exercise 8 of Section 4.2. But because the Poisson distribution is so important, you might want to do them again now for review and practice.
   a. Find EX. **b.** Find $E(X(X-1))$.
   c. Find EX^2. **d.** Find VX.

7. Find EX, EX^2, and VX if X has the density $f(x) = 3/x^4$ for $x > 1$ (and 0 otherwise).

8. Repeat Exercise 7 with $3/x^4$ replaced by $2/x^3$.

9. Repeat Exercise 7 with $3/x^4$ replaced by a/x^{a+1}, assuming a is a number (not necessarily an integer) greater than 2. What goes wrong if $a \le 2$?

10. **The "gamma formula."** This useful and important formula says that for a positive integer n and a positive real number λ,

$$\int_0^\infty x^{n-1} e^{-\lambda x} dx = \frac{(n-1)!}{\lambda^n}.$$

To prove it, define $g_n(\lambda)$ to be the integral on the left; we want to show that $g_n(\lambda)$ equals the quantity on the right.
   a. Show that $g_1(\lambda) = 1/\lambda$.
   b. Show that for $n = 2, 3, 4, \ldots, g_n(\lambda) = (n-1)g_{n-1}(\lambda)/\lambda$. [Integrate by parts with $u = x^{n-1}$.]
   c. Conclude that for any n and λ, $g_n(\lambda) = (n-1)!/\lambda^n$. [Use induction.]

We call this the "gamma formula" because of its close relation to the gamma *function* $\Gamma(z)$, which is defined for all positive real numbers z by

$$\Gamma(z) = \int_0^\infty x^{z-1} e^{-x} dx.$$

The gamma *formula,* with $\lambda = 1$, shows that if z is a positive integer, then $\Gamma(z) = (z-1)!$; so the gamma *function* extends the definition of factorials to arbitrary positive numbers. We will see the gamma function in connection with the gamma family of distributions in Sections 8.3 and 8.4. But we will find many uses for the gamma formula before then.

11. Let n be a nonnegative integer and λ a positive real number. Consider the function

$$f_n(x) = \frac{\lambda^n}{(n-1)!} x^{n-1} e^{-\lambda x} \quad \text{for } x > 0 \quad \text{(and 0 otherwise).}$$

Use the gamma formula derived in Exercise 10 to do the following.
a. Show that $f_n(x)$ is a probability density function.
b. Suppose X has this density function. Find EX, EX^2, and VX.
c. Find the kth moment μ_k, for $k = 0, 1, 2, \ldots$.

The distribution with this density function is a member of the gamma family, called the *Erlang* distribution with parameters n and λ; as we will see later, it is the distribution of the time of the nth arrival in a Poisson process.

12. a. Can the variance of a random variable be negative? Under what conditions?
b. Can the variance of a random variable equal 0? Under what conditions? [*Hint:* See Exercise 6 of Section 4.2.]

13. The *standard deviation* of a random variable is the square root of its variance. Suppose $EX = \mu$ and the standard deviation of X equals σ. (This is a lowercase sigma, the usual symbol for a standard deviation.) Let $Z = (X - \mu)/\sigma$. Find EZ and VZ.

Subtracting the expected value and dividing by the standard deviation is called *standardizing* the random variable X. This is an important operation; we will see it again in Sections 6.2 and 8.2.

14. If $EX = \mu$ and $VX = b$, find the following.
a. $E(2X - 3)$
b. $V(2X - 3)$
c. The second moment of X
d. The second moment of X about $a = 1$
e. $V(5 - X)$
f. $E((X - 2)(X + 1))$

15. Let Z be a standard normal random variable and let $X = aZ + b$. Find EX and VX. [See Example (4.3.9).]

16. Suppose $EX = 4$ and $VX = 3$. Find, if you can:
a. EX^2
b. $V(2X - 246)$
c. The variance of $Y = X^2$

17. Find a formula for the variance of $Y = X^2$ in terms of the second and fourth moments μ_2 and μ_4 of X.

18. Find an expression for the variance of XY if X and Y are independent.

19. In a Bernoulli process let X be the number of failures before the first success.
a. Give the mass function and the name of the distribution of X.
b. Find VX. [See Exercise 9 of the previous section.]

ANSWERS

1. $\frac{1}{2}(b + a), \frac{1}{12}(b - a)^2$
2. $\frac{1}{2}(n + 1), \frac{1}{12}(n^2 - 1)$
3. $2/\lambda$; $6/\lambda^2$; $2/\lambda^2$
4. **a.** $\frac{2}{3}$; $\frac{1}{18}$ **b.** $2/(k + 2)$
5. 1
6. **a.** λ **b.** λ^2 **c.** $\lambda^2 + \lambda$ **d.** λ
7. $\frac{3}{2}$; 3; $\frac{3}{4}$
8. $EX = 2$; EX^2 and VX do not exist.
9. $a/(a - 1)$; $a/(a - 2)$; $a/(a - 2)(a - 1)^2$
11. **b.** n/λ; $n(n + 1)/\lambda^2$; n/λ^2
    c. $n(n + 1)(n + 2) \cdots (n + k - 1)/\lambda^k$
12. **a.** No **b.** It is possible.
13. $EZ = 0$, $VZ = 1$
14. **a.** $2\mu - 3$ **b.** $4b$ **c.** $b + \mu^2$
    d. $b + (\mu - 1)^2$ **e.** b **f.** $b + \mu^2 - \mu - 2$
15. b, a^2
16. **a.** 19 **b.** 12 **c.** Cannot be found
17. $\mu_4 - \mu_2^2$
18. $V(XY) = E(X^2) \cdot E(Y^2) - (EX)^2 \cdot (EY)^2$
19. **a.** Geometric with parameter p **b.** q/p^2

4.4 COVARIANCE AND THE VARIANCE OF A SUM OF RANDOM VARIABLES

We begin by working out a formula for $V(X + Y)$. As we remarked at the end of the previous section, it is not simply $VX + VY$ except under special circumstances.

According to the standard formula (4.3.4) for computing variances,

$$V(X + Y) = E[(X + Y)^2] - [E(X + Y)]^2.$$

Look at these two terms separately. We did the first one in Example (4.2.21); it is

$$E[(X + Y)^2] = E(X^2 + Y^2 + 2XY) = EX^2 + EY^2 + 2E(XY),$$

and the second is

$$[E(X + Y)]^2 = [EX + EY]^2 = (EX)^2 + (EY)^2 + 2 \cdot EX \cdot EY.$$

When we subtract them we get

$$V(X + Y) = \{EX^2 - (EX)^2\} + \{EY^2 - (EY)^2\} + 2\{E(XY) - EX \cdot EY\}.$$

This is $VX + VY$ plus an additional quantity. So $V(X + Y)$ is not equal to $VX + VY$ unless $E(XY) - EX \cdot EY$ is 0. Remember Theorem (4.2.19) and Example (4.2.20), which show that $E(XY) = EX \cdot EY$ when X and Y are independent, and sometimes when they are not, but not always. Their difference is the *covariance* of X and Y. The usual definition of covariance looks different, but they turn out to be equal (see Note 2 after the definition):

(4.4.1) **Definition.** The ***covariance*** of two random variables X and Y is the number $\mathrm{Cov}(X, Y)$ defined by

$$\mathrm{Cov}(X, Y) = E[(X - EX)(Y - EY)],$$

provided that EX and EY are defined and this expected value exists.

If $\mathrm{Cov}(X, Y)$ is positive, then we say that X and Y are ***positively correlated;*** if it is negative, they are ***negatively correlated.*** If $\mathrm{Cov}(X, Y) = 0$ (as happens when X and Y are independent, and also in some other cases), then we say that X and Y are ***uncorrelated.***

The covariance of X and Y measures, in a relative and somewhat imperfect sense, the degree of dependence of X and Y. It is 0 when they are independent; as we will see shortly, it is positive when X and Y tend to be large together or small together, and it is negative when small values of one tend to be observed along with large values of the other.

Note 1. It can be proved that if VX and VY both exist, then EX, EY, and the expected value in the definition of covariance all exist. An equivalent assumption is that EX^2 and EY^2 both exist. For more details, see (4.3.16) on existence of moments.

Note 2. The covariance is usually computed not from the definition, but from the formula

(4.4.2)
$$\mathrm{Cov}(X, Y) = E(XY) - EX \cdot EY,$$

which we can prove easily as follows. For simplicity write a for EX and b for EY. Then according to Definition (4.4.1),

$$\begin{aligned}\mathrm{Cov}(X, Y) &= E[(X - a)(Y - b)] = E[XY - aY - bX + ab] \\ &= E(XY) - aEY - bEX + ab = E(XY) - ab - ba + ab \\ &= E(XY) - ab = E(XY) - EX \cdot EY.\end{aligned}$$ ∎

Note 3. A variance is a special kind of covariance: according to Definition (4.4.1), the covariance of X with itself is

(4.4.3)
$$\mathrm{Cov}(X, X) = E(X^2) - EX \cdot EX = VX.$$

Note 4. As we showed at the beginning of this section,

(4.4.4)
$$V(X + Y) = VX + VY + 2 \cdot \mathrm{Cov}(X, Y),$$

and consequently,

(4.4.5) if X and Y are uncorrelated, then $V(X + Y) = VX + VY$.

Note 5. Independent random variables are uncorrelated, because they satisfy $E(XY) = EX \cdot EY$ and therefore $\text{Cov}(X, Y) = 0$. But uncorrelated random variables need not be independent. See Example (4.2.20) above, in which X and Y are not independent even though they are uncorrelated.

(4.4.6) **Example: The meaning of positive or negative covariances.** Here we show two different joint distributions for two discrete random variables, one pair positively correlated and one negatively correlated.

Suppose first that X and Y have the joint mass function $p(x, y)$ as given in the following table.

	$p(x, y)$	y = 1	2	3	
x	0	.02	.05	.15	.22
	1	.08	.24	.05	.37
	2	.23	.15	.03	.41
		.33	.44	.23	

Notice that it is unlikely that both X and Y will be at the same end of their ranges; if X is 0, Y is much more likely to be 3 than 1, whereas if X is 2, Y is more likely to be 1. Small values of X will tend to be observed along with large values of Y, and vice versa.

It is a simple matter to compute EX, EY, and $E(XY)$:

$$\begin{aligned}
EX &= 0 \cdot .22 + 1 \cdot .37 + 2 \cdot .41 = 1.19; \\
EY &= 1 \cdot .33 + 2 \cdot .44 + 3 \cdot .23 = 1.90; \\
E(XY) &= 0 \cdot 1 \cdot .02 + 0 \cdot 2 \cdot .05 + 0 \cdot 3 \cdot .15 \\
&\quad + 1 \cdot 1 \cdot .08 + 1 \cdot 2 \cdot .24 + 1 \cdot 3 \cdot .05 \\
&\quad + 2 \cdot 1 \cdot .23 + 2 \cdot 2 \cdot .15 + 2 \cdot 3 \cdot .03 = 1.95.
\end{aligned}$$

Consequently,

$$\text{Cov}(X, Y) = 1.95 - 1.19 \cdot 1.90 = -.3110.$$

Now here is a different pair of discrete random variables, Z and W, which are likely to be large together or small together. The tendency is not quite so strong as it was in the other direction with X and Y, but it is still there:

$p(z, w)$		$w = 1$	$w = 2$	$w = 3$	
z	0	.22	.05	.05	.32
	1	.18	.14	.05	.37
	2	.07	.10	.14	.31
		.47	.29	.24	

Now as an exercise, and to be sure you understand the process, find EZ, EW, $E(ZW)$, and $\text{Cov}(Z, W)$. The results are: $EZ = .99$, $EW = 1.77$, $E(ZW) = 1.99$, and $\text{Cov}(Z, W) = +.2377$.

Indeed, the positive covariance of Z and W is are not quite as large in magnitude as the negative covariance of X and Y. However, any comparison by these two numbers would be deceptive; as we will see later in (4.4.15), they need to be scaled before they can be used as measures of correlation. ∎

The preceding example, although it demonstrates what is indicated by positive and negative covariances, gives us no sense of *why* they should work in this way. To see why, let Z denote the random variable $(X - EX)(Y - EY)$. Then $\text{Cov}(X, Y) = EZ$. Now when X and Y are both large or both small, then $X - EX$ and $Y - EY$ will have the same sign, and Z will be positive. (See Figure 4.4a.) But if one is large and the other is small, then Z will be negative because $X - EX$ and $Y - EY$ will be of opposite signs.

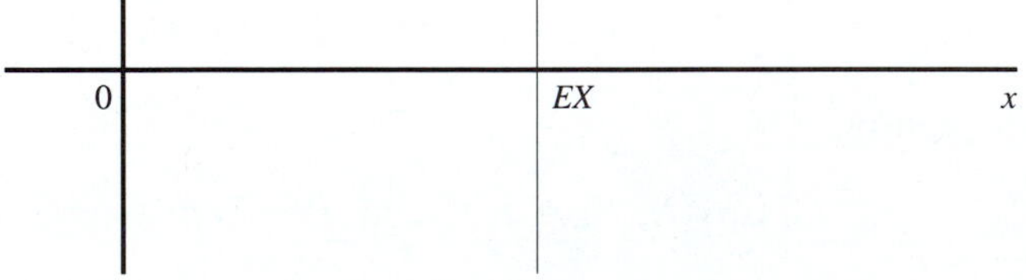

y

$(X - EX)(Y - EY) < 0$ $(X - EX)(Y - EY) > 0$

EY

$(X - EX)(Y - EY) > 0$ $(X - EX)(Y - EY) < 0$

0 EX x

FIGURE 4.4a **The lines $x = EX$ and $y = EY$ divide the plane into four regions; the sign of $(x - EX)(y - EY)$ is constant throughout each region, positive in two and negative in two.**

If Z tends on the average to be positive rather than negative, then $X - EX$ and $Y - EY$ will tend to have the same sign, which says that X and Y will tend to be near the tops or bottoms of their ranges at the same time. And vice versa.

(4.4.7) Example. Let us find $\text{Cov}(X, Y)$ if X and Y have the joint density function

$$f(x, y) = \begin{cases} x + y & \text{if } 0 \le x \le 1 \text{ and } 0 \le y \le 1, \\ 0 & \text{otherwise.} \end{cases}$$

We have to find EX, EY, and $E(XY)$. We find $E(XY)$ directly from the density by integrating, using Theorem (4.2.12b):

$$E(XY) = \int_0^1 \int_0^1 xy \cdot (x + y) dy\, dx = \tfrac{1}{3},$$

as you can check. As for EX and EY, probably the simplest way to proceed in this example is to find them also as double integrals as we did $E(XY)$. Check for yourself that

$$EX = \int_0^1 \int_0^1 x \cdot (x + y) dy\, dx = \tfrac{7}{12}.$$

Similarly check that $EY = \frac{7}{12}$. Consequently,

$$\text{Cov}(X, Y) = \tfrac{1}{3} - \tfrac{7}{12} \cdot \tfrac{7}{12} = -\tfrac{1}{144}.$$

We can see that X and Y are not independent; if they were, their covariance would be 0. (Of course, a covariance of 0 would not imply independence, as we know from Note 5 under the definition of covariance. But a nonzero covariance does imply nonindependence.) ∎

In some examples it might happen that we already know the marginal densities of X and Y as well as their joint density. In such cases it will probably be simpler to find EX and EY as single integrals, $\int x f_X(x) dx$ and $\int y f_Y(y) dy$. If you want to do it that way in the preceding example, you will find that

$$f_X(x) = \tfrac{1}{2} + x \quad \text{for } 0 \le x \le 1 \quad \text{(and 0 otherwise),}$$

and therefore,

$$EX = \int_0^1 x \cdot \left(\tfrac{1}{2} + x\right) dx = \tfrac{7}{12}.$$

The calculations for Y will be similar.

(4.4.8) Covariances of linear functions and of sums. As with expected values and variances, there is a simple connection between $\text{Cov}(aX + b, cY + d)$ and $\text{Cov}(X, Y)$:

$$\text{Cov}(aX + b, cY + d) = ac\text{Cov}(X, Y).$$

As with variances, all that matters are the multipliers; adding a constant to a random variable changes neither its variance nor its covariance with any other random variable.

In addition, the covariance acts linearly on sums of random variables:

$$\text{Cov}(X + Y, Z) = \text{Cov}(X, Z) + \text{Cov}(Y, Z)$$

and

$$\text{Cov}(X, Y + Z) = \text{Cov}(X, Y) + \text{Cov}(X, Z).$$

Exercise 1 at the end of this section asks you to check these formulas. They can of course be combined in numerous ways to find covariances such as

$$\text{Cov}(aX + bY, cZ + d) = ac\text{Cov}(X, Z) + bc\text{Cov}(Y, Z).$$

(4.4.9) **Example.** Suppose $VX = 5$, $VY = 8$, and $\text{Cov}(X, Y) = -3$. Let us find $V(X + Y)$, $V(X - Y)$, and $V(X + 5Y)$. We will use (4.4.4) and (4.4.8), as well as (4.3.11), which says that $V(aX + b)$ equals a^2VX.

We can find $V(X + Y)$ directly:

$$V(X + Y) = VX + VY + 2\text{Cov}(X, Y) = 5 + 8 + 2(-3) = 7.$$

To find $V(X - Y)$, first use (4.4.4):

$$V(X + (-Y)) = VX + V(-Y) + 2\text{Cov}(X, -Y);$$

then use (4.3.11) to get $V(-Y) = VY$ and (4.4.8) to get $\text{Cov}(X, -Y) = -\text{Cov}(X, Y)$:

$$V(X - Y) = VX + VY - 2\text{Cov}(X, Y) = 5 + 8 - 2 \cdot (-3) = 19.$$

We get $V(X + 5Y)$ similarly:

$$\begin{aligned} V(X + 5Y) &= VX + V(5Y) + 2\text{Cov}(X, 5Y) \\ &= VX + 25VY + 2 \cdot 5\text{Cov}(X, Y) \\ &= 5 + 25 \cdot 8 + 2 \cdot 5 \cdot (-3) = 175. \end{aligned}$$ ∎

(4.4.10) **Variance of a sum of more than two random variables.** The formula (4.4.4) extends naturally to the sum of any finite number of random variables. The result is easier to say than it is to write in symbols.

> The variance of a sum is the sum of the variances, plus twice the sum of all the covariances.

For three random variables, it reads as follows:

$$V(X_1 + X_2 + X_3)$$
$$= VX_1 + VX_2 + VX_3 + 2\mathrm{Cov}(X_1, X_2) + 2\mathrm{Cov}(X_2, X_3) + 2\mathrm{Cov}(X_1, X_3).$$

For four random variables there will be four variance terms VX_i and six covariance terms $2\mathrm{Cov}(X_i, X_j)$, because there are $\binom{4}{2}$, or six, pairs among four objects. In general

$$V\left(\sum_{i=1}^{n} X_i\right) = \sum_{i=1}^{n} VX_i + 2\sum_{i<j}\mathrm{Cov}(X_i, X_j);$$

the "$i < j$" under the sum of covariances indicates that each pair of distinct indices is included just once.

If the X_j's are uncorrelated, then all the covariances equal 0, and the result, given in the following theorem, is simple.

(4.4.11) **Theorem.** If $X_1, X_2, \ldots, X_n$ are uncorrelated random variables and VX_j exists for each j, then

$$V(X_1 + X_2 + \cdots + X_n) = VX_1 + VX_2 + \cdots + VX_n.$$

Of course independent random variables are uncorrelated.

(4.4.12) **Example: The variance of a binomially distributed random variable is *npq*.** Now that we know how to find the variance of a sum, we can use the method of indicators to find the variance of X, the number of successes in n Bernoulli trials. You might want to recall Example (4.2.17) and (4.2.18) on indicator random variables. The trick is to write

$$X = X_1 + X_2 + \cdots + X_n,$$

where X_i is the indicator of the event "success on trial i"; that is, X_i is 1 if the ith trial results in success and 0 if not. What we did in Example (4.2.17) was to notice that each X_i has expected value $1 \cdot p + 0 \cdot q = p$, and therefore that

$$EX = EX_1 + EX_2 + \cdots + EX_n = p + p + \cdots + p = np.$$

Now we can do the same trick with the variances. Because the X_i's are associated with different trials of a Bernoulli process, we can assume that they are independent random variables, and so

$$VX = VX_1 + VX_2 + \cdots + VX_n.$$

And the variance of an individual X_i is almost as easy as the expected value:

$$EX_i^2 = 1^2 \cdot p + 0^2 \cdot q = p;$$

therefore,

$$VX_i = EX_i^2 - (EX_i)^2 = p - p^2 = p(1 - p) = pq,$$

and so

$$VX = pq + pq + \cdots + pq = npq. \qquad \blacksquare$$

You might be curious to know what calculations we have been spared by using the method of indicators to find $VX = npq$ for a binomial random variable. They may not seem pleasant, but they are not beyond us, and it is a useful skill to be able to handle such things. Here we give the method, without going through the calculations.

First recall Calculation 1 in Example (4.1.22), which showed directly that

$$EX = \sum_{k=0}^{n} k \cdot \binom{n}{k} p^k q^{n-k} = np.$$

The trick was to notice that the first term (for $k = 0$) equals 0, and that in all the other terms the k cancels with the $k!$ in the denominator of $\binom{n}{k}$, leaving $(k - 1)!$ in the denominator. If we then factor np out of the sum, what is left is the sum of binomial probabilities for parameters $n - 1$ and p, and so it adds to 1, leaving np.

To find the variance, we first find $E(X(X - 1))$, then add $EX = np$ to get EX^2, and finally subtract $(EX)^2 = (np)^2$ to get VX. The last two steps are easy once we have done the first step, in which we find that

$$E(X(X - 1)) = \sum_{k=0}^{n} k(k - 1) \cdot \binom{n}{k} p^k q^{n-k} = n(n - 1)p^2.$$

If you want to try this, notice that the first two terms (for $k = 0$ and $k = 1$) are 0, and in all the others the $k(k - 1)$ cancels with the $k!$ in the denominator of $\binom{n}{k}$, leaving $(k - 2)!$ in the denominator. Then factor out $n(n - 1)p^2$ from every term, and notice that what is left is a sum of binomial probabilities for parameters $n - 2$ and p.

(4.4.13) Example: The numbers of successes in overlapping sets of Bernoulli trials. In a Bernoulli process with success probability p, let X be the number of successes in the first five trials, and let Y be the number of successes in trials numbered 3–7. There is an overlap of three trials in the successes counted by X and by Y, and we might guess that they are positively correlated. How can we compute their covariance?

Each of X and Y is the number of successes in five Bernoulli trials, so they are binomially distributed, each with $n = 5$. So $EX = np = 5p$, and similarly

$EY = 5p$. But how can we find $E(XY)$? We could try to find the possible values of XY and their probabilities, but that would be very tedious. It happens that there is a way to find $\text{Cov}(X, Y)$ from the formula (4.4.4). That formula says that $V(X + Y) = VX + VY + 2\text{Cov}(X, Y)$ and therefore

(4.4.14)
$$\text{Cov}(X, Y) = \tfrac{1}{2}\big(V(X + Y) - VX - VY\big).$$

The trick works because VX and VY are easy to find, and $V(X + Y)$ is obtainable if we notice the right thing about $X + Y$.

We know VX and VY, because the variance of a binomial random variable is npq; so $VX = VY = 5pq$. (As always, $q = 1 - p$.) What about $V(X + Y)$? The trick is to notice that

$$X + Y = S + 2T,$$

where S is the number of successes in trials 1, 2, 6, and 7, and T is the number of successes in trials 3, 4, and 5.

What does this get us? S and T are independent, because they count the successes in nonoverlapping sets of trials. So S and $2T$ are also independent, and therefore uncorrelated; therefore,

$$V(X + Y) = V(S + 2T) = VS + V(2T) = VS + 4VT.$$

And $VS = 4pq$ and $VT = 3pq$, because S and T are binomial random variables based on 4 and 3 trials. So $V(X + Y) = 4pq + 4 \cdot 3pq = 16pq$. Consequently, from (4.4.14),

$$\text{Cov}(X, Y) = \tfrac{1}{2}(16pq - 5pq - 5pq) = 3pq.$$

Notice that 3 is the number of overlapping trials. It is natural to wonder whether the covariance in such a situation is always kpq, where k is the number of trials that X and Y count in common. Exercise 7 at the end of the section will lead you to the answer. ∎

(4.4.15) **The correlation coefficient.** As we remarked, the covariance of X and Y is only a relative measure of their degree of positive or negative correlation. It would seem, for example, that if X and Y have a certain degree of positive correlation, then X and $2Y$ should have the same correlation; otherwise a change of scale would change the amount of correlation of two random quantities. But because of (4.4.8), the covariance of X and $2Y$ is twice that of X and Y.

Fortunately there is a natural way to normalize $\text{Cov}(X, Y)$ so that it gives an absolute measure of correlation—by dividing it by $\sqrt{VX \cdot VY}$. The square root of the variance of a random variable is its *standard deviation,* usually denoted by σ or σ_X (this is a lowercase Greek sigma).

(4.4.16) Definition. Let X and Y be random variables. The ***correlation coefficient*** of X and Y is the number $\rho = \rho(X, Y)$ (ρ is a lowercase Greek rho), defined by

$$\rho = \frac{\mathrm{Cov}(X, Y)}{\sigma_X \sigma_Y},$$

where $\sigma_X = \sqrt{VX}$ and $\sigma_Y = \sqrt{VY}$ are the ***standard deviations*** of X and Y.

The correlation coefficient is used a great deal in statistical inference, where it is used to measure in some sense how nearly the random variables X and Y can be regarded as being linear functions of each other.

As we mentioned earlier, the correlation coefficient is unaffected by scaling or by addition of constants: the correlation coefficient of $aX + b$ and $cY + d$ is the same as that of X and Y. This is true because if we switch from X and Y to $aX + b$ and $cY + d$, then we multiply the covariance by ac, but we also multiply the variances by a^2 and c^2; consequently, we multiply the standard deviations by a and c, respectively. (Actually, we multiply them by $|a|$ and $|c|$, but this is the right thing to do: it produces the appropriate sign for the correlation coefficient.)

(4.4.17) Example. Let us find the correlation coefficient for each of the two discrete joint distributions in Example (4.4.6). In that example, the mass function tables seemed to show that the correlation of the negatively correlated pair (X, Y) was stronger than that of the positively correlated pair (Z, W); indeed $|\mathrm{Cov}(X, Y)|$ is larger than $|\mathrm{Cov}(Z, W)|$. But that is not a fair comparison; it is not the covariance, but the correlation coefficient, that best measures correlation.

To find the two correlation coefficients, we divide the covariances by the standard deviations—that is, by the square roots of the variances. We begin by restating the calculations that we did in Example (4.4.6) and then we will proceed to the standard deviations. First X and Y:

$EX = 1.19$, $EY = 1.90$, and $E(XY) = 1.95$, and therefore $\mathrm{Cov}(X, Y) = 1.95 - 1.19 \cdot 1.90 = -.311$.

Now to find VX: Remember that X has possible values 0, 1, and 2, with probabilities .22, .37, and .41, respectively. So you can check that $EX^2 = 2.01$ and therefore $VX = 2.01 - 1.19^2 = .5939$, and so

$$\sigma_X = \sqrt{.5939} \doteq .7706.$$

As for VY: Y has possible values 1, 2, and 3, with probabilities .33, .44, and .23, respectively, so $EY^2 = 4.16$ and $VY = 4.16 - 1.90^2 = .55$. Thus,

$$\sigma_Y = \sqrt{.55} \doteq .7416.$$

Therefore,

$$\rho(X, Y) = \frac{-.311}{.7706 \cdot .7416} \doteq -.5442.$$

Now you can work out the correlation coefficient of Z and W as an exercise. So that you will not have to look back, here is the pertinent information: In Example (4.4.6) we found that $EZ = .99$, $EW = 1.77$, and $E(ZW) = 1.99$, and therefore $\text{Cov}(Z, W) = 1.99 - .99 \cdot 1.77 = .2377$. To find the standard deviations, use the facts that Z has possible values 0, 1, and 2, with corresponding probabilities .32, .37, and .31, and W has possible values 1, 2, and 3, with corresponding probabilities .47, .29, and .24. The result, which you should be sure you know how to get, is

$$\rho(Z, W) \doteq .3695.$$

Comparing $-.5542$ with .3695 shows the extent to which the negative correlation of X and Y is stronger than the positive correlation of Z and W. ▮

(4.4.18) **The correlation coefficient of a joint normal distribution.** In Example (3.5.14) we introduced the joint density

$$f(x, y) = \frac{1}{2\pi\sqrt{1-\rho^2}} e^{-(x^2+y^2-2\rho xy)/2(1-\rho^2)}.$$

We saw that although X and Y are not independent if they have this joint density, their marginal densities are both standard normal. Here we see that the parameter ρ is actually the correlation coefficient (which is why it was called ρ in the first place).

Since X and Y are standard normal, we know that $EX = EY = 0$ and $VX = VY = 1$; nothing about their joint distribution is needed to get this. Consequently, we need only find $E(XY)$, because the covariance is

$$\text{Cov}(X, Y) = E(XY) - EX \cdot EY = E(XY),$$

and the correlation coefficient is

$$\rho(X, Y) = \frac{\text{Cov}(X, Y)}{\sqrt{VX \cdot VY}} = \text{Cov}(X, Y) = E(XY).$$

Now

$$E(XY) = \int_{-\infty}^{\infty}\int_{-\infty}^{\infty} xy \cdot \frac{1}{2\pi\sqrt{1-\rho^2}} e^{-(x^2+y^2-2\rho xy)/2(1-\rho^2)} dx\, dy,$$

and to find this we proceed much as in Example (3.5.14). It is not pleasant when one first encounters this kind of integral, but here is an outline of the evaluation. First do a little rearranging, as in Example (3.5.14), to get

$$E(XY) = \int_{-\infty}^{\infty} y \cdot \frac{1}{\sqrt{2\pi}} e^{-y^2/2} \left[\int_{-\infty}^{\infty} x \cdot \frac{1}{\sqrt{2\pi}\sqrt{1-\rho^2}} e^{-(x-\rho y)^2/2(1-\rho^2)} dx \right] dy.$$

Next, work on the inner integral, in brackets. Begin by making the substitution $z = (x - \rho y)/\sqrt{1-\rho^2}$. The inner integral becomes

$$\int_{-\infty}^{\infty} \frac{1}{\sqrt{2\pi}} (z\sqrt{1-\rho^2} + \rho y) e^{-z^2/2} dz.$$

This is the sum of two integrals; the first equals 0 because it is $\sqrt{1-\rho^2}$ times the expected value of a standard normal random variable, and the second equals ρy because it is ρy times the integral of the standard normal density. So the whole inner integral reduces to ρy. Putting this in gives

$$E(XY) = \int_{-\infty}^{\infty} \rho y^2 \cdot \frac{1}{\sqrt{2\pi}} e^{-y^2/2} dy = \rho \cdot E(Y^2) = \rho.$$

Summarizing the results of (3.5.14) and this example: If X and Y have the joint density given above, then the marginal distributions of X and Y are both the standard normal distribution; X and Y are independent if and only if $\rho = 0$, and

$$\text{Cov}(X, Y) = \rho(X, Y) = \rho.$$

Remember that, in general, independent random variables are uncorrelated, but uncorrelated random variables are not necessarily independent. However, if X and Y have this joint density, then they are independent if and only if they are uncorrelated—that is, if and only if $\rho = 0$.

We will return to this joint density, and the more general class of *bivariate normal* densities, in (5.4.10).

The next fact is perhaps the most important feature of the correlation coefficient, along with the related fact that it is impervious to scaling.

(4.4.19) The correlation coefficient is always between 1 and -1.

There is actually more: The only way that $\rho(X, Y)$ can equal ± 1 is for X and Y to be in a *perfect linear relationship*—that is, for there to exist constants c and d such that $Y = cX + d$ with probability 1.

The fact that $-1 \le \rho \le 1$ follows from the following important inequality, which appears in different forms in many places in mathematics.

(4.4.20) **Theorem: The Cauchy–Schwarz inequality.** If S and T are random variables and ES^2 and ET^2 exist, then

$$[E(ST)]^2 \leq ES^2 \cdot ET^2.$$

The proof involves a wonderful trick; we will give it in (4.4.22), after we show how this inequality implies that the correlation coefficient is between 1 and -1. Then we will show how a covariance of ± 1 implies a linear relationship.

(4.4.21) Corollary. If X and Y are random variables and VX and VY exist, then

$$\text{Cov}(X, Y)^2 \leq VX \cdot VY,$$

and consequently $|\rho(X, Y)| \leq 1$.

Proof. Let $S = X - EX$ and $T = Y - EY$. Then $\text{Cov}(X, Y) = E(ST)$, $VX = ES^2$, and $VY = ET^2$; so the Cauchy–Schwarz inequality gives the stated result. ∎

(4.4.22) Proof of the Cauchy–Schwarz inequality. This gem of a proof is based on a little high-school algebra and the observation (4.2.23) that the expected value of a nonnegative random variable is a nonnegative number. Let z be any real number and consider the random variable $(zS + T)^2$. It is nonnegative, so

$$0 \leq E(zS + T)^2 = z^2 ES^2 + 2zE(ST) + ET^2 \quad \text{for all real numbers } z.$$

This tells us that the quadratic polynomial $az^2 + bz + c$, where $a = ES^2$, $b = 2E(ST)$, and $c = ET^2$, is never negative for any value of z. So the graph of $az^2 + bz + c$, as a function of z, may touch the horizontal axis but will never go below it. That is, the quadratic polynomial $az^2 + bz + c$ has at most one real root. So its discriminant $b^2 - 4ac$ is either 0 (if there is one real root) or negative (if there are none). But $b^2 - 4ac = 4[E(ST)]^2 - 4ES^2 \cdot ET^2$, and consequently

$$[E(ST)]^2 - ES^2 \cdot ET^2 \leq 0.$$

∎

(4.4.23) **Theorem.** If $\rho(X, Y) = 1$ or $\rho(X, Y) = -1$, then there are constants c and d such that $Y = cX + d$ with probability 1.

Proof. The hypothesis says the same thing as $\rho^2 = 1$—that is, that

$$\text{Cov}(X, Y)^2 = VX \cdot VY,$$

which is the same as

$$[E(X - EX)(Y - EY)]^2 = E(X - EX)^2 \cdot E(Y - EY)^2.$$

This is the same as

$$[E(ST)]^2 = ES^2 \cdot ET^2,$$

where, as in the proof of Corollary (4.4.21), $S = X - EX$ and $T = Y - EY$. This says that the Cauchy–Schwarz inequality is satisfied as an exact equality. In the language of (4.4.22), the discriminant $b^2 - 4ac$ is exactly 0, which implies that the quadratic polynomial $E(zS + T)^2 = az^2 + bz + c$ has one real root. Call this root r; that means that $E(rS + T)^2 = 0$.

But the only way the expected square of a random variable can be 0 is for the random variable to be constantly 0. That is, with probability 1 we have

$$0 = rS + T = r(X - EX) + (Y - EY);$$

and thus

$$Y = -rX + (rEX + EY).$$

So if we set $c = -r$ and $d = rEX + EY$, we get $Y = cX + d$. ▯

EXERCISES FOR SECTION 4.4

1. Verify the following properties of the covariance. (Here X, Y, and Z are random variables and a, b, c, and d are constants.)
   a. $\text{Cov}(aX + b, cY + d) = ac\text{Cov}(X, Y)$
   b. $\text{Cov}(X + Y, Z) = \text{Cov}(X, Z) + \text{Cov}(Y, Z)$
   c. $\text{Cov}(Y, X) = \text{Cov}(X, Y)$

2. Suppose $EX = a$, $EY = b$, $EX^2 = c$, $EY^2 = d$, and $E(XY) = e$. Find the following.
   a. $\text{Cov}(X, Y)$ b. $\text{Cov}(-X, 3Y + 2)$
   c. σ_X and σ_Y d. $\rho(X, Y)$

3. Let X and Y have the joint distribution shown in the table. Find their covariance and their correlation coefficient. [Begin by finding EX, EY, EX^2, EY^2, and $E(XY)$. Then use the results of Exercise 2a and 2d.]

	$p(x, y)$	y = 1	2	3	
x	1	.12	.08	.11	.31
	2	.18	.14	.07	.39
	3	.17	.05	.08	.30
		.47	.27	.26	

4. Let X and Y have the following joint density function:

$$f(x, y) = \tfrac{6}{5}(x^2 + y) \quad \text{for } 0 < x < 1 \text{ and } 0 < y < 1 \quad \text{(and 0 otherwise).}$$

Find the covariance and the correlation coefficient.

5. Let $X_1, X_2, \ldots, X_n$ be independent random variables, each with expected value μ and variance σ^2. Let S denote their sum $X_1, + X_2 + \cdots + X_n$.
   a. Find the expected value, variance, and standard deviation of S.
   b. Find the expected value, variance, and standard deviation of the random variable S/n.

6. Find the correlation coefficient of the two random variables X and Y in Example (4.4.13). There X is the number of successes in trials 1–5, and Y the number in trials 3–7, of a Bernoulli process with success probability p.

7. We want to find out whether the covariance in Example (4.4.13) is a coincidence. In a Bernoulli process with success probability p, consider three adjacent sequences of trials: the first j trials, the next k after those j, and the next m after those $j + k$. Let X be the number of successes in the first $j + k$ trials, and let Y be the number in the $k + m$ trials after the first j. [In the example, j equals 2, k equals 3, m equals 2, and $\mathrm{Cov}(X, Y) = 3pq$.] Mimic the trick used in Example (4.4.13) to find whether $\mathrm{Cov}(X, Y)$ is kpq in this general situation.

8. Find the correlation coefficient of the two random variables in Exercise 7.

9. Let the joint distribution of the pair (X, Y) of random variables be the uniform distribution on the diamond-shaped region in the plane whose four corners are the points $(1, 0)$, $(-1, 0)$, $(0, 1)$, and $(0, -1)$.
   a. Sketch this region. It is bounded by four lines; write the equations of those lines. Notice that this region is $\{(x, y): |x| + |y| \le 1\}$.
   b. Before doing any calculations, decide whether X and Y are independent.
   c. Write down the joint density of X and Y.
   d. Find the marginal density of X.
   e. Find EX and VX.
   f. Without doing any further calculation, write down the marginal density, expected value, and variance of Y.
   g. Find $E(XY)$, $\mathrm{Cov}(X, Y)$, and $\rho(X, Y)$. Why does the value of the covariance not contradict the answer to part b?

   [If you have difficulty with parts b–f, see Section 3.5. In any case, do part g even if you have to look at the answers for parts b–f.]

10. Suppose $EX = 3$, $EY = 2$, $EX^2 = 13$, $EY^2 = 7$, and $E(XY) = 3$. Find the following.
    a. $\mathrm{Cov}(X, Y)$
    b. $\mathrm{Cov}(X + Y, X - Y)$ [*Hint:* Use (4.4.2) or Exercise 1 above.]

11. Let X and Y be random variables. Show the following.
a. If X and Y are independent, then $\text{Cov}(X, X + Y) = VX$.
b. If Y is a constant, then X and Y are uncorrelated.

12. Show that for any two random variables X and Y, $\text{Cov}(X + Y, X - Y) = VX - VY$. Conclude that if X and Y have the same variance, then $X + Y$ and $X - Y$ are uncorrelated.

13. Suppose $EX = 3$, $EX^2 = 14$, $EY = 5$, $EY^2 = 31$, and $E(XY) = 13$. Find the following.
a. VX **b.** $\text{Cov}(X, Y)$ **c.** $V(X + Y)$

14. Let Y be the number of failures preceding the second success in a Bernoulli process with success probability p. We want to show that $EY = 2q/p$ and $VY = 2q/p^2$. To do this, write $Y = X_1 + X_2$, where X_1 is the number of failures preceding the first success and X_2 is the number between the first and second successes. [All of the following have been done already. You will find everything you need in Example (4.1.3), Exercise 19 of Section 4.3, Exercise 9 of Section 4.2, and rule B4 for Bernoulli processes as restated in Example (2.4.9).]
a. Give the mass function of X_1 and the name of its distribution.
b. Do the same for X_2.
c. Find EX_1 and EX_2.
d. Conclude that $EY = 2q/p$.
e. Find VX_1 and VX_2. [See Exercise 9 of Section 4.2 or Exercise 19 of Section 4.3.]
f. Conclude that $VY = 2q/p^2$. (Be sure to justify your conclusion.)

There are two ways to find EY and VY; this exercise shows the simpler of the two. The other way is to find them directly from the mass function of Y, which is $p(k) = (k + 1)p^2q^k$ for $k = 0, 1, 2, \ldots$. You found EY in this way in Exercise 16 of Section 4.1; probably the best way to find EY^2 is first to find $E[Y(Y + 2)]$. But wait for Chapter 7, where we will study a powerful tool called the probability generating function. It enables us not only to find moments quickly, but also in many cases to identify the distribution of a random variable.

ANSWERS

2. a. $e - ab$ **b.** $-3(e - ab)$
c. $\sqrt{c - a^2}$; $\sqrt{d - b^2}$
d. $(e - ab)/\sqrt{(c - a^2)(d - b^2)}$

3. Cov $= -.0821$; $\rho = -.1269$

4. Cov $= -1/100$; $\rho = -.130558$

5. a. $n\mu$; $n\sigma^2$; $\sigma\sqrt{n}$ **b.** μ; σ^2/n; $\sigma/\sqrt{n}$

6. $\frac{3}{5}$

7. $\text{Cov}(X, Y) = kpq$; the result of (4.4.13) is no coincidence.

8. $k/\sqrt{(j + k)(k + m)}$

9. b. They are not independent.
c. $f(x, y) = \frac{1}{2}$ if $|x| + |y| \leq 1$ (0 otherwise)

d. $f_X(x) = 1 - |x|$ if $-1 \le x \le 1$ (0 otherwise) **e.** $EX = 0$; $VX = \frac{1}{6}$ **g.** All are 0.

10. a. -3 **b.** 1

13. a. 5 **b.** -2 **c.** 7

4.5 LAWS OF LARGE NUMBERS

At the beginning of this chapter we gave a definition of EX that was intended to produce a number representing the *expected long-run average value* of X. How would we restate this intent in the form of a proposition that could be proved? That is, is there a theorem which provides that the long-run average of X is likely to be close to the number EX?

If the experiment that produces X is repeated a large number of times, say n times, and we let $X_1, X_2, \ldots, X_n$ denote the values of X that come out, then these X_j's are independent, identically distributed random variables, and their average is also a random variable, a function of $X_1, X_2, \ldots, X_n$, denoted by $\overline{X}_n$:

(4.5.1)
$$\overline{X}_n = \frac{1}{n}(X_1 + X_2 + \cdots + X_n).$$

Our definition of EX was motivated by the desire to produce a number that the random variable $\overline{X}_n$ is likely to be close to if n is large. And one of the remarkable features of probability theory is that it produces a theorem which says just that: in some sense, the random variable $\overline{X}_n$ has a high probability of being close to the number EX if n is large.

There are actually several theorems to this effect; they are called *laws of large numbers*. They all deal with a sequence $X_1, X_2, X_3, \ldots$ of independent, identically distributed (i.i.d.) random variables, representing the values of some random variable X on independent repetitions of the experiment. Subject to various additional hypotheses on the distribution of X, they make precise in different ways the notion that $\overline{X}_n$ is likely to be close to EX if EX is large. We will discuss two of these theorems in this section—the Kolmogorov Strong Law and the Chebyshev Weak Law. (The terms "strong" and "weak" actually have a technical meaning; the difference lies in how the law provides that $\overline{X}_n$ is "likely to be close to" EX.)

An implication of the laws of large numbers is that an unknown average of a population of numbers can be estimated by the average of a sample from the population, as indicated in the following example.

(4.5.2) **Example: Estimating a population mean.** Suppose we are interested in the average cholesterol level of teenagers in the United States. Call it μ. This is a quantity that we cannot know exactly; we would need to do cholesterol tests on tens of millions of people.

But if X is the cholesterol level of a randomly chosen teenager, then X is a random variable whose expected value EX equals μ. The law of large numbers says that if $X_1, X_2, \ldots, X_n$ are a large number of independent observations of X—that is, the cholesterol levels of a large sample of randomly chosen people—then the average $\overline{X}_n$ will be close to μ. We do not know μ, but we can have some confidence that we are close to it if we take a random sample of teenagers, measure their cholesterol levels, and average them. ∎

Even the theoretical implications of the laws of large numbers contain some surprising insights—for example, the following.

(4.5.3) **Probability as expected long-run relative frequency.** Long before we defined EX and interpreted it as expected long-run average value, we interpreted the probability of an event A as its expected long-run relative frequency. A side benefit of the laws of large numbers is that they can be used to confirm that the theory supports this interpretation as well. The reason is that the probability $P(A)$ is really the expected value of a certain random variable X—the indicator function of the event A.

We discussed indicators earlier, in Example (4.2.17) and (4.2.18). Remember that the indicator of A is the random variable X defined to be 1 if A occurs and 0 if not; more precisely, X is the function on the outcome space Ω defined by

$$X(\omega) = \begin{cases} 1 & \text{if } \omega \in A, \\ 0 & \text{if not.} \end{cases}$$

The expected value of X is $P(A)$:

$$\begin{aligned} EX &= 1 \cdot P(X = 1) + 0 \cdot P(X = 0) \\ &= 1 \cdot P(A) + 0 \cdot P(A^c) = P(A). \end{aligned}$$

Now suppose we do the experiment n times, and let $X_1, X_2, \ldots, X_n$ be the values of the indicator of A that come out. Then the laws of large numbers say that the average $\overline{X}_n$ will be close to $P(A)$ if n is large. But what is $\overline{X}_n$? Each X_i is 1 or 0, so the sum of the X_i's is the number of times A occurs in the n trials; that is, the frequency of occurrence of A in n trials. And the average $\overline{X}_n$, which is the sum divided by n, is the relative frequency. So the statement $\overline{X}_n \to EX$ implies that the relative frequency of A approaches $P(A)$.

Thus laws of large numbers also imply that probabilities can indeed be interpreted as expected long-run relative frequencies. This gives a satisfying sort of inner consistency to the theory.

(4.5.4) **Kolmogorov's (Strong) Law of Large Numbers.** This is the strongest of the laws of large numbers. It needs no hypothesis at all about the distribution of X; it tells us what the sequence $\overline{X}_1, \overline{X}_2, \overline{X}_3, \ldots, \overline{X}_n, \ldots$ of averages will do whether EX exists or not. Its proof requires measure theory and is usually given in beginning graduate level probability courses.

Theorem (Kolmogorov's Strong Law). Let $X_1, X_2, X_3, \ldots$ be a sequence of i.i.d. random variables, all having the same distribution as a random variable X. For each n let $\overline{X}_n = \frac{1}{n}(X_1 + X_2 + \cdots + X_n)$.

a. If EX exists, then $\lim_{n\to\infty} \overline{X}_n = EX$ with probability 1.

b. If EX does not exist, then with probability 1 the sequence $\overline{X}_1, \overline{X}_2, \overline{X}_3, \ldots$ fails to have a limit.

Interpreting the statement "$\lim_{n\to\infty} \overline{X}_n = EX$ with probability 1" requires a little imagining. It means that if we do a great many repetitions of the experiment of observing a sequence $X_1, X_2, X_3, \ldots$ and finding the limit of the sequence of averages $\overline{X}_1, \overline{X}_2, \overline{X}_3, \ldots$, then the fraction of times in which that limit is EX approaches 1.

Of course, there is no way we can do even one repetition of this experiment in reality, because one repetition of the experiment involves observing infinitely many X_j's, computing the averages $\overline{X}_n$, and finding the limit of that sequence. But the theorem still gives us insight as to what we will actually see if we repeat some chance experiment a large number of times.

This kind of conclusion, a "probability-1" statement about the limit of a sequence, is what distinguishes a strong law of large numbers from a weak one.

A. N. Kolmogorov was born in 1903 and died in 1987. He spent his entire career at Moscow University and was for years the leading member of the brilliant Russian school of probabilists. His contributions to probability are monumental; it was he who in 1933 first formulated the definition of a probability space (the same one given at the beginning of Chapter 2) and showed how to base the theory on it. The English translation of his 1933 book (which was in German) is entitled *Foundations of the Theory of Probability* and is only 70 pages long. See the bibliography for a reference.

Now we turn to Chebyshev's Law of Large Numbers, a weak law whose proof is easily within our reach. It requires the additional assumption that the random variable X has not only a finite expected value, but also a finite variance VX. The essence of Chebyshev's law will be apparent when we look at the expected value and variance of $\overline{X}_n$ under these assumptions and remember what variances tell us.

As we have done before, let us use the letter μ to denote EX and σ^2 to denote VX. (A variance is customarily viewed as a square, because it is never negative, and because its square root, the standard deviation σ, is a common quantity in statistics.)

So let $X_1, X_2, \ldots, X_n$ be independent random variables, each having $EX_i = \mu$ and $VX_i = \sigma^2$, and define

$$S_n = X_1 + X_2 + \cdots + X_n$$

and

$$\overline{X}_n = \frac{1}{n} \cdot S_n.$$

Then it follows from the formulas for the expected value and variance of a sum that

$$ES_n = n\mu \quad \text{and} \quad VS_n = n\sigma^2.$$

Notice that we need the assumption of independence for the variance, although the expected value would equal $n\mu$ even if the X_i's were dependent. Now when we divide S_n by n, we divide its expected value by n and its variance by n^2; then

$$E\overline{X}_n = \mu \quad \text{and} \quad V\overline{X}_n = \frac{\sigma^2}{n}.$$

Now you can see why this implies that $\overline{X}_n$ will be close to μ—its expected value *is* μ, and its variance is small when n is large. A random variable with small variance is likely to be close to its expected value; once we have made this statement precise, we will have a theorem.

It is *Chebyshev's inequality* that makes precise the idea that a small variance means a high probability of being close to the expected value.

(4.5.5) **Chebyshev's inequality.** Let Y be a random variable with a finite expected value and a finite variance. Then for any positive number a,

$$P(|Y - EY| \geq a) \leq \frac{VY}{a^2}.$$

That is, the probability that Y *fails* to be within a of its expected value is no bigger than VY/a^2; so the smaller the variance of Y, the smaller this probability. We will prove this later in Theorem (4.5.9) and (4.5.10).

Chebyshev's inequality is just what we need to prove a law of large numbers; we simply apply it to $\overline{X}_n$. Remember that $\overline{X}_n$ has expected value μ and variance σ^2/n. In a law of large numbers, we think of n as increasing to ∞, so $V\overline{X}_n$ approaches 0. That is, for any positive number a,

$$P(|\overline{X}_n - \mu| \geq a) \leq \frac{\sigma^2}{na^2},$$

and this approaches 0 as $n \to \infty$. We have proved the following theorem.

(4.5.6) **Chebyshev's (Weak) Law of Large Numbers.** If $X_1, X_2, X_3, \ldots$ are i.i.d. random variables with finite expected value μ and finite variance σ^2, then for any positive number a, $P(|\overline{X}_n - \mu| \geq a) \to 0$ as $n \to \infty$. That is, for any positive a, no matter how small, the event that $\overline{X}_n$ deviates from μ by a or more becomes vanishingly improbable as n increases.

Notice the difference between the conclusions of a weak law and a strong law. Both involve the sequence $\{\overline{X}_n\}$ of random variables; but a strong law implies something about the limit of that sequence, whereas a weak law involves only the limit of a sequence of probabilities associated with the random variables. It takes a stronger hypothesis to prove that something is true with probability 1 about the limit of a sequence of random variables, than it does to prove something about the limit of a sequence of numbers.

Let us summarize informally the steps in our thinking that led to Chebyshev's Law of Large Numbers. Then we will give an example that should also help in understanding the logic.

1. Laws of large numbers say that if $X_1, X_2, X_3, \ldots$ are i.i.d. random variables with expected value μ, then $\overline{X}_n$ is likely to be close to μ when n is large. (That is, they say that expected values really are expected long-run observed average values.)
2. If the X_i's have a finite variance σ^2, then $\overline{X}_n$ has expected value μ and variance σ^2/n, and this variance approaches 0 as n increases.
3. A random variable with a small variance is likely to be close to its expected value. Chebyshev's inequality shows that if the variance approaches 0, the probability of *not* being within a given distance of the expected value approaches 0.

(4.5.7) Example: **How Chebyshev's inequality leads to the law of large numbers.** Suppose we repeatedly perform the experiment of rolling a die until the first 6 appears and recording X, the number of non-6's we get before the first 6. We know that X has the geometric distribution with $p = \frac{1}{6}$ $\left(\text{and } q = \frac{5}{6}\right)$; the mass function is $P(X = k) = pq^k = \frac{1}{6} \cdot \left(\frac{5}{6}\right)^k$ for $k = 0, 1, 2, 3, \ldots$. In Example (4.1.3) and in Exercise 19 of Section 4.3, we saw that $EX = q/p = 5$ and $VX = q/p^2 = 30$.

Let X_i be the value of X on the ith performance of this experiment. Then $\overline{X}_n$ is the average number of non-6's before a 6, on the first n performances. Because $EX = 5$, the law of large numbers says that $\overline{X}_n$ should be close to 5 if n is large. Let us find, for various values of n, what Chebyshev's inequality guarantees about the probability that $\overline{X}_n$ *fails* to be between 4.8 and 5.2.

This probability is $P(|\overline{X}_n - E\overline{X}_n| \geq .2)$; so by Chebyshev's inequality it is less than or equal to $V\overline{X}_n/(.2)^2 = 25 \cdot V\overline{X}_n$. We know that $V\overline{X}_n = VX/n = 30/n$. Thus the Chebyshev bound is $25 \cdot 30/n = 750/n$. Putting this all together, we get the following result.

$$\text{For any } n,\ P(\overline{X}_n \text{ is } \textit{not} \text{ between 4.8 and 5.2}) \leq \frac{750}{n}.$$

Now this may not seem like much of a bound. It tells us nothing at all about the probability when $n \leq 750$, because $750/n$ is at least 1, and all probabilities are less than or equal to 1. But it does provide information as soon as n exceeds 750; for $n = 1{,}500$, for example, we are guaranteed that the probability is no greater than $\frac{1}{2}$. (But even that information is crude; the true probability when $n = 1{,}500$ is approximately .16, which is quite a bit less than $\frac{1}{2}$.)

But as crude as it is, Chebyshev's inequality is enough to guarantee that the probability tends to 0 as $n \to \infty$ (because $750/n \to 0$). And that is Chebyshev's Law of Large Numbers. ▮

Another example will show how Chebyshev's inequality applies to random variables other than $\overline{X}_n$. It will also show us why, although its importance in the theory is great, it is not much used in practice: the number VY/a^2 is often much greater than the true probability.

(4.5.8) **Example.** Suppose customers are arriving according to a Poisson process with a rate of ten customers per minute. Let Y be the arrival time of the 50th customer. What can we say about Y? It would seem that Y should be somewhere near 5 minutes—in fact, we will see that EY equals 5. But what can we say about how close it is likely to be? For example, what is the probability that Y fails to be within 48 seconds (.8 minute) of 5 minutes? That is, what is $P(|Y - EY| \geq .8)$?

In a later chapter we will learn what the density of Y is, and so we might be able to find this probability by integrating that density from 4.2 to 5.8. However, doing that integral turns out to be rather complicated, and we may not care to go to the trouble. Chebyshev's inequality will not give us the probability, but it will give us an upper bound on the probability. All we need to know are EY and VY, because the inequality says that $P(|Y - EY| \geq .8) \leq VY/(.8)^2$.

Now Y is the sum of 50 independent waiting times, each having the exponential distribution with $\lambda = 10$. Each of these waiting times has expected value $1/\lambda = .1$ and variance $1/\lambda^2 = .01$, so the expected value and variance of Y are 50 times these:

$$EY = 5 \quad \text{and} \quad VY = .5.$$

So according to Chebyshev's inequality,

$$P(|Y - EY| \geq .8) \leq .5/(.8)^2 = .78125.$$

Now as it happens, the true probability in this example is about .26, which is indeed less than or equal to .78125. So while Chebyshev's inequality is correct, it may not give a very useful upper bound.

In fact, in some cases VY/a^2 turns out to be greater than 1; then Chebyshev's inequality is of no use at all, because all probabilities are 1 or less. We will see this in some of the exercises at the end of this section; see in particular Exercises 9c and 10a. On the other hand, see Exercise 5 for a case in which no improvement on Chebyshev's inequality is possible. ▮

Of course, we have not yet proved Chebyshev's inequality (4.5.5), so we cannot properly say we have proved Chebyshev's Law of Large Numbers. But we will do this now. It turns out that there is a simpler inequality than (4.5.5) that has an easy (and quite clever) proof, and it implies (4.5.5):

(4.5.9) **Theorem.** If W is a nonnegative random variable with a finite expected value, then for any positive number b,

$$P(W \geq b) \leq \frac{EW}{b}.$$

Proof. Define a new random variable Z to be the following function of W:

$$Z = \begin{cases} b & \text{if } W \geq b, \\ 0 & \text{if } W < b. \end{cases}$$

A little thought will convince you that $W \geq Z$. Consequently, by (4.2.24), $EW \geq EZ$. But EZ is easy to find, because Z has only two possible values:

$$EZ = b \cdot P(Z = b) + 0 \cdot P(Z = 0) = b \cdot P(W \geq b).$$

Putting this together with $EW \geq EZ$, we get $EW \geq b \cdot P(W \geq b)$, which is just what the theorem says. ∎

(4.5.10) **How Chebyshev's inequality follows from Theorem (4.5.9).** Chebyshev's inequality refers to $P(|Y - EY| \geq a)$, which is the same as $P((Y - EY)^2 \geq a^2)$. All we do is apply (4.5.9) to the nonnegative random variable $W = (Y - EY)^2$, with $b = a^2$; we get

$$\begin{aligned} P(|Y - EY| \geq a) &= P((Y - EY)^2 \geq a^2) \\ &= P(W \geq b) \leq \frac{EW}{b} = \frac{E(Y - EY)^2}{a^2} = \frac{VY}{a^2}. \end{aligned}$$

A more appropriate name for the inequality of this section is the *Chebyshev–Bienaymé inequality*. Irenée-Jules Bienaymé (1796–1878) was a French mathematician who, in a paper published in 1853, gave the inequality and the weak law of large numbers that appear in this section. Chebyshev did the same in 1867, but publicly acknowledged Bienaymé's prior publication. The two corresponded on various mathematical topics, and each obtained membership for the other in his own country's academy of sciences.

Pafnuty L. Chebyshev lived from 1821 to 1894 and was an influential Russian mathematician in the University of St. Petersburg. He made many contributions to probability and analysis in addition to what we have seen in this section.

EXERCISES FOR **SECTION 4.5**

1. Let X be the number of heads in 100 tosses of a fair coin.
   a. What does Chebyshev's inequality say about the probability that X is not within 10% of its expected value—that is, that it is not between 45 and 55?
   b. Write the actual probability in part a as 1 minus the sum of nine binomial probabilities. If you can, calculate this number. [It is about .37. We will

learn how to approximate it using the Central Limit Theorem in a later chapter.]

c. Repeat part a for 1,000 tosses instead of 100. [The true value is about .0014, which is considerably less than what Chebyshev's inequality says.]

d. Repeat part a for n tosses.

2. Let $X_1, X_2, \ldots, X_n$ be i.i.d. random variables, each with the uniform distribution on (0, 1).
   a. What are $E\overline{X}_n$ and $V\overline{X}_n$?
   b. What does Chebyshev's inequality say about the probability that $\overline{X}_n$ is not between .4 and .6?
   c. How large must n be in order for Chebyshev's inequality to guarantee that this probability is less than .01?

3. Let $X_1, X_2, \ldots, X_{100}$ be i.i.d. random variables, each having the exponential distribution with parameter $\lambda = .25$.
   a. What are $E\overline{X}_{100}$ and $V\overline{X}_{100}$?
   b. What does Chebyshev's inequality say about the probability that $\overline{X}_{100}$ is not within .5 of its expected value?

4. Suppose $EX = 20$ and $VX = 9$. What does Chebyshev's inequality say about the probability that X is between 15 and 25?

5. **Although Chebyshev's inequality is often crude, it sometimes cannot be improved on.** Let X have the possible values $-a$, $+a$, and 0, with probabilities p, p, and $1 - 2p$, respectively, where p is some number in the interval $\left(0, \frac{1}{2}\right)$.
   a. What are EX and VX?
   b. What does Chebyshev's inequality say about $P(|X - EX| \geq a)$?
   c. What is the actual value of $P(|X - EX| \geq a)$?

   This example shows that there are cases in which the Chebyshev upper bound is exactly equal to the probability that it bounds. Therefore, although the Chebyshev bound is usually much greater than the true probability, there is no point in looking for an inequality that gives a better bound in general.

6. **Estimating the unknown arrival rate of a Poisson process.** Consider a Poisson process with a rate of λ arrivals per minute; suppose λ is unknown. Let X_1 be the number of arrivals in the first minute, X_2 the number in the second minute, and so on. One way to estimate λ is to observe a number of these X_i's and average them. Accordingly, let $\overline{X}_n$ be the average of $X_1, X_2, \ldots, X_n$.
   a. What is the distribution of a single X_i? What are EX_i and VX_i? [See Exercise 6 of Section 4.3.]
   b. What are $E\overline{X}_n$ and $V\overline{X}_n$?
   c. What does Chebyshev's inequality say about the probability that $\overline{X}_n$ fails to be within 5% of λ (that is, fails to be between $\lambda - .05\lambda$ and $\lambda + .05\lambda$)?

   Another way to estimate λ makes use of the fact that if Y_1 is the time of the first arrival, Y_2 the time between the first and second, and so on, then each of the Y_i's has the exponential distribution with parameter λ, and so their common expected value is $1/\lambda$. Thus the average of a large number of Y_i's will likely be close to $1/\lambda$. This method is usually not as efficient as the one described above, for reasons having to do with the number of observations that must be taken to get a given degree of certainty that the estimator is tolerably close to the true value. But if λ is extremely small, it may be preferable to count times between arrivals than to count arrivals in fixed time periods. Questions like this, concerning how to decide between competing estimators, are covered in courses in statistical inference.

[a]7. In Exercise 6, suppose that (unknown to us) λ equals 19.63, and that our sample size n is 30. Then the answer to Exercise 6c is that the probability is less than or equal to .6792. But we can find the actual probability in this case by considering $S_{30} = 30 \cdot \overline{X}_{30}$, which is the number of arrivals in the first 30 minutes.

a. What is the distribution of S_{30}?

b. Translate the statement "$\overline{X}_{30}$ is not within 5% of 19.63" to a statement about S_{30}.

c. Write the probability of the event in part b as 1 minus the sum of a set of Poisson probabilities. If you have access to a computer, find it. [It is about .22.]

8. Let X be the number of 1's in six rolls of a fair die.

a. Find the probability that X is not within 1 of 1; that is, find the probability that $|X - 1| \geq 1$.

b. Find the Chebyshev upper bound on the probability in part a.

9. **Will the relative frequency of an event be close to its probability?** Suppose A is an event associated with some chance experiment and $P(A)$ equals .35. Suppose the experiment is repeated 45 times, and on the ith repetition X_i equals 1 if A occurs and 0 if not.

a. What is the distribution of S_{45}?

b. What are $E\overline{X}_{45}$ and $V\overline{X}_{45}$?

c. What does Chebyshev's inequality say about the probability that $\overline{X}_{45}$ is not between .3 and .4?

As pointed out in (4.5.3), $S_{45} = X_1 + X_2 + \cdots + X_{45}$ is the number of occurrences of A in 45 trials and $\overline{X}_{45}$ is the relative frequency of occurrence of A. Since probability represents expected long-run relative frequency, we expect $\overline{X}_{45}$ to be close to .35; so the probability that $\overline{X}_{45}$ is not between .3 and .4 should be small. The answer does not look promising. However, remember two things: (i) The answer is only an upper bound; the true probability, as we will find in Supplementary Exercise 13 for Chapter 6, is about .44. (ii) If the number of repetitions were 4,500 instead of 45, the result would be more satisfying.

10. Let X have the Poisson distribution with parameter $\lambda = 3.5$.

a. What does Chebyshev's inequality say about the probability that X is not within .4 of its expected value?

b. What is the true probability that X is not within .4 of its expected value?

11. The weight of a bag of a certain brand of potato chips is given as 6 ounces,

but in fact it is a random variable whose expected value is 6.2 ounces and whose standard deviation is .17 ounce. Nothing else about its distribution is known.

a. What can you say about the probability that the weight of a randomly chosen bag is between 6 ounces and 6.4 ounces?

b. Find an interval in which you can be sure that 90% of the bags' weights will be found.

12. (*Computer exercise*) If you did Exercise 24 in Section 4.1, you wrote a program that prints out the average of a large number, n, of independent observations of a random variable having the exponential distribution with $\lambda = 3.2$. The law of large numbers says that as n increases, these averages will approach 1/3.2, the expected value of the distribution. Modify the program so that it generates 100,000 observations, and computes and prints the average after every 1,000 observations. Run it and see whether the 100 averages look like terms in a sequence whose limit is 1/3.2.

13. (*Computer exercise*) This exercise involves simulating observations from a Cauchy distribution, which has no expected value.

a. Write a program that generates and prints (on the screen) 100 observations of a random variable having the standard Cauchy distribution. [Such a random variable X can be generated by letting U have the uniform distribution on (0, 1), letting $V = \pi \cdot \left(U - \frac{1}{2}\right)$, so that V has the uniform distribution on $\left(-\frac{\pi}{2}, \frac{\pi}{2}\right)$, and letting $X = \arctan(V)$.]

b. Run it a few times to see what standard Cauchy observations look like. (They cluster around 0, but with noticeably more observations outside the interval $[-3, 3]$ than the standard normal, and occasional very large observations.)

c. Modify it so that it generates 1,000 observations, but prints only the average instead of printing all 1,000 numbers. Run this several times to see whether the averages seem to be close to anything.

d. Kolmogorov's Law of Large Numbers says that as the number of observations increases, the sequence of averages will fail to have a limit. Modify your program so that it generates 100,000 observations, and computes and prints the current average after each 1,000. Run it and see whether the 100 numbers look like terms in a sequence that has no limit.

[a]**14.** **Monte Carlo approximation of an integral.** The method, which we outlined in Exercise 23 of Section 4.2, is as follows. To approximate $\int_0^1 h(x)dx$, let X have the uniform distribution on (0, 1), make a large number of observations of X, and average the values of $Y = h(X)$. Because EY equals the desired integral, the expected long-run average of Y should be equal to that integral.

In symbols, let $X_1, X_2, \ldots, X_n$ be independent observations of X and let $Y_j = h(X_j)$. Then $\overline{Y}_n$ is the Monte Carlo estimator of the integral based on a sample of size n.

a. What does Kolmogorov's Strong Law say about the situation?

b. Chebyshev's inequality can be used to make a statement about the goodness of the approximation, provided that the variance of Y is finite. Find VY in terms of the function $h(x)$. What assumption about $h(x)$ needs to be made to guarantee that VY is finite?

c. Unfortunately we would need to know more than we do about $h(x)$ to find VY. But show that if we know an upper bound for the absolute value of h, say $|h(x)| \leq A$ for all $x \in (0, 1)$, then $VY \leq A^2$.

d. Suppose $VY \leq A^2$. For a given tolerance $\epsilon > 0$, what does Chebyshev's inequality say about the probability that the estimator $\overline{Y}_n$ is not within ϵ of the true value of the integral? (The symbol ϵ is a lowercase Greek epsilon.)

e. Suppose $|h(x)| \leq 1$ for all $x \in (0, 1)$. How large a sample must we take to get a 95% chance that $\overline{Y}_n$ is within .01 of that true integral?

The result makes the Monte Carlo method look very inefficient. But remember that we have required the Chebyshev bound to be less than .05; the real probability may be less than .05 with far fewer evaluations of the function. Still, for functions of one variable, standard numerical integration techniques usually produce comparable accuracy with far fewer evaluations of the function than this. The real power of the Monte Carlo method is in evaluating integrals of functions of several variables. As the number of variables increases, the number of evaluations needed by nonrandom techniques grows considerably faster than that needed by Monte Carlo methods. There are also techniques for improving the accuracy of the estimator well beyond the Chebyshev bound discussed in this exercise.

For further reading, see Sheldon Ross's book *A Course in Simulation*. Also, many undergraduate numerical analysis textbooks contain material on Monte Carlo methods.

15. **A possible gambling strategy: bet a constant proportion of current holdings.** Suppose you repeatedly play a certain game of chance; you may bet any amount on each play, and you win that amount with probability p and lose that amount with probability q. Successive plays are independent. You adopt the strategy of always betting a certain proportion α (this is a lowercase Greek alpha) of the amount you currently have. If at any point you find yourself with x dollars, then on the next play you bet αx dollars, where α is some constant between 0 and 1. You start with C dollars; let C_n be the amount you have after the nth play. It is natural to wonder if there is an optimal value of α that maximizes your chance of success in some sense.

a. Show that $C_n = C \cdot (1 + \alpha)^{S_n}(1 - \alpha)^{n-S_n}$, where S_n is the number of wins in the first n plays.

b. Let $Y_n = (1/n)\ln(C_n/C)$. Use Kolmogorov's Strong Law to show that as n increases, Y_n approaches the limit $p \ln(1 + \alpha) + q \ln(1 - \alpha)$ with probability 1.

It follows from part b that C_n is asymptotic to $C \cdot r^n$, where r is the constant $(1 + \alpha)^p \cdot (1 - \alpha)^q$. [The reason is that $C_n = C \cdot (e^{Y_n})^n$ and e^{Y_n} approaches r.] Thus, as the number of plays increases, the amount you have will grow exponentially if $r > 1$, shrink to 0 exponentially if $r < 1$, and will approach your initial amount C if $r = 1$. Therefore, given the probabilities p and q for the game you are playing, you would like to choose an α so that $r > 1$, if there is such a (nonnegative) α. In any case, you would like to know the α for which r is largest.

c. To this end think of r as a function of α; that is, $r(\alpha) = (1 + \alpha)^p(1 - \alpha)^q$. Show that $r(\alpha)$ is increasing on the interval $(-\infty, p - q)$ and decreasing on $(p - q, \infty)$. [*Hint:* It is much easier to show this for $\ln r(\alpha)$ than for $r(\alpha)$.]

d. Conclude that if $p \le q$, then the maximum of $r(\alpha)$ for $0 \le \alpha$ is $r(0) = 1$, and $r(\alpha) < 1$ for any $\alpha > 0$. [Thus, if $p \le q$, then your holdings will shrink exponentially to 0 no matter what proportion α you use. It is somewhat surprising that this is true even if $p = q$.]

e. Show that if $p > q$ then the maximum value is $r(p - q) = 2p^pq^q$. [This number is greater than 1, but not by much unless p is quite large. If $p =$ .55, for example, it is only about 1.005. Still, after 1,000 plays you can expect your initial holdings to have been multiplied by approximately $r^{1,000} \doteq 149.66$.]

ANSWERS

1. **a.** Less than or equal to 1 **c.** Less than or equal to .1 **d.** Less than or equal to $100/n$
2. **a.** $\frac{1}{2}$, $1/12n$ **b.** Less than or equal to $100/12n$ **c.** At least 834
3. **a.** 4, .16 **b.** Less than or equal to .64
4. It is greater than $\frac{16}{25}$.
5. **b.** It is less than or equal to $2p$. **c.** It equals $2p$.
6. **c.** It is less than or equal to $400/\lambda n$.
7. **a.** Poisson with parameter 588.9 **b.** S_{30} is not within 5% of 588.9, i.e., not between 559.455 and 618.345, i.e., not between 560 and 618, inclusive. **c.** $1 - \sum_{k=560}^{618} \frac{588.9^k e^{-588.9}}{k!}$
8. **a.** .5981 **b.** $\frac{5}{6} \doteq .8333$
9. **a.** Binomial with $n = 45$ and $p = .35$ **b.** .35, .005056 **c.** Less than or equal to 2.02
10. **a.** It is less than or equal to 21.875. **b.** 1
11. **a.** It is greater than .2775. **b.** (5.66, 6.74)
14. **a.** $\lim_{n\to\infty} \overline{Y}_n = \int_0^1 h(x)dx$ with probability 1. **b.** $VY = \int_0^1 h(x)^2dx - \left(\int_0^1 h(x)dx\right)^2$. It is enough to assume that $\int_0^1 h(x)^2dx$ is finite. **d.** $P(\overline{Y}_n$ is not within ϵ of $\int_0^1 h(x)dx) \le \frac{A^2}{n\epsilon^2}$. **e.** At least 200,000

SUPPLEMENTARY EXERCISES FOR **CHAPTER 4**

1. If X is a discrete random variable whose possible values are 1, 2, 3, and 4, and if the mass function $p(x)$ is proportional to $1/x$, find EX, EX^2, and VX.

2. Let X have the density $f(x) = 1/(1 + x)^2$ for $x > 0$ (and 0 otherwise). What is EX?

3. Let X have the density $f(x) = \frac{3}{26}x(3 + x)$ for $0 < x < 2$ (and 0 otherwise). Find EX, EX^2, and VX.

4. If the first two moments of X are $\mu_1 = a$ and $\mu_2 = a^2$, what is the distribution of X?

5. Using the gamma formula found in Exercise 10 of Section 4.3, find EX^k if X has the exponential distribution with parameter λ. [The density is $\lambda e^{-\lambda x}$ for $x > 0$.]

6. Suppose $EX = a$, $EX^2 = b$, $EX^3 = c$, and $EX^4 = d$. Find the following in terms of a, b, c, d, and r.
   a. VX **b.** $E(X - r)^3$
   c. $E(X - r)^4$ **d.** $V(2X + 3)$

7. Find the expected value and variance of the sum of n independent random variables if each of them has the geometric distribution with parameter p. Recall that the mass function of this distribution is $p(k) = pq^k$ for $k = 0, 1, 2, \ldots$. [This is the distribution of the number of failures before the nth success in Bernoulli trials.]

8. Let X be the smaller and Y the larger of the two numbers that show when a pair of fair dice is rolled. (If the two numbers that show are equal, then $X = Y$.)
   a. Make a table of the joint probability mass function of X and Y. Include the marginal mass functions.
   b. Find EX, EX^2, EY, EY^2, and $E(XY)$.
   c. Find VX and VY.
   d. Find the covariance of X and Y. [Guess first whether it is positive, negative, or 0.]
   e. Find the correlation coefficient of X and Y.

9. Suppose $EX = 4$, $VX = 3$, $EY = 2$, $VY = 5$, and $\rho(X, Y) = .32$. Find the following.
   a. EX^2 **b.** $\text{Cov}(X, Y)$
   c. $E(XY)$ **d.** $\text{Cov}(2X + 3, 1 - Y)$
   e. $V(X + Y)$ **f.** $V(3X - Y)$
   g. $\text{Cov}(X, X + Y)$

10. Is it possible for a random variable X to have $EX = 2$ and $EX^2 = 3$? Why or why not?

11. Let $X_1, X_2, X_3, \ldots$ be independent, each having $EX_j = 5$ and $VX_j = 10$. For any n let $\overline{X}_n$ denote the average of $X_1, X_2, \ldots, X_n$. For each of the following values of n, what does Chebyshev's inequality say about the probability that $\overline{X}_n$ is *not* between 4.5 and 5.5?
    a. $n = 1$ [$\overline{X}_1$ is just X_1.] **b.** $n = 10$
    c. $n = 100$ **d.** $n = 1{,}000$
    e. $n = 1{,}000{,}000$

12. Let X have the uniform distribution on $(0, 1)$.
    a. What does Chebyshev's inequality say about the probability that X is not within .2 of its expected value?

b. What is the true probability that X is not within .2 of its expected value?
c. Repeat parts a and b, replacing .2 by .5.

13. Let μ_k be the kth moment $E(X^k)$ of X. Show that $\mu_{2n} \geq (\mu_n)^2$ for any positive integer n.

14. In Exercise 7 above you found the expected value and variance of the number of failures preceding the nth success in Bernoulli trials. Using those results, find (with very little additional work) the expected value and variance of the number of trials needed to produce the nth success.

15. Suppose X and Y are independent random variables and that $EX = r$, $EX^2 = s$, $EY = t$, and $EY^2 = u$. Find the following in terms of r, s, t, and u.
a. $E[(X - Y)^2]$ **b.** $V(2X - 1)$

16. You plan to observe 1,000 random variables that have the density $f(x) = 160/x^6$ for $x > 2$ (and 0 otherwise). You will record the squares of these numbers, and finally you will add these 1,000 squares. What number do you expect the sum to be close to?

17. The possible values of X are the positive integers, and $P(X = k) = kp^2q^{k-1}$ for $k = 1, 2, 3, \ldots$. (As always, p and q are positive constants and $p + q = 1$.) Find EX. [*Hint:* It is difficult to find EX directly; try finding $E(X + 1)$ or $E(X - 1)$.]

18. A random variable X has expected value 5 and variance 7. Someone makes n independent observations of X and averages the n numbers.
a. What can you say about the probability that this average is not between 4.2 and 5.8?
b. How large must n be in order that the probability in part a is guaranteed to be less than .01?
[a]**c.** What does Chebyshev's Law of Large Numbers say about the average?

ANSWERS

1. 1.92, 4.8, 1.1136
2. Does not exist
3. 1.3846, 2.1231, .2059
4. $X = a$ with probability 1
5. $k!/\lambda^k$
6. **a.** $b - a^2$ **b.** $c - 3br + 3ar^2 - r^3$ **c.** $d - 4cr + 6br^2 - 4ar^3 + r^4$ **d.** $4(b - a^2)$
7. nq/p; nq/p^2
8. **b.** $EX = \frac{91}{36}$; $EX^2 = \frac{301}{36}$; $EY = \frac{161}{36}$; $EY^2 = \frac{791}{36}$; $E(XY) = \frac{441}{36}$ **c.** Both 1.97145 **d.** +.9452 **e.** .47945
9. **a.** 19 **b.** 1.23935 **c.** 9.23935 **d.** −2.4787 **e.** 10.4787 **f.** 24.5639 **g.** 4.23935
10. No
11. **a.** ≤ 40 **b.** ≤ 4 **c.** $\leq .4$ **d.** $\leq .04$ **e.** $\leq .00004$

12. **a.** It is less than or equal to 2.0833. . . . **b.** .6 **c.** Chebyshev's inequality says it is less than or equal to $\frac{1}{3}$; true value is 0.

14. n/p; nq/p^2

15. **a.** $s + u - 2rt$ **b.** $4(s - r^2)$

16. $6{,}666\frac{2}{3}$

17. $\dfrac{2}{p} - 1$

18. **a.** It is less than or equal to $7/.64n$. **b.** At least 1,094 **c.** For any positive a, the probability that the average differs from 5 by more than a approaches 0 as $n \to \infty$.

CHAPTER

5

Functions of Random Variables

5.1 Introduction: Finding Distributions

5.2 Using Cumulative Distribution Functions

5.3 The Change-of-Variable Theorem for Densities

5.4 The Multivariate Change-of-Variable Theorem

Supplementary Exercises

5.1 INTRODUCTION: FINDING DISTRIBUTIONS

The central question of this chapter is: If X is a random variable with a known distribution, and if $Y = h(X)$ is some function of X, what is the distribution of Y? Or, more generally, if Y is a random variable that is given as a function of several random variables, or as a function on a probability space, what is the distribution of Y? Most generally of all, if $X_1, X_2, \ldots$ are random variables with a known joint distribution, and $Y_1, Y_2, \ldots$ are functions of them, what is the joint distribution of the Y_j's?

The method we give in this chapter, called the CDF method, is the foundation for all the theorems and formulas in the chapter. It is most useful in dealing with absolutely continuous distributions—and all our examples will be of this type. We will also assume throughout the chapter that the set of possible values of any distribution is an interval, as opposed, say, to a union of disjoint intervals or some more complicated set. This is not much of a restriction, because essentially all the absolutely continuous distributions encountered in practice have this property.

First let us recall what is meant by "knowing the distribution of a random variable." The distribution of a random variable X is a recipe for finding $P(X$ is in $I)$ for any interval I, and hence for finding $P(X$ is in $B)$ for any Borel set B. But "knowing the distribution of X" can take several forms, because a distribution is specified completely by giving any of the following:

its cumulative distribution function;

its density function, if it is an absolutely continuous distribution;

its mass function, if it is a discrete distribution;

any recipe that gives $P(X$ is in $I)$ for interval I [for example, the recipe $P(a < X \leq b) = e^{-\lambda a} - e^{-\lambda b}$ for $0 \leq a \leq b$, or the recipe $P(X \in I) = \frac{1}{2} \cdot$ length of I for subintervals of $(0, 2)$];

a name for the distribution, assuming that the name given is unambiguous to our audience.

The point is that if we know any one of the above for a given random variable, then (at least in principle) we can find the others.

(5.1.1) Examples: How distributions can be specified.

a. If we learn that $P(X > a) = e^{-\lambda a}$ for $a \geq 0$, then we know that for $0 \leq a \leq b$,

$$P(a < X \leq b) = P(X > a) - P(X > b) = e^{-\lambda a} - e^{-\lambda b}.$$

We recognize this as the integral $\int_a^b \lambda e^{-\lambda x} dx$. Thus we know that a density for X is $f(x) = \lambda e^{-\lambda x}$ for $x \geq 0$ (and 0 for $x < 0$), and that X has the exponential distribution with parameter λ. We also know the CDF, of course, because it is just

$$\begin{aligned} F(x) &= P(X \leq x) = 1 - P(X > x) \\ &= 1 - e^{-\lambda x} \quad \text{for } x \geq 0 \quad (\text{and } 0 \text{ for } x < 0). \end{aligned}$$

Alternatively, given that $P(X > a) = e^{-\lambda a}$ for $a > 0$, we can see that the CDF of the distribution is $F_X(a) = P(X \leq a) = 1 - e^{-\lambda a}$ for $a > 0$, and differentiating this tells us that $f(x) = \lambda e^{-\lambda x}$ is a density.

b. If we know that the CDF of X is

$$F(x) = \begin{cases} 0 & \text{if } x < 1, \\ 1 - 1/x^2 & \text{if } x \geq 1, \end{cases}$$

then we can find the probabilities of intervals directly from this. We can also find the density by differentiating; it is $f(x) = 2/x^3$ for $x > 1$.

c. If we are told that X has the Poisson distribution with parameter λ, then by convention we know that X is a discrete random variable whose mass function is $p(k) = P(X = k) = (\lambda^k e^{-\lambda})/k!$ for $k = 0, 1, 2, 3, \ldots$.

d. Suppose we are told that the density of X is proportional to xe^{-3x} for $x > 0$, and 0 otherwise. That is, we know that $f(x) = cxe^{-3x}$ $(x > 0)$ is a density for X, but we are not given the value of the constant c. We still know the distribution of X, because we can use the requirement

$$1 = \int_{-\infty}^{\infty} f(x)dx = \int_0^{\infty} cxe^{-3x}dx,$$

evaluate the integral, and solve for c. In general, if we are given the nonconstant part of a density for X—that is, if we are given the fact that some constant times $r(x)$ is a density for X—then we can find the constant; so the distribution is specified.

e. In a later chapter we will learn yet another way to specify a distribution—by giving a *generating function* for it. A generating function does not give probabilities directly, but it enables us to find other things, such as moments. In addition, it is useful as a way, different from the methods we study in this chapter, of finding the distribution of a function of one or more random variables. ∎

Notice that some distributions have names while others do not. For instance, in Example (5.1.1a), we would most likely communicate this distribution to someone simply by saying that it is the exponential distribution with parameter λ, rather than by giving its density or its CDF. On the other hand, the distribution in Example (5.1.1b) has no commonly-agreed-upon name. Anyone who wanted to refer to it would probably give its density, because for that distribution the density is simpler than the CDF.

The distributions with names are of course the ones that arise most frequently in practice. We will summarize some of them in Chapters 7 and 8. Just for reference, the names we have encountered to this point are the following.

Discrete distributions: binomial, Bernoulli, geometric, Poisson

Absolutely continuous distributions: uniform, exponential, normal, Cauchy

More details of these and other "brand-name" distributions are on the endpapers at the back of the book. Still more details are in Appendix B.

Now recall the question we asked first. If we know the distribution of X, and Y is defined as $h(X)$ for some given function h, what is the distribution of Y? In this chapter we will study the following general method, called the **CDF method,** for answering this question when X has an absolutely continuous distribution.

> The CDF of Y is $F_Y(y) = P(Y \leq y) = P(h(X) \leq y)$. If we can solve the inequality $h(X) \leq y$ by transforming it algebraically into an inequality or inequalities for X, then we can use what we know about the distribution of X to find $P(h(X) \leq y)$, and we will know the CDF of Y.

The next section begins with an example of this method.

5.2 USING CUMULATIVE DISTRIBUTION FUNCTIONS

We begin with a simple example. It is essentially the same as Example (1.3.11), except that in Chapter 1 we were not using the language of random variables and distributions.

(5.2.1) Example. Let X have the uniform distribution on $(0, 1)$, and let $Y = X^2$. [That is, $Y = h(X)$ where h is the function defined by $h(x) = x^2$.] What is the distribution of Y? We solve this by finding the CDF of Y, $F_Y(y) = P(Y \leq y)$.

It is essential to think first about the set of possible values of Y. The set of possible values of X is the interval $(0, 1)$, and Y is the square of X, so the possible values of Y are the squares of the numbers in $(0, 1)$—that is, the numbers in $(0, 1)$ again. So we can say before doing any calculations that $F_Y(y)$ equals 0 if $y \leq 0$ and $F_Y(y)$ equals 1 if $y \geq 1$.

In connection with this first step, it is always a good idea to begin by sketching the graph of the function $y = h(x)$ on the set of possible values of X. This set will be the domain of h, and since $Y = h(X)$, the range of h will be the set of possible values of Y. In this example, we sketch the graph of $y = x^2$ on the interval $(0, 1)$; this is shown in Figure 5.2a. You can see that the range of this function is also $(0, 1)$. This graph will be useful in the next step of the method also.

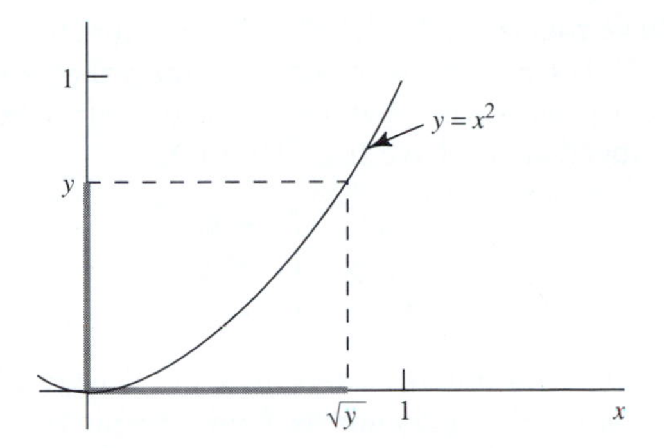

FIGURE 5.2a Sketch for finding the distribution of $Y = X^2$ if X has possible values in $(0, 1)$

Now we find $F_Y(y)$ when y is one of the possible values of Y—that is, when $0 < y < 1$. We proceed as follows:

$$F_Y(y) = P(Y \leq y) = P(X^2 \leq y).$$

We have to transform the event $(X^2 \leq y)$ into an event whose probability we can find using the distribution of X, which we were given. There are two ways to approach this, and it is good to look at them both.

Solution a. In this example we can do it simply by solving the inequality $X^2 \leq y$ for X, and that is easy to do: $X^2 \leq y$ is the same as the pair of inequalities $-\sqrt{y} \leq X \leq \sqrt{y}$. Therefore,

$$\begin{aligned} F_Y(y) &= P(-\sqrt{y} \leq X \leq \sqrt{y}) \\ &= P(0 < X \leq \sqrt{y}) \qquad \text{(because } P(X \leq 0) = 0\text{)} \\ &= \sqrt{y} - 0 = \sqrt{y} \qquad \text{(because } X \text{ has the uniform dist. on } (0, 1)\text{).} \end{aligned} \tag{5.2.2}$$

Solution b. Alternatively (and in nearly all examples this is necessary because the algebra is not so simple), we can make the translation with the help of the graph of $y = h(x)$, as follows. (Look at Figure 5.2a in connection with this procedure.)

i. Pick a typical y on the vertical axis in the set of possible values of Y. Identify the set of possible values of Y that are less than or equal to this y. This is the heavily marked interval on the vertical axis in the figure. Let us call it the "y-interval."

ii. Now identify on the graph the set of possible values of X for which $h(x)$ is in the y-interval. This is the marked interval on the horizontal axis. Call it the "x-set." (It is an interval in this example, but it will sometimes be a union of two or more disjoint intervals.)

iii. Now do the algebra to find, in terms of y, the endpoints of the x-set. In this case they are 0 and $\sqrt{y}$.

iv. $F_Y(y)$ equals the probability that X is in the x-set. In this example, that is $P(0 \leq X \leq \sqrt{y}) = \sqrt{y}$.

At this point we have found $F_Y(y) = P(Y \le y)$, as a function of y, for a typical possible value y of Y. For a y that is outside the interval of possible values of Y, the CDF is 0 or 1, depending on whether y is to the left or to the right of that interval. Putting it together, we have the CDF of Y:

$$F_Y(y) = \begin{cases} 0 & \text{if } y < 0, \\ \sqrt{y} & \text{if } 0 \le y < 1, \\ 1 & \text{if } y \ge 1. \end{cases}$$

Now, if we need it, we can differentiate to get the density $f(y) = 1/(2\sqrt{y})$ for $0 < y < 1$ (and 0 otherwise). We do not recognize the distribution of Y as one of our named distributions, but it is a member of the beta family. We will encounter this family later.

In Example (1.3.11) we found $F_Y(y)$ only for $y = .5$. You may recall the warning given in that example about why and how the step marked with the hazard sign **Z** must be made. The graphical method in Solution b above takes care of this hazard more or less automatically, whereas in Solution a it is up to us to be aware of it and to handle it correctly. ▮

(5.2.3) **The CDF method.** If X is an absolutely continuous random variable with a known distribution and Y is defined to be $h(X)$ for some function h, then the CDF of Y can be found as follows.

1. Sketch the graph of $y = h(x)$. Let the domain be the set of possible values of X, and identify the range for this domain, which will be the set of possible values of Y. $F_Y(y)$ is of course 0 for any y that is less than all members of this range, and 1 for any y that is greater than all members of this range.
2. Choose a typical y in the set of possible values of Y, and identify the "y-interval"—the interval on the vertical axis consisting of all numbers in the range that are less than or equal to y. Then find the "x-set"—the set on the horizontal axis consisting of the numbers x in the domain for which $h(x)$ is in the y-interval.
3. Do the algebra needed to identify, in terms of y, the endpoints of the interval or intervals that make up the x-set.
4. Now $F_Y(y) = P(Y \le y)$ equals the probability that X is in the x-set.

This is the method of Solution b in Example (5.2.1). Solution a corresponds to replacing steps 2–4 by the following algebraic manipulations:

> The inequality $Y \le y$ is the inequality $h(X) \le y$. Solve this inequality for X, getting one or more inequalities involving X and y. Then find, in terms of y, the probability that X satisfies these inequalities. That probability is $F_Y(y)$.

Many students resist the idea of graphing a function and working from the graph, and would prefer to use an algebraic solution. This is certainly simple enough to do in Example (5.2.1), but it is much more difficult in most examples.

Solving inequalities is always easier if you have a graph to look at. Without a graph you are doing blind algebra.

You are strongly advised to overcome any resistance you may have to graphing functions. The ability to sketch a graph quickly is invaluable in nearly all areas of mathematics, and it is arguably the only way to develop a feeling for the functions one works with. Furthermore, the graphing procedures you may have learned in calculus courses, involving first and second derivatives, critical points, and inflection points, rarely need to be used in full in solving the problems of this chapter. The goal is simply to get an idea of the shape of the function—in particular, where it increases and where it decreases. As you gain experience, you will find yourself able to do this more quickly.

In addition, it is usually not difficult to use a computer, if you have access to one, to get a quick graph of a given function on a given domain. The best advice is to try to see the graph without using a computer; but it would be foolish not to make use of one if you need it.

Our next example shows how useful the graph can be.

(5.2.4) **Example.** Let X have the uniform distribution on $(0, \pi)$. What is the distribution of $y = \sin X$? We will find the CDF of Y. We can then differentiate the CDF to find a density.

First recall how we find probabilities for X. The density is $f(x) = 1/\pi$ for $0 < x < \pi$. Because this density is a constant, the probability that X is in any subinterval of $(0, \pi)$ is $1/\pi$ times the length of the subinterval.

We begin by sketching the graph of $y = \sin x$. This is shown in Figure 5.2b (page 310). The domain we use is $(0, \pi)$, the set of possible values of X, and as you can see from the graph, the range is $(0, 1]$. This is the set of possible values of Y; so $F_Y(y)$ will equal 0 when $y \leq 0$ and will equal 1 when $y \geq 1$.

We pick a typical y in the range, and identify the y-interval of possible values that are less than or equal to y. This is the interval shaded on the vertical axis in Figure 5.2b. Then, looking at the graph, we identify the x-set, those values of x for which $h(x)$ is in the y-interval. This is the union of the two intervals that are shaded on the horizontal axis in Figure 5.2b.

Next we find, in terms of y, the endpoints of the two intervals that make up the x-set. Here is where the algebra comes in, but there is not much in this case: the left interval's endpoints are 0 and $\sin^{-1}y$, and the right interval's are $\pi - \sin^{-1}y$ and π.

As you can see, there are two numbers in $(0, \pi)$ whose sines are y, one in $(0, \pi/2)$ and one in $(\pi/2, \pi)$. We are using $\sin^{-1}y$ to denote the one in $(0, \pi/2)$; this is called the "principal value" of the inverse sine function. The other one happens to be $\pi - \sin^{-1}y$.

So $P(Y \leq y)$ equals the probability that X is in the x-set; that is,

$$\begin{aligned} F_Y(y) = P(Y \leq y) &= P(\sin X \leq y) \\ &= P(0 < X \leq \sin^{-1}y) + P(\pi - \sin^{-1}y \leq X \leq \pi). \end{aligned}$$

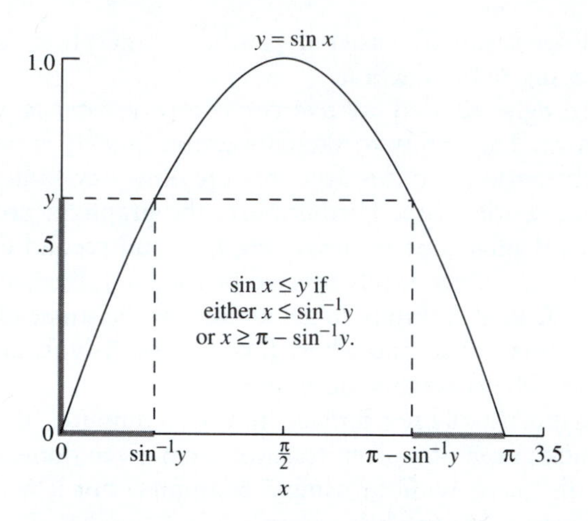

FIGURE 5.2b Sketch for finding the distribution of $Y = \sin X$ if X has possible values in $(0, \pi)$

Since the probability that X is in a subinterval of $(0, \pi)$ is $1/\pi$ times its length, the above is just

$$F_Y(y) = \frac{\sin^{-1}y - 0}{\pi} + \frac{\pi - (\pi - \sin^{-1}y)}{\pi} = \frac{2}{\pi}\sin^{-1}y.$$

So the CDF of Y is

$$F_Y(y) = \begin{cases} 0 & \text{if } y \le 0, \\ \dfrac{2}{\pi}\sin^{-1}y & \text{if } 0 < y < 1, \\ 1 & \text{if } y \ge 1. \end{cases}$$

As a partial check, notice that this function increases from 0 when $y = 0$, to 1 when $y = 1$.

If we want a density, we can get it by differentiating:

$$f_Y(y) = \begin{cases} 2/\pi\sqrt{1 - y^2} & \text{if } 0 < y < 1, \\ 0 & \text{otherwise.} \end{cases}$$ ∎

If you insisted on doing this example without a graph, your blind algebra would proceed as follows.

1. The set of possible values of X is $(0, \pi)$, and the range of $h(X) = \sin X$ on this set is $(0, 1]$. This is the set of possible values of Y.
2. For a typical y in $(0, 1]$, the inequality $\sin X \le y$ is satisfied when $X \le \sin^{-1}y$ or when $X \ge \pi - \sin^{-1}y$.
3. Consequently,

$$\begin{aligned} F_Y(y) &= P(Y \le y) = P(X \le \sin^{-1}y) + P(X \ge \pi - \sin^{-1}y) \\ &= P(0 < X \le \sin^{-1}y) + P(\pi - \sin^{-1}y \le X \le \pi) \\ &= \frac{\sin^{-1}y - 0}{\pi} + \frac{\pi - (\pi - \sin^{-1}y)}{\pi} = \frac{2}{\pi}\sin^{-1}y. \end{aligned}$$

If you can be sure of step 2 without looking at, or at least imagining, the graph, you are remarkably good at doing algebra with trigonometric functions.

(5.2.5) **Example.** Let X have the exponential distribution with parameter λ, and let $Y = 1/(X + 1)$. What is a density for Y? We will find the CDF and then differentiate it to get a density.

The set of possible values of X is $(0, \infty)$, and so we start by sketching the graph of $y = 1/(x + 1)$ on this domain. This is in Figure 5.2c. We see from this graph that the set of possible values of Y is the interval $(0, 1)$. Therefore, $F_Y(y) = 0$ when $y \le 0$ and $F_Y(y) = 1$ when $y \ge 1$. So we need only to find $F_Y(y)$ for an arbitrary y between 0 and 1.

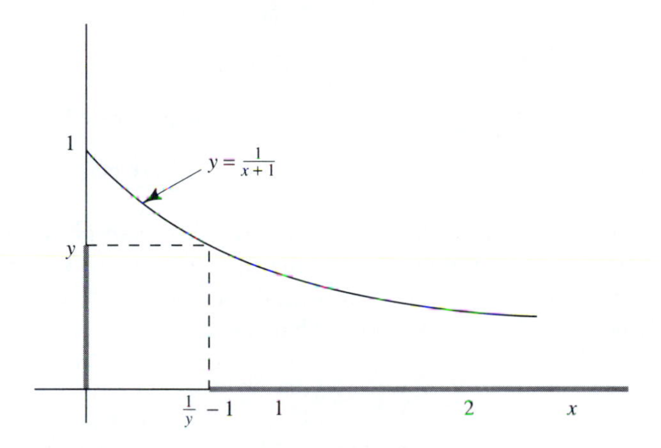

FIGURE 5.2c Sketch for finding the distribution of $Y = \dfrac{1}{X+1}$ if X has possible values in $(0, \infty)$

We pick a typical y in $(0, 1)$ and mark the y-interval, the shaded interval on the vertical axis in Figure 5.2c. We then identify the x-set corresponding to this interval; it is the shaded interval on the horizontal axis. Then we identify the endpoint of this interval in terms of y, by doing a little algebra:

$$\frac{1}{x+1} = y \quad \text{implies that} \quad x = \frac{1}{y} - 1.$$

Therefore, for a typical y in the interval $(0, 1)$,

(5.2.6)
$$F_Y(y) = P(Y \le y) = P(X \ge \tfrac{1}{y} - 1).$$

Now X has the exponential distribution with parameter λ, so for any $a \geq 0$ we know that $P(X \geq a) = e^{-\lambda a}$. (You might have remembered this directly, or you might need to evaluate the integral $\int_a^\infty \lambda e^{-\lambda x} dx$.) With $(1/y) - 1$ for a, we get

$$F_Y(y) = e^{-\lambda(1/y-1)} \quad \text{for } 0 < y < 1.$$

Putting this together with the values for y outside $(0, 1)$, we get the CDF:

$$F_Y(y) = \begin{cases} 0 & \text{if } y \leq 0, \\ e^{-\lambda(1/y-1)} & \text{if } 0 < y < 1, \\ 1 & \text{if } y \geq 1. \end{cases}$$

Finally, we differentiate $F_Y(y)$ to get a density. The derivative exists everywhere except perhaps at $y = 0$ and $y = 1$; to the left of $y = 0$ and to the right of $y = 1$ it is 0, and between 0 and 1 it is

$$F_Y'(y) = \frac{\lambda}{y^2} e^{-\lambda(1/y-1)} \quad \text{for } 0 < y < 1.$$

So a density for Y is

$$f_Y(y) = \begin{cases} \dfrac{\lambda}{y^2} e^{-\lambda(1/y-1)} & \text{if } 0 < y < 1, \\ 0 & \text{otherwise.} \end{cases}$$

∎

(5.2.7) **A shortcut to the density of Y.** In the previous example we were asked for the density of Y, and we found it by finding the CDF and differentiating. But it turns out that there is no need to find the CDF of Y explicitly if we do not want it, as long as we know a density for X. That density is

(5.2.8)
$$f_X(x) = \begin{cases} \lambda e^{-\lambda x} & \text{if } x > 0, \\ 0 & \text{otherwise.} \end{cases}$$

Start back at (5.2.6), which says that $F_Y(y) = P(X \geq 1/y - 1)$. This means that

(5.2.9)
$$F_Y(y) = 1 - F_X\left(\tfrac{1}{y} - 1\right).$$

Instead of evaluating this, we differentiate it to get a density for Y. Using the chain rule (the one from calculus, not the one for conditional probabilities), we get

$$F_Y'(y) = -F_X'\left(\frac{1}{y} - 1\right) \cdot \left(-\frac{1}{y^2}\right) = \frac{1}{y^2} \cdot F_X'\left(\frac{1}{y} - 1\right).$$

But F_X' is a density for X, and so is (5.2.8). Therefore, for $0 < y < 1$,

$$f_Y(y) = F_Y'(y) = \frac{1}{y^2} F_X'\left(\frac{1}{y} - 1\right) = \frac{1}{y^2} f_X\left(\frac{1}{y} - 1\right) = \frac{1}{y^2} \cdot \lambda e^{-\lambda(1/y-1)},$$

which agrees with the result of the calculation in Example (5.2.5). ∎

The point of the shortcut is that we do not need to write the CDF of X explicitly in order to differentiate it; we just write F_X' and then use the fact that $F_X' = f_X$. The key (for this example) was (5.2.9), which shows that $F_Y(y)$ is simply expressed in terms of $F_X(g(y))$ for a certain function $g(y)$, which is in fact $h^{-1}(y)$.

You might be wondering if this approach will work all the time, and if there is a formula that gives us a density for $Y = h(X)$ directly from a density for X. Well, there almost is, and we will investigate the situation in the next section. But things get a little messy without some restrictions on the function $h(x)$, as we will see.

For one thing, in the example above, the function $h(x) = 1/(x + 1)$ is strictly decreasing, and therefore one-to-one, and that made (5.2.9) simple. It is even simpler if $h(x)$ is strictly increasing. But in the next example $h(x)$ is not one-to-one, and we will see what happens.

(5.2.10) Example. Let X have the standard normal distribution and let $Y = X^2/2$. What is a density for Y?

Again we use the CDF method, but with the shortcut of not writing down any CDFs explicitly. Recall that a density for X is the famous

$$f_X(x) = \varphi(x) = \frac{1}{\sqrt{2\pi}} e^{-x^2/2} \quad \text{for } -\infty < x < \infty.$$

First, as always, we sketch the graph of $h(x) = x^2/2$ on the set of possible values of X, which is $(-\infty, \infty)$. Figure 5.2d shows it, with a y-interval and the corresponding x-set shaded. We see that the set of possible values of Y is the set $[0, \infty)$ of nonnegative numbers. [We can ignore 0 and consider just $(0, \infty)$ if we like, because X and Y are absolutely continuous and so $P(Y = 0)$ is 0.]

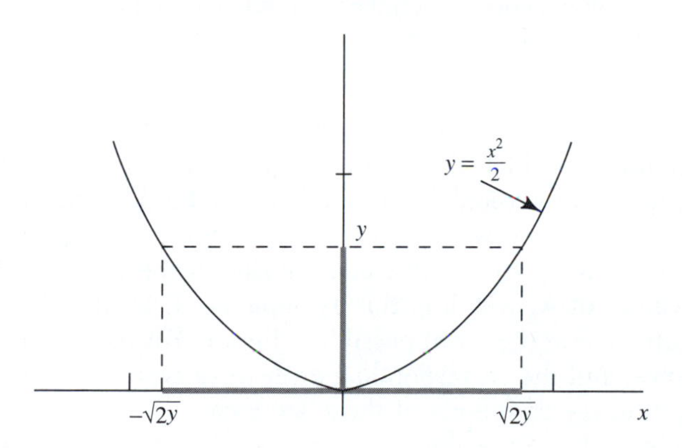

FIGURE 5.2d Sketch for finding the distribution of $Y = \dfrac{X^2}{2}$ if X has possible values in $(-\infty, \infty)$

For a typical y in $(0, \infty)$, the y-interval and the corresponding x-set are shaded on the axes in Figure 5.2d. You can see that in terms of y, the two endpoints of the x-set are $\pm\sqrt{2y}$. Therefore,

(5.2.11)
$$F_Y(y) = P(Y \le y) = P\left(\frac{X^2}{2} \le y\right) = P(-\sqrt{2y} \le X \le \sqrt{2y}) = F_X(\sqrt{2y}) - F_X(-\sqrt{2y}).$$

Now, without worrying about what this is, we differentiate it to get a density for Y:

$$\begin{aligned} F_Y'(y) &= F_X'(\sqrt{2y}) \cdot \left(\frac{1}{\sqrt{2y}}\right) - F_X'(-\sqrt{2y}) \cdot \left(-\frac{1}{\sqrt{2y}}\right) \\ &= \frac{1}{\sqrt{2y}} \cdot \left(f_X(\sqrt{2y}) + f_X(-\sqrt{2y})\right) = \frac{1}{\sqrt{2y}} \cdot \left(\frac{1}{\sqrt{2\pi}}e^{-y} + \frac{1}{\sqrt{2\pi}}e^{-y}\right) \\ &= \frac{e^{-y}}{\sqrt{\pi y}}. \end{aligned}$$

That is, a density for Y is

$$f(y) = \begin{cases} \dfrac{e^{-y}}{\sqrt{\pi y}} & \text{if } y > 0, \\ 0 & \text{otherwise.} \end{cases}$$

Again we do not recognize this as a distribution with a name, although it is a member of the important gamma family, which we will study later. ∎

Compare the key expression (5.2.11) in this example with (5.2.9) in the previous one, and you see that in (5.2.11) the expression for F_Y in terms of F_X involved two appearances of F_X, with different functions of y in them. This happens because $h(x)$ does not have a unique inverse; for a typical y, there are two values of x for which $h(x) = y$.

(5.2.12) Example. Let X have the uniform distribution on $(0, 1)$, let λ be a given positive number, and let $Y = -(1/\lambda)\ln X$. What is the CDF of Y?

This time we are asked for the CDF and not a density, so we will not use the shortcut; that is, when we get an expression for $F_Y(y)$ we will evaluate it rather than differentiate it. As always, first we sketch the graph of $h(x)$ on the set of possible values of X, which in this example is $(0, 1)$. It is shown in Figure 5.2e.

We can see that the set of possible values of Y is $(0, \infty)$. The figure also shows the y-interval and the corresponding x-set for a typical y in this set. The algebra needed to find the endpoints of the x-set is easy:

$$y = -\frac{1}{\lambda}\ln x \quad \text{implies that} \quad x = e^{-\lambda y}.$$

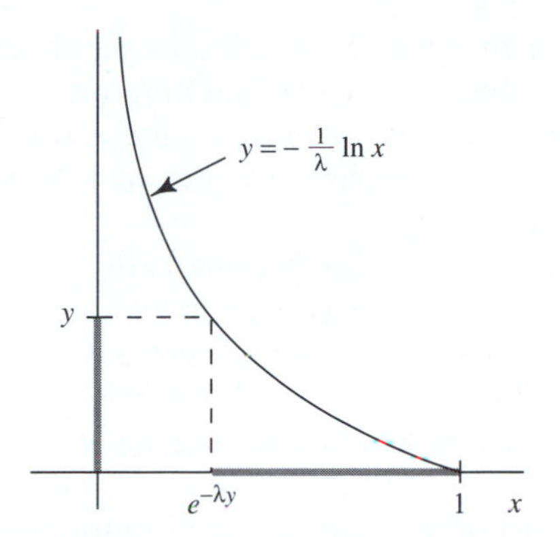

FIGURE 5.2e Sketch for finding the distribution of $Y = -\frac{1}{\lambda} \ln X$ if X has possible values in $(0, \infty)$

So

(5.2.13)
$$F_Y(y) = P(Y \le y) = P\left(-\frac{1}{\lambda}\ln X \le y\right) = P(\ln X \ge -\lambda y) = P(X \ge e^{-\lambda y}).$$

[You might have been able to do (5.2.13) without looking at the graph, but it is hoped that you will feel more comfortable having it.]

Now this equals $1 - F_X(e^{-\lambda y})$, but because we want the CDF of Y, we must evaluate this function of y rather than taking the shortcut of differentiating it. We can use what we know about the distribution of X [uniform on (0, 1)] to see quickly that $P(X \ge e^{-\lambda y})$ equals $1 - e^{-\lambda y}$. Thus we have

$$F_Y(y) = \begin{cases} 0 & \text{if } y \le 0, \\ 1 - e^{-\lambda y} & \text{if } y > 0. \end{cases}$$

You may recognize this CDF; Y has the exponential distribution with parameter λ. ∎

Thus we have discovered a way to generate exponentially distributed random variables with a computer: ask the computer for a random number, which has the uniform distribution on (0, 1); then take the natural logarithm, change the sign, and divide by λ.

This also gives us a second way to simulate a Poisson process with a computer. [In (1.4.11) we saw how to simulate one approximately by using a Bernoulli process.] Recall that the times between arrivals of a Poisson process are independent exponentially distributed random variables. So to simulate a Poisson process, we can just generate a sequence $X_1, X_2, X_3, \ldots$ of such random variables, and let the arrival times of the process be X_1, $X_1 + X_2$, $X_1 + X_2 + X_3$, etc.

(5.2.14) **Functions of two or more random variables.** We can also use the CDF method to find the distribution of a function of several variables such as $W = h(X, Y)$. The basic idea of the CDF method still works: $F_W(w) = P(W \le w) = P(h(X, Y) \le w)$, and we translate the statement "$h(X, Y) \le w$" into the description of an event involving X and Y whose probability we can find in terms of w. The difference is that the graph of the function h is not useful, because it is a surface in three (or more) dimensions. Instead, we have to visualize the event "$h(X, Y) \le w$" and find its probability as best we can.

We have actually done similar calculations earlier—see the calculations involving R in Example (1.3.13), Exercise 14 in Section 1.3, Supplementary Exercise 4 in Chapter 1, and Exercise 12 in Section 3.3. And our first example here is simply a reworking of Example (3.5.12), where we used the CDF method without knowing it. We do it this time in the language of the CDF method.

(5.2.15) **Example: The distribution of the sum of two independent, exponentially distributed random variables.** Let X and Y be independent random variables, each having the exponential distribution with parameter λ, and let $W = X + Y$. We want the CDF of W. Here the function $h(X, Y)$ is of course $X + Y$.

The set of possible values of W is $(0, \infty)$, because X and Y have the positive numbers as possible values, and therefore so does $X + Y$. So we take a typical possible value w and seek $P(W \le w) = P(X + Y \le w)$.

We do not graph the function $h(x, y)$, but we look instead at the set of possible values of the pair (X, Y)—the positive quadrant—and identify the set of points (x, y) for which $x + y \le w$. (This is shown back in Figure 3.5d, except that we are now using w where we used a in the earlier example.) It is the set of points in the positive quadrant below the line $x + y = w$; that is, the triangle whose vertices are at $(0, 0)$, $(0, w)$, and $(w, 0)$. We need to find its probability, as a function of w.

We will not repeat the calculation; it is the double integral of the joint density of X and Y over that triangle, and you might want to look back at it to remember how it goes. The answer is $1 - e^{-\lambda w} - \lambda w e^{-\lambda w}$, and so the CDF of W is

$$F_W(w) = \begin{cases} 0 & \text{if } w \le 0, \\ 1 - e^{-\lambda w} - \lambda w e^{-\lambda w} & \text{if } w > 0. \end{cases}$$

Of course, we can differentiate this (except possibly at $w = 0$) to get a density for W:

$$f_W(w) = \lambda^2 w e^{-\lambda w} \quad \text{for } w > 0.$$

This is called the Erlang density with parameters 2 and λ; it is a member of the gamma family, which we will study in Chapter 8. In general, as we will see there, the sum of n independent random variables having the exponential distribution with parameter λ has the Erlang distribution with parameters n and λ. Its density is a constant times $x^{n-1}e^{-\lambda x}$. ∎

(5.2.16) **Example: The distribution of the larger of two independent uniform random variables.** Let the pair (X, Y) have the uniform distribution on the unit square. This is the square whose corners are at the points $(0, 0)$, $(0, 1)$, $(1, 0)$, and $(1, 1)$ in the plane. For any subset A of the square, the probability that the point (X, Y) falls in A is equal to the area of A (divided by the area of the square, which is 1). This is of course equivalent to saying that X and Y are independent random variables, each having the uniform distribution on (0, 1). Notice that the square is the Cartesian product $(0, 1) \times (0, 1)$.

Let W be the random variable defined by saying that W is the larger of X and Y. Thus, if $X = .28$ and $Y = .63$, then $W = .63$. If X and Y should happen to be equal, then W is their common value (although the probability of this event is 0). That is, $W = h(X, Y)$ where $h(x, y) = \max(x, y)$. What is the CDF of W?

Again, rather than graph the function $w = h(x, y)$, we take a typical possible value w of W; and, looking at a picture of the set of possible values of the pair (X, Y), we identify the set of points (x, y) for which $h(x, y) \leq w$.

The possible values of W are not difficult to see: If X and Y range between 0 and 1, then the larger of the two also ranges between 0 and 1. So the set of possible values of W is the interval (0, 1); $F_W(w)$ is 0 if $w \leq 0$ and 1 if $w \geq 1$.

Now take a typical w between 0 and 1, and look at the set of possible values of (X, Y), which is the larger square in Figure 5.2f. Where is the set of (x, y) for which the larger of x and y is less than or equal to w? The key is to notice that for any (x, y), saying that the larger of x and y is less than or equal to w is exactly the same as saying that *both* x and y are less than or equal to w. In Figure 5.2f, the shaded region is the set of points whose larger coordinate is less than or equal to w, and it is clearly the same as the region in which both coordinates are less than or equal to w.

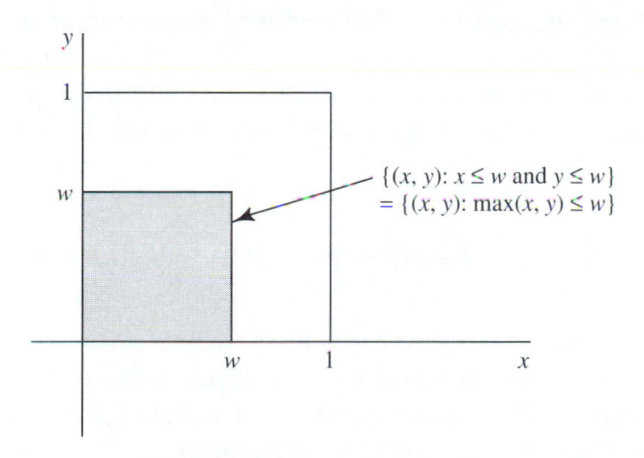

FIGURE 5.2f The set of (x, y) in $(0, 1) \times (0, 1)$ for which the larger of x and y is less than or equal to w (i.e., both x and y are less than or equal to w)

The probability of this event is just w^2, the area of the shaded square in Figure 5.2f. So the CDF of W is

$$F_W(w) = \begin{cases} 0 & \text{if } w \leq 0, \\ w^2 & \text{if } 0 < w < 1, \\ 1 & \text{if } w \geq 1. \end{cases}$$

Of course we can get a density by differentiating this (except at $w = 1$):

$$f_W(w) = \begin{cases} 2w & \text{if } 0 < w < 1, \\ 0 & \text{otherwise.} \end{cases}$$ ∎

(5.2.17) **Convolutions for densities of sums of independent random variables.** In Example (5.2.15) we found a density for $X + Y$ where X and Y were independent, each with an exponential density. If we go through that calculation with an arbitrary, unspecified pair of densities, we get a theorem that sometimes saves a little work.

Suppose X and Y are independent, with densities $f_X(x)$ and $f_Y(y)$, and $W = X + Y$. For any real number w, whether it is a possible value of W or not, the CDF $F_W(w)$ equals $P(X + Y \leq w)$, and this is the probability of the region of the xy-plane below the line $x + y = w$. This is the integral of the joint density of X and Y over that region, which is

$$\int_{-\infty}^{\infty}\int_{-\infty}^{w-x} f_X(x) \cdot f_Y(y)dy\, dx = \int_{-\infty}^{\infty} f_X(x)\int_{-\infty}^{w-x} f_Y(y)dy\, dx.$$

That is,

(5.2.18)
$$F_W(w) = \int_{-\infty}^{\infty} f_X(x) \cdot F_Y(w - x)dx.$$

Notice that the right side of (5.2.18) is indeed a function of w; it gives the correct value of $P(W \leq w)$ for any real number w. To find a density for W, we want to differentiate it with respect to w. To do this we need to do what is called "differentiating under the integral sign," which is valid if certain conditions are satisfied:

(5.2.19)
$$f_W(w) = F_W'(w) = \int_{-\infty}^{\infty} f_X(x) \cdot \frac{d}{dw}F_Y(w - x)dx.$$

Now there are two questions: First, what does it take to justify this differentiation under the integral sign? Second, what is $(d/dw)F_Y(w - x)$? We know the answer to the second question: it equals $f_Y(w - x)$, at least at those points $w - x$ where the density f_Y is continuous.

Luckily, that is essentially the same condition that we need to justify the differentiation under the integral sign. The theorem, usually proved in advanced calculus courses, implies that such a differentiation is valid as long as x is in an interval on which f_X is continuous and $w - x$ is in an interval on which f_Y is continuous.

Now we look again at the integral (5.2.19), and ask: for which x is the integrand positive? The answer is easy—for those x that are possible values of X, and for which $w - x$ is a possible value of Y. So if f_X and f_Y are continuous on their intervals of possible values, then the differentiation is justified.

If we put these considerations together, we have proved the following theorem.

(5.2.20) **Convolution theorem for densities.** Suppose X and Y are independent random variables with densities $f_X(x)$ and $f_Y(y)$. Suppose their sets of possible values are intervals, and these densities are continuous on those intervals. Then a density for $W = X + Y$ is

$$f_W(w) = \int_{-\infty}^{\infty} f_X(x) \cdot f_Y(w - x)dx. \qquad \square$$

This integral is called the ***convolution*** of the functions f_X and f_Y. Notice that in a given example, the limits of integration will not really be $-\infty$ and ∞; the integral will be over only those x that are possible values of X, and for which $w - x$ is a possible value of Y. It is always easy to write down the integrand; the tricky step is to get the limits of integration right.

(5.2.21) **Example.** Let us use the convolution theorem to rederive the result of Example (5.2.15). There X and Y are independent, each having the exponential density:

$$f_X(x) = \lambda e^{-\lambda x} \text{ for } x > 0 \quad \text{and} \quad f_Y(y) = \lambda e^{-\lambda y} \text{ for } y > 0.$$

For a given w, where is the integrand $f_X(x) \cdot f_Y(w - x)$ positive? The answer is that both x and $w - x$ must be positive. That is, w must be positive and x must be between 0 and w. So a density for $W = X + Y$ is

$$\begin{aligned} f_W(w) &= \int_0^w \lambda e^{-\lambda x} \cdot \lambda e^{-\lambda(w-x)} dx \\ &= \lambda^2 e^{-\lambda w} \int_0^w dx \\ &= \lambda^2 w e^{-\lambda w} \quad \text{for } w > 0. \end{aligned}$$

Notice how much quicker this calculation is than the (essentially similar) CDF-method calculation in Example (5.2.15).

Do not leave this example without looking again at the limits of integration. It is very important to understand why we integrate only from 0 to w in this example, although the theorem says to integrate from $-\infty$ to ∞. It is because the function $f_X(x) \cdot f_Y(w - x)$ equals $\lambda e^{-\lambda x} \cdot \lambda e^{-\lambda(w-x)}$ only when x and $w - x$ are both positive; it equals 0 elsewhere. This will not happen in all examples. When the densities are given by the same formulas for all values of the variables, the limits will be $-\infty$ to ∞. As we have said repeatedly, the important thing to remember is that ***when you replace the name of a function, such as f(x), by a formula, you must pay attention to the set of values for which the formula is valid.*** ∎

In Chapter 6 we will need to find the distribution of the sum of independent normally distributed random variables. The result is simple, but the calculation involves some messy (but essential) algebra. One way to find it uses the convolution theorem, and another uses the multivariate change-of-variable theorem of Section 5.4. There is yet a third, and quicker, way, using generating functions, which we will cover in Chapter 7.

EXERCISES FOR SECTION 5.2

1. Use the CDF method to find a density for $Y = aX$ if X has the exponential distribution with parameter λ and a is a positive constant. What is the name of the distribution of Y?

2. The (standard) *lognormal* distribution is the distribution of a random variable Y for which $X = \ln Y$ has a standard normal distribution. Find a density for Y.

   Y is not $\ln X$, X is $\ln Y$. Y is called "lognormal" because its logarithm is normal, not because it is the logarithm of a normal random variable. This may seem perverse, but it is standard terminology. Random variables whose logarithms are normal (or approximately so) arise more often than logarithms of normal random variables.

3. Let X have the uniform distribution on $(0, 2)$; the density is $f(x) = \frac{1}{2}$ for $0 < x < 2$. Find the CDF of $Y = (X - 1)^2$.

4. Find a density for $Y = X^{1/r}$ if X has the exponential distribution with parameter λ and r is some positive number.

5. If X has the standard normal distribution, find a density for $Y = X^2$. [This is one of the chi-squared densities, which make up an important subfamily of the gamma family.]

6. Let X have the uniform distribution on $(0, 1)$, and let $Y = a + (b - a)X$, where a and b are numbers with $a < b$. Find the distribution of Y. What is its name?

7. Let X have the uniform distribution on $(-\pi/2, \pi/2)$ (probability of an interval is its length divided by π), and let $Y = \tan X$. Find the CDF and a density for Y. What is the name of the distribution of Y?

8. **The normal family of distributions.**

   This important exercise involves a fairly simple use of the CDF method. If you do it now, not only will you get a little practice with the method, but you will also be more familiar with the results when they reappear. They will reappear in Example (5.3.11) in the next section, in a slightly different way. A summary can be found in (5.3.14), which you may want to read even if you are going to omit Section 5.3. The results will be needed in Chapter 6. The early part of Section 6.2 collects the basic facts needed to work with normal distributions.

   Let Z have the standard normal distribution, whose density we first saw in (1.3.5):

$$\varphi(z) = \frac{1}{\sqrt{2\pi}} e^{-z^2/2} \quad \text{for } z \in \mathbb{R}.$$

a. Use the CDF method to show that if $Y = aZ + b$ where a is nonzero, then a density for Y is

$$f(y) = \frac{1}{\sqrt{2\pi a^2}} e^{(y-b)^2/2a^2} \quad \text{for } y \in \mathbb{R}.$$

[*Hint:* There are two cases to consider, $a > 0$ and $a < 0$. Note that $\sqrt{a^2} = |a|$ in either case.]

The distribution of Y is called the ***normal distribution with parameters b and a²***, or the $N(b, a^2)$ distribution for short. Notice that the $N(0, 1)$ distribution is the standard normal distribution. Graphs of some normal densities are shown in Figure 5.3c.

b. Show that $EY = b$ and $VY = a^2$. [You do not need to do any integrals; just remember from Examples (4.1.19) and (4.3.9) that $EZ = 0$ and $VZ = 1$, and use the algebra of expected values and variances.] Because of this result we often use the symbols μ for b and σ^2 for a^2.

c. Suppose X has the $N(\mu, \sigma^2)$ distribution and $Y = aX + b$, where again a is nonzero. Use the CDF method to show that Y has the $N(a\mu + b, a^2\sigma^2)$ distribution.

d. Conclude that if X has the $N(\mu, \sigma^2)$ distribution and $Z = (X - \mu)/\sigma$, then Z has the standard normal distribution.

[Despite the order in which the results above are derived, it is worth noting that part c is the most general result, and that parts a and d are special cases of it.]

9. Let the CDF of X be

$$F_X(x) = \begin{cases} 0 & \text{if } x < 0, \\ x^2/4 & \text{if } 0 \le x < 2, \\ 1 & \text{if } x \ge 2. \end{cases}$$

Let $Y = X^2/4$. Find the distribution of Y. [It is a familiar distribution. You should find the CDF, and then the density, and recognize the distribution by name.]

10. Let the CDF of X be

$$F_X(x) = \begin{cases} 0 & \text{if } x < 1, \\ 1 - 1/x^2 & \text{if } x \ge 1. \end{cases}$$

Let $Y = 1 - 1/X^2$. Find the distribution of Y.

11. Do you see a pattern in Exercises 9 and 10? Suppose X has the CDF $F_X(x)$ and let $Y = F_X(X)$. If the function $F_X(x)$ has an inverse $F_X^{-1}(y)$ on the set of possible values of X, then you can use the CDF method to show that Y has the uniform distribution on $(0, 1)$.

12. Let X and Y be the coordinates of a randomly chosen point having the uniform distribution on the unit disk. That is, the probability that (X, Y) falls in any Borel subset of the disk is $1/\pi$ times the area of the subset. Let $Z = h(X, Y)$ be the distance of (X, Y) from the center: $h(x, y) = \sqrt{x^2 + y^2}$. Find the distribution of Z. [You did it already in Exercise 14 of Section 1.3.]

13. In Example (5.2.15), X and Y, are independent, each having the exponential distribution with parameter λ. Their joint density is $f(x, y) = \lambda^2 e^{-\lambda(x+y)}$ (for $x > 0$ and $y > 0$).

a. Define V to be the larger of X and Y. Find the CDF of V. [Use the method of Example (5.2.16).]

b. Define U to be the smaller of X and Y. For a typical possible value u, find $P(U > u)$. Then find the CDF of U, and differentiate it to find a density. What is the name of the distribution of U?

This repeats Exercise 6 in Section 3.5.

14. Let X and Y be independent, each with the uniform distribution on $(0, 1)$. Let U be the smaller of X and Y. Find a density for U. [Proceed as in Exercise 13b.]

15. Let the joint density of X and Y be

$$f(x, y) = \begin{cases} 6e^{-3x-2y} & \text{if } x > 0 \text{ and } y > 0, \\ 0 & \text{otherwise.} \end{cases}$$

Find a density for the random variables defined as follows.

a. $Z = X + Y$

b. $W =$ the larger of X and Y

c. $U =$ the smaller of X and Y

16. In Exercise 15, X and Y are independent random variables, so in part a we can use the convolution theorem if we wish. Redo Exercise 15a by first finding the marginal densities of X and Y, and then applying the convolution theorem.

17. Let X and Y be independent, each with the same density $f(x) = \lambda^2 x e^{-\lambda x}$ for $x > 0$.

a. Use the convolution theorem to find a density for $W = X + Y$.

b. Use the CDF method to get the same result.

18. **The rejection method for generating random variables with a desired density on (0, 1).** Suppose we need to generate random variables whose density is $f(z)$, but our computer has only a uniform random number generator. If $f(z)$ is nonzero only on the interval $(0, 1)$, and if $f(z)$ is bounded, so that there exists a constant a for which $af(z) \leq 1$ for all z, then we can proceed as follows. Let

X and Y be a pair of independent random variables, each with the uniform distribution on (0, 1). If $Y \le af(X)$, then let $Z = X$. Otherwise reject X and Y and generate a new pair (X, Y). Continue, stopping and taking $Z = X$ when first $Y \le af(X)$.

a. To show that the Z produced by this method actually has $f(z)$ as its density, show that the CDF of Z is $F(z) = \int_0^z f(x)dx$. Do this using the fact that $P(Z \le z)$ is really a conditional probability: $F(z) = P(Z \le z) = P(X \le z | Y \le af(X))$. [*Hint:* In the unit square draw the graph of the function $y = af(x)$, and find the probabilities (areas) of the two events "$X \le z$ and $Y \le af(X)$" and "$Y \le af(X)$."]

b. What is the probability mass function of the number of pairs that are rejected before $Y \le af(X)$?

This method is restricted by two requirements on the generated random variable Z: it must have (0, 1) as its set of possible values, and it must have a bounded density. The boundedness restriction is difficult to get around, but the interval (0, 1) is not necessary. The method can easily be modified to generate random variables with any given bounded density. See Ross's book *A Course in Simulation,* pp. 63–68.

ANSWERS

1. Exponential with parameter λ/a
2. $\dfrac{1}{y\sqrt{2\pi}} e^{-(\ln y)^2/2}$, $y > 0$
3. $F_Y(y) = \sqrt{y}$ for $0 < y < 1$ (0 for $y \le 0$ and 1 for $y \ge 1$)
4. $\lambda r y^{r-1} e^{-\lambda y^r}$ for $y > 0$
5. $\dfrac{1}{\sqrt{2\pi y}} e^{-y/2}$ $(y > 0)$
6. Uniform on (a, b)
7. $F_Y(y) = \dfrac{1}{2} + \dfrac{1}{\pi}\tan^{-1} y$ for $-\infty < y < \infty$; $f_Y(y) = \dfrac{1}{\pi(1 + y^2)}$ for $y \in \mathbb{R}$; standard Cauchy

9. Uniform on (0, 1)
10. Uniform on (0, 1)
12. Density is $2z$ for $0 < z < 1$.
13. **a.** $F_V(v) = (1 - e^{-\lambda v})^2$ for $v \ge 0$
    b. Exponential with parameter 2λ
14. $2(1 - u)$ for $0 < u < 1$
15. **a.** $6e^{-2z} - 6e^{-3z}$ for $z > 0$
    b. $3e^{-3w} + 2e^{-2w} - 5e^{-5w}$ for $w > 0$
    c. $5e^{-5u}$ for $u > 0$
17. $f_W(w) = \dfrac{\lambda^4 w^3}{6} e^{-\lambda w}$ for $w > 0$
18. **b.** $P(k \text{ pairs rejected}) = a \cdot (1 - a)^k$ for $k = 0, 1, 2, \ldots$ (the geometric mass function with success probability a)

5.3 THE CHANGE-OF-VARIABLE THEOREM FOR DENSITIES

The "shortcut" version of the CDF method described in (5.2.7) finds a density for $Y = h(X)$, given a density for X. It involves the CDFs of X and of Y, but not explicitly. It turns out that this method produces a simple formula that gives us

$f_Y(y)$ in terms of $f_X(x)$ and $h(x)$, as long as the function h is of the right type. The formula is given in Theorem (5.3.8).

The main requirements on h are that it be invertible, and that its inverse be differentiable everywhere on the interval of possible values of X. A condition sufficient to guarantee this is that $h'(x)$ exist and be nonzero everywhere on that interval. This implies that h is either an increasing function or a decreasing function on that interval.

It is possible to use the method to get general formulas that work under less restrictive conditions (see (5.3.17)), but they tend to be complicated. Rather than extend Theorem (5.3.8) to these cases, it is usually simpler to rely on the CDF method of the previous section.

Another extension of Theorem (5.3.8) *is* very useful, however—the multidimensional version, which we will cover in the next section. Suppose $X_1, X_2, \ldots, X_n$ are random variables with a known joint density, and we have n random variables $Y_1, Y_2, \ldots, Y_n$ that are functions of the X_j's:

(5.3.1)
$$Y_1 = h_1(X_1, X_2, \ldots, X_n), \quad Y_2 = h_2(X_1, X_2, \ldots, X_n), \quad \ldots, \\ Y_n = h_n(X_1, X_2, \ldots, X_n).$$

Then under certain conditions on the functions $h_i(x_1, \ldots, x_n)$, there is a formula that gives a joint density for $Y_1, Y_2, \ldots, Y_n$. The conditions on the h_j's amount to differentiability and invertibility—we have to be able to solve (5.3.1) uniquely for the X_j's in terms of the Y_i's, and the inverse has to be differentiable. Sufficient for this is that the transformation (5.3.1) have continuous partial derivatives, and that its Jacobian matrix be nonzero on the set of possible values of $(X_1, X_2, \ldots, X_n)$.

If you are going to omit the next section, you may want to omit this one also. Examples (5.3.11) and (5.3.14) on normal distributions are important, but that material was covered briefly in Exercise 8 of the previous section, and it will be reviewed at the beginning of Section 6.2.

In this section we deal only with the one-dimensional case. Perhaps the easiest way to get to the theorem is to imagine that we are solving one of the problems of the previous section using the "shortcut" CDF method, but with no specific density $f_X(x)$ or function $h(x)$ in mind. If we do it once for an increasing $h(x)$ and then again for a decreasing $h(x)$, the theorem will emerge.

(5.3.2) **The change-of-variable theorem for densities, if $h(x)$ is increasing.** Suppose X has a known density $f_X(x)$, which is nonzero on some interval I_X, the set of possible values of X. Let $Y = h(X)$, where the function $h(x)$ is differentiable, with $h'(x) > 0$, for all x in I_X. Let Y be defined as $h(X)$; we want a density for Y.

Figure 5.3a shows this situation. The set of possible values of Y is an interval I_Y on the vertical axis. For a typical y in this set, the "x-set" corresponding to the "y-interval" (to use the language of the previous section) is the interval of points to the left of $h^{-1}(y)$. It will be helpful to give the inverse function $h^{-1}(y)$ a name; let us define $g(y) = h^{-1}(y)$. So for a typical y in the set of possible values of Y,

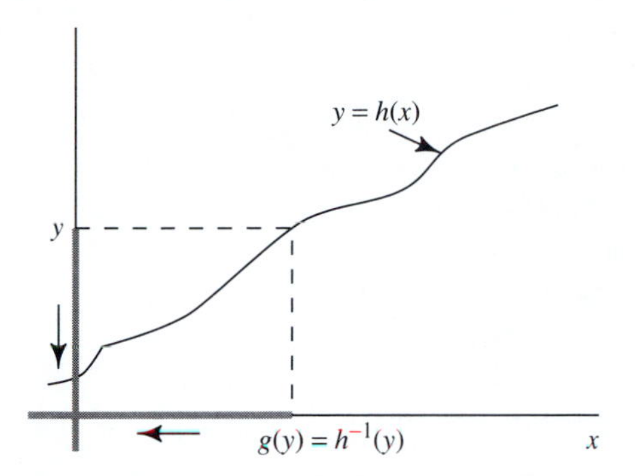

FIGURE 5.3a Sketch illustrating proof of change-of-variable theorem when $h(x)$ is increasing

(5.3.3)
$$F_Y(y) = P(Y \le y) = P(X \le g(y)) = F_X(g(y)).$$

Now to find a density for Y we want to differentiate this function of y. We can do so using the chain rule, because if $h'(x)$ exists and is positive on I_X, then $g'(y)$ exists and is positive on I_Y. When we differentiate the left and right members of (5.3.3) we get

$$F_Y'(y) = F_X'(g(y)) \cdot g'(y).$$

Since the derivative of a CDF is a density, this says

(5.3.4)
$$f_Y(y) = f_X(g(y)) \cdot g'(y),$$

provided, of course, that $Y = h(X)$, h is differentiable, h' is positive, and $g = h^{-1}$.

(5.3.5) The change-of-variable theorem for densities, if $h(x)$ is decreasing. Now we repeat the same steps, but this time we assume that $h'(x)$ is negative for all x in the set of possible values of X. Now $h(x)$ is a decreasing function, as shown in Figure 5.3b (page 326). Notice that for a typical y in the set of possible values of Y, the corresponding x-set is now the interval of points to the *right* of $g(y) = h^{-1}(y)$. So instead of (5.3.3) we have in this case

(5.3.6)
$$F_Y(y) = P(Y \le y) = P(X \ge g(y)) = 1 - F_X(g(y)).$$

Consequently, when we differentiate to get a density, a negative sign comes in:

$$F_Y'(y) = -F_X'(g(y)) \cdot g'(y).$$

That is,

(5.3.7)
$$f_Y(y) = -f_X(g(y)) \cdot g'(y),$$

provided that $Y = h(X)$, h is differentiable, h' is negative, and $g = h^{-1}$.

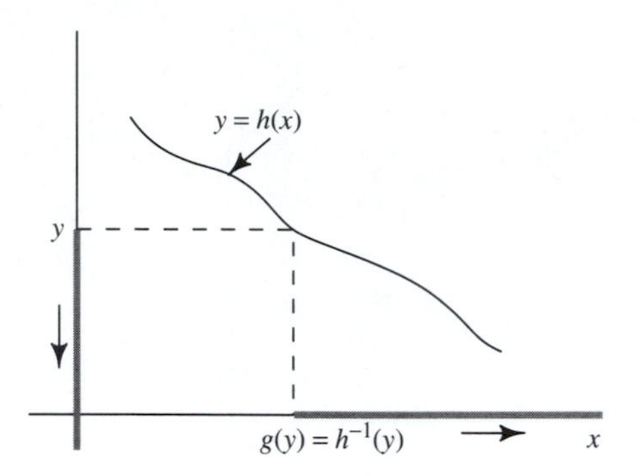

FIGURE 5.3b Sketch illustrating proof of change-of-variable theorem when $h(x)$ is decreasing

The negative sign in (5.3.7) is disconcerting until we realize that $g'(y)$ is also negative in this case. That is true because $h(x)$ is decreasing, and so its inverse $g(y)$ is also decreasing.

We appear to be left with two different formulas for a density for Y, (5.3.4) if h is increasing and (5.3.7) if h is decreasing. But they are easily combined. In (5.3.4) $g'(y)$ is positive, and in (5.3.7) it is negative, but there is a negative sign that makes it positive. So in both cases the formula reads $f_X(g(y)) \cdot |g'(y)|$. And that is our theorem.

(5.3.8) **Theorem: Change of variable for densities.** Let X be a random variable with density $f_X(x)$; suppose the set of possible values of X is an interval I_X. Let $Y = h(X)$, where h is a function that is invertible on I_X and whose inverse $g(y)$ is differentiable. Then a density for Y is

$$f_Y(y) = f_X(g(y)) \cdot |g'(y)|.$$

(The hypothesis that h has a differentiable inverse is equivalent to assuming that h is differentiable and its derivative is either positive throughout I_X or negative throughout I_X. As is shown in texts on advanced calculus, this in turn is equivalent to assuming that $h'(x)$ exists and is nonzero for all X in I_X.)

(5.3.9) Applying the theorem. When we use this theorem we are given two things—a density for X, including the interval of possible values of X, and the function $h(x)$ on that interval. To use the theorem to find a density for Y we need to do three things:

1. Check that $h'(x)$ is strictly positive or strictly negative throughout I_X (otherwise the theorem does not apply), and find the range of $h(x)$. This is the set of possible values of Y.
2. Find the inverse $g(y)$ of the function $h(x)$, and the derivative $g'(y)$.
3. Apply the formula.

You will often find, when you do this, that you have essentially done all the hard work of the CDF method, and that the formula is not really such a work-saver after all. In particular, you should sketch the graph of $h(x)$ on the set of possible values of X, no matter which method you use. Furthermore, as we have noted, the formula applies only for very special types of functions $h(x)$—those whose derivatives are nonzero throughout I_X.

Because the CDF method as given in the previous section always works, and because using the formula of Theorem (5.3.8) often involves essentially the same work, many people simply use the CDF method in all problems involving a one-dimensional function of one random variable. This is probably a good practice.

So why, you ask, have we included Theorem (5.3.8) at all? The reason, as we mentioned at the beginning of the section, is that its multidimensional version, Theorem (5.4.3) in the next section, becomes much easier to understand after a little practice with the one-dimensional version. And the multidimensional version really is useful in many instances.

(5.3.10) Example. Let X have the density $f_X(x) = \frac{4}{27}x^2(3 - x)$ for $0 < x < 3$, and let $Y = \sqrt{1/(1 + X)}$. We find a density for Y.

Step 1. The function $h(x) = \sqrt{1/(1 + x)}$ is strictly decreasing on the domain $0 < x < 3$, and its range on that domain—the set of possible values of Y—is the interval $\left(\frac{1}{2}, 1\right)$. Be sure you understand how to see this. You may be able to visualize it without sketching a graph, but it would not hurt to make the sketch.

Step 2. The inverse of

$$y = h(x) = \sqrt{\frac{1}{1 + x}}$$

is

$$x = g(y) = \frac{1}{y^2} - 1.$$

This is simple algebra, and it is valid for any y in the interval $\left(\frac{1}{2}, 1\right)$. (If $y = 0$ were in the set of possible values of Y, we would have a problem.) Differentiating is also mechanical:

$$g'(y) = -\frac{2}{y^3}.$$

Step 3. Apply the formula. For $\frac{1}{2} < y < 1$,

$$\begin{aligned} F_Y(y) &= f_X(g(y)) \cdot |g'(y)| \\ &= \frac{4}{27}\left(\frac{1}{y^2} - 1\right)^2\left(3 - \left(\frac{1}{y^2} - 1\right)\right) \cdot \frac{2}{y^3}, \end{aligned}$$

which simplifies after a little algebra to

$$f_Y(y) = \frac{8(1 - y^2)^2(4y^2 - 1)}{27y^9} \quad \text{for } \tfrac{1}{2} < y < 1.$$

It's not pleasant, but it is the answer. ∎

(5.3.11) Example: A linear function of a standard normal random variable. Let X have the standard normal distribution; so a density for X is the well-known

(5.3.12)
$$f_X(x) = \varphi(x) = \frac{1}{\sqrt{2\pi}} e^{-x^2/2} \quad \text{for } x \in \mathbb{R}.$$

Let $Y = aX + b$. We assume a is nonzero, so that Y is not a constant, but we are not saying whether a is positive or negative. We want a density for Y. In this case everything is quite simple.

Step 1. The function $h(x) = ax + b$ is strictly increasing if $a > 0$ and strictly decreasing if $a < 0$. On the domain $\mathbb{R}$, its range is $\mathbb{R}$.
Step 2. The inverse of $y = h(x) = ax + b$ is $x = g(y) = (y - b)/a$. The derivative is $g'(y) = 1/a$.
Step 3. Therefore, a density for Y is

$$\begin{aligned} f_Y(y) &= f_X(g(y)) \cdot |g'(y)| \\ &= \frac{1}{\sqrt{2\pi}} e^{-(y-b)^2/2a^2} \cdot \frac{1}{|a|} \quad \text{for } -\infty < y < \infty. \end{aligned}$$

Notice the absolute value; if you forget this you risk calling a negative quantity a density. A convenient way to make the density look a little neater is to use the fact that $|a| = \sqrt{a^2}$; this works whether a is positive or negative. Thus a density for Y is

(5.3.13)
$$f_Y(y) = \frac{1}{\sqrt{2\pi a^2}} e^{-(y-b)^2/2a^2} \quad (-\infty < y < \infty).$$ ∎

(5.3.14) The normal family of distributions. The density (5.3.13) found in the previous example is called the ***normal density with parameters b and a^2***. Figure 5.3c shows the graphs of some of these densities for a few different settings of the parameters. Regardless of the values of a and b, the density is bell shaped. Changing b shifts the curve left or right (it is always "centered" at b), whereas changing a^2 makes it wider or narrower (if a^2 is larger, the curve is wider).

This family of distributions is so important that it has a standard abbreviation. The normal distribution with parameters b and a^2, whose density is (5.3.13), is referred to as the $N(b, a^2)$ distribution. Notice that if $b = 0$ and $a^2 = 1$, (5.3.13) reduces to the standard normal density (5.3.12); that is, the standard normal density is the $N(0, 1)$ density.

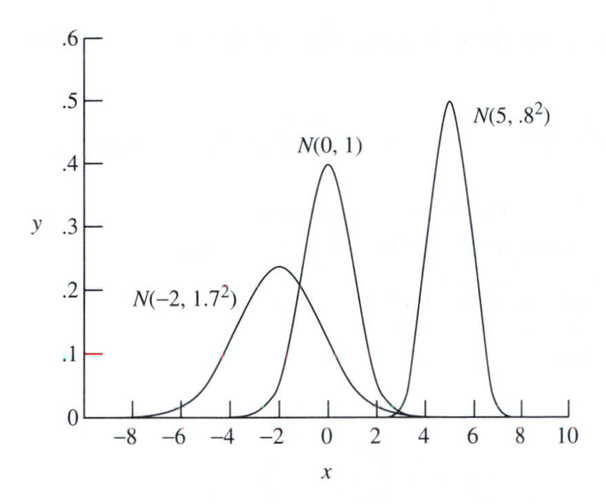

FIGURE 5.3c Some normal densities

We will encounter this very important family of distributions many times in following chapters. We can restate what we learned in the previous example as follows:

> For any number b and any positive number a^2, the $N(b, a^2)$ distribution (the normal distribution with parameters b and a^2) is the one whose density is given by (5.3.13).
>
> If X has the $N(0, 1)$ (standard normal) distribution, then for any number b and any nonzero number a (positive or negative), $Y = aX + b$ has the $N(b, a^2)$ distribution.

Notice also that we can quickly find the expected value and variance of the normal distribution with parameters b and a^2, without having to do any integrations:

X has the standard normal distribution, so $EX = 0$ and $VX = 1$,

and

$$Y = aX + b;$$

consequently,

$$EY = aEX + b = b \quad \text{and} \quad VY = a^2VX = a^2.$$

Notice again that we do not care whether a is positive or negative; only its square appears in the density, expected value, and variance. That is why we are content to use a^2 as the parameter of the density rather than a. Restating these results:

If Y has the $N(b, a^2)$ distribution, then $EY = b$ and $VY = a^2$.

For this reason, we usually use the letters μ and σ^2 in place of b and a. We will do this in later encounters with normal densities.

Even if we had not been able to make use of the fact that $Y = aX + b$ where X is standard normal, we could still have found the expected value and variance by integrating. We would need to find

(5.3.15)

$$EY = \int_{-\infty}^{\infty} y \cdot \frac{1}{\sqrt{2\pi a^2}} e^{-(y-b)^2/2a^2} dy,$$

which comes out to be b, and

(5.3.16)

$$EY^2 = \int_{-\infty}^{\infty} y^2 \cdot \frac{1}{\sqrt{2\pi a^2}} e^{-(y-b)^2/2a^2} dy,$$

which comes out to be $a^2 + b^2$.

These integrals look formidable, but they yield very nicely to the substitution $x = (y - b)/a$. These kinds of integrals come up often in dealing with normal distributions. It would be a good idea to be certain that you can do them both, which is what Exercise 5 at the end of the section asks you to do.

As a further exercise at this point, you might want to verify the following, using the change-of-variable theorem. We did it also in Exercise 8 of Section 5.2 using the CDF method:

If X has the $N(\mu, \sigma^2)$ distribution and $Y = aX + b$ where a is nonzero, then Y has the $N(a\mu + b, a^2\sigma^2)$ distribution.

The two most used special cases of this result are the following.

If Z has the $N(0, 1)$ distribution and $Y = \sigma Z + \mu$, then Y has the $N(\mu, \sigma^2)$ distribution.

If X has the $N(\mu, \sigma^2)$ distribution and $Z = (X - \mu)/\sigma$, then Z has the $N(0, 1)$ distribution. ▮

(5.3.17) **The one-dimensional change-of-variable theorem for functions $h(x)$ that are not invertible.** As we remarked at the beginning of this section, it is possible to say something in general along the lines of Theorem (5.3.8) even though the function $h(x)$ is neither strictly increasing nor strictly decreasing. It may not be worthwhile to do so, because the CDF method is always reliable. But let us look briefly at what is involved, to get an idea of what the complete theorem would look like.

To do this we return to Example (5.2.4). There X has the uniform distribution on $(0, \pi)$ and $Y = \sin X$. The function $h(x) = \sin x$ increases on the first half of the interval $(0, \pi)$ and decreases on the second half, so we cannot use Theorem (5.3.8).

Had we used the CDF method with the shortcut in that example, we would have gone through the following steps:

The set of possible values of Y is $(0, 1]$.

For a typical y in $(0, 1]$, $P(Y \leq y) = P(X \leq \sin^{-1}y) + P(X \geq \pi - \sin^{-1}y)$.

That is, $F_Y(y) = F_X(\sin^{-1}y) + 1 - F_X(\pi - \sin^{-1}y)$.

Differentiating gives

(5.3.18)
$$f_Y(y) = f_X(\sin^{-1}y) \cdot \sqrt{\frac{1}{1 - y^2}} - f_X(\pi - \sin^{-1}y) \cdot \left(-\sqrt{\frac{1}{1 - y^2}}\right).$$

Now in Example (5.2.4), $f_X(x)$ is the constant $1/\pi$ on the set of possible values of X, because the distribution of X is uniform; so (5.3.18) simplifies further. But that is not what we want to notice here. The point now is that there are two "inverses" of the function $h(x) = \sin x$ on $(0, \pi)$. More properly, there are two solutions of the equation $y = \sin x$ for $0 < y \leq 1$; they are

$$g_1(y) = \sin^{-1}y \quad \text{and} \quad g_2(y) = \pi - \sin^{-1}y.$$

In terms of these, (5.3.18) says

(5.3.19)
$$f_Y(y) = f_X(g_1(y)) \cdot |g_1'(y)| + f_X(g_2(y)) \cdot |g_2'(y)|.$$

This shows us what happens in general. Instead of the single term $f_X(g(y)) \cdot |g'(y)|$ in Theorem (5.3.8), we need a separate term for each different solution of $y = h(x)$. [Sometimes these solutions are called the different "branches" of the inverse of $h(x)$.]

But we will not pursue the matter further. You can ignore the theorem of this section and use the CDF method of the previous section in any example involving a single function $Y = h(X)$ of a single random variable. As we pointed out earlier, the main reason for covering it is to lead into the multivariate theorem of the next section, in which we deal with one or more functions of several random variables.

EXERCISES FOR **SECTION 5.3**

1. Let X have the exponential distribution with parameter λ. Use Theorem (5.3.8) to find a density for Y in each of the following.
   a. $Y = X^2$
   b. $Y = aX + b$, where a and b are given constants and a is positive

c. $Y = 1 - e^{-\lambda X}$ [Compare with Exercises 9, 10, and 11 of Section 5.2.]
d. $Y = 1/(1 + X^2)$

2. Let X have the uniform distribution on (0, 1). Use Theorem (5.3.8) to find a density for Y in each of the following.
a. $Y = \tan\left(\pi\left(X - \frac{1}{2}\right)\right)$ b. $Y = 1/(1 - X)^2$ c. $Y = -(1/\lambda)\ln X$.

3. Do Exercises 2 and 6 of Section 5.2 using Theorem (5.3.8). Compare the calculations with the ones you made using the CDF method.

4. Let X have the $N(b, a^2)$ distribution, as described in (5.3.14). [The density is (5.3.13).] Use Theorem (5.3.8) to find the following.
a. A density for $Z = (X - b)/a$. Is the result the same whether a is positive or negative?
b. A density for $Y = cX + d$. What is the name of the distribution of Y?

5. Evaluate the integrals (5.3.15) and (5.3.16) to show directly that if Y has the $N(b, a^2)$ distribution, then $EY = b$ and $EY^2 = a^2 + b^2$. [Use the substitution $z = (y - b)/a$.]

6. **One way to generate random variables having a desired density.** Suppose that $f(y)$ is a density, and that we would like to generate observations of a random variable Y having this density using a computer. The computer gives us observations of a random variable X having the uniform distribution on (0, 1).
a. Show, using Theorem (5.3.8), that if $F(y)$ is the CDF corresponding to the density $f(y)$, and $y = h(x)$ is the inverse of the function $x = F(y)$, then the random variable $Y = h(X)$ has the desired density.
b. Let the desired density be $f(y) = 2/y^3$ for $y > 1$ (and 0 otherwise). Find the transformation $Y = h(X)$ that produces a Y with this density.

Compare the result of part a with Exercises 9, 10, and 11 of Section 5.2. This is a sort of converse to the discovery made there. We will discuss these results in Section 8.5 on uniform distributions. For now we note that, although this is sometimes a simple way to get random variables with a desired distribution, it is not always the most efficient way.

ANSWERS

1. a. $\dfrac{\lambda}{2\sqrt{y}} e^{-\lambda\sqrt{y}}$ $(y > 0)$

b. $\dfrac{\lambda}{a} e^{-(\lambda/a)(y-b)}$ $(y > b)$

c. 1 $(y > 0)$

d. $\dfrac{1}{2y^2\sqrt{\dfrac{1}{y} - 1}} e^{-\lambda\sqrt{(1/y)-1}}$ $(0 < y < 1)$

2. a. $\dfrac{1}{\pi(1 + y^2)}$ $(y \in \mathbb{R})$ b. $\dfrac{1}{2y\sqrt{y}}$ $(1 < y)$
c. $\lambda e^{-\lambda y}$ $(y > 0)$

4. a. The $N(0, 1)$ (standard normal) density in either case b. $N(cb + d, c^2a^2)$

6. b. $Y = 1/\sqrt{1 - X}$

5.4 THE MULTIVARIATE CHANGE-OF-VARIABLE THEOREM

In this section we give (without proof) the extension of the one-dimensional change-of-variable theorem to the case of several random variables. We suppose that we have n random variables $X_1, X_2, \ldots, X_n$ with a known joint density, and n random variables $Y_1, Y_2, \ldots, Y_n$ that are functions of them; and we want a joint density for $Y_1, Y_2, \ldots, Y_n$. As in the one-variable case, if we can invert the functions and express the X_j's in terms of the Y_i's, and if the inverse transformation has a derivative, then there is a formula giving a joint density for the Y_i's.

As an example of the need for such a formula, there are many problems in statistical inference where we need to know the distribution of a ratio of two random variables. For example, the famous "Student's t statistic," the basis of the t-test common in statistics, is the ratio X_1/X_2 of two independent random variables, where X_1 has the standard normal distribution and X_2 has another known distribution (it is the square root of a constant multiple of a chi-squared random variable; we will discuss this in Section 8.4).

We can find a density for X_1/X_2 in two steps. First, define $Y_1 = X_1/X_2$ and $Y_2 = X_2$ and use the change-of-variable theorem to find a joint density for Y_1 and Y_2. Second, integrate to find the marginal density for Y_1. This looks as though we have made extra work for ourselves by adding an unneeded Y_2 and then immediately getting rid of it. But it is in fact the easiest approach. In Example (5.4.4) we will apply the theorem to find a density for a ratio similar to this one.

Exercise 6 at the end of this section gives an important formula that can be used to shorten the work of finding a density for a ratio when the denominator is positive.

Another use of the multivariate change-of-variable theorem is to find a joint density for the polar coordinates of two random variables. Suppose we know a joint density for X_1 and X_2, and we think of (X_1, X_2) as a random point in the plane. The polar coordinates of that point are another pair of random variables, which we call R and Θ (uppercase theta) instead of Y_1 and Y_2:

$$R = \sqrt{X_1^2 + X_2^2} \quad \text{and} \quad \Theta = \tan^{-1}(X_2/X_1).$$

See Figure A.9f in Appendix A. The inverse of this pair of functions is simpler and more familiar than the functions themselves:

$$X_1 = R\cos\Theta \quad \text{and} \quad X_2 = R\sin\Theta.$$

It is sometimes useful to know the distributions of R and Θ, and the multivariate change-of-variable theorem will give them to us.

For an example involving more than two X_j's, the method can be used to find a density for the sum of n random variables whose joint density is known. We define

$$Y_1 = X_1 + X_2 + \cdots + X_n, \quad Y_2 = X_2, \quad Y_3 = X_3, \quad \ldots, \quad Y_n = X_n.$$

Next, use the theorem to find a joint density for the Y_i's, and then integrate $n - 1$ times to find the marginal density of Y_1.

More generally, if $X_1, X_2, \ldots, X_n$ have any known joint distribution, and $Y_1, Y_2, \ldots Y_n$ are any linear combinations of them, then as long as the matrix of the coefficients of these linear combinations is invertible, the method works simply to find a joint density for the Y_i's. This is especially useful in dealing with multivariate normal distributions.

In this section we will state the theorem for n functions of n random variables, but our examples will be confined to the case $n = 2$.

To state the theorem we need to introduce some notation. Denote the joint density of the X_j's, which we know, by $f_X(x_1, x_2, \ldots, x_n)$, and the joint density of the Y_i's, which we want, by $f_Y(y_1, y_2, \ldots, y_n)$. [Notation like $f_{X_1, X_2, \ldots, X_n}(x_1, x_2, \ldots, x_n)$ might be more proper, but seems too cumbersome. Just remember that the subscript X stands for the vector $(X_1, X_2, \ldots, X_n)$.]

As for the function h, there are n functions now:

$$
\begin{aligned}
Y_1 &= h_1(X_1, X_2, \ldots, X_n),\\
Y_2 &= h_2(X_1, X_2, \ldots, X_n),\\
&\vdots\\
Y_n &= h_n(X_1, X_2, \ldots, X_n).
\end{aligned}
$$

We suppose that for every possible value $(y_1, y_2, \ldots, y_n)$ of $(Y_1, Y_2, \ldots, Y_n)$, the equations

$$
\begin{aligned}
y_1 &= h_1(x_1, x_2, \ldots, x_n)\\
y_2 &= h_2(x_1, x_2, \ldots, x_n)\\
&\vdots\\
y_n &= h_n(x_1, x_2, \ldots, x_n)
\end{aligned}
\tag{5.4.1}
$$

can be solved uniquely for the x_j's in terms of the y_i's—that is, that the functions $h_1, h_2, \ldots h_n$ can be inverted. Let the solution be

$$
\begin{aligned}
x_1 &= g_1(y_1, y_2, \ldots, y_n)\\
x_2 &= g_2(y_1, y_2, \ldots, y_n)\\
&\vdots\\
x_n &= g_n(y_1, y_2, \ldots, y_n).
\end{aligned}
\tag{5.4.2}
$$

Finally, suppose the inverse functions $g_j(x_1, x_2, \ldots, x_n)$ have continuous partial derivatives. Let $J(y_1, y_2, \ldots, y_n)$ denote the Jacobian determinant of these functions:

$$J(y_1, y_2, \ldots, y_n) = \begin{vmatrix} \frac{\partial x_1}{\partial y_1} & \frac{\partial x_1}{\partial y_2} & \cdots & \frac{\partial x_1}{\partial y_n} \\ \frac{\partial x_2}{\partial y_1} & \frac{\partial x_2}{\partial y_2} & \cdots & \frac{\partial x_2}{\partial y_n} \\ \cdot & \cdot & \cdot & \cdot \\ \cdot & \cdot & \cdot & \cdot \\ \cdot & \cdot & \cdot & \cdot \\ \frac{\partial x_n}{\partial y_1} & \frac{\partial x_n}{\partial y_2} & \cdots & \frac{\partial x_n}{\partial y_n} \end{vmatrix}.$$

(5.4.3) **Theorem: Change of variable for joint densities.** Under the suppositions above, the joint density of $Y_1, Y_2, \ldots, Y_n$ is given by

$$f_Y(y_1, y_2, \ldots, y_n) = f_X(g_1(y_1, \ldots, y_n), \ldots, g_n(y_1, \ldots, y_n)) \cdot |J(y_1, y_2, \ldots, y_n)|.$$

Although this is notationally complicated, it is just an extension of the formula in Theorem (5.3.8), which is $f_Y(y) = f_X(g(y)) \cdot |g'(y)|$. The Jacobian determinant $J(y_1, y_2, \ldots, y_n)$ of the inverses $g_1, g_2, \ldots, g_n$ takes the place of the derivative $g'(y)$. Notice also that this determinant may be positive or negative; the formula calls for its absolute value. Furthermore, notice that the factor involving f_X on the right side of the formula is just the joint density of the X_j's, but with the variables x_j expressed in terms of the y_i's.

Perhaps a good way to remember how to use Theorem (5.4.3) is as follows. Express the X_j's in terms of the Y_i's and find the Jacobian of these inverse functions. Find the set of possible values of the Y_i's. Write down a joint density for the X_j's, involving the variables $x_1, x_2, \ldots, x_n$. In this density express the x_j's in terms of the y_i's; then multiply by the absolute value of the Jacobian.

We will not prove this theorem; its proof involves what is sometimes called the change-of-variable theorem for multiple integrals, which is usually proved in advanced calculus or real analysis courses. To make this theorem applicable, we added an additional hypothesis that is not in Theorem (5.3.8): the partial derivatives of the g_j's are assumed to be continuous, whereas in Theorem (5.3.8) we simply assume that $g(y)$ is differentiable. The real assumption needed is that the inverse transformation (5.4.2), as a function from $\mathbb{R}^n$ to $\mathbb{R}^n$ that takes $(y_1, y_2, \ldots, y_n)$ to $(x_1, x_2, \ldots, x_n)$, be differentiable. The theory of differentiability of a function of several variables is also discussed in advanced calculus or real analysis courses: one of the important theorems is that if all the partial derivatives are continuous, then the function is differentiable. (This is not necessary for differentiability, but it is sufficient. The mere existence of the partial derivatives is not.) Thus the additional hypothesis—continuity of the partial derivatives—is just a simple way of assuring, without going into the theory of the derivative of a function of several variables, that the inverse function is differentiable.

(5.4.4) Example. Let X_1 and X_2 be independent, each having the exponential distribution with parameter λ. We will use the theorem to find a density for X_1/X_2, by finding a joint density for

$$Y_1 = X_1/X_2 \quad \text{and} \quad Y_2 = X_2,$$

and then integrating to find the marginal density of Y_1.

Before thinking about the set of possible values of the pair (Y_1, Y_2), it is helpful to invert the functions. The functions to invert are

$$\begin{aligned} y_1 &= x_1/x_2 \quad [=h_1(x_1, x_2)], \\ y_2 &= x_2 \qquad\ \ [=h_2(x_1, x_2)]. \end{aligned}$$

The inverses are

$$\begin{aligned} x_1 &= y_1y_2 \quad [=g_1(y_1, y_2)], \\ x_2 &= y_2 \qquad [=g_2(y_1, y_2)]. \end{aligned}$$

Now we can see the set of possible values. If X_1 and X_2 range independently over all positive numbers, then Y_1 and Y_2 also can be any pair of positive numbers. One way to look at this is to notice that any pair of positive numbers, y_1 and y_2, can be realized as x_1/x_2 and x_2, for some pair (x_1, x_2) of positive numbers; we just have to let $x_1 = y_1y_2$ and $x_2 = y_2$. So the set of possible values of the pair (Y_1, Y_2) is the positive quadrant.

Next we need the Jacobian of the inverse:

$$J(y_1, y_2) = \begin{vmatrix} \dfrac{\partial x_1}{\partial y_1} & \dfrac{\partial x_1}{\partial y_2} \\ \dfrac{\partial x_2}{\partial y_1} & \dfrac{\partial x_2}{\partial y_2} \end{vmatrix} = \begin{vmatrix} y_2 & y_1 \\ 0 & 1 \end{vmatrix} = y_2.$$

And of course we need a joint density for X_1 and X_2. They are independent, each having the exponential distribution with parameter λ, so a joint density is

$$f_X(x_1, x_2) = \lambda^2 e^{-\lambda(x_1+x_2)} \quad \text{for } x_1 > 0 \quad \text{and} \quad x_2 > 0.$$

Now we use Theorem (5.4.3) to get a joint density for Y_1, and Y_2. We just express the x_i's in terms of the y_i's in this density, and multiply by the absolute value of the Jacobian:

$$\begin{aligned} f_Y(y_1, y_2) &= f_X(y_1y_2, y_2) \cdot |y_2| \\ &= \lambda^2 e^{-\lambda(y_1y_2+y_2)} \cdot y_2 \quad \text{for } 0 < y_1 \text{ and } 0 < y_2. \end{aligned}$$

That completes the application of the theorem. We finish the problem at hand by integrating with respect to y_2 to find a density for $Y_1 = X_1/X_2$:

$$f_{Y_1}(y_1) = \int_0^\infty f_Y(y_1, y_2)dy_1 = \lambda^2 \int_0^\infty y_2 e^{-\lambda(y_1+1)y_2}\, dy_2.$$

This is the same as

$$\lambda^2 \int_0^\infty x e^{-\lambda a x}\, dx,$$

where $a = y_1 + 1$ and the variable of integration y_2 is replaced by x. An integration by parts (or the gamma formula given in Exercise 10 of Section 4.3) shows that this equals $1/a^2$. Thus a density for Y_1 is

$$f_{Y_1}(y_1) = \frac{1}{(1 + y_1)^2} \quad \text{for } y_1 > 0.$$

Notice that this density has no expected value. ∎

(5.4.5) Example. Let X_1 and X_2 be independent random variables, each having the $N(0, 1)$ distribution (the standard normal distribution). We will find a joint density for the polar coordinates of the pair X_1, X_2. Then, by integrating, we will be able to find the marginal densities.

Remember that the polar coordinates of a point (x_1, x_2) in the plane r and θ (see Figure A.9f in Appendix A) are defined by

$$\begin{aligned} r &= \sqrt{x_1^2 + x_2^2} \quad [=h_1(x_1, x_2)], \\ \theta &= \tan^{-1}(x_2/x_1) \quad [=h_2(x_1, x_2)]; \end{aligned}$$

the inverse of this pair of functions of x_1 and x_2 is

$$\begin{aligned} x_1 &= r\cos\theta \quad [=g_1(r, \theta)], \\ x_2 &= r\sin\theta \quad [=g_2(r, \theta)]. \end{aligned}$$

So we denote the polar coordinates of a pair (X_1, X_2) of random variables by R and Θ (uppercase theta) instead of Y_1 and Y_2. The set of possible values of (R, Θ) is the set of points (r, θ) for which r is nonnegative and $0 \le \theta < 2\pi$. The Jacobian of the inverses is familiar from calculus courses:

$$J(r, \theta) = \begin{vmatrix} \dfrac{\partial x_1}{\partial r} & \dfrac{\partial x_1}{\partial \theta} \\ \dfrac{\partial x_2}{\partial r} & \dfrac{\partial x_2}{\partial \theta} \end{vmatrix} = \begin{vmatrix} \cos\theta & -r\sin\theta \\ \sin\theta & r\cos\theta \end{vmatrix} = r\cos^2\theta + r\sin^2\theta = r.$$

A joint density for X_1 and X_2 is

$$f_X(x_1, x_2) = \frac{1}{2\pi} e^{-(x_1^2+x_2^2)/2} \quad \text{for } -\infty < x_1 < \infty \text{ and } -\infty < x_2 < \infty.$$

When we express the x_i's in terms of r and θ in this density, $x_1^2 + x_2^2$ becomes simply r^2, and so a joint density for R and Θ is

$$f_{R,\Theta}(r,\theta) = \frac{1}{2\pi}e^{-r^2/2} \cdot |r| = \frac{1}{2\pi}re^{-r^2/2} \quad \text{for } r \geq 0 \text{ and } 0 \leq \theta < 2\pi.$$

That ends the application of Theorem (5.4.3). We go on to find the marginal densities; they are not at all difficult:

(5.4.6)
$$f_R(r) \int_0^{2\pi} \frac{1}{2\pi}re^{-r^2/2}d\theta = re^{-r^2/2} \quad \text{for } r \geq 0;$$

and

$$f_\Theta(\theta) = \int_0^\infty \frac{1}{2\pi}re^{-r^2/2}dr = \frac{1}{2\pi}\Big[-e^{-r^2/2}\Big]_0^\infty = \frac{1}{2\pi} \quad \text{for } 0 \leq \theta < 2\pi.$$

We see that the distribution of Θ is the uniform distribution on $[0, 2\pi)$; this is not surprising when we realize that the joint density $f_X(x_1, x_2)$ of X_1 and X_2 involves only $x_1^2 + x_2^2 = r^2$, and not the angle θ of the vector (x_1, x_2). So the joint distribution of X_1 and X_2 is radially symmetric about the origin.

We have seen the density of R in examples and exercises earlier, but now for the first time we see how a random variable with this density could arise—as the distance from the origin of a random vector (X_1, X_2) whose coordinates are independent standard normal random variables.

Notice one other interesting thing. The joint density of R and Θ is the product of the marginal densities; that is, R and Θ are independent.

Although the main purpose for this example is to illustrate the use of Theorem (5.4.3), its results are important enough for us to summarize them:

> If X_1 and X_2 are independent standard normal random variables, then their polar coordinates R and Θ are independent; Θ has the uniform distribution on $[0, 2\pi)$, and R has the density given by (5.4.6). ▮

(5.4.7) Example. Let X_1 and X_2 be independent, each with the density

$$f(x) = \frac{1}{\sqrt{2\pi x}}e^{-x/2} \quad \text{for } x > 0.$$

In Exercise 5 of Section 5.2 we saw that this is the density of the square of a standard normal random variable; in Chapter 8 we will see that it is the chi-squared density with 1 degree of freedom. Now we are interested in the distribution of $X_1/(X_1 + X_2)$.

Define $Y_1 = X_1/(X_1 + X_2)$ and $Y_2 = X_1 + X_2$. The inverses of the functions

$$y_1 = x_1/(x_1 + x_2) \quad \text{and} \quad y_2 = x_1 + x_2$$

are

$$x_1 = y_1 y_2 \quad \text{and} \quad x_2 = y_2(1 - y_1).$$

Because X_1 and X_2 range independently over the positive numbers, we see that Y_1 will be between 0 and 1 and Y_2 will be a positive number, and also that *any* pair (y_1, y_2) with $0 < y_1 < 1$ and $y_2 > 0$ is a possible value of the pair (Y_1, Y_2). The Jacobian of the inverse functions is

$$J(y_1, y_2) = \begin{vmatrix} y_2 & y_1 \\ -y_2 & 1 - y_1 \end{vmatrix} = y_2.$$

A joint density for X_1 and X_2 is

$$f_X(x_1, x_2) = \frac{1}{2\pi\sqrt{x_1 x_2}} e^{-(x_1+x_2)/2} \quad \text{for } x_1 > 0 \text{ and } x_2 > 0,$$

and so by Theorem (5.4.3) a joint density for Y_1 and Y_2 is

$$f_Y(y_1, y_2) = \frac{1}{2\pi\sqrt{y_1 y_2^2(1 - y_1)}} e^{-y_2/2} \cdot |y_2|;$$

that is,

(5.4.8)
$$f_Y(y_1, y_2) = \frac{1}{2\pi\sqrt{y_1(1 - y_1)}} e^{-y_2/2} \quad \text{for } 0 < y_1 < 1 \text{ and } y_2 > 0.$$

To find the marginal density of Y_1 we integrate with respect to y_2:

$$f_{Y_1}(y_1) = \frac{1}{\pi\sqrt{y_1(1 - y_1)}} \int_0^\infty \tfrac{1}{2} e^{-y_2/2} dy_2.$$

Notice that we moved everything involving y_1 outside the integral, as well as the constant $1/\pi$, but left the constant $\frac{1}{2}$ inside. We did that because the resulting integrand is precisely the exponential density with parameter $\lambda = \frac{1}{2}$, and so the integral is 1. Thus

$$f_{Y_1}(y_1) = \frac{1}{\pi\sqrt{y_1(1 - y_1)}} \quad \text{for } 0 < y_1 < 1.$$

This is a beta density. The general beta density is a constant times $x^{\alpha-1}(1 - x)^{\beta-1}$ for $0 < x < 1$, and this one has $\alpha = \beta = \frac{1}{2}$. The symbols α and β are lowercase Greek alpha and beta.) We will discuss the beta family in Chapter 8.

Finally, we find the distribution of $Y_2 = X_1 + X_2$. The result may be a little surprising. Finding it also reveals a useful fact about joint densities, which we will take up in (5.4.9). We find a density for Y_2 by integrating with respect to y_1:

$$f_{Y_2}(y_2) = \tfrac{1}{2} e^{-y_2/2} \int_0^\infty \frac{1}{\pi\sqrt{y_1(1 - y_1)}} dy_1.$$

We moved everything involving y_2 outside the integral. We moved the $\frac{1}{2}$ outside and left the $1/\pi$ inside because that makes the integrand a density. (We just learned that it was a density, in fact, when we saw that it is a density for Y_1.) So the integral is 1, and

$$f_{Y_2}(y_2) = \tfrac{1}{2}e^{-y_2/2} \quad \text{for } y_2 > 0.$$

That is, $Y_2 = X_1 + X_2$ has the exponential distribution with $\lambda = \frac{1}{2}$. ∎

(5.4.9) **When a joint density factors.** Let us switch notation for simplicity and refer to Y and Z instead of Y_1 and Y_2. We know what happens when the joint density of Y and Z factors into the product of the marginal densities—that is, when $f_{Y,Z}(y, z) = f_Y(y)f_Z(z)$. The answer is that Y and Z are independent.

But what if we know only that $f_{Y,Z}(y, z)$ factors as *some* function of y times *some* function of z, but we do not know that these functions are the densities? For example, (5.4.8) is the same as

$$f(y, z) = \frac{1}{2\pi\sqrt{y(1-y)}}e^{-z/2} \quad \text{for } 0 < y < 1 \text{ and } z > 0,$$

which clearly factors as a function of y times a function of z. But it factors in many different ways, depending on how we arrange the constants:

$$\begin{aligned} f(y, z) &= \frac{1}{2\pi\sqrt{y(1-y)}} \cdot e^{-z/2} = \frac{1}{\pi\sqrt{y(1-y)}} \cdot \frac{1}{2}e^{-z/2} \\ &= \frac{3}{\sqrt{y(1-y)}} \cdot \frac{1}{6\pi}e^{-z/2}, \quad \text{etc.} \end{aligned}$$

When we integrated we found out that the second of these factorizations shows the densities—that is, $1/\pi\sqrt{y(1-y)}$ is a density for Y and $\frac{1}{2}e^{-z/2}$ is a density for Z—and, consequently, Y and Z are independent.

The question is: if the joint density of Y and Z factors into a function of Y times a function of Z, then is there always a way to arrange the constants so that one factor is a density for Y and the other is a density for Z? If so, then just knowing that the density factors will be enough to show that Y and Z are independent.

The answer is yes. Suppose $f_{Y,Z}(y, z) = r(y) \cdot s(z)$ for some functions r and s. Then $r(y)$ and $s(z)$ are not necessarily the densities of Y and Z, but some constant times $r(y)$ is a density for Y, and some constant times $s(z)$ is a density for Z.

This is easy to see by integrating. A density for Y is found by integrating with respect to z:

$$f_Y(y) = \int_{-\infty}^{\infty} f_{Y,Z}(y, z)dz = \int_{-\infty}^{\infty} r(y) \cdot s(z)dz = r(y)\int_{-\infty}^{\infty} s(z)dz.$$

Now this does not look like much progress, but the integral $\int_{-\infty}^{\infty} s(z)dz$ is just a constant, and so $f_Y(y)$ is this constant times $r(y)$. A similar argument shows that $f_Z(z)$ is some constant times $s(z)$. Furthermore, the product of these two correcting constants must equal 1, and therefore the product of the two densities will equal $r(y) \cdot s(z)$.

In general, if the joint density $f_{Y,Z}(y, z)$ factors as $r(y) \cdot s(z)$, look at the nonconstant parts of the functions $r(y)$ and $s(z)$; they are the nonconstant parts of the densities of Y and Z. If you recognize one or both of them as the nonconstant part of some familiar density, then just use the correct constant along with that density, and adjust the constant of the other density so that the product is the same as the original product.

Had we known this before the previous example, we could have proceeded from (5.4.8) as follows. The joint density factors into

$$r(y_1) = \frac{1}{2\pi\sqrt{y_1(1 - y_1)}} \quad \text{for } 0 < y_1 < 1$$

and

$$s(y_2) = e^{-y_2/2} \quad \text{for } y_2 > 0.$$

Now look at $s(y_2)$. We recognize it as the nonconstant part of the exponential density with $\lambda = \frac{1}{2}$. We know that the constant multiplying $e^{-\lambda y}$ in an exponential density is λ. So a density for Y_2 is $\frac{1}{2}s(y_2)$, the exponential density, and this tells us that a density for Y_1 must be $2r(y_1)$, so that the two correcting constants will multiply to 1. This agrees with the results we got by integrating in the example. ∎

(5.4.10) **Bivariate normal densities.** Here we give a general class of joint densities, the *bivariate normal* densities, for which the marginal densities are normal. We have looked at a subclass of these twice before, in (3.5.14) and (4.4.18); now we are able to work with the full class.

Given any two real numbers μ_1 and μ_2, any two positive numbers σ_1^2 and σ_2^2, and a number ρ in the interval $(-1, 1)$, there is a bivariate normal density so that, if X_1 and X_2 have that joint density, then

$$EX_1 = \mu_1, \quad EX_2 = \mu_2, \quad VX_1 = \sigma_1^2, \quad VX = \sigma_2^2, \quad \text{and} \quad \text{Cov}(X, Y) = \rho.$$

The density is

(5.4.11)
$$f(x_1, x_2) = \frac{1}{2\pi\sigma_1\sigma_2\sqrt{1 - \rho^2}} \cdot \exp\left\{-\frac{1}{2(1 - \rho^2)}\left[\left(\frac{x_1 - \mu_1}{\sigma_1}\right)^2 + \left(\frac{x_2 - \mu_2}{\sigma_2}\right)^2 - 2\rho\frac{(x_1 - \mu_1)(x_2 - \mu_2)}{\sigma_1\sigma_2}\right]\right\},$$

where $\exp\{a\}$ means e^a. This looks frightening, but we can notice two things about it.

First, regardless of the μ's and σ's, X_1 and X_2 are independent if $\rho = 0$, because the joint density then factors into a function of x_1 times a function of x_2. Second, when the expected values μ_1 and μ_2 equal 0 and the variances equal 1, (5.4.11) reduces to the joint density discussed in (3.5.14) and (4.4.18). That joint density is

(5.4.12)
$$f(z_1, z_2) = \frac{1}{2\pi\sqrt{1 - \rho^2}} e^{-(z_1^2 + z_2^2 - 2\rho z_1 z_2)/2(1-\rho^2)}.$$

Remember also that if Z_1 and Z_2 have this joint density, then they each have the standard normal distribution, and their correlation coefficient is ρ. The density in (5.4.12) is the most general bivariate normal density for standard normal random variables. (We have switched from x's to z's because this is customary with standard normal random variables.)

Notice also that just as a single normal distribution is completely determined once we know two parameters—its expected value and variance—so a bivariate normal density is completely determined by five parameters—the two μ's, the two σ^2's, and ρ. But there are two things we have not proved yet:

a. These five numbers actually are the expected values, variances, and correlation coefficient of X_1 and X_2.
b. The marginal densities of X_1 and X_2 are actually normal densities.

We can prove these things using what is summarized above about random variables with the joint density (5.4.12), as follows.

Suppose X_1 and X_2 have the bivariate normal joint density (5.4.11). Let

(5.4.13)
$$Z_1 = \frac{X_1 - \mu_1}{\sigma_1} \quad \text{and} \quad Z_2 = \frac{X_2 - \mu_2}{\sigma_2}.$$

These are the standardizations of X_1 and X_2. It is a relatively easy exercise, an application of the multivariate change-of-variable theorem, to show that the joint density of Z_1 and Z_2 is (5.4.12). Consequently, $Z_1 \sim N(0, 1)$, $Z_2 \sim N(0, 1)$, and $\rho(Z_1, Z_2) = \rho$.

But then it follows immediately from Exercise 8 in Section 5.2, or equivalently from (5.3.11), that $X_1 \sim N(\mu_1, \sigma_1^2)$ and $X \sim N(\mu_2, \sigma_2^2)$. Furthermore, it follows from (4.4.15) that the correlation coefficient of Z_1 and Z_2 is the same as that of X_1 and X_2 (in general, the correlation coefficient of $aX + b$ and $cY + d$ is the same as that of X and Y). Since we know that the correlation coefficient of Z_1 and Z_2 is ρ, it follows that ρ is the correlation coefficient of X_1 and X_2 also. Thus (a) above is proved.

Also, it follows from what we did in (3.5.14) that the marginal densities of Z_1 and Z_2 are standard normal, and thence that the densities of X_1 and X_2 are $N(\mu_1, \sigma_1^2)$ and $N(\mu_2, \sigma_2^2)$, respectively. Thus (b) is proved.

It can be checked also that the above result goes in the other direction as well. That is, if Z_1 and Z_2 have the joint density (5.4.12)—so that they are standard normal and their correlation coefficient is ρ—then the random variables

$$X_1 = \sigma_1 X_1 + \mu_1 \quad \text{and} \quad X_2 = \sigma_2 X_2 + \mu_2$$

have the general bivariate normal density (5.4.11).

EXERCISES FOR **SECTION 5.4**

1. Let X_1 and X_2 be independent, each having the exponential distribution with parameter λ. Define $Y_1 = X_1/(X_1 + X_2)$ and $Y_2 = X_1 + X_2$. The main point of this exercise is to find the distribution of Y_1, but we can learn other things as well.
   a. Write down the set of possible values of the pair (Y_1, Y_2).
   b. Find a joint density for Y_1 and Y_2.
   c. Find the marginal densities for Y_1 and Y_2. What is the name of the distribution of Y_1? [Notice that we found the distribution of Y_2 in Example (5.2.15).]
   d. Are Y_1 and Y_2 independent?

2. Let X_1 and X_2 be independent, each having the standard normal distribution. Let $Y_1 = X_1 + X_2$ and $Y_2 = X_1 - X_2$.
   a. Find a joint density for Y_1 and Y_2.
   b. Find the marginal densities of Y_1 and Y_2. Identify the distributions by name.
   c. Are Y_1 and Y_2 independent?

3. Let X_1 and X_2 be independent, each with an exponential distribution. Let X_1 have parameter λ and X_2 have parameter μ. Find a density for $\lambda X_1 + \mu X_2$. [*Hint:* Let this equal Y_1 and let $Y_2 = \mu X_2$.]

4. Redo Exercise 3 without using the multivariate change-of-variable theorem, in the following steps.
   a. Use either the one-dimensional change-of-variable theorem or the CDF method to find densities for $Z_1 = \lambda X_1$ and $Z_2 = \mu X_2$.
   b. Use the convolution theorem (at the end of Section 5.2) to find a density for $Z_1 + Z_2$.

5. In Exercise 17 of Section 5.2, X and Y are independent, each with the density $f(x) = \lambda^2 x e^{-\lambda x}$ for $x > 0$. We found a density for $W = X + Y$ in two ways: the CDF method and the convolution theorem. Do it again using the multivariate change-of-variable theorem.

6. **The density of a ratio.** Let X and Y be independent random variables with densities $f_X(x)$ and $f_Y(y)$, and suppose $Y > 0$ with probability 1. If $Z = X/Y$, show that a density for Z is

$$f_Z(z) = \int_0^\infty x f_X(zx) f_Y(x)\,dx.$$

7. **The Box–Muller transformation for generating normally distributed random variables.** Let X_1 and X_2 be independent, each with the uniform distribution on (0, 1). Define

$$Z_1 = \cos(2\pi X_1)\sqrt{-2 \ln X_2} \quad \text{and} \quad Z_2 = \sin(2\pi X_1)\sqrt{-2 \ln X_2}.$$

Show that Z_1 and Z_2 are independent, each with the standard normal distribution.

This remarkable observation provides a way to generate independent standard normal random variables, two at a time. They are not easy to generate one at a time from uniform random variables, and so the Box–Muller transformation is more than just a curiosity. For more on generating normal and other random variables, see Sheldon Ross's book *A Course in Simulation.*

ANSWERS

1. **a.** $0 < y_1 < 1$, $y_2 > 0$ **b.** $f_Y(y_1, y_2) = \lambda^2 y_2 e^{-\lambda y_2}$ for y_1 and y_2 as given in part a. **c.** $f_{Y_1}(y_1) = 1$ for $0 < y_1 < 1$ [uniform on (0, 1)]; $f_{Y_2}(y_2) = \lambda^2 y_2 e^{-\lambda y_2}$ for $y_2 > 0$ **d.** Yes

2. **a.** $f_Y(y_1, y_2) = \dfrac{1}{4\pi} e^{-(y_1^2+y_2^2)/4}$ $(y_1 \in \mathbb{R}, y_2 \in \mathbb{R})$ **b.** Each is $\dfrac{1}{\sqrt{4\pi}} e^{-y^2/4}$ $(y \in \mathbb{R})$, i.e., $N(0, 2)$. **c.** Yes

3. $f_{Y_1}(y_1) = y_1 e^{-y_1}$ for $y_1 > 0$

SUPPLEMENTARY EXERCISES FOR **CHAPTER 5**

1. Suppose X has the density $f(x) = 2/x^3$ for $x > 1$ (and 0 for $x \leq 1$). Find a density for $Y = X^2 - 1$, in three ways.
   a. Use the CDF method without the "shortcut."
   b. Use the CDF method with the "shortcut."
   c. Use the change-of-variable theorem.

2. Find a density (again, in three ways) for the square root of the absolute value of a standard normal random variable.

3. Suppose $X \sim \exp(\lambda)$ and $Y = aX + b$, where a is positive. Find a density for Y (three ways).

4. Suppose $X \sim \text{unif}(0, 1)$ and $Y = \cos \pi X$. Find a density for Y.

5. Let X have the density $\frac{2}{3}x$ for $1 < x < 2$ (and 0 otherwise). Find a density for $Y = 1 + (X - 1)^2$.

6. Let X and Y be independent, each having the uniform distribution on (0, 1). Find a density for $Z = XY$ in two different ways.

a. Use the CDF method (find the probability of the subset of the unit square consisting of the points below the curve $xy = z$).
b. Use the multivariate change-of-variable theorem, letting $Z_1 = XY$ and $Z_2 = Y$. [Be careful about the set of possible values of the pair (Z_1, Z_2).]

7. Let X and Y be independent, with densities $f_X(x) = \lambda^2 x e^{-\lambda x}$ for $x > 0$ and $f_Y(y) = (\lambda^3/2)y^2 e^{-\lambda y}$ for $y > 0$. Find a density for $W = X + Y$ in three different ways.
a. Use the CDF method.
b. Use the convolution theorem.
c. Use the multivariate change-of-variable theorem.

8. Find a density for the reciprocal of a random variable that has the uniform distribution on $(0, 1)$.

9. Let X have the density $\lambda^2 x e^{-\lambda x}$ for $x > 0$. Find a density for $Y = \sqrt{X}$.

10. Find a density for $Y = X^2$ if X has the standard Cauchy density $1/\pi(1 + x^2)$ $(-\infty < x < \infty)$.

11. Let X have the exponential distribution with parameter λ; recall that a density is $f_X(x) = \lambda e^{-\lambda x}$ for $x > 0$. Find a density for $Y = 1/\sqrt{X}$.

12. Let X have the standard normal distribution. Find a density for $Y = X^4$.

ANSWERS

1. $f_Y(y) = 1/(1 + y)^2$ for $y > 0$ (and 0 for $y \leq 0$)

2. $f(y) = \dfrac{4y}{\sqrt{2\pi}} e^{-y^4/2}$ for $y > 0$ (and 0 for $y \leq 0$)

3. $f(y) = (\lambda/a)e^{-(\lambda/a)(y-b)}$ for $y > b$ (and 0 for $y \leq b$)

4. $f(y) = \dfrac{1}{\pi\sqrt{1 - y^2}}$ for $-1 < y < 1$ (and 0 otherwise)

5. $f(y) = \dfrac{1 + \sqrt{y - 1}}{3\sqrt{y - 1}}$ for $1 < y < 2$ (and 0 otherwise)

6. $f(z) = -\ln z$ for $0 < z < 1$

7. $\dfrac{\lambda^5}{24} w^4 e^{-\lambda w}$ for $w > 0$

8. $f_Y(y) = 1/y^2$ for $y > 1$

9. $f(y) = 2\lambda^2 y^3 e^{-\lambda y^2}$ for $y > 0$

10. $f(y) = \dfrac{1}{\pi\sqrt{y}(1 + y)}$ for $y > 0$

11. $f(y) = \dfrac{2\lambda}{y^3} e^{-\lambda/y^2}$ for $y > 0$

12. $f(y) = \dfrac{1}{2y^{3/4}\sqrt{2\pi}} e^{-\sqrt{y}/2}$ for $y > 0$

CHAPTER

Normal Distributions and the Central Limit Theorem

6.1 Sums and Averages of i.i.d. Random Variables

6.2 Normal Distributions

6.3 The Central Limit Theorem

6.4 The Poisson and Normal Approximations for Binomial Distributions

Supplementary Exercises

6.1 SUMS AND AVERAGES OF I.I.D. RANDOM VARIABLES

Recall the laws of large numbers that we discussed in Section 4.5. They deal with a sequence $X_1, X_2, X_3, \ldots$ of i.i.d. (independent, identically distributed) random variables, which we think of as repeated independent observations of some random variable X. The random variable X_j represents the value of X observed on the jth repetition of the experiment. For any positive integer n, the random variable $\overline{X}_n = (X_1 + X_2 + \cdots + X_n)/n$ is the average of the first n observed values of X.

Laws of large numbers are theorems which say that under certain conditions, the random variable $\overline{X}_n$ has a high probability of being close to the number EX if n is large. The strongest of the laws of large numbers, Kolmogorov's, says that if EX exists then (with probability 1) the limit of the sequence $\overline{X}_1, \overline{X}_2, \ldots, \overline{X}_n, \ldots$ equals EX; and if EX does not exist then that sequence has no limit.

The Central Limit Theorem involves an additional assumption—namely, that VX exists as well as EX—and it has a stronger conclusion: *If n is large, then $\overline{X}_n$ has approximately a normal distribution.* Thus not only do we know that $\overline{X}_n$ is likely to be close to EX; we can also find approximate probabilities for X_n. The theorem as proved actually says that a certain linear function of the average, $Z_n = (\overline{X}_n - E\overline{X}_n)/\sqrt{V\overline{X}_n}$, has approximately the *standard* normal distribution. But since linear functions of normal random variables are also normal, it follows that $\overline{X}_n$ itself is approximately normally distributed. So is the sum $S_n = X_1 + X_2 + \cdots + X_n$, which is a linear function of $\overline{X}_n$ and of Z_n.

This theorem is of enormous importance for the practical applications of probability and statistical theory. It is important because it implies that even if the distribution of X is unknown, we can still find approximate probabilities for sums and averages of large numbers of observations of X, as long as we can assume that EX and VX are finite.

In this section we will look at sums and averages of i.i.d. random variables, and we will notice what some of our previous results on expected values and variances say about them. In Section 6.2 we will study the normal family of distributions. In Section 6.3, we will state the Central Limit Theorem—not proving it but giving examples of its use. We will discuss its proof later, in Chapter 8, after we have studied a powerful tool (generating functions) for finding the distributions of certain functions of random variables. Finally, Section 6.4 deals with an important special case of the Central Limit Theorem that gives a useful approximation for binomial probabilities.

We begin with a sequence $X_1, X_2, X_3, \ldots$ of i.i.d. random variables, representing repeated independent observations of a random variable X. For each positive integer n, the random variables $X_1, X_2, \ldots, X_n$ represent the values of X that are observed on the first n trials. Together, these n random variables are called a ***sample of size n*** from the distribution.

For any positive integer n, we define the following random variables associated with the sample $X_1, X_2, \ldots, X_n$:

(6.1.1) $$\text{the } \textbf{\textit{sample sum}}, \quad S_n = X_1 + X_2 + \cdots + X_n,$$

and

(6.1.2) $$\text{the } \textbf{\textit{sample mean}} \text{ or } \textbf{\textit{sample average}}, \quad \overline{X}_n = \frac{S_n}{n} = \frac{X_1 + X_2 + \cdots + X_n}{n}.$$

In addition, suppose the distribution has a finite expected value, $EX = \mu$, and a finite variance, $VX = \sigma^2$. Then it is easy to see that

(6.1.3) $$ES_n = n\mu \quad \text{and} \quad VS_n = n\sigma^2,$$

because the expected value of any sum (of finitely many random variables) is the sum of the expected values, and the same holds for the variance if the random variables are independent. In addition, for $\overline{X}_n$ we have

(6.1.4) $$E\overline{X}_n = \mu \quad \text{and} \quad V\overline{X}_n = \frac{\sigma^2}{n},$$

because for any random variable Y the expected value of aY is a aEY, and the variance is a^2VY. Recall that (6.1.4) is what makes Chebyshev's Law of Large Numbers work; if n is large, then $V\overline{X}_n$ is small, so $\overline{X}_n$ is likely to be close to μ.

The above is all valid regardless of the distribution of X that we start with—except, of course, that we cannot proceed beyond (6.1.2) unless X has finite expected value and variance. But nothing has been said about the *distributions* of S_n and $\overline{X}_n$. These do depend on the original distribution of X. As we have said, the Central Limit Theorem says that they all have *approximate* normal distributions, but we cannot say anything about their actual distributions unless we know something about the distribution of X.

In Chapter 8 we will learn how generating functions can be used in many cases to find the actual, not just the approximate, distribution of S_n and $\overline{X}_n$, given the distribution of X.

We have already studied the distribution of a sum S_n of random variables in a few special cases, mostly in Chapter 5. In these cases we knew the distribution of the X_j's and used the methods of Chapter 5 to find the distribution of S_n. In Example (5.2.15), for instance, we used the CDF method to find a density for the sum of two independent random variables, each having the exponential distribution with parameter λ. We did the same thing with the convolution theorem in Example (5.2.21). And in Exercises 1, 2, and 5 of Section 5.4 we found densities for sums using the multivariate change-of-variable theorem.

There is one other case (so far) in which we know the distribution of a sum of i.i.d. random variables—the case in which the X_j's have a Bernoulli distribution. If X_j equals 1 with probability p and 0 with probability $q = 1 - p$, then X_j can be thought of as the number of successes on the jth Bernoulli trial; it is the indicator function of the event "success on trial j." See Example (4.2.17) and (4.2.18) for a discussion of this. Accordingly, S_n is the number of successes in the first n trials, and we know that S_n has the binomial distribution with parameters n and p.

But this chapter is concerned with what we can say about the distributions of S_n and $\overline{X}_n$ even if we do *not* know the distributions of the X_j's. What we can say is that they have approximate normal distributions as long as the X_j's have finite variance—that is the crux of the Central Limit Theorem. That is important, because if you know something has a normal distribution and you know its expected value and variance, then you can find probabilities for it, as we will see in the next section.

EXERCISES FOR SECTION 6.1

[a]**1.** In each part let $X_1, X_2, \ldots, X_n$ be a sample from the distribution described. Find the expected value and variance of S_n and $\overline{X}_n$. [Very little calculation is necessary, as long as you know the expected value and variance of one of the X_j's. You can look these up in the examples or exercises indicated.]

a. The exponential distribution with parameter λ [See Example (4.3.6).]
b. The uniform distribution on (0, 1) [See Example (4.3.8).]
c. The geometric distribution with parameter p—that is, the distribution of the number of failures before the first success in Bernoulli trials [See Exercise 9 of Section 4.2 and Exercise 19 of Section 4.3.]
d. The $N(b, a^2)$ distribution [See (5.3.14).]
e. The Poisson distribution with parameter λ [See Exercise 6 of Section 4.3.]
f. The binomial distribution with parameters k and p [See Example 4.4.12).]
g. The distribution with density $f(x) = 2/x^3$ for $x > 1$ [Find EX and VX yourself.]

2. If $X_1, X_2, \ldots, X_n$ are a sample from a distribution whose expected value is 3 and whose variance is 2.4, how large must n be in order that the variance of $\overline{X}_n$ be less than .01?

3. If $EX = \mu$ and $VX = \sigma^2$, find two different linear functions of X, of the form $Z = cX + d$, such that $EZ = 0$ and $VZ = 1$.

4. The vehicles that cross a certain bridge are found to have weights whose average is 4,675 pounds and whose standard deviation is 345 pounds. If there are 40 vehicles on the bridge at one time, what are the expected value and standard deviation of the total weight?

5. Scores on a certain standardized test average 500, with a standard deviation of 100. If we take a random sample of 200 of these scores and average them, what are the expected value and standard deviation of our average?

ANSWERS

1.

	ES_n	VS_n	$E\overline{X}_n$	$V\overline{X}_n$
a.	n/λ	n/λ^2	$1/\lambda$	$1/n\lambda^2$
b.	$n/2$	$n/12$	$1/2$	$1/12n$
c.	nq/p	nq/p^2	q/p	q/np^2
d.	nb	na^2	b	a^2/n
e.	$n\lambda$	$n\lambda$	λ	λ/n
f.	nkp	$nkpq$	kp	kpq/n
g.	$2n$	DNE	2	DNE

2. At least 241.
3. $c = \pm 1/\sigma$; $d = -\mu/\sigma$
4. Expected value 187,000; standard deviation 2,181.97
5. Expected value 500; standard deviation 7.0711

6.2 NORMAL DISTRIBUTIONS

We have been working with normal distributions since Section 1.3. In this section we summarize, in (6.2.1)–(6.2.9), the basic facts needed to work with the family of one-dimensional normal distributions.

If you did Exercise 8 in Section 5.2, or if you covered the material in Example (5.3.11) and (5.3.14), this will be familiar. If not, it might be a good idea to do Exercise 8 in Section 5.2 now. We will also use the theorem of Section 5.4 in the proof (6.2.11) of one of the main results; but if you did not cover Section 5.4, you can omit (6.2.11); a simpler proof will appear later, in (8.2.9).

Let μ be any number (positive, negative, or zero) and let σ^2 be any positive number. We say that X has the ***normal distribution with parameters μ and σ^2***, or the $N(\mu, \sigma^2)$ distribution for short, if a density for X is

(6.2.1)
$$f(x) = \frac{1}{\sqrt{2\pi\sigma^2}} e^{-(x-\mu)^2/2\sigma^2} \quad \text{for } -\infty < x < \infty.$$

It is customary in probability to use the symbol $\sim$ to indicate the distribution of a random variable; so we can abbreviate the phrase "X has the $N(\mu, \sigma^2)$ distribution" by writing just "$X \sim N(\mu, \sigma^2)$."

The expected value and variance of the $N(\mu, \sigma^2)$ distribution are

(6.2.2)
$$EX = \mu \quad \text{and} \quad VX = \sigma^2,$$

as we saw in (5.3.14). This is important since these two parameters, which identify which normal distribution we are talking about, are precisely the expected value and the variance of the distribution. If you know that X has *some* normal distribution, then you know exactly *which* normal distribution if you know EX and VX.

The positive square root of σ^2, which is obviously denoted by σ, is the standard deviation of X. It may seem strange that we think of σ^2 as the parameter rather than σ. The reason is that some of the facts about normal distributions are more easily stated in terms of variances than standard deviations. (For example, the variance of a sum is the sum of the variances plus twice the covariances; to express this using standard deviations is not so simple.) You may think, consequently, that it would be simpler to denote the variance by some single letter, say $\alpha = VX$, and just remember that α is always a positive parameter. The reason we do not is that there are many formulas that involve the standard deviation rather than the variance; we would have to use $\sqrt{\alpha}$ in these, making them more complicated.

The things to remember about the symbol σ^2 and the abbreviation $N(\mu, \sigma^2)$ are the following: (i) The second parameter of a normal distribution is the variance, not the standard deviation. If $X \sim N(3, 5)$, for example, then $VX = 5$, and the standard deviation of X is $\sqrt{5}$. (ii) The standard deviation σ is the *positive* square root of the parameter σ^2.

The special case $\mu = 0$ and $\sigma^2 = 1$ is of particular importance: The ***standard normal***, or $N(0, 1)$, distribution is the one whose density is

(6.2.3)
$$\varphi(z) = \frac{1}{\sqrt{2\pi}} e^{-z^2/2} \quad \text{for } -\infty < z < \infty.$$

If Z has this distribution, then $EZ = 0$ and $VZ = 1$. It is customary to denote a standard normal random variable by Z, and this is why we use the letter z for the variable in the density function.

The next important fact about normal distributions is one that we saw in Exercise 8 of Section 5.2:

(6.2.4)
If $X \sim N(\mu, \sigma^2)$ and $Y = aX + b$, then $Y \sim N(a\mu + b, a^2\sigma^2)$.

We got this by using the change-of-variable theorem, although it could have been done just as easily with the ordinary CDF method of Section 5.2. In Chapter 8 we will see an even simpler way to obtain it using generating functions.

Two particular cases of (6.2.4) are important:

(6.2.5)
If $Z \sim N(0, 1)$ and $X = \sigma Z + \mu$, then $X \sim N(\mu, \sigma^2)$.

(6.2.6)
If $X \sim N(\mu, \sigma^2)$ and $Z = \dfrac{X - \mu}{\sigma}$, then $Z \sim N(0, 1)$.

(Even though σ refers to the positive square root of σ^2, it is worth noting that these two statements are valid for any nonzero number σ, positive or negative.)

Notice the relationship of Z to X in (6.2.6); we can think of Z as being the deviation of X from its expected value μ, expressed in multiples of its standard deviation σ. What (6.2.6) says is that any normal random variable becomes standard normal when we express it in this way. The transformation $(X - \mu)/\sigma$ is called the ***standardization*** of X.

The last of our basic facts about normal densities is one we have not seen yet.

(6.2.7) If X_1 and X_2 are independent and $X_1 \sim N(\mu_1, \sigma_1^2)$ and $X_2 \sim N(\mu_2, \sigma_2^2)$, then

$$X_1 + X_2 \sim N(\mu_1 + \mu_2, \sigma_1^2 + \sigma_2^2).$$

It comes as no surprise that $E(X_1 + X_2) = \mu_1 + \mu_2$ and $V(X_1 + X_2) = \sigma_1^2 + \sigma_2^2$. The former is true of any pair of random variables, and the latter is true of any pair of independent random variables. What is new here is that $X_1 + X_2$ has a normal distribution. We will prove this in (6.2.11) below.

As a corollary of (6.2.7), we can prove by induction that if $X_1, X_2, \ldots, X_n$ are independent, each having the $N(\mu, \sigma^2)$ distribution, then

(6.2.8) $S_n = X_1 + X_2 + \cdots + X_n$ has the $N(n\mu, n\sigma^2)$ distribution.

From (6.2.8) it follows, because of (6.2.4), that

(6.2.9) $\overline{X}_n = \frac{1}{n} S_n$ has the $N\left(\mu, \frac{\sigma^2}{n}\right)$ distribution.

(6.2.10) Finding normal probabilities. Tables such as the one in Appendix C give the values of the function $\Phi(z)$, the CDF of the standard normal distribution. If Z is a random variable with this distribution, then for any a and b, $P(Z \in (a, b]) = \Phi(b) - \Phi(a)$. Because normal distributions are absolutely continuous, this is also the probability of the other intervals with the same endpoints, (a, b), $[a, b]$, and $[a, b)$.

If X has a normal distribution $N(\mu, \sigma^2)$ other than the standard normal, then we can find probabilities for X using (6.2.5), as follows:

$$\begin{aligned} P(a < X \leq b) &= P\left(\frac{a-\mu}{\sigma} < \frac{X-\mu}{\sigma} \leq \frac{b-\mu}{\sigma}\right) \\ &= P\left(\frac{a-\mu}{\sigma} < Z \leq \frac{b-\mu}{\sigma}\right) \qquad \text{where } Z = \frac{X-\mu}{\sigma} \sim N(0, 1) \\ &= \Phi\left(\frac{b-\mu}{\sigma}\right) - \Phi\left(\frac{a-\mu}{\sigma}\right). \end{aligned}$$

Example a. Suppose that the heights of adult males are normally distributed with expected value $\mu = 70.3$ inches and variance $\sigma^2 = 9.5$. What fraction of the adult males are between 72 and 75 inches tall?

Solution. This is the same as asking for $P(72 < X < 75)$, where X is the height of a randomly chosen adult male. We find it as indicated above:

$$\begin{aligned} P(72 < X < 75) &= P\left(\frac{72 - 70.3}{\sqrt{9.5}} < \frac{X - 70.3}{\sqrt{9.5}} < \frac{75 - 70.3}{\sqrt{9.5}}\right) \\ &\doteq P(.55 < Z < 1.52) \qquad \text{where } Z = \frac{X - 70.3}{\sqrt{9.5}} \sim N(0, 1) \\ &= \Phi(1.52) - \Phi(.55) \doteq .9357 - .7088 = .2269. \end{aligned}$$

Approximately 23% of the population are between 6 feet and 6 feet 3 inches tall (assuming that 70.3 and 9.5 are the correct expected value and variance of heights of adult males in the population under study). ∎

Notice the dots over two of the equality signs in this example, indicating that the equality is not exact because of rounding. In the first one, the numbers .55 and 1.52 have been rounded to two decimal places, because we cannot use the normal table exactly with numbers more accurate than that. In the second, the dot indicates that the normal table gives numbers only to four decimal places of accuracy.

It is a mistake to think that the final answer, .2269, is accurate to four decimal places, because we entered the table with numbers that were accurate to only two decimal places. If more accuracy were needed, we would have to use more decimal places in the two ratios .55 and 1.52, and then use either interpolation in the table, a more accurate table, or some other means of computing $\Phi(z)$. Computer approximations for $\Phi(z)$ are available in most mathematical software libraries.

Using such an approximation in this problem shows that the true probability, to four decimal places, is .2270, and so we have not made much of an error. In most applications, probabilities are needed to only two or three significant digits, and so we will not pursue further the question of achieving more accuracy than we got in this example. That is, it is satisfactory for most purposes to round off numbers to two decimal places before entering the normal table, as long as it is recognized that the probabilities obtained from the table are accurate to only two or three decimal places.

Example b. Scores on a certain standardized test are assumed to be normally distributed with $\mu = 100$ and $\sigma^2 = 100$. What fraction of the scores are over 120?

Solution. This is $P(X > 120)$ where $X \sim N(100, 10^2)$, and it equals

$$\begin{aligned} P(X > 120) &= P\left(\frac{X - 100}{10} > \frac{120 - 100}{10}\right) \\ &= P(Z > 2) \qquad \text{where } Z = \frac{X - 100}{10} \sim N(0, 1) \\ &= 1 - P(Z \le 2) = 1 - \Phi(2) \doteq 1 - .9772 = .0228. \end{aligned}$$ ∎

Example c. In a normal population of numbers, what fraction of the numbers are within 3 standard deviations of the mean?

Solution. This is a disguised version of the question: If $X \sim N(\mu, \sigma^2)$, what is $P(\mu - 3\sigma < X < \mu + 3\sigma)$? The answer is

$$\begin{aligned} &P\left(\frac{\mu - 3\sigma - \mu}{\sigma} < \frac{X - \mu}{\sigma} < \frac{\mu + 3\sigma - \mu}{\sigma}\right) \\ &= P(-3 < Z < 3) \qquad \text{where } Z = \frac{X - \mu}{\sigma} \sim N(0, 1) \\ &= 2 \cdot P(0 < Z < 3) \qquad \text{by symmetry: } P(-3 < Z < 0) = P(0 < Z < 3) \\ &= 2 \cdot (\Phi(3) - \Phi(0)) \doteq 2 \cdot (.9987 - .5000) = .9974. \end{aligned}$$

Over 99% of observations from any normal population are within 3 standard deviations of the mean. That is, a normally distributed random variable has better than a 99% chance of falling within 3σ of μ.

In the previous calculation, the step from $P(-3 < Z < 3)$ to $2 \cdot P(0 < Z < 3)$ is valid because the standard normal distribution is symmetric about $z = 0$, and consequently $P(-3 < Z < 0) = P(0 < Z < 3)$. It could have been done alternatively as follows:

$$\begin{aligned} P(-3 < Z < 3) & \\ &= P(Z < 3) - P(Z < -3) \\ &= P(Z < 3) - P(Z > 3) \qquad [P(Z < -3) = P(Z > 3) \text{ by symmetry}] \\ &= P(Z < 3) - (1 - P(Z \leq 3)) \\ &= 2 \cdot P(Z \leq 3) - 1 \doteq 2 \cdot .9987 - 1 = .9974. \end{aligned}$$

∎

Example d. The weights of a certain type of laboratory animal are normally distributed with $\mu = 185$ grams and $\sigma = 16$ grams. What is the number w such that 80% of the animals weigh less than w? This number is called the 80th *percentile* of the weights of the population.

Solution. Let X be the weight of a randomly chosen animal from this population. We want the w such that $P(X \leq w) = .80$. The way to find this is as follows:

$$\begin{aligned} P(X \leq w) &= P\left(\frac{X - 185}{16} \leq \frac{w - 185}{16}\right) \\ &= P\left(Z \leq \frac{w - 185}{16}\right) \qquad \text{where } Z \sim N(0, 1). \end{aligned}$$

We want this to be .80. Looking at the normal table, we see that $P(Z \leq z)$ is closest to .80 when z is .84; therefore, we should choose w so that

$$\frac{w - 185}{16} = .84; \quad \text{that is, } w = 198.44.$$

∎

Example e. Suppose $X \sim N(3, 5)$, and suppose $X_1, X_2, \ldots, X_{30}$ are independent observations of X—that is, they are a sample of size 30 from the $N(3, 5)$ distribution. Let $\overline{X}_n$ be the sample mean. We ask two questions: (i) What is the probability that X is between 2.5 and 3.5? (ii) What is the probability that $\overline{X}_{30}$ is between 2.5. and 3.5?

Solution. Answering (i) is just like the calculations in Examples a and b above:

$$\begin{aligned} P(2.5 < X < 3.5) &= P\left(\frac{2.5 - 3}{\sqrt{5}} < \frac{X - 3}{\sqrt{5}} < \frac{3.5 - 3}{\sqrt{5}}\right) \\ &\doteq P(-.22 < Z < .22) \qquad \text{where } Z \sim N(0, 1) \\ &\doteq 2 \cdot (.5871 - .5) = .1742. \end{aligned}$$

Answering (ii) is similar, except that we must first use (6.2.9) to find that $V\overline{X}_{30}$ equals $\frac{5}{30}$. Consequently,

$$
\begin{aligned}
P(2.5 < \overline{X}_{30} < 2.5) &= P\left(\frac{2.5-3}{\sqrt{5/30}} < \frac{\overline{X}_{30}-3}{\sqrt{5/30}} < \frac{3.5-3}{\sqrt{5/30}}\right) \\
&\doteq P(-1.22 < Z < 1.22) \qquad \text{where } Z \sim N(0, 1) \\
&\doteq 2 \cdot (.8888 - .5) = .7776.
\end{aligned}
$$

Notice that both X, a single observation, and $\overline{X}_{30}$, the average of 30 independent observations, have 3 as their expected values. But because the variance of $\overline{X}_{30}$ is only $\frac{1}{30}$ that of X, it has a much greater probability of being within .5 of its expected value. ▮

Example f. Let X have the $N(100, 20)$ distribution. We want to find two numbers on either side of the expected value 100, equidistant from 100, such that the probability is 95% that X is between these two numbers.

Solution. Call these numbers $100 - a$ and $100 + a$. What we want is

$$
\begin{aligned}
.95 &= P(100 - a \leq X \leq 100 + a) \\
&= P\left(\frac{100-a-100}{\sqrt{20}} < \frac{X-100}{\sqrt{20}} < \frac{100+a-100}{\sqrt{20}}\right) \\
&= P\left(\frac{-a}{\sqrt{20}} < Z < \frac{a}{\sqrt{20}}\right) \qquad \text{where } Z \sim N(0, 1).
\end{aligned}
$$

So we need to find the two numbers, one positive and one negative, that have 95% of the standard normal probability between them. This implies that 2.5% of the probability must be to the right of the positive one, and 2.5% to the left of the negative one. The positive one must therefore have 97.5% of the probability to the left of it. Looking at the table in Appendix C, we see that the positive one must be 1.96. Therefore, the negative one must be -1.96. That is,

$$\frac{a}{\sqrt{20}} = 1.96,$$

and, therefore, $a = 1.96 \cdot \sqrt{20} \doteq 8.77$. So the two numbers asked for are approximately $100 - 8.77$ and $100 + 8.77$—that is, 91.23 and 108.77.

As an exercise, do Example f again, assuming that the variance is 40 instead of 20. The numbers are approximately 100 ± 12.40. Notice that whatever the variance σ^2 is, the two numbers will be $100 \pm 1.96\sigma$. ▮

(6.2.11) **Proof of (6.2.7).** Let X_1 and X_2 be independent and suppose $X_1 \sim N(\mu_1, \sigma_1^2)$ and $X_2 \sim N(\mu_2, \sigma_2^2)$. Let $W = X_1 + X_2$. Then (6.2.7) says that $W \sim N(\mu_1 + \mu_2, \sigma_1^2 + \sigma_2^2)$, and this is what we want to prove here.

As we have mentioned, there is a simple and short proof of this result using generating functions; we will see it in (8.2.9). We are giving a proof here using the multivariate change-of-variable theorem of Section 5.4, in part to demonstrate some useful techniques.

But we will not do it directly—getting a density for W from the densities for X_1 and X_2—because the algebra involved is very messy. Instead, we rely on a little trickery.

First, we make a slight reduction in what needs to be proved: Let $Y_1 = X_1 - \mu_1$ and $Y_2 = X_2 - \mu_2$; then $Y_1 \sim N(0, \sigma_1^2)$ and $Y_2 \sim N(0, \sigma_2^2)$. Let $U = Y_1 + Y_2$, which equals $W - (\mu_1 + \mu_2)$. If we can show that $U \sim N(0, \sigma_1^2 + \sigma_2^2)$, then it will follow that W, which equals $U + (\mu_1 + \mu_2)$, will be $N(\mu_1 + \mu_2, \sigma_1^2 + \sigma_2^2)$. So we have reduced the proof to showing that

> If Y_1 and Y_2 are independent, $Y_1 \sim N(0, \sigma_1^2)$, $Y_2 \sim N(0, \sigma_2^2)$, and $U = Y_1 + Y_2$, then $U \sim N(0, \sigma_1^2 + \sigma_2^2)$.

To show this with the multivariate change-of-variable theorem, we let

$$U_1 = Y_1 + Y_2 \quad \text{and} \quad U_2 = \sigma_2^2 Y_1 - \sigma_1^2 Y_2.$$

We will get a joint density $f_U(u_1, u_2)$ for U_1 and U_2. Then we will see that the joint density factors into a function of u_1 times a function of u_2.

Now in (5.4.9) we looked at what happens when a joint density factors:

> If a joint density $f_U(u_1, u_2)$ for U_1 and U_2 factors as $r(u_1) \cdot s(u_2)$, then some constant times $r(u_1)$ is a density for U_1.

In addition, U_1 and U_2 are independent, and some constant times $s(u_2)$ is a density for U_2; but we will not need this part of the result. We will need to show only that the nonconstant part of the function of u_1 is exactly the nonconstant part of the $N(0, \sigma_1^2 + \sigma_2^2)$ density. As we saw in (5.4.9), this implies that U_1 has this density.

The inverses of

$$u_1 = y_1 + y_2 \quad \text{and} \quad u_2 = \sigma_2^2 y_1 - \sigma_1^2 y_2$$

are

$$y_1 = \frac{1}{\sigma_1^2 + \sigma_2^2}(\sigma_1^2 u_1 + u_2) \quad \text{and} \quad y_2 = \frac{1}{\sigma_1^2 + \sigma_2^2}(\sigma_2^2 u_1 - u_2).$$

You can check that the Jacobian of this inverse equals $-1/(\sigma_1^2 + \sigma_2^2)$, a constant.

Now a joint density for Y_1 and Y_2 is

(6.2.12)
$$f_Y(y_1, y_2) = \frac{1}{\sqrt{2\pi\sigma_1^2}\sqrt{2\pi\sigma_2^2}} e^{-A/2}, \quad \text{where } A = \frac{y_1^2}{\sigma_1^2} + \frac{y_2^2}{\sigma_2^2}.$$

So a joint density for U_1 and U_2 is

(6.2.13)
$$f_U(u_1, u_2) = \frac{1}{2\pi\sigma_1\sigma_2} e^{-B/2} \cdot \left|\frac{-1}{\sigma_1^2 + \sigma_2^2}\right|,$$

where

$$B = \frac{1}{(\sigma_1^2 + \sigma_2^2)^2} \cdot \left(\frac{(\sigma_1^2 u_1 + u_2)^2}{\sigma_1^2} + \frac{(\sigma_2^2 u_1 - u_2)^2}{\sigma_2^2}\right).$$

A little manipulation shows that

(6.2.14)
$$B = \frac{u_1^2}{\sigma_1^2 + \sigma_2^2} + \frac{u_2^2}{\sigma_1^2\sigma_2^2(\sigma_1^2 + \sigma_2^2)}.$$

Now looking at (6.2.13) and (6.2.14), we see that $f_U(u_1, u_2)$ factors as $r(u_1) \cdot s(u_2)$, where

$$r(u_1) = (\text{some constant}) \cdot \exp\left(-\frac{1}{2} \cdot \frac{u_1^2}{\sigma_1^2 + \sigma_2^2}\right).$$

[Here we are using $\exp(x)$ for e^x.] But the nonconstant part of this density is the same as the nonconstant part of the $N(0, \sigma_1^2 + \sigma_2^2)$ density, and it follows from the result of (5.4.9) that U_1 has this density. □

EXERCISES FOR **SECTION 6.2**

1. Find the indicated probabilities, assuming that $X \sim N(\mu, \sigma^2)$ for the given values of μ and σ^2.
   a. $P(X \leq 8)$ for $\mu = 5$ and $\sigma^2 = 10$
   b. $P(40 < X < 60)$ for $\mu = 50$ and $\sigma^2 = 25$
   c. $P(40 < X < 60)$ for $\mu = 50$ and $\sigma^2 = 37.5$
   d. $P(X \leq 200)$ for $\mu = 300$ and $\sigma^2 = 2{,}500$

2. Find the indicated probabilities, assuming that $X \sim N(\mu, \sigma^2)$ for the given values of μ and σ. [*Note:* In each case you are given σ, not σ^2.] Compare this problem with Supplementary Exercise 2 for Chapter 1.
   a. $P(X \leq 28)$ for $\mu = 20$ and $\sigma = 5$
   b. $P(X > 10)$ for $\mu = 6.2$ and $\sigma = 1.4$
   c. $P(X$ is within 8 of 12) for $\mu = 12$ and $\sigma = 4$
   d. $P(X$ is within 2σ of $\mu)$ for arbitrary μ and σ

3. Let $X \sim N(86, 25)$. Find the following.
   a. The probability that X is not within 10% of its expected value (that is, not between $\mu - .1 \cdot \mu$ and $\mu + .1 \cdot \mu$)

b. The 75th percentile of the distribution of X
c. The 20th percentile of the distribution of X
d. The number a so that the probability is 90% that X is between $86 - a$ and $86 + a$

4. Let $X_1, X_2, \ldots, X_{40}$ be a sample of size 40 from the $N(50, 36)$ distribution. Find the following.
a. The probability that $\overline{X}_{40}$ is between 50 and 51
b. The probability that S_{40} is less than 1,940
c. The 98th percentile of the distribution of $\overline{X}_{40}$
d. Two numbers, equidistant from the expected value of S_{40}, so that the probability is .75 that S_{40} is between them

5. A company sells sugar in 5-pound bags; it is subject to a penalty for each bag it sells that is found to contain less than 5 pounds of sugar. The packaging machines cannot guarantee an exact weight of sugar per bag; in fact, the weights are normally distributed with an average weight of $\mu = 5.13$ pounds and a standard deviation of $\sigma = .08$ pound.
a. What proportion of bags weigh less than 5 pounds?
b. The machine can be adjusted to increase the average weight μ. Assume that the standard deviation will not change. To what value should μ be set so that only 1% of the bags will weigh less than 5 pounds?

6. Heights of children in a certain age group average 58.4 inches, with a standard deviation of 2.9 inches. Assume that the heights are normally distributed.
a. What proportion of children are between 57 and 61 inches tall?
b. What is the 90th percentile of the children's heights?
c. What is the probability that the average of a sample of 20 heights will be greater than 60 inches?

7. Let $X_1, X_2, \ldots, X_{20}$ be a sample from the $N(45, 15)$ distribution.
a. What is the probability that $\overline{X}_{20}$ is less than 47?
b. What is the probability that S_{20} is greater than 925?

8. Let $X_1, X_2, \ldots, X_n$ be a sample from the $N(\mu, \sigma^2)$ distribution, where σ^2 is known to be equal to 5 but μ is unknown. Find, for each of the following sample sizes n, the probability that $\overline{X}_n$ is within .2 of the unknown μ (that is, between $\mu - .2$ and $\mu + .2$).
a. $n = 25$ **b.** $n = 50$ **c.** $n = 500$ **d.** $n = 1{,}500$

9. Let X and Y be independent random variables; $X \sim N(2.4, 3.5)$ and $Y \sim N(3.3, 2.6)$. What is the probability that $X + Y$ is between 5 and 6?

10. Cholesterol levels in a certain population are normally distributed with mean equal to 187 and standard deviation equal to 24.
a. Find two numbers equally distant from 187, say $187 + a$ and $187 - a$, so that the probability is 95% that the cholesterol level of a random member of the population is between these two numbers.

b. Now find two such numbers, so that the probability is 95% that the sample mean $\overline{X}_{30}$ of a sample of size 30 will be between these two numbers.

11. Let X be a random variable with finite expected value and variance.
    a. Given that X is normally distributed, find the probability that X is not within 2 standard deviations of its expected value. [See Exercise 2d above.]
    b. If you are not given that X has a normal distribution, what can you say about the probability found in part a?

12. A certain brand of candy bar has an advertised weight of 4 ounces, but the weights are actually random, with expected value 4.25 ounces and standard deviation .15 ounce. The manufacturer wants to know an interval of weights that will contain 95% of the weights of these candy bars.
    a. If we can assume that the weights are normally distributed, then there are many such intervals. Find the one centered at 4.25.
    b. Find the one whose left endpoint is $-\infty$ (still assuming the weights are normally distributed).
    c. Now suppose we cannot assume anything about the distribution of weights, other than the given expected value and standard deviation. Find an interval that is sure to contain 95% of the weights. [*Hint:* Use Chebyshev's inequality.]

13. In Exercise 4 of the previous section, we assumed that the vehicles crossing a certain bridge have an average weight of 4,675 pounds, with a standard deviation of 345 pounds. There we found the expected value and standard deviation of the combined weight of 40 vehicles on the bridge. Now, assuming in addition that the weights are normally distributed, find the probability that the combined weight exceeds 190,000 pounds.

ANSWERS

1. a. .8289 **b.** .9544 **c.** .8968 **d.** .0228

2. a. .9452 **b.** .0034 **c.** .9544 **d.** .9544

3. a. .0854 **b.** 89.35 **c.** 81.8 **d.** 8.225

4. a. .3531 **b.** .0571 **c.** 51.94
d. 2,000 ± 43.64

5. a. .0516 **b.** 5.19 pounds

6. a. .5003 **b.** 62.112 inches **c.** .0068

7. a. .9896 **b.** .0749

8. a. .3472 **b.** .4714 **c.** .9544 **d.** .9994

9. .1581

10. a. 187 ± 47.04, i.e., 139.96 and 234.04
b. 187 ± 8.59, i.e., 178.41 and 195.59

11. a. .0456 **b.** It is less than or equal to $\frac{1}{4}$.

12. a. (3.956, 4.544) [i.e., 4.25 ± .294]
b. $(-\infty, 4.49675)$ **c.** (3.8, 4.7)
[i.e., 4.25 ± .45]

13. .0853

6.3 THE CENTRAL LIMIT THEOREM

An informal statement of this amazing theorem is the following.

> **If $X_1, X_2, X_3, \ldots$ are i.i.d. random variables with finite expected value μ and variance σ^2, then when n is large, the sample sum, $S_n = X_1 + X_2 + \cdots + X_n$, and the sample average, $\overline{X}_n = S_n/n$, have distributions that are approximately normal.**

Which normal distributions do they have? To specify them we need only give their expected values and variances. As we saw in the previous section,

$$ES_n = n\mu \quad \text{and} \quad VS_n = n\sigma^2,$$

and

$$E\overline{X}_n = \mu \quad \text{and} \quad V\overline{X}_n = \frac{\sigma^2}{n}.$$

> So **S_n is approximately $N(n\mu, n\sigma^2)$ and $\overline{X}_n$ is approximately $N(\mu, \sigma^2/n)$.**

Notice that if the individual random variables X_j themselves have a normal distribution, then, as we saw in the previous section, S_n and $\overline{X}_n$ also have normal distributions. What makes the Central Limit Theorem so amazing is that no matter what distribution the X_j's have, S_n and $\overline{X}_n$ have distributions that are approximately normal, as long as n is large and the X_j's have finite expected value and variance.

A natural question is: how large? How big must n be for the approximation to be very good? This question is nearly impossible to answer satisfactorily in general. How good the approximation is for a given n depends on the nature of the true distribution of the X_j's, and in many cases that distribution is unknown. A rule of thumb that has little mathematical justification, but is based on experience with normal approximations, is that if the distribution of the X_j's is itself not too unusual, then if n is 30 or more, the normal approximation is fairly reliable. The more bell-shaped the distribution of the X_j's is, the better the approximation is.

We will present a rigorous statement of the Central Limit Theorem, and discuss how it is proved, in Chapter 8. In this section we will give several examples showing how it is used, and then make a few comments on the theorem and its history.

(6.3.1) Example. Suppose we use a computer to generate 100 independent observations, $X_1, X_2, \ldots, X_{100}$, from the uniform distribution on (0, 1), and we let

$S_{100} = X_1 + X_2 + \cdots + X_{100}$. What is the probability that S_{100} is between 45 and 55?

Solution. It is possible, but difficult, to find this probability exactly. The distribution of S_{100} is quite complicated; it has (0, 100) as its set of possible values, and there is no simple formula for a density.

But because S_{100} is the sum of a large number of i.i.d. random variables with finite expected value and variance, it has an approximately normal distribution. So to find approximate probabilities for S_{100}, we need only to work out its expected value and its variance.

For each individual X_j, we have

$$\mu = EX_j = \tfrac{1}{2} \quad \text{and} \quad \sigma^2 = VX_j = \tfrac{1}{12};$$

so for their sum S_{100}, we have

$$ES_{100} = 100 \cdot \mu = 50 \quad \text{and} \quad VS_{100} = 100 \cdot \sigma^2 = \tfrac{25}{3}.$$

Therefore, the distribution of S_{100} is approximately $N(50, 25/3)$. So

$$\begin{aligned} P(45 \le X \le 55) &= P\left(\frac{45 - 50}{\sqrt{25/3}} \le \frac{X - 50}{\sqrt{25/3}} \le \frac{55 - 50}{\sqrt{25/3}}\right) \\ &\approx P(-1.73 \le Z \le 1.73) \qquad \text{where } Z \sim N(0, 1) \\ &\doteq .9164. \end{aligned}$$

∎

Notice the two types of approximate equality signs in this example. The symbol $\doteq$, of course, means that rounding has taken place; the normal table gives only a rounded answer. The symbol $\approx$ indicates not rounding, but an approximate probability; the distribution of X is not exactly normal, so neither is that of Z, but it is approximately so, according to the Central Limit Theorem. (Of course, the number 1.73 is rounded to the two decimal places as well, so perhaps the symbol $\risingdotseq$ would be appropriate. But such pedantry in the name of precision about imprecision is probably best avoided.)

(6.3.2) Example. In a certain town the yearly household incomes have an average of \$33,700 and a standard deviation of \$3,521. What is the probability that the average income of 40 randomly chosen households is under \$33,000?

Solution. In the language of probability theory, what we are given is as follows: if X denotes the income of a randomly chosen household, then $EX = 33{,}700$ and $VX = 3{,}521^2$. We have a sample $X_1, X_2, \ldots, X_{40}$ of independent observations of X and we want $P(\overline{X}_{40} \le 33{,}000)$. There is no way we can find this probability exactly, because we do not know the distribution of X; consequently, we cannot find the distribution of $\overline{X}_{40}$. All we can find are its expected value and variance:

$$E\overline{X}_{40} = 33{,}700 \quad \text{and} \quad V\overline{X}_{40} = \frac{3{,}521^2}{40}.$$

And because $\overline{X}_{40}$ is the average of a large number of i.i.d. random variables with finite expected value and variance, it has approximately a normal distribution. Thus,

$$\begin{aligned} P(\overline{X}_{40} \le 33{,}000) &= P\left(\frac{\overline{X}_{40} - 33{,}700}{3{,}521/\sqrt{40}} \le \frac{33{,}000 - 33{,}700}{3{,}521/\sqrt{40}}\right) \\ &\approx P\left(Z \le \frac{-700}{556.72}\right) \qquad \text{where } Z \sim N(0, 1) \\ &\doteq P(Z \le -1.26) \doteq .1038. \end{aligned}$$

This gives us an idea of how well we can expect the sample average to predict the population average; in this example, there is only about a 10% chance that the sample average will be under the population average by \$700 or more. In other words, only about 10% of samples of size 40 from this population will have their averages under \$33,000. Of course, in another example the probability will be different; it depends on the sample size n and on the population standard deviation.

There is something important to notice about this example: We can find an approximate probability for the average of 40 incomes even though we have no idea what the distribution of incomes in the population is. Income distributions do not tend to be normal, so without further information we have no way of finding the probability, say, that a *single* randomly chosen income is under \$33,000. But we can approximate without difficulty the same probability for the average of a number of randomly chosen incomes.

Another point needs to be made about this example before we leave it. Although the Central Limit Theorem enables us to find approximate probabilities for deviations of the sample mean from the expected value of the distribution, in practice the expected value of an income distribution is not much used as a population measure. The reason is simple; in a large set of incomes the high ones tend to pull the average up unduly, giving a false impression of the "typical" income in the population. For distributions that are skewed in this way, the *median*—the number m such that $P(X > m)$ equals $\frac{1}{2}$—is usually thought of as being a better indicator of a "typical" value. See Exercise 11 of Section 4.1 or Exercise 2 of Section 8.3. ∎

(6.3.3) Example. Suppose we have a Poisson process with $\lambda = 5$ arrivals per minute. We are interested in the time of the 50th arrival. Before asking the questions that we will answer in this example, we note that the expected time of the 50th arrival is 10 minutes after the start of observations.

This is easy to see intuitively: at 5 arrivals per minute, we should get 50 arrivals in 10 minutes. To see it rigorously, notice that the time of the 50th arrival is the sum of 50 waiting times—the time before the first arrival, the time between the first and the second, and so on. In a Poisson process, these waiting times are independent random variables, each having the exponential distribution with parameter λ. So the expected value of each waiting time is $1/\lambda$, and the expected sum of 50 of them is $50/\lambda$, which in this example is $50/5 = 10$.

One other comment: the distribution of the sum of 50 i.i.d. exponential random variables with parameter λ is called the *Erlang* (or *gamma*) distribution with parameters $n = 50$ and λ. In the next chapter we will get a density for it using generating functions. It is also possible to get this density using the multivariate change-of-variable theorem, although we did not do so in the last chapter. But even this density is not very useful for finding probabilities, because it equals a constant times $x^{49}e^{-\lambda x}$, and to find its integral from a to b we need 49 successive integrations by parts, each adding a new term to the answer. The most convenient thing to do is to use the Central Limit Theorem to get approximate probabilities.

We consider two questions: (i) What is the probability that the 50th arrival comes between 9 and 11 minutes after the start of observations? (ii) Find two numbers on either side of 10, say $10 - a$ and $10 + a$, such that the probability is 99% that the 50th arrival is between these two numbers.

Solution to (i). We want the probability that the sum of 50 independent waiting times differs from its expected value by 1 minute or less. In symbols, let $X_1, X_2, \ldots, X_{50}$ be these waiting times and let S_{50} be their sum. For each X_j, we have [see Examples (4.1.6) and (4.3.6)]

$$EX_j = \frac{1}{\lambda} = \frac{1}{5} \quad \text{and} \quad VX_j = \frac{1}{\lambda^2} = \frac{1}{25};$$

and so

$$ES_{50} = 50 \cdot \frac{1}{5} = 10 \quad \text{and} \quad VS_{50} = 50 \cdot \frac{1}{25} = 2.$$

Therefore,

$$\begin{aligned} P(9 \le S_{50} \le 11) &= P\left(\frac{9 - 10}{\sqrt{2}} \le \frac{S_{50} - 10}{\sqrt{2}} \le \frac{11 - 10}{\sqrt{2}}\right) \\ &\approx P(-.71 \le Z \le +.71) \qquad \text{where } Z \sim N(0, 1) \\ &\doteq .5222. \end{aligned}$$

For the record, the correct answer, found by using numerical analysis techniques to integrate the true density of S_{50} from 9 to 11, is .5210. So the normal approximation is quite good here.

Solution to (ii). We want a so that

$$\begin{aligned} .99 &= P(10 - a \le S_{50} \le 10 + a) \\ &= P\left(\frac{10 - a - 10}{\sqrt{2}} \le \frac{S_{50} - 10}{\sqrt{2}} \le \frac{10 + a - 10}{\sqrt{2}}\right) \\ &\approx P\left(\frac{-a}{\sqrt{2}} \le Z \le \frac{a}{\sqrt{2}}\right). \end{aligned}$$

So we want a such that 99% of the standard normal probability is between $-a/\sqrt{2}$ and $+a/\sqrt{2}$. This means that .5%, or .005, is to the right of $+a/\sqrt{2}$, and so .995 is to the left of $a/\sqrt{2}$. Looking at the table, we see that $a/\sqrt{2}$ must be 2.58, and so a must equal $\sqrt{2} \cdot 2.58$, or approximately 3.65. The two numbers we want are 10 ± 3.65. The 50th arrival has a 99% chance of coming between 6.35 and 13.65 minutes after the start of observations. ∎

(6.3.4) **Why many random quantities appear to be normally distributed.** The normal distribution appears very often in probability models of real-world random phenomena. Many physical measurements on living organisms, for example, are shown by experience to have bell-shaped distributions for which normal densities are good approximations—for example, heights of people, blood cholesterol measurements, and scores on tests of physical strength. Deviations from desired values in manufacturing processes, also, are often assumed to be normally distributed. In addition, errors in measurement of physical quantities, or "random noise" added to a transmitted signal, are often modeled as normally distributed random variables.

Why should normal distributions be appropriate for such quantities? The Central Limit Theorem gives us some insight. First we should remark that, although we stated the theorem for i.i.d. random variables, it applies to random variables that are not identically distributed as well—as long as they are independent and certain restrictions on their variances are satisfied.

Now a quantity such as the height of a person, or the error made by some measuring device, can perhaps be thought of as the accumulation of many independent influences, all of which are uncontrollable, or not well understood, and therefore can be viewed as random. The height of an adult, for example, might be thought of as the sum of a random birth height and several independent random increments of growth, due to different genetic and environmental factors.

Although such thinking is surely an oversimplification, it might nevertheless make it plausible that a complicated, imperfectly understood quantity might be the sum of many independent quantities, and therefore would have approximately a normal distribution.

(6.3.5) **Poisson random variables with large λ are approximately normal.** Suppose X has the Poisson distribution with parameter λ. We abbreviate the name of this distribution by Pois(λ). Suppose λ is large. If we could show that the distribution of X is that of a sum of many i.i.d. random variables, then it would follow that X has approximately a normal distribution. Since $EX = \lambda$ and $VX = \lambda$ (we found these in Exercise 6 of Section 4.3), X is approximately $N(\lambda, \lambda)$.

Now this seems strange when we realize that the normal distribution is an absolutely continuous one, with $\mathbb{R}$ as its set of possible values, whereas a Poisson distribution is discrete, with the nonnegative integers as possible values. Still, for a given pair of numbers a and b, both distributions assign a probability to the interval (a, b), and if λ is large, these two probabilities are close to each other.

[The Poisson probability of (a, b) is a sum of probabilities of the integers in (a, b); and the normal probability of (a, b) is the integral of some normal density from a to b; but that does not mean they cannot be nearly equal.]

So why should a Poisson random variable have the same distribution as the sum of a large number of i.i.d. random variables? Think of a Poisson process with arrival rate λ per hour. Let X be the number of arrivals in the first hour; then X has the Pois(λ) distribution. For some large n, divide the hour into n equal time intervals; let X_j be the number of arrivals in the jth interval. Then the X_j's are independent and each has the Pois(λ/n) distribution; and X is their sum.

You might ask: why does λ have to be large for this to work? We chose n arbitrarily, and no matter what λ is, we can simply choose a very large n, and then X will be the sum of n i.i.d. Pois(λ/n) random variables. By this reasoning, it would seem that even the Poisson distribution with parameter $\lambda = 1$, or $\lambda = .1$, should be approximately normal. But it is not. If λ is small, the Pois(λ) distribution is far from bell shaped.

Where is the catch? Remember what we said earlier about how good the normal approximation is: it is good if the distribution of the X_j's is not too far from normal, but the less bell shaped the distribution is, the larger n needs to be for a good approximation.

And if λ is small, then the larger n gets, the further from bell shaped is the Pois(λ/n) distribution, and the larger n needs to be to make the approximation good. No value of n produces a good approximation. But if λ is large, then we can choose n to be near λ, so that λ/n is near 1; and the Pois(1) distribution is not so far from bell shaped as to ruin the approximation.

(6.3.6) **Binomial random variables with large n and moderate p are approximately normal.** This result is actually more important than the previous result on the Poisson distribution. It was discovered before the Central Limit Theorem itself and is called the De Moivre–Laplace limit theorem in honor of Abraham De Moivre (1667–1754) and Pierre Simon de Laplace (1749–1827). These two French mathematicians made early contributions to the understanding of the fact that binomial probabilities can be approximated by integrating the function that we now call the standard normal density.

We will not cover this result in detail here, but it will occupy a large part of the next section. For now, we just recall Example (4.2.17) and (4.2.18); there we saw that if X has the binomial distribution with parameters n and p, then X can be regarded as the number of successes in n Bernoulli trials, and thus $X = X_1 + X_2 + \cdots + X_n$, where X_j equals 1 if the jth trial results in success and 0 if not. So X is the sum of n i.i.d. random variables (each having the Bernoulli distribution with parameter p). Therefore, as long as n is large and the Bernoulli distribution is not too far from bell shaped, the normal approximation is reliable.

There is an easy rule of thumb that is often used to decide whether the approximation will be valid: if np and nq are both 5 or more, then n is large enough and the Bernoulli distribution is close enough to bell shaped for the approximation to be valid.

Which normal distribution should be used for a given n and p? Again that is easy. We know that $EX = np$ and $VX = npq$, so if X has approximately a normal distribution, then it must be the $N(np, npq)$ distribution.

(6.3.7) **The Central Limit Theorem in statistical inference.** One of the most common situations in statistical inference is that of using a random sample to get information about the average of a large population of scores. To take an example, suppose we want to know something about the average income of households in a certain city. The natural thing to do is to take a sample of households and find the average of their incomes. The resulting number, the sample average, is called, for obvious reasons, an *estimator* of the unknown population average. But how good an estimator is it?

We have seen that the probability model for this situation is a collection $X_1, X_2, \ldots, X_n$ of i.i.d. random variables. Each X_j is the income of a randomly chosen household. The distribution of X_j is unknown; to know it would require knowing all the incomes in the city. But the expected value of X_j is exactly the unknown population average—call it μ, as we customarily do. In addition, VX_j is the variance of the collection of all incomes in the city, another unknown number. We call it σ^2.

As we saw in Example (4.5.2), the law of large numbers says that $\overline{X}_n$ is a random variable that is likely to be close to the unknown μ if n is large. But it is reasonable to ask: how likely to be how close if n is how large? The Central Limit Theorem allows us to take a step toward an answer, for even though the distribution of the individual X_j's is unknown, the distribution of $\overline{X}_n$ is approximately $N(\mu, \sigma^2/n)$.

Suppose for the moment, hypothetically, that we know σ^2. Then, for any desired tolerance w, we can find (approximately) the probability that $\overline{X}_n$ is within w of the unknown μ:

$$\begin{aligned} P(\mu - w \leq \overline{X}_n \leq \mu + w) &= P\left(\frac{\mu - w - \mu}{\sigma/\sqrt{n}} \leq \frac{\overline{X}_n - \mu}{\sigma/\sqrt{n}} \leq \frac{\mu - w + \mu}{\sigma/\sqrt{n}}\right) \\ &\approx P\left(-\frac{w\sqrt{n}}{\sigma} \leq Z \leq \frac{w\sqrt{n}}{\sigma}\right) \quad \text{where } Z \sim N(0, 1) \\ &= 2 \cdot \Phi\left(\frac{w\sqrt{n}}{\sigma}\right) - 1. \end{aligned}$$

So if, for example, we know $\sigma = \$5{,}000$, we took a sample of size $n = 75$, and we wanted the probability that $\overline{X}_{75}$ is within $w = \$1{,}000$ of the unknown μ, then that probability would be approximately

$$2 \cdot \Phi\left(\frac{1{,}000 \cdot \sqrt{75}}{5{,}000}\right) - 1 \doteq .9164.$$

Interpreting this requires a little careful thought. Suppose you take the sample of 75 households and find that the average income in the sample is \$29,173. There is about a 92% chance that the sample average is within \$1,000 of the unknown population average; consequently statisticians would call the interval $[\overline{X}_{75} - 1{,}000, \overline{X}_{75} + 1{,}000]$ a "92% confidence interval" for μ. Does this mean that there is a 92% chance that the unknown average is between \$28,173 and \$30,173? No; it makes no sense to talk about the probability that μ is in some interval, because although μ is unknown, it is not a randomly varying quantity. When we repeat the experiment of taking a sample, μ does not come up differently each time; $\overline{X}_{75}$ does. The proper interpretation is that if we took many samples of size 75, we would get many different sample averages, and about 92% of these sample averages would be within \$1,000 of the unknown μ.

Actually, there is a branch of statistical inference that gets information about an unknown number μ by modeling it as a random variable. This branch of inference is called "Bayesian inference," because the calculations involved make heavy use of Bayes's theorem, in a form that we will see in Chapter 9.

But there is a more serious difficulty. The preceding analysis is of no use if σ is not known, and it is difficult to imagine a situation in which we will know the standard deviation but not the average of a population. Fortunately, there is a way out. We can use the same sample to estimate σ and use that estimate in the calculation. This use of the same sample twice in one calculation, however, changes the probabilities, and the standard normal probabilities must be "corrected" a little to take this into account. The correction depends on the sample size—the bigger the sample, the smaller the correction. The proper distribution to use is called a t distribution, and the calculations involve a quantity called "Student's t-statistic." This subject is covered fully in courses in statistical inference. We will have more to say about t distributions in Section 8.4.

(6.3.8) **How the Central Limit Theorem improves on Chebyshev's inequality.** In several examples and exercises in Section 4.5, we worked out what Chebyshev's inequality tells about the probability that $\overline{X}_n$ is not within some given distance of μ. If n is large, we can use the Central Limit Theorem to get an approximation to the true probability, not just a number that we know to be larger than it.

For example, suppose we have $n = 100$ independent random variables, $X_1, X_2, \ldots, X_{100}$, each having the exponential distribution with parameter $\lambda = .25$. Then each X_j has $EX_j = 1/\lambda = 4$ and $VX_j = 1/\lambda^2 = 16$. So $E\overline{X}_{100} = 4$ and $V\overline{X}_{100} = 16/100 = .16$. Let us consider the probability that $\overline{X}_{100}$ is not within .5 of 4 (its expected value).

By Chebyshev's inequality (4.5.5), this probability is less than or equal to $V\overline{X}_{100}/(.5)^2 = .64$.

By the Central Limit Theorem, $\overline{X}_{100}$ has approximately a normal distribution, so we can use its expected value (4) and its variance (.16) to approximate the probability that it is not within .5 of 4. That is,

$$\begin{aligned} &P(\overline{X}_{100} \le 3.5) + P(\overline{X}_{100} \ge 4.5) \\ &\quad = P\left(\frac{\overline{X}_{100} - 4}{\sqrt{.16}} \le \frac{3.5 - 4}{\sqrt{.16}}\right) + P\left(\frac{\overline{X}_{100} - 4}{\sqrt{.16}} \ge \frac{4.5 - 4}{\sqrt{.16}}\right) \\ &\quad \approx P(Z \le -1.25) + P(Z \ge 1.25) = 2 \cdot (1 - P(Z \le 1.25)) = .2112. \end{aligned}$$

The probability, which Chebyshev's inequality bounds by .64, is actually close to .2112. Again we see the crudeness of Chebyshev's inequality. (However, remember Exercise 5 in Section 4.5, which shows that there is no better inequality that works for *all* distributions.)

(6.3.9) A historical note. The Central Limit Theorem has a long history. As we mentioned in (6.3.6), the special case that we call the De Moivre–Laplace limit theorem seems to have been the first appearance of normal distributions. By the late eighteenth century, De Moivre and Laplace had shown that the probability of getting between a and b successes in n Bernoulli trials can be approximated by the area under the standard normal curve between $(a - np)/\sqrt{npq}$ and $(b - np)/\sqrt{npq}$.

The most important name in the history of the Central Limit Theorem and of normal distributions is that of Karl Friedrich Gauss (1777–1855), who was among the first to realize that areas under the normal curve could be used to approximate probabilities in a very wide range of contexts, not merely for Bernoulli trials. Normal distributions are often called *Gaussian* distributions in his honor. Gauss has been called by some people the greatest mathematician of all time.

But the story of the Central Limit Theorem is far from finished; there is much current research on various generalizations of it. In Chapter 8 we will discuss the proof of the theorem, and in (8.2.22) is a little more of the history, as well as some discussion of the kinds of questions on which research continues.

EXERCISES FOR **SECTION 6.3**

1. Suppose you generate 1,000 independent observations from the uniform distribution on (0, 1) and average them.
   a. What are the expected value and variance of the average?
   b. What is the approximate probability that the average is less than .48?
   c. What is the approximate probability that the average is between .49 and .51?

2. Suppose $X_1, X_2, \ldots, X_n$ are i.i.d., each with some unknown distribution that has an expected value of $\mu = 25.7$ and a standard deviation of $\sigma = 4.3$.
   a. If $n = 50$, find the approximate probability that the average $\overline{X}_n$ is less than 25. Repeat for $n = 100$ and $n = 1{,}000$.
   [a]b. What additional knowledge about the distribution of the X_i's would make the probabilities in part a exact instead of approximate?

3. Find the approximate probability that 100 tosses of a fair coin produce between 40 and 60 heads. [Use the fact, mentioned in (6.3.6), that the number of successes in n Bernoulli trials has approximately a normal distribution. In the next section we will see a way to improve the approximation a little.]

4. Let S_{45} be the number of failures preceding the 45th success in a Bernoulli process with success probability $p = .36$. We can write $S_{45} = X_1 + X_2 + \cdots + X_{45}$, where X_1 is the number of failures preceding the first success, X_2 is the number between the first and second successes, and so forth. The X_j's are independent.
   a. Give the name of the distribution of a single X_j, also give its expected value μ and its variance σ^2. [See Exercise 9 of Section 4.2.]
   b. What are the expected value and variance of S_{45}?
   c. Approximate the probability that S_{45} is within 20 of its expected value.
   d. What does Chebyshev's inequality say about the probability that you approximated in part c?

5. In a Poisson process with $\lambda = 500$ arrivals per day, find the approximate probability that there will be between 475 and 525 arrivals tomorrow. [See (6.3.5).]

6. In a Poisson process with $\lambda = 500$ arrivals per day, find the approximate probability that the thousandth arrival comes between 47.5 and 48.5 hours after the start of observations. [See Solution to (i) in (6.3.3).]

7. At a certain bank, the deposits made by customers in the past have averaged \$638, with a standard deviation of \$126; but apart from this, nothing is known about the distribution of deposits. As part of a study, 75 deposits are chosen independently and at random, and they are found to average \$671. If the expected value and standard deviation have not changed, what is the approximate probability that the average of 75 independent deposits is \$671 or greater?

   This is typical of statistical hypothesis testing. If the probability is too small, then we could reject the hypothesis that the expected value and standard deviation have not changed, because if they have *not* changed, then something rare has occurred, and rare events do not often occur.

8. **Will a football team break a freight elevator?** A freight elevator has a safe load of 10,000 pounds. Members of a certain professional football league have an average weight of 215 pounds, with a standard deviation of 26 pounds. If 44 football players get on this elevator at once, what is the approximate probability that their combined weight exceeds 10,000 pounds? (Assume the players are a random sample from the league.)

9. The vehicles that cross a certain bridge are found to have an average weight of 4,675 pounds, with a standard deviation of 345 pounds. If there are 40 vehicles on the bridge at once, find the number a such that the (approximate) probability is 99% that their combined weight does not exceed a. [You found the expected value and standard deviation of their combined weight in Exercise 4 of Section 6.1, and you found a probability for it in Exercise 13 of Section 6.2, assuming the weights were normally distributed. Notice that in this exercise we do not assume normality, but we still get an approximate answer.]

10. **An experiment in personal taste for money, after Daniel Bernoulli.** You have won a contest and you get to choose between two prizes. Prize 1 is \$95 in cash. For prize 2, someone will toss a fair coin; if heads show you will get \$200, if tails, you will get nothing. Which prize will you choose? It is easy to see that your expected winnings are \$100 with prize 2 and only \$95 with prize 1. Nevertheless, as Daniel Bernoulli pointed out in his St. Petersburg paper (see Exercise 23 of Section 4.1), a poor person who desperately needed money would be well advised to take prize 1. There is no right choice; it depends on the value of a dollar to you—the "utility" of money.

    But suppose you are offered 50 prize 1's or 50 prize 2's; you must make the choice once for all 50 prizes, but if you choose prize 2 there will be 50 coin tosses. Now to make the choice rationally you would want the probability that 50 prize 2's will yield more than 50 prize 1's. Find the approximate value of this probability.

    [If you had no trouble deciding which prize you would take in the case of just one play, consider the following choices. In each, the expected values of the two prizes are exactly the same.

Prize 1: \$100	Prize 2: \$1,000 with probability .1, \$0 with probability .9
Prize 1: \$1,000	Prize 2: \$1,000,000 with probability .001, \$0 with probability .999
Prize 1: \$1	Prize 2: \$1,000,000 with probability .000001, \$0 with probability .999999

    The point of this exercise is that—although decisions involving many plays are often based on expected gains and probabilities of positive gains—if there is to be only one play, the choice can be quite subjective, and the one with lesser expected value may be more attractive. Moreover, as the last three choices show, the expected value sometimes has little if anything to do with what you would choose.]

ANSWERS

1. **a.** $\frac{1}{2}$; $\frac{1}{12{,}000}$ **b.** .0143 **c.** .7286
2. **a.** .1251; .0516; essentially 0 [actually about $\Phi(-5.15)$, or about .0000001302] **b.** Knowing that the X_i's have a normal distribution
3. .9544
4. **a.** $\mu = q/p = 16/9$; $\sigma^2 = q/p^2 \doteq 4.938$ **b.** $ES_{45} = 80$; $VS_{45} = 222.22 \ldots$ **c.** .8198 **d.** It is greater than .44444 . . . .
5. .7372
6. .2586
7. .0116
8. .0009
9. 192,083.99
10. .6368 [50 prize 2's are better than 50 prize 1's with this probability.]

6.4 THE POISSON AND NORMAL APPROXIMATIONS FOR BINOMIAL DISTRIBUTIONS

If X is the number of successes in n Bernoulli trials with success probability p, then X has the binomial distribution with parameters n and p. We say for short that X has the bin(n, p) distribution. The mass function is of course

(6.4.1)
$$p(k) = P(X = k) = \binom{n}{k} p^k q^{n-k} \quad \text{for } k = 0, 1, 2, \ldots, n,$$

where as always q denotes the failure probability $1 - p$. We also know that

$$EX = np \quad \text{and} \quad VX = npq.$$

Now the probabilities (6.4.1) are quite easy to find with a hand calculator if n is small. If n is large, it can be more difficult. For example, finding $\binom{750}{25}(.03)^{25}(.97)^{725}$ on a calculator is tedious at best, and in any case involves us in questions of accuracy and roundoff error. It is the product of a very large number with a very small number, and even with a computer it must be done carefully if we want to avoid serious inaccuracies due to roundoff.

The need for a computer rather than a hand calculator is even more apparent when we ask, say, for the probability that 600 rolls of a fair die produce between 90 and 110 6's (inclusive). This is a sum of 21 terms,

(6.4.2)
$$P(90 \leq X \leq 110) = \binom{600}{90}\left(\tfrac{1}{6}\right)^{90}\left(\tfrac{5}{6}\right)^{510} + \binom{600}{91}\left(\tfrac{1}{6}\right)^{91}\left(\tfrac{5}{6}\right)^{509} + \cdots + \binom{600}{110}\left(\tfrac{1}{6}\right)^{110}\left(\tfrac{5}{6}\right)^{490},$$

and it would take a good deal of time to get this with a hand calculator. Besides the roundoff problem, the likelihood of hitting a wrong key is quite high with such a large calculation.

In (6.4.15) we will say a few words about efficient ways to program the computation of a binomial probability. But it seems clear that even with a computer at hand, we would find it useful to have a quick way to approximate a sum of probabilities like (6.4.2), or even a single probability like $\binom{750}{25}(.03)^{25}(.97)^{725}$. In this section we will look at two different approximations that are reliable and quick.

(6.4.3) **The Poisson approximation to the binomial distribution.** This approximation is very easy to state:

If n is large, p is small, and np is moderate, then

$$\binom{n}{k} p^k q^{n-k} \quad \text{is approximately equal to} \quad \frac{(np)^k e^{-np}}{k!}.$$

(6.4.4) **Example.** Consider $n = 30$ Bernoulli trials with success probability $p = .02$. What is the probability that the 30 trials produce three successes? The Poisson approximation is simple: $np = 30 \cdot .02 = .6$, and so the probability is approximately $(.6)^3e^{-.6}/3!$, which to four significant digits is .01976.

The exact answer is $\binom{30}{3}(.02)^3(.98)^{27}$, which equals .01882 (to four significant digits). The error in the approximate answer is about 5% of the exact answer.

Here are a few of the other exact probabilities for this binomial distribution, along with the Poisson approximations. You might want to compute a few of these numbers yourself as an exercise. In each case the true probability is $\binom{30}{k}(.02)^k(.98)^{30-k}$ and the Poisson approximation is $(.6)^ke^{-.6}/k!$.

bin(30, .02) probabilities and Pois(.6) approximations to them

k	0	1	2	3	4	5	6
True binomial prob.	.54548	.33397	.09883	.01882	.00259	.00028	.00002
Poisson approx.	.54881	.32929	.09879	.01976	.00296	.00036	.00004

You can see that the approximation is quite good, especially for the larger probabilities; the error is less than 1.5% of the true probability for $k = 0, 1$, and 2. ∎

(6.4.5) **Precise statement and proof of the Poisson approximation.** It is difficult to imagine proving the informal statement we made in (6.4.3). It is easy to use, but it is too vague to be proved.

Most statements involving an approximation, valid when something is large or small, are made rigorous by changing them into statements about limits. To make n large, we let it tend to ∞; to make p small, we let it tend to 0; and (in this case) to make np moderate, we keep it constant. Thus we have the following.

Poisson approximation theorem. If $n \to \infty$ and $p \to 0$ in such a way that np is always equal to λ (that is, if $n \to \infty$ and $p = \lambda/n$), then for any fixed nonnegative integer k,

$$\lim_{n\to\infty} \binom{n}{k}p^kq^{n-k} = \frac{\lambda^ke^{-\lambda}}{k!}.$$

Now in this form, it may look familiar. In fact, we saw it in (1.4.11), when we discussed how to simulate a finite time interval of a Poisson process. The method is to divide the hour into n subintervals, and do n Bernoulli trials with success probability λ/n. Each success counts as an arrival in the corresponding subinterval. The probability of k arrivals in the simulated hour is $\binom{n}{k}p^kq^{n-k}$, and this, we said, is close to the probability $\lambda^ke^{-\lambda}/k!$ that a Poisson process gives to that event.

So proving the Poisson approximation theorem will also prove the validity of the simulation method given in (1.4.11), because they are really the same thing. We actually did prove that result in (1.4.11), for the special case $k = 4$. Here we will do it for an arbitrary nonnegative integer k.

Before we do, we recall the following fact: for any real number λ,

(6.4.6)
$$\lim_{n\to\infty}\left(1 - \frac{\lambda}{n}\right)^n = e^{-\lambda}.$$

This is often proved in advanced calculus courses. It is easy to believe if you notice that the logarithm of $(1 - (\lambda/n))^n$ is $n \cdot \ln(1 - \lambda/n)$, expand the latter logarithm using the Maclaurin series for $\ln(1 - x)$, multiply each term in the series by n, and let n tend to ∞. The result is that the logarithms of both sides of (6.4.6) are equal. We will simply take (6.4.6) as given.

Proof of the Poisson approximation theorem. First we will rewrite the binomial probability in such a way that we can see where the various factors in the Poisson probability come from. First,

$$\binom{n}{k}p^k q^{n-k} = \frac{n(n-1)(n-2)\cdots(n-k+1)}{k!}\left(\frac{\lambda}{n}\right)^k\left(1 - \frac{\lambda}{n}\right)^{n-k};$$

we have done nothing here but expand the binomial coefficient and write λ/n for p. If we rearrange this it becomes

$$\left[\frac{n(n-1)(n-2)\cdots(n-k+1)}{n^k}\right]\cdot\left[\left(1 - \frac{\lambda}{n}\right)^{-k}\right]\cdot\left[\frac{\lambda^k}{k!}\left(1 - \frac{\lambda}{n}\right)^n\right].$$

Now as $n \to \infty$, the third bracketed quantity approaches the Poisson probability $\lambda^k e^{-\lambda}/k!$ because of (6.4.6). So the proof consists of showing that the first two bracketed quantities approach 1 as $n \to \infty$.

The first bracketed quantity is the same as

$$\frac{n}{n}\cdot\frac{n-1}{n}\cdot\frac{n-2}{n}\cdot\cdots\cdot\frac{n-k+1}{n},$$

a product of k factors, each of which approaches 1 as $n \to \infty$. Since k is constant, the product approaches 1.

And the second bracketed quantity is just the reciprocal of the kth power of a quantity that approaches 1; again, because k is constant, the power approaches 1. ∎

Now we turn to the other approximation of this section.

(6.4.7) **The normal approximation to the binomial distribution** The basis for the approximation is the De Moivre–Laplace limit theorem: if $X \sim \text{bin}(n, p)$, then

X has approximately the $N(\mu, \sigma^2)$ distribution with $\mu = np$ and $\sigma^2 = npq$; the approximation is good if n is large and neither p nor q is close to 0.

The usual rule of thumb for a good approximation is that both np and nq should be at least 5. Thus for $n = 100$, for example, we expect a good approximation as long as p is between .05 and .95. If p were less than .05, we would use the Poisson approximation for the binomial probabilities. If p were greater than .95, then q would be less than .05, and we could let Y be the number of *failures* and use the Poisson approximation for probabilities for Y.

We mentioned the De Moivre–Laplace limit theorem in (6.3.6), where we showed how it follows as a special case of the Central Limit Theorem. We are still relying on informal statements of both of these limit theorems; we will discuss the precise statement and proof of the Central Limit Theorem in Chapter 8. Here we will describe how to use the normal approximation effectively.

(6.4.8) **Example.** In 100 tosses of a fair coin, what is the probability that the number of heads is between 40 and 60? This was Exercise 3 of the previous section; we do it here so that we can comment on the method.

The random variable X has the $\text{bin}\left(100, \frac{1}{2}\right)$ distribution; np and nq are both equal to 50, and so the distribution of X ought to be well approximated by the $N(\mu, \sigma^2)$ distribution with $\mu = np = 50$ and $\sigma^2 = npq = 25$. We proceed as though X had this normal distribution:

$$\begin{aligned} P(40 < X < 60) &= P\left(\frac{40 - 50}{\sqrt{25}} < \frac{X - 50}{\sqrt{25}} < \frac{60 - 50}{\sqrt{25}}\right) \\ &\approx P(-2 < Z < 2) \qquad \text{where } Z \sim N(0, 1) \\ &= 2 \cdot P(0 < Z < 2) \doteq 2 \cdot .4772 = .9544. \end{aligned}$$

We observe also that this approximation, as we used it here, is incapable of distinguishing between $P(40 \le X \le 60)$ and $P(40 < X < 60)$. If X really had the $N(50, 25)$ distribution these two probabilities would be the same, because the singletons 40 and 60 would have probability 0; but because X really has a binomial distribution, the possible values 40 and 60 have positive probabilities, and they are not the same.

The true probabilities, to six decimal places, are

$$P(40 \le X \le 60) \doteq .964800 \quad \text{and} \quad P(40 < X < 60) \doteq .943112.$$

The difference between them is of course $P(X = 40) + P(X = 60)$, which (to six places) is $.010844 + .010844 = .021688$.

The normal approximation to these two probabilities, .9544, is in error by only a little over 1% of either of the two probabilities. If so small an error is still intolerable in a given situation, the "continuity correction," covered in (6.4.10), is a means of improving the approximation.

Finally, notice how likely it is that the number of heads in 100 tosses is within 10 of its expected value. When we compute the normal approximation we see that

$\sigma = \sqrt{npq} = 5$, and so $10 = 2\sigma$. Any normally distributed random variable has probability .9544 of being within 2 standard deviations of its expected value.

As an exercise here (Exercise 6 at the end of this section), compute the probability that the number of successes is between 40 and 60 for a number of different Bernoulli trial situations in which $np = 50$. (For example, $n = 200$, $p = .25$.) When you do, you will see that what is crucial is the standard deviation $\sigma = \sqrt{npq}$, or more accurately, the value of $10/\sigma$, which is the number 10 measured in standard deviations. ▮

(6.4.9) Example. Let X be the number of 6's in 30 rolls of a fair die. We expect X to average 5 in the long run. What is the probability that on a given trial it is 3 or less? Let us find the true value and the normal approximation. Here $n = 30$ and $p = \frac{1}{6}$, so that np is just 5, which is right on the border of what the usual rule of thumb recommends for a good approximation.

The true value of $P(X \le 3)$ is

$$\begin{aligned} &P(X = 0) + P(X = 1) + P(X = 2) + P(X = 3) \\ &= \binom{30}{0}\left(\tfrac{1}{6}\right)^0\left(\tfrac{5}{6}\right)^{30} + \binom{30}{1}\left(\tfrac{1}{6}\right)^1\left(\tfrac{5}{6}\right)^{29} + \binom{30}{2}\left(\tfrac{1}{6}\right)^2\left(\tfrac{5}{6}\right)^{28} + \binom{30}{3}\left(\tfrac{1}{6}\right)^3\left(\tfrac{5}{6}\right)^{27} \\ &\doteq .004213 + .025276 + .073301 + .136829 = .239619. \end{aligned}$$

For the normal approximation we have $\mu = np = 5$ and $\sigma^2 = npq = \frac{25}{6}$, so

$$\begin{aligned} P(X \le 3) &= P\left(\frac{X - 5}{\sqrt{25/6}} \le \frac{3 - 5}{\sqrt{25/6}}\right) \approx P(Z \le -.98) \qquad \text{where } Z \sim N(0, 1) \\ &\doteq .1635. \end{aligned}$$

The approximation is not so good here; in practice, it would be a good idea to use the "continuity correction" discussed in (6.4.10) and described in (6.4.12). ▮

(6.4.10) Approximating the probability of a single possible value and the "continuity correction." Because a normal distribution is absolutely continuous, it gives a probability of 0 to any single possible value. You might think that this renders it useless for approximating a probability like $P(X = k)$ for a single number k of successes.

But there is something that can be done. The probability $P\left(k - \frac{1}{2} \le X \le k + \frac{1}{2}\right)$ is positive with a normal distribution, and gives a good approximation to the binomial probability $P(X = k)$. To see why this should be so, look at Figure 6.4a, which shows the bin(30, 4) probabilities as a bar graph along with the $N(12, 7.2)$ density, which is the appropriate normal approximation. The bar graph is set up so that the width of each bar is 1 and its height is the binomial probability of the integer at the center of the bar; this makes the *area* of the bar equal to the probability. But because the curve fits the bar graph so well, the area of a given bar is fairly close to the area under the curve for the same x values. Notice that each bar differs from the curve area by the difference between the areas of two "triangles" (they have one curved side) of nearly equal size.

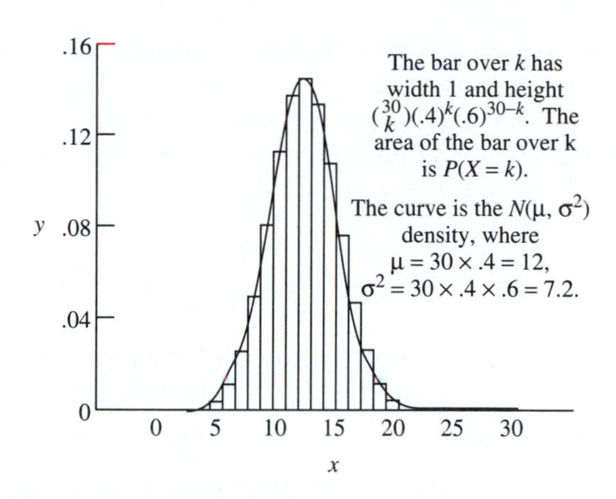

FIGURE 6.4a bin(30, .4) probabilities

(6.4.11) **Example.** Suppose $n = 30$ and $p = \frac{1}{6}$ as in Example (6.4.9). Let us approximate $P(X = 3)$ and then $P(X \le 3)$.

Solution. The true value of $P(X = 3)$ is

$$\binom{30}{3}\left(\tfrac{1}{6}\right)^3\left(\tfrac{5}{6}\right)^{27} = .136829,$$

and if we want a normal approximation to this we can use

$$\begin{aligned} P(X = 3) = P(2.5 \le X \le 3.5) &= P\left(\frac{2.5 - 5}{\sqrt{25/6}} \le \frac{X - 5}{\sqrt{25/6}} \le \frac{3.5 - 5}{\sqrt{25/6}}\right) \\ &\approx P(-1.22 \le Z \le -.73) \qquad \text{where } Z \sim N(0, 1) \\ &= P(.73 \le Z \le 1.22) \doteq .8888 - .7673 = .1215. \end{aligned}$$

This is still not an extremely good approximation; it is in error by more than 11%. We really need to have both np and nq well above 5 for the approximation to be very good. However, it may be more than adequate for many purposes.

We turn now to $P(X \le 3)$. If we approximate each of the four probabilities $P(X = k)$ $(k = 0, 1, 2, 3)$ in the same way, we get

$$\begin{aligned} P(X \le 3) &= P(X = 0) + P(X = 1) + P(X = 2) + P(X = 3) \\ &= P(-.5 \le X \le .5) + P(.5 \le X \le 1.5) + P(1.5 \le X \le 2.5) \\ &\quad + P(2.5 \le X \le 3.5). \end{aligned}$$

There is no need to evaluate these four separately; we can combine them as $P(-.5 \le X \le 3.5)$, and then use the normal approximation:

$$\begin{aligned} P(-.5 \le X \le 3.5) &= P\left(\frac{-.5 - 5}{\sqrt{25/6}} \le \frac{X - 5}{\sqrt{25/6}} \le \frac{3.5 - 5}{\sqrt{25/6}}\right) \\ &\approx P(-2.69 \le Z \le -.73) \doteq .2291. \end{aligned}$$

This approximation is much closer to the true probability (to four places, .2396) than the one we got from $P(0 \le X \le 3)$.

Notice that we get nearly the same approximation (as it happens, a little better) if we ignore the left endpoint of the interval [0, 3] and simply calculate as follows:

$$P(X \le 3) = P(X \le 3.5) = P\left(\frac{X-5}{\sqrt{25/6}} \le \frac{3.5-5}{\sqrt{25/6}}\right) \approx P(Z \le -.73) \doteq .2327. \quad \blacksquare$$

In general, the normal approximation for binomial probabilities is improved if we use the following device, which is usually called the ***continuity correction*** for the approximation:

(6.4.12) If $X \sim \text{bin}(n, p)$, then, using $\mu = np$ and $\sigma^2 = npq$, approximate $P(X \le k)$ by using normal tables to find $P\left(X \le k + \frac{1}{2}\right)$, and for $P(X \ge k)$ use $P\left(X \ge k - \frac{1}{2}\right)$.

Thus we approximate $P(k \le X \le l)$ by finding $P\left(k - \frac{1}{2} \le X \le l + \frac{1}{2}\right)$, and $P(X = k)$ by finding $P(k - \frac{1}{2} \le X \le k + \frac{1}{2})$.

(6.4.13) **Example.** If 51% of the voters in a large population favor candidate A over candidate B, what is the probability that in a sample of 150 voters, a majority favor candidate B? (This is the probability that the poll makes the wrong prediction.)

We assume the sampling is with replacement. [Real sampling is done without replacement, but the probabilities are practically the same if the population is large; we will discuss this phenomenon later in Example (7.5.4).] So we have $n = 150$ Bernoulli trials with success probability $p = .49$ (we are counting a voter who favors candidate B as a success), and we want the probability that the number of successes is 76 or more.

The quantities np and nq are both well over 5, so according to (6.4.12) we approximate this by finding the normal probability that X, the number of successes, is 75.5 or more. The parameters are $\mu = np = 150 \cdot .49 = 73.5$ and $\sigma^2 = npq = 37.485$; so $\sigma = \sqrt{37.485} \doteq 6.1225$. We get

$$P(X \ge 76) = P(X \ge 75.5) = P\left(\frac{X - 73.5}{6.1225} \ge \frac{75.5 - 73.5}{6.1225}\right)$$
$$\approx P(Z \ge .33) \doteq .3707.$$

This is quite a sizeable probability of error; if the election is close, a poll based on 150 voters has less than $\frac{2}{3}$ chance of predicting it correctly. At this point you ought to do Exercise 5, which asks you to find the same probability, but assuming the sample size is 1,500 instead of 150. Polling agencies typically use sample sizes of 1,200 or more. Even with 1,500, there is only about an 80% chance of a correct call in an election where the population is 51% for one candidate and 49% for another. But see also Exercise 9, in which we see that the probability of a correct call is much greater if the election is not so close.

Let us see whether the continuity correction was worth the effort, by computing the same probability without it:

$$P(X \geq 76) = P\left(\frac{X - 73.5}{6.1225} \geq \frac{76 - 73.5}{6.1225}\right) \approx P(Z \geq .41) \doteq .3409.$$

For many purposes we might not be concerned about the difference between .3707 and .3409; in either case, we have a pretty good idea of the chance that the poll fails us. But in some situations more accuracy may be important.

Of course, if it is crucial to know the probability of a polling error as exactly as possible, then we would use a computer to find the actual probability as a sum of binomial probabilities:

$$\begin{aligned} P(X \geq 76) &= p(76) + p(77) + \cdots + p(150) \\ &= \sum_{k=76}^{150} \binom{150}{k} (.49)^k (.51)^{150-k} = .3718. \end{aligned}$$

▮

(6.4.14) Example. Should we be suspicious if we roll a pair of dice 100 times and get nine 11's? The probability of an 11 with a fair pair of dice is $\frac{1}{18}$, so the expected number of 11's in 100 rolls is $100 \cdot \frac{1}{18} \doteq 5.556$. We are wondering if the observed number of 11's, nine, deviates suspiciously from the expected number, 5.556.

The statistical approach to this question is to compute the probability that 100 rolls of a fair pair of dice will produce nine or more 11's. If this probability is small, then we can say that *either* the dice are not fair, *or* we have just witnessed a rare occurrence.

Notice that we do not ask for the probability of *exactly* nine 11's; that is going to be fairly small in any case, even if nine were the expected number of 11's. We would be suspicious only if it would be rare for fair dice to produce a deviation *equal to or greater than* the observed deviation.

So our problem is this. The probability of an 11 is $p = \frac{1}{18}$. What is the probability that $X \geq 9$, where X is the number of successes in $n = 100$ trials? Both np and nq are greater than 5, so we can use the normal approximation. We use $\mu = np \doteq 5.556$ and $\sigma = \sqrt{npq} \doteq 2.291$. We will also use the continuity correction. We want the probability of the possible values 9, 10, 11, 12, and so on; so we find the normal probabilities of the intervals [8.5, 9.5], [9.5, 10.5], and so on. That is, we find

$$\begin{aligned} P(X \geq 8.5) &= P\left(\frac{X - 5.556}{2.291} \geq \frac{8.5 - 5.556}{2.291}\right) \\ &\approx P(Z \geq 1.29) \qquad \text{where } Z \sim N(0, 1) \\ &\doteq .0985. \end{aligned}$$

This is rather small; we can say that either the dice are not fair, or else the dice are fair and something has occurred whose probability is only about 10%. This is the way inferences are made in the branch of statistical inference called *hypothesis testing*. ∎

(6.4.15) **Computing binomial probabilities.** Here we consider the computation of true—not approximate—binomial probabilities using a computer or a calculator. It is often most efficient to compute them recursively, starting with either $p(0)$ or $p(n)$. Starting with $p(0)$, for example, we would compute $p(0) = q^n$ and then use

$$\frac{p(1)}{p(0)} = \frac{npq^{n-1}}{q^n} = \frac{np}{q}, \quad \text{and so} \quad p(1) = p(0) \cdot n \cdot \frac{p}{q};$$

$$\frac{p(2)}{p(1)} = \frac{\binom{n}{2}p^2q^{n-2}}{\binom{n}{1}pq^{n-1}} = \frac{(n-1)p}{2q}, \quad \text{and so} \quad p(2) = p(1) \cdot \frac{n-1}{2} \cdot \frac{p}{q};$$

and, in general,

$$p(k) = p(k-1) \cdot \frac{n-k+1}{k} \cdot \frac{p}{q}.$$

Or it may be desirable to start with $p(n) = p^n$ and use

$$p(n-k) = p(n-k+1) \cdot \frac{n-k+1}{k} \cdot \frac{q}{p}.$$

Typically we would start with $p(0) = q^n$ if q is larger than p and with $p(n) = p^n$ if p is larger.

(6.4.16) **Example.** Let us find a few probabilities for the bin(45, .16) distribution. Because q is larger than p, $p(0) = q^n$ is larger than $p(n) = p^n$, and so we start with $p(0)$:

$$p(0) = (.84)^{45} \doteq .00039135228.$$

If we were using a computer we would not worry about how many decimal places we were carrying. We would simply take what the computer gave us, using double precision arithmetic if we were able and wanted to do so, and round off the answers to the desired number of places. With a hand calculator, we can also leave each probability in the calculator to start the calculation of the next; this is preferable. The calculator that was used for this example carries twelve significant digits, and although we are reporting eight, twelve were carried throughout the calculations.

For the recursion we will be multiplying at each step by p/q, which is $.16/.84 = 4/21$. This is a repeating decimal, .190476190476. . . ; if we are using a computer we can just store this number, but with a hand calculator it is simpler to multiply by 4 and divide by 21 each time.

$$p(1) = p(0) \cdot 45 \cdot \tfrac{4}{21} \doteq .0033544481.$$

Continuing,

$$p(2) = p(1) \cdot \tfrac{44}{2} \cdot \tfrac{4}{21} \doteq .014056735;$$

$$p(3) = p(2) \cdot \tfrac{43}{3} \cdot \tfrac{4}{21} \doteq .038377118;$$

$$p(4) = p(3) \cdot \tfrac{42}{4} \cdot \tfrac{4}{21} \doteq .076754326;$$

and so on. ∎

EXERCISES FOR **SECTION 6.4**

1. For n Bernoulli trials with success probability p, find the true value and the Poisson approximation for each of the following probabilities.
   a. For $n = 20$ and $p = .1$, the probability of three successes
   b. For $n = 40$ and $p = .05$, the probability of three successes
   c. For $n = 100$ and $p = .02$, the probability of three successes

2. For n Bernoulli trials with success probability p, find the true value and the Poisson approximation for each of the following probabilities.
   a. For $n = 10$ and $p = .1$, the probability of at least two successes
   b. For $n = 20$ and $p = .015$, the probability of no successes
   c. For $n = 35$ and $p = .2$, the probability of five successes

3. Verify several of the numbers in the table in Example (6.4.4).

4. The people who patronize a certain restaurant have a probability of .001 (one chance in a thousand) of getting up and singing with the band that is playing. On a certain evening there are 150 people in the restaurant. What is the probability that at least one person gets up and sings with the band? (Assume that each person in the restaurant makes the decision independently of the others.) Find the true value and the Poisson approximation.

5. In a large population, 51% of the voters favor candidate A and 49% favor candidate B. What is the probability that in a sample of 1,500 voters, a majority (751 or more) favor candidate B? Use the normal approximation, both with and without the continuity correction.

   Notice that the continuity correction makes only a slight difference when n is large.

6. Find the normal approximation to the probability that the number of successes in n Bernoulli trials is between 40 and 60, for each of the following sets of

parameters. In each case the expected number of successes is $np = 50$. Do not bother with the continuity correction.
a. $n = 100, p = .5$
b. $n = 200, p = .25$
c. $n = 500, p = .1$
d. $n = 5{,}000, p = .01$

Notice that the probability of a deviation of 10 or less from the expected value of 50 depends on the standard deviation σ—or, more precisely, on $10/\sigma$, which is 10 measured in units of one standard deviation.

7. Use the normal approximation to the binomial distribution to find the approximate probability that in 79 tosses of a fair coin there are fewer than 30 heads. Use a continuity correction.

8. The event whose probability is found in Exercise 7, "fewer than 30 heads in 79 tosses," can be restated as "80 or more trials are needed to produce the 30th head"—that is, "50 or more tails before the 30th head." You can approximate this using the Central Limit Theorem as follows.
a. What are the expected value and variance of the number of tails before the first head, or between the first and second, etc.?
b. What are the expected value and variance of the number of tails before the 30th head?
c. Use the Central Limit Theorem to approximate the probability that this number is 50 or more. Use 49.5 instead of 50; this amounts to a "continuity correction."

In Exercises 7 and 8 we found two approximations to the same probability, and they are very different. The approximation in Exercise 7 is much closer to the true probability. There are two reasons for this. First, the approximation in Exercise 8 is based on a sum of only 30 independent random variables, whereas the one in Exercise 7 uses 79. Second, the 30 random variables in Exercise 8 have a geometric distribution, which is highly skewed, whereas the 79 in Exercise 7 have the Bernoulli distribution with $p = \frac{1}{2}$, which, while not very bell shaped, is at least symmetric.

For the record, the true value of the probability that we approximated in Exercises 7 and 8 is .0119 (to three significant digits).

9. What is the approximate probability that a poll of 1,500 randomly chosen voters makes the correct prediction in an election in which 47% of the voters favor candidate B and 53% favor candidate A?

10. Event A has probability .4. This of course means that we expect the relative frequency of A to be near .4 in a large number of trials of the experiment being modeled. What is the probability that in 1,000 trials the relative frequency of A is between .38 and .42 (inclusive)? Use an appropriate approximation.

The relative frequency "should" be near .4, because probabilities are supposed to represent long-run relative frequencies. But in fact it only has a reasonably high probability of being within .02 of .4. See the comment after Exercise 12 also.

11. **Can a .329 hitter have a .400 season?** What is the probability that a baseball player whose "real" batting average is .329 will bat .400 or better for a given season? Assume that .329 is the player's true probability of getting a hit in a given time at bat, that successive times at bat are Bernoulli trials, and that the player has 600 times at bat in the season.

Daniel Seligman, in his "Keeping Up" column in *Fortune* magazine for September 22, 1980, discussed the possibility of George Brett's batting .400 that year. He pointed out that .329 is a generous estimate of Brett's "real" batting average, since although it was his previous season's average it was considerably higher than his lifetime average. Consequently, he argued, Brett would be very lucky to hit .400. (In fact, he did not.) What a statistician would say, had Brett hit .400, is that either Brett was very lucky, or the model of Bernoulli trials with success probability .329 was not appropriate for Brett in 1980.

We should point out that in September 1980 Brett was hitting nearly .400 and there were fewer than 100 times at bat to go. So his probability at that time of finishing with a .400 season was considerably higher than the answer to this problem. This problem is more appropriately thought of in connection with a completed season, or one not yet begun.

12. When five fair dice are rolled, the chance of getting five of a kind (a "yacht") is $1/6^5$, or 1/7,776. If five fair dice are rolled 7,776 times, what is the probability of getting five of a kind at least once? Find the approximate value using the appropriate approximation. [The true value is $1 - \left(\frac{7{,}775}{7{,}776}\right)^{7{,}776} \doteq .632144$, and your calculator might give you this without difficulty. But the approximation is quite close, and is not subject to roundoff error as is a high power of a number near 1.]

In Exercises 10 and 12 we see two things that many people think "should" happen, but that in fact only have a reasonably high probability of happening. The relative frequency of an event "should" be close to its probability, but might not be. An event with probability $1/n$ "should" occur once in n trials, but the probability that it does not is approximately 37%.

13. **Skepticism about successful prognosticators.** If you predict in advance the outcomes of ten tosses of a fair coin, the probability that you get them all right is $1/2^{10}$, which equals 1/1,024. Show that if 2,000 people try to predict the ten outcomes, the chance that at least one of them succeeds is better than 85%.

In his book *A Random Walk down Wall Street,* Burton Malkiel points out that we should not be overly impressed when someone correctly predicts the action of the stock market over some period of time. As this exercise shows, when a large number of people are making such predictions it is not surprising that one or more will be correct, purely by chance. If one person out of 2,000 correctly predicted the outcome of ten coin tosses, a prudent observer would probably ask that person to repeat the performance before deciding that the person has special abilities.

14. Someone offers a prize to anyone who can get 125 or more 6's in 600 rolls of a fair die. What is the probability that if 300 people try, at least two will win prizes? [*Hint:* Two approximations are needed.]

15. Compute the binomial probabilities in the table in Example (6.4.4) using the recursive method described in (6.4.15).

ANSWERS

1. **a.** .1901; .1804 **b.** .1851; .1804 **c.** .1823; .1804

2. **a.** .2639; .2642 **b.** .7391; .7408 **c.** .1286; .1277

4. .13936; .13929

5. With: .2119; without: .2033

6. **a.** .9544 **b.** .8968 **c.** .8638 **d.** .8444

7. .0122

8. **a.** 1; 2 **b.** 30; 60 **c.** .0059

9. Without continuity correction: .9887; with continuity correction: .9893

10. .8132 (with continuity correction)

11. About .0001 (continuity correction used)

12. .632121

13. It is about .8582.

14. .3046

SUPPLEMENTARY EXERCISES FOR CHAPTER 6

1. X has a normal distribution; $EX = 23$ and $VX = 15$. Find the following.
   a. $P(X \leq 30)$
   b. $P(X$ fails to be within 5 of its expected value)
   c. The 80th percentile of the distribution of X
   d. The number a such that $P(23 - a \leq X \leq 23 + a) = .98$

2. A factory makes ball bearings; the specifications call for bearings whose diameter is .6 cm. The actual diameters of the bearings are random, having a normal distribution with an expected value of .6 cm and a variance of .00012. A bearing is useless to the customer if its diameter is less than .58 cm or greater than .615 cm. What proportion of bearings are useless?

3. Suppose we take a random sample of 65 of the ball bearings in Exercise 2, and we compute the average of their diameters.
   a. What is the distribution of the average?
   b. What is the probability that the average is within .001 cm of .6 cm?

4. Suppose that the noontime temperature on a random winter day in a certain resort city can be modeled as a normally distributed random variable, with expected value 78 (degrees Fahrenheit) and standard deviation 4.
   a. What proportion of winter days have noontime temperatures above 75°?
   b. The local tourist bureau wants to be able to say that on 90% of the winter days the noontime temperature is above T. What should T be?

5. The demand for regular gasoline at a filling station averages 6,430 gallons per week, with a standard deviation of 421 gallons. The station receives gasoline

weekly from its supplier. How many gallons of regular gasoline should the manager order weekly, so as to have a 98% probability of meeting the demand in a given week? Assume that the demand is a random variable with a normal distribution.

6. In Bernoulli trials with success probability .75, find the approximate probability that the number of successes in the first thousand trials is
   a. Less than 780.
   b. Between 740 and 760 (inclusive).
   c. Within 5% of its expected value.

7. In Bernoulli trials with success probability .24, let X be the number of failures preceding the 100th success.
   a. What are the expected value and variance of X?
   b. Find the approximate probability that X is less than 300. Why is the normal approximation appropriate?

8. We want the probability that we get exactly three double 6's in 90 rolls of a fair pair of dice.
   a. Find the exact probability, to four significant digits.
   b. Find the Poisson approximation.
   c. Find the normal approximation.
   d. Why should we expect the Poisson approximation to be better?

9. Let X be the number of heads in n tosses of a fair coin. For each of the following values of n, find the approximate probability that X differs from its expected value by 10 or less. Use the continuity correction.
   a. $n = 100$ b. $n = 500$ c. $n = 1{,}500$ d. $n = 5{,}000$

Notice that these probabilities decrease as n increases. This may seem to contradict the fact that X, the frequency of heads, is supposed to be close to $n/2$ when n is large (because the *relative* frequency of heads is supposed to approach $\frac{1}{2}$ as the number of trials increases). It does get close, but only proportionately, as the next exercise shows.

10. Repeat Exercise 9, but find in each case the probability that X differs from its expected value by 10% of the expected value or less.

11. Let X have the $N(8, 5)$ distribution and let Y have the $N(-4, 3)$ distribution. Suppose X and Y are independent. Find the following.
   a. $P(X + Y \leq 1)$
   b. $P(X - Y \leq 20)$
   c. $P(3X + 2Y \geq 18)$
   d. $P(X \leq -2Y)$

12. An event A has probability .15. So in a large number of trials, we expect the relative frequency of A to be about .15. If the number of trials is 300, find the approximate probability that the relative frequency is strictly between .13 and .17. Use the continuity correction.

13. Let an event A have probability .35. If we do 45 trials of the experiment, we

expect the relative frequency of A to be near .35; approximate the probability that it is between .3 and .4 (inclusive).

In Exercise 9 of Section 4.5 we used Chebyshev's inequality to bound the probability that the relative frequency of A in 45 trials is not between .3 and .4, and at that time we asserted that the probability bounded was around .44. This is where that answer came from.

ANSWERS

1. **a.** .9649 **b.** .1970 **c.** 26.25 **d.** 9.02
2. .1189
3. **a.** $N(.6, .000001846)$ **b.** .5408
4. **a.** .7734 **b.** 72.88
5. 7,293.05
6. **a.** With continuity correction: .9842; without: .9857 **b.** With: .5588; without: .5346 **c.** .9938
7. **a.** 316.666 . . . ; 1,319.444 . . . **b.** .3228
8. **a.** .2171 **b.** .2138 **c.** .2389
9. **a.** .9642 **b.** .6528 **c.** .4108 **d.** .2358
10. **a.** .7286 **b.** .9774 **c.** About .9999 d. $2 \cdot \Phi(7.085) - 1$, which is essentially 1
11. **a.** .1446 **b.** .9977 **c.** .3974 **d.** .5000
12. .6266
13. .5631

CHAPTER 7

Some Important Distributions on the Nonnegative Integers

7.1 Probability Generating Functions

7.2 Binomial and Bernoulli Distributions

7.3 Geometric and Negative Binomial Distributions

7.4 Poisson Distributions

7.5 Hypergeometric Distributions

7.6 Multinomial Distributions

Supplementary Exercises

7.1 PROBABILITY GENERATING FUNCTIONS

In this chapter we want to cover the important families of discrete distributions more systematically than we have done so far. We have already seen all but two of these families at various times in examples and exercises. The ones we have seen are the Bernoulli, binomial, geometric, and Poisson families. The new ones in this chapter are the negative binomial and the hypergeometric families.

These are all discrete distributions, but they have an additional feature in common: they have only nonnegative integers as possible values. There are of course many discrete distributions that have other numbers as possible values—if, for example, X has a binomial distribution with $n = 4$, then the distribution of $X/4$ assigns positive probabilities to the numbers $0, \frac{1}{4}, \frac{1}{2}, \frac{3}{4}$, and 1. But the discrete distributions that arise most often in applications have possible values that are nonnegative integers.

The distributions we cover in this chapter and the next are summarized in Appendix B. You might find it useful to have those summaries at hand while going through Sections 2–5 of this chapter and Sections 2–6 of Chapter 8.

The main difference between the discussions in this chapter and previous discussions of the same distributions is the use of a very important tool for dealing with distributions on nonnegative integers—the probability generating function, or PGF. This is a function associated with a distribution on the nonnegative integers that provides information about it in a way different from the CDF or the mass function. The rest of this section deals with this tool.

The important properties of PGFs are as follows.

1. The PGF determines the distribution uniquely; that is, there is just one distribution on the nonnegative integers for a given PGF. So if we find the PGF of a random variable and recognize it as the PGF of a known distribution, then that is the distribution of the random variable.
2. The PGF can be used to find the mass function and the expected value, variance, and other moments of the distribution.
3. The PGF of a sum of independent random variables is the product of their PGFs. We can use this together with property 1 to find distributions of sums.
4. If the PGF of X is the limit of some sequence of PGFs, then the distribution of X can be found as a certain kind of limit of the distributions associated with the PGFs in the sequence.

We begin with an example intended to illustrate the concept.

(7.1.1) Example. In Example (4.1.22) we made two different calculations showing that $EX = np$ if X is the number of successes in n Bernoulli trials with success probability p. Here we look again at the second calculation. The task was to show that

$$\sum_{k=0}^{n} k \cdot \binom{n}{k} p^k q^{n-k} = np. \tag{7.1.2}$$

We began by pulling the following identity out of the air; it appeared as (4.1.23) in Section 4.1.

(7.1.3)
$$\sum_{k=0}^{n} \binom{n}{k} p^k t^k q^{n-k} = (pt + q)^n.$$

It it easy to see that this is true; it comes from the binomial theorem (A.4.7) with pt in place of a and q in place of b. But it involves a new variable t that seems to have nothing to do with the problem; we gave no hint as to how anyone could know in advance that it would be useful. However, it is true for all t, and we can differentiate it and set t to anything we want. Differentiating (7.1.3) gives

$$\sum_{k=0}^{n} k \cdot \binom{n}{k} p^k t^{k-1} q^{n-k} = np(pt + q)^{n-1}$$

(where the first term on the left, the one for $k = 0$, is 0 for all t). And when we take $t = 1$, the left side becomes the same as the left side of (7.1.2) and the right side becomes $np(p + q)^{n-1} = np$. ∎

Now (7.1.3) is a function of t that we will define shortly as the probability generating function of the binomial distribution. The question is: where did it come from? If we wanted to try a similar trick for some other distribution, how would we know what identity to pull out of the air? The answer comes from looking at the left side of (7.1.3). It equals

(7.1.4)
$$\sum_{k} t^k p(k),$$

where $p(k)$ is the mass function of the binomial distribution. What happens is that we write down this sum and notice that it sums to a simple function of t, namely $(pt + q)^n$. Then we are in business.

Let us try the same trick with another familiar distribution, and see if we can get a quick calculation of its expected value.

(7.1.5) Example. Let X have the Poisson distribution with parameter λ. The mass function is

$$p(k) = \begin{cases} \dfrac{\lambda^k e^{-\lambda}}{k!} & \text{if } k = 0, 1, 2, \ldots, \\ 0 & \text{otherwise,} \end{cases}$$

and we want to show that

$$\sum_{k=0}^{\infty} k \cdot \frac{\lambda^k e^{-\lambda}}{k!} = \lambda.$$

What do we get if we write down (7.1.4) and try to evaluate it? It is not difficult:

$$\sum_k t^k p(k) = \sum_{k=0}^{\infty} t^k \cdot \frac{\lambda^k e^{-\lambda}}{k!} = e^{-\lambda} \sum_{k=0}^{\infty} \frac{(\lambda t)^k}{k!} = e^{-\lambda} e^{\lambda t} = e^{\lambda t - \lambda}.$$

The part of this that corresponds to (7.1.3) is

(7.1.6)
$$\sum_{k=0}^{\infty} t^k \cdot \frac{\lambda^k e^{-\lambda}}{k!} = e^{\lambda t - \lambda}.$$

Differentiating with respect to t (without simplifying the left side) gives

$$\sum_{k=0}^{\infty} k \cdot t^{k-1} \cdot \frac{\lambda^k e^{-\lambda}}{k!} = \lambda e^{\lambda t - \lambda}.$$

So it works again; setting $t = 1$ gives

$$\sum_{k=0}^{\infty} k \cdot \frac{\lambda^k e^{-\lambda}}{k!} = \lambda e^{\lambda - \lambda} = \lambda.$$ ∎

Motivated by these successes, we make the definition.

(7.1.7) **Definition.** Let X be a discrete random variable whose possible values are the nonnegative integers 0, 1, 2, 3, . . . or some subset of that set. Let the mass function of X be $p(k)$. The ***probability generating function*** (PGF) of X is the function $\pi(t)$ [or $\pi_X(t)$ if we want to emphasize which random variable we are dealing with] defined by

(7.1.8)
$$\pi(t) = \sum_k t^k p(k).$$

This is either a finite sum or an infinite series. If it is an infinite series, then we need to consider whether it converges or not. But it is easy to see that it converges, in fact converges absolutely, at least for all t in the interval $[-1, 1,]$, because if $|t| \leq 1$, then we have

$$\sum_k |t^k p(k)| = \sum_k |t|^k p(k) \leq \sum_k p(k) = 1.$$

Notice also that $\pi(1)$ always equals 1, because it is just the sum of all the values of the mass function of X.

Now from the elementary theory of power series, covered in first-year calculus courses, it follows that any power series $\sum_{k=0}^{\infty} a_k t^k$ with nonnegative coefficients a_k has the following properties.

Property A. If $\sum_{k=0}^{\infty} a_k t^k$ converges for any one positive value of t, say for $t = r$, then it converges for all t in the interval $[-r, r]$ and thus defines a function of t on that interval.

Property B. For any t in $(-r, r)$ this function is differentiable at t, and the series $\sum_{k=0}^{\infty} k a_k t^{k-1}$ converges to the derivative.

This derivative formula is called the "term-by-term derivative" of the power series. From these facts we get the following theorem.

(7.1.9) Theorem. If the possible values of X are 0, 1, 2, . . . or some subset of this set, and $\pi(t)$ is its PGF, then

the power series $\sum_{k=0}^{\infty} t^k p(k)$ for $\pi(t)$ converges for all t in $[-1, 1]$.

If it also happens to converge for any value of t *greater* than 1, then $\pi(t)$ is differentiable at $t = 1$, EX exists, and

$$\pi'(1) = EX.$$

Proof. The series converges at $t = 1$, because there it is just $\sum_{k=0}^{\infty} p(k)$, which equals 1. So it converges for all t in $[-1, 1]$ by Property A above.

Suppose the series converges for some $t > 1$. The differentiability of $\pi(t)$ at $t = 1$ follows from Property B, and when we find the term-by-term derivative we get

$$\pi'(1) = \sum_{k=0}^{\infty} kp(k)1^{k-1} = \sum_{k=0}^{\infty} kp(k).$$

Since this is finite, EX exists and equals it. □

Theorem (7.1.9) is not the whole truth about finding expected values from PGFs; some PGFs do not converge for any values of t greater than 1, but they can still be used to find expected values. We will discuss this briefly in (7.1.23). But this theorem will serve us in all the examples we deal with.

At the risk of repetition, let us summarize what we learned in the two examples at the beginning of this section, in the language of PGFs.

If X has the binomial distribution with parameters n and p, then the PGF of X is $\pi(t) = (pt + q)^n$ for all t. Consequently,

$$EX = \pi'(1) = [np(pt + q)^{n-1}]_{t=1} = np.$$

If X has the Poisson distribution with parameter λ, then the PGF of X is $\pi(t) = e^{\lambda t - \lambda}$ for all t. Consequently,

$$EX = \pi'(1) = [\lambda e^{\lambda t - \lambda}]_{t=1} = \lambda.$$

Another note is in order here. There is no point in looking for any significance in the variable t or the value of $\pi(t)$. We rarely put any t into $\pi(t)$ to see what the value is, and we seldom are interested in a graph of the function π. It is simply that $\pi(t)$ is the sum of a power series whose coefficients have meaning.

The useful thing to know about a PGF is simply a formula for it. For example, the left sides of (7.1.3) and (7.1.6) are not of much use; it is only when we realize that they sum to $(pt + q)^n$ and $e^{\lambda t - \lambda}$ that we can make use of them.

(7.1.10) Example. Let X be the number of tails before the second head in repeated tosses of a fair coin. We will find the PGF of X. Then we will differentiate and set $t = 1$ to get EX.

As we know from (1.2.7), the mass function of X is $p(k) = (k + 1)\left(\frac{1}{2}\right)^{k+2}$ for $k = 0, 1, 2, \ldots$. So the probability generating function is

$$\begin{aligned}\pi(t) &= \sum_{k=0}^{\infty}(k + 1)t^k\left(\tfrac{1}{2}\right)^{k+2} = 1 \cdot \left(\tfrac{1}{2}\right)^2 + 2t \cdot \left(\tfrac{1}{2}\right)^3 + 3t^2 \cdot \left(\tfrac{1}{2}\right)^4 + 4t^3 \cdot \left(\tfrac{1}{2}\right)^5 + \cdots \\ &= \left(\tfrac{1}{2}\right)^2(1 + 2x + 3x^2 + 4x^3 + \cdots) \quad \left(\text{where } x = \tfrac{t}{2}\right) \\ &= \left(\tfrac{1}{2}\right)^2 \frac{1}{\left(1 - \frac{t}{2}\right)^2} = \frac{1}{(2 - t)^2}.\end{aligned}$$

For which t is this valid? The formula

$$1 + 2x + 3x^2 = \cdots = \frac{1}{(1 - x)^2}$$

is valid for $|x| < 1$, because it is the derivative of the geometric series formula, which in turn is valid for $|x| < 1$. So, since $x = t/2$, our calculation is valid for $|t| < 2$. So the PGF converges for at least one value of t greater than 1, and, consequently, we can differentiate and set $t = 1$ to get EX:

$$EX = \pi'(1) = \left[\frac{2}{(2 - t)^3}\right]_{t=1} = 2. \qquad \blacksquare$$

As an exercise at this point (Exercise 4 at the end of this section), you should do the above for Bernoulli trials with success probability p instead of $\frac{1}{2}$.

(7.1.11) Finding higher moments of X from the probability generating function. We can find the other moments of X as well if we know its PGF $\pi(t)$, by taking higher derivatives. To keep things simple, we continue to assume that $\pi(t)$ exists for some $t > 1$, so that $\pi(t)$ is differentiable the required number of times in a neighborhood of $t = 1$.

We start as before with

$$\pi(t) = \sum_k t^k p(k).$$

and the first derivative,

$$\pi'(t) = \sum_k k t^{k-1} p(k).$$

If we differentiate again, we get

(7.1.12) $$\pi''(t) = \sum_k k(k-1) t^{k-2} p(k),$$

where the terms for $k = 0$ and $k = 1$ are both 0 for all t. If we now set $t = 1$ we get

(7.1.13) $$\pi''(1) = \sum_k k(k-1)p(k) = E(X(X-1)) = EX^2 - EX.$$

Of course, from this we can get the second moment $\mu_2 = EX^2$, if we know EX.

Continuing, if we differentiate (7.1.12) and set $t = 1$ we get

(7.1.14) $$\pi'''(1) = \sum_k k(k-1)(k-2)p(k) = E(X(X-1)(X-2)) = EX^3 - 3EX^2 + 2EX.$$

From this we can get the third moment $\mu_3 = EX^3$, if we know EX and EX^2.

In general, it is not difficult to see that the nth derivative at $t = 1$ is

(7.1.15) $$\begin{aligned}\pi^{(n)}(1) &= \sum_k k(k-1)(k-2)\cdots(k-n+1)p(k)\\ &= E(X(X-1)(X-2)\cdots(X-n+1)),\end{aligned}$$

at least provided that $\pi(t)$ can be differentiated n times in some neighborhood of $t = 1$.

Notice that we do not get the moments of X directly by putting $t = 1$ in the derivatives of the PGF. Instead we get what are called the ***factorial moments.*** The expected value in (7.1.15) is called the nth factorial moment of X, and to find the nth moment EX^n from it we need to know the first $n - 1$ moments as well. There exist formulas for finding the moments EX^n if the factorial moments $E(X(X-1)(X-2)\cdots(X-n+1))$ are known. They involve some coefficients, called Stirling numbers, that arise in various contexts in combinatorial mathematics. We will not go into them here. If n is small, it is a simple matter to find the moments, as we noted in the remarks following (7.1.13) and (7.1.14).

(7.1.16) **Example: PGF and moments of the geometric distribution.** In repeated Bernoulli trials, let X be the number of failures preceding the first success, assuming the success probability is p. In Example (4.1.3) we found that $EX = q/p$, and in Exercise 9 of Section 4.2 we found that $E(X(X-1)) = 2q^2/p^2$. Each of these was found by summing a series. Now we see that we need to sum just one series—the PGF—and then we can get these two quantities, and more, from it.

From EX and $E(X(X - 1))$, we can proceed to find EX^2 as $E(X(X - 1)) + EX$, and then to find VX as $EX^2 - (EX)^2$. There is no shortcut for these last two steps; but finding EX and $E(X(X - 1))$ can be shortened—and we can find higher moments if we want them—by finding the PGF first.

We know the mass function well:

$$p(k) = pq^k \quad \text{for } k = 0, 1, 2, \ldots .$$

The PGF is simple to find from the mass function:

(7.1.17)
$$\pi(t) = \sum_{k=0}^{\infty} t^k \cdot pq^k = p \sum_{k=0}^{\infty} (qt)^k = p(1 + qt + (qt)^2 + (qt)^3 + \cdots),$$

which converges, as long as $|qt| < 1$, to

(7.1.18)
$$\pi(t) = \frac{p}{1 - qt}.$$

Since $q < 1$, there is a value of t greater than 1 for which this converges, and so we are justified in differentiating to get moments. Differentiating (7.1.18), we get

(7.1.19)
$$\pi'(t) = \frac{pq}{(1 - qt)^2},$$

and putting in $t = 1$ gives us

$$EX = \pi'(1) = \frac{pq}{(1 - q)^2} = \frac{q}{p}.$$

Now we differentiate (7.1.19) and set $t = 1$ to get $E(X(X - 1))$:

(7.1.20)
$$\pi''(t) = \frac{2pq^2}{(1 - qt)^3},$$

so

$$E(X(X - 1)) = \pi''(1) = \frac{2pq^2}{(1 - q)^3} = \frac{2q^2}{p^2}.$$

As we remarked above, if we need VX we still have to do the following steps: First find EX^2,

$$EX^2 = E(X(X-1)) + EX = \frac{2q^2}{p^2} + \frac{q}{p} = \frac{2q^2 + pq}{p^2},$$

and then find VX,

$$VX = EX^2 - (EX)^2 = \frac{2q^2 + pq}{p^2} - \frac{q^2}{p^2} = \frac{q^2 + pq}{p^2} = \frac{q}{p^2}.$$

Now we see how we can proceed to find further factorial moments of the geometric distribution. Further differentiation, starting from where we left off at (7.1.20), gives

(7.1.21)

$$\pi'''(t) = \frac{6pq^3}{(1-qt)^4},$$

$$\pi^{(\text{iv})}(t) = \frac{24pq^4}{(1-qt)^5},$$

and, in general,

(7.1.22)

$$\pi^{(n)}(t) = \frac{n!pq^n}{(1-qt)^{n+1}}.$$

Putting $t = 1$ in this gives us all the factorial moments of the geometric distribution:

$$E(X(X-1)(X-2)\cdots(X-n+1)) = \frac{n!q^n}{p^n} \text{ for } n = 1, 2, 3, \ldots. \quad \blacksquare$$

(7.1.23) **Note on the hypothesis of existence of $\pi(t)$ for some $t > 1$.** We included this hypothesis in Theorem (7.1.9) and (7.1.11) only because the proofs are very simple in that case. In fact, there are some PGFs that do not exist for any $t > 1$, but that can still be used because they have derivatives from the left at $t = 1$. The full truth of the situation can be stated as follows (although we omit its proof, which requires some advanced calculus).

Any PGF $\pi(t)$ is differentiable at least for $-1 < t < 1$. As for differentiability at $t = 1$, *either*

$$\pi'(1-) = \lim_{t \uparrow 1} \pi'(t) = EX$$

(that is, all three quantities are finite and equal), *or*

$\pi'(1-)$ does not exist, $\lim_{t\uparrow 1} \pi'(t) = +\infty$, and the series for EX diverges to $+\infty$.

Consequently, one way to check convergence is to find $\pi'(t)$ for $-1 < t < 1$, which always exists, and then to look at the limit as t increases to 1. If this limit is finite then it equals EX [and also $\pi'(1)$]; if not, EX does not exist.

The same approach works for higher moments as well. The kth derivative of $\pi(t)$ always exists for $-1 < t < 1$, and the three quantities $\pi^{(k)}(1)$, $\lim_{t\uparrow 1} \pi^{(k)}(t)$, and $E(X(X-1)(X-2)\cdots(X-k+1))$ are either all finite or all nonexistent (divergent to $+\infty$).

Of course, if it happens that $\pi(t)$ does exist for some $t > 1$, then we have no need of these stronger theorems.

At the beginning of the section we listed four important properties of generating functions, and so far we have concentrated only on the second. We turn now to two other important properties.

(7.1.24) **The probability generating function determines the distribution.** Suppose random variables X and Y (whose possible values are nonnegative integers) have the same probability generating function. Could they possibly have different distributions? No, because the two power series

$$\pi_X(t) = \sum_{k=0}^{\infty} t^k P(X = k) \quad \text{and} \quad \pi_Y(t) = \sum_{k=0}^{\infty} t^k P(Y = k)$$

are equal for all t in the interval $(-1, 1)$; one of the basic facts about power series is that if two power series agree on an interval, then the corresponding coefficients of these power series are the same. That is, $P(X = k) = P(Y = k)$ for all k, which says that X and Y have the same distribution.

In fact, we can find the probabilities $P(X = k)$ from the PGF $\pi_X(t)$, because they are the coefficients in the Maclaurin series (Taylor series about 0) for $\pi_X(t)$, and therefore they are related in a simple way to the derivatives of $\pi(t)$ at $t = 0$. Suppose that $p(k)$ is the mass function and $\pi(t)$ is the PGF of X. Then

$$\pi(t) = p(0) + tp(1) + t^2p(2) + t^3p(3) + t^4p(4) + \cdots, \quad \text{and so} \quad \pi(0) = p(0),$$
$$\pi'(t) = p(1) + 2tp(2) + 3t^2p(3) + 4t^3p(4) + \cdots \quad \text{and so} \quad \pi'(0) = p(1),$$
$$\pi''(t) = 2p(2) + 6tp(3) + 12t^2p(4) + \cdots \quad \text{and so} \quad \frac{\pi''(0)}{2} = p(2),$$
$$\pi'''(t) = 6p(3) + 24tp(4) + \cdots \quad \text{and so} \quad \frac{\pi'''(0)}{3!} = p(3),$$

and, in general,

(7.1.25)

$$p(k) = P(X = k) = \frac{\pi^{(k)}(0)}{k!}.$$

This is why $\pi(t)$ is called a *probability generating function:* it is a function whose Maclaurin series has the probabilities as its coefficients, and the probabilities can accordingly be found from the derivatives of $\pi(t)$ at 0, or, equivalently, from the power series expansion of $\pi(t)$.

Using (7.1.25) is usually a tedious way to find the coefficients of a power series. We would use it in general only as a last resort. It is much better, if possible, to use known power series expansions to get the coefficients, as in the next example.

(7.1.26) **Example.** Suppose we learn through some means or other that X is a discrete random variable, whose possible values are nonnegative integers, and whose PGF is

$$\pi(t) = \frac{p}{1 - qt^2},$$

for some p and q with $q = 1 - p$. What is the probability mass function of X?

The first thing to try is to expand this function in a power series. Using the geometric series formula (A.5.8), we get

$$\begin{aligned}\pi(t) &= p(1 + qt^2 + (qt^2)^2 + (qt^2)^3 + (qt^2)^4 + \cdots) \\ &= p + pqt^2 + pq^2t^4 + pq^3t^6 + pq^4t^8 + \cdots.\end{aligned}$$

The pattern in the coefficients of powers of t is discernible, so we can read off the mass function:

$$\begin{aligned}p(0) &= p \quad \text{(the coefficient of } t^0\text{)},\\ p(1) &= 0 \quad \text{(the coefficient of } t^1\text{)},\\ p(2) &= pq \quad \text{(the coefficient of } t^2\text{)},\\ p(3) &= 0,\\ p(4) &= pq^2,\end{aligned}$$

and, in general,

$$p(k) = \begin{cases} pq^{k/2} & \text{if } k \text{ is an even integer,} \\ 0 & \text{otherwise.}\end{cases}$$

If this power series expansion had not been apparent to us, we would have had to use (7.1.25), as follows. You can see how tedious it will get.

$$\pi(0) = p, \quad \text{so} \quad p(0) = p;$$

$$\pi'(t) = \frac{2pqt}{(1 - qt^2)^2}, \quad \text{so} \quad \pi'(0) = 0, \quad \text{so} \quad p(1) = 0;$$

$$\pi''(t) = \frac{2pq(1 - qt^2)^2 - 2pqt \cdot 2(1 - qt^2)(-2qt)}{(1 - qt^2)^4},$$

$$\text{so} \quad \pi''(0) = 2pq, \quad \text{so} \quad p(2) = \frac{\pi''(0)}{2!} = pq,$$

and so forth. ∎

Of course, in many cases we will not even need to expand the PGF; instead, we will simply recognize it as one we have seen before. In this case, we will know the distribution. The next example shows how this works.

(7.1.27) **Example.** Suppose we learn somehow that the PGF of X is $\pi(t) = e^{3t-3}$. We could expand this as a power series if we wanted; but we remember from Example (7.1.5) that the Poisson distribution with parameter λ has the PGF $e^{\lambda t - \lambda}$. Since two different distributions cannot have the same PGF, our X must have the Poisson distribution with $\lambda = 3$. No more work needs to be done. ∎

As an exercise, write down the power series expansion of e^{3t-3} and notice that the coefficients really are the probabilities of the Poisson distribution with $\lambda = 3$. (To expand it, write it as $e^{-3} \cdot e^{3t}$ and use the familiar Maclaurin series for e^x with $3t$ in place of x.)

(7.1.28) **The probability generating function as an expected value.** If X is a random variable whose possible values are nonnegative integers, then by definition the probability generating function of X is

$$\pi(t) = \sum_k t^k P(X = k).$$

But now think back to Theorem (4.2.2), which tells us how to get the expected value of a function of a discrete random variable. For any function $h(X)$,

$$Eh(X) = \sum_k h(k)P(X = k).$$

It is clear that these two expressions are the same if $h(k) = t^k$—that is, if $h(X) = t^X$. We see that the PGF of X is

$$\pi(t) = E(t^X). \tag{7.1.29}$$

This formula makes it easy to remember the definition of the probability generating function. It also makes it easy to reconstruct the way in which the moments can be found from the PGF, by differentiating (7.1.29) formally ("formally" means "according to form, but not necessarily justified"):

$$\pi'(t) = \frac{d}{dt}E(t^X) \stackrel{*}{=} E\left(\frac{d}{dt}t^X\right) = E(Xt^{X-1}),$$

and, therefore,

$$\pi'(1) = EX;$$

$$\pi''(t) = \frac{d}{dt}E(Xt^{X-1}) \stackrel{*}{=} E\left(\frac{d}{dt}Xt^{X-1}\right) = E(X(X-1)t^{X-2}),$$

and, therefore,

$$\pi''(1) = E(X(X-1));$$

and so forth. We emphasize that these calculations are only *formally* correct; in particular, the equality signs marked with asterisks (*) are not justified. We would need to develop more theory to justify this kind of differentiation of expected values.

Notice that (7.1.29) makes sense even if X is not a discrete random variable with nonnegative integer values. The probability generating function can be defined for any random variable X whatever. However, there are other types of generating functions that are more useful if X is something other than nonnegative-integer-valued. We will cover this in the next chapter.

(7.1.30) **The probability generating function of a sum of independent random variables.** Let X and Y be random variables having nonnegative integer possible values, with mass functions

$$p_X(k) = P(X = k) \quad \text{and} \quad p_Y(k) = P(Y = k) \quad \text{for } k = 0, 1, 2, \ldots,$$

and probability generating functions

$$\pi_X(t) = \sum_{k=0}^{\infty} t^k p_X(t) \quad \text{and} \quad \pi_Y(t) = \sum_{k=0}^{\infty} t^k p_Y(t).$$

(Notice that we have to subscript the PGFs as well as the mass functions to indicate which random variables they pertain to.)

Now suppose that X and Y are independent and $Z = X + Y$. It then happens that the PGF of Z is the product of the PGFs of X and Y, as given by the following theorem.

(7.1.31) **Theorem.** If X and Y are independent random variables with nonnegative integer possible values, and $Z = X + Y$, then

$$\pi_Z(t) = \pi_X(t)\pi_Y(t) \quad \text{for all } t \text{ in the interval } [-1, 1].$$

Proof 1. A proof that is formally correct is easy if we use (7.1.29):

$$\pi_Z(t) = E(t^Z) = E(t^{X+Y}) = E(t^X \cdot t^Y).$$

Because X and Y are independent, so are t^X and t^Y; therefore, by (4.2.19) the expected value of their product is the product of their expected values:

$$\pi_Z(t) = E(t^X) \cdot E(t^Y) = \pi_X(t)\pi_Y(t).$$ □

Proof 2. The preceding proof is only formally correct because we did not justify the existence of all the expected values in it. A more mundane, but completely justified, proof of Theorem (7.1.31) considers $\pi_X(t)\pi_Y(t)$ as the product of two power series:

$$\pi_X(t)\pi_Y(t) = \left(p_X(0) + tp_X(1) + t^2p_X(2) + \cdots\right) \cdot \left(p_Y(0) + tp_Y(1) + t^2p_Y(2) + \cdots\right).$$

Now because both series converge in $(-1, 1)$ the product does also, and the coefficients of the various powers of t in the product are found as though we were multiplying polynomials:

the coefficient of t^0 in $\pi_X(t)\pi_Y(t)$ is $p_X(0)p_Y(0)$;

the coefficient of t^1 in $\pi_X(t)\pi_Y(t)$ is $p_X(0)p_Y(1) + p_X(1)p_Y(0)$;

the coefficient of t^2 in $\pi_X(t)\pi_Y(t)$ is $p_X(0)p_Y(2) + p_X(1)p_Y(1) + p_X(2)p_Y(0)$;

the coefficient of t^3 in $\pi_X(t)\pi_Y(t)$ is $p_X(0)p_Y(3) + p_X(1)p_Y(2) + p_X(2)p_Y(1) + p_X(3)p_Y(0)$;

and so forth.

But **because X and Y are independent,** these coefficients are precisely the probabilities of the various possible values of $Z = X + Y$:

$$\begin{aligned}
P(Z = 0) &= P(X = 0 \text{ and } Y = 0) = p_X(0)p_Y(0); \\
P(Z = 1) &= P(X = 0 \text{ and } Y = 1) + P(X = 1 \text{ and } Y = 0) \\
&= p_X(0)p_Y(1) + p_X(1)p_Y(0); \\
P(Z = 2) &= P(X = 0 \text{ and } Y = 2) + P(X = 1 \text{ and } Y = 1) \\
&\quad + P(X = 2 \text{ and } Y = 0) \\
&= p_X(0)p_Y(2) + p_X(1)p_Y(1) + p_X(2)p_Y(0); \\
P(Z = 3) &= P(X = 0 \text{ and } Y = 3) + P(X = 1 \text{ and } Y = 2) \\
&\quad + P(X = 2 \text{ and } Y = 1) + P(X = 0 \text{ and } Y = 3) \\
&= p_X(0)p_Y(3) + p_X(1)p_Y(2) + p_X(2)p_Y(1) + p_X(3)p_Y(0);
\end{aligned}$$

and so forth. □

Before giving an example, we note that although the independence of X and Y implies the product formula of Theorem (7.1.31), the converse is false; there are dependent random variables for which the product formula holds. Exercise 17 at the end of this section gives an example.

The following example illustrates the use of (7.1.24) and Theorem (7.1.31).

(7.1.32) Example: A sum of independent Poisson random variables has a Poisson distribution. Let X and Y be independent random variables; let X have the Poisson distribution with parameter λ_1 and let Y have the Poisson distribution with parameter λ_2. Let $Z = X + Y$. We will use Theorem (7.1.31) to find the PGF of Z. Then we will recognize this function as the PGF of the Poisson distribution with parameter $\lambda_1 + \lambda_2$, and so by (7.1.24) we will conclude that Z must have that distribution.

Recall that in Example (7.1.5) we found the PGF of the Poisson distribution with parameter λ; it is $e^{\lambda t - \lambda}$. Thus,

$$\pi_X(t) = e^{\lambda_1 t - \lambda_1} \quad \text{and} \quad \pi_Y(t) = e^{\lambda_2 t - \lambda_2},$$

and so by Theorem (7.1.31)

$$\pi_Z(t) = \pi_X(t) \cdot \pi_Y(t) = e^{\lambda_1 t - \lambda_1} \cdot e^{\lambda_2 t - \lambda_2} = e^{(\lambda_1 + \lambda_2)t - (\lambda_1 + \lambda_2)}.$$

This is exactly the Poisson PGF with $\lambda_1 + \lambda_2$ in place of λ. Since there can be only one distribution with a given PGF, Z can have no other distribution than the Poisson distribution with parameter $\lambda_1 + \lambda_2$. ∎

Summary of Section 7.1. Because this section is long, we will summarize the main results here, slightly rearranged from the order in which we encountered them.

1. The probability generating function of a random variable X is the function $\pi_X(t)$ of a real variable t defined by $\pi_X(t) = E(t^X)$. If the possible values of X are nonnegative integers, then

$$\pi_X(t) = \sum_{k=0}^{\infty} t^k P(X = k);$$

this power series converges for $-1 \le t \le 1$ and is differentiable for $-1 < t < 1$.

2. No two different distributions on the nonnegative integers can have the same probability generating function. Thus, if we find the PGF of X, and we recognize it as the PGF of a known distribution, then X has that distribution. Alternatively, if we can expand it in a power series, then the coefficients of that power series are the values of the mass function of the distribution.
3. If X and Y are independent random variables, then the PGF of their sum is the product of their PGFs.

4. $EX = \pi'(1)$, and more generally, for any positive integer n, we can find $E(X(X-1)(X-2)\cdots(X-n+1))$ by differentiating the PGF n times and setting $t = 1$ (or taking the limit as $t \uparrow 1$), always provided that the expected value exists (which is equivalent to saying that the limit as $t \uparrow 1$ of the nth derivative exists).

EXERCISES FOR **SECTION 7.1**

1. If we choose a random permutation of the digits 1, 2, 3, and 4 and let S be the number of digits in their proper places, then as we saw in Exercise 6d of Section 1.1, the possible values of S are 0, 1, 2, and 4, and the probabilities are 9, 8, 6, and 1, all divided by 24. Write down the PGF of S and differentiate it to find ES. [Notice how closely your calculation follows the ordinary calculation of ES from the mass function of S. Also see Exercise 19 below.]

2. Let X be the number of spots that show when a fair die is rolled. Find the PGF of X. Differentiate it and set $t = 1$ to show that $EX = 3.5$.

3. Suppose the PGF of X is $\pi(t) = e^{t^2-1}$.
   a. Find EX, EX^2, and VX.
   b. Expand $\pi(t)$ in a power series [$\pi(t) = (1/e)e^{t^2}$; use (A.5.11)] and from that find the mass function. [See Example (7.1.26) for a similar calculation.]

4. Let the mass function of X be $p(k) = (k+1)p^2q^k$ for $k = 0, 1, 2, \ldots$, where q denotes $1 - p$. [This is the distribution of the number of failures before the second success in Bernoulli trials with success probability p. This exercise was done for $p = \frac{1}{2}$ in Example (7.1.10).]
   a. Find the PGF of X.
   b. Find EX, EX^2, and VX.

5. If Y is the number of trials needed for the first success in repeated Bernoulli trials (not the number of failures preceding it), then $P(Y = k) = pq^{k-1}$ for $k = 1, 2, 3, \ldots$. Here as usual p is the success probability and $q = 1 - p$.
   a. Find the PGF of Y.
   b. Find EY and VY.
   c. If X is the number of *failures* preceding the first success, then we already know about the distribution of X; see Example (7.1.16). Compare EY with EX, and also compare VY with VX. Can you explain these connections?

6. In Example (7.1.1) we found that if X has the binomial distribution with parameters n and p, then the PGF of X is $\pi(t) = (pt + q)^n$, where $q = 1 - p$. We also saw again in that example that $EX = np$. Now use the PGF to show that $VX = npq$. [This will be done in Section 7.2; it appears here simply for more practice with PGFs.]

7. You do n Bernoulli trials with success probability p; X is the number of

successes. Independently of those trials, a friend of yours does m trials with the same success probability; Y is the number of successes on these. Now let $Z = X + Y$, your combined number of successes. You would think that Z ought to have the binomial distribution with parameters $n + m$ and p. Does it? Write down the PGFs of X and Y, multiply them together, and see if the result is the PGF of the binomial distribution with parameters $n + m$ and p.

8. Let X and Y be independent random variables, each having the distribution of the number of failures before the first success in Bernoulli trials—that is, the geometric distribution with parameter p, whose mass function is $p(k) = pq^k$ ($k = 0, 1, 2, \ldots$), where $q = 1 - p$. So the PGF of X and that of Y are the same, i.e., $\pi(t) = p/(1 - qt)$. Find the PGF of the random variable $Z = X + Y$. Compare with the answer to Exercise 4a above. Of course, this is not surprising. Why?

9. **A "back-door" derivation of the binomial probabilities.** Suppose we had somehow come this far in the subject without ever thinking about the binomial distribution. We know what Bernoulli trials are, but we have never seen the formula $\binom{n}{k}p^kq^{n-k}$ for the probability of k successes in n Bernoulli trials. We can now derive this formula using PGFs, without even thinking of the outcome space (the set of 2^n strings made of n H's and T's), as follows.
   a. Let Z be the number of successes in *one* Bernoulli trial, with success probability p. Then Z has only two possible values, 0 and 1. Find its PGF. [The distribution of Z is called the *Bernoulli distribution with parameter* p; notice that it is really the binomial distribution with parameters $n = 1$ and p.]
   b. Let $Z_1, Z_2, \ldots, Z_n$ be independent, each having the Bernoulli distribution with parameter p. Let $X = Z_1 + Z_2 + \cdots + Z_n$. What is the PGF of X?
   c. Use the binomial theorem [(A.4.7) in Appendix A] along with the definition of PGF, (7.1.8), to read off the binomial probabilities.

This kind of rederivation of a known result by other methods looks like "reinventing the wheel," but in mathematics it is not only illuminating but satisfying to see different ideas leading to the same conclusion. This instance is particularly satisfying because the rederivation completes an *Ideenkreis,* or circle of ideas, as follows. We started by deriving the binomial mass function from elementary considerations in Section 1.2, and then (in Section 4.1 and again in this section) we computed the associated PGF. Here we see that the PGF can be derived without knowing the mass function, and then it can be used to find the mass function.

Section 7.3 contains a less transparent but similar *Ideenkreis*; see especially (7.3.18).

[a]10. $\pi(t) = \frac{1}{3}t^2(1 + 2t^3)$ is a probability generating function. What is the mass function? [*Hint:* Look at the coefficients of the various powers of t.]

11. If $\pi(t)$ is the PGF of a nonnegative integer-valued random variable X, what is $\pi(1)$? What is $\pi(0)$?

12. If $\pi(t)$ is the PGF of a nonnegative integer-valued random variable X, what are $\frac{1}{2}(\pi(1) + \pi(-1))$ and $\frac{1}{2}(\pi(1) - \pi(-1))$?

13. If $0 < p < 1$ and $q = 1 - p$, then $p/(1 - qt)$ is a probability generating function. [See Example (7.1.16).] But $\pi(t) = p/(1 + qt)$ is not a PGF; for one thing, $\pi(1)$ does not have the right value (see Exercise 11 above). However, $\pi(t) = \alpha p/(1 + qt)$ does have the right value at $t = 1$ if α is chosen correctly. Why is it still not a PGF?

14. The function $h(x) = 1/(5 - x)^{3/2}$ has a power series expansion, $h(x) = a_0 + a_1x + a_2x^2 + a_3x^3 + \cdots$, valid for $-1 \leq x \leq 1$.

a. What is the sum of the series $a_0 + a_1 + a_2 + a_3 + \cdots$?

b. What must be the value of the constant c in order that $p(k) = c \cdot a_k$ (for $k = 0, 1, 2, \ldots$) defines a probability mass function?

c. If X is a discrete random variable with this mass function, what is EX?

15. The function $\pi(t) = C(e^t + e^{-t})$ happens to be a probability generating function if the constant C is chosen correctly. Its Maclaurin series is

$$\pi(t) = 2C\left(1 + \frac{t^2}{2!} + \frac{t^4}{4!} + \frac{t^6}{6!} + \cdots\right).$$

a. Find $P(X = 1)$ and $P(X = 10)$ in terms of C.

b. Find EX in terms of C.

c. Find C.

16. In combinatorial mathematics, the *generating function* of an arbitrary sequence $q_0, q_1, q_2, \ldots$ of numbers (whether they are probabilities or not) is the function of t defined by

$$Q(t) = \sum_{n=0}^{\infty} q_n t^n.$$

[Thus the PGF of X is really the generating function of the sequence $p_0, p_1, p_2, \ldots$ where $p_n = P(X = n)$.] Show that if $\pi(t)$ is the PGF of X and q_n is defined to be $P(X \leq n)$, then the generating function of the sequence $q_0, q_1, q_2, \ldots$ is $\pi(t)/(1 - t)$.

17. **The product formula for PGFs does not imply independence.** Let X and Y have the joint mass function shown in the following table; c is an arbitrary constant with $|c| \leq \frac{1}{9}$.

		y		
		0	1	2
	0	$\frac{1}{9}$	$\frac{1}{9}+c$	$\frac{1}{9}-c$
x	1	$\frac{1}{9}-c$	$\frac{1}{9}$	$\frac{1}{9}+c$
	2	$\frac{1}{9}+c$	$\frac{1}{9}-c$	$\frac{1}{9}$

Show that X and Y are not independent unless $c = 0$, but that the product formula $\pi_{X+Y}(t) = \pi_X(t)\pi_Y(t)$ holds for any c in $\left[-\frac{1}{9}, \frac{1}{9}\right]$.

This example, due to Wassily Hoeffding, shows that although independence implies the product formula of Theorem (7.1.31), the converse is false.

18. **Not all functions are PGFs.** We noted after (7.1.8) that if $\pi(t)$ is a PGF, then $\pi(1)$ must equal 1. Here we consider further restrictions on the functions that can be PGFs.

a. Show that if $\pi(t)$ is a PGF, then $\pi(0)$ is nonnegative, and furthermore that all derivatives at 0 [$\pi'(0)$, $\pi''(0)$, and so on] are nonnegative.

b. Show that $\pi'(1)$ must also be nonnegative.

c. Show that, although $\pi''(1)$ can be negative, $\pi'(1) + \pi''(1)$ cannot.

19. **The number of items in their proper places in a random permutation: the PGF and factorial moments.** Let S_n be the number of items in their proper places in a random permutation of the integers $1, 2, \ldots, n$. Let p_{nk} denote $P(S_n = k)$ $(k = 0, 1, 2, \ldots, n)$. In Exercise 23 of Section 2.2 we saw that $p_{nk} = (1/k)p_{n-1,k-1}$ for $n \geq 2$ and $k \geq 1$. This recurrence does not apply to give p_{n0}, the probability of a derangement. This exercise concerns the PGF of S_n, which we denote $\pi_n(t)$.

a. Show that $\pi_n'(t) = \pi_{n-1}(t)$ for all $n > 1$.

b. Conclude that $\pi_n'(1) = 1$ for all n—that is, that $ES_n = 1$ for all n. [We showed this using indicators in Exercise 19 of Section 4.2, and by a direct computation using the recurrence in Exercise 26 of Section 4.1.]

c. Show that for any positive integer $k \leq n + 1$, the kth derivative at 1, $\pi_n^{(k)}(1)$, equals 1. That is, all factorial moments of S_n (up to and including the nth, $E[S_n(S_n - 1)(S_n - 2)(S_n - n + 1)]$, equal 1.

This result can also be obtained using indicators, but it is not nearly so simple to get that way.

It is not too difficult to show that the distribution of S_n is the only distribution whose possible values are $0, 1, 2, \ldots, n$ and whose first n factorial moments equal 1. Are there any other distributions with a property like this? It can be shown that there is just one distribution on the set of all nonnegative integers whose factorial moments all equal 1—the Poisson distribution with parameter $\lambda = 1$. This is also the limiting distribution of S_n as n increases.

ANSWERS

1. $ES = 1$

3. **a.** 2; 8; 4; **b.** $p(k) = 1/(e \cdot (k/2)!)$ for $k = 0, 2, 4, 6, \ldots$

4. **a.** $(p/(1 - qt))^2$
b. $2q/p$; $(6q^2 + 2pq)/p^2$; $2q/p^2$

5. **a.** $pt/(1 - qt)$ **b.** $1/p$; q/p^2

7. Yes

9. **b.** $(pt + q)^n$

10. There are only two possible values, 2 and 5.

11. $\pi(1) = 1$; $\pi(0) = P(X = 0)$

12. One is $P(X \text{ is even})$; the other is $P(X \text{ is odd})$.

13. Probabilities cannot be negative.

14. **a.** $\frac{1}{8}$ **b.** 8 **c.** $\frac{3}{8}$

15. **a.** 0; $2C/10!$ **b.** $C(e^2 - 1)/e$ **c.** $e/(e^2 + 1)$

7.2 BINOMIAL AND BERNOULLI DISTRIBUTIONS

We have done so many examples and exercises involving binomial distributions that there is little more to say about them in this section. The main new thing about them will be the use of PGFs to derive some of the results more simply.

In addition we will spend some time looking at Bernoulli distributions, which are binomial distributions with $n = 1$. They may seem trivial, but what is important is that *indicator random variables* have Bernoulli distributions, and there are some very useful tricks involving indicators.

(7.2.1) **Mass function of a binomial distribution.** For a positive integer n and a number p between 0 and 1, the ***binomial distribution with parameters n and p*** has the mass function

$$p(k) = \binom{n}{k} p^k q^{n-k} \quad \text{for } k = 0, 1, 2, \ldots, n,$$

where q denotes $1 - p$. The abbreviated name for this distribution is $\text{bin}(n, p)$.

This distribution arises in connection with a Bernoulli process (see Section 1.4) having success probability p. If X is the number of successes in any set of n trials, then X has the $\text{bin}(n, p)$ distribution. This is true because the experiment of doing n Bernoulli trials has 2^n outcomes, corresponding to the n-letter words from the alphabet {S, F}; the event "k successes" consists of the words made of k S's and $n - k$ F's; there are $\binom{n}{k}$ of these, and each has probability $p^k q^{n-k}$.

One important experiment for which Bernoulli trials are an appropriate model is *sampling with replacement* from a population of objects of two types. If the proportion of type 1 objects in the population is p and we draw a random sample of size n with replacement, then the individual draws are Bernoulli trials, and the number of type 1 objects in the sample has the $\text{bin}(n, p)$ distribution.

(7.2.2) **PGF, expected value, and variance of a binomial distribution.** If $X \sim \text{bin}(n, p)$, then the PGF of X is found from the binomial theorem, after which the distribution is named; that is,

$$\pi(t) = \sum_{k=0}^{n} t^k \cdot \binom{n}{k} p^k q^{n-k} = \sum_{k=0}^{n} \binom{n}{k} (pt)^k q^{n-k},$$

which implies that

$$\pi(t) = (pt + q)^n.$$

Because this is a polynomial (and not an infinite series), we can differentiate it for all t. So we quickly get the expected value,

$$EX = \pi'(1) = [np(pt + q)^{n-1}]_{t=1} = np,$$

and the second factorial moment,

$$E(X(X - 1)) = \pi''(1) = [n(n - 1)p^2(pt + q)^{n-2}]_{t=1} = n(n - 1)p^2.$$

Then it is just a matter of algebra to get the second moment,

$$EX^2 = E(X(X - 1)) + EX = n(n - 1)p^2 + np = n^2p^2 + npq,$$

and finally the variance,

$$VX = EX^2 - (EX)^2 = npq.$$

It is also possible to find EX and $E(X(X - 1))$ without using the PGF; we can evaluate the sums

$$EX = \sum_{k=1}^{n} k\binom{n}{k}p^k q^{n-k} \quad \text{and} \quad E(X(X - 1)) = \sum_{k=2}^{n} k(k - 1)\binom{n}{k}p^k q^{n-k}$$

using the facts that

$$k\binom{n}{k} = n\binom{n-1}{k-1} \quad \text{and} \quad k(k - 1)\binom{n}{k} = n(n - 1)\binom{n-2}{k-2}.$$

(7.2.3) **Higher moments of a binomial distribution.** The moments EX^k and the central moments $E(X - np)^k$ are complicated, but, as often happens with distributions on the nonnegative integers, the factorial moments are easy to get using the PGF. And it is always possible, using just algebraic manipulation, to get moments and central moments from factorial moments. The factorial moments are

$$\pi'(t) = np(pt + q)^{n-1}, \quad \text{so} \quad EX = \pi'(1) = np;$$
$$\pi''(t) = n(n - 1)p^2(pt + q)^{n-2}, \quad \text{so} \quad E(X(X - 1)) = \pi''(1) = n(n - 1)p^2;$$
$$\pi'''(t) = n(n - 1)(n - 2)p^3(pt + q)^{n-3}, \quad \text{so} \quad E(X(X - 1)(X - 2)) = n(n - 1)(n - 2)p^3;$$

and in general, for any $j = 1, 2, 3, \ldots,$

$$E(X(X - 1)(X - 2) \cdots (X - j + 1)) = n(n - 1)(n - 2) \cdots (n - j + 1)p^j.$$

Notice that if $j > n$, then this equals 0. Only the first n factorial moments of the bin(n, p) distribution are nonzero.

(7.2.4) **Sums of independent binomial random variables with the same p are binomial.** Suppose X and Y are independent, binomially distributed random variables *with the same* p; say $X \sim \text{bin}(n, p)$ and $Y \sim \text{bin}(m, p)$. Then $X + Y$ has the bin($n + m, p$) distribution. This is most easily proved by noticing that the PGF of $X + Y$ is the product of the PGFs of X and Y, which is $(pt + q)^{n+m}$, and this is the PGF of the bin($n + m, p$) distribution. This was Exercise 7 in the previous section.

More generally, let $X_1, X_2, \ldots, X_k$ be independent, binomially distributed random variables with the same p; say $X_j \sim \text{bin}(n_j, p)$ for $j = 1, 2, \ldots, n$. Then a similar PGF argument shows that the sum $X_1 + X_2 + \cdots + X_k$ has the bin($n_1 + n_2 + \cdots + n_k, p$) distribution.

If two or more independent random variables have binomial distributions with different p's, however, it can be shown that their sum does *not* have a binomial distribution. Exercise 11 at the end of this section shows how to see this in a simple case.

(7.2.5) Computing and approximating binomial probabilities. In computing the values of the binomial probabilities for a given n and p, if n is large, it is often most efficient to compute them recursively, as described in (6.4.15).

If n is large enough that both np and nq are 5 or more, binomial probabilities can be well approximated using the normal distribution with $\mu = np$ and $\sigma^2 = npq$, as described beginning at (6.4.7).

If n is large, p is small, and np is moderate, then the values of the bin(n, p) mass function are well approximated by the Poisson probabilities with $\lambda = np$, as described beginning at (6.4.3).

(7.2.6) Bernoulli distributions, a special case of binomial distributions. If $n = 1$, the bin($1, p$) distribution has only two possible values, 0 and 1, and

$$p(0) = \binom{1}{0}p^0q^1 = q \quad \text{and} \quad p(1) = \binom{1}{1}p^1p^0 = p.$$

This is the ***Bernoulli distribution with parameter p***, or, for short, the Bern(p) distribution. If X has this distribution, then $EX = p$ and $EX^2 = p$, and so $VX = p - p^2 = pq$. All the moments are equal to p, in fact—for any k, $EX^k = 0^k \cdot q + 1^k \cdot p = p$.

The PGF is simply $\pi(t) = pt + q$.

(7.2.7) Binomial random variables can be viewed as sums of independent Bernoulli random variables. This can be seen in two different ways.

First, if X is the number of successes in n Bernoulli trials with success probability p, then $X = X_1 + X_2 + \cdots + X_n$, where X_j equals 1 if the jth trial results in success and 0 if failure. The X_j's are independent, and each obviously has the Bern(p) distribution, because it equals 1 with probability p and 0 with probability q. This situation was discussed in (4.2.18).

Alternatively, let $X_1, X_2, \ldots, X_n$ be independent random variables, each with the Bern(p) distribution. If X is their sum, then the PGF of X is the product of their PGFs, which is $(pt + q)^n$, and this is the PGF of the bin(n, p) distribution.

This is the same *Ideenkreis,* or circle of ideas, that we saw in Exercise 9 of Section 7.1. A similar but bigger circle, involving geometric distributions, is completed in (7.3.18) below. Although such "reinventing the wheel" is sometimes seen as wasteful in business or government, in mathematics and other areas of thought it often provides valuable insights and new techniques. But even if it had no such use, as a form of play it is pleasant and satisfying, a way to stop and smell the flowers.

(7.2.8) **Indicator random variables have Bernoulli distributions.** If A is any event in any outcome space Ω, then the random variable X that equals 1 if A occurs and 0 if not is the *indicator random variable* of A. It has the Bernoulli distribution with parameter $p = P(A)$. Its expected value is $P(A)$ and its variance is $P(A) - P(A)^2$. It is sometimes denoted by I_A; for any outcome ω, $I_A(\omega) = 1$ if $\omega \in A$ and $I_A(\omega) = 0$ if $\omega \notin A$.

Indicators are sometimes very useful in finding expected values. If a random variable can be written as a sum of indicators, then its expected value is simply the sum of the p's that are their parameters—that is, the sum of the probabilities of the events they indicate. This was discussed in (4.2.17) and (4.2.18). An example will show what we mean.

(7.2.9) **Example: An expected value found by indicators.** In a group of n people, what is the expected number of pairs of people who share a birthday? In Exercise 17 of Section 2.1 we found the probability that there is at least one such pair; that was the famous "birthday problem," in which it is surprising to see how high the probability is, even when n is only moderately large.

But here we are asking for EX, where X is the number of pairs in the group who share birthdays. Notice that if there are three people in the group with the same birthday, we count them as three pairs; if there are four with the same birthday, then we count them as six pairs; if five, then ten pairs. [There are three subsets of size 2 in a set of size 3, six in a set of size 4, and ten in a set of size 5. That is, $\binom{3}{2} = 3$, $\binom{4}{2} = 6$, and $\binom{5}{2} = 10$.]

To find EX we notice that there are a total of $\binom{n}{2} = n(n-1)/2$ pairs in the group. Call this number k for short. For each pair of people there is an indicator random variable X_j ($j = 1, 2, \ldots, k$), which equals 1 if that pair share a birthday and 0 if not. Then $X = X_1 + X_2 + \cdots + X_k$, and so $EX = EX_1 + EX_2 + \cdots + EX_k$.

But EX_j is the same for all j; it is the probability of the event that the jth pair have the same birthday, and this is $\frac{1}{365}$. (Why? Because there are 365^2 assignments of birthdays to two people, and among these are 365 in which they have the same birthday.) Consequently, we have

$$EX = k \cdot \frac{1}{365} = \frac{n(n-1)}{2} \cdot \frac{1}{365} = \frac{n(n-1)}{730}.$$

This number is greater than 1 when n is 28 or more. ∎

In some problems involving sums of indicators, we can also find quickly the variance of X as $VX = VX_1 + VX_2 + \cdots + VX_k$, *provided* that $X_1, X_2, \ldots, X_k$ are independent. We did this in (4.4.12) to show that the variance of a random variable with the bin(n, p) distribution is npq.

In Example (7.2.9), however, they are not all independent; X_i and X_j are independent if the ith and jth pairs consist of four different people, but not if there is a person who is in both pairs.

When random variables are not independent, the variance of their sum is not the sum of their variances; as we saw in (4.4.10), we have to add twice the sum of the covariances of all the pairs among the X_i's. The covariance of two indicators is often not difficult to find, and this sometimes turns out to be a useful way to find a variance. We will look at the covariance of two indicators in (7.2.11).

(7.2.10) **The essential properties of indicators.** The usefulness of indicators for various purposes rests essentially on three simple facts:

The probability of a set is the expected value of its indicator.

The indicator of an intersection of sets is the product of their indicators.

The indicator of the complement of a set is 1 minus the indicator of the set.

We have already noted the first of these. The other two are quite simple to prove.

Proof. $I_A \cdot I_B$ is the product of two random variables each of which equals 1 or 0; so it equals 1 if both I_A and I_B equal 1, and 0 otherwise. That is, $I_A \cdot I_B$ equals 1 if both A and B occur and 0 otherwise. Therefore, it is the indicator of $A \cap B$.

$1 - I_A$ equals $1 - 1 = 0$ if A occurs, and $1 - 0 = 1$ if not; so $1 - I_A$ is the indicator of A^c. □

(7.2.11) **The covariance of two indicators.** Let X be the indicator of event A and Y the indicator of B. We know that $EX = P(A)$ and $EY = P(B)$. We also know the variances, $VX = P(A) - P(A)^2$ and $VY = P(B) - P(B)^2$.

The covariance is also not difficult to find. Since $\text{Cov}(X, Y) = E(XY) - (EX)(EY)$, all we need to find is $E(XY)$. But XY is the indicator of $A \cap B$, as we saw in (7.2.10), and so $E(XY) = P(A \cap B)$. Consequently,

$$\begin{aligned}\text{Cov}(X, Y) &= P(A \cap B) - P(A)P(B)\\ &= P(X = 1 \text{ and } Y = 1) - P(X = 1)P(Y = 1).\end{aligned}$$

This may not look like much of a formula, but it makes it possible to find variances using indicators. In particular, it will enable us to find the variance of the hypergeometric distributions of Section 7.5. Notice also that it equals 0 if and only if A and B are independent; that is,

Events A and B are independent if and only if their indicators are uncorrelated.

(7.2.12) The principle of inclusion and exclusion via indicators. Indicators can be used to give a clever algebraic proof of the principle of inclusion and exclusion. We do it for three sets, although the proof for any number is similar. Recall that the principle for three sets can be stated as follows:

$$P(A \cup B \cup C)$$
$$= P(A) + P(B) + P(C) - P(A \cap B) - P(A \cap C) - P(B \cap C) + P(A \cap B \cap C).$$

Now $A \cup B \cup C = (A^c \cap B^c \cap C^c)^c$ by De Morgan's Law, and so the principle can be restated as follows:

(7.2.13) $$P(A^c \cap B^c \cap C^c) = 1 - P(A) - P(B) - P(C) + P(A \cap B) + P(A \cap C) + P(B \cap C) - P(A \cap B \cap C).$$

To prove this algebraically, let X, Y, and Z be the indicators of the events A, B, and C. Because of the three essential properties mentioned in (7.2.10), the left side of (7.2.13) is

$$E[(1 - X)(1 - Y)(1 - Z)],$$

which equals

$$E[1 - X - Y - Z + XY + XZ + YZ - XYZ]$$
$$= 1 - EX - EY - EZ + E(XY) + E(XZ) + E(YZ) - E(XYZ);$$

but this is exactly the right side of (7.2.13).

EXERCISES FOR SECTION 7.2

1. For practice with binomial distributions, look back at the following:
   Section 1.2, Exercises 11, 12, and 13;
   Section 1.4, Exercises 1, 2 (except part c), and 5;
   Supplementary Exercises 9a–d and 11 for Chapter 1;
   Section 2.1, Exercise 1;
   Section 2.2, Exercise 4;
   Section 2.4, Exercises 5 and 10;
   Supplementary Exercises 6 and 12 for Chapter 2;
   Section 6.4 on approximating binomial probabilities.

2. For practice with indicators, look back at the following:
   Section 2.4, Exercise 10;
   Section 4.2, Exercise 19.

3. Let A be an event whose probability is .32. What is the probability that the relative frequency of occurrence of A in 1,000 trials is less than .29? Use the normal approximation. [*Hint:* Translate to a question involving the number of occurrences of A in 1,000 trials.]

4. In a Poisson process with an arrival rate of $\lambda = 4$ per minute, let X be the number of arrivals among the first 100 that come less than 5 seconds after the previous one. Give the name of the distribution of X, and its expected value and variance.

5. In ten Bernoulli trials with success probability p, let X equal the number of times a success is followed immediately by a failure. Thus, for example, if the outcome sequence is SSFSFFSSFS, then X equals 3, because SF appears on trials 2–3, 4–5, and 8–9. Find EX.

   It is not so simple to find the probability that X equals k for various values of k—that is, to find the distribution of X—but its expected value is easy to find, using a method that should be familiar by now.

6. Compare the normal and Poisson approximations to the probability that 600 rolls of a pair of fair dice produce exactly 20 double 6's. The true probability is about .0654.

7. Let X be the number of successes in n Bernoulli trials; suppose the success probability is $p = .28$. Of course $EX = .28 \cdot n$. Find the approximate probability that X is within 10% of $.28 \cdot n$ (that is, between $.28 \cdot n - .10 \cdot .28 \cdot n$ and $.28 \cdot n + .10 \cdot .28 \cdot n$), for each of the following values of n. Do not bother with the continuity correction.
   a. $n = 100$ **b.** $n = 1{,}000$ **c.** $n = 10{,}000$

8. **Chevalier de Méré's "paradox."** Let X be the number of 1's in four rolls of a fair die, and Y the number of double 1's in 24 rolls of two fair dice.
   a. Find the probability that X is at least 1.
   b. Find the probability that Y is at least 1.
   [a]**c.** The Chevalier de Méré was a high roller who (according to William Feller, in Section II.10 of *An Introduction to Probability Theory and Its Applications,* Volume 1), thought the answers to parts a and b should be equal, and he knew the answer to part a was greater than $\frac{1}{2}$. Accordingly, he bet on Y being positive and blamed mathematics when he lost money over a long sequence of bets. What might have led him to think wrongly that they would be equal? That is, what quantity *do* X and Y have in common?

   The Chevalier made a more significant contribution to probability theory, however. See Exercise 14 of Section 1.4.

9. A military train consisting of 100 boxcars sits motionless on a track. Enemy airplanes drop 60 bombs on this train; each bomb lands on one car at random, or on a place where a car was; all 100 places are equally likely. Each

bomb destroys three cars, the car it lands on and the two cars adjacent to it, if they happen to be still there when the bomb lands. What is the expected number of cars *not* destroyed? [*Hint:* Let X_j be the indicator of the event "car j is not destroyed," for $j = 1, 2, \ldots, 100$. Then EX_j is the probability that car j is not destroyed, which is the probability that none of the 60 bombs hits it or a car adjacent to it. The "interior" X's, X_2 through X_{99}, all have the same expected value; X_1 and X_{100} have a different expected value.]

10. **Edges on a random chessboard.** An ordinary chessboard has 32 black squares and 32 white squares arranged in an alternating pattern; you can check that there are 112 line segments that are borders between black and white squares. But what if the squares are colored black and white at random; how many of the segments will separate differently colored squares? Find the expected number:
   a. If the squares are colored according to 64 Bernoulli trials with $p = q = \frac{1}{2}$
   b. If they are colored using Bernoulli trials with arbitrary p and q
   c. If there are 32 black squares and 32 white ones, but the locations are chosen at random
   d. If there are b black squares and $w = 64 - b$ white squares, but the locations are chosen at random

This exercise is less frivolous than you might think. It turns out that if X is the number of segments that separate black–white pairs in part b, then the PGF of X is closely related to an important function in statistical mechanics, the *Ising partition function*, having to do with the Ising model for ferromagnetism. From the partition function most of the important properties of the two-dimensional lattice Ising model can be derived. Lars Onsager first found it in 1944. Notice that we have found only EX, not the PGF.]

11. **The sum of independent Bernoulli random variables is not binomial unless the success probabilities are the same.** Let X and Y be independent Bernoulli random variables, but suppose $X \sim \text{Bern}(p_1)$ and $Y \sim \text{Bern}(p_2)$ where $p_1 \neq p_2$. Let $Z = X + Y$.
   a. Find the PGF of Z.
   b. Expand the PGF of Z in powers of t and read off the mass function of Z.
   c. Show that if $p_1 = \frac{1}{2}$ and $p_2 = \frac{1}{3}$, this cannot be the bin$(2, p)$ mass function for any p.
   [a]d. Show that no matter what p_1 and p_2 are, unless $p_1 = p_2$ this cannot be the bin$(2, p)$ mass function for any p.

We know from (7.2.4) that the sum of independent binomial random variables *with the same p* has a binomial distribution; this exercise shows that the requirement that they have the same p is essential.

12. Let $p(k)$ be the bin$(n = 100, p = .03)$ mass function. Starting with $p(0) = (.97)^{100} \doteq .0475525$, compute several of the probabilities $p(1), p(2), p(3), \ldots$ using the recursive formula of (6.4.15). [Notice that the chance of eight or fewer successes is over 99%.]

13. **The probability of getting exactly the expected number of heads, approximated by Stirling's formula.** Let X be the number of heads in $2n$ tosses of a fair coin; then the expected value of X, n, is also its most likely value. Its probability is of course $\binom{2n}{n}\left(\frac{1}{2}\right)^{2n}$. We can approximate this using Stirling's formula, which says that if n is large, $n!$ is approximated by

$$n! \sim n^n e^{-n}\sqrt{2\pi n},$$

where "$\sim$" indicates that the ratio of the two sides of the expression tends to 1 as n increases. Show that $P(X = n)$ is approximately $1/\sqrt{\pi n}$.

The most likely value becomes increasingly unlikely as the number of trials increases. For $n = 100$, the probability of 50 heads and 50 tails is about $1/\sqrt{50\pi} \doteq .0798$. For $n = 1{,}000$, the probability of 500 heads and 500 tails is .0798 divided by the square root of 10.

As we know by now, of course, probability theory does not predict in general that the expected value, or even the most likely value, is highly probable. In this situation the law of large numbers says only that as the number of trials increases, the number of successes is increasingly likely to be within a given fixed percentage of the expected number. This was illustrated in Supplementary Exercises 9 and 10 in Chapter 6, and also in Exercise 7 above. The next exercise shows the result a little more generally.

14. **The probability of being within 1% of the expected number of heads, using the normal approximation.** Let X be the number of heads in $2n$ tosses of a fair coin. So the expected value of X, which is also its most likely value in this case, is n. Show that the probability that X is within 1% of n (that is, between $n - .01n$ and $n + .01n$) is approximately $2\Phi(.01\sqrt{2n}) - 1$. [Do not bother with the continuity correction.]

This probability approaches 1 as n increases. Of course, there is nothing special about 1%; if it were replaced by any percentage r, the probability would be $2\Phi(r\sqrt{2n}) - 1$, which would still approach 1.

15. **The expected number of shared birthdays.** In example (7.2.9) we used indicators to find the expected number of pairs of people, in a group of n, who share a birthday. But to do it conveniently we had to agree that if three share a birthday there are three pairs; if four, six pairs, and so on. Perhaps a more interesting quantity is the expected number of days on which two or more people in the group have a birthday. (If we count days instead of pairs of people, then when three or more share a birthday it still counts only once.) Use indicators to find this expected value. Continue to assume that there are 365 days in the year and that all are equally likely to be a birthday, and that the birthdays of the n people are chosen independently.

ANSWERS

3. With continuity correction: .0192; without: .0212
4. $EX = 28.35$; $VX = 20.31$
5. $9pq$
6. Normal approximation: .0709; Poisson approximation: .0650

7. **a.** .4648 **b.** .9512
c. $2 \cdot \Phi(6.24) - 1$ ($\doteq 1$)

8. **a.** .5177 **b.** .4914 **c.** $EX = EY$

9. $2 \cdot (.98)^{60} + 98 \cdot (.97)^{60} \doteq 16.35$

10. **a.** 56 **b.** $224pq$ **c.** 56.89 **d.** $224 \cdot \frac{b}{64} \cdot \frac{w}{63}$

11. **c.** p would have to equal $\sqrt{1/6}$ and q would be $\sqrt{2/6}$. These do not add to 1. **d.** If the probabilities found in part b were bin(2, p) probabilities, then we would have $p^2 = p_1p_2$ and $(1 - p)^2 = (1 - p_1)(1 - p_2)$. Expand the latter and put in p_1p_2 for p^2 and $\sqrt{p_1p_2}$ for p, and simplify to get $2\sqrt{p_1p_2} = p_1 + p_2$. Square both sides to get $(p_1 - p_2)^2 = 0$.

12. .147070, .225153, .227474, .170606, .101308, .049610, .020604, .007408, .002342

15. $365 - (n + 364) \cdot \left(\frac{364}{365}\right)^{n-1}$. This exceeds 1 when n is 29 or more.

7.3 GEOMETRIC AND NEGATIVE BINOMIAL DISTRIBUTIONS

Geometric distributions are familiar to us; they are waiting-time distributions for Bernoulli processes. Specifically, the number of failures before the first success has the geometric distribution with parameter p. In this section we will begin by reviewing these distributions, using PGFs to get the results quickly.

Then we will introduce negative binomial distributions. In a Bernoulli process, if X is the number of failures preceding the nth success, then X is the sum of n independent random variables, each of which has the geometric distribution. The distribution of X is the negative binomial distribution with parameters n and p. If $n = 1$, of course, we have a geometric distribution.

So there is really only one family of distributions here; the negative binomial family includes the geometric family.

(7.3.1) **Definition.** For any number p between 0 and 1, the ***geometric distribution with parameter p***, which we abbreviate as the geom(p) distribution, is the one whose mass function is

$$p(k) = pq^k \quad \text{for } k = 0, 1, 2, \ldots,$$

where $q = 1 - p$.

This is the distribution of the number of failures before the first success in a Bernoulli process with success probability p. It is also the distribution of the number of failures before the next success after any given trial; and, although we have not proved it, it is the distribution of the number of failures *between* any success and the next success.

One of the important properties of geometric distributions is the "lack of memory" that Bernoulli processes have. Let X be a waiting time that has a geometric distribution; then $P(X \geq a + k \mid X \geq a) = P(X \geq k)$ for any nonnegative integers a and k. Given that there have been a failures since a success (or since the start of observations), the conditional probability that there will be at least k more failures, does not depend on how large a is. This was shown in Supplementary Exercise 15 at the end of Chapter 2.

(7.3.2) PGF, expected value, and variance of a geometric distribution. If $X \sim \text{geom}(p)$, then we find the PGF of X by summing a geometric series (whence the name of the distribution):

$$\pi(t) = \sum_{n=0}^{\infty} t^k \cdot pq^k = p \sum_{n=0}^{\infty} (qt)^k = \frac{p}{1 - qt}, \quad \text{for all } t \text{ such that } |t| < 1/q.$$

Because $1/q > 1$, this converges for at least one value of $t > 1$, and so we can get factorial moments by differentiating and setting $t = 1$. In particular,

$$EX = \pi'(1) = \left[\frac{pq}{(1 - qt)^2}\right]_{t=1} = \frac{q}{p},$$

and

$$E(X(X - 1)) = \pi''(1) = \left[\frac{2pq^2}{(1 - qt)^3}\right]_{t=1} = \frac{2q^2}{p^2}. \tag{7.3.3}$$

Algebraic manipulation then gives us

$$EX^2 = E(X(X - 1)) + EX = \frac{pq + 2q^2}{p^2}$$

and then

$$VX = EX^2 - (EX)^2 = \frac{pq + 2q^2 - q^2}{p^2} = \frac{q}{p^2}.$$

We can also find EX and $E(X(X - 1))$ without using the PGF, if we want to, by computing

$$EX = \sum_{k=1}^{\infty} k \cdot pq^k \quad \text{and} \quad E(X(X - 1)) = \sum_{k=2}^{\infty} k(k - 1) \cdot pq^k$$

using familiar operations with derivatives of geometric series. However, it should be clear that PGFs enable us to find these and other moments by successive differentiation of one function.

(7.3.4) Higher moments of a geometric distribution. As with most distributions on the nonnegative integers, the factorial moments $E(X(X - 1)(X - 2) \cdots (X - j + 1))$ are much simpler than the moments EX^j or the central moments $E(X - EX)^j$. We can always get the jth moment of one kind from the first j moments of another kind if we need to.

It is easy to continue differentiating from (7.3.3) to see, for any $j = 1, 2, 3, \ldots$, that the jth factorial moment of the geom(p) distribution is

$$E(X(X-1)(X-2)\cdots(X-j+1)) = \frac{j!q^j}{p^j}.$$

(7.3.5) **The number of *trials* needed for the first success.** It is obvious that if Y is the number of *trials* needed for the first success, then $Y = X + 1$ where X is the number of *failures* before the first success. The mass function of Y is

$$P(Y = k) = pq^{k-1} \quad \text{for } k = 1, 2, 3, \ldots .$$

This distribution is different from the geometric distribution whose mass function is given in (7.3.1), but it is sometimes called a geometric distribution. Confusion is possible, and it is well, when you read about a geometric distribution, to be sure which one is meant. In this text we use the name "geometric" only for the distribution of the number of failures, whose mass function is in (7.3.1).

It is not difficult to find the PGF of Y and thence its expected value and variance, or to find EY and VY directly from the mass function; you might want to do these for practice. But we can also find them quickly from those of X, because $Y = X + 1$ (we did this, in fact, in Exercise 5 of Section 7.1):

$$EY = E(X+1) = EX + 1 = \frac{q}{p} + 1 = \frac{1}{p};$$

$$VY = V(X+1) = VX = \frac{q}{p^2};$$

as for the PGF,

$$\pi_Y(t) = E(t^Y) = E(t^{X+1}) = E(t \cdot t^X) = tE(t^X) = t\pi_X(t) = \frac{pt}{1-qt}.$$

(7.3.6) **Negative binomial distributions.** Now let X be the number of failures before the nth success in a Bernoulli process with success probability p. (This is meant to include all the failures before the first success, between the first and the second, and so on.) The distribution of X is called the negative binomial distribution with parameters n and p. It is important to notice from the start that X is the sum of independent random variables, each with the geom(p) distribution; that is, $X = X_1 + X_2 + \cdots + X_n$, where X_1 is the number of failures before the first success, X_2 is the number of failures between the first and second successes, and so on.

We will do two things:

1. Find the mass function of this distribution by a simple probability argument; and

2. Find the PGF of the distribution, not from the mass function but quickly by noting that X is a sum of independent geometric random variables.

Then we will proceed to discuss the expected value, variance, and other moments, as well as to give some examples of calculations with negative binomial distributions.

(7.3.7) **The mass function of a negative binomial distribution.** The possible values of X are $0, 1, 2, \ldots$. What is $P(X = k)$ for one of these possible values? It is the probability that there are k failures before the nth success, which says that the nth success comes on trial number $n + k$, after $n - 1$ successes and k failures in the first $n + k -$ trials. So

$$\begin{aligned} P(X = k) &= p(n - 1 \text{ S's in the first } n + k - 1 \text{ trials and S on the } (n + k)\text{th trial}) \\ &= \binom{n + k - 1}{n - 1} p^{n-1} q^k \cdot p; \end{aligned}$$

that is, the mass function is

$$p(k) = \binom{n + k - 1}{n - 1} p^n q^k \quad \text{for } k = 0, 1, 2, \ldots.$$

This is the mass function of the ***negative binomial distribution with parameters n and p,*** or the neg bin(n, p) distribution for short.

Notice the special case in which $n = 1$: then X is the number of failures before the first success, and we already know that in this case X has the geom(p) distribution. So the mass function above had better turn out to be the geom(p) mass function when we put in $n = 1$. You can check that it does.

The negative binomial mass function is one of those formulas that are difficult to memorize and easy to get wrong in solving problems. It is usually preferable to find probabilities by a direct argument, as in the following examples.

(7.3.8) **Example.** What is the probability that in rolling a fair die repeatedly, ten non-6's precede the third 6? If X is the number of non-6's preceding the third 6, then X has the negative binomial distribution with parameters $n = 3$ and $p = \frac{1}{6}$, and we want $P(X = 10)$. The formula in (7.3.7) gives

$$\binom{3 + 10 - 1}{3 - 1}\left(\tfrac{1}{6}\right)^3\left(\tfrac{5}{6}\right)^{10} = \binom{12}{2}\left(\tfrac{1}{6}\right)^3\left(\tfrac{5}{6}\right)^{10} \doteq .0493.$$

But as we mentioned, it is better to find this probability by thinking as follows:

> The probability that ten failures precede the third success is the probability of ten failures and two successes in the first 12 trials, followed by a success on trial 13. This equals $\binom{12}{2}\left(\frac{1}{6}\right)^2\left(\frac{5}{6}\right)^{10} \cdot \frac{1}{6}$, which equals $\binom{12}{2}\left(\frac{1}{6}\right)^3\left(\frac{5}{6}\right)^{10}$. ∎

Notice also that a negative binomial probability is not a binomial probability. In a binomial probability $\binom{n}{j}p^j q^k$, the two exponents j and k always add to n, and the other number j in the binomial coefficient is the exponent of p. In a negative binomial probability there is always an extra factor of p. That is, if we take a negative binomial probability and reduce the exponent of p by 1, it becomes a binomial probability. This can serve as a partial check that you have written a negative binomial probability correctly.

(7.3.9) **Example.** If the success probability is p, what is the probability that the 8th success comes on the 20th trial? Arguing directly, we say the following:

> This is the probability that the first 19 trials include 7 successes and 12 failures, followed by a success on trial 20. This equals $\binom{19}{7}p^7 q^{12} \cdot p = \binom{19}{7}p^8 q^{12}$.

If we had to use the formula, we would say the following:

> This is the probability that $X = k = 12$, where X is the number of failures preceding the $n = 8$th success. By (7.3.7) it equals
>
> $$\binom{n+k-1}{n-1}p^n q^k = \binom{8+12-1}{8-1}p^8 q^{12} = \binom{19}{7}p^8 q^{12}.$$ ∎

(7.3.10) **Converting negative binomial probabilities to binomial probabilities.** In the preceding two examples we found probabilities of equalities such as $P(X = k)$ for negative binomial random variables. But probabilities of *inequalities,* like $P(X \leq k)$ or $P(X \geq k)$, involving negative binomial random variables can be converted to inequalities involving binomial random variables. These can then be approximated with the normal approximation described in Section 6.4.

Example a. If the success probability is .25, what is the probability that the 10th success comes after the 50th trial?

Solution. Do not be tempted to identify n and k for a negative binomial random variable; it is nearly always simpler when finding such probabilities to proceed as we did in Examples (7.3.8) and (7.3.9).

$$\begin{aligned} P(\text{10th success after 50th trial}) &= P(\text{9 or fewer successes in the first 50 trials}) \\ &= P(Y \leq 9) \quad \text{where } Y \sim \text{bin}(50, .25). \end{aligned}$$

To approximate this we use the fact that Y is approximately normally distributed, with $\mu = 50 \cdot .25 = 12.5$ and $\sigma^2 = 50 \cdot .25 \cdot .75 = 9.375$. Using the continuity correction, we find

$$\begin{aligned} P(Y \le 9) = P(Y \le 9.5) &= P\left(\frac{Y - 12.5}{\sqrt{9.375}} \le \frac{9.5 - 12.5}{\sqrt{9.375}}\right) \\ &\approx P(Z \le -.98) \quad \text{where } Z \sim N(0, 1) \\ &\doteq .1635. \end{aligned}$$

∎

Example b. What is the probability, in a Bernoulli process with success probability .6, that 25 or fewer failures precede the 35th success?

Solution. Again we translate this into a statement about the number of successes in a fixed number of trials. We are looking for the probability that the 35th success comes on or before the 60th trial—that is, that there are 35 or more successes in 60 trials. If Y is the number of successes in 60 trials, we will find $P(Y \ge 34.5)$ using the normal approximation with $\mu = 60 \cdot .6 = 36$ and $\sigma^2 = 60 \cdot .6 \cdot .4 = 14.4$. This is

$$\begin{aligned} P\left(\frac{Y - 36}{\sqrt{14.4}} \ge \frac{34.5 - 36}{\sqrt{14.4}}\right) &\approx P(Z \ge -.40) \quad \text{where } Z \sim N(0, 1) \\ &\doteq .6554. \end{aligned}$$

∎

(7.3.11) **The PGF of a negative binomial distribution.** Rather than sum the series $\Sigma t^k p(k)$ to get the PGF of X, we do it by remembering that $X = X_1 + X_2 + \cdots + X_n$, where the X_j's are independent random variables having the geom(p) distribution. So the PGF of X is the product of the PGFs of the X_j's, and each of these PGFs is equal to $p/(1 - qt)$. Therefore, the PGF of the neg bin(n, p) distribution is

$$\pi_X(t) = \left(\frac{p}{1 - qt}\right)^n.$$

This is valid as long as $|t| < 1/q$, because the series for the PGFs of the X_j's converge for these values of t.

(7.3.12) **A formula that follows from the two preceding calculations.** In (7.3.7) we showed that if X has the neg bin(n, p) distribution, then the mass function of X is

$$p(k) = \binom{n + k - 1}{n - 1} p^n q^k \quad \text{for } k = 0, 1, 2, \ldots;$$

and in (7.3.11) we showed that the probability generating function of X is

$$\sum_{k=0}^{\infty} t^k p(k) = \left(\frac{p}{1 - qt}\right)^n.$$

Putting these together, we see that the following formula must be valid even though we have never proved it directly:

(7.3.13)
$$\left(\frac{p}{1-qt}\right)^n = \sum_{k=0}^{\infty} t^k \cdot \binom{n+k-1}{n-1} p^n q^k.$$

In (7.3.18), to complete the circle of ideas, we will prove (7.3.13) directly by expanding the left side in a power series. It is not necessary, because the arguments in (7.3.7) and (7.3.11) are correct; but it is perhaps instructive, or at least a good exercise, to do so.

(7.3.14) **Expected value and variance of a negative binomial distribution.** Let X have the neg bin(n, p) distribution. Three ways to find EX and VX are available to us.

- **i.** Evaluate the sums $EX = \sum_k k \cdot p(k)$ and $E(X(X-1)) = \sum_k k(k-1) \cdot p(k)$, where $p(k)$ is the mass function found in (7.3.7). Then find VX from these by first finding $EX^2 = E(X(X-1)) + EX$.
- **ii.** Differentiate the PGF (7.3.11) twice and use $\pi'(1) = EX$ and $\pi''(1) = E(X(X-1))$; then proceed to find VX as in (i).
- **iii.** Use the fact that X is the sum of n independent random variables $X_1, X_2, \ldots, X_n$ that have the geom(p) distribution. As we saw in (7.3.2), each of the X_j's has expected value q/p and variance q/p^2, and, therefore,

(7.3.15)
$$EX = \frac{nq}{p} \quad \text{and} \quad VX = \frac{nq}{p^2}.$$

It is apparent that (iii) is the simplest method; it produces the results immediately. When we find higher moments, next, we will use method (ii).

(7.3.16) **Higher moments of a negative binomial distribution.** As with the other distributions on the nonnegative integers, the moments and the central moments are complicated, but the factorial moments are not so bad, and it is possible to find a given moment or central moment if the factorial moments are known.

We get the factorial moments in the usual manner. Starting with

$$\pi(t) = \left(\frac{p}{1-qt}\right)^n = p^n(1-qt)^{-n},$$

we get

$$EX = \pi'(1) = [nqp^n(1-qt)^{-n-1}]_{t=1} = nq/p;$$
$$E(X(X-1)) = \pi''(1) = [n(n+1)q^2p^n(1-qt)^{-n-2}]_{t=1} = n(n+1)q^2/p^2;$$

and, in general, you can check that

$$E(X(X-1)(X-2)\cdots(X-j+1)) = n(n+1)(n+2)\cdots(n+j-1)q^j/p^j.$$

(7.3.17) **Sums of independent negative binomial random variables with the same p are negative binomial.** We have already seen that a sum of independent *geometric* random variables with the same p—that is, of independent neg bin$(1, p)$ random variables—has the neg bin(n, p) distribution. In fact, that is how we came across negative binomial distributions in the first place.

But the individual random variables do not have to be neg bin$(1, p)$ for this to hold; they can be neg bin(n_i, p), each with its own n_i. The result is as follows.

> If $Y = X_1 + X_2 + \cdots + X_j$, where $X_1, X_2, \ldots, X_j$ are independent and $X_1 \sim$ neg bin(n_1, p), $X_2 \sim$ neg bin(n_2, p), $\ldots$, and $X_j \sim$ neg bin(n_j, p), then $Y \sim$ neg bin(n, p), where $n = n_1 + n_2 + \cdots + n_j$.

We can see this immediately by looking at the PGF of Y, which is the product of the PGFs of the X_i's:

$$\pi_Y(t) = \left(\frac{p}{1-qt}\right)^{n_1} \cdot \left(\frac{p}{1-qt}\right)^{n_2} \cdot \cdots \cdot \left(\frac{p}{1-qt}\right)^{n_j} = \left(\frac{p}{1-qt}\right)^{n}.$$

(7.3.18) Proof that the PGF calculation and the mass function calculation agree. As we pointed out above in (7.3.12), we found the mass function

$$p(k) = \binom{n+k-1}{n-1} p^n q^k \quad \text{for } k = 0, 1, 2, \ldots$$

using a probability argument, and the PGF

$$\pi(t) = \left(\frac{p}{1-qt}\right)^{n}$$

using the fact that X is the sum of independent geometrically distributed random variables. Although it is not necessary to do so, because both arguments are correct, we now show the consistency of these two arguments by expanding $\pi(t)$ in a power series and showing that the coefficient of t^k *is* $p(k)$.

As we have said, efficiency-minded people think of this as "reinventing the wheel." Mathematicians think of it as a circle of ideas, an *Ideenkreis*; such a thing is valued because it provides both insight and fun. We saw a similar one involving the binomial and Bernoulli distributions in Exercise 9 of Section 7.1; it was mentioned again in (7.2.7).

To expand the PGF we notice what happens when we repeatedly differentiate the geometric series

$$\frac{1}{1-x} = 1 + x + x^2 + x^3 + \cdots = \sum_{k=0}^{\infty} x^k.$$

The first derivative is

$$\frac{1}{(1-x)^2} = 1 + 2x + 3x^2 + 4x^3 + \cdots = \sum_{k=0}^{\infty} (k+1)x^k, \tag{7.3.19}$$

which is the same as

$$\frac{1}{(1-x^2)} = \sum_{k=0}^{\infty} \binom{k+1}{1} x^k.$$

Differentiating (7.3.19) gives us

(7.3.20)

$$\begin{aligned}\frac{2}{(1-x)^3} &= 2 \cdot 1 + 3 \cdot 2x + 4 \cdot 3x^2 + 5 \cdot 4x^3 + \cdots \\ &= \sum_{k=0}^{\infty} (k+2) \cdot (k+1)x^k,\end{aligned}$$

and dividing by 2 gives

$$\frac{1}{(1-x)^3} = \sum_{k=0}^{\infty} \binom{k+2}{2} x^k.$$

If we continue, we get for any positive integer n,

(7.3.21)

$$\frac{1}{(1-x)^n} = \sum_{k=0}^{\infty} \binom{k+n-1}{n-1} x^k.$$

This is also derived in Appendix A; it is really the same formula as (A.5.10).

Now all we have to do is to write down (7.3.21) with qt in place of x, and then multiply by p^n:

$$p^n \left(\frac{1}{1-qt} \right)^n = p^n \sum_{k=0}^{\infty} \binom{k+n-1}{n-1} (qt)^k;$$

that is,

(7.3.22)

$$\left(\frac{p}{1-qt} \right)^n = \sum_{k=0}^{\infty} t^k \cdot \binom{k+n-1}{n-1} p^n q^k.$$

But (7.3.22) says just what we mentioned in (7.3.12); that is, the mass function given in (7.3.7) and the PGF given in (7.3.11) belong to each other. □

(7.3.23) **A further digression: Newton's binomial theorem and the reason for the name "negative binomial."** Let us start by recalling the familiar binomial theorem (A.4.7) in a slightly different form:

(7.3.24)

$$(1+x)^r = \sum_k \binom{r}{k} x^k.$$

We know this is true when r is a positive integer; in this case, the sum extends over $k = 0, 1, \ldots, r$ and amounts to a polynomial function of x. Recall also that

(7.3.25)
$$\binom{r}{k} = \begin{cases} \dfrac{r(r-1)(r-2)\cdots(r-k+1)}{k!} & \text{if } k = 1, 2, 3, \ldots \\ 1 & \text{if } k = 0 \end{cases}$$

[which equals $r!/k!(r-k)!$ if r is a positive integer].

Now one of Isaac Newton's many discoveries was that even if r is not a positive integer—it can be a fraction or an irrrational number, positive or negative—(7.3.24) is still true. The only difference is that if r is not a nonnegative integer, then the sum becomes an infinite series, with nonzero terms for $k = 0, 1, 2, 3, \ldots$. Notice that (7.3.25) defines a binomial coefficient no matter what r is (as long as k is a nonnegative integer), and it is nonzero for all k if r is not a nonnegative integer.

Because (7.3.24) is a power series if r is not a nonnegative integer, we need to know for which x it converges. The answer is that it converges for all x in the interval $(-1, 1)$. (Convergence for $x = 1$ and absolute convergence for $x = -1$ depend on the value of r.) We can summarize this by stating

Newton's binomial theorem: The binomial formula (7.3.24) holds for all real numbers r, if we define binomial coefficients by (7.3.25). If r is a nonnegative integer, (7.3.25) gives 0 for $k > r$ and so the sum (7.3.24) has nonzero terms only for $k = 0, 1, 2, \ldots, r$. Otherwise, there are nonzero terms for all $k = 0, 1, 2, 3, \ldots$, and the sum is a power series; in this case, it converges absolutely for $-1 < x < 1$.

Now what does this have to say about the negative binomial distribution? The answer is that Newton's binomial theorem gives us another power series expansion of the PGF: if in (7.3.24) we use $-n$ for r and $-qt$ for x, and if we then multiply by p^n, we get

$$\left(\frac{p}{1-qt}\right)^n = p^n(1-qt)^{-n} = p^n \sum_{k=0}^{\infty} \binom{-n}{k}(-qt)^k = \sum_{k=0}^{\infty} t^k \cdot \binom{-n}{k} p^n (-q)^k.$$

Comparing this with (7.3.22), we see that the coefficients of the two power series must be the same, and thus we have two apparently different formulas for the mass function:

(7.3.26)
$$p(k) = \binom{n+k-1}{n-1} p^n q^k = \binom{-n}{k} p^n (-q)^k \quad \text{for } k = 0, 1, 2, \ldots.$$

If this is true, then it must be the case that

$$\binom{n+k-1}{n-1} = (-1)^k \binom{-n}{k}$$

for any positive integer n and nonnegative integer k. You can check that this is true by noting that the left side is the same as $\binom{n+k-1}{k}$ and by using (7.3.24).

At any rate, now we see why the name "negative binomial" is used for this distribution. Its PGF involves a negative power of a binomial, and can be written in terms of binomial coefficients $\binom{-n}{k}$ with a negative number in them.

EXERCISES FOR **SECTION 7.3**

1. In a Bernoulli process with success probability $p = .65$, find the probabilities of the following events.
a. The first success comes on the fourth trial.
b. The third success comes after four failures.
c. The second success comes after at least four failures.
d. The first success comes after no more than three failures.

2. In a Bernoulli process with success probability $p = .65$, find the following.
a. The expected number of failures before the first success
b. The expected number of failures before the tenth success
c. The expected value of the number of the trial on which the third success occurs
d. The expected value and variance of the number of failures before the 20th success

3. Let X have the geometric distribution with parameter p; that is, X has the distribution of the number of failures preceding the first success in Bernoulli trials with success probability p.
a. Show that no matter how large or small p is, the most probable value of X is 0.
b. Find the probability that X is even and show that it is greater than $\frac{1}{2}$ regardless of the value of p.

4. Let X have the geometric distribution with parameter p. [The mass function is $p(k) = pq^k$ for $k = 0, 1, 2, \ldots$.]
a. Show in two ways that for any nonnegative integer k, $P(X \geq k) = q^k$ (where q denotes $1 - p$):
i. By evaluating the sum $p(k) + p(k + 1) + p(k + 2) + \cdots$.
ii. By noting that $X \geq k$ if and only if the first k trials result in failure.
b. Show that for any nonnegative integers k and j, $P(X \geq k + j \mid X \geq k) = P(X \geq j)$.

That is, for any j the conditional probability of having an additional j failures before a success, given that no success has occurred in k trials, is the same no matter how large k is. This property is called the *lack of memory* of Bernoulli trials.

5. We have seen that in tossing a fair coin, the probability that the second head comes on the kth toss is $(k - 1)\left(\frac{1}{2}\right)^k$ for $k = 2, 3, 4, \ldots$. Derive this result again using the negative binomial mass function with $n = 2$ and $p = \frac{1}{2}$, by noticing that the second head comes on the kth toss if and only if $k - 2$ tails precede the second head.

6. A certain experiment has probability $\frac{1}{3}$ of success.
   a. What is the probability that the fourth success comes after exactly ten failures (that is, there are ten failures interspersed among the first three successes)?
   b. What is the probability that the third success comes on the eighth trial?

7. You roll a die until you have seen six 6's.
   a. How many non-6's do you expect to roll on the average?
   b. How many rolls do you expect to make on the average?
   c. What is the probability that you see exactly the expected number of non-6's?
   d. What is the approximate probability that it takes more than 50 rolls to get the sixth 6?

[a]8. Let $p(k)$ be the mass function of the neg bin(n, p) distribution, given in (7.3.7). Find the ratios $p(1)/p(0)$, $p(2)/p(1)$, $p(3)/p(2)$, and, in general, $p(k)/p(k-1)$. Thus, get a recursive formula for computing the probabilities $p(k)$, starting with $p(0) = p^n$.

9. Use the recursive formula developed in Exercise 8 to compute the first several values of $p(k)$ for the neg bin(5, .2) distribution, starting with $p(0) = (.2)^5 = .000320$.

10. Let X be the number of failures before the 15th success in a Bernoulli process with success probability .32. We want the probability that X is 36 or less. Convert this to a probability involving a binomially distributed random variable, and approximate it using the normal approximation. Use the continuity correction.

11. In repeated rolls of a fair pair of dice:
    a. What is the expected number of *trials* needed to produce the 20th double 6?
    b. Should you be suspicious if, in repeated rolls of a pair of dice, the 20th double 6 has not appeared within the first 800 trials? [Approximate the probability that this would happen with a fair pair of dice.]

12. In a Bernoulli process with success probability p, a success is counted as a "great success" if it is immediately preceded by at most one failure.
    a. What is the distribution of the number of great successes among the first 50 successes?
    b. What is the distribution of the number of non-great successes preceding the first great success?

13. Consider Bernoulli trials with success probability p (and, of course, $q = 1 - p$).
    a. What is the probability that the fifth success comes on the 12th trial?
    b. What is the probability that there are no failures between the fourth and fifth successes?

c. Let X denote the total number of failures before the 35th success (including failures before the first success, those between the first and second successes, and so on). What are the expected value and variance of X?

14. In repeated rolls of a fair die, let X be the number of rolls needed for all six faces to come up. Find EX. [*Hint:* $X = X_1 + X_2 + \cdots + X_6$ where $X_1 = 1$ (the first roll); X_2 is the number of rolls needed for the next new face after the one that appeared on the first roll; X_3 is the number needed for the next new face after the first two, and so on.]

15. **Should we spend money on asteroid-deflection technology?** There is a positive probability that at some time in the future the earth will be hit by an asteroid large enough and speedy enough to wipe out human life. The probability does not seem large—according to one estimate, the chance that such an asteroid will hit the earth in a given year is one in a million. But some argue that the potential loss is so great that the development of anti-asteroid systems would be worth the cost. Is there a rational way to think about these things?

Probably not, depending on your view of humanity and its place in the universe. But if you are willing to take a strictly monetary view, Daniel Seligman, in his *Fortune* magazine column "Keeping Up" for June 1, 1992, suggests a way to estimate the dollar value of the expected loss of human lives due to an asteroid, as follows. Suppose there are N people alive today, and each life is valued at V dollars. Then the dollar value of human lives today is NV. If we assume an inflation rate of 4% and a population growth rate of .6%, then the value of human lives n years from now will be $NV(1.04)^n(1.006)^n$. To see this in terms of today's dollars we discount it at 7%; that is, we multiply it by $(.93)^n$, producing $NV \cdot r^n$, where $r = 1.04 \cdot 1.006 \cdot .93 \doteq .973$. This is the present value of the human lives n years from now, and, consequently, it is the present value of the loss, should human life be eradicated in that year. (You could substitute your own value for r if you do not agree with these rates of inflation, population growth, and discount.)

Now let p be the probability that an asteroid large enough to destroy human life hits the earth in a given year. Let X be the (random) year in which such an asteroid first hits the earth, counting the current year as year 0. Then the present value (on January 1 of this year) of the loss in year X is $Y = NV \cdot r^X$. Show that the expected value of Y is $NVrp/(1 - rq)$.

The numbers suggested by Seligman are the following: $p = .000001$ (one chance in a million); $r = .973$, as indicated above; and $NV = 7.5 \times 10^{14}$. This is 250 million times 3 million; it is a conservative figure, taking into account only the population of the United States; it is based on the current U.S. population and some evidence that the average American values his or her life at \$3 million. Seligman discussed this evidence in an earlier article, "How Much Money Is Your Life Worth?", in *Fortune,* March 3, 1986.

Putting these values together gives the expected loss EY as about \$$2.7 \times 10^{10}$, or just over \$27 billion. Seligman suggests that a \$50 million program to develop the technology to detect and deflect an asteroid might be well worth the money.

See Supplementary Exercise 19 at the end of this chapter for a closely related but different threat to life on earth—Comet Swift–Tuttle, which at one time was thought to have one chance in 10,000 of striking the earth on August 14, 2126. It is interesting to compare the expected magnitudes of the two threats—a one-in-a-million chance every year, or a one-time, one-in-ten-thousand chance over 130 years from now.

We must also mention that the very idea of putting a value on the existence of the human race is in some sense absurd. A good case can be made for saying that the loss would be infinite, or without price. There is also a philosophical, or theological, problem involved. If all human life disappears, who suffers the loss? Whoever it is will not likely measure it in dollars.

16. **Will allowing each family only one male child change the proportion of males to females?** A developing country wants to control its population, but recognizes that its culture puts a high premium on male children. Accordingly, each family may continue to have children until they have had their first boy; then they may have no more. [This model depicts a practice that is at variance with modern attitudes in the United States, but it has been used by public-health practitioners in considering third-world population problems.] It has been argued by opponents of such a policy that it would soon produce a population in which females outnumbered males, because no family would have more than one boy, while many would have large numbers of girls. Is this argument valid? Show that the expected ratio of females to males in a random family, chosen from among families with children, is q/p. Assume for simplicity that every family with children actually continues to have children until they have a male child, and ignore the effect of differing death rates for males and females. [In an unrestricted population, the ratio of females to males would be q/p, the ratio of female births to male births. So the policy does not change the proportion.]

ANSWERS

1. a. .0279 **b.** .0618 **c.** .0540 **d.** $1 - (.35)^4 \doteq .9850$

2. a. .5385 **b.** 5.385 **c.** 4.615 **d.** 10.769; 16.568

3. b. $1/(1 + q)$

6. a. .0612 **b.** .1024

7. a. 30 **b.** 36 **c.** .02931 **d.** .1401

8. $p(k) = p(k - 1) \cdot q \cdot (n + k - 1)/k$

9. .001280; .003072; .005734; .009175; .013212; .017616

10. .7088

11. a. 720 **b.** The probability is .2776, not small enough for suspicion.

12. a. $\text{bin}(50, p + pq)$ **b.** $\text{geom}(p + pq)$

13. a. $\binom{11}{4}p^5q^7$ **b.** p **c.** $35q/p$; $35q/p^2$

14. 14.7

7.4 POISSON DISTRIBUTIONS

Poisson distributions, like binomial and geometric distributions, are familiar to us by now. In this section we will have little to say that has not been covered before; we will mostly summarize the essential facts.

(7.4.1) **Mass function of a Poisson distribution.** For any positive number λ, the ***Poisson distribution with parameter*** λ is the one who mass function is

$$p(k) = \frac{\lambda^k e^{-\lambda}}{k!} \quad \text{for } k = 0, 1, 2, \ldots.$$

This arises in connection with Poisson processes. In a Poisson process with arrival rate λ per unit time, let X be the number of arrivals in a given time interval whose length is s units. Then X has the Poisson distribution with parameter λs. This was covered in Section 1.4.

In addition, if X has the bin(n, p) distribution and n is large and p is small, then (as we saw in Section 6.4) the distribution of X is approximately the Poisson distribution with parameter np.

(7.4.2) **PGF, expected value, and variance of a Poisson distribution.** These are particularly simple. The PGF is

$$\pi(t) = \sum_{k=0}^{\infty} t^k \cdot \frac{\lambda^k e^{-\lambda}}{k!} = e^{-\lambda} \sum_{k=0}^{\infty} \frac{(\lambda t)^k}{k!} = e^{-\lambda} e^{\lambda t};$$

and this is valid for all real t; that is,

$$\pi(t) = e^{\lambda t - \lambda} \quad \text{for all } t.$$

From this it is easy to get EX and VX if $X \sim \text{Pois}(\lambda)$:

$$EX = \pi'(1) = [\lambda e^{\lambda t - \lambda}]_{t=1} = \lambda,$$

(7.4.3)
$$E(X(X-1)) = \pi''(1) = [\lambda^2 e^{\lambda t - \lambda}]_{t=1} = \lambda^2;$$

and, therefore,

(7.4.4)
$$EX^2 = E(X(X-1)) + EX = \lambda^2 + \lambda,$$

and so

$$VX = EX^2 - (EX)^2 = \lambda.$$

That is,

If $X \sim \text{Pois}(\lambda)$, then $EX = \lambda$ and $VX = \lambda$.

It is also not difficult to find EX and $E(X(X-1))$ directly from the mass function, without using the PGF:

$$EX = \sum_{k=0}^{\infty} k \cdot \frac{\lambda^k e^{-\lambda}}{k!} = \sum_{k=1}^{\infty} \frac{\lambda^k e^{-\lambda}}{(k-1)!} = \lambda \sum_{k=1}^{\infty} \frac{\lambda^{k-1} e^{-\lambda}}{(k-1)!} = \lambda \sum_{k=0}^{\infty} \frac{\lambda^k e^{-\lambda}}{k!} = \lambda,$$

and

$$E(X(X-1)) = \sum_{k=0}^{\infty} k(k-1) \cdot \frac{\lambda^k e^{-\lambda}}{k!} = \sum_{k=2}^{\infty} \frac{\lambda^k e^{-\lambda}}{(k-2)!} = \lambda^2 \sum_{k=2}^{\infty} \frac{\lambda^{k-2} e^{-\lambda}}{(k-2)!} = \lambda^2 \sum_{k=0}^{\infty} \frac{\lambda^k e^{-\lambda}}{k!} = \lambda^2.$$

(7.4.5) Higher moments of a Poisson distribution. Using the PGF and continuing from (7.4.3), we get the factorial moments:

$$E(X(X-1)(X-2)) = \pi'''(1) = [\lambda^3 e^{\lambda t - \lambda}]_{t=1} = \lambda^3,$$

and, in general, the jth factorial moment is

$$E(X(X-1)(X-2)\cdots(X-j+1)) = \lambda^j \quad \text{for } j = 1, 2, 3, \ldots. \tag{7.4.6}$$

As with the other distributions on the nonnegative integers, the moments and central moments are complicated, though less so for the Poisson distribution than the others. For example, (7.4.6) implies for $j = 3$ that

$$EX^3 - 3EX^2 + 2EX = \lambda^3.$$

Putting this together with (7.4.4) and the fact that $EX = \lambda$, we get the third moment,

$$EX^3 = \lambda^3 + 3(\lambda^2 + \lambda) - 2\lambda = \lambda^3 + 3\lambda^2 + \lambda,$$

and the third central moment,

$$\begin{aligned} E(X-\lambda)^3 &= EX^3 - 3\lambda EX^2 + 3\lambda^2 EX - \lambda^3 \\ &= \lambda^3 + 3\lambda^2 + \lambda - 3\lambda(\lambda^2 + \lambda) + 3\lambda^2 \cdot \lambda - \lambda^3 = \lambda. \end{aligned}$$

The third central moment, as well as the second central moment (the variance), both equal the expected value, λ.

Could it be that all the central moments equal λ? Unfortunately not. You can check, using computations similar to the preceding ones, that $E(X-\lambda)^4 = \lambda + 3\lambda^2$.

(7.4.7) Sums of independent Poisson random variables are Poisson. Here the parameters of the individual random variables do not need to be the same. We can see this immediately using the PGF:

If $Y = X_1 + X_2 + \cdots + X_n$, where $X_1, X_2, \ldots, X_n$ are independent and $X_1 \sim \text{Pois}(\lambda_1)$, $X_2 \sim \text{Pois}(\lambda_2)$, . . . , and $X_n \sim \text{Pois}(\lambda_n)$, then $Y \sim \text{Pois}(\lambda)$, where $\lambda = \lambda_1 + \lambda_2 + \cdots + \lambda_n$.

The reason is that PGF of Y is the product of the PGFs of the X_j's, which equals

$$e^{\lambda_1 t - \lambda_1} \cdot e^{\lambda_2 t - \lambda_2} \cdot \cdots \cdot e^{\lambda_n t - \lambda_n} = e^{\lambda t - \lambda}.$$

(7.4.8) **Computing Poisson probabilities.** As with binomial probabilities, it is often most efficient to compute Poisson probabilities recursively, starting with $p(0)$. The recursion is very simple:

$$p(0) = \frac{\lambda^0 e^{-\lambda}}{0!} = e^{-\lambda};$$

$$\frac{p(1)}{p(0)} = \frac{\lambda e^{-\lambda}}{e^{-\lambda}} = \lambda, \quad \text{so} \quad p(1) = p(0) \cdot \lambda;$$

$$\frac{p(2)}{p(1)} = \frac{\lambda^2 e^{-\lambda}/2!}{\lambda e^{-\lambda}} = \frac{\lambda}{2}, \quad \text{so} \quad p(2) = p(1) \cdot \frac{\lambda}{2};$$

$$\frac{p(3)}{p(2)} = \frac{\lambda^3 e^{-\lambda}/3!}{\lambda^2 e^{-\lambda}/2!} = \frac{\lambda}{3}, \quad \text{so} \quad p(3) = p(2) \cdot \frac{\lambda}{3};$$

and, in general,

$$p(k) = p(k-1) \cdot \frac{\lambda}{k}.$$

That is, to get $p(k)$ from $p(k-1)$, just multiply by λ and divide by k.

EXERCISES FOR **SECTION 7.4**

1. For a Poisson process with an arrival rate of 6.2 per minute, find the probabilities of the following events.
   a. There are 6.2 arrivals in the first minute.
   b. There are six arrivals in the first minute.
   c. There are more than two arrivals in the first 30 seconds.
   d. There are four arrivals in the first minute and three in the next 45 seconds.

2. In a Poisson process with λ arrivals per minute, if we observe a time interval of length $1/\lambda$ minutes, we expect one arrival.
   a. What is the probability that we see exactly one arrival?
   b. What is the probability that we see no arrivals?

3. We do 80 Bernoulli trials; the success probability is $\frac{1}{20}$. Find the approximate probability that we get exactly the expected number of successes.

4. Suppose customers arrive at a bank each day in accordance with a Poisson process, with an arrival rate of 1.6 customers per minute.
   a. Let Z be the sum of the numbers of customers that arrive in the first minute on each of two successive days. Find the probability that Z equals 4.
   b. The number of customers arriving in the first minute is observed on five different days, and W is the average of the five observations. What is the probability that W equals 1.6? [*Hint:* Translate "$W = 1.6$" into a statement about the sum of the five numbers.]

5. In Exercise 12 of Section 7.2 you were asked to compute recursively the first several binomial probabilities for $n = 100$ and $p = .03$. Now compute the Poisson approximation to those binomial probabilities, using the recursive formula in (7.4.8). Start with $p(0)$, the Poisson approximation to which is $e^{-3} \doteq .049787$.

6. **Fisher's insect egg distribution.** It sometimes happens that a random variable with a Poisson distribution can be observed only if it is not 0. For example, suppose a certain species of insect lays its eggs on leaves and the number of eggs deposited on a leaf by an insect is a Poisson random variable. Then empty leaves cannot be counted as observations of $X = 0$, because they might be leaves never visited by an insect. In this case we model the number of eggs on a leaf as a random variable X whose possible values are $1, 2, 3, \ldots$; the probabilities are the conditional probabilities for a Poisson random variable, given that it is nonzero. Write down the mass function and find the expected value. (A common name for this distribution is "truncated Poisson.")

7. **The mode of a Poisson distribution.** Which integer is given the largest probability by the Poisson distribution with parameter λ? In Exercise 17 of Section 1.2 we found the answer, for several particular values of λ, simply by inspecting the probabilities. It can be done more systematically by computing the ratio $r_k = p(k)/p(k-1)$ for $k = 1, 2, 3, \ldots$ and noting when r_k is greater than, equal to, or less than 1.
   a. Find r_k in terms of λ and k.
   b. Suppose first that λ is not an integer. Show that $p(k) > p(k-1)$ if $k < \lambda$ and $p(k) < p(k-1)$ if $k > \lambda$. For which k is $p(k)$ largest in this case?
   c. Now suppose λ is a positive integer. Then $p(k)$ is largest for two adjacent integers. What are they?

   The *mode* of a discrete distribution is its most probable value—that is, the number at which the mass function is largest, if there is such a number. For an absolutely continuous distribution, the mode is defined as the value at which the density function is largest, if there is such a value.

8. **Stirling's formula to approximate the probability of the expected value.** Stirling's formula is

$$n! \sim n^n e^{-n} \sqrt{2\pi n}.$$

The symbol "~" indicates that the expression is valid for large n in the sense that the ratio of the two sides approaches 1 as $n \to \infty$. Use this to find the approximate probability that $X = \lambda$ if X has the Pois(λ) distribution and λ is a large integer.

The expected value is (tied for) the most probable value if λ is an integer, but we see that its probability is quite small if λ is large. We saw a similar phenomenon for the binomial distribution with $p = .5$, in Exercise 13 in Section 7.2. These are instances of the general truth that *the expected value is only an expected average;* it is not likely that X will equal its expected value, only that it will be within a fixed percentage of it.

ANSWERS

1. a. 0 **b.** .1601 **c.** .5988 **d.** .0200

2. a. $e^{-1} \doteq .3679$ **b.** e^{-1}

3. .1954

4. a. .1781 **b.** .1396

5. .149361, .224042, .224042, .168031, .100819, .050409, .021604, .008102, .002701

6. $EX = \lambda/(1 - e^{-\lambda})$

7. b. $k = \lfloor \lambda \rfloor$, the greatest integer less than or equal to λ. **c.** $k = \lambda$ and $k = \lambda - 1$

8. $1/\sqrt{2\pi\lambda}$

7.5 HYPERGEOMETRIC DISTRIBUTIONS

Hypergeometric distributions arise in the following situation: Suppose we have a finite population consisting of items of two types. For convenience we label the two types S and F. For example, we might have an urn containing balls of two colors, or we might have a population of voters, each of whom favors one of two candidates. We draw a sample of n items from the population without replacing any sampled item, so that no item can be sampled twice. If X is the number of S's in the sample, what is the probability that $X = k$ for various k?

Notice that if the sampling were done *with* replacement instead of without, then the successive draws would be independent trials, and the probability of an S on each trial would be the proportion of S's in the population. So the sampling would amount to a Bernoulli process, and the distribution of X would be bin(n, p), where n is the sample size and p is the proportion of S's in the population.

But if the sampling is done *without* replacement, as seems more sensible, then the trials are not independent: drawing an S reduces the proportion of S's in the population and thus lowers the conditional probability that the next draw produces an S. The distribution of X is no longer binomial; it has a *hypergeometric* distribution. Naturally, the probabilities are not the same as those given by the binomial distribution, but as we will see, the difference is not very noticeable if the population is large.

(7.5.1) **The mass function of a hypergeometric distribution.** Suppose we have a finite population consisting of S's and F's. Let us say that the number of S's in the population is σ and the number of F's is ϕ, and the total size of the population is $\lambda = \sigma + \phi$. (These are not the usual uses of these Greek letters, but we choose them so that it will be easier to remember what they stand for: σ for the number of S's in the population, ϕ for the number of F's, and π for population size.) We take a random sample of n items from the population, without replacement, and let X be the number of S's in the sample. What is the mass function of X? That is, what is $p(k) = P(X = k)$, as a function of σ, ϕ, n, and k?

To find the mass function of X we go back to Section 1.2 and think of the outcome space Ω as the set of all possible samples of size n from the population. It is just the set of all subsets of size n from a set of size π; so the number of possible outcomes is $\binom{\pi}{n}$. Now the event "$X = k$" is the set of all subsets of size n consisting of k S's and $n - k$ F's, taken from the σ S's and ϕ F's in the population. The number of such subsets is $\binom{\sigma}{k}\binom{\phi}{n-k}$. Since all members of Ω are equally likely (random sampling means that no possible sample is more likely than any other), we see that the mass function of X is

(7.5.2)
$$p(k) = \frac{\binom{\sigma}{k}\binom{\phi}{n-k}}{\binom{\pi}{n}} \quad (k = 0, 1, 2, \ldots, n).$$

This is the ***hypergeometric* mass function with parameters n, σ, and π.** Only three parameters are needed to determine the distribution: the sample size, the number of S's in the population, and the population size. But it is a difficult mass function to memorize, because it involves five different quantities (the parameters n, σ, and π, $\phi = \pi - \sigma$, and the possible value k). It is much easier to work out the probabilities directly, as in the following example.

(7.5.3) **Example.** An urn contains 8 blue balls and 12 white balls. If five are drawn at random, without replacement, what is the probability that the sample contains two blues and three whites? The answer is

$$\frac{\binom{8}{2}\binom{12}{3}}{\binom{20}{5}}.$$

This is because the number of possible samples is the number of subsets of size 5 in a set of 20 balls; this $\binom{20}{5}$. The number of samples containing two blue balls and three white balls is the number of subsets of size 2 from the eight blues, times the number of subsets of size 3 from the 12 whites; this is $\binom{8}{2}\binom{12}{3}$.

It is much easier to write down such a probability just by looking at the problem than by trying to use the formula. The fraction above has a property that all hypergeometric probabilities have, which you can use as a partial check—in this one, $8 + 12 = 20$ and $2 + 3 = 5$. The first sum, $8 + 12 = 20$, concerns the numbers of S's and F's in the population, and the other one, $2 + 3 = 5$, shows the numbers of S's and F's in the sample.

If we *had* to use the formula (7.5.2) for this problem, we would say that we have $\sigma = 8$ S's (blue balls), $\phi = 12$ F's (white balls), $\pi = 20$ balls in the population, the sample size is $n = 5$, and we want the probability of $k = 2$ S's in the sample. But you are most strongly advised not to try to find hypergeometric probabilities in this way.

What about finding the numerical value of the probability above? Probably the best way is to write out the binomial coefficients using the formula

$$\binom{r}{k} = \frac{r(r-1)(r-2)\cdots(r-k+1)}{k!}.$$

For the probability in question we get

$$\frac{\binom{8}{2}\binom{12}{3}}{\binom{20}{5}} = \frac{8\cdot 7}{2\cdot 1}\cdot\frac{12\cdot 11\cdot 10}{3\cdot 2\cdot 1}\cdot\frac{5\cdot 4\cdot 3\cdot 2\cdot 1}{20\cdot 19\cdot 18\cdot 17\cdot 16};$$

we can do several cancellations before we go to our calculator. The result is about .379317.

Even if you do not trust yourself to do the cancellations and want to do it all on your calculator, it is good to write it out as we did above first, and then to do alternately one multiplication and one division. That keeps the numbers from getting too big or too small for the calculator's capacity. ∎

(7.5.4) **Example: Sampling with replacement vs. without.** It will not surprise you to learn that if the population is large compared to the size of the sample, then the probabilities for sampling without replacement are very close to the ones for sampling with replacement. The reason is that even if we return sampled items to the population for possible resampling, it is unlikely that a member of a large population will be chosen more than once in a small sample.

For example, suppose we have a population of 1,000 items, of which 400 are S's and 600 are F's. What is the probability that a sample of size 12 contains 4 S's and 8 F's?

Solution. If the sampling is *without* replacement, the answer is the hypergeometric probability

$$\frac{\binom{400}{4}\binom{600}{8}}{\binom{1{,}000}{12}}$$
$$= \frac{400 \cdot 399 \cdot 398 \cdot 397}{4!} \cdot \frac{600 \cdot 599 \cdot \cdots \cdot 593}{8!} \cdot \frac{12!}{1{,}000 \cdot 999 \cdot \cdots \cdot 989}$$
$$\doteq .213768.$$

If it is *with* replacement, then we are doing $n = 12$ Bernoulli trials with success probability $p = \frac{400}{1{,}000}$, and consequently the answer is the binomial probability

$$\binom{12}{4}\left(\frac{400}{1{,}000}\right)^4\left(\frac{600}{1{,}000}\right)^8 \doteq .212841.$$

The difference between these two is less than half a percent. You might want to study the two calculations; by looking hard at the hypergeometric one you can see why it is close to the binomial one. The $\binom{12}{4}$ is easy to spot, and a little rearranging of the rest of the factors will show you why they are close to the powers of $\frac{400}{1{,}000}$ and $\frac{600}{1{,}000}$ that are found in the binomial probability.

Now suppose the population has the same mix of S's and F's, but only 100 items instead of 1,000. That is, it contains 40 S's and 60 F's. Now what is the probability that a sample of size 12 contains 4 S's and 8 F's?

If the sampling is *without* replacement, then the answer is the hypergeometric probability

$$\frac{\binom{40}{4}\binom{60}{8}}{\binom{100}{12}} = \frac{40 \cdot 39 \cdot 38 \cdot 37}{4!} \cdot \frac{60 \cdot 59 \cdot \cdots \cdot 53}{8!} \cdot \frac{12!}{100 \cdot 99 \cdot \cdots \cdot 89} \doteq .222608.$$

If it is *with* replacement, then the answer is the same binomial probability computed above—about .212841; in sampling *with* replacement we have Bernoulli trials, and the success probability depends only on the *proportions* of S's and F's in the population, not on the size. You can see that the two answers are still close together, but not as close as they are for a larger population. The difference now is about 4.5%.

The other thing you will have noticed is that the binomial probabilities are much easier to compute than the hypergeometric ones. For this reason, they are often used as an approximation to hypergeometric probabilities when the population is large compared to the sample. ▮

Rather than prove an approximation theorem as we did with the Poisson approximation to the binomial probabilities, we will simply restate what we have seen above.

(7.5.5) Binomial approximation to the hypergeometric distribution. For samples of size n without replacement from a population consisting of σ S's and ϕ F's: if the population size $\pi = \sigma + \phi$ is large compared to the sample size n, then the (hypergeometric) probability of k S's in the sample,

$$p(k) = \frac{\binom{\sigma}{k}\binom{\phi}{n-k}}{\binom{\pi}{n}},$$

is well approximated by the binomial probability

$$p(k) = \binom{n}{k} p^k q^{n-k}$$

where $p = \sigma/\pi$ and $q = \phi/\pi$ are the proportions of S's and F's in the population.

(7.5.6) Expected value of a hypergeometric distribution. Let X have the hypergeometric distribution based on a sample of size n from a population of size π with σ S's and $\phi = \pi - \sigma$ F's. That is, X is the number of S's in a sample of size n without replacement from this population. We want to find EX.

After our experience with other discrete distributions, the inclination is to try

$$EX = \sum_{k=1}^{n} k \cdot \frac{\binom{\sigma}{k}\binom{\phi}{n-k}}{\binom{\pi}{n}},$$

or perhaps the PGF

$$\pi(t) = \sum_{k=0}^{n} t^k \cdot \frac{\binom{\sigma}{k}\binom{\phi}{n-k}}{\binom{\pi}{n}}.$$

The sum for EX can be done with considerable manipulation, but the PGF has no simple sum. It is what mathematicians call a "special function," along with Bessel functions, Legendre functions, and the like, which means that it arises in trying to solve certain types of differential equations. In the theory of special functions this PGF is called a *hypergeometric function,* and this is where the family of distributions gets its name.

However, it is not difficult to find EX using indicator functions. The result is a bit of a surprise:

(7.5.7)
$$EX = n \cdot \frac{\sigma}{\pi}.$$

What is surprising about this? If the sampling were *with* replacement, X would have the binomial distribution with parameters n (sample size) and $p = \sigma/\pi$

(proportion of S's in population), and so EX would equal $np = n\sigma/\pi$. Now we see that the expected value is the same even though the sampling is *without* replacement.

The indicators are as follows. Let X_1 equal 1 if the first item sampled is an S and 0 if it is an F. Then X_1 has the Bern(p) distribution, where $p = P(X_1 = 1) = \sigma/\pi$. Therefore, $EX_1 = \sigma/\pi$. Now let X_2 equal 1 if the second item sampled is an S and 0 if an F. Then $EX_2 = P(X_2 = 1)$, and this also equals σ/π. [See the remarks in small print below if you are suspicious of this.]

Similarly, let X_j equal 1 if the jth item sampled is an S and 0 if it is an F. Then $EX_j = P(X_j = 1)$, and this equals σ/π no matter what j is. Finally, because $X = X_1 + X_2 + \cdots + X_n$, it follows that $EX = EX_1 + EX_2 + \cdots + EX_n = n \cdot \sigma/\pi$.

If you are uneasy about $P(X_2 = 1) = \sigma/\pi$ because you think the probabilities should be different on the second draw, remember that we are looking for the *unconditional* probability that the second item sampled is an S, not the conditional probability given something about the first item. We can find this by conditioning on the first item drawn:

$$\begin{aligned} &P(\text{second is S}) \\ &\quad = P(\text{first is S}) \cdot P(\text{second is S} \mid \text{first is S}) + P(\text{first is F}) \cdot P(\text{second is S} \mid \text{first is F}) \\ &\quad = \frac{\sigma}{\pi} \cdot \frac{\sigma - 1}{\pi - 1} + \frac{\phi}{\pi} \cdot \frac{\sigma}{\pi - 1} = \frac{\sigma \cdot (\sigma - 1 + \phi)}{\pi \cdot (\pi - 1)} = \frac{\sigma \cdot (\pi - 1)}{\pi \cdot (\pi - 1)} = \frac{\sigma}{\pi}. \end{aligned}$$

You might want to look back at Exercise 12i of Section 2.2, where we found that the (unconditional) probability that the second card drawn from a deck is an ace equals the probability that the first card drawn is an ace.

(7.5.8) **The variance of a hypergeometric distribution.** We use the same indicators as in (7.5.7). The variance of any one of these indicators is $(\sigma/\pi) \cdot (\phi/\pi)$, because the variance of a Bern(p) random variable is pq. But because these indicators are not uncorrelated, the variance of their sum is not just the sum of their variances; as we saw in (4.4.10), we have to add twice the sum of the covariances. (If the X_j's *were* independent, VX would be the sum of their variances, which equals $n \cdot (\sigma/\pi) \cdot (\phi/\pi)$, the same as the npq for a binomial distribution.)

The result turns out to be

(7.5.9)
$$VX = n \cdot \frac{\sigma}{\pi} \cdot \frac{\phi}{\pi} \cdot \frac{\pi - n}{\pi - 1}.$$

Before we prove this, notice that if the population size π is very large compared to the sample size n, then the fraction $(\pi - n)/(\pi - 1)$ is close to 1, and the variance is close to $npq = n \cdot (\sigma/\pi) \cdot (\phi/\pi)$. This is consistent with what we said above in (7.5.5), that a hypergeometric distribution is approximately a binomial distribution when π is large compared to n.

Proof. To prove (7.5.9), recall the formula for the variance of a sum, which we found in (4.4.10):

(7.5.10)
$$V\left(\sum_{j=1}^{n} X_j\right) = \sum_{j=1}^{n} VX_j + 2\sum_{i<j}\text{Cov}(X_i, X_j).$$

The first sum on the right, the sum of the variances, is $n \cdot \sigma/\pi \cdot \phi/\pi$, as we saw above. The sum of the covariances is the sum of $\binom{n}{2}$ terms, one for each pair of X_j's; so we need to find the covariance of one of these pairs.

As we saw in (7.2.11), the covariance of a pair of indicators is

$$\text{Cov}(X_i, X_j) = P(X_i = 1 \text{ and } X_j = 1) - P(X_i = 1) \cdot P(X_j = 1).$$

We found $P(X_i = 1)$ and $P(X_j = 1)$ in (7.5.6) above; they both equal σ/π. As for $P(X_i = 1 \text{ and } X_j = 1)$, it is the probability that two draws without replacement produce two successes; this is a hypergeometric probability:

$$P(X_i = 1 \text{ and } X_j = 1) = \frac{\binom{\sigma}{2}\binom{\phi}{0}}{\binom{\pi}{2}} = \frac{\sigma(\sigma - 1)}{\pi(\pi - 1)}.$$

So the covariance of one pair is

$$\text{Cov}(X_i, X_j) = \frac{\sigma(\sigma - 1)}{\pi(\pi - 1)} - \frac{\sigma^2}{\pi^2},$$

which simplifies to

$$\text{Cov}(X_i, X_j) = -\frac{\sigma\phi}{\pi^2(\pi - 1)}.$$

The X_j's are negatively correlated, which is not surprising when we consider that because the drawing is without replacement, if X_i is known to equal 1, then X_j becomes slightly less likely to equal 1.

Putting it together, we see that the variance of X is

$$VX = n \cdot \frac{\sigma}{\pi} \cdot \frac{\phi}{\pi} - 2 \cdot \binom{n}{2} \cdot \frac{\sigma\phi}{\pi^2(\pi - 1)},$$

and a little algebra shows that this equals the formula given in (7.5.9). □

(7.5.11) **The normal approximation to the hypergeometric distribution.** A hypergeometric random variable is not a sum of independent random variables; it is a sum of indicators, but as we saw in computing the variance just above, the indicators are not independent. Nevertheless, we have mentioned that if the population size is large, then the binomial distribution is a good approximation to the hypergeometric; and we already know about the normal approximation to the binomial.

As a result, it is common to use a normal approximation to the hypergeometric distribution. As with all normal approximations, it is easy to use: just subtract the expected value and divide by the standard deviation of the hypergeometric random variable, and use the standard normal table. An example will illustrate the method.

(7.5.12) Example. A population of size 1,000 contains 500 S's and 500 F's. We take a sample of size 100, without replacement. What is the probability that the sample contains between 40 and 60 S's (inclusive)?

Solution. If X is the number of S's in the sample, then

$$EX = n \cdot \frac{\sigma}{\pi} = 100 \cdot \frac{500}{1{,}000} = 50$$

and

$$VX = n \cdot \frac{\sigma}{\pi} \cdot \frac{\phi}{\pi} \cdot \frac{\pi - n}{\pi - 1} = 100 \cdot \frac{500}{1{,}000} \cdot \frac{500}{1{,}000} \cdot \frac{900}{999} = 22.5225.$$

So the probability we want is approximately (using the continuity correction as with the binomial distribution)

$$\begin{aligned} P(39.5 \le X \le 60.5) &\approx P\left(\frac{39.5 - 50}{\sqrt{22.5225}} \le Z \le \frac{60.5 - 50}{\sqrt{22.5225}}\right) \quad \text{where } Z \sim N(0, 1) \\ &\doteq P(-2.21 \le Z \le 2.21) \doteq .9728. \end{aligned}$$

It is interesting to notice that this probability is greater than the corresponding probability for sampling without replacement, because the variance of X is smaller, by a factor of $(\pi - n)/(\pi - 1)$, in sampling without replacement. In this example, if the sampling were with replacement, the variance of X would be $100 \cdot \frac{500}{1{,}000} \cdot \frac{500}{1{,}000} = 25$, and the probability would be

$$P\left(\frac{39.5 - 50}{\sqrt{25}} \le Z \le \frac{60.5 - 50}{\sqrt{25}}\right) = P(-2.1 \le Z \le 2.1) \doteq .9642.$$

See Exercise 6 for a similar comparison.

In general, the proportion of successes in the sample is more likely to be close to the proportion in the population if the sampling is without replacement. For this reason, polling agencies use sampling without replacement in estimating population proportions, even though sampling with replacement is conceptually simpler. This is in good accord with our intuition: it would seem foolish, in sampling voters to ascertain public opinion, to allow voters to be repeated in the sample. ∎

In Appendix B.6, instead of a summary of hypergeometric distributions, there is a summary and comparison of the distributions for sampling with and without replacement from a finite two-type population.

EXERCISES FOR **SECTION 7.5**

1. A committee consists of 15 women and 9 men. A subcommittee of size 6 is chosen and there are only two women on it.
   a. What is the probability that a randomly chosen subcommittee consists of two women and four men?
   b. What is the probability that there are two or fewer women on a randomly chosen subcommittee?
   c. What is the expected number of women on the subcommmittee?

   If you were suspicious because a committee of six contained only two women, the probability in part b is the one you would be interested in—not the one in part a, which is fairly small for any single number of women. If the probability in part b is very small, then either the selection was not really random, or something very rare occurred.

2. An urn contains 5 blue balls and 15 white balls. A sample of size 5 is drawn without replacement.
   a. What is the probability that the sample contains four blue balls? Five blue balls?
   b. Give the expected value and the variance of the number of blue balls in the sample.
   c. What would the expected value and variance have been if the sampling were with replacement?

3. A shipment of 100 garlic presses happens to contain four defective presses. A quality control inspector chooses five presses at random from the shipment and tests them.
   a. What is the probability that the sample contains no defective presses? One? Two? Three? Four?
   b. What is the probability that the sample contains five defective presses? [Notice that you can see the answer without any calculation. What happens when you try to use the formula (7.5.2)?]

4. Find the true (hypergeometric) probability and the binomial approximation for each of the following.
   a. A population contains 10 S's and 20 F's; what is the probability that a sample of size 10 contains three S's and seven F's?
   b. Same as part a, but population contains 20 S's and 40 F's.
   c. Same as part a, but population contains 100 S's and 200 F's.

5. Find EX and VX if X is the number of S's in a sample of size 10 for each of the following populations. Find the values for sampling with replacement, and also for sampling without replacement.

a. Population has 20 S's and 30 F's.
b. 200 S's and 300 F's
c. 8 S's and 12 F's
d. 4 S's and 6 F's (What is X for sampling without replacement in this case?)

6. A population consists of S's and F's; the proportion of S's is .3. We want the normal approximation to the probability that a sample of size 50 (in which the expected number of S's is 15) contains between 12 and 18 S's (inclusive). To use the continuity correction, we find the probability that the number of S's is between 11.5 and 18.5. Find this probability, assuming:
a. That the sampling is with replacement.
b. That the sampling is without replacement and the population size is 500.
c. That the sampling is without replacement and the population size is 5,000. [Notice that sampling without replacement gives a higher probability of being close to the expected value, although the difference is less with a larger population.]

7. **Seligman's corrupt-senator problem.** The United States Senate has 100 members. A random sample of two senators is taken, and it is found that both of them are engaged in corruption or fraud of some kind. What inference, if any, can we draw about the number of corrupt people in the Senate as a whole? One statistical approach to this question is through the *likelihood principle*. For each possible number, σ, of corrupt senators, we compute the likelihood of the observed data—that is, the probability that a sample of size 2 contains two corrupt senators, under the hypothesis that σ is the true number. [We might then use the σ that gives the greatest likelihood as an estimate of the number of corrupt people in the Senate; this is called *maximum likelihood estimation*.]
a. Find this likelihood as a function of σ.
b. In statistical inference, one regards a hypothesis as untenable if the likelihood of the observed data under that hypothesis is very small. Show that the likelihood is less than .05 if σ is 22 or less, and is less than .01 if σ is 10 or less.

Loosely speaking, the observation of two corrupt senators, in a random sample of size 2, is quite a rare occurrence (probability less than 1 in 100) unless there are at least 11 corrupt senators; and it is fairly rare (less than 1 in 20) unless there are 23 or more. In his *Fortune* magazine column "Keeping Up" for November 6, 1978, Daniel Seligman noted that something like this actually happened when it was discovered that two senators, in the news for other reasons, both turned out to be involved in large-scale corruption. As Seligman points out, though, the inference is suspect in this case: the two were in the news because of other activities that might or might not be independent of corruption. Thus the sample of size 2 may not have been random.

ANSWERS

1. **a.** .09829 **b.** .11296 **c.** $3\frac{3}{4}$

2. **a.** .004837; .0000645 **b.** $\frac{5}{4}$; $\frac{15}{16} \cdot \frac{15}{19}$ **c.** $\frac{5}{4}$; $\frac{15}{16}$

3. **a.** .8119; .1765; .01139; .000242; .000001 **b.** 0

4. Bin. approx.: .260123 in all parts. True values: **a.** .309615 **b.** .281901 **c.** .264107

5. **a.** With: 4; 2.4; without: 4; 1.9592
   b. With: 4; 2.4; without: 4; 2.3567
   c. With: 4; 2.4; without: 4; 1.263
   d. With: 4; 2.4; without: 4; 0

6. **a.** .7198 **b.** .7458 **c.** .7242

7. **a.** $\sigma(\sigma - 1)/9{,}900$

7.6 MULTINOMIAL DISTRIBUTIONS

A multinomial distribution is a certain kind of joint distribution of several discrete random variables, not the distribution of a single random variable. It arises often in statistical work, and so we include a brief description of it here.

Suppose we do n independent trials that are like Bernoulli trials, except that each trial has k different outcomes instead of 2. That is, instead of two outcomes S and F with probabilities p and q that add to 1, we have k outcomes, say $\mathcal{O}_1, \mathcal{O}_2, \ldots, \mathcal{O}_k$, with probabilities $p_1, p_2, \ldots, p_k$ that add to 1. Let X_1 be the number of $\mathcal{O}_1$'s, X_2 the number of $\mathcal{O}_2$'s, and so on.

Now the distribution of one of the individual X_j's is easy, when you think about it: X_j has the bin(n, p_j) distribution. To see this, think of outcome $\mathcal{O}_j$ as a success and any of the others as a failure; then X_j just counts the number of successes.

But the X_j's are not independent; if X_1 is large, for example, then we would expect the other X_j's to be small. In fact, the X_j's are constrained by

$$X_1 + X_2 + \cdots + X_k = n,$$

because each X_j counts a number of trials, and these numbers have to add to the total number of trials made, which is n.

The joint distribution of the X_j's in this situation is what is called a multinomial distribution. The marginal distributions are all binomial distributions, as we said. All we want to do here is to find the joint mass function, and then to give a few examples, including some that illustrate how multinomial distributions arise in statistical inference.

(7.6.1) **The multinomial joint mass function.** As described above, consider independent trials, each trial having k possible outcomes $\mathcal{O}_1, \mathcal{O}_2, \ldots, \mathcal{O}_k$, with probabilities $p_1, p_2, \ldots, p_k$ that add to 1. Let X_j be the number of $\mathcal{O}_j$'s that occur in the n trials.

Then a possible value of the k-tuple $(X_1, X_2, \ldots, X_k)$ is a k-tuple $(m_1, m_2, \ldots, m_k)$ of nonnegative integers having the property that $m_1 + m_2 + \cdots + m_k = n$. What is the probability that $(X_1, X_2, \ldots, X_k)$ equals this k-tuple; that is, what is $P(X_1 = m_1, X_2 = m_2, \ldots, X_k = m_k)$?

The possible outcomes of the n-trial experiment are n-letter "words" from the "alphabet" $\{\mathcal{O}_1, \mathcal{O}_2, \ldots, \mathcal{O}_k\}$. There are k^n of these, and the probability of any one of them is $p_1^{m_1} \cdot p_2^{m_2} \cdots p_k^{m_k}$, where m_1 is the number of $\mathcal{O}_1$'s in the word, and so on. So $P(X_1 = m_1, X_2 = m_2, \ldots, X_k = m_k)$ must be this probability times the number of n-letter words that happen to contain m_1 $\mathcal{O}_1$'s, m_2 $\mathcal{O}_2$'s, and so on. This number is a "multinomial coefficient"; it is usually shown in discrete mathematics courses that

> the number of n-letter words containing m_1 $\mathcal{O}_1$'s, m_2 $\mathcal{O}_2$'s, . . . , and m_k $\mathcal{O}_k$'s is $\dfrac{n!}{m_1!m_2!\cdots m_k!}$.

Consequently, the joint mass function of $X_1, X_2, \ldots, X_k$ is

(7.6.2)

$$p(m_1, m_2, \ldots, m_k) = \frac{n!}{m_1!m_2!\cdots m_k!} \cdot p_1^{m_1} \cdot p_2^{m_2} \cdots p_k^{m_k}$$

if $m_1, \ldots, m_k$ are nonnegative integers whose sum is n.

This is the joint mass function of the ***multinomial distribution with parameters $n, p_1, p_2, \ldots, p_k$***. For each $j = 1, 2, \ldots, k$, the random variable X_j has the $\text{bin}(n, p_j)$ distribution.

(7.6.3) Example. If we toss two fair coins there are three possible outcomes: no heads, with probability $\frac{1}{4}$; one head, with probability $\frac{1}{2}$; and two heads, with probability $\frac{1}{4}$. If we do this 20 times, what is the probability that we get no heads 4 times, one head 9 times, and two heads 7 times? The answer is easy to write down:

$$\frac{20!}{4!9!7!}\left(\frac{1}{4}\right)^4\left(\frac{1}{2}\right)^9\left(\frac{1}{4}\right)^7 \doteq .0258.$$

What we have are 20 independent three-outcome trials; the outcomes are $\mathcal{O}_0$, "no heads," whose probability is $p_0 = \frac{1}{4}$; $\mathcal{O}_1$, "one head," with $p_1 = \frac{1}{2}$; and $\mathcal{O}_2$, "two heads," with $p_2 = \frac{1}{4}$. We have found the probability that the triple (X_1, X_2, X_3) equals $(4, 9, 7)$.

Notice the following pitfall. What is the probability that $(X_1, X_2, X_3) = (4, 8, 6)$? The answer is 0, because $4 + 8 + 6$ is not equal to 20. The formula (7.6.2) will not automatically give 0 here. ▮

(7.6.4) Example. What is the probability that 36 rolls of a fair die produce exactly six of each face? The answer is

$$\frac{36!}{6!6!6!6!6!6!}\left(\frac{1}{6}\right)^6\left(\frac{1}{6}\right)^6\left(\frac{1}{6}\right)^6\left(\frac{1}{6}\right)^6\left(\frac{1}{6}\right)^6\left(\frac{1}{6}\right)^6 \doteq .0002589.$$

Notice how unlikely this is, even though it is the most likely outcome of 36 rolls of a fair die, and even though 6 is the expected number of times each face appears. There is no reason to think the most likely outcome is very likely if there are a lot of possible outcomes. And there is no reason to think that the expected outcome is likely; it is only the expected average outcome over the long run. ▮

(7.6.5) **Bernoulli trials as a special case of multinomial sampling.** Suppose $k = 2$. Then we have only two outcomes, $\mathcal{O}_1$ and $\mathcal{O}_2$, with probabilities p_1 and p_2 adding to 1, and we do n trials. Let X_1 be the number of $\mathcal{O}_1$'s and X_2 the number of $\mathcal{O}_2$'s. What is the probability that $X_1 = m_1$ and $X_2 = m_2$, where $m_1 + m_2 = n$?

The multinomial formula (7.6.2) gives us the answer,

$$\frac{n!}{m_1!m_2!}p_1^{m_1}p_2^{m_2}.$$

But we could have found this as a binomial probability also. We can write S for $\mathcal{O}_1$ and F for $\mathcal{O}_2$, and we just want the probability of m_1 successes in n trials. The success probability p equals p_1 and the failure probability q equals p_2, so the answer is

$$\binom{n}{m_1}p_1^{m_1}p_2^{n-m_1}.$$

Needless to say, these two answers are the same.

(7.6.6) **The covariance of two of the variables in a multinomial distribution.** Suppose X_1, $X_2, \ldots, X_k$ have the multinomial distribution with parameters $n, p_1, p_2, \ldots,$ p_k. What is the covariance of X_1 and X_2? Remember that X_1 is the number of times $\mathcal{O}_1$ occurs in the n trials, and X_2 is the number of times $\mathcal{O}_2$ occurs. A high value of X_1 slightly decreases the chance that X_2 is high, because if there are many $\mathcal{O}_1$'s, then there is a little less chance of getting many $\mathcal{O}_2$'s. So we would guess that the covariance will be slightly negative.

We know from (4.4.2) that

$$\text{Cov}(X_1, X_2) = E(X_1X_2) - (EX_1)(EX_2).$$

Finding EX_1 and EX_2 is easy, because X_1 and X_2 are binomially distributed:

$$EX_1 = np_1 \quad \text{and} \quad EX_2 = np_2;$$

but $E(X_1X_2)$ is not so easy to get directly.

Instead we turn to indicators. We have two sets of indicators to deal with—the indicators for $\mathcal{O}_1$ on the various trials and those for $\mathcal{O}_2$. For each trial number $j = 1, 2, \ldots, n$, let

$$I_j^{(1)} = \begin{cases} 1 & \text{if } \mathcal{O}_1 \text{ occurs on trial } j, \\ 0 & \text{otherwise,} \end{cases} \quad \text{and} \quad I_j^{(2)} = \begin{cases} 1 & \text{if } \mathcal{O}_2 \text{ occurs on trial } j, \\ 0 & \text{otherwise.} \end{cases}$$

Then

$$X_1 = \sum_{j=1}^{n} I_j^{(1)} \quad \text{and} \quad X_2 = \sum_{l=1}^{n} I_l^{(2)},$$

and so

(7.6.7)
$$\mathrm{Cov}(X_1, X_2) = \mathrm{Cov}\left(\sum_{j=1}^{n} I_j^{(1)}, \sum_{l=1}^{n} I_l^{(2)}\right) = \sum_{j=1}^{n}\sum_{l=1}^{n} \mathrm{Cov}(I_j^{(1)}, I_l^{(2)}).$$

This comes from the linearity of the covariance that we saw in (4.4.8). It is a sum of n^2 covariances, one for each pair (j, l) of trial numbers.

Now in (7.2.11) we saw how to find the covariance of two indicators. It is the probability that they are both 1 together, minus the product of the probabilities that they equal 1:

$$\mathrm{Cov}(I_j^{(1)}, I_l^{(2)}) = P(I_j^{(1)} = 1 \text{ and } I_l^{(2)} = 1) - P(I_j^{(1)} = 1) \cdot P(I_l^{(2)} = 1).$$

The two individual probabilities are easy:

$$P(I_j^{(1)} = 1) = P(\mathbb{O}_1 \text{ occurs on trial } j) = p_1$$

and

$$P(I_l^{(2)} = 1) = P(\mathbb{O}_2 \text{ occurs on trial } l) = p_2.$$

As for $P(I_j^{(1)} = 1$ and $I_l^{(2)} = 1)$, it depends on whether trials j and l are the same trial. If $j = l$, it is impossible that both $\mathbb{O}_1$ and $\mathbb{O}_2$ occur on the same trial, so the probability is 0. Otherwise, the two trials are independent, so the probability is p_1p_2. Putting this together, we get that

$$\mathrm{Cov}(I_j^{(1)}, I_l^{(2)}) = \begin{cases} -p_1p_2 & \text{if } j = l, \\ 0 & \text{if } j \neq l. \end{cases}$$

So, looking back at (7.6.7), we see that although it is a sum of n^2 covariances, one for each pair (j, l) of trial numbers, only the n covariances for which $j = l$ are nonzero, and each of these is $-p_1p_2$. Therefore,

(7.6.8)
$$\mathrm{Cov}(X_1, X_2) = -np_1p_2.$$

Of course, the same argument applies to any two of the outcomes: the covariance is $-n$ times the product of the single-trial probabilities of those outcomes.

EXERCISES FOR **SECTION 7.6**

1. The experiment of tossing a fair coin three times and counting the number of heads has four possible outcomes: 0, 1, 2, 3. Suppose we do this experiment 8 times. What is the probability that we get exactly the expected numbers of 0's, 1's, 2's, and 3's? [It is disconcertingly small, even though it is the "expected" outcome and (as can be shown) the most probable outcome. But as we have

pointed out repeatedly, the most probable outcome does not have to be very probable when there are many outcomes.]

2. Runners A, B, C, and D have probabilities .3, .25, .1, and .35 of winning when they race against each other. Suppose they race 15 times, independently. Find the following.
   a. The probability that A wins four times, B wins three times, C wins twice, and D wins six times
   b. The probability that A wins four times, B wins three times, C wins three times, and D wins four times
   c. The probability that C wins more than once.
   d. The expected number of times B wins
   e. The probability that there are exactly eight races that are won by either A or B [*Hint:* Consider two outcomes: one is (A or B), the other is (C or D).]
   f. The probability that A wins five times and B wins three times [*Hint:* Consider three outcomes.]

3. If X_1, X_2, X_3, and X_4 have the multinomial distribution with parameters 50, .1, .2, .3, and .4, find the following.
   a. The probability that X_2 is 2 or less
   b. The expected values of the X_i's
   c. The distribution of $X_2 + X_4$
   d. The joint distribution of X_1, X_2, and $X_3 + X_4$

4. Let X equal the number of successes and Y the number of failures in n Bernoulli trials with success probability p. Find the following.
   a. The joint distribution of X and Y
   b. $\mathrm{Cov}(X, Y)$ [You can find it most easily using (7.6.8); but for practice with covariances you can also find it using the definition of covariance, the fact that $Y = n - X$, and a little algebra.]

5. **Feller's encounter with parapsychologists.** A classical experiment to detect extrasensory perception involves a deck of cards, in which each card contains one of five symbols. [Subjects are asked to determine which symbol is on a randomly chosen card without looking at it.] Assume the deck contains equal numbers of each of the five symbols. Suppose we make 25 independent draws, with replacement, from such a deck, shuffling carefully before each draw. Find the probability that each symbol is drawn five times.

This probability is small enough that, if someone claimed to have obtained equal numbers of the five symbols in 25 trials, we might suspect that the draws were not done independently and at random. William Feller argued in this way in criticizing a research paper in the *Journal of Parapsychology*. Feller comments on the response he received in Volume 1 of his *An Introduction to Probability Theory and Its Applications* (see the bibliography). The comments, in footnotes, can be found by looking up "parapsychology" in Feller's index.

Students of trivia in the history of science will be interested to know that in the second edition of his book, Feller called the parapsychologists' alleged randomness "extraordinary" and their response to his critique "amusing," whereas in the third edition he used the words "miraculous" and "amazing" instead. In both editions, he remarks of the parapsychologists that "both their arithmetic and their experiments have a distinct tinge of the supernatural."

6. Let $X_1, X_2, \ldots, X_k$ have the multinomial distribution with parameters $n, p_1, p_2, \ldots, p_k$. Let $S = X_1 + X_2 + \cdots + X_k$.
   a. Use the formulas for the expected value and variance of a sum given in (4.2.14) and (4.4.10), along with the variances and covariances of the X_j's found in this section, to find the expected value and variance of S.
   b. Explain how you could have known these results without doing any calculations.

ANSWERS

1. .04867
2. a. .01467 b. 0 c. .45096 d. 3.75 e. .2013 f. .05113
3. a. .001285 b. 5, 10, 15, 20 c. bin(50, .6) d. Multinomial with parameters 50, .1, .2, .7
4. a. Multinomial with parameters n, p, q b. $-npq$
5. .00209
6. a. $ES = n$, $VS = 0$

SUPPLEMENTARY EXERCISES FOR **CHAPTER 7**

1. Let X be a nonnegative integer-valued random variable whose PGF is $\pi(t) = (4/(11 - 7t))^2$.
   a. Expand this in a power series to find the mass function. [*Hint:* Divide numerator and denominator by 11 and use the power series expansion for $1/(1 - x)^2$.]
   b. What is the name of the distribution of X?

2. The probability mass function of X is $p(k) = \alpha q^k/k$ for $k = 1, 2, 3, \ldots$. The parameter q is in the interval (0, 1), and α is the appropriate constant.
   a. Find the PGF of X. [*Hint:* Use (A.5.12).]
   b. Find EX (in terms of α and q) in two ways—from the definition and using the PGF.
   c. Find the value of α in terms of q in two ways—by summing the probabilities and by using the fact that $\pi_X(1) = 1$.

   [The point of doing these calculations two ways is to illustrate how, in a sense, finding the PGF really does all such calculations at once.]

3. If X has possible values 0, 1, 2, . . . and if for each of these numbers we have $P(X \geq k) = \left(\frac{1}{3}\right)^k$, what is the distribution of X? [*Hint:* If you know $P(X \geq k)$ for each k, you can find $P(X = k)$ for any k.]

4. Suppose you can hit free throws in basketball with probability .64. You are required to attempt free throws until you have hit ten. For simplicity assume successive throws are independent. (This is not really a good assumption, but may be valid as a first approximation for some purposes.)
   a. Find the probability that your tenth hit comes on the 15th throw.
   b. Find the approximate probability that you miss ten or more before the tenth hit. Use a continuity correction.
   c. What is the expected number of free throws you must attempt to make ten?

5. Suppose $\pi(t) = t^r$ is the PGF of X, where r is a given positive integer.
   a. By looking at the power series expansion of $\pi(t)$, determine the distribution of X.
   b. Confirm this by using the PGF to find EX and VX.

6. The PGF of X is $\pi(t) = \frac{1}{125}(2t + 3)^3$. Find the following.
   a. EX b. VX c. $P(X = 0)$ d. $P(X = 3)$ e. $P(X = 4)$ f. $P(X = 5)$

7. If the probability that you win a certain sweepstakes is one in a million, how many such sweepstakes must you enter to have a 50% chance of winning at least one? Do this two ways—using the exact probability of no successes in n trials and using the Poisson approximation.

8. You roll a fair die until a 6 appears, and you win a dollar if it happens on or before the third trial. Otherwise you lose a dollar. Find the following.
   a. The probability of winning on one play of this game
   b. The expected value and variance of your number of wins in 100 plays
   c. Your expected net gain after 100 plays
   d. The approximate probability that you will be ahead (51 or more wins) after 100 plays

9. If you bet a dollar on a typical "even-bet" gambling game in a casino, then you will win a dollar with probability .48 and lose a dollar with probability .52. (The actual probabilities are different for different types of games, and .48 is really a better chance than the player has in most games.)
   a. Find the expected value and variance of your number of wins in 1,000 such bets.
   b. Find the expected value of your net gain in 1,000 bets.
   c. Find the approximate probability that you win 500 or more times in 1,000 bets.
   d. Find the approximate probability that you win 2,500 or more times in 5,000 bets.

The answers to parts c and d show again why it is so profitable to run a gambling house. From the house's point of view, it is nearly impossible to be behind if as few as 5,000 bets are made. In a typical large casino on a typical evening, tens of thousands of bets are made, and on most of them the house's probability of winning is greater than .52.

10. To study the efficiency of a telephone switchboard system, an analyst observes successive minutes, classifying each minute according to the number of calls that come in. It is assumed that the calls arrive according to a Poisson process with $\lambda = 2.6$ per minute. The minutes are classified into four categories: C_1, no calls or one call; C_2, two calls: C_3, three calls; and C_4, four or more calls.
    a. Find the probability that over the next 8 minutes, there are equal numbers of minutes in each of the four categories.
    b. Find the expected value and variance of the number of minutes classified C_4 in the next hour.
    c. What is the distribution of X, the number of minutes classified C_2 in the next 2 hours?

11. You toss a fair coin until the second head has come up, and you regard the experiment as successful if four or fewer tosses are needed. You do this experiment repeatedly, until you have been successful ten times. What is the expected number of times you have to do the experiment?

12. In a class of 50 students there are 35 seniors and 15 juniors. If five members of the class are chosen at random (without replacement), what is the probability that either four or five of them are juniors?

13. An urn contains 15 red balls and 25 green balls. Ten balls are drawn at random, without replacement.
    a. What is the expected number of red balls in the sample?
    b. What is the probability that six of them are red and the other four green?

14. An urn contains 10 red balls, 15 white balls, and 20 blue balls.
    a. If eight are drawn at random, with replacement, what is the probability that there are two reds, three whites, and three blues in the sample?
    b. If ten are drawn at random, with replacement, what is the probability that four are red?
    c. Repeat part b if the drawing is without replacement.

15. The probability that a random five-card poker hand is a royal flush is $\frac{4}{2{,}598{,}960}$. Find the approximate probability that, in 100,000 independent poker hands, there will be two or more royal flushes.

16. In repeated rolls of a fair die:
    a. What is the probability that the third 6 appears on the 12th roll?
    b. What is the expected number of non-6's preceding the third 6?
    c. What is the expected number of non-6's between the second and third 6's?

17. An urn contains 6 blue balls and 11 white balls. We take a sample of size 4, without replacement. We then replace the sample, remix the balls, and repeat the process, until a total of ten such samples have been taken.
    a. What is the expected number of samples that contain at least one blue ball?
    b. What is the expected number of blue balls in the ten samples?

18. The probability generating function of X is $\pi(t) = \frac{1}{16}(t^2 + 3t)^2$.
    a. Find EX.
    b. Find $P(X = 3)$.
    c. Find $E(2^X)$.
    d. Find the PGF of $X_1 + X_2$, if X_1 and X_2 are independent observations of X.

19. **Should we spend money on technology to deflect Comet Swift–Tuttle?** In October 1992, some astronomers estimated that there was approximately one chance in 10,000 that Comet Swift–Tuttle will strike the earth on August 14, 2126. The same scientists estimate this comet's size and speed to be sufficient that the impact and its aftermath would likely eradicate human life on the planet (*New York Times,* November 3, 1992). Subsequently, at least one of these scientists withdrew the prediction (*Washington Post*, December 8, 1992), because additional data led to a different projected path for the comet, one that does not come dangerously close to Earth.

    Still, it is interesting to consider the expected loss from such an event. One approach is to use Seligman's estimate of the present monetary value of human life n years from now, as described in Exercise 15 of Section 7.3. This value is NVr^n, where N is today's population, V is today's monetary value of a human life, and r is a number combining the rates of inflation, population growth, and present value discount. In terms of these parameters, find the expected present value of the loss due to the possible impact of Swift–Tuttle.

The values suggested by Seligman, $NV = 7.5 \times 10^{14}$ and $r = .973$, lead to an expected 1993 present value of almost \$2 billion. In Exercise 15 of Section 7.3 we got the figure \$27 billion as the expected present value of the loss due to the impact of an asteroid in some unspecified future year. It is interesting to compare these two calculations. The chance of a calamitous asteroid impact in a given year is only one in a million, as opposed to what was thought to be one in ten thousand for Swift–Tuttle. Yet the fact that an asteroid could hit in *any* future year, as opposed to Swift–Tuttle's one chance, makes the expected loss from an asteroid much greater—by a factor of about 14. (This factor is independent of the human-life estimate NV; it depends only on the rate r, the two probabilities, and the number of years before the one-time threat.)

It is also interesting that the press and the public took immediate notice of the Swift–Tuttle threat, but paid much less attention to the asteroid threat, even though the latter is much greater, in Seligman's sense of expected loss. One explanation for this is the concreteness of the Swift–Tuttle threat: the object had been seen, it has a name, and the year and even the day were given, whereas the asteroid threat, though more likely, involves an unknown object and date. An unspecified future disaster seems infinitely remote; a disaster in August of 2126 feels somehow imminent.

ANSWERS

1. **b.** neg bin$\left(n = 2, p = \frac{4}{11}\right)$
2. **a.** $-\alpha \ln(1 - qt)$
   b. $\alpha q/p$ (where $p = 1 - q$) **c.** $-1/\ln p$
3. Geometric with $p = \frac{2}{3}$
4. **a.** .1396 **b.** .1020 **c.** 15.625

6. a. $\frac{6}{5}$ **b.** $\frac{18}{25}$ **c.** $\frac{27}{125}$ **d.** $\frac{8}{125}$ **e.** 0 **f.** 0

7. Using exact probability: at least 693,147; using approximation: at least 693,148

8. a. .4213 **b.** 42.13; 24.38 **c.** −$15.74 **d.** .0446

9. a. 480; 249.6 **b.** −$40.00 **c.** .1093 **d.** .0024

10. a. .0375 **b.** 15.8400; 11.6582 **c.** bin(120, .2510)

11. $\frac{160}{11}$

12. .0240

13. a. $3\frac{3}{4}$ **b.** .0747

14. a. .0899 **b.** .1134 **c.** .1068

15. .0107

16. a. .0493 **b.** 15 **c.** 5

17. a. 8.6134 **b.** $40 \cdot \frac{6}{17} = 14.1176$

18. a. $\frac{40}{16}$ **b.** $\frac{6}{16}$ **c.** $\frac{100}{16}$ **d.** $\frac{1}{256}(t^2 + 3t)^4$

19. $NVr^n/10{,}000$

CHAPTER 8

Some Important Absolutely Continuous Distributions

8.1 Moment Generating Functions

8.2 Normal Distributions

8.3 Exponential and Erlang Distributions

8.4 Gamma Distributions

8.5 Uniform Distributions

8.6 Beta Distributions

Supplementary Exercises

8.1 MOMENT GENERATING FUNCTIONS

In this chapter we cover the important families of absolutely continuous distributions in a systematic way. As we know, an absolutely continuous distribution is determined by its density function; the probability it assigns to any interval is the integral of the density over that interval. But there are other functions that serve as useful tools in dealing with these distributions as well.

One is the CDF, or cumulative distribution function, which we introduced in Section 3.2 and used heavily in Chapter 5. The CDF of a discrete distribution is not usually of much use, but for an absolutely continuous distribution it is often more convenient than the density function.

The other important tool is the moment generating function, or MGF. This function plays the same role with absolutely continuous distributions as the PGF (probability generating function) plays with discrete distributions on the nonnegative integers. Its important properties and uses are similar to those that we gave for PGFs at the beginning of the previous chapter:

1. The MGF determines the distribution uniquely; that is, it is impossible for two different distributions to have the same MGF. So, if we find the MGF of a random variable and recognize it as the MGF of a known distribution, then that is the distribution of the random variable.
2. The MGF can be used to find the moments of the distribution. Also, in principle but not often in practice, it can be used to find the density.
3. The MGF of the sum of independent random variables is the product of their MGFs. In addition, the MGF of a function $h(X)$ can often be found if the MGF of X is known. Combined with property 1 above, this gives us a powerful method for finding distributions of sums or other functions of random variables.
4. If the MGF of X is the limit of a sequence of MGFs, then the distribution of X is a certain kind of limit of the distributions. This fact makes a proof of the Central Limit Theorem possible.

(8.1.1) Definition. Let X be a random variable. The ***moment generating function*** **(MGF)** of X is the function $m_X(s)$, or simply $m(s)$, defined by

$$m(s) = E(e^{sX}),$$

for those s for which this expected value exists.

If X is an absolutely continuous random variable with density function $f(x)$, then this definition says that

$$m(s) = \int_{-\infty}^{\infty} e^{sx} f(x)dx \tag{8.1.2}$$

for all s for which this integral converges. Note that the integrand is nonnegative, so absolute convergence is the same as ordinary convergence. In addition, the integral is not always an improper integral: if, as often happens, the density is 0 except on a bounded interval and bounded on that interval, then the integral is proper and convergence is not in question.

Definition (8.1.1) makes sense even if X is a discrete random variable. In that case it gives

(8.1.3) $$m(s) = \sum_x e^{sx} p_X(x),$$

again just for those s for which the series converges, if there are infinitely many possible values. As we mentioned, though, MGFs are most often used with absolutely continuous random variables; with a discrete random variable, especially one whose possible values are nonnegative integers, calculations involving the PGF are generally simpler.

(8.1.4) **Example: The MGF of an exponential distribution.** Let X have the exponential distribution with parameter λ. We will find its MGF. The density function is

$$f(x) = \begin{cases} \lambda e^{-\lambda x} & \text{if } x > 0, \\ 0 & \text{otherwise.} \end{cases}$$

and so the MGF is

$$m(s) = \int_0^\infty e^{sx} \lambda e^{-\lambda x} dx = \left[\frac{\lambda}{s - \lambda} e^{(s-\lambda)x} \right]_{x=0}^{x=\infty} = \frac{\lambda}{\lambda - s}.$$

We have been careless, though, in not bothering to check whether the integral converges. The calculation really should go as follows:

$$m(s) = \int_0^\infty e^{sx} \lambda e^{-\lambda x} dx = \lim_{A \to \infty} \int_0^A e^{sx} \lambda e^{-x} dx.$$

If $s = \lambda$ this is $\lim_{A\to\infty} \int_0^A \lambda dx = \lim_{A\to\infty} \lambda A$, which does not exist (diverges to $+\infty$); and if $s \neq \lambda$ it is

$$m(s) = \lim_{A\to\infty} \left[\frac{\lambda}{s - \lambda} e^{(s-\lambda)x} \right]_{x=0}^{x=A} = \lim_{A\to\infty} \frac{\lambda}{s - \lambda} \left[e^{(s-\lambda)A} - 1 \right].$$

Now the limit of $e^{(s-\lambda)A}$ does not exist (diverges to $+\infty$) if $s - \lambda$ is positive, but equals 0 if $s - \lambda$ is negative. So the MGF exists only when $s - \lambda < 0$—that is, when $s < \lambda$. Summarizing:

$$\text{If } X \sim \exp(\lambda), \text{ then } m_X(s) = \frac{\lambda}{\lambda - s} \quad \text{for } s < \lambda.$$

If this calculation seems familiar, remember that we did it in Exercise 7b of Section 4.2.

∎

(8.1.5) **Example: The MGF of the uniform distribution on (0, 1).** If X has this distribution, then its MGF is

$$m(s) = \int_0^1 e^{sx} \cdot 1dx = \left[\frac{1}{s}e^{sx}\right]_{x=0}^{x=1} = \frac{e^s - 1}{s}$$

for any real s, except that this expression makes no sense if $s = 0$. However, for $s = 0$ the integral is even simpler; it is

$$m(0) = \int_0^1 e^{0 \cdot x} \cdot 1dx = \int_0^1 1dx = 1.$$

So we can use the formula $(e^s - 1)/s$ for $m(s)$ as long as we agree that $(e^0 - 1)/0$ is to be interpreted as 1. [The limit of $(e^s - 1)/s$ as s approaches 0 is indeed 1, as you can check using L'Hôpital's rule. See (A.6).] ∎

8.16 **Example: The MGF compared with the PGF for a discrete distribution.** In the previous chapter we found the PGF of the binomial distribution with parameters n and p; it was $\pi(t) = (pt + q)^n$, where $q = 1 - p$. What is the MGF? The calculation is almost the same:

$$m(s) = E(e^{sX}) = \sum_{k=0}^{n} e^{sk}\binom{n}{k}p^k q^{n-k} = \sum_{k=0}^{n} \binom{n}{k}(pe^s)^k q^{n-k} = (pe^s + q)^n.$$

It is instructive to compare the PGF and the MGF:

$$\pi(t) = (pt + q)^n \quad \text{and} \quad m(s) = (pe^s + q)^n.$$

In the preceding example, if we put e^s into $\pi(t)$ in place of t, we get $m(s)$; that is, $m(s) = \pi(e^s)$. This is no coincidence: for any random variable X,

(8.1.7)
$$m(s) = E(e^{sX}) = E((e^s)^X) = \pi(e^s),$$

because $\pi(t) = E(t^X)$. The connection (8.1.7) is valid, of course, only at those s where the MGF exists. ∎

(8.1.8) **Example: The MGF of the standard normal distribution.** Let Z have the standard normal distribution; we will find the moment generating function. It is

$$m(s) = E(e^{sZ}) = \int_{-\infty}^{\infty} e^{sz} \cdot \frac{1}{\sqrt{2\pi}}e^{-z^2/2}dz = \int_{-\infty}^{\infty} \frac{1}{\sqrt{2\pi}}e^{-(z^2-2sz)/2}dz.$$

The easiest way to evaluate this integral is to complete the square in the exponent; if we add s^2 to the quantity in parentheses it becomes $(z - s)^2$, and we can compensate for adding that s^2 by multiplying the whole integral by $e^{s^2/2}$. That is,

$$m(s) = e^{s^2/2} \int_{-\infty}^{\infty} \frac{1}{\sqrt{2\pi}} e^{-(z^2-2sz+s^2)/2}\, dz = e^{s^2/2} \int_{-\infty}^{\infty} \frac{1}{\sqrt{2\pi}} e^{-(z-s)^2/2} dz.$$

Now if we make the change of variable $y = z - s$, the integral becomes simply $\int_{-\infty}^{\infty} \frac{1}{\sqrt{2\pi}} e^{-y^2/2} dy$, which equals 1 because it is the integral of the standard normal density, and we are left with just $e^{s^2/2}$. Therefore, the MGF exists for all real s, and we have the following.

(8.1.9) If $X \sim N(0, 1)$, then $m(s) = e^{s^2/2}$. ▮

(8.1.10) Existence of $m(s)$. Notice that $m(0)$ always equals 1 because it equals $E(e^{0 \cdot X}) = E(1)$. However, there is no guarantee of the existence of $m(s)$ for any other values of s, unless X happens to satisfy certain conditions.

Suppose, for example, that X is nonnegative with probability 1. Then $m(s)$ exists for all negative values of s, because if $s < 0$ and $X > 0$ then $e^{sX} \leq 1$ and so $E(e^{sX}) \leq E(1) = 1$. But even for a nonnegative random variable, $m(s)$ may fail to be finite for any positive values of s.

More generally, if for some real number a we have $X \geq a$ with probability 1, then for any negative value of s we get $e^{sX} \leq e^{sa}$ and so $m_X(s)$ exists (it is no greater than e^{sa}) for all $s < 0$. Similarly, if for some a we have $X \leq a$ with probability 1, then $m_X(s)$ exists for all positive values of s.

Some distributions have MGFs that do not exist for any value of s other than $s = 0$. The most famous example is the Cauchy distribution, whose density is $1/\pi(1 + x^2)$. The MGF is

$$m(s) = \int_{-\infty}^{\infty} \frac{e^{sx}}{\pi(1 + x^2)} dx,$$

and this improper integral can be shown to diverge to $+\infty$ for all $s \neq 0$, although it equals 1 for $s = 0$. This is closely related to the fact that the Cauchy distribution has no moments: if X has this distribution, then EX, EX^2, and all the EX^n fail to exist.

However, in most cases the situation is not so bleak. If it happens that $m(s)$ exists for at least one positive value of s, say s_+, and at least one negative value, say s_-, then it can be shown that $m(s)$ is finite for all s in the interval (s_-, s_+), which contains 0. Then nice things happen, as we will see in Theorem (8.1.15) below.

There is a third kind of generating function, called a ***characteristic function.*** It is a complex-valued function:

$$\phi(u) = E(e^{iuX}) = E(\cos X) + iE(\sin X).$$

As you can see, it is really just $m(iu)$, the result of putting an imaginary number iu instead of a real number s into the moment generating function. It has the slight disadvantage of requiring some knowledge of complex functions, but it has one great advantage: no matter what distribution X has, its characteristic function $\phi(u)$ exists for all real u. It is the generating function best suited to general, abstract probability theory. But we will not consider it further here.

(8.1.11) **Moment generation.** You are right if the name leads you to expect the moment generating function to provide a way to generate moments. Recall that we get moments from the PGF by differentiating it and then setting $t = 1$. Recall also that although this works well for the first moment EX, when we use it for higher moments we get factorial moments $E(X(X-1)(X-2)\cdots(X-k+1))$ instead of ordinary moments EX^k.

With MGFs the situation is much simpler. If we differentiate the equation in Definition (8.1.1) formally with respect to s, we get

(8.1.12)
$$m'(s) = \frac{d}{ds}E(e^{sX}) \overset{*}{=} E\left(\frac{d}{ds}e^{sX}\right) = E(Xe^{sX}),$$

although we have not justified the equality sign marked with the asterisk. Now it is $s = 0$ that gives us what we want:

$$m'(0) = EX.$$

What about higher derivatives and higher moments? If we continue and differentiate (8.1.12) repeatedly with respect to s, we get

(8.1.13)
$$\begin{aligned} m''(s) &= E(X^2e^{sX}), \quad \text{so} \quad m''(0) = E(X^2); \\ m'''(s) &= E(X^3e^{sX}), \quad \text{so} \quad m'''(0) = E(X^3); \end{aligned}$$

and, in general, provided the derivative exists,

(8.1.14)
$$m^{(n)}(s) = E(X^n e^{sX}), \quad \text{so} \quad m^{(n)}(0) = E(X^n).$$

The moment generating function really generates moments—that is, expected values of the powers X^n, not factorial moments as the PGF does.

But what about rigor? Can we justify the equality with the asterisk in (8.1.12)? [There are similar unjustified steps in (8.1.13) and (8.1.14) as well. In all of them, we are differentiating an expected value by differentiating what is inside.] Let us look at what this differentiation involves in the case of an absolutely continuous distribution:

$$m(s) = \int_{-\infty}^{\infty} e^{sx}f(x)dx,$$

and we want to know if we can justify the equality marked with the asterisk in the following:

$$m'(s) = \frac{d}{ds}\int_{-\infty}^{\infty} e^{sx}f(x)dx \overset{*}{=} \int_{-\infty}^{\infty}\frac{d}{ds}e^{sx}f(x)dx = \int_{-\infty}^{\infty} xe^{sx}f(x)dx = E(Xe^{sX}).$$

This is called a differentiation under the integral sign; the conditions under which it is allowable are studied in advanced calculus courses. The simplest kind of assumption that allows it is that the derivative with respect to s of $e^{sx}f(x)$ be a continuous function of x and s, and that the improper integral converge uniformly in s.

We will not go any further into these questions, but will simply state the theorem on the relation between the MGF of a random variable and its moments. The theorem says that if the MGF exists for all s in any open interval (s_-, s_+) containing 0 (called a *neighborhood* of 0), then all the moments of X are finite and can be found from the MGF. Remember that if the MGF exists for one positive value s_+ and one negative value s_-, then it exists in the neighborhood (s_-, s_+).

(8.1.15) Theorem. If the MGF $m(s)$ of X exists in any neighborhood of 0, then for any nonnegative integer n, the nth derivative of $m(s)$ at $s = 0$ and the nth moment of X are finite and equal. In other words:

> If $m(s)$ exists in any neighborhood of 0, then for any $n = 0, 1, 2, \ldots,$ $m^{(n)}(0) = EX^n$.

Furthermore, in this case $m(s)$ has a Maclaurin expansion: for all s in some neighborhood of 0, the following series converges:

$$m(s) = c_0 + c_1 s + c_2 s^2 + c_3 s^3 + \cdots,$$

where

$$c_n = \frac{m^{(n)}(0)}{n!} = \frac{EX^n}{n!}.$$

So if we can see the power series expansion for $m(s)$ about $s = 0$, then we can read off the moments of X from the coefficients (remembering that we must multiply the coefficient of s^n by $n!$ to get EX^n).

(8.1.16) Example: Moments of the standard normal distribution. In Example (8.1.8) we found that the MGF of the standard normal distribution exists for all real s, and equals

$$m(s) = e^{s^2/2}.$$

We can expand this in a power series using (A.5.11):

$$\begin{aligned} m(s) &= 1 + \frac{1}{1!}\left(\frac{s^2}{2}\right) + \frac{1}{2!}\left(\frac{s^2}{2}\right)^2 + \frac{1}{3!}\left(\frac{s^2}{2}\right)^3 + \frac{1}{4!}\left(\frac{s^2}{2}\right)^4 + \cdots \\ &= 1 + \frac{1}{2}s^2 + \frac{1}{2^2 \cdot 2!}s^4 + \frac{1}{2^3 \cdot 3!}s^6 + \frac{1}{2^4 \cdot 4!}s^8 + \cdots. \end{aligned}$$

Now according to Theorem (8.1.15), EX^k is $k!$ times the coefficient of s^k. So

$$\begin{aligned}
EX^1 &= 0;\\
EX^2 &= 2! \cdot \frac{1}{2} = 1;\\
EX^3 &= 0;\\
EX^4 &= 4! \cdot \frac{1}{2^2 \cdot 2!} = 3;\\
EX^5 &= 0;\\
EX^6 &= 6! \cdot \frac{1}{2^3 \cdot 3!} = 15;
\end{aligned}$$

and so forth.

This may seem complicated, but it is much simpler than finding the moments by evaluating

$$EX^k = \int_{-\infty}^{\infty} x^k \cdot \frac{1}{\sqrt{2\pi}} e^{-x^2/2} dx$$

using repeated integrations by parts. ∎

(8.1.17) **Summary and comparison of moment generation and power series expansions for PGFs and MGFs.** It is easy to be confused by the ways in which the two different kinds of generating functions generate moments and probabilities, the ways in which these quantities appear in the power series expansions, and the connections between the various quantities and the derivatives of the functions. One way to keep it all straight is to remember the two definitions in the form

(8.1.18)
$$m(s) = E(e^{sX}) \quad \text{and} \quad \pi(t) = E(t^X).$$

For any X, we can formally use the Maclaurin series for e^x to get

(8.1.19)
$$\begin{aligned}
m(s) = E(e^{sX}) &= E\left(1 + sX + \frac{(sX)^2}{2!} + \frac{(sX)^3}{3!} + \cdots\right)\\
&= 1 + s \cdot EX + s^2 \cdot \frac{EX^2}{2!} + s^3 \cdot \frac{EX^3}{3!} + \cdots.
\end{aligned}$$

From this we can see that the coefficients are the moments divided by factorial integers, provided the moments exist. If $m(s)$ is finite in a neighborhood of 0, then all the moments do exist.

On the other hand, the PGF is usually used only for nonnegative integer-valued random variables, and for these we can write down a power series using the definition of expected value:

(8.1.20)
$$\pi(t) = E(t^X) = t^0 \cdot P(X = 0) + t^1 \cdot P(X = 1) + t^2 \cdot P(X = 2) + \cdots .$$

From this we can see that the coefficients are the probabilities.

As for the derivatives, we can differentiate the basic formulas (8.1.18) and then see what values of t or s to use for moments:

$$m'(s) = E(Xe^{sX}), \quad \text{so} \quad m'(0) = EX;$$
$$m''(s) = E(X^2e^{sX}), \quad \text{so} \quad m''(0) = E(X^2);$$
etc.

$$\pi'(t) = E(Xt^{X-1}), \quad \text{so} \quad \pi'(1) = EX;$$
$$\pi''(t) = E(X(X-1)t^{X-2}), \quad \text{so} \quad \pi''(1) = E(X(X-1));$$
etc.

Or we can differentiate the power series (8.1.19) and (8.1.20) and see what putting in 0 for s or 1 for t will give.

(8.1.21) Example: Moments of the exponential distribution. In Example (8.1.4) we found that if X has the exponential distribution with parameter λ, then the MGF of X is

$$m(s) = \frac{\lambda}{\lambda - s} \quad \text{for } s < \lambda.$$

To expand this in a power series, we divide numerator and denominator by λ and get a simple geometric series:

$$m(s) = \frac{1}{1 - (s/\lambda)} = 1 + \frac{s}{\lambda} + \left(\frac{s}{\lambda}\right)^2 + \left(\frac{s}{\lambda}\right)^3 + \left(\frac{s}{\lambda}\right)^4 + \cdots .$$

The coefficient of s^k is $1/\lambda^k$, and from (8.1.19) we know that this is $EX^k/k!$, so we see that

(8.1.22)
$$EX^k = \frac{k!}{\lambda^k} \quad \text{for } k = 1, 2, 3, \ldots .$$

In Exercise 7 at the end of this section you are asked to get the same result by differentiating $m(s)$ repeatedly and setting $s = 0$ in each derivative. You are also asked, in the same exercise, to confirm it again by integrating x^k times the density of X. ∎

(8.1.23) The moment generating function determines the distribution. MGFs have the same uniqueness property that PGFs have: it is impossible for two different distributions to have MGFs that agree for all values of s. This theorem is proved using facts from measure theory and from the theory of Laplace transforms (or Fourier transforms)—and we will not say any more about its proof here.

But the theorem is important and much used. As with PGFs, it is often possible to find the MGF of a random variable X in terms of the MGFs of other random variables; if we then recognize this as the MGF of a known distribution, then we can conclude that X has that distribution.

(8.1.24) **The moment generating function of a sum of independent random variables.** It is true for MGFs, as for PGFs, that the generating function of a sum of independent random variables is the product of their generating functions:

If X and Y are independent and $Z = X + Y$, then $m_Z(s) = m_X(s)m_Y(s)$,

for all s for which both $m_X(s)$ and $m_Y(s)$ exist.

We can see why this is true in the same way that we saw the result for PGFs, that is,

$$m_Z(s) = E(e^{sZ}) = E(e^{s(X+Y)}) = E(e^{sX}e^{sY}).$$

Because X and Y are independent, so are e^{sX} and e^{sY}; therefore, by (4.2.19) the expected product is the product of the expected values, and we get

$$m_Z(s) = E(e^{sX})E(e^{sY}) = m_X(s)m_Y(s).$$

This of course extends to more than two random variables:

If $X_1, X_2, \ldots, X_n$ are independent, and $Z = X_1 + X_2 + \cdots + X_n$, then $m_Z(s) = m_{X_1}(s)m_{X_2}(s) \cdots m_{X_n}(s)$.

(8.1.25) **Example.** What is the MGF of the sum of two independent random variables if each of them has the exponential distribution with parameter λ? The answer is $(\lambda/(\lambda - s))^2$, because each of the two random variables has MGF $\lambda/(\lambda - s)$.

More generally, suppose $X_1, X_2, \ldots, X_n$ are independent and all have the exponential distribution with parameter λ. Then the sum $X = X_1 + X_2 + \cdots + X_n$ has the MGF $(\lambda/(\lambda - s))^n$.

We have not yet encountered the distribution that has this MGF, but it is an important one, called the Erlang distribution with parameters n and λ. Erlang distributions form a subfamily of the gamma family of distributions, and we will cover these in Sections 8.3 and 8.4. Also see Exercise 8 at the end of this section. ▮

(8.1.26) **The moment generating function of $aX + b$.** Suppose X has a known MGF and $Y = aX + b$. Can we find the MGF of Y from that of X? Easily; we see that

$$m_Y(s) = E(e^{sY}) = E(e^{s(aX+b)}) = E(e^{asX}e^{bs}) = e^{bs}E(e^{asX}); \tag{8.1.27}$$

that is,

$$m_Y(s) = e^{bs} m_X(as). \tag{8.1.28}$$

This is another case in which you should remember the proof of the theorem rather than try to memorize the statement of the theorem. It is easy to make a mistake if you try to memorize (8.1.28); but if you remember that $m_Y(s) = E(e^{sY})$ you can finish the derivation in (8.1.27).

(8.1.29) **Example.** Let X have the exponential distribution with parameter λ and let $Y = aX$, where a is some positive number. What is the MGF of Y?

Solution. The MGF of X is of course $\lambda/(\lambda - s)$, and so by (8.1.28) the MGF of Y is just $\lambda/(\lambda - as)$. Do we recognize this as the MGF of a known distribution? Perhaps not, but if we divide the numerator and denominator by a we do:

$$m_Y(s) = \frac{\lambda/a}{(\lambda/a) - s}.$$

This is the MGF of the exponential distribution with parameter λ/a. ∎

Notice the power of this method. Because there is only one distribution with a given MGF, we have proved in a very simple way that if X has the exponential distribution with parameter λ, then aX must have the exponential distribution with parameter λ/a. We obtained this same result using the CDF method in Exercise 1 of Section 5.2.

We will use this MGF technique throughout this chapter.

(8.1.30) **MGFs of sample sums and sample averages.** Let $X_1, X_2, \ldots, X_n$ be a sample from some distribution; that is, let them be independent, each with that distribution. We view them as independent observations of some random variable X. We are then interested in the distributions of the sample sum,

$$S_n = X_1 + X_2 + \cdots + X_n,$$

and the sample average,

$$\overline{X}_n = \frac{1}{n} S_n.$$

It is often difficult to find the CDFs or densities for these two important random variables, but it is easy to find their MGFs if we know the MGF of X. Suppose $m(s)$ is that MGF. Then by (8.1.24), the MGF of S_n is

$$m_{S_n}(s) = [m(s)]^n, \tag{8.1.31}$$

and hence by (8.1.28),

(8.1.32)
$$m_{\overline{X}_n}(s) = \left[m\left(\frac{s}{n}\right)\right]^n.$$

These formulas do not look like much, but they are useful. The next example illustrates their use, and in the next section we will see how they are used in proving the Central Limit Theorem.

(8.1.33) Example. If X has the exponential distribution with parameter λ, then by the preceding two formulas, the MGFs of the sample sum and sample average are

$$m_{S_n}(s) = \left(\frac{\lambda}{\lambda - s}\right)^n,$$
$$m_{\overline{X}_n}(s) = \left(\frac{\lambda}{\lambda - s/n}\right)^n.$$

That ends the application of the theorem. But it is interesting to note that this can be used to show that $\overline{X}_n$ approaches $1/\lambda$ in the same sense given by the weak law of large numbers, as follows. The MGF of $\overline{X}_n$ has a limit as $n \to \infty$, because a little manipulation gives

$$m_{\overline{X}_n}(s) = \frac{1}{(1 - s/\lambda n)^n}.$$

Remembering that $(1 - x/n)^n$ approaches e^{-x} as $n \to \infty$, we see that

$$\lim_{n \to \infty} m_{\overline{X}_n}(s) = e^{s/\lambda}.$$

Do we recognize this as the MGF of a known distribution? After you have done Exercise 4 at the end of this section, you will: this is the MGF of a constant random variable that equals $1/\lambda$ with probability 1.

What does this tell us? Remember that $EX = 1/\lambda$ when X has an exponential distribution; we have just seen that the MGF of $\overline{X}_n$ approaches that of the constant $1/\lambda$. It can be shown that this implies that $\overline{X}_n$ approaches $1/\lambda$ in the same sense that is given by the weak law of large numbers. So (except for proving this limiting theorem) we have rederived the weak law for the case when the X_k's have an exponential distribution. ∎

Summary of Section 8.1. As we did for PGFs at the end of Section 7.1, we summarize here the main facts about moment generating functions.

1. The moment generating function of a random variable X is the function $m_X(s)$ of a real variable s defined by $m_X(s) = E(e^{sX})$. It is defined only for those s for which the expected value exists.

2. No two different distributions can have the same moment generating function. That is, if we find the MGF of X and recognize it as the MGF of a known distribution, then X must have that distribution.
3. If X and Y are independent random variables, then the MGF of $X + Y$ is the product of their MGFs. If X is any random variable and the MGF of X is $m(s)$, then for any a and b the MGF of $Y = aX + b$ is $e^{bs}m(as)$.
4. If $m(s)$ exists in any neighborhood (s_-, s_+) of 0, then all moments of X exist, and can be found from the power series expansion of $m(s)$,

$$m(s) = 1 + s \cdot EX + s^2 \cdot \frac{EX^2}{2!} + s^3 \cdot \frac{EX^3}{3!} + \cdots,$$

or from the derivatives of $m(s)$ at 0,

$$EX^k = m^{(k)}(0).$$

EXERCISES FOR **SECTION 8.1**

1. Suppose the MGF of X is $m(s) = 2/\sqrt{4 - s}$.
   a. Find EX, EX^2, and VX.
   b. If X and Y are independent, both with this MGF, find the MGF, and identify the distribution, of $X + Y$.

2. Let X have the $N(\mu, \sigma^2)$ distribution. Find the MGF of X by using the fact that $X = \sigma Z + \mu$ where Z has the standard normal distribution. [See (8.1.26).] Then use the MGF to confirm that $EX = \mu$ and $VX = \sigma^2$.

3. If the moment generating function of X is $m(s) = a^2/(a^2 - s^2)$, what is the kth moment EX^k? [*Hint:* Expand $m(s)$ as a power series and use (8.1.19).]

4. If X is a constant random variable that equals a with probability 1, what is the MGF of X?

5. Suppose the MGF of X has the Maclaurin series $m(s) = 1 + a_1 s + a_2 s^2 + a_3 s^3 + \cdots$. Find the variance and the third central moment of X in terms of a_1, a_2, and a_3.

6. Let $X_1, X_2, \ldots, X_n$ be i.i.d., each having the normal distribution with parameters μ and σ^2. Find the MGFs of the sample sum and the sample average, using the result of Exercise 2a. What are the distributions of these two random variables?

7. The MGF of the exponential distribution with parameter λ is $m(s) = \lambda/(\lambda - s)$. We found all the moments in Example (8.1.21) by expanding this in a power series.
   a. Confirm these values by differentiating $m(s)$ repeatedly and setting $s = 0$ in each derivative.

b. Confirm them again without using the MGF, by using the "gamma formula,"

$$\int_0^\infty x^{n-1}e^{-\lambda x}dx = \frac{(n-1)!}{\lambda^n},$$

which we derived in Exercise 10 of Section 4.3.

8. For a positive integer n and a positive number λ, let

$$f(x) = \frac{\lambda^n}{(n-1)!}x^{n-1}e^{-\lambda x}$$

for $x > 0$.

a. Using the gamma formula (see the previous exercise), show that $f(x)$ is a density.

b. Show that the MGF associated with this density is $(\lambda/(\lambda - s))^n$.

c. Compare the result of part b with Example (8.1.25). What can you conclude?

d. Let X have this density. Find EX and EX^2 using the MGF, and then find VX.

e. Confirm the values of EX and EX^2 using the gamma formula.

This exercise anticipates the Erlang distributions studied in Section 8.3.

9. Let X have the uniform distribution on $(0, 1)$; the density is of course $f(x) = 1$ for $0 < x < 1$ and 0 otherwise. In Example (8.1.5) we found that the MGF of X is given by $m(s) = (e^s - 1)/s$ for $s \neq 0$ and $m(0) = 1$. We also know that $EX = \frac{1}{2}$ for this distribution, so $m'(0)$ must equal $\frac{1}{2}$; but we cannot find $m'(0)$ by simply differentiating the formula $(e^s - 1)/s$ and setting $s = 0$.

a. Show that $m'(0) = \frac{1}{2}$ by using the definition of the derivative:

$$m'(0) = \lim_{s \to 0} \frac{m(s) - m(0)}{s}.$$

b. Now expand $m(s)$ in a power series by writing down the Maclaurin series for e^s, subtracting 1, and dividing each term by s. Now you can read off *all* the moments, remembering that EX^k is $k!$ times the coefficient of s^k.

This is an example in which it is easier to find the moments directly from the integrals $EX^k = \int_0^1 x^k f(x)dx$, than it is to use the MGF; but it is instructive to see how the MGF gives the same results.

10. Find the MGF of the Poisson distribution with parameter λ. Do it in two ways:

a. Use $m(s) = E(e^{sX}) = \sum_{k=0}^{\infty} e^{sk} \cdot \lambda^k e^{-\lambda}/k!$.

b. Use $m(s) = \pi(e^s)$ along with the formula for $\pi(t)$ that we found in Example (7.1.5), $\pi(t) = e^{\lambda t - \lambda}$.

c. Confirm by differentiating the MGF and setting $s = 0$ that $EX = \lambda$ and $EX^2 = \lambda + \lambda^2$.

Compare the MGF calculations with the PGF calculations and note how much simpler the PGF is for nonnegative integer-valued random variables.

11. Suppose the kth moment $E(X^k)$ of X equals ka^k for each $k = 1, 2, 3, \ldots$.
a. What is the coefficient of s^k (for $k \geq 1$) in the power series for the MGF of X?
b. What is the coefficient of $s^0 = 1$?
c. What is the MGF of X? [*Hint:* Sum the series.]

12. **Not all functions are MGFs.**
a. Show that if $m(s)$ is the MGF of a random variable, then $m(0)$ must equal 1.
b. Show that even though $\cos(0) = 1$, $m(s) = \cos(s)$ cannot be the MGF of any distribution. [*Hint:* Look at EX^2.]
c. Show that if $m(s) = 1 + a_1 s + a_2 s^2 + \cdots$ is the MGF of some distribution, then $2a_2$ cannot be less than a_1^2.
d. If $m(s) = 1 + a_1 s + a_2 s^2 + \cdots$ is the MGF of some nonconstant random variable, what must be true of the coefficients $a_2, a_4, a_6, \ldots$? [*Hint:* What property do VX, EX^2, EX^4, EX^6, etc. all have?]
e. Can a polynomial be the MGF of a nonconstant random variable?

[Exercise 18 of Section 7.1 considered some restrictions of a similar type on PGFs. In Exercises 1, 2, and 10 of the same section we saw examples of polynomials that are PGFs.]

13. Suppose the MGF of X can be expressed as a power series,

$$m(s) = \sum_{k=1}^{\infty} a_k s^k = a_0 + a_1 s + a_2 s^2 + a_3 s^3 + \cdots,$$

and suppose further that $a_0 = 1$, $a_1 = 3$, and $a_2 = 7$. Find EX and VX.

ANSWERS

1. a. $EX = \frac{1}{8}$; $EX^2 = \frac{3}{64}$; $VX = \frac{1}{32}$
b. $X + Y \sim \exp(4)$

3. 0 if k is odd, $k!/a^k$ if k is even

4. $m(s) = e^{as}$ for all s

5. $VX = 2a_2 - a_1^2$; $E(X - EX)^3 = 6a_3 - 6a_1 a_2 + 2a_1^3$

6. $m_{S_n}(s) = e^{n\mu s + n\sigma^2 s^2/2}$, $N(n\mu, n\sigma^2)$; $m_{\overline{X}_n}(s) = e^{\mu s + \sigma^2 s^2/2n}$, $N(\mu, \sigma^2/n)$

8. c. The sum of n independent $\exp(\lambda)$ random variables has density $f(x)$.
d. n/λ; $n(n + 1)/\lambda^2$; n/λ^2

9. b. $EX^k = 1/(k + 1)$

10. $m(s) = e^{\lambda(e^s - 1)}$

11. a. $a^k/(k - 1)!$ **b.** $m(0) = 1$ **c.** $1 + ase^{as}$ (for all s)

12. e. No

13. $EX = 3$; $VX = 5$

8.2 NORMAL DISTRIBUTIONS

We have already seen the essential facts about normal distributions; Section 6.2 was devoted to them. In this section we will briefly summarize the facts again; the only new feature will be the use of moment generating functions to derive some of the results very easily.

At the end of the section we will return to the Central Limit Theorem, which we discussed in Section 6.3. We will indicate how it is proved, and will add a few comments about its history and about the research that continues on various aspects of it.

(8.2.1) Standard normal density and CDF. The ***standard normal*** density is the well-known

$$\varphi(z) = \frac{1}{\sqrt{2\pi}} e^{-z^2/2} \quad \text{for } z \in \mathbb{R}.$$

(It is customary to use Z to denote a standard normal random variable; hence we use z as the variable in the density.)

The CDF is

$$\Phi(z) = \int_{-\infty}^{z} \varphi(t)dt.$$

Although there is no formula for this function of z in terms of standard functions, its values have been tabulated; the table in Appendix C gives $\Phi(z)$ for z from .00 to 3.69 in steps of .01. Such tables are constructed from formulas that produce good approximations to the true value of $\Phi(z)$.

(8.2.2) Standard normal MGF and moments. The moment generating function of the standard normal distribution is

$$m(s) = e^{s^2/2} \quad \text{for } s \in \mathbb{R}.$$

We showed this in Example (8.1.8) by evaluating the integral $m(s) = \int_{-\infty}^{\infty} e^{sz}\varphi(z)dz$ using the change of variable $y = z - s$. With the MGF we can get the moments of a standard normal random variable Z by repeatedly differentiating and setting $s = 0$, or, more simply, by expanding the MGF in a power series and multiplying the coefficient of x^k by $k!$. That is what we did in Example (8.1.16); the result can be summarized as follows.

If k is odd, then $EZ^k = 0$;

If k is even, say $k = 2j$, then $EZ^k = k!/(2^j \cdot j!)$.

In particular, the first two moments of the standard normal distribution are

$EZ = 0$ and $EZ^2 = 1$; and therefore $VZ = 1$.

It is also possible to get these first two moments without using the MGF, simply using the definition of expected value; we did these in (4.1.19) and (4.2.9). It is even possible, though tedious, to get all the moments in this way, using repeated integration by parts.

(8.2.3) General normal distributions: densities and moments. The ***normal distribution with parameters μ and σ^2***, or the $N(\mu, \sigma^2)$ distribution, is the one determined by the density

$$f(x) = \frac{1}{\sqrt{2\pi\sigma^2}} e^{-(x-\mu)^2/2\sigma^2} \quad \text{for } x \in \mathbb{R}.$$

If we put in 0 for μ and 1 for σ^2, we see that the $N(0, 1)$ distribution is the standard normal distribution.

The fundamental connection among different normal distributions is this:

(8.2.4) If $X \sim N(\mu, \sigma^2)$ and $Y = aX + b$, then $Y \sim N(a\mu + b, a^2\sigma^2)$.

We proved this in (5.3.11) using change-of-variable techniques. Two particular cases of (8.2.4) are especially important:

(8.2.5) If $X \sim N(\mu, \sigma^2)$ and $Z = \dfrac{X - \mu}{\sigma}$, then $Z \sim N(0, 1)$;

and

(8.2.6) If $Z \sim N(0, 1)$ and $X = \sigma Z + \mu$, then $X \sim N(\mu, \sigma^2)$.

We use (8.2.5) to find probabilities for any normally distributed random variable; we studied this in (6.2.10) and have used it often.

We can use (8.2.6) to find quickly the expected value and variance of a random variable with the $N(\mu, \sigma^2)$ distribution:

If $X \sim N(\mu, \sigma^2)$, then $EX = \mu$ and $VX = \sigma^2$.

That is, the two parameters of a normal distribution are its expected value and its variance.

(8.2.7) The MGF of an arbitrary normal distribution. We can also use (8.2.6) to find the MGF of the $N(\mu, \sigma^2)$ distribution: Let $Z \sim N(0, 1)$ and let $X = \sigma Z + \mu$. Then from (8.2.2)

$$m_Z(s) = e^{s^2/2},$$

and as we saw in (8.1.26),

$$\begin{aligned} m_X(s) &= E(e^{sX}) = E(e^{s(\sigma Z+\mu)}) \\ &= e^{\mu s}E(e^{\sigma s Z}) = e^{\mu s}m_Z(\sigma s) \\ &= e^{\mu s}e^{(\sigma s)^2/2}; \end{aligned}$$

that is,

(8.2.8) If $X \sim N(\mu, \sigma^2)$, then $m_X(s) = e^{\mu s+\sigma^2 s^2/2}$ for $s \in \mathbb{R}$.

You could also get (8.2.8) directly from the definition of the MGF; you would need to evaluate the integral

$$m_X(s) = \int_{-\infty}^{\infty} e^{sx} \cdot \frac{1}{\sqrt{2\pi\sigma^2}} e^{-(x-\mu)^2/2\sigma^2} dx.$$

This looks unwieldy, but the substitution $z = (x - \mu)/\sigma$ makes it less so.

(8.2.9) **Sums of independent normal random variables are normal.** We first proved this important fact in (6.2.11); the proof was complicated, using the multivariate change-of-variable theorem, which was all we had at our disposal then. Now we can give a really short proof using MGFs. The fact is:

> If $X_1 \sim N(\mu_1, \sigma_1^2)$ and $X_2 \sim N(\mu_2, \sigma_2^2)$ are independent, and $Y = X_1 + X_2$, then $Y \sim N(\mu_1 + \mu_2, \sigma_1^2 + \sigma_2^2)$.

Proof. According to (8.1.24), the MGF of the sum of two independent random variables is the product of their MGFs. Therefore,

$$\begin{aligned} m_Y(s) &= m_{X_1}(s) \cdot m_{X_2}(s) = e^{\mu_1 s+\sigma_1^2 s^2/2} \cdot e^{\mu_2 s+\sigma_2^2 s^2/2} \\ &= e^{(\mu_1+\mu_2)s+(\sigma_1^2+\sigma_2^2)s^2/2}, \end{aligned}$$

and comparing this with (8.2.8), we recognize it as the moment generating function of the $N(\mu_1 + \mu_2, \sigma_1^2 + \sigma_2^2)$ distribution. ∎

The generalization to more than two random variables is as follows:

> If $Y = X_1 + X_2 + \cdots + X_n$, where $X_j \sim N(\mu_j, \sigma_j^2)$ for $j = 1, 2, \ldots, n$, and if $X_1, X_2, \ldots, X_n$ are independent, then $Y \sim N(\mu, \sigma^2)$, where $\mu = \mu_1 + \mu_2 + \cdots + \mu_n$ and $\sigma^2 = \sigma_1^2 + \sigma_2^2 + \cdots + \sigma_n^2$.

The proof is essentially the same as for two random variables; alternatively, it follows by induction from that result. ∎

(8.2.10) **Sums and averages of normal samples.** An important special case of (8.2.9) is that in which the X_j's all have the same normal distribution. They are then i.i.d. (independent, identically distributed) random variables, and together they are called a sample from the normal distribution that they share. In this case their sum is denoted by S_n, and their average, S_n/n, is denoted by $\overline{X}_n$. The results are

> If $X_1, X_2, \ldots, X_n$ are independent, each with the $N(\mu, \sigma^2)$ distribution, then their sum S_n has the $N(n\mu, n\sigma^2)$ distribution, and their average $\overline{X}_n$ has the $N(\mu, \sigma^2/n)$ distribution.

The distribution of S_n comes directly from (8.2.9), and that of $\overline{X}_n$ comes from the distribution of S_n if we use (8.2.4) with $b = 0$ and $a = 1/n$.

This result was obtained, in essentially the same way, in Exercise 6 of the previous section.

(8.2.11) Uses for normal distributions. Normal distributions are certainly the most frequently used absolutely continuous distributions. They are used to model, at least approximately, a very wide range of quantities that vary symmetrically about their averages. What is perhaps surprising is that so many quantities that vary randomly as a result of many uncontrollable factors should have symmetric distributions that can be modeled by normal ones.

As we saw in Section 6.3, the Central Limit Theorem gives a plausible explanation for this. It says (in one of its forms) that if a random variable Y can be regarded as a sum of a large number of independent random variables, say $Y = X_1 + X_2 + \cdots + X_n$, then no matter what the distributions of the X_j's, that of Y is approximately normal if certain conditions on the variances of the X_j's are satisfied.

An important special case of the Central Limit Theorem is the De Moivre–Laplace limit theorem, which we discussed at length in Section 6.4. It says that a binomially distributed random variable with large n has approximately a normal distribution, because it can be regarded as the sum of n Bernoulli random variables. Of course, we know *which* normal distribution, because we know the expected value, np, and the variance, npq, of a binomial random variable.

(8.2.12) Rigorous statement of the Central Limit Theorem. We will be able to use MGFs to give an indication of how the theorem is proved, but before we can prove it we need to state it rigorously. The only statement of it that we have made so far is that if S_n and $\overline{X}_n$ are the sum and average of a sample from some distribution that has a finite expected value and variance, then *approximately*

$$S_n \sim N(n\mu, n\sigma^2) \quad \text{and} \quad \overline{X}_n \sim N(\mu, \sigma^2/n).$$

That is, we can find approximate probabilities for S_n from a standard normal table if we first subtract $n\mu$ and divide by $\sqrt{n\sigma^2}$; we can do the same for $\overline{X}_n$ by subtracting μ and dividing by $\sqrt{\sigma^2/n}$. To be a little more precise: for any number s,

(8.2.13)
$$P(S_n \le s) = P\left(\frac{S_n - n\mu}{\sigma\sqrt{n}} \le \frac{s - n\mu}{\sigma\sqrt{n}}\right) \approx \Phi\left(\frac{s - n\mu}{\sigma\sqrt{n}}\right);$$

and for any x,

(8.2.14)
$$P(\overline{X}_n \le x) = P\left(\frac{\overline{X}_n - \mu}{\sigma/\sqrt{n}} \le \frac{x - \mu}{\sigma/\sqrt{n}}\right) \approx \Phi\left(\frac{x - \mu}{\sigma/\sqrt{n}}\right).$$

Here of course Φ denotes the standard normal CDF.

But now notice that the two "standardized" random variables in (8.2.13) and (8.2.14) are the same random variable, which we denote by Z_n. That is,

(8.2.15)
$$Z_n = \frac{S_n - n\mu}{\sigma\sqrt{n}} = \frac{\overline{X}_n - \mu}{\sigma/\sqrt{n}}.$$

This random variable Z_n, the standardized version of S_n and $\overline{X}_n$, has expected value 0 and variance 1. Now we see that (8.2.13) and (8.2.14) are really the same statement; they say that

(8.2.16) $P(Z_n \le z)$ is approximately $\Phi(z)$ for any number z

[whether we call that number z, $(s - n\mu)/(\sigma\sqrt{n})$, or $(x - \mu)/(\sigma/\sqrt{n})$].

It is easy to make (8.2.16) into a statement about limits. That statement is the Central Limit Theorem, which follows.

Central Limit Theorem. If $X_1, X_2, X_3, \ldots$ are i.i.d. random variables whose distribution has a finite expected value μ and a finite variance σ^2, then for any real number z

$$\lim_{n\to\infty} P(Z_n \le z) = \Phi(z)$$

where Φ is the standard normal CDF.

In other words:

If $F_n(z)$ is the CDF of Z_n, then for every z, $F_n(z)$ approaches $\Phi(z)$ as $n \to \infty$.

(8.2.17) **How the Central Limit Theorem is proved.** The proof is not at all apparent, because the distribution of Z_n is usually fairly complicated, and so we have little hope of finding its CDF and showing that its limit for large n is Φ.

However, there happens to be another way to show that the limit of a sequence of CDFs is Φ: show that the limit of the associated MGFs is $e^{s^2/2}$, which is the MGF of Φ. The fact that this works is one of the important properties of generating functions that we mentioned at the beginning of Section 8.1. We have not proved or used it since then; it is usually covered in graduate courses in probability theory.

Actually, what is used in the full-fledged proof is not the MGF, but the characteristic function, which you may remember is a generating function closely related to the MGF. We mentioned it briefly in (8.1.10). It has the minor disadvantage of being a complex-valued function, but it has the great advantage of existing for every distribution, regardless of whether the distribution has moments or not. However, since we are giving only a brief indication of how the proof goes, we will use MGFs.

At this point you might want to look at Exercise 9 at the end of this section. There it is shown how to prove that if the X_j's have a particular distribution—namely, the Pois(λ) distribution—then the MGF of Z_n approaches $e^{s^2/2}$. The general proof that we will indicate here shows that this happens no matter what the original distribution is, as long as it has finite expected value and variance.

Outline of proof. First, we make a simplification. We assume that the X_j's have expected value 0. This turns out not to be a serious restriction, though, because for any X_j's with nonzero expected value μ, the random variables $X_j - \mu$ have expected value 0 and the same variance. You can check that the Z_n's for the sequence of X_j's are the same as the Z_n's for the sequence of $(X_j - \mu)$'s.

So here is where we are. We have a sequence $X_1, X_2, X_3, \ldots$ of i.i.d. random variables with expected value 0 and finite variance σ^2, and we want to show that the MGF of Z_n—call it $m_n(s)$—tends to $e^{s^2/2}$, which is the MGF of the $N(0, 1)$ distribution.

We will need two basic facts for the proof. First, you remember that if $m(s)$ is the MGF of X, and all the moments EX^k of X are finite, then $m(s)$ has a Maclaurin series

$$m(s) = 1 + EX \cdot s + \frac{EX^2}{2!} \cdot s^2 + \frac{EX^3}{3!} \cdot s^3 + \cdots + \frac{EX^k}{k!} \cdot s^k + \cdots .$$

Therefore, if we denote the MGF of the X_j's by $m(s)$, then since $EX = 0$ and $VX = E(X - 0)^2 = EX^2 = \sigma^2$, then *if the other moments of the X_j's existed* we would have the Maclaurin series

$$m(s) = 1 + \frac{\sigma^2}{2} \cdot s^2 + \frac{EX^3}{3!} \cdot s^3 + \cdots + \frac{EX^k}{k!} \cdot s^k + \cdots .$$

Now what can be shown is that even if the other moments do not all exist, we can still (under certain other mild assumptions) write

$$m(s) = 1 + \frac{\sigma^2}{2} \cdot s^2 + r(s), \tag{8.2.18}$$

where $r(s)$, instead of being $\frac{EX^3}{3!} \cdot s^3 + \cdots + \frac{EX^k}{k!} \cdot s^k + \cdots$, is some remainder function which is unknown except for the important property that

$$\frac{r(s)}{s^2} \to 0 \quad \text{as } s \to 0. \tag{8.2.19}$$

This—that is, (8.2.18) and (8.2.19)—is our first basic fact. The second is also a generalization of something you remember, namely that

$$\lim_{n\to\infty}\left(1+\frac{x}{n}\right)^n = e^x \quad \text{for any real } x.$$

The generalization is that if w_n is any sequence of real numbers that tend to 0 as $n\to\infty$, then we still have

(8.2.20)
$$\lim_{n\to\infty}\left(1+\frac{x+w_n}{n}\right)^n = e^x \quad \text{for any real } x.$$

We will omit the proof of this generalization. It is at the level of an advanced calculus course.

Now what is the MGF of Z_n? We did something like this in (8.1.30):

The MGF of X_j is $m(s)$, so the MGF of S_n is $[m(s)]^n$, and

$$\text{since } Z_n = \frac{S_n - n\mu}{\sigma\sqrt{n}} = \frac{S_n}{\sigma\sqrt{n}}, \quad \text{its MGF is } \left[m\left(\frac{s}{\sigma\sqrt{n}}\right)\right]^n.$$

Now putting (8.2.18) into this, we see that the MGF of Z_n is

$$\left[m\left(\frac{s}{\sigma\sqrt{n}}\right)\right]^n = \left(1+\frac{s^2}{2n}+r\left(\frac{s}{\sigma\sqrt{n}}\right)\right)^n.$$

This can be rewritten as

(8.2.21)
$$m_{Z_n}(s) = \left(1+\frac{(s^2/2)+t_n(s)}{n}\right)^n,$$

where

$$t_n(s) = n\cdot r\left(\frac{s}{\sigma\sqrt{n}}\right).$$

Now because of (8.2.20), we see that if we can only show that $t_n(s)\to 0$ as $n\to\infty$ for any s, then (8.2.21) shows that the MGF of Z_n approaches $e^{s^2/2}$ as $n\to\infty$. That will complete the proof.

But $t_n(s)$ can be written

$$t_n(s) = \left(\frac{s}{\sigma}\right)^2\cdot\frac{r\left(\dfrac{s}{\sigma\sqrt{n}}\right)}{\left(\dfrac{s}{\sigma\sqrt{n}}\right)^2},$$

and because $s/(\sigma\sqrt{n})\to 0$ as $n\to\infty$ for any s, we get by (8.2.19) that $t_n(s)\to 0$. ☐

(8.2.22) **More on the history and future of the Central Limit Theorem.** We made a brief historical note on the Central Limit Theorem in (6.3.9). Here we want to mention two interesting directions in which the Central Limit Theorem has been extended:

1. The Central Limit Theorem we gave has the hypothesis that the X_j's are i.i.d. and have a finite expected value and variance. Is it possible to weaken this hypothesis? That is, can S_n, $\overline{X}_n$, and Z_n ever be approximately normally distributed even if the X_j's are not independent, or not identically distributed, or both?
2. The Central Limit Theorem says that a certain kind of standardization of a sum of random variables (subtracting the mean and dividing by the standard deviation) produces a random variable with approximately a normal distribution. Are there other distributions that arise from similar situations as a result of other kinds of standardizations?

Much work has been done through the years on the first question. Although Karl Friedrich Gauss (1777–1855) knew the Central Limit Theorem essentially as we stated it here, work has continued on generalizing and extending it up to the present. In the early part of the 20th century the Russian mathematician Lyapounov proved a Central Limit Theorem for the following situation: for each n, we take a new sample of size n from a possibly different distribution instead of just sampling one more random variable from the same distribution. The set of samples is called a "triangular array," because of the way it looks when we list the sample of size 1, the sample of size 2 below it, the sample of size 3 below that, and so on. Lyapounov gave conditions on the moments of the distributions of the samples, under which the sample sums and averages still have approximately normal distributions.

Later, the Swedish mathematician Lindeberg introduced some conditions that were weaker than Lyapounov's but still implied convergence to a normal distribution. Following that, William Feller proved that Lindeberg's conditions could not be improved on: in any triangular array in which the distribution of the Z_n's converges to $N(0, 1)$, Lindeberg's conditions must hold.

Continuing research involves a wide range of other types of situations in which something or other has an approximate normal distribution—sums of independent vectors of random variables, for example, or random variables defined in terms of certain types of stochastic processes.

On the second question there has also been much work in the 20th century, continuing to the present. It turns out that there is a large family of distributions which are the limiting distributions for triangular arrays in which Lindeberg's conditions do not necessarily hold. The most important distributions in this family are the normal and the Poisson distributions, but they are only a small subfamily. Work continues on identifying the properties of various other members of the family. This topic is usually covered, in connection with what are called "infinitely divisible distributions," in graduate probability courses.

Although William Feller made very important theoretical contributions, he may be better known for his influential textbooks, Volumes 1 and 2 of *An Introduction to Probability*

Theory and Its Applications. See the bibliography for more details on these books; Volume 1 in particular is highly recommended to readers of this text.

Feller was born in 1906 in Zagreb and received his Ph.D. at Göttingen at the age of 20. He spent time at the universities in Kiel and Stockholm before coming to the United States in 1939. He spent 5 or 6 years each at Brown and Cornell Universities before going to Princeton in 1950. He died there in 1970. He was a colorful and lively teacher, a prolific researcher, and an active consultant to industry and science who was unafraid of confrontation. (See the note following Exercise 5 in Section 7.6.) In Princeton he lived in a house on Random Road.

EXERCISES FOR **SECTION 8.2**

1. If $X \sim N(15, 30)$ and $Y \sim N(-12, 15)$, find the following.
   a. The distribution of $X + Y$
   b. The distribution of $X - 2Y$
   c. The probability that $2X + 3Y \leq 5$
   d. The probability that $X > Y$

2. Let $X_1, X_2, \ldots, X_{30}$ be a sample from the $N(20, 25)$ distribution. Find the following:
   a. The probability that $\overline{X}_{30}$ is between 19 and 21
   b. The probability that S_{30} is greater than 650
   c. Z_{30} in terms of $\overline{X}_{30}$ and S_{30}

3. Find the normal approximation to the probability that a random variable with the Pois($\lambda = 32$) distribution is less than 25.

4. Find the number a so that a $N(\mu, \sigma^2)$ random variable has probability 99% of being less than $\mu + a\sigma$—that is, less than a standard deviations above its expected value. Repeat for 95% and 90%.

5. Find the number a so that the probability is 99% that a normally distributed random variable is within a standard deviations of its expected value. Repeat for 95% and 90%.

The numbers found in Exercises 4 and 5 come up often in introductory courses in statistical inference. If something that has been observed has a probability of 5% or less under some hypothesis, then we say that we can reject that hypothesis at the 5% level. Such probabilities are most often computed using normal approximations. Thus, for example, if we observe a sample mean that is more than 1.96 standard deviations away from the hypothesized population mean (the standard deviation being that of the sample mean $\overline{X}_n$, and assuming that n is large enough that $\overline{X}_n$ is approximately normal), then we can reject that hypothesized mean at the 5% level.

6. If you got 25 7's in 100 rolls of a pair of dice, would you have reason to suspect that the dice were loaded to produce 7's? Find the approximate probability that 100 rolls of a fair pair produce 25 or more 7's.

(In the language of statistical inference, you can reject the hypothesis of fairness at the .05 level if this probability is less than .05. Statisticians sometimes regard the .05 level as "significant" and the .01 level as "highly significant." In this problem, we can reject the hypothesis of fairness at the 5% level, but not the 1% level. That is, 25 is a significant, but not highly significant, deviation from $16\frac{2}{3}$, the expected number of 7's.

7. The weights of children in a certain age group are normally distributed with expected value 65 pounds and standard deviation 15 pounds.
   a. What is the 80th percentile?
   b. What is the distribution of the average of the weights of 20 children chosen independently and at random from this group?

8. Let $X_1, X_2, \ldots, X_{50}$ be independent observations of some random variable X whose expected value is 100 and whose variance is 25. Let $\overline{X}_{50}$ be their average.
   a. What are the expected value and variance of $\overline{X}_{50}$?
   b. What does Chebyshev's inequality say about the probability that $\overline{X}_{50}$ fails to be between 98 and 102?
   c. What does the Central Limit Theorem say about the probability described in part b?

9. This exercise illustrates how the proof of the Central Limit Theorem works by proving it under the additional hypothesis that the X_j's have the Pois(λ) distribution. Let $X_1, X_2, \ldots, X_n$ be a sample from this distribution. We want to show that the MGF of Z_n approaches $e^{s^2/2}$. We do this by writing down the logarithm of the MGF and noticing that it approaches $s^2/2$.
   [a]**a.** Write down the distribution, expected value, and variance of S_n.
   b. Show that the MGF of S_n is $e^{n\lambda(e^s-1)}$. [See Exercise 10 of Section 8.1.]
   [a]**c.** Write down Z_n in terms of S_n.
   d. Show that the MGF of Z_n is $\exp(n\lambda(e^{s/\sqrt{n\lambda}} - 1 - s/\sqrt{n\lambda}))$. [Use (8.1.28).]
   e. To show that the logarithm of this MGF approaches $\sigma^2/2$, expand $e^{s/\sqrt{n\lambda}}$ using the Maclaurin series, subtract 1 and $s/\sqrt{n\lambda}$, and multiply by $n\lambda$. Notice that as n approaches ∞, all the terms approach 0 except the term $s^2/2$.

ANSWERS

1. **a.** $N(3, 45)$ **b.** $N(39, 90)$ **c.** .7549 **d.** Essentially 1
2. **a.** .7286 **b.** .0336
3. .0918, with continuity correction; .1075, without
4. 99%: 2.33; 95%: 1.645 is usually used; 90%: 1.28
5. 99%: 2.58; 95%: 1.96; 90%: 1.645
6. .0179, with continuity correction; .0125, without
7. **a.** 77.6 **b.** $N\left(65, \frac{225}{20}\right)$
8. **a.** 100; $\frac{1}{2}$ **b.** It is less than or equal to $\frac{1}{8}$. **c.** It is approximately equal to .0046.
9. **a.** Pois($n\lambda$); $n\lambda$; $n\lambda$ **c.** $(S_n - n\lambda)/\sqrt{n\lambda} = \frac{1}{\sqrt{n\lambda}}S_n - \sqrt{n\lambda}$

8.3 EXPONENTIAL AND ERLANG DISTRIBUTIONS

We are very familiar with exponential distributions by now. They were one of our first examples of absolutely continuous distributions, they play an important part in Poisson processes, and we have used them in many examples and exercises. We will summarize them briefly here.

Then we will discuss the Erlang distributions. In a Poisson process with rate λ, if X is the time of the nth arrival, then X is the sum of n independent random variables, each having the exponential distribution with parameter λ. The distribution of X is called the Erlang distribution with parameters n and λ. If $n = 1$, of course, this is the exponential distribution.

So we are really dealing with only one family of distributions—the Erlang family includes the exponential family. This is like the situation discussed in Section 7.3, where we saw that the negative binomial family includes the geometric distributions; that is, neg bin$(1, p)$ is the same as geom(p) and neg bin(k, p) is the distribution of the sum of k independent geom(p) random variables.

In the next section we will see an even larger family that includes the Erlang family, in which the parameter n is allowed to be any positive number instead of just an integer. This larger family is called the gamma family.

(8.3.1) Definition. For any positive number λ, the ***exponential distribution with parameter λ***, which we abbreviate as the exp(λ) distribution, is the one whose density function is

$$f(x) = \lambda e^{-\lambda x} \quad \text{for } x > 0.$$

The corresponding CDF is

$$F(x) = \begin{cases} 0 & \text{if } x \leq 0, \\ 1 - e^{-\lambda x} & \text{if } x > 0. \end{cases}$$

Probably the most convenient recipe for finding probabilities is the familiar

$$P(X \in I) = e^{-\lambda a} - e^{-\lambda b} \quad \text{if } I \text{ is any subinterval of } (0, \infty) \text{ with endpoints } a \text{ and } b.$$

This is the distribution of the waiting time, in a Poisson process with arrival rate λ, from the start of observations to the first arrival; it is also the distribution of the waiting time from any given time point to the next arrival, and also of the waiting time from any one arrival to the next.

One of the important properties of an exponential distribution is its "lack of memory." As we saw in Exercise 9 of Section 2.2, if X has an exponential distribution, then for any positive numbers a and t,

$$P(X > a + t \mid X > a) = P(X > t) = e^{-\lambda t}.$$

The conditional probability that we have to wait t more time units, given that there has been no arrival in a time units, is the same no matter how big or small a is. This is a characteristic property of Poisson processes, as well as of Bernoulli processes.

(8.3.2) **MGF and moments of an exponential distribution.** The MGF is

$$m(s) = \frac{\lambda}{\lambda - s} \quad \text{for } s < \lambda.$$

We found this in Example (8.1.4). From it we found all the moments in Example (8.1.21), by expanding this MGF in a power series. The result is

$$EX^k = \frac{k!}{\lambda^k} \quad \text{for } k = 0, 1, 2, \ldots .$$

The first two moments and the variance are, as we have known for a long time,

$$EX = \frac{1}{\lambda}, \quad EX^2 = \frac{2}{\lambda^2}, \quad \text{and} \quad VX = \frac{1}{\lambda^2}.$$

(8.3.3) **Distribution of a multiple of an exponentially distributed random variable.** If $X \sim \exp(\lambda)$ and $y = aX$ for some positive number a, then $Y \sim \exp(\lambda/a)$. This is easy to see using MGFs, as in Example (8.1.29), or using the CDF method, as in Exercise 1 of Section 5.2.

(8.3.4) **The gamma formula.** This formula is the key to a great many computations involving exponential, Erlang, and general gamma distributions. We first saw it in Exercise 10 of Section 4.3, and we will see it again, in more generality, in (8.4.10). The version we want here is

$$\int_0^\infty x^{n-1} e^{-\lambda x} dx = \frac{(n-1)!}{\lambda^n} \quad \text{for any } \lambda > 0 \text{ and } n = 1, 2, 3, \ldots .$$

This is most easily proved by first making the change of variable $y = \lambda x$ and then integrating repeatedly by parts. In the next section we will see that there is a version of it even for values of n that are not integers; the factorial is replaced by the gamma function in that case.

It may seem a little strange to use $n - 1$ in place of n in this formula; after all, it is equivalent to

$$\int_0^\infty x^n e^{-\lambda x} dx = \frac{n!}{\lambda^{n+1}} \quad \text{for any } \lambda > 0 \text{ and } n = 0, 1, 2, \ldots ,$$

and you may wonder why we did not present it in this form. When we extend it to noninteger values of n in the next section, though, the other form will be a little more convenient.

We can use the gamma formula to find the moments of an exponentially distributed random variable quickly without using the MGF,

$$EX^k = \int_0^\infty x^k \cdot \lambda e^{-\lambda x} dx = \lambda \cdot \frac{k!}{\lambda^{k+1}} = \frac{k!}{\lambda^k}.$$

(8.3.5) **Erlang distributions** (gamma distributions with integer parameter). Now let X be the time of the nth arrival in a Poisson process with arrival rate λ. The distribution of X is called the Erlang distribution with parameters n and λ. It is clear that X is the sum of n independent random variables, each having the $\exp(\lambda)$ distribution: $X = X_1 + X_2 + \cdots + X_n$, where X_1 is the waiting time to the first arrival, X_2 is the waiting time between the first and second arrivals, and so on.

We will do two things, which have actually been done before, in Exercise 8 of Section 8.1.

1. Present a certain density, define it as the Erlang density with parameters n and λ, and show by the gamma formula that it *is* a density.
2. Find the MGF of this density (also by using the gamma formula), and recognize it as the MGF of the sum of n independent $\exp(\lambda)$ random variables.

It will follow that the Erlang density is indeed the correct density for the time of the nth arrival. But this seems a little unsatisfactory, because you might wonder how anyone arrived at it in the first place. So we will then show, using the convolution theorem, how to get this density for the sum of n independent $\exp(\lambda)$ random variables.

Finally, we will find the moments of an Erlang distribution. Surprisingly, they are not best found using the MGF; rather, the gamma formula makes it just as simple to get them directly. (Neither calculation is particularly difficult, however.)

(8.3.6) **Density of an Erlang distribution.** Let n be a positive integer and let λ be a positive number. The ***Erlang distribution with parameters n and λ*** [Erl(n, λ)] is the one whose density function is

$$f(x) = \frac{\lambda^n}{(n-1)!} x^{n-1} e^{-\lambda x} \quad \text{for } x > 0.$$

The gamma formula (8.3.4) shows immediately that this integrates to 1.

There is no closed form for the CDF of this density; the most convenient expression is

$$F(x) = \begin{cases} 0 & \text{if } x \le 0, \\ 1 - \sum_{k=0}^{n-1} \frac{(\lambda x)^k e^{-\lambda x}}{k!} & \text{otherwise.} \end{cases}$$

This can be proved by integrating the density, using repeated integrations by parts. It is probably easier to derive it using the argument given in (8.3.14). We will discuss it briefly at that point. Notice the Poisson probabilities in the sum. We will see in (8.3.14) that it is no coincidence that they are there.

Also notice the special case in which $n = 1$. In this case, $f(x)$ is simply $\lambda e^{-\lambda x}$ for $x > 0$, which is the $\exp(\lambda)$ distribution. And of course we know that this is the distribution of the first arrival.

Graphs of some of the Erlang densities are shown in the next section in Figure 8.4b.

(8.3.7) **The MGF of an Erlang distribution.** The gamma formula also gives the MGF to us:

$$m(s) = \int_0^\infty e^{sx} \cdot \frac{\lambda^n}{(n-1)!} x^{n-1} e^{-\lambda x} dx$$
$$= \frac{\lambda^n}{(n-1)!} \int_0^\infty x^{n-1} e^{-(\lambda - s)x} dx.$$

As long as $\lambda - s$ is positive, the gamma formula tells us that this equals

$$\frac{\lambda^n}{(n-1)!} \cdot \frac{(n-1)!}{(\lambda - s)^n};$$

that is,

$$m(s) = \left(\frac{\lambda}{\lambda - s}\right)^n \quad \text{for } s < \lambda.$$

Now we know that $\lambda/(\lambda - s)$ is the MGF of the $\exp(\lambda)$ distribution; so by (8.1.24) this is the MGF of the sum of n independent $\exp(\lambda)$ random variables. As a result, we now have proved that the time of the nth arrival has the $\text{Erl}(n, \lambda)$ distribution.

(8.3.8) **Direct derivation of the Erlang density for the time of the *n*th arrival.** Here we will show, without using MGFs, that the density of the sum of n i.i.d. $\exp(\lambda)$ random variables is the Erlang density defined in (8.3.6). We do it by induction, using the convolution theorem (5.2.20).

Proof. For $n = 1$, the assertion is that the density of one $\exp(\lambda)$ random variable is obtained by putting $n = 1$ in the Erlang density. We have already noted this in (8.3.6).

So suppose that X is the sum of $k - 1$ independent $\exp(\lambda)$ random variables, and that a density for X is

$$f_X(x) = \frac{\lambda^{k-1}}{(k-2)!} x^{k-2} e^{-\lambda x} \quad \text{for } x > 0,$$

which is the result of putting $n = k - 1$ into (8.3.6). Let $Y \sim \exp(\lambda)$ be independent of X, so that $Z = X + Y$ is the sum of k i.i.d. $\exp(\lambda)$ random variables. We want to show that a density for Z is

(8.3.9)
$$f_Z(x) = \frac{\lambda^k}{(k-1)!} z^{k-1} e^{-\lambda z} \quad \text{for } z > 0.$$

Of course we know that a density for Y is

$$f_Y(y) = \lambda e^{-\lambda y} \quad \text{for } y > 0,$$

and that the set of possible values of Z is $(0, \infty)$. By the convolution theorem (5.2.20), for any $z > 0$

$$f_Z(z) = \int_{-\infty}^{\infty} f_X(x) \cdot f_Y(z - x) dx$$
$$= \int_0^z \frac{\lambda^{k-1}}{(k-2)!} x^{k-2} e^{-\lambda x} \cdot \lambda e^{-\lambda(z-x)} dx;$$

the integral goes from 0 to z because either $f_X(x)$ or $f_Y(z - x)$ is 0 if x is outside that range. But now we get a pleasant surprise: the x's cancel each other in the exponent, and everything in the integral except x^{k-2} is a constant, so that

$$f_Z(z) = \frac{\lambda^k}{(k-2)!} e^{-\lambda z} \int_0^z x^{k-2} dx = \frac{\lambda^k}{(k-2)!} e^{-\lambda z} \cdot \frac{z^{k-1}}{k-1},$$

which equals (8.3.9). □

(8.3.10) **Expected value and variance of an Erlang distribution.** The simplest way to find these is to use the fact that if $X \sim \text{Erl}(n, \lambda)$, then X has the distribution of $X_1 + X_2 + \cdots + X_n$, where the X_j's are i.i.d. $\exp(\lambda)$ random variables. Each X_j has expected value $1/\lambda$ and variance $1/\lambda^2$; therefore,

$$EX = \frac{n}{\lambda} \quad \text{and} \quad VX = \frac{n}{\lambda^2}.$$

(8.3.11) **General moments of an Erlang distribution.** It is interesting to note that the MGF is not the best way to find these. Instead, the gamma formula makes it quite simple to find them directly. If $X \sim \text{Erl}(n, \lambda)$, then

$$EX^k = \int_0^\infty x^k \cdot \frac{\lambda^n}{(n-1)!} x^{n-1} e^{-\lambda x} dx = \frac{\lambda^n}{(n-1)!} \int_0^\infty x^{k+n-1} e^{-\lambda x} dx,$$

and by the gamma formula this is

$$\frac{\lambda^n}{(n-1)!} \cdot \frac{(k+n-1)!}{\lambda^{k+n}};$$

that is,

$$EX^k = \frac{n(n+1)(n+2)\cdots(n+k-1)}{\lambda^k}.$$

It is also not difficult to get this same result by differentiating the MGF k times and then setting $s = 0$.

(8.3.12) **Distribution of a constant multiple of an Erlang random variable.** If $X \sim \text{Erl}(n, \lambda)$ and $Y = aX$ for some positive number a, then it is easy to use (8.1.28) to find the MGF of Y and recognize it as another Erlang MGF:

$$m_X(s) = \left(\frac{\lambda}{\lambda - s}\right)^n, \quad \text{so} \quad m_Y(s) = m_X(as) = \left(\frac{\lambda}{\lambda - as}\right)^n = \left(\frac{\lambda/a}{\lambda/a - s}\right)^n.$$

Thus, Y has the $\text{Erl}(n, \lambda/a)$ distribution.

(8.3.13) **Distribution of a sum of independent Erlang random variables with the same λ.** Suppose $Y = X_1 + X_2 + \cdots + X_k$, where $X_j \sim \text{Erl}(n_j, \lambda)$. Notice that they do not have to have the same n, but the λ must be the same for all of them. Then by (8.1.24) the MGF of Y is

$$m_Y(s) = m_{X_1}(s)m_{X_2}(s)\cdots m_{X_k}(s) = \left(\frac{\lambda}{\lambda - s}\right)^{n_1+n_2+\cdots+n_k}$$

So Y has the $\text{Erl}(n_1 + n_2 + \cdots + n_k, \lambda)$ distribution. This result is not surprising when we realize that Y can be thought of as the time of the $(n_1 + n_2 + \cdots + n_k)$th arrival.

(8.3.14) **Finding probabilities for Erlang distributions: converting Erlang probabilities to Poisson probabilities.** Unfortunately, although we can use the gamma formula to show that the integral of an Erlang density from 0 to ∞ equals 1, it is not easy to integrate the $\text{Erl}(n, \lambda)$ density between two given positive numbers. It can always be done using $n - 1$ integrations by parts, but if n is large this is inconvenient.

However, it is often possible to translate statements about Erlang random variables, which are waiting times, to statements about Poisson random variables, which are numbers of arrivals. The key is the following simple fact:

$$P(n\text{th arrival comes before time } x) = P(n \text{ or more arrivals in the interval } (0, x)).$$

We have already done this kind of translation in several examples and exercises in Section 1.4, and it is similar to the kind of conversions we made in (7.3.10) for negative binomial random variables.

As an important example of this, we can verify that the CDF of the $\text{Erl}(n, \lambda)$ distribution is the function that we gave without proof in (8.3.6) above. If X has this distribution, then we can view X as the time of the nth arrival in a Poisson process. So the CDF of X, for $x > 0$, is

$$\begin{aligned} F(x) &= P(X \leq x) = P(\text{the } n\text{th arrival comes before time } x) \\ &= P(n \text{ or more arrivals in interval } (0, x)) \\ &= 1 - \sum_{k=0}^{n-1} P(k \text{ arrivals in } (0, x)) \\ &= 1 - \sum_{k=0}^{n-1} \frac{(\lambda x)^k e^{-\lambda x}}{k!}. \end{aligned}$$

(8.3.15) **Example.** In a Poisson process with three arrivals per minute, what is the probability that the sixth arrival comes within the first 2 minutes?

Solution. This could be found as

(8.3.16)
$$\int_0^2 \frac{3^6}{5!} x^5 e^{-3x} dx,$$

because the time of the sixth arrival has the Erl(6, 3) density, and what we want is the integral of that density from 0 to 2. You can do that integral using five integrations by parts if you like.

Alternatively, if you have access to (or can write) a computer program that does numerical integration, it is possible to get the value of this integral to a desired degree of accuracy. [The use of one such program shows that the value of (8.3.16), to four significant digits, is .5543.]

But there is a relatively easy way to get the answer with just a hand calculator. We are asking for the probability that there are six or more arrivals in the first 2 minutes. The number of arrivals in the first 2 minutes has the Poisson distribution with parameter $2 \cdot \lambda = 2 \cdot 3 = 6$; so the answer is

(8.3.17)
$$1 - \left(\frac{6^0 e^{-6}}{0!} + \frac{6^1 e^{-6}}{1!} + \frac{6^2 e^{-6}}{2!} + \frac{6^3 e^{-6}}{3!} + \frac{6^4 e^{-6}}{4!} + \frac{6^5 e^{-6}}{5!} \right) \doteq .5543.$$

We have replaced five integrations by parts with Poisson probabilities. Note that this is not an approximation; it gives the exact probability by translating the statement about waiting times into a statement about numbers of arrivals.

You might find it instructive to evaluate (8.3.16) using five integrations by parts. Each integration by parts produces a new term; you will notice that the terms are exactly the probabilities in (8.3.17). (The sixth term comes from the integral that remains after the fifth integration by parts.) ∎

(8.3.18) **Example.** If $X \sim$ Erl(20, 2), then we know $EX = 10$. What is the probability that X is between 9 and 11?

Solution. There are at least three ways to approach this, two exact and one approximate. As in the previous example, the direct approach, using an Erlang density, is to evaluate

$$\int_9^{11} \frac{2^{20}}{19!} x^{19} e^{-2x} dx.$$

This can be done using 19 integrations by parts, or using a numerical integration program to get the answer to a desired degree of accuracy. To four significant figures, the answer is .3449.

The second approach is to translate the question into a question about numbers of arrivals in a Poisson process. What we are asking for is the probability that in a Poisson process with two arrivals per minute, the 20th arrival comes between the 9-minute mark and the 11-minute mark. We can translate this as follows: it is the probability that there are 20 or more arrivals in the first 11 minutes, but that there are fewer than 20 arrivals in the first 9 minutes. That is, it equals

$$P(\text{20 or more in the first 11 minutes}) - P(\text{20 or more in the first 9 minutes}).$$

This is still fairly messy, but at least it involves Poisson probabilities for which we need no numerical software, but only a simple programming language, to get the answer:

$$\left(1 - \sum_{k=0}^{19} \frac{22^k e^{-22}}{k!}\right) - \left(1 - \sum_{k=0}^{19} \frac{18^k e^{-18}}{k!}\right).$$

The third approach, normal approximation, works much better. Because the time of the 20th arrival is a sum of 20 i.i.d. random variables [each having the exp(2) distribution], the Central Limit Theorem provides that it has approximately a normal distribution. Now 20 is not overly large, and the exp(2) distribution is not very bell shaped, so the approximation might be suspect; but let us do it anyway. Let X denote the time of the 20th arrival. Then $EX = 20/2 = 10$, $VX = 20/2^2 = 5$, and, since X is approximately normally distributed,

$$\begin{aligned} P(9 < X < 11) &= P\left(\frac{9-10}{\sqrt{5}} < \frac{X-10}{\sqrt{5}} < \frac{11-10}{\sqrt{5}}\right) \\ &\approx P(-.45 < Z < .45) \doteq .3472. \end{aligned}$$

This is by far the simplest of the three methods, although we pay for the simplicity by getting an answer that is only approximate. However, notice that it differs from the correct probability (which to four significant digits is .3449) by less than 1%. ∎

EXERCISES FOR **SECTION 8.3**

1. Consider the CDF $F(x) = 1 - e^{-2x} - 2xe^{-2x}$ for $x > 0$ (and 0 for $x \le 0$). Which Erlang distribution is this?

2. Let X have the exponential distribution with parameter λ.

a. What is the probability that X exceeds its expected value? [Notice that the answer does not involve λ; but more importantly, it does not equal $\frac{1}{2}$.]
b. What is the number m such that $P(X \geq m) = \frac{1}{2}$? [This number is called the *median* of X.]
c. What is the number M at which the density of X is largest? [This is the *mode*.]

Exponential distributions provide a rather extreme counterexample to common misconceptions about expected values. For them, and for other nonsymmetric distributions, EX is neither the median value, which has half the probability above it and half below, nor the modal value, where the density is highest. They are an extreme case because they are so *skewed*—that is, so far from symmetric. For such distributions—for example, income distributions, though not usually exponential, are often heavily skewed—the median is often regarded as a better indication of a "typical" value.

3. In Exercise 9c of Section 1.4 we assumed that telephone calls are coming in according to a Poisson process with arrival rate $\lambda = 9.3$ calls per minute. We found the probability that the fourth call comes in more than 30 seconds after the start of observations. Now we know the distribution of the time of the fourth call. Give its name, expected value, and variance.

4. In Exercise 3 of Section 4.3 we found $EX = 2/\lambda$ and $VX = 2/\lambda^2$ for the density $\lambda^2 x e^{-\lambda x}$ $(x > 0)$. Identify this density as a member of the Erlang family and use (8.3.10) to confirm the values.

5. What density on $(0, \infty)$ is proportional to $x^4 e^{-3x}$? Write down the density and give its name.

6. Customers arrive according to a Poisson process with an average of eight arrivals per minute. Let X be the arrival time of the 100th customer, in minutes after the start of observations. We are interested in the probability that X is less than 10 minutes.
   [a]a. Write down (but do not try to evaluate) the integral that expresses this probability.
   [a]b. Translate it to a probability involving a Poisson random variable Y, and write down (but do not evaluate) the sum that expresses it.
   c. Approximate it using the fact that X is approximately normally distributed.
   d. Why is X approximately normally distributed?
   e. Approximate it using the fact that Y is approximately normally distributed.
   f. Why is Y approximately normally distributed?

As in Example (8.3.18), the two approximations in parts c and e are quite different, and the one in part e is closer to the true probability.

7. Let $X_1, X_2, \ldots, X_n$ be a sample from the $\exp(\lambda)$ distribution.
   a. Find the CDF and thence a density for U_n, the smallest of $X_1, X_2, \ldots, X_n$. Identify the distribution of U_n by name. [*Hint:* $U_n > u$ if and only if every X_j is greater than u.]

b. Find the CDF and a density for V_n, the largest of X_1 X_2, . . . , X_n. [This distribution does not have a standard name.]

8. Find the Erlang distribution whose expected value is 16 and whose variance is 32/3.

9. The lightbulbs used in a certain factory have lifetimes that are exponentially distributed; the expected lifetime is 3,000 hours. If there are 5,000 bulbs in the factory, what is the distribution of the time of the next burnout? [Use the result of Exercise 7a.] What are the expected value and standard deviation?

10. The variance of a Poisson random variable always equals its expected value. What about an exponential random variable?

11. You buy four new tires for your car. Assume that the "lifetime" of one of these tires—the number of miles it will go before wearing out or blowing out—is a random variable having the exponential distribution with parameter λ. Also assume, for simplicity, that the lifetimes of the four tires are independent.
    a. What is the probability that your left rear tire goes at least a miles? (a is some positive number.)
    b. Let X be the number of miles you go before a tire wears out or blows out. What is $P(X > a)$?
    c. What is the name of the distribution of X?

12. Let X and Y be independent random variables with different exponential distributions: suppose $X \sim \exp(3)$ and $Y \sim \exp(7)$.
    a. Find the MGF of $X + Y$ and show that it is not an Erlang MGF. [*Hint:* Write it as a fraction with numerator 1 and show that it cannot equal $1/(1 - s/\lambda)^2$ for any positive λ.]
    b. Find a value of a so that $X + aY$ has an Erlang distribution.
    [If X and Y had the *same* exponential distribution and were independent, their sum would have an Erlang distribution; part a shows the necessity of having the same distribution.]

13. **Should banks have a separate waiting line for each teller or one line?** Which is better, a bank in which customers choose a teller and join that teller's waiting line, or one in which all customers join one line and the customer at the head of the line goes to the next available teller? Banks have been switching to the latter scheme in recent years. We looked at one possible advantage of this "new" scheme in Exercise 12 of Section 3.5. Here we want to look at the expected value and variance of your waiting time in each of the two schemes.

    We consider a concrete but typical example. The bank has six tellers, and when you arrive there are 22 people already there, six being served and 16 waiting. Suppose that once a customer reaches a teller, the time it takes to serve that customer has the $\exp(\lambda)$ distribution. Because of the lack of memory of this distribution, the time to finish the service of one of the customers already being served also has the same distribution.

a. In the "old" scheme, the 22 people are in six lines. It seems safe to assume that all lines have at least three people, and naturally at least some of them have no more than three. So you choose a line with three people—one being served and two waiting. Your waiting time is then the sum of the three independent service times of the people ahead of you. Find its expected value and variance.

b. In the "new" scheme, six people are being served and the waiting line has 16 people ahead of you, each going to the next available teller. But it follows from the result of Exercise 7a that tellers become available according to a Poisson process whose arrival rate is 6λ. Find the expected value and variance of your waiting time in this case.

c. Now suppose there are n people there when you arrive, six being served and the rest in line. Write n as $6k + j$, where the remainder j is one of 0, 1, . . . , 5. Then in the "old" scheme you will join a line with k people ahead of you, including the one being served; in the "new" scheme there will be 6 being served and $6(k - 1) + j$ in line ahead of you. Show that your expected waiting time, and the variance, are less in the "new" scheme.

ANSWERS

1. Erl(2, 2)
2. a. .3679 b. $(\ln 2)/\lambda$ c. 0
3. Erl(4, 9.3); .4301; .04625
4. Erl(2, λ)
5. Erl(5, 3)
6. a. $\int_0^{10} (8^{100}/99!)x^{99}e^{-8x}dx$

   b. $1 - \sum_{k=0}^{99} \frac{80^k e^{-80}}{k!}$ c. .0228 e. .0146
7. a. $\exp(n\lambda)$
   b. Density is $n\lambda e^{-\lambda v}(1 - e^{-\lambda v})^{n-1}$ for $v > 0$.
8. Erl($n = 24$, $\lambda = 1.5$)
9. $\exp\left(\frac{5}{3}\right)$; expected value is $\frac{3}{5}$ hour; standard deviation is $\frac{3}{5}$
10. The standard deviation equals the expected value.
11. a. $e^{-\lambda a}$ b. $e^{-4\lambda a}$ c. $\exp(4\lambda)$
12. a. $21/(3 - s)(7 - s)$ b. $\frac{7}{3}$
13. a. Expected value, $3/\lambda$; variance, $3/\lambda^2$
    b. Expected value, $16/6\lambda$; variance, $16/36\lambda^2$ c. Old: expected value, k/λ; variance, k/λ^2. New: expected value, $(6(k - 1) + j)/6\lambda$; variance, $(6(k - 1) + j)/36\lambda^2$.

8.4 GAMMA DISTRIBUTIONS

Remember that the Erlang density with parameters α and λ is

$$f(x) = \frac{\lambda^\alpha}{(\alpha - 1)!} x^{\alpha - 1} e^{-\lambda x} \quad \text{for } x > 0.$$

In the previous section we studied this distribution in the case where α is a positive integer, but there is also a way to make sense of it for any positive α.

In this section we will introduce the gamma function $\Gamma(\alpha)$, defined for all positive α, and show that it equals $(\alpha - 1)!$ when α is a positive integer. We will then be able to define the gamma density with parameters α and λ,

$$f(x) = \frac{\lambda^\alpha}{\Gamma(\alpha)} x^{\alpha - 1} e^{-\lambda x} \quad \text{for } x > 0,$$

once we have shown that it integrates to 1.

The gamma densities with noninteger α do not arise directly in connection with Poisson processes; there is no such thing as the time of the αth arrival if α is not an integer. But they are not just a curiosity. In particular, when $\lambda = \frac{1}{2}$ and α is a half-integer $\left(\frac{1}{2}, 1, \frac{3}{2}, 2, \frac{5}{2}, \ldots\right)$, we get one of the important chi-squared distributions.

In this section we begin by defining the gamma function and showing its important properties. Then we will define the gamma densities and derive the usual facts about these distributions: moments, sums of independent gamma random variables, and so on. Finally, we will discuss the chi-squared distributions.

(8.4.1) The gamma function. First consider the following fact; it is a special case ($\lambda = 1$) of the gamma formula (8.3.4) that we used in the previous section.

(8.4.2)
$$\int_0^\infty x^{n-1} e^{-x} dx = (n - 1)! \quad \text{for any positive integer } n.$$

This formula shows that the factorials of the integers can be expressed as improper integrals of certain functions of x from 0 to ∞. But it happens that this improper integral converges when the integer n is replaced by any positive number α, integer or not. (It diverges for $\alpha \leq 0$.) Its value for α is denoted by $\Gamma(\alpha)$, and this is the definition of the ***gamma function:***

(8.4.3)
$$\Gamma(\alpha) = \int_0^\infty x^{\alpha - 1} e^{-x} dx \quad \text{for any positive number } \alpha.$$

We take (8.4.3) as our starting point. The only thing we have not justified is the fact that this improper integral converges for all $\alpha > 0$. It is easy to justify for $\alpha \geq 1$, but needs a little work, at the level of advanced calculus, for $0 < \alpha < 1$.

(It is improper at both ends in that case.) But if you accept (8.4.3), you can justify all of the following steps using the hints given.

Putting 1 for α in (8.4.3) gives immediately

(8.4.4) $$\Gamma(1) = 1.$$

If we put $\alpha + 1$ in place of α in (8.4.3) and then integrate by parts, we get

(8.4.5) $$\Gamma(\alpha + 1) = \alpha\Gamma(\alpha).$$

If we use (8.4.4) as the first step and (8.4.5) as the induction step, we can prove that

(8.4.6) $$\Gamma(\alpha) = (\alpha - 1)! \quad \text{when } \alpha \text{ is a positive integer,}$$

which is the same thing as (8.4.2).

We now have a function $\Gamma(\alpha)$, defined for all positive numbers α, which gives factorials when integers are plugged in. Thus we can regard the gamma function as an extension of the factorial function. Specifically, (8.4.6) says the same thing as

(8.4.7) $$r! = \Gamma(r + 1) = \int_0^\infty x^r e^{-x} dx,$$

when r is a nonnegative integer; but (8.4.7) can be used to define $r!$ for all numbers $r > -1$. Indeed, some calculators with factorial buttons will give a value for $r!$ even when r is not an integer; what they are giving is (8.4.7). To see whether yours does, try to get $(-.5)!$, which is $\Gamma(.5)$; it should equal the square root of π, which is about 1.77245, as we will see at (8.4.8). If that doesn't work, try $(.5)!$, which is $\Gamma(1.5) = \frac{1}{2}\sqrt{\pi} \doteq .886227$.

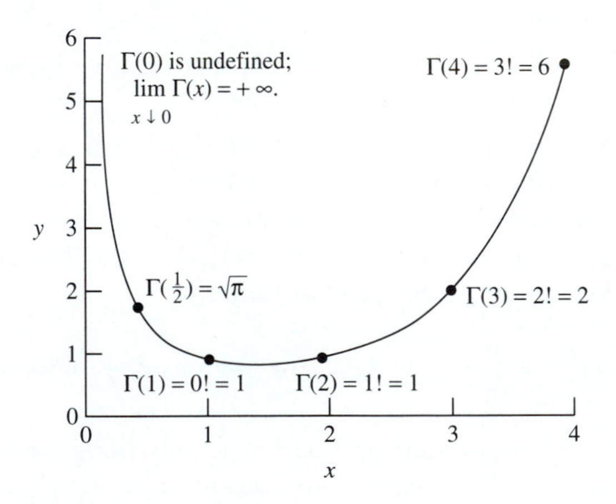

FIGURE 8.4a Gamma function

The gamma function is one of those functions [like $\Phi(z)$, the normal CDF] whose values we do not know except by approximation. That is, there is no closed formula for $\Gamma(\alpha)$ except for certain values of α; we must resort to tables that have been constructed, or to an approximation formula (which is what a calculator uses). A graph of the gamma function is shown in Figure 8.4a. It shows that the gamma function gives factorials when 1, 2, 3, . . . are plugged in; but its behavior is a bit unexpected between $\alpha = 0$ and $\alpha = 2$.

There happen to be closed formulas for certain important values of α, however. These include the positive integers, of course, but also the *half-integers*—numbers $n/2$ for odd integers n. First of all,

(8.4.8)
$$\Gamma\left(\tfrac{1}{2}\right) = \sqrt{\pi}.$$

This strange result can be proved in a rather enjoyable way. The definition (8.4.3) says that

$$\Gamma\left(\tfrac{1}{2}\right) = \int_0^\infty x^{-1/2}e^{-x}dx;$$

make the substitution $x = z^2/2$ (that is, $z = \sqrt{2x}$), and adjust the constants to get

$$\Gamma\left(\tfrac{1}{2}\right) = 2\sqrt{\pi}\int_0^\infty \frac{1}{\sqrt{2\pi}}e^{-z^2/2}dz.$$

But the integral equals $\frac{1}{2}$ (be sure you see why!), which cancels the 2 and leaves $\sqrt{\pi}$.

Now we can use (8.4.8) along with (8.4.5) to get $\Gamma\left(\frac{n}{2}\right)$ for odd integers n:

$$\Gamma\left(\tfrac{3}{2}\right) = \tfrac{1}{2}\Gamma\left(\tfrac{1}{2}\right) = \tfrac{1}{2}\sqrt{\pi};$$
$$\Gamma\left(\tfrac{5}{2}\right) = \tfrac{3}{2}\Gamma\left(\tfrac{3}{2}\right) = \tfrac{3}{2}\cdot\tfrac{1}{2}\sqrt{\pi};$$
$$\Gamma\left(\tfrac{7}{2}\right) = \tfrac{5}{2}\Gamma\left(\tfrac{5}{2}\right) = \tfrac{5}{2}\cdot\tfrac{3}{2}\cdot\tfrac{1}{2}\sqrt{\pi};$$

and so forth. In general, odd integers are of the form $2k + 1$ for $k = 0$, 1, 2, . . . , and

(8.4.9)
$$\Gamma\left(\frac{2k+1}{2}\right) = \frac{1\cdot 3\cdot 5\cdot\cdot\cdot(2k-1)}{2^k}\sqrt{\pi} \quad \text{for } k = 1, 2, 3, \ldots.$$

Finally, check the following, by making the substitution $y = \lambda x$ on the left and using (8.4.3):

(8.4.10)
$$\int_0^\infty x^{\alpha-1}e^{-\lambda x}dx = \frac{\Gamma(\alpha)}{\lambda^\alpha} \quad \text{for any positive numbers } \alpha \text{ and } \lambda.$$

This is the **gamma formula** in its full generality.

There are many formulas in the preceding development, from (8.4.2) to (8.4.10), and even if you wanted to have them all available without reference (something that is not essential), you would not have to memorize them all. You can reconstruct all the important facts if you remember three basic formulas:

the gamma formula (8.4.10) (from which you can get the definition of $\Gamma(\alpha)$ by setting $\lambda = 1$);

(8.4.6), the connection between the gamma function and factorials;

and the recursion (8.4.5).

These suffice for most calculations; the only other thing that might be needed is the fact that $\Gamma\left(\frac{1}{2}\right) = \sqrt{\pi}$.

(8.4.11) Gamma distributions. Because of the gamma formula (8.4.10), the following function integrates to 1; since it is obviously positive for $x > 0$, it is a density for any positive values of α and λ. It is called the ***gamma density with parameters α and λ***:

$$f(x) = \frac{\lambda^\alpha}{\Gamma(\alpha)} x^{\alpha-1} e^{-\lambda x} \quad \text{for } x > 0 \quad \text{(and 0 otherwise).}$$

The abbreviation for this distribution is gamma(α, λ). Notice that when α is an integer, say $\alpha = n$, this is the Erl(n, λ) density; and of course, when $\alpha = 1$, it is the exp(λ) density.

There is no convenient formula for the CDF; when α is an integer, the CDF is best expressed as a sum of Poisson probabilities, and when α is not an integer, it is even more complicated.

We have not yet seen graphs of any of the gamma densities. Figure 8.4b shows five of them. In that figure, the density with most of its mass near 0 is the gamma$\left(\frac{1}{2}, \frac{1}{2}\right)$ density. All the gamma densities with $\alpha < 1$ look roughly like this one, tending to ∞ as $x \to 0$.

The two densities with their peaks near (but not at) $x = 5$ each have $EX = \alpha/\lambda = 5$; the wider one has $\alpha = 5$ and $\lambda = 1$, so that its variance is $\alpha/\lambda^2 = 5$, and the taller one has $\alpha = 10$ and $\lambda = 2$, so that its variance is 2.5. [The formulas for EX and VX are derived in (8.4.12).] You can see that for any desired expected value and variance (as long as they are positive) you can find α and λ so that α/λ is the expected value and α/λ^2 is the variance. This makes the gamma family a flexible family that can be used to model many real populations.

The "rightmost" density in Figure 8.4b has $\alpha = 27$ and $\lambda = 3$, giving it an expected value of 9 and a variance of 3. The "lowest" one is the gamma$\left(\frac{8}{2}, \frac{1}{2}\right)$ density.

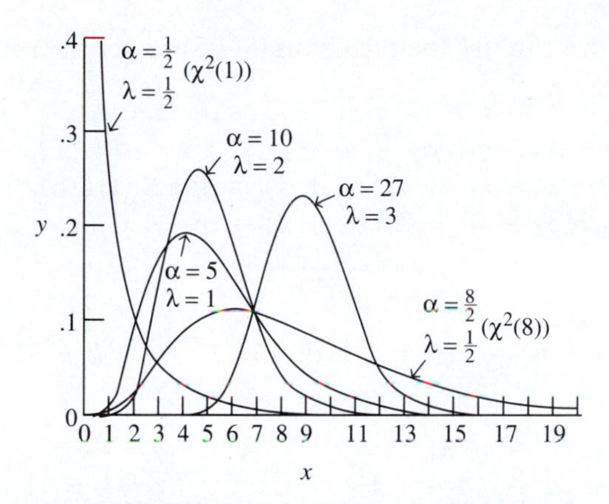

FIGURE 8.4b Some gamma densities

(8.4.12) **Moments and MGF of a gamma distribution.** To illustrate the usefulness of the gamma formula, we will indicate the calculations that produce the expected value and moment generating function of this distribution. You will notice that all the results are exactly the same as for Erlang distributions. There are no new formulas to remember, except that the gamma function replaces factorials.

For the expected value we get

$$EX = \int_0^\infty x \cdot \frac{\lambda^\alpha}{\Gamma(\alpha)} x^{\alpha-1} e^{-\lambda x} dx = \frac{\lambda^\alpha}{\Gamma(\alpha)} \int_0^\infty x^\alpha e^{-\lambda x} dx,$$

and by the gamma formula we get

$$EX = \frac{\lambda^\alpha}{\Gamma(\alpha)} \cdot \frac{\Gamma(\alpha+1)}{\lambda^{\alpha+1}} = \frac{\lambda^\alpha}{\Gamma(\alpha)} \cdot \frac{\alpha\Gamma(\alpha)}{\lambda^{\alpha+1}} = \frac{\alpha}{\lambda}.$$

The MGF is just as simple:

$$m(s) = \int_0^\infty e^{sx} \frac{\lambda^\alpha}{\Gamma(\alpha)} x^{\alpha-1} e^{-\lambda x} dx = \frac{\lambda^\alpha}{\Gamma(\alpha)} \int_0^\infty x^{\alpha-1} e^{-(\lambda-s)x} dx,$$

and as long as $\lambda - s$ is positive we can use the gamma formula on this to get

$$m(s) = \frac{\lambda^\alpha}{\Gamma(\alpha)} \cdot \frac{\Gamma(\alpha)}{(\lambda-s)^\alpha} = \left(\frac{\lambda}{\lambda-s}\right)^\alpha.$$

We can get all the moments EX^k from this, or we can get them directly from the definition, again using the gamma formula:

$$EX^k = \int_0^\infty x^k \cdot \frac{\lambda^\alpha}{\Gamma(\alpha)} x^{\alpha-1} e^{-\lambda x} dx = \frac{\lambda^\alpha}{\Gamma(\alpha)} \int_0^\infty x^{\alpha+k-1} e^{-\lambda x} dx = \frac{\lambda^\alpha}{\Gamma(\alpha)} \cdot \frac{\Gamma(\alpha+k)}{\lambda^{\alpha+k}}.$$

To simplify this, we can use the recursion (8.4.5) to get the ratio of $\Gamma(\alpha + k)$ and $\Gamma(\alpha)$:

$$\begin{aligned}\Gamma(\alpha + 1) &= \alpha\Gamma(\alpha),\\ \Gamma(\alpha + 2) &= (\alpha + 1)\Gamma(\alpha + 1) = \alpha(\alpha + 1)\Gamma(\alpha),\\ \Gamma(\alpha + 3) &= (\alpha + 2)\Gamma(\alpha + 2) = \alpha(\alpha + 1)(\alpha + 2)\Gamma(\alpha),\end{aligned}$$

and in general

$$\Gamma(\alpha + k) = \alpha(\alpha + 1)(\alpha + 2) \cdots (\alpha + k - 1)\Gamma(\alpha).$$

Thus we get

$$EX^k = \frac{\alpha(\alpha + 1)(\alpha + 2) \cdots (\alpha + k - 1)}{\lambda^k}.$$

In particular, if $X \sim$ gamma(α, λ), then

(8.4.13) $$EX = \frac{\alpha}{\lambda}, \quad EX^2 = \frac{\alpha(\alpha + 1)}{\lambda^2}, \quad \text{and therefore,} \quad VX = \frac{\alpha}{\lambda^2}.$$

(8.4.14) **Sums of independent gamma random variables.** It is easy to use MGFs to show that the sum of independent gamma-distributed random variables ***with the same λ*** also has a gamma distribution. Suppose $X \sim$ gamma(α, λ) and $Y \sim$ gamma(β, λ) are independent and $Z = X + Y$. Then the MGF of Z is the product of the MGFs of X and Y:

$$m_Z(s) = \left(\frac{\lambda}{\lambda - s}\right)^{\alpha}\left(\frac{\lambda}{\lambda - s}\right)^{\beta} = \left(\frac{\lambda}{\lambda - s}\right)^{\alpha+\beta},$$

which is the MGF of the gamma$(\alpha + \beta, \lambda)$ distribution.

Of course, this generalizes to sums of more than two independent gamma-distributed random variables, as long as they all have the same λ: if $X_1, X_2, \ldots, X_k$ are independent and $X_j \sim$ gamma(α_j, λ) for each j, and $Y = X_1 + X_2 + \cdots + X_k$, then $Y \sim$ gamma$(\alpha_1 + \alpha_2 + \cdots + \alpha_k, \lambda)$.

(8.4.15) **Distribution of a constant multiple of a gamma random variable.** The MGF shows quickly the distribution of aX if $X \sim$ gamma(α, λ):

$$m_Y(s) = m_X(as) = \left(\frac{\lambda}{\lambda - as}\right)^{\alpha} = \left(\frac{\lambda/a}{\lambda/a - s}\right)^{\alpha},$$

so Y has the gamma$(\alpha, \lambda/a)$ distribution.

(8.4.16) **Special case: Chi-squared distributions.** In Exercise 5 of Section 5.2, we saw that if $Z \sim N(0, 1)$ and $X = Z^2$, then a density for X is

$$f(x) = \frac{1}{\sqrt{2\pi x}} e^{-x/2} \quad \text{for } x > 0.$$

A little rewriting shows that this is a disguised gamma density, and fits into the form of (8.4.11):

$$f(x) = \frac{\left(\frac{1}{2}\right)^{1/2}}{\Gamma\left(\frac{1}{2}\right)} x^{1/2-1} e^{-x/2} \quad \text{for } x > 0;$$

that is, the square of a $N(0, 1)$ random variable has the gamma$\left(\frac{1}{2}, \frac{1}{2}\right)$ distribution. The MGF of this distribution is

$$m(s) = \left(\frac{\frac{1}{2}}{\frac{1}{2} - s}\right)^{1/2} = \frac{1}{\sqrt{1 - 2s}} \quad \text{for } s < \tfrac{1}{2}.$$

By (8.4.14), we can see that if $Z_1, Z_2, \ldots, Z_n$ are independent $N(0, 1)$ random variables, then their squares are independent gamma$\left(\frac{1}{2}, \frac{1}{2}\right)$ random variables, and, consequently, $Z_1^2 + Z_2^2 + \cdots + Z_n^2$ has the gamma$\left(\frac{n}{2}, \frac{1}{2}\right)$ distribution. This distribution is called the ***chi-squared distribution with parameter n.*** In statistical contexts, for reasons that we will not go into here, it is called the *chi-squared distribution with n degrees of freedom.* The abbreviation for this distribution is $\chi^2(n)$.

We can summarize as follows.

(8.4.17) If $Z_1, Z_2, \ldots, Z_n$ are independent $N(0, 1)$ random variables, then $X = Z_1^2 + Z_2^2 + \cdots + Z_n^2$ has the $\chi^2(n)$ distribution, which is the same as the gamma$\left(\frac{n}{2}, \frac{1}{2}\right)$ distribution.

From (8.4.13) we get immediately:

(8.4.18) If $X \sim \chi^2(n)$, then $EX = n$ and $VX = 2n$.

We see also from (8.4.12) that the MGF is

$$m(s) = \left(\frac{\frac{1}{2}}{\frac{1}{2} - s}\right)^{n/2} = \left(\frac{1}{1 - 2s}\right)^{n/2} \quad \text{for } s < \tfrac{1}{2}.$$

Of course, for $n = 1$, the square of a single $N(0, 1)$ random variable has the $\chi^2(1)$ distribution.

The chi-squared distributions are encountered often in statistical inference, where an important class of hypothesis-testing procedures is based on sums of squares of independent normal random variables. Partial tables of the CDFs of these distributions are used in conducting the chi-squared tests.

It is not important to know the densities of the $\chi^2(n)$ distributions, although you can easily write them down using (8.4.11) with $\alpha = \frac{n}{2}$ and $\lambda = \frac{1}{2}$. If you remember the basic facts about gamma distributions, and remember that $\chi^2(n)$ means gamma$\left(\frac{n}{2}, \frac{1}{2}\right)$, you can reconstruct what is needed about chi-squared distributions.

(8.4.19) **_F_ distributions.** In certain statistical testing problems it is important to have an idea of whether the variances of two distributions can safely be regarded as equal or not. The standard test for this purpose is based on the F statistic, which is a constant times the ratio of two independent chi-squared random variables.

Specifically, let $X \sim \chi^2(n)$ and $Y \sim \chi^2(m)$ be independent, and let $F = (X/n)/(Y/m)$. The set of possible values of F is $(0, \infty)$; we can use the integral formula that we found in Exercise 6 of Section 5.4 to find a density for F.

The densities of X and Y are

$$f_X(x) = \frac{1}{2^{n/2}\Gamma(n/2)} x^{(n/2)-1} e^{-x/2} \quad \text{for } x > 0;$$

$$f_Y(y) = \frac{1}{2^{m/2}\Gamma(m/2)} y^{(m/2)-1} e^{-y/2} \quad \text{for } y > 0.$$

So a density for $Z = X/Y$ is (according to Exercise 6 in Section 5.4)

$$f_Z(z) = \int_0^\infty x f_X(zx) f_Y(x)\,dx,$$

and you can check (using the gamma formula) that this equals

$$f_Z(z) = \frac{\Gamma((n+m)/2)}{\Gamma(n/2)\Gamma(m/2)} \cdot \frac{z^{(n/2)-1}}{(1+z)^{(m+n)/2}} \quad \text{for } z > 0.$$

The last step is to notice that $F = (m/n) \cdot Z$, and therefore a density for F is $(n/m) \cdot f_Z((n/m)z)$.

We will not bother to write down this density; there is little need to know it in practice. The F test is carried out using tables of the CDF for this density; those tables are constructed by using numerical integration techniques.

The distribution whose density we did not quite write down above is called the *F distribution with parameters n and m.* The parameters n and m are called the "numerator degrees of freedom" and the "denominator degrees of freedom." The density was apparently first found by the American statistician George

Snedecor, who used the letter F in honor of R. A. Fisher, the British biologist of the early 20th century who was a pioneer in statistical inference.

(8.4.20) **Student's t distributions.** Another important quantity that arises in statistical inference is $T = Z/\sqrt{X/n}$, where Z and X are independent, $Z \sim N(0, 1)$, and $X \sim \chi^2(n)$. The distribution of T is called *Student's t distribution with parameter n*; the number n is called the number of "degrees of freedom."

To find a density for T, we first find a density for $Y = \sqrt{X/n}$, and then use the result of Exercise 6 of Section 5.4 to find a density for $T = Z/Y$. The result, which you can check, is

$$f_T(t) = \frac{\Gamma((n+1)/2)}{\sqrt{n}\Gamma(1/2)\Gamma(n/2)} \cdot \frac{1}{(1 + t^2/n)^{(n+1)/2}} \quad \text{for } t > 0.$$

"Student" was the pseudonym used by William S. Gosset (1876–1937), who worked for the Guinness brewery in Dublin. Recognizing the need for statistical expertise in the brewing industry, he went to London in 1906, where he worked under Karl Pearson, who was one of the leaders (along with R. A. Fisher) in the developing area of statistical inference. A common, but false, legend about Gosset is that he published his work pseudonymously because his employers did not approve of his statistical work. In fact, Guinness sent him to London.

The famous Student's t test arises in connection with inference about the unknown expected value μ of a normal distribution, which is estimated by $\bar{X}_n$, the average of a sample of size n from that distribution. The formulas that arise necessarily involve the standard deviation σ of the same distribution, and of course σ is also unknown in most cases. The usual procedure, now often called "Studentizing," is to use the same sample to estimate σ, and then use that estimate in the formulas. When this is done, the formulas then involve the ratio whose density we found above.

EXERCISES FOR **SECTION 8.4**

1. What density on $(0, \infty)$ is proportional to $\sqrt{x}e^{-2x}$? Write down the density and give its name.

2. Let X and Y be independent, each having the chi-squared distribution with parameter 1. Find the MGF of the average of X and Y. What is the distribution of the average?

3. If the set of possible values of X is $(0, \infty)$ and the density is some constant times $x^{3.7}e^{-2.3x}$, give the name of the distribution, the value of the constant, EX, and VX.

4. Let $f(x)$ be the gamma(α, λ) density. Suppose $\alpha > 1$. For which value of x is this density largest? This number is called the *mode* of the distribution. Is the mode larger or smaller than the expected value of the distribution? [*Hint:* You can ignore the constant in the density, and the x that maximizes the logarithm of the nonconstant part also maximizes the density.]

5. Repeat Exercise 4 if $\alpha = 1$, and also if $\alpha < 1$.

6. Do the calculations leading to the density $f_Z(z)$ given in (8.4.19) and the t density given in (8.4.20).

7. **Finding the gamma distribution with desired expected value and variance.** If X has a gamma distribution, $EX = 21.3$, and $VX = 12.1$, which gamma distribution is it? More generally, find the gamma distribution whose expected value and variance are given numbers μ and σ^2 (assuming that μ, as well as σ^2, is positive).

   Exercise 8 in the previous section was similar. Note that there is not always an Erlang distribution with given expected value and variance, because the parameter $\alpha = n$ of an Erlang distribution must be an integer. But in the general gamma family there is no such restriction; we can tailor the parameters to get any (positive) expected value and variance we want.

8. **a.** The $\chi^2(2)$ distribution is also known by another name. What is it?
   b. Let the coordinates X and Y of a random point in the plane be independent $N(0, 1)$ random variables, and let W be the squared distance of the random point from the origin. Without doing any calculations, name the distribution of W.

9. **The logarithm of an exponentially distributed random variable.** Let X have the $\exp(\lambda)$ distribution and let $Y = \ln X$.
   a. Using the definition of MGF and the gamma formula, show that the MGF of Y is $\Gamma(s + 1)/\lambda^s$.
   b. Find a density for Y using the CDF method (see Section 5.2).

10. **Alternative parametrization for gamma distributions.** It is often convenient to use $\beta = 1/\lambda$ instead of λ as the parameter when working with exponential, Erlang, and gamma distributions. The densities are less convenient to write, but expected values and MGFs are simpler in terms of β. Let $X \sim \exp(\lambda)$, $Y \sim \text{Erl}(n, \lambda)$, and $Z \sim \text{gamma}(\alpha, \lambda)$. Write down the densities, MGFs, expected values, and variances of X, Y, and Z in terms of $\beta = 1/\lambda$.

11. **The Gauss–Legendre duplication formula.** Use (8.4.9) to show that if $n = 2k + 1$ is a positive odd integer, then

$$\Gamma\left(\tfrac{n}{2}\right) = \frac{(2k)!}{k!2^{2k}}\sqrt{\pi}.$$

   Then use (8.4.5) and (8.4.6) to derive the Gauss–Legendre duplication formula:

$$\Gamma(2k) = \Gamma(k)\Gamma\left(k + \tfrac{1}{2}\right)\frac{2^{2k-1}}{\sqrt{\pi}}.$$

   This exercise is here simply for practice in algebraic manipulation involving the gamma function. The Gauss–Legendre formula turns out to be valid whether k is an integer or not; it is useful in constructing tables of the gamma function, or in recursive algorithms for computing it, as well as in theory.

12. In Supplementary Exercise 9 for Chapter 5 we found a density for $Y = \sqrt{X}$ when X has the Erl(2, λ) distribution. Now we want EY. Write down two integrals for this number, one using the density of X and one using the density of Y. One of them can be evaluated using the gamma formula. [The best way to evaluate the other one is the make the substitution that turns it into the first one.]

ANSWERS

1. Gamma$\left(\frac{3}{2}, 2\right)$
2. Exponential with $\lambda = 1$
3. Constant is $(2.3)^{4.7}/\Gamma(4.7)$; $EX = 2.0435$; $VX = .8885$
4. $(\alpha - 1)/\lambda$ (smaller than the expected value)
5. If $\alpha = 1$ the mode is 0; if $\alpha < 1$ there is no mode.
7. Gamma$(\mu^2/\sigma^2, \mu/\sigma^2)$
8. **a.** $\exp\left(\frac{1}{2}\right)$ **b.** $\exp\left(\frac{1}{2}\right)$
9. **b.** $f_Y(y) = \lambda e^{(y - \lambda e^y)}$ for $y \in \mathbb{R}$
12. $\Gamma\left(\frac{5}{2}\right)/\lambda^{1/2}\left[= \frac{3}{4}\sqrt{\pi/\lambda}\right]$

8.5 UNIFORM DISTRIBUTIONS

In some sense uniform distributions are the simplest of all absolutely continuous distributions, and yet there is still a good deal to be said about them. For one thing, most computer random number generators give numbers that behave as though they were independent observations from the uniform distribution on (0, 1). If we want to generate any other kind of random variables, we have to start with these.

We have already seen most of these ideas, and in this section we will just collect and summarize what we know.

(8.5.1) Definition. For any interval (a, b) of real numbers, the ***uniform distribution on*** $(\boldsymbol{a}, \boldsymbol{b})$ [the abbreviation is unif(a, b)] is the one whose density is

$$f(x) = \frac{1}{b - a} \quad \text{for } a < x < b \quad \text{(and 0 otherwise).}$$

The corresponding CDF is

$$F(x) = \begin{cases} 0 & \text{if } x < a, \\ \dfrac{x - a}{b - a} & \text{if } a \le x \le b, \\ 1 & \text{if } x > b. \end{cases}$$

The case in which $a = 0$ and $b = 1$ is especially common; the unif(0, 1) density is

$$f(x) = 1 \quad \text{if } 0 < x < 1 \quad \text{(and 0 otherwise),}$$

and the CDF is

$$F(x) = \begin{cases} 0 & \text{if } x < 0, \\ x & \text{if } 0 \le x \le 1, \\ 1 & \text{if } x > 1. \end{cases}$$

The unif(0, 1) distribution is the one simulated by the "random number generators" of computer programming languages. This distribution can be regarded in some sense as fundamental to much of probability theory, in that any random variables, with any distributions whatever, can (at least in principle) be regarded as functions of random variables that have the unif(0, 1) distribution. It is also important in statistical theory, because [as we will see in (8.5.6)], any random variable with a continuous distribution can be transformed into a random variable with the unif(0, 1) distribution.

(8.5.2) **Moments and MGF of a uniform distribution.** The moments of X, if $X \sim \text{unif}(a, b)$, can be found quickly just by integrating:

$$EX^k = \int_a^b x^k \cdot \frac{1}{b-a}\,dx = \frac{1}{k+1} \cdot \frac{b^{k+1} - a^{k+1}}{b-a},$$

which can also be written

$$EX^k = \frac{b^k + b^{k-1}a + b^{k-2}a^2 + \cdots + a^k}{k+1}.$$

In particular,

$$EX = \frac{b+a}{2} \quad \text{and} \quad EX^2 = \frac{b^2 + ba + a^2}{3},$$

and from this we get

$$VX = \frac{(b-a)^2}{12}.$$

We found the MGF for the unif(0, 1) distribution in Example (8.1.5). For the general unif(a, b) distribution, the derivation is similar and the result is

$$m(s) = \int_a^b e^{sx} \cdot \frac{1}{b-a}\,dx = \begin{cases} \dfrac{e^{bs} - e^{as}}{s(b-a)} & \text{if } s \ne 0, \\ 1 & \text{if } s = 0. \end{cases}$$

You might want to find the first two moments using the MGF, just to see how much easier it is to get them directly for this distribution. (See Exercise 2 at the end of this section.)

(8.5.3) **Connections between uniform distributions.** You can check, using the CDF method described in Section 5.2, that

If $X \sim \text{unif}(a, b)$ and $Y = cX + d$, then $Y \sim \text{unif}(ca + d, cb + d)$.

Two particular cases of this are noteworthy:

(8.5.4)

$$\text{If } X \sim \text{unif}(a, b) \text{ and } U = \frac{X - a}{b - a}, \text{ then } U \sim \text{unif}(0, 1).$$

$$\text{If } U \sim \text{unif}(0, 1) \text{ and } X = a + (b - a)U, \text{ then } X \sim \text{unif}(a, b).$$

You can also use (8.5.4) to reconstruct the expected value and variance of the unif(a, b) distribution if you forget them, if you remember that the unif(0, 1) distribution has expected value $\frac{1}{2}$ and variance $\frac{1}{12}$.

(8.5.5) Sums of i.i.d. uniformly distributed random variables. Such sums have complicated distributions, and we do not often deal with them explicitly. For example, if X_1 and X_2 are independent, each having the unif(0, 1) distribution, and $Y = X_1 + X_2$, then the CDF of Y is

$$F(y) = \begin{cases} 0 & \text{if } y < 0, \\ \frac{1}{2}y^2 & \text{if } 0 \le y < 1, \\ 2y - \frac{1}{2}y^2 - 1 & \text{if } 1 \le y < 2, \\ 1 & \text{if } y \ge 2. \end{cases}$$

This was done in parts b and c of Supplementary Exercise 4 in Chapter 1, and again in parts d and e of Exercise 8 in Section 3.5. Notice that the cases $0 \le y < 1$ and $1 \le y < 2$ have to be treated separately. Differentiating the CDF gives us a density for Y:

$$f(y) = \begin{cases} y & \text{if } 0 < y < 1, \\ 2 - y & \text{if } 1 \le y < 2, \\ 0 & \text{otherwise.} \end{cases}$$

This is a "tent-shaped" function. Things get more complicated as we add more X_j's; the sum of three independent unif(0, 1) random variables, for example, has (0, 3) as its set of possible values, and there are three cases to consider in finding the CDF. The density looks like a tent with a flat top.

If you worked out the densities for the sums of three, four, and five independent unif(0, 1) random variables, you would see increasingly complicated functions, but you would notice that they start to become bell shaped. Of course, it is the Central Limit Theorem that we would use if we had to find a probability for a sum of more than a few such random variables.

(8.5.6) The CDF of X transforms X to a unif(0, 1) random variable. Let X have any distribution at all, and let $F(x)$ be its CDF. Define $Y = F(X)$. This seems like a strange thing to do—to plug a random variable into its own CDF—but if $F(x)$ is a continuous function, the result is interesting:

If the CDF $F(x)$ of X is continuous, and $Y = F(X)$, then $Y \sim \text{unif}(0, 1)$.

We can prove this easily if we make an additional assumption on $F(x)$—that it has an inverse on the set of possible values of X. (Not every continuous CDF has this property, but one that is strictly increasing on the set of possible values does.) The result is still true if F is not invertible, but the proof is a little more complicated, and we will omit it here.

Proof (for continuous invertible F). The set of possible values of Y is the range of F, which is the interval (0, 1). We will show that $F_Y(y) = y$ for each y between 0 and 1; this is the CDF of the unif(0, 1) distribution, so the proof will be complete.

Let $x = F^{-1}(y)$ denote the inverse of $y = F(x)$. For any $y \in (0, 1)$,

$$F_Y(y) = P(Y \leq y) = P(F(X) \leq y) = P(X \leq F^{-1}(y)) = F(F^{-1}(y)) = y. \qquad \square$$

The full truth of the matter, which we are not proving here, is that, if F is the CDF of X and $Y = F(X)$, then $Y \sim \text{unif}(0, 1)$ if and only if F is continuous.

(8.5.7) **Example.** We saw instances of (8.5.6) in Exercises 9, 10, and 11 of Section 5.2. In Exercise 9, for example, we saw that if X has the CDF

$$F(x) = \begin{cases} 0 & \text{if } x \leq 0, \\ x^2/4 & \text{if } 0 < x < 2, \\ 1 & \text{if } x \geq 2, \end{cases}$$

and $Y = X^2/4$, then $Y \sim \text{unif}(0, 1)$. If you find the proof of (8.5.6) confusing, do that exercise again using the CDF method; the proof of (8.5.6) is just a terse generalization of how it is done. ▮

You might be wondering how (8.5.6) could be of any use. As an example of how it is used in statistical inference, suppose we wanted to decide whether a set of numbers $x_1, x_2, \ldots, x_n$ could be regarded as independent observations of a random variable X whose CDF is $F(x)$. Well, if they are, then the numbers $F(x_1)$, $F(x_2), \ldots, F(x_n)$ can be regarded as independent observations of a unif(0, 1) random variable. And there are many ways to test this hypothesis statistically—the kinds of tests used to evaluate random number generators. If we can reject the hypothesis that the $F(x_j)$'s come from the unif(0, 1) distribution, then we can reject the hypothesis that the x_j's come from the distribution whose CDF is $F(x)$.

(8.5.8) **The function $F^{-1}(x)$ transforms a unif(0, 1) random variable to one whose CDF is F.** This is the reverse of (8.5.6). Just as F transforms X to a uniform random variable, so F^{-1} transforms a uniform random variable to one whose CDF is F. More precisely:

If $F(y)$ is a CDF that is invertible on its set of possible values, and if $Y = F^{-1}(X)$ where $X \sim \text{unif}(0, 1)$, then $F(y)$ is the CDF of Y.

Proof. The set of possible values of Y is the range of $F^{-1}(x)$ for $0 < x < 1$, which is the set of possible values of the distribution whose CDF is F. (Look at the graph of a CDF to see this.) For one of these numbers, say y,

$$F_Y(y) = P(Y \le y) = P(F^{-1}(X) \le y) = P(X \le F(y)) = F(y).$$

That is, F is the CDF of Y. □

It is pretty clear how this could be used. Suppose we want to generate observations of a random variable with a given distribution. A computer generates only unif(0, 1) observations. If F is the CDF of the distribution we want, and we are able to find F^{-1}, then we just apply F^{-1} to the uniform observations, and we get observations from the desired distribution. (In some situations, as for example with normal distributions, we cannot find a formula for F^{-1} and other methods have to be used.)

(8.5.9) Example: Generating exponentially distributed random variables. Suppose we want to transform X, a unif(0, 1) random variable, to Y, an $\exp(\lambda)$ random variable. According to the above, we just write down the CDF of the $\exp(\lambda)$ distribution, find its inverse, and apply that to X.

The $\exp(\lambda)$ CDF is familiar:

$$F(x) = \begin{cases} 0 & \text{if } x \le 0, \\ 1 - e^{-\lambda x} & \text{if } x > 0. \end{cases}$$

We invert this on the set of possible values—that is, on $(0, \infty)$—simply by solving $x = 1 - e^{-\lambda y}$ for y:

$$\text{If } x = 1 - e^{-\lambda y}, \text{ then } e^{-\lambda y} = 1 - x, \text{ so } y = -\frac{1}{\lambda}\ln(1 - x).$$

So the inverse of the $\exp(\lambda)$ CDF is $F^{-1}(x) = -(1/\lambda)\ln(1 - x)$. Consequently, if $X \sim \text{unif}(0, 1)$, then

$$Y = F^{-1}(X) = -\frac{1}{\lambda}\ln(1 - X)$$

has the $\exp(\lambda)$ distribution.

Now if you look back at Example (5.2.12), you will see that there we found another transformation of a unif(0, 1) random variable that also produces an $\exp(\lambda)$ random variable: $Y = -(1/\lambda)\ln X$. You can generate X and then take $-1/\lambda$ times either X or $1 - X$; either will produce an $\exp(\lambda)$ random variable.

Of course, $Y = -(1/\lambda)\ln X$ is simpler, and it is the one to use. The two methods are really not different, because if X has the unif(0, 1) distribution, then so does $1 - X$. ∎

We remark in closing that, while (8.5.8) provides one way to generate random variables with a given distribution, for some distributions there are other ways that are more efficient. For more on the subject, see Sheldon Ross's book *A Course in Simulation*.

EXERCISES FOR **SECTION 8.5**

1. In (8.5.2) we saw that if $X \sim \text{unif}(a, b)$, then $EX = (b + a)/2$ and $VX = (b - a)^2/12$. Confirm this again as suggested after (8.5.4). [That is, use the fact that if $X \sim \text{unif}(a, b)$, then $X = a + (b - a)Z$ for a random variable $Z \sim \text{unif}(0, 1)$.]

2. The MGF of the unif(0, 1) distribution is

$$m(s) = \begin{cases} \dfrac{e^s - 1}{s} & \text{if } s \neq 0, \\ 1 & \text{if } s = 0. \end{cases}$$

Use this to show that the expected value of the distribution is $\frac{1}{2}$ by showing that $m'(0) = \frac{1}{2}$. Notice that you cannot do this by simply differentiating $(e^s - 1)/s$ and setting $s = 0$. Do it in the following two ways.

   a. Use the definition of derivative; that is, evaluate the limit

$$\lim_{h \to 0} \frac{m(h) - m(0)}{h}.$$

   [You will need L'Hôpital's rule.]

   b. Get a power series for $(e^s - 1)/s$ (and note that it gives the correct value for $s = 0$ as well). Now you can read off all the moments of the distribution.

   [This exercise repeats Exercise 9 of Section 8.1.]

3. You need to generate a random variable Y whose CDF is $F(y) = 1 - 1/(1 + y)^2$ for $x > 0$ (and 0 for $x \leq 0$). The computer gives you a random variable $X \sim \text{unif}(0, 1)$.

   a. What function of X should you use?

   b. Explain why $Y = -1 + 1/\sqrt{X}$ works as well as your answer to part a.

4. Consider the density $f(y) = 2ye^{-y^2}$ for $y > 0$ (and 0 for $y \leq 0$).

   a. Let $X \sim \text{unif}(0, 1)$; find a function of X that has the density $f(y)$.

   b. Let Y have the density $f(y)$. Find a function of Y whose distribution is unif(0, 1).

5. Let $X \sim \text{unif}(0, 1)$. Find a function of X whose density is $f(y) = 6(1 - y)^5$ for $0 < y < 1$.

6. **Not all digits have the same probability of being a first digit.** It is surprising that, if a number is chosen at random from a table of data (say from an almanac or a book of census data), the first significant digit is more likely to be 1 than 2, more likely to be 2 than 3, and so on. This is not true for all tables, but if the numbers in a table have a very wide range it can be justified as follows.
   a. Let x be a positive number, let $y = \log_{10}x$, and let $z = y - \lfloor y \rfloor$, the fractional part of y. Show that the first significant digit of x is k if and only if z is between $\log_{10}(k)$ and $\log_{10}(k + 1)$.
   b. If X is a random variable whose interval of possible values is very large, then that of $\log_{10}X$ is also large, and it is not unreasonable to assume that the fractional part of $\log_{10}X$ has the uniform distribution on $(0, 1)$. Under this assumption show that the mass function of the first digit of X is as given in the following table.

k	1	2	3	4	5	6	7	8	9
$p(k)$	.3010	.1761	.1249	.0969	.0792	.0669	.0580	.0512	.0458

Note: The probability that the first digit is 1, 2, or 3 is more than 60%, whereas one would think it should be $\frac{1}{3}$. The probability that it is 4 or less is not $\frac{4}{9}$ but almost 70%. Before trying to make any money by betting on this, however, you should be warned that the assumption does not likely hold for mathematical tables; be sure to use a table of data.

William Feller, in Volume 2 of *An Introduction to Probability Theory and Its Applications* (on page 63 of the second edition), attributes the argument in part b to R. S. Pinkham.

7. **Breaking a stick at two random places.** Let X and Y be independent, each with the uniform distribution on $(0, 1)$. Then X and Y divide the interval into three subintervals, and we are interested in the (marginal, not joint) distributions of the lengths of these subintervals. One way to deal with this is to let U be the smaller and V the larger of X and Y; then the three lengths are $Z_1 = U$, $Z_2 = V - U$, and $Z_3 = 1 - V$.
   a. Find the CDF and thence a density for Z_1. [*Hint:* $Z_1 > z$ if and only if both X and Y are greater than z.]
   b. Find the CDF and thence a density for Z_3. [*Hint:* $Z_3 > z$ if and only if both X and Y are less than something.]
   c. Find the CDF and thence a density for Z_2. [*Hint:* $Z_2 \le z$ if and only if the point (X, Y) lies in a certain region of the unit square.]

The fact that the three lengths have the same distribution is perhaps surprising; one might think that the middle one would be more likely to be longer, while the two end ones would be shorter. The interesting thing is that three different arguments lead to the same density.

William Feller, in Volume 2 of *An Introduction to Probability Theory and Its Applications*, gives an appealing intuitive argument from symmetry, in which we start with a circle of circumference 1, choose three points at random—now it is obvious that the three intervals have the same distribution—and then cut the circle at one of the points to make a stick, using the other two points as the random break points.

Using conditioning arguments, as discussed in Chapter 9, it is possible to give a proof that generalizes to the $n + 1$ intervals formed by n independent uniform random variables.

ANSWERS

3. a. $Y = 1/\sqrt{1 - X} - 1$
b. If $X \sim \text{unif}(0, 1)$, then $1 - X \sim \text{unif}(0, 1)$ also.

4. a. $Y = \sqrt{-\ln(1 - X)}$ **b.** $X = 1 - e^{-Y^2}$

5. $Y = 1 - (1 - X)^{1/6}$ [$Y = 1 - X^{1/6}$ also has the same density.]

7. $f(z) = 2(1 - z)$ for $0 < z < 1$ in all three parts.

8.6 BETA DISTRIBUTIONS

Here we introduce a family of distributions on the interval (0, 1) that we have not seen before, except incidentally in an example or two. It is a two-parameter family of distributions; the parameters, denoted by α and β, can be any positive numbers, and the density of the beta(α, β) distribution is proportional to $x^{\alpha-1}(1 - x)^{\beta-1}$. We will see shortly what the constant of proportionality has to be.

This gives a very flexible family of distributions on (0, 1); by proper choice of α and β we can put the expected value of the distribution anywhere we like in the interval, and we can make the variance large or small as we please. (See Exercise 3 at the end of this section.) In addition, various members of the beta family arise individually in different contexts in probability.

Another very important context in which beta distributions arise is that of *order statistics*. If $X_1, X_2, \ldots, X_n$ are a sample from the uniform distribution on (0, 1), and we arrange them in increasing order, we get n random variables $Y_1, Y_2, \ldots, Y_n$ that are no longer independent (because $Y_1 \leq Y_2 \leq \cdots \leq Y_n$); but their marginal distributions are all beta distributions. We will see this in (8.6.6).

We begin with the constant of proportionality.

(8.6.1) **The beta formula.** This formula is the key to computations involving beta distributions. For any $\alpha > 0$ and $\beta > 0$,

$$\int_0^1 x^{\alpha-1}(1 - x)^{\beta-1}dx = \frac{\Gamma(\alpha)\Gamma(\beta)}{\Gamma(\alpha + \beta)}.$$

The symbol Γ denotes the same gamma function that we discussed in Section 8.4. The proof is quite clever; we will give it in (8.6.4). The value of this integral is denoted by $B(\alpha, \beta)$; this function of two variables is called the *beta function*. (The

B is a capital beta, despite its appearance.) Its reciprocal is of course the constant of proportionality we need to make the function $x^{\alpha-1}(1-x)^{\beta-1}$ into a density.

When α and β are positive integers, the gammas become factorials:

$$\int_0^1 x^{i-1}(1-x)^{j-1}dx = \frac{(i-1)!(j-1)!}{(i+j-1)!},$$

which is the same as

$$\int_0^1 x^k(1-x)^l dx = \frac{k!l!}{(k+l+1)!}.$$

(8.6.2) **Definition.** For positive numbers α and β, the ***beta distribution with parameters α and β*** (the abbreviation is beta(α, β)) is the one whose density is

$$f(x) = \frac{\Gamma(\alpha+\beta)}{\Gamma(\alpha)\Gamma(\beta)} x^{\alpha-1}(1-x)^{\beta-1} \quad \text{for } 0 < x < 1.$$

An important special case is the case $\alpha = 1$ and $\beta = 1$; the density reduces to $f(x) = 1$ for $0 < x < 1$. That is,

the beta(1, 1) distribution is the unif(0, 1) distribution.

Another case that arises in connection with the theory of repeated coin tossing is the beta$\left(\frac{1}{2}, \frac{1}{2}\right)$ distribution; the density is

$$f(x) = \frac{\Gamma(1)}{\Gamma\left(\frac{1}{2}\right)\Gamma\left(\frac{1}{2}\right)} x^{-1/2}(1-x)^{-1/2} = \frac{1}{\pi\sqrt{x(1-x)}}.$$

This is called the "arc sine" distribution, because its CDF is $(2/\pi)\sin^{-1}\sqrt{x}$ for $0 < x < 1$.

Graphs of several different beta densities are shown in Figure 8.6a.

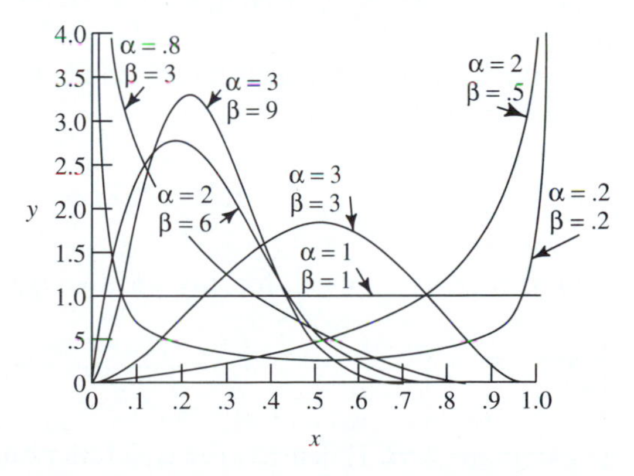

FIGURE 8.6a Some beta densities

(8.6.3) Moments of beta distributions. The beta formula, along with the recurrence $\Gamma(\alpha + 1) = \alpha\Gamma(\alpha)$, which we proved at (8.4.5), makes it easy to find moments of beta-distributed random variables. If $X \sim \text{beta}(\alpha, \beta)$, then

$$EX = \int_0^1 x \cdot \frac{\Gamma(\alpha+\beta)}{\Gamma(\alpha)\Gamma(\beta)} x^{\alpha-1}(1-x)^{\beta-1}dx = \frac{\Gamma(\alpha+\beta)}{\Gamma(\alpha)\Gamma(\beta)} \int_0^1 x^{\alpha}(1-x)^{\beta-1}dx$$
$$= \frac{\Gamma(\alpha+\beta)}{\Gamma(\alpha)\Gamma(\beta)} \cdot \frac{\Gamma(\alpha+1)\Gamma(\beta)}{\Gamma(\alpha+1+\beta)} = \frac{\Gamma(\alpha+1)}{\Gamma(\alpha)} \cdot \frac{\Gamma(\alpha+\beta)}{\Gamma(\alpha+\beta+1)};$$

and because of the recurrence, we see that

$$EX = \frac{\alpha}{\alpha+\beta}.$$

A similar argument (try it) shows that

$$EX^2 = \frac{\alpha(\alpha+1)}{(\alpha+\beta)(\alpha+\beta+1)};$$

some algebra then shows that

$$VX = \frac{\alpha\beta}{(\alpha+\beta)^2(\alpha+\beta+1)}.$$

The MGF of a beta density is a little complicated and less useful than are the MGFs of other distributions. The reason is twofold. First, as we saw above, the beta formula gives us the moments without much difficulty. Second, the distribution of a sum of independent beta random variables is fairly complicated. [See (8.5.5), where we saw this for uniform random variables.]

(8.6.4) Proof of the beta formula. We begin with something seemingly irrelevant: if X and Y are independent, and $X \sim \text{gamma}(\alpha, \lambda)$ and $Y \sim \text{gamma}(\beta, \lambda)$, then $W = X + Y \sim \text{gamma}(\alpha + \beta, \lambda)$. We proved this using MGFs in (8.4.14). We will use it here only in the case $\lambda = 1$.

Recall the convolution theorem for densities (5.2.20); it says that if X and Y are independent and $W = X + Y$, then

$$f_W(w) = \int_{-\infty}^{\infty} f_X(x) \cdot f_Y(w-x)dx.$$

Putting these two facts together, we get, for any positive w,

(8.6.5)
$$\frac{1}{\Gamma(\alpha+\beta)} w^{\alpha+\beta-1}e^{-w} = \int_0^w \frac{1}{\Gamma(\alpha)} x^{\alpha-1}\, e^{-x} \cdot \frac{1}{\Gamma(\beta)}(w-x)^{\beta-1}e^{-(w-x)}dx.$$

The left side is the gamma($\alpha + \beta$, 1) density; the two functions inside the integral are the gamma(α, 1) density and the gamma(β, 1) density. The integral goes only from 0 to w, because one or the other of the two densities is 0 for any x outside that range.

Now the exponentials in (8.6.5) cancel out; when we do this and bring all the gammas together, we get

$$\frac{\Gamma(\alpha)\Gamma(\beta)}{\Gamma(\alpha+\beta)} w^{\alpha+\beta-1} = \int_0^w x^{\alpha-1}(w-x)^{\beta-1}dx;$$

and when we set $w = 1$ we get the beta formula. □

This proof is a nice illustration of the use of two different probability techniques (namely, the convolution theorem and the MGF method for getting the distribution of a sum of random variables), to prove an integral formula that is not really probabilistic at all.

(8.6.6) **Order statistics of samples from unif(0, 1) have beta distributions.** The ***order statistics*** of a set $X_1, X_2, \ldots, X_n$ of random variables are the same random variables arranged in increasing order. We denote them by Y_j's; so

$$\begin{aligned} Y_1 &= \text{the smallest of } X_1, X_2, \ldots, X_n; \\ Y_2 &= \text{the second smallest of } X_1, X_2, \ldots, X_n; \\ &\vdots \\ Y_n &= \text{the largest of } X_1, X_2, \ldots, X_n. \end{aligned}$$

Another notation that is often used is $X_{(1)}$ for the smallest, $X_{(2)}$ for the next smallest, and so on.

Notice that even if the X_i's are independent, the Y_j's cannot be, because

$$Y_1 \le Y_2 \le \cdots \le Y_n \text{ with probability 1.}$$

It may seem strange at first that the Y_j's do not have the same distribution as the X_j's. After all, each Y_j is one of the X_i's, and so you might think it would have the same distribution as one of them. But this is wrong because it does not take into account that Y_j will be a different X_i on different trials of the experiment. Intuitively, notice that if j is large, then Y_j is more likely to be close to 1 than to 0, so its distribution will not be uniform.

We have seen instances of order statistics before, and in those cases it was clear that they did not have the same distributions as the random variables they came from. In Supplementary Exercise 4 for Chapter 1, V was the larger of the two coordinates of a point chosen at random from the unit square; this is the second order statistic Y_2 of a sample (X_1, X_2) from the unif(0, 1) distribution. You may recall that we found $P(V \le z) = z^2$ for $0 < z < 1$, because this is the probability that both coordinates are less than or equal to z. And in Exercise 6 of Section 3.5, we looked at the order statistics of a sample of size 2 from the exp(λ) distribution; they did not have the exp(λ) distribution.

What we want to do here is to find the density of Y_k, the kth smallest among $X_1, X_2, \ldots, X_n$, in the case where $X_1, X_2, \ldots, X_n$ are independent, each with the uniform distribution on (0, 1). We begin by first finding the CDF of Y_k and then differentiating. We will state the result here, in case you do not want to bother with the derivation:

(8.6.7) If $X_1, X_2, \ldots, X_n$ are a sample from unif(0, 1) and Y_k is the kth order statistic, then $Y_k \sim \text{beta}(k, n + 1 - k)$.

Proof. Before proving this general result, we prove it first for Y_1 and Y_n, because these two particular cases are easier to prove than the general case.

The proof for Y_n, the largest, is the easiest one; we have seen the technique several times. The set of possible values of Y_n is of course (0, 1). For $0 < y < 1$,

$$F_{Y_n}(y) = P(Y_n \leq y) = P(\text{all of } X_1, X_2, \ldots, X_n \text{ are} \leq y) = y^n.$$

Therefore, a density for Y_n is

$$f_{Y_n}(y) = ny^{n-1} \quad \text{for } 0 < y < 1 \quad (\text{and } 0 \text{ otherwise}).$$

This is indeed the beta(n, 1) density. You can see that without checking the constant by noticing that the density is proportional to $y^{n-1}(1 - y)^{1-1}$ on (0, 1).

Next let us prove the result for Y_1, the smallest. It is almost as simple. Again the set of possible values is (0, 1), and for $0 < y < 1$,

$$\begin{aligned} F_{Y_1}(y) &= 1 - P(Y_1 > y) \\ &= 1 - P(\text{all of } X_1, X_2, \ldots, X_n \text{ are} > y) \\ &= 1 - (1 - y)^n. \end{aligned}$$

Therefore, a density for Y_1 is

$$f_{Y_1}(y) = n(1 - y)^{n-1} \quad \text{for } 0 < y < 1 \quad (\text{and } 0 \text{ otherwise}).$$

This is indeed the beta(1, n) density; it is proportional to $y^{1-1}(1 - y)^{n-1}$.

The proof for other values of k involves more computation. The set of possible values of Y_k is (0, 1), and for $0 < y < 1$,

$$F_{Y_k}(y) = P(Y_k \leq y) = P(k \text{ or more among } X_1, X_2, \ldots, X_n \text{ are} \leq y).$$

Now think of y as a constant, and think of the n events $(X_j \leq y)$ as successes in n Bernoulli trials. The success probability is y, because this is the probability that a single unif(0, 1) random variable is less than or equal to y. So for $0 < y < 1$, $F_{Y_k}(y)$ is the probability of k or more successes in n trials with success probability y; that is,

$$F_{Y_k}(y) = \sum_{j=k}^{n} \binom{n}{j} y^j (1 - y)^{n-j}.$$

Now we have to differentiate this sum and show that the result is the beta$(k, n + 1 - k)$ density. The derivative is

$$f_{Y_k}(y) = \sum_{j=k}^{n} j\binom{n}{j} y^{j-1}(1 - y)^{n-j} - \sum_{j=k}^{n-1} (n - j)\binom{n}{j} y^j (1 - y)^{n-j-1};$$

we can drop the term for $j = n$ from the second sum because the $n - j$ makes it 0. Now notice that

$$j\binom{n}{j} = n\binom{n-1}{j-1} \quad \text{and} \quad (n - j)\binom{n}{j} = n\binom{n-1}{j},$$

so this can be rewritten

$$f_{Y_k}(y) = n \cdot \sum_{j=k}^{n} \binom{n-1}{j-1} y^{j-1}(1 - y)^{n-j} - n \cdot \sum_{j=k}^{n-1} \binom{n-1}{j} y^j (1 - y)^{n-j-1}.$$

Now rewrite the first sum, using i for $j - 1$:

$$f_{Y_k}(y) = n \cdot \sum_{i=k-1}^{n-1} \binom{n-1}{i} y^i (1 - y)^{n-i-1} - n \cdot \sum_{j=k}^{n-1} \binom{n-1}{j} y^j (1 - y)^{n-j-1}.$$

Looking at these two sums, we see that their terms are exactly the same, except that the first sum contains a term for $i = k - 1$ that the second sum does not. So all that survives is that one term:

$$f_{Y_k}(y) = n \cdot \binom{n-1}{k-1} y^{k-1}(1 - y)^{n-k}.$$

This is indeed the beta$(k, n - k + 1)$ density. □

EXERCISES FOR **SECTION 8.6**

1. What density on (0, 1) is proportional to $x^4(1 - x)^5$? Give the constant of proportionality, the name of the distribution, the expected value, and the variance.

2. Suppose $X \sim$ beta(α, β). Find EX and VX for each of the following sets of parameters.
   a. $\alpha = 1, \beta = 2$
   b. $\alpha = 5, \beta = 10$
   c. $\alpha = 50, \beta = 100$

3. **Finding a beta distribution with desired expected value and variance.**
   a. Show that if $X \sim$ beta(α, β) and $EX = \mu$ and $VX = \sigma^2$, then

$$\alpha = \mu\left(\frac{\mu(1-\mu)}{\sigma^2} - 1\right) \quad \text{and} \quad \beta = (1-\mu)\left(\frac{\mu(1-\mu)}{\sigma^2} - 1\right).$$

[a]**b.** If μ and σ^2 are given numbers, what must be true in order that there be a beta distribution with expected value μ and variance σ^2?

4. Let $X \sim \text{beta}(\alpha, \beta)$ and let $Y = (1/X) - 1$. Find a density for Y.

The distribution of Y is called the *Pareto* distribution (or sometimes the "beta distribution of type II") with parameters α and β; distributions of incomes are sometimes modeled by Pareto distributions. Notice that if α and β are half-integers, we get the distribution of Z in (8.4.19), which is a constant times an F statistic.

5. The following densities, which we have seen in examples and exercises, are actually beta densities. Identify which beta densities they are; that is, give α and β for each. Each density is defined as shown only for x in the interval (0, 1), and is 0 otherwise.
a. $f(x) = 2x$ [The density of the square root of a unif(0, 1) random variable]
b. $f(x) = 12x^2(1 - x)$ [Example (1.3.8)]
c. $f(x) = ax^{a-1}$ where a is a positive constant
d. $f(x) = 1/2\sqrt{x}$

6. What density on (0, 1) is proportional to $\sqrt{x(1-x)}$? Give the name of the distribution and the density.

7. Find $E(1/X)$ if $X \sim \text{beta}(\alpha, \beta)$, provided it exists. For which α and β does it exist?

8. The expected value of the beta(α, β) distribution is $\alpha/(\alpha + \beta)$. What is the mode—that is, the x for which the density is largest? [*Hint:* It is easier to maximize $\ln f(x)$ than $f(x)$.]

9. Let Y_k be the kth order statistic of a sample of size n from the unif(0, 1) distribution.
a. Find EY_k and VY_k.
b. Suppose $n \to \infty$ and $k \to \infty$, but $k/(n + 1)$ is always equal to a constant α. Show that $Y_k \to \alpha$ in the sense of the Chebyshev Weak Law of Large Numbers; that is, that for any positive number c, $P(|Y_k - \alpha| > c) \to 0$ as $k \to \infty$. [*Hint:* Use the result of part a.]

The result of part b is unnecessarily restrictive; it can be shown that there is no need for $k/(n + 1)$ to be constant, as long as it approaches some constant α as a limit. Thus, for example, if k is $(n + 1)/2$ when n is odd and $n/2$ when n is even, then the kth order statistic (the *sample median* in this case) approaches $\frac{1}{2}$ in the Chebyshev sense as the sample size grows.

10. Let $X_1, X_2, \ldots, X_n$ be a sample from an arbitrary distribution; suppose their CDF is $F(x)$. Let Y be the kth order statistic of $F(X_1), F(X_2), \ldots, F(X_n)$. What is the distribution of Y?

11. Let $X_1, X_2, \ldots, X_{10}$ be a sample from the standard normal distribution. Find the probability that the third order statistic of the sample (that is, the third smallest of the X_j's) is less than 0. [*Hint:* As in the proof of (8.6.7), consider ten Bernoulli trials, with success on trial j occurring if X_j is negative.]

12. Find density functions for the 1st and the nth order statistics of a sample of size n from the $\exp(\lambda)$ distribution. [This repeats Exercise 7 of Section 8.3.]

13. Consider a Poisson process with arrival rate λ. Suppose we are given that there were exactly n arrivals in the interval $(0, 1)$.

a. Show that the conditional CDF of the last arrival in $(0, 1)$ is the beta$(n, 1)$ CDF. [*Hint:* This is essentially the same as Exercise 11 of Section 2.4.]

b. Show that the conditional CDF of the first arrival in $(0, 1)$ is the beta$(1, n)$ CDF.

c. Show that the conditional CDF of the second arrival is the beta$(2, n - 1)$ CDF. [*Hint:* The second arrival time is less than or equal to a if and only if there are 2 or more arrivals in $(0, a)$.]

In Exercise 11 of Section 2.4 we mentioned the following informal statement: In a Poisson process, given that there were n arrivals in time interval I, their conditional distribution is the same as that of n points chosen independently from the uniform distribution on I. A more accurate statement is the following: *Given that there were n arrivals in I, the joint conditional distribution of the arrival times is the same as the joint distribution of the order statistics of a sample of size n from the uniform distribution on I.* We have still not proved this statement, but this exercise indicates that the marginal distributions of the arrival times are the same as those of the order statistics.

ANSWERS

1. Beta(5, 6); $EX = 5/11$; $VX \doteq .02066$

2. $EX = \frac{1}{3}$ in all parts; $VX = \frac{1}{18}$ in part a, .01389 in part b, .001472 in part c

3. b. μ must be between 0 and 1, and σ^2 must be less than $\mu(1 - \mu)$.

4. $f(y) = \dfrac{\Gamma(\alpha + \beta)}{\Gamma(\alpha)\Gamma(\beta)} \cdot \dfrac{y^{\alpha-1}}{(1 + y)^{\alpha+\beta}}$ for $y > 0$

5. a. $\alpha = 2, \beta = 1$ **b.** $\alpha = 3, \beta = 2$ **c.** $\alpha = a, \beta = 1$ **d.** $\alpha = \frac{1}{2}, \beta = 1$

6. Density is $\dfrac{8}{\pi}\sqrt{x(1 - x)}$ $(0 < x < 1)$

7. $EX = \dfrac{\alpha + \beta - 1}{\alpha - 1}$ if $\alpha > 1$ (and $\beta > 0$); EX does not exist if $\alpha \le 1$.

8. $(\alpha - 1)/(\alpha + \beta - 2)$

9. a. $\dfrac{k}{n + 1}, \dfrac{k(n + 1 - k)}{(n + 1)^2(n + 2)}$

10. Beta$(k, n + 1 - k)$. [The random variables $F(X_j)$ are a sample from the unif(0, 1) distribution.]

11. .9453

12. $f_{Y_1}(y) = n\lambda e^{-n\lambda y}$ for $y > 0$;
$f_{Y_n}(y) = n\lambda e^{-\lambda y}(1 - e^{-\lambda y})^{n-1}$ for $y > 0$

SUPPLEMENTARY EXERCISES FOR **CHAPTER 8**

1. There are densities proportional to the following functions of x. In each case identify the distribution by name and give the density. [You should not have to do any integrals.]
   a. x^4e^{-2x} for $x > 0$
   b. $e^{-x/2}$ for $x > 0$
   c. $e^{-x^2/2}$ for $x \in \mathbb{R}$
   d. $x^4(1-x)$ for $0 < x < 1$
   e. $\sqrt{1-x}$ for $0 < x < 1$
   f. 1 for $2 < x < 6$
   g. $e^{-(x-1)^2/3}$ for $x \in \mathbb{R}$
   h. x^3 for $0 < x < 1$

2. In a Poisson process, if the expected time of the tenth arrival is 4 minutes after the start of observations, what is the arrival rate? What is the distribution of the time of the tenth arrival?

3. Let X be the sum of 50 independent random variables, each having the Erlang distribution with parameters $n = 2$ and $\lambda = 1.4$.
   a. What are EX and VX?
   b. Find the approximate probability that X is between 60 and 80. [What approximation is appropriate, and why?]
   c. What is the name of the distribution of X?

4. If $X \sim \chi^2(2)$, find and identify by name the distribution of $X/2\lambda$, where λ is a positive constant. [Use the MGF.]

5. Suppose $EX^k = k!$ for $k = 0, 1, 2, \ldots$. What is the moment generating function of X? What distribution has this MGF?

6. Let a be a real number. Suppose $m(s) = e^{as}$ is the moment generating function of X.
   a. Expand $m(s)$ in a power series and from it find the moments of X.
   b. Find VX.
   c. What is the distribution of X?

7. Let X have the density

$$f(x) = \begin{cases} \frac{1}{2} & \text{if } 0 < x < 2, \\ 0 & \text{otherwise.} \end{cases}$$

   [This is the uniform density on (0, 2).]
   a. Find the moment generating function of X.
   b. Use the MGF to find EX, EX^2, and VX.
   c. Find EX^k directly by integrating x^k times the density.

8. Suppose $EX^k = a^k$ for $k = 0, 1, 2, \ldots$, where a is a constant.
   a. What is the MGF of X?
   b. What is the distribution of X?
   c. Confirm the result of part b by using $EX^k = a^k$ to find EX and VX.

9. Suppose $X \sim$ unif(0, 1). Find the distribution of X^n (the CDF method of Section 5.2 works well) and identify it by name.

10. Let α be an arbitrary positive number. Write down the beta(α, 1) density and simplify the constant as much as possible.

11. Let $X_1, X_2, \ldots, X_{45}$ be independent random variables having the unif(0, 1) distribution. Find the approximate probability that the sum of the squares of the X_j's is less than 17.

12. Let $X_1, X_2, \ldots, X_n$ be independent random variables with an unknown distribution; n is large, but is not given. Find the approximate probability that the random variable $Y = \sum_{j=1}^{n} \sqrt{\sin(X_j)}$ is within two standard deviations of its expected value.

13. Let X and Y be independent, each with the same distribution, whose MGF is $m(s) = \sqrt{1/(1-4s)}$ $\left(\text{for } x < \frac{1}{4}\right)$. What is the name of the distribution of $X + Y$?

14. Let $X_1, X_2, \ldots, X_n$ be independent random variables, each having the uniform distribution on (0, 1). Let $Y = -2\ln(X_1 X_2 \cdot \cdot \cdot X_n)$. Show that Y has the $\chi^2(2n)$ distribution. [*Hint:* What is the distribution of $-2\ln X_j$ for a single j?]

ANSWERS

1. **a.** $f(x) = \frac{4}{3}x^4 e^{-2x}$ $(x > 0)$
   b. $f(x) = \frac{1}{2}e^{-x/2}$ $(x > 0)$
   c. $f(x) = \dfrac{1}{\sqrt{2\pi}} e^{-x^2/2}$ $(x \in \mathbb{R})$
   d. $f(x) = 30x^4(1 - x)$ $(0 < x < 1)$
   e. $f(x) = \frac{3}{2}\sqrt{1 - x}$ $(0 < x < 1)$
   f. $f(x) = \frac{1}{4}$ $(2 < x < 6)$
   g. $f(x) = \dfrac{1}{\sqrt{3\pi}} e^{-(x-1)^2/3}$ $(x \in \mathbb{R})$
   h. $f(x) = 4x^3$ $(0 < x < 1)$
2. Time of tenth arrival ~ Erl(10, 2.5)
3. **a.** $EX = 71.43$; $VX = 51.02$ **b.** .7698 **c.** Erl(100, 1.4)
4. Exp(λ)
5. Exponential with $\lambda = 1$
6. **a.** $EX^k = a^k$ **b.** 0 **c.** $X = a$ with probability 1
7. **a.** $\dfrac{e^{2s} - 1}{2s}$ (for $s \neq 0$; $m(0) = 1$)
   b. 1; $\frac{4}{3}$; $\frac{1}{3}$ **c.** $\dfrac{2^{k+1}}{2(k+1)}$
8. **a.** e^{as} for all $s \in \mathbb{R}$ **b.** $X = a$ with probability 1 **c.** $EX = a$; $VX = 0$
9. Beta($1/n$, 1)
10. $f(x) = \alpha x^{\alpha-1}$ for $0 < x < 1$
11. $\Phi(1) = .8413$
12. .9544
13. exp$\left(\lambda = \frac{1}{4}\right)$

CHAPTER 9

Conditioning and Bayes's Theorem for Random Variables

9.1 Conditioning on a Discrete Random Variable

9.2 Conditioning on an Absolutely Continuous Random Variable

Supplementary Exercises

9.1 CONDITIONING ON A DISCRETE RANDOM VARIABLE

We saw the idea of conditioning—the law of total probability, as well as Bayes's theorem—in Section 2.3. The formulas there deal with a partition of the outcome set Ω into events $A_1, A_2, \ldots$, which represent the "conditions" under which an event B might or might not happen. The two formulas were the law of total probability, which gives the probability of B in terms of its conditional probabilities under the various conditions,

$$P(B) = P(B|A_1)P(A_1) + P(B|A_2)P(A_2) + \cdots,$$

and its corollary, Bayes's theorem, which gives the "reverse" conditional probability of one of the conditions, given B,

$$P(A_k|B) = \frac{P(B|A_k)P(A_k)}{P(B|A_1)P(A_1) + P(B|A_2)P(A_2) + \cdots}.$$

In this chapter the conditions A_n are replaced by the events $(X = x)$ for various values of a random variable X, and the event B is some event described in terms of another random variable Y. For example, if X is discrete with possible values $x_1, x_2, \ldots$, and if B is the event $(Y = y)$, then the law of total probability says

$$P(Y = y) = P(Y = y|X = x_1)P(X = x_1) + P(Y = y|X = x_2)P(X = x_2) + \cdots.$$

Such a restatement may not seem to add much to our understanding. But because we are dealing with random variables and not just events, two additional things happen. First, as always, the language of random variables allows the possibility of more efficient and suggestive notation. The previous identity, for example, will appear at (9.1.17) in the form

$$p_Y(y) = p_Y(y|X = x_1)p_X(x_1) + p_Y(y|X = x_2)p_X(x_2) + \cdots,$$

which, as we will see in (9.1.31), is actually the expected value of a certain function of X.

Second, not only new notations but also new concepts will become available. In particular, the notions of conditional expected values, variances, CDFs, and densities give rise to additional formulas analogous to the law of total probability. We will see this in (9.1.1).

When X is discrete the restatements are easy, as we saw in the paragraphs above. When X is not discrete, however, the events $(X = x)$ have probability 0. So even though they form a partition of Ω, it seems as though conditional probabilities given $X = x$ will be meaningless. We will cover this situation in Section 9.2. The pleasant surprise is that the notation we introduce in this section gives us a clue as to how to define things so that meaningful formulas come out in the next section.

This section is long, with many formulas, but there is really only one fundamental principle, the "principle of conditioning." We will state it here for a discrete random variable, even though it might not be clear yet what it means.

To find a quantity by conditioning on a discrete random variable X, multiply the conditional quantity, given $X = x$, by the mass function of X, and sum the result over all possible values of X.

We will see several forms of this principle, for finding the probability of an event B, the mass function or density function of a random variable Y, and the expected value of Y. The principle does not work for the variance of Y, but there is a conditioning formula that works. In addition, we will discuss Bayes's theorem in the context of a pair of discrete random variables.

We begin by defining a number of conditional quantities, given a value of a discrete random variable X. But we emphasize at the start that although we will need to know these definitions, we rarely use them to compute the quantities. Usually we find that we are *given* one or more of them, as assumptions about the situation being modeled.

(9.1.1) **Conditional probabilities, mass functions, expected values, and densities, given a value of a discrete random variable.** Some of what we do here has already been done in Section 3.4, and you may want to look back at (3.4.21) and (3.4.22).

If X is a discrete random variable and x is one of its possible values, and if B is any event, then the ***conditional probability of B, given X = x***, is

$$P(B|X = x) = \frac{P(B \cap (X = x))}{P(X = x)}. \tag{9.1.2}$$

This looks more familiar in the special case in which B is the event $(Y = y)$, where Y is another discrete random variable and y is one of its possible values. We get $P(Y = y|X = x)$, which, seen as a function of y, is denoted $p_Y(y|X = x)$ and called the ***conditional mass function of Y, given X = x:***

$$p_Y(y|X = x) = P(Y = y|X = x) = \frac{P((Y = y) \cap (X = x))}{P(X = x)} = \frac{p(x, y)}{p_X(x)} \tag{9.1.3}$$

where $p(x, y)$ denotes the joint mass function of X and Y, as discussed in Section 3.4.

The ***conditional expected value of Y, given X = x***, is (if Y is discrete)

$$E(Y|X = x) = \sum_y y p_Y(y|X = x), \tag{9.1.4}$$

where the sum is over all the possible values y of Y. As with unconditional expected values, we define this only when the sum converges absolutely. It can be proved, although we will not go into it, that if the sum for EY converges absolutely, then so does the sum in (9.1.4) for any random variable X.

In the same way, we can define the ***conditional variance of Y***, given that $X = x$:

(9.1.5)
$$V(Y|X = x) = \sum_y (y - E(Y|X = x))^2 p_Y(y|X = x).$$

Again, this is defined only if the sum converges absolutely; since all terms are positive, this is equivalent to assuming that it converges. If we denote $E(Y|X = x)$ by c, we get that $V(Y|X = x) = E((Y - c)^2|X = x)$, and with a little algebra we can see that

(9.1.6)
$$V(Y|X = x) = E(Y^2|X = x) - [E(Y|X = x)]^2.$$

Now what if X is discrete but Y is not? We can still define the ***conditional CDF of Y, given X = x,*** by taking $(Y \leq y)$ for our event B in (9.1.2):

(9.1.7)
$$F_Y(y|X = x) = P(Y \leq y|X = x) = \frac{P((Y \leq y) \cap (X = x))}{P(X = x)}.$$

If this function of Y has a derivative on the set of possible values of Y, we call that derivative a ***conditional density for Y, given X = x:***

(9.1.8)
$$f_Y(y|X = x) = \frac{d}{dy} F_Y(y|X = x).$$

And then there is nothing to stop us from defining the ***conditional expected value of Y, given X = x,*** for an absolutely continuous Y, as

(9.1.9)
$$E(Y|X = x) = \int_{-\infty}^{\infty} y f_Y(y|X = x)dy,$$

and the ***conditional variance*** as

(9.1.10)
$$V(Y|X = x) = \int_{-\infty}^{\infty} (y - c)^2 f_Y(y|X = x)dy$$

where c again denotes $E(Y|X = x)$. Again we have that $V(Y|X = x) = E((Y - c)^2|X = x)$, and it follows that

(9.1.11)
$$V(Y|X = x) = E(Y^2|X = x) - [E(Y|X = x)]^2.$$

Notice that the two definitions of conditional variance (9.1.6) and (9.1.11), for discrete and absolutely continuous Y, are the same.

Before we proceed, remember the two kinds of calculations with conditional probabilities that we did in Chapter 2. One was to compute them from the definition. That would be analogous to computing the conditional quantities above using the definitions and formulas (9.1.2) through (9.1.11).

But the more common use of conditional objects (probabilities, mass functions, etc.), as in Chapter 2, arises in situations where we take one or more of them as given, and go on to use them in formulas similar to the law of total probability. The following examples show typical situations.

(9.1.12) **Example.** In Exercise 2 of Section 2.3 we had the following situation. We roll a fair die, and then we put one blue ball into an urn, along with as many white balls as there are spots showing on the die. Then we mix the balls and draw one; we are interested in the event—call it B—that the blue ball is drawn.

In this problem we took the following conditional probabilities as given (that is, we did not compute them from the definition of conditional probability): $P(B \mid \text{die shows } 1) = \frac{1}{2}$, $P(B \mid \text{die shows } 2) = \frac{1}{3}$, and so on. If we want to let X be the number of spots showing on the die, we can rewrite these, $P(B \mid X = 1) = \frac{1}{2}$, etc. We can be more succinct and say

$$P(B \mid X = x) = \frac{1}{x + 1} \quad \text{for } x = 1, 2, 3, 4, 5, 6.$$

Notice that we did *not* find this from (9.1.2); that would have involved finding the two probabilities in that formula and computing their ratio. Instead, we simply noticed that part of what we know about this situation is expressed in the form of conditional probabilities.

The other thing we know in this situation is not a conditional probability; it is simply the unconditional mass function of X:

$$p_X(x) = \tfrac{1}{6} \quad \text{for } x = 1, 2, 3, 4, 5, 6.$$

What we did in Section 2.3 was to put these two functions together using the law of total probability; we will see this again shortly. ∎

(9.1.13) **Example: The number of successes in a random number of Bernoulli trials.** Suppose we do Bernoulli trials with success probability p, but the number of trials we do is determined by tossing a fair coin until we get the first H and counting the number of T's. We are interested in the number of successes.

This arises in connection with a population model that was first given in Exercise 16 of Section 7.3. In this model every couple is allowed to have children until they have had their first male child; then they may have no more children. Let us assume for simplicity now that male and female children are equally likely, so that the number of births in a family is modeled by the number of fair-coin tosses needed to get the first H. The number of female births is the number of T's that precede the first H. In addition, suppose that this country has a high infant mortality rate, and children have probability p of surviving the first year. Then in a given family, the number of female children who survive their first year is the number of successes in a random number of Bernoulli trials as described.

We condition on the number of female births in a family; let this be X. Let Y be the number of surviving females; that is, Y is the number of successes in X Bernoulli trials. What do we know? First of all, the mass function of X is

$$p_X(n) = \left(\tfrac{1}{2}\right)^{n+1} \quad \text{for } n = 0, 1, 2, \ldots :$$

this is the probability that n T's precede the first H. And *given* that $X = n$, Y is *conditionally* the number of successes in n Bernoulli trials. That is, the conditional mass function of Y, given $X = n$, is the bin(n, p) distribution. So we can immediately write down the conditional mass function, expected value, and variance:

$$p_Y(k|X = n) = \binom{n}{k} p^k q^{n-k} \quad \text{for } k = 0, 1, 2, \ldots, n \quad \text{and} \quad n = 0, 1, 2, \ldots,$$
$$E(Y|X = n) = np, \quad \text{and} \quad V(Y|X = n) = npq.$$

Again, we did not find these from any of the definitions or formulas given above; we simply noticed that given $X = n$, Y is conditionally bin(n, p).

[Notice that we have to think a bit about the case $X = 0$. The above says that in this case Y has conditionally the bin$(0, p)$ distribution. We have not considered this a proper binomial distribution, but the above is all valid if we agree that when $X = 0$, Y is the constant random variable that equals 0 with probability 1. This is reasonable—the number of successes in 0 Bernoulli trials equals 0.]

We will return to this model in Examples (9.1.18), (9.1.23), and (9.1.41). ∎

(9.1.14) Example. In a Poisson process with arrival rate λ per hour, suppose that with each arrival is associated a trial in a Bernoulli process, which results in success with probability p and failure with probability $q = 1 - p$. Assume the two processes are independent of each other. We are interested in Y, the time of the first arrival that produces a success. The way to do it is to condition on X, the number of trials needed to produce the first success. We know that the mass function of X is

$$p_X(n) = pq^{n-1} \quad \text{for } n = 1, 2, 3, \ldots.$$

[This is not quite the geom(p) mass function, because X is the number of *trials* needed for the first success, not the numbers of failures before it.]

Now Y is the time of the first arrival that produces a success. We do not know which arrival it is that comes at time Y; but given that $X = n$, we know that Y is conditionally the time of the nth arrival. The time of the nth arrival has the Erl(n, λ) distribution, whose density, expected value, and variance are, for $n = 1, 2, 3, \ldots,$

$$f_Y(y|X = n) = \frac{\lambda^n}{(n-1)!} y^{n-1} e^{-\lambda y} \quad \text{for } y > 0,$$
$$E(Y|X = n) = \frac{n}{\lambda}, \quad \text{and} \quad V(Y|X = n) = \frac{n}{\lambda^2}.$$

Again, we did not compute these from the definitions. They simply come from the fact that, given $X = n$, Y is conditionally the nth arrival time and, consequently, has the Erl(n, λ) distribution. ∎

Now we turn to the conditioning formulas in which we use the conditional quantities (mass functions, densities, expected values, etc.) that were defined above. We begin by recalling the law of total probability from (2.3.1):

If a finite or countable collection of events $A_1, A_2, A_3, \ldots$ forms a partition of Ω, and if $P(A_n) > 0$ for each A_n, then for any event B,

(9.1.15)
$$\begin{aligned} P(B) &= \sum_n P(B|A_n)P(A_n) \\ &= P(B|A_1)P(A_1) + P(B|A_2)P(A_2) + P(B|A_3)P(A_3) + \cdots . \end{aligned}$$

(9.1.16) **Finding the mass function of *Y* by conditioning on *X*.** Suppose X is a discrete random variable with possible values $x_1, x_2, x_3, \ldots$, and suppose we know the mass function of X, $p_X(x) = P(X = x)$. Suppose also that Y is discrete, and that we know its conditional mass function $p_Y(y|X = x_n)$ for each possible value x_n of X. Then we can let A_n denote the event $(X = x_n)$, and for a given y we can let B denote the event $(Y = y)$. Then (9.1.15) becomes

$$\begin{aligned} p_Y(y) &= \sum_n p_Y(y|X = x_n)p_X(x_n) \\ &= p_Y(y|X = x_1)p_X(x_1) + p_Y(y|X = x_2)p_X(x_2) + p_Y(y|X = x_3)p_X(x_3) + \cdots . \end{aligned}$$

This is more conveniently written

(9.1.17)
$$p_Y(y) = \sum_x p_Y(y|X = x)p_X(x),$$

the sum extending over all the possible values x of X.

(9.1.18) **Example.** In Example (9.1.13), X is the number of female children in a family, with mass function

$$p_X(n) = \left(\tfrac{1}{2}\right)^{n+1} \quad \text{for } n = 0, 1, 2, \ldots,$$

and Y is the number of females who survive the first year. Let us find the unconditional mass function of Y. We know that, conditional on $X = n$, Y has the bin(n, p) distribution:

$$p_Y(k|X = n) = \binom{n}{k}p^k q^{n-k} \quad \text{for } k = 0, 1, 2, \ldots, n \quad \text{and} \quad n = 1, 2, 3, \ldots .$$

By (9.1.17), the unconditional mass function of Y is

$$p_Y(k) = \sum_{n=k}^{\infty} \binom{n}{k}p^k q^{n-k} \cdot \left(\tfrac{1}{2}\right)^{n+1} \quad \text{for } k = 0, 1, 2, \ldots .$$

Notice that the sum goes from k, not 1, to ∞, because $p_Y(k|X = n)$ equals 0 unless $n \geq k$.

To sum this, we do a little judicious factoring, and see that what is left inside is a kth derivative:

$$p_Y(k) = \frac{1}{2} \cdot \left(\frac{p}{2}\right)^k \cdot \frac{1}{k!} \cdot \sum_{n=k}^{\infty} n(n-1)(n-2)\cdots(n-k+1)\left(\frac{q}{2}\right)^{n-k}.$$

Remembering (see "geometric series" in Section A.5 of Appendix A) that

$$\sum_{n=k}^{\infty} n(n-1)(n-2)\cdots(n-k+1)x^{n-k} = \frac{d^k}{dx^k}\left(\frac{1}{1-x}\right) = \frac{k!}{(1-x)^{k+1}},$$

we see then that

$$p_Y(k) = \frac{1}{2} \cdot \left(\frac{p}{2}\right)^k \cdot \frac{1}{k!} \cdot \frac{k!}{\left(1 - \frac{q}{2}\right)^{k+1}} = \frac{p^k}{(2-q)^{k+1}}.$$

This does not look like a familiar mass function, but remember that $2 - q = 1 + p$, and so

$$p_Y(k) = \left(\frac{p}{1+p}\right)^k \cdot \frac{1}{1+p} \quad \text{for } k = 0, 1, 2, \ldots.$$

The unconditional distribution of Y is the geometric distribution with parameter $1/(1 + p)$. ▮

(9.1.19) **Finding a density for Y by conditioning on X.** If Y is absolutely continuous and X is discrete, and we know the conditional density $f_Y(y|X = x)$ for each possible value x of X, and if we also know the mass function $p_X(x)$ of X, then we can find a density for Y by a formula similar to (9.1.17):

(9.1.20)
$$f_Y(y) = \sum_x f_Y(y|X = x)p_X(x).$$

To prove this we first find the CDF of Y, $F_Y(y) = P(Y \leq y)$. In (9.1.15) let B be the event $(Y \leq y)$; then we get

$$F_Y(y) = P(Y \leq y) = \sum_x P(Y \leq y|X = x)p_X(x) = \sum_x F_Y(y|X = x)p_X(x).$$

Differentiating this term by term produces (9.1.20), because of (9.1.8). Such term-by-term differentiation is not valid without some extra assumptions. In this case, we need to assume something about the convergence of the right side of (9.1.20)—namely, that it converges uniformly as a function of y. We will not go into this here; uniform convergence is usually covered in advanced calculus or real analysis courses. The necessary assumptions are valid in any of the cases we will encounter.

(9.1.21) **Example.** Let X have the Pois(λ) distribution, and let Y be the largest observation in a sample of size $X + 1$ from the unif(0, 1) distribution. What is the unconditional distribution of Y? We know from (8.6.6) that for $n = 0, 1, 2, 3, \ldots$ we have

$$f_Y(y|X = n) = (n + 1)y^n \text{ for } 0 < y < 1.$$

This is the beta($n + 1$, 1) density, which as we saw in (8.6.6) is the density of the largest among $n + 1$ i.i.d. unif(0, 1) random variables.

By (9.1.20), then, a density for Y is

$$f_Y(y) = \sum_{n=0}^{\infty} (n + 1)y^n \cdot \frac{\lambda^n e^{-\lambda}}{n!}.$$

To sum this we break it into two sums:

$$f_Y(y) = \sum_{n=0}^{\infty} ny^n \cdot \frac{\lambda^n e^{-\lambda}}{n!} + \sum_{n=0}^{\infty} y^n \cdot \frac{\lambda^n e^{-\lambda}}{n!}.$$

The second sum is easy; factor out $e^{-\lambda}$ and the rest is the Maclaurin series for $e^{\lambda y}$, so the second sum is $e^{-\lambda(1-y)}$. In the first sum, factor out $e^{-\lambda}$, notice that the $n = 0$ term is 0, cancel the n into the $n!$, factor out λy, and what is left is again $e^{\lambda y}$. The first sum is therefore $\lambda y e^{-\lambda(1-y)}$, and so

$$f_Y(y) = (1 + \lambda y)e^{-\lambda(1-y)} \quad \text{for } 0 < y < 1.$$

This is not a density we have seen before, but it is a density for Y. ∎

(9.1.22) **Finding EY by conditioning on X.** We have to derive the formula separately for discrete Y and absolutely continuous Y, but the result is the same in both cases; it is sometimes called the ***law of total expectation***:

$$EY = \sum_x E(Y|X = x)p_X(x).$$

Compare this with (9.1.17), and you see what is going to become a familiar pattern: to find an unconditional quantity, write down the conditional version of it, and then multiply by the mass function of the conditioning random variable and sum. (In the next section we will see the same thing happen when X is absolutely continuous—we multiply the conditional quantity by the density of X and integrate.)

Before proving this we will give two examples.

(9.1.23) Example. In Example (9.1.13) we had $E(Y|X = n) = np$, and so we quickly get EY:

$$EY = \sum_{n=1}^{\infty} np \cdot \left(\frac{1}{2}\right)^{n+1}.$$

You could sum this directly, or you could notice that it is p times the expected value of the geometric distribution with parameter $\frac{1}{2}$. That expected value is 1, so $EY = p$.

Notice that you can also get this expected value from the last sentence of Example (9.1.18). ▮

(9.1.24) Example. In Example (9.1.14), the arrivals of a Poisson process (with parameter λ) have Bernoulli trials associated with them, and Y is the time of the first arrival that produces a success. We conditioned on X, the number of trials needed to get that first success, and we saw that, given $X = n$, the conditional expected value of Y is

$$E(Y|X = n) = \frac{n}{\lambda}.$$

The mass function of X is

$$p_X(n) = pq^{n-1} \quad \text{for } n = 1, 2, 3, \ldots;$$

so by the law of total expectation

$$EY = \sum_{n=0}^{\infty} \frac{n}{\lambda} \cdot pq^{n-1}.$$

To sum this, factor out p/λ, and what is left inside is a familiar derivative. The result is

$$\begin{aligned} EY &= \frac{p}{\lambda} \cdot \frac{1}{(1-q)^2} \\ &= \frac{1}{\lambda p}. \end{aligned}$$

▮

(9.1.25) Proof of the law of total expectation. We have to give two separate proofs—one for a discrete random variable Y and one for an absolutely continuous Y. In both cases we are trying to prove (9.1.22), which says

$$EY = \sum_x E(Y|X = x)p_X(x),$$

but $E(Y|X = x)$ is different in the two cases.

If Y is discrete, we use (9.1.4) to see that the right side of (9.1.22) is

$$\sum_x \sum_y y p_Y(y|X = x) p_X(x) = \sum_y y \sum_x p_Y(y|X = x) p_X(x).$$

(The reversal of the order of summing can be justified because the series can be proved to converge absolutely.) But look at the inner sum here; by the law of total probability it equals the unconditional mass function $p_Y(y)$. So the right side of (9.1.22) is just $\sum_y y p_Y(y)$, which is EY.

If Y is absolutely continuous, we have to be a little more roundabout. We use (9.1.9) to see that the right side of (9.1.22) is

$$\sum_x \int_{-\infty}^{\infty} y f_Y(y|X = x) dy \cdot p_X(x) = \int_{-\infty}^{\infty} y \sum_x f_Y(y|X = x) \cdot p_X(x) dy,$$

and by (9.1.20) this equals

$$\int_{-\infty}^{\infty} y f_Y(y) dy,$$

which equals EY.

Again, we need something to justify the interchange of summation and integration in this proof. It turns out that the same uniform convergence that we needed to justify (9.1.20) is sufficient. □

(9.1.26) **The law of total expectation as an expected value.** Now we come to one of those places in mathematics where the right notation produces a marvelous simplification of formulas—if we can keep in mind what the notation means. Look back at (9.1.22), the law of total expectation, but in it let $h(x)$ denote $E(Y|X = x)$:

$$EY = \sum_x E(Y|X = x) p_X(x) = \sum_x h(x) p_X(x).$$

This last sum is just the expected value of the function $h(X)$. That is,

(9.1.27) $$EY = Eh(X) \quad \text{where} \quad h(x) = E(Y|X = x).$$

Now, with a little more notation, we can make this even more succinct. If $h(x)$ denotes $E(Y|X = x)$, then we denote $h(X)$ by $E(Y|X)$. Then (9.1.27) becomes the remarkable formula

(9.1.28) $$EY = E(E(Y|X)).$$

But to make sense of this, we need to keep in mind what the right side of (9.1.28) represents.

> $E(Y|X)$ is a random variable that is a function of X; its value when $X = x$ is $E(Y|X = x)$. Its expected value is EY.

That is, finding EY by conditioning on X amounts to the following procedure:

> Find $E(Y|X = x)$ as a function of x. Then make it a function of X, and find its expected value.

Sometimes this procedure can be very quick, as a few examples will show.

(9.1.29) **Example.** You roll a fair die, and then you do as many Bernoulli trials as there are spots showing on the die. If the success probability in the Bernoulli trials is p, what is the expected number of successes?

Solution. We condition on the number of spots showing; let it equal X. Recall before beginning that $EX = 3.5$.

Now let Y equal the number of successes. Given that $X = x$, Y is the number of successes in x Bernoulli trials, so Y has conditionally the bin(x, p) distribution. Therefore,

$$E(Y|X = x) = xp;$$

that is,

$$E(Y|X) = Xp.$$

So

$$EY = E(Xp) = p \cdot EX = 3.5 \cdot p.$$

Just in case this seems too much like magic, let us do it again the long way, using (9.1.22). Starting again with $E(Y|X = x) = xp$, and using (9.1.22), we have

$$EY = \sum_x E(Y|X = x)p_X(x) = \sum_{x=1}^{6} xp \cdot \tfrac{1}{6} = p \sum_{x=1}^{6} x \cdot \tfrac{1}{6}.$$

Now we recognize this as $p \cdot EX$; but even if we did not, we could continue and get

$$EY = p \cdot \tfrac{1}{6} \cdot (1 + 2 + 3 + 4 + 5 + 6) = p \cdot 3.5.$$ ∎

(9.1.30) **Example.** Let $X \sim$ Pois(λ) and let Y have the $\chi^2(X)$ distribution—the chi-squared distribution with a random number, X, of "degrees of freedom." What is EY?

Even if you have not looked at the chi-squared distributions in (8.4.16), all you need to know is that the expected value of the $\chi^2(n)$ distribution is n. [It is n because the $\chi^2(n)$ distribution is the gamma $\left(\frac{n}{2}, \frac{1}{2}\right)$ distribution, and the expected value of the gamma(α, λ) distribution is α/λ.]

Solution. Given that $X = x$, Y has conditionally the $\chi^2(x)$ distribution, whose expected value is x. That is, $E(Y|X = x) = x$. That is, $E(Y|X) = X$. So $EY = E(E(Y|X)) = EX$. Since $X \sim \text{Pois}(\lambda)$, $EY = \lambda$. ∎

(9.1.31) **Summary of conditioning formulas as expected values.** The law of total expectation is one of three conditioning formulas that we have seen so far. All of them can be written as expected values of certain functions of X. We will give each of them here in three different forms; you should use the one you are most comfortable with in the problem at hand.

The first one is the law of total expectation, which we saw in (9.1.26):

$$EY = \sum_x E(Y|X = x)p_X(x),$$

which can be written as

$$EY = Eh(X), \quad \text{where} \quad h(x) = E(Y|X = x),$$

or, more briefly,

$$EY = E(E(Y|X)).$$

The next is the formula (9.1.17) for the mass function of a discrete random variable Y:

$$p_Y(y) = \sum_x p_Y(y|X = x)p_X(x),$$

which can be written as

$$p_Y(y) = Eh(X), \quad \text{where} \quad h(x) = p_Y(y|X = x),$$

or, more briefly,

$$p_Y(y) = E(p_Y(y|X)).$$

The third conditioning formula is (9.1.20), which gives the density of an absolutely continuous random variable Y:

$$f_Y(y) = \sum_x f_Y(y|X = x)p_X(x),$$

which reads

$$f_Y(y) = Eh(X) \quad \text{where} \quad h(x) = f_Y(y|X = x),$$

or, more briefly,

$$f_Y(y) = E(f_Y(y|X)).$$

Although we did not state it, there is of course a formula for the conditional CDF of any random variable Y in terms of the conditional CDF that we defined in (9.1.7). We get it from the law of total probability (9.1.15) by using $(Y \le y)$ for B:

$$F_Y(y) = \sum_x F_Y(y|X = x)p_X(x),$$

which reads

$$F_Y(y) = Eh(X), \quad \text{where} \quad h(x) = F_Y(y|X = x),$$

or, more briefly,

$$F_Y(y) = E(F_Y(y|X)).$$

In the next section we will develop formulas for conditioning on an absolutely continuous random variable. The concept of a conditioning formula as an expected value will get us around the difficulty of defining a conditional quantity like $f_Y(y|X = x)$ when $P(X = x)$ equals 0.

(9.1.32) Example. Suppose X is the number of failures before the second success in a Bernoulli process, and Y has (conditionally) the uniform distribution on the interval $(0, X + 1)$. We will find EY and a density $f_Y(y)$ using the conditioning formulas. The expected value is very easy to find; the density requires a little care, but it works out nicely.

First, it helps to understand the distribution of X. The mass function, as we saw in Exercise 15 of Section 1.2, is

$$p_X(x) = (x + 1)p^2q^x \quad \text{for } x = 0, 1, 2, \ldots.$$

We saw in Exercise 16 of Section 4.1 that

$$EX = \frac{2q}{p}.$$

[If you have been through Section 7.3, you know that $X \sim$ neg bin$(2, p)$, and you can get the mass function and the expected value from there.]

Finding EY is very simple. First, we notice that

$$E(Y|X) = \frac{X + 1}{2},$$

because that is the expected value of the uniform distribution on $(0, X + 1)$; it follows that

$$EY = E\left(\frac{X + 1}{2}\right) = \frac{1}{2}EX + \frac{1}{2} = \frac{q}{p} + \frac{1}{2}.$$

As for a density for Y, we can see that

$$f_Y(y|X) = \begin{cases} \dfrac{1}{X+1} & \text{if } 0 < y < X + 1, \\ 0 & \text{otherwise.} \end{cases}$$

We need to find the expected value of this function of X to get a density for Y, and that is difficult in this form. It is simpler to use the notation that refers to a specific possible value x of X:

$$f_Y(y|X = x) = \begin{cases} \dfrac{1}{x+1} & \text{if } 0 < y < x + 1, \\ 0 & \text{otherwise.} \end{cases}$$

and we want to multiply this by $p_X(x)$ and add. Notice that y (some positive number) is thought of as constant in this calculation, and the x's we will sum over are those for which $x + 1 > y$. That is, x will range from $\lfloor y \rfloor$ to ∞. We get

$$f_Y(y) = \sum_{x=\lfloor y \rfloor}^{\infty} \frac{1}{x+1} \cdot (x+1)p^2q^x = p^2q^{\lfloor y \rfloor}\frac{1}{1-q};$$

that is,

$$f_Y(y) = \begin{cases} pq^{\lfloor y \rfloor} & \text{if } y > 0, \\ 0 & \text{if } y \le 0. \end{cases}$$

This is a little confusing, because it looks like a geometric mass function, but it is a density, not a mass function. It is nonzero for all positive y, and if we integrated it from 0 to ∞ we would get 1. Notice that this density is constant on each of the intervals between integers. ∎

Now we turn to Bayes's theorem, which, as you may remember from Chapter 2, goes along with the law of total probability and the other conditioning formulas.

(9.1.33) Theorem (Bayes's theorem for two discrete random variables). Remember Bayes's theorem for a partition $A_1, A_2, \ldots$ and an event B with $P(B) > 0$:

$$P(A_j|B) = \frac{P(B|A_j)P(A_j)}{\sum_n P(B|A_n)P(A_n)}.$$

Suppose X and Y are discrete random variables, the possible values of X are

$x_1, x_2, \ldots,$ and y is a possible value of Y. Then let B be the event $(Y = y)$, and for each n let A_n be the event $(X = x_n)$; Bayes's theorem then reads

$$p_X(x_j|Y = y) = \frac{p_Y(y|X = x_j)p_X(x_j)}{\sum_n p_Y(y|X = x_n)p_X(x_n)}.$$

A slightly more convenient way to state the same thing is as follows: for any possible value x of X and any possible value y of Y,

$$p_X(x|Y = y) = \frac{p_Y(y|X = x)p_X(x)}{\sum_z p_Y(y|X = z)p_X(z)}.$$

Because the denominator is just $p_Y(y)$ by (9.1.17), we can also remember this as

(9.1.34)
$$p_X(x|Y = y) = \frac{p_Y(y|X = x)p_X(x)}{p_Y(y)}.$$

There is really nothing new to learn here; this formula is the same as

$$P(X = x|Y = y) = \frac{P(Y = y|X = x)P(X = x)}{\sum_z P(Y = y|X = z)P(X = z)},$$

which is obviously the same Bayes's theorem that we saw in Section 2.3.

(9.1.35) **Example.** Let $X \sim \text{Pois}(\lambda)$, and let Y be the number of successes in X Bernoulli trials, with success probability p. Given that there were 10 successes, what is the probability that $X = 15$?

Solution. We want $P(X = 15|Y = 10)$, and by Bayes's theorem

$$P(X = 15|Y = 10) = \frac{P(Y = 10|X = 15)P(X = 15)}{\sum_{k=0}^{\infty} P(Y = 10|X = k)P(X = k)}$$

$$= \frac{\binom{15}{10}p^{10}q^5 \cdot \frac{\lambda^{15}e^{-\lambda}}{15!}}{\sum_{k=10}^{\infty}\binom{k}{10}p^{10}q^{k-10} \cdot \frac{\lambda^k e^{-\lambda}}{k!}}.$$

This is not quite as bad as it looks. A fair amount of simplification takes place, and the result is a pleasant surprise:

$$P(X = 15|Y = 10) = \frac{(\lambda q)^5 e^{-\lambda q}}{5!}.$$

You can check by a similar argument that the general result is

$$P(X = n | Y = k) = \frac{(\lambda q)^{n-k} e^{-\lambda q}}{(n-k)!}$$
$$\text{for } k = 0, 1, 2, \ldots \quad \text{and} \quad n = k, k+1, k+2, \ldots .$$

Given that $Y = k$, the conditional distribution of X can be described by saying that $X - k \sim \text{Pois}(\lambda q)$. ▮

(9.1.36) Bayesian inference: Prior and posterior distributions. In the preceding example, we were given the distribution of X and the conditional distribution of Y, given a value of X. We used Bayes's theorem to find the conditional distribution of X, given a value of Y.

There is a statistical context in which conditional distributions found in this way from Bayes's theorem are used; it is called Bayesian inference. Here is how the preceding example might arise:

Suppose that n Bernoulli trials with success probability p have been done, and we know the value of p but not the value of n, and we would like to estimate n based on the number of successes. In Bayesian analysis the approach is to regard the unknown parameter n as a random variable and call it X instead of n; then we assume that it has some appropriate distribution, called the *prior distribution* of the random parameter X. In this example the prior distribution is the Pois(λ) distribution.

The number of successes is also a random variable, Y, and we know the conditional distribution of Y, given any value of X: given that $X = n$, Y is conditionally bin(n, p). But what happens is that we observe a value of Y; in the preceding example we observed $Y = 10$ or, in general, $Y = k$. We then use Bayes's theorem, as we did above, to find the conditional distribution of X, given the observed value of Y. This conditional distribution is called the *posterior distribution* of X, given the observed value of Y. In the example above, it is the distribution given by the mass function at the end of the example.

Notice the reasons for the names *prior* and *posterior*. The prior distribution reflects our belief about likely values of the parameter X, *before* making any observations, and the posterior distribution reflects our changed belief about X *after* observing Y.

Once we have the posterior distribution of the unknown parameter X, there are several ways we could use it. For example, we might use the expected value of this posterior distribution, or perhaps the most likely value (mode), as an estimator of the true value of X.

In Example (9.1.35), given that $Y = k$, the expected value of the posterior distribution is $k + \lambda q$, and the mode is $k + \lfloor \lambda q \rfloor$ (although there is a tie for most likely value if λq is an integer). These are different kinds of *Bayesian estimators* of the unknown parameter X.

We will discuss Bayesian inference a little more in the next section; see (9.2.38).

Bayesian techniques are much used in statistics; they often yield results in situations where standard methods are less successful. But the subject is not free of controversy. Many statisticians consider themselves "Bayesians" or "non-Bayesians," and have more or less strong feelings about whether the Bayesian approach is valid, or preferable to other approaches, in various situations.

The controversy is not over the correctness of Bayes's theorem, or of the probabilities that result from its application; it is more philosophical than that. Detractors take issue with the idea of considering an unknown number as a random variable and assigning a prior distribution to it; they point out the difficulty of knowing which prior distribution to use. Bayesians counter by pointing out the practical results that the method can produce in some problems where other methods—sometimes called "classical statistics"—are less satisfactory.

(9.1.37) **Finding VY by conditioning on X.** The usual conditioning procedure—multiply the conditional quantity by the mass function of X and sum—does not work to produce the variance. Let us see first what it does produce. From (9.1.6) we have

$$V(Y|X = x) = E(Y^2|X = x) - [E(Y|X = x)]^2,$$

and if we multiply by $p_X(x)$ and sum over all x we get

$$\sum_x V(Y|X = x)p_X(x) = E(Y^2) - \sum_x [E(Y|X = x)]^2 p_X(x).$$

The $E(Y^2)$ term on the right comes from the law of total expectation, but that law does not simplify the sum on the right. If we restate this equality in the expected-value notation of (9.1.31), it reads

$$E(V(Y|X)) = E(Y^2) - E([E(Y|X)]^2).$$

Remember that $E(Y|X)$ is a random variable, a function of X. It will help to give it a name, so let us call it Z. Then what we have so far is $E(V(Y|X)) = E(Y^2) - E(Z^2)$; that is,

(9.1.38)
$$E(Y^2) = E(V(Y|X)) + E(Z^2).$$

What we want is $VY = E(Y^2) - (EY)^2$. However, we can get EY by conditioning and then square it:

$$EY = E(E(Y|X)) = EZ,$$

and, therefore,

(9.1.39)
$$(EY)^2 = (EZ)^2.$$

Putting (9.1.38) and (9.1.39) together, we get

$$\begin{aligned} VY &= E(Y^2) - (EY)^2 = E(V(Y|X)) + E(Z^2) - (EZ)^2 \\ &= E(V(Y|X)) + VZ. \end{aligned}$$

Since $Z = E(Y|X)$, we have derived the amazing formula

(9.1.40)
$$\mathbf{VY = E(V(Y|X)) + V(E(Y|X)).}$$

The variance of Y is the expected value of the conditional variance plus the variance of the conditional expected value.

To make sense of this formula, remember that $E(Y|X)$ and $V(Y|X)$ are random variables that are functions of X; their values when $X = x$ are $E(Y|X = x)$ and $V(Y|X = x)$.

Another way to understand it is to think about the procedure for using it:

> To find VY by conditioning on X, first find $E(Y|X = x)$ and $V(Y|X = x)$ as functions of x. Make them functions of X, take the variance of one and the expected value of the other, and add them.

(9.1.41) **Example.** Let $X \sim \text{geom}(p)$, and given $X = x$, let Y have conditionally the $\text{bin}(x + 1, p)$ distribution. This is somewhat similar to the situation in Example (9.1.13). In this case, imagine doing Bernoulli trials with success probability p until the first success; let X be the number of failures, so that $X + 1$ is the number of trials. Then we do $X + 1$ more trials, and count the number of successes; this is Y. Let us find EY and VY.

This is not quite the same as Example (9.1.13), though; there, the success probability was $\frac{1}{2}$ in the first series of trials and p in the second; here it is p in both.

In problems like this, it is always wise to begin by finding EX and VX so that they will be ready to use when we need them. In this case, because $X \sim \text{geom}(p)$, we know that

$$EX = q/p \quad \text{and} \quad VX = q/p^2.$$

Now we need to find $E(Y|X)$ and $V(Y|X)$. These will be functions of X. Then we need to take expected values and variances of them as indicated in the two formulas

$$EY = E(E(Y|X)) \quad \text{and} \quad VY = E(V(Y|X)) + V(E(Y|X)).$$

Given that $X = x$, Y is binomial with parameters $x + 1$ and p, so

$$E(Y|X = x) = (x + 1)p \quad \text{and} \quad V(Y|X = x) = (x + 1)pq.$$

That is,

$$E(Y|X) = (X + 1)p \quad \text{and} \quad V(Y|X) = (X + 1)pq.$$

Consequently,

$$EY = E((X + 1)p) = p(EX + 1),$$

and, because $EX = q/p$, we get

$$EY = p\left(\frac{q}{p} + 1\right) = 1.$$

As for the variance,

$$VY = E((X + 1)pq) + V((X + 1)p) = pq(EX + 1) + p^2VX,$$

and since $EX = q/p$ and $VX = q/p^2$, we get

$$VY = pq\left(\frac{q}{p} + 1\right) + p^2\frac{q}{p^2} = 2q.$$

It should come as no surprise that $EY = 1$. It simply says that if we do Bernoulli trials until we get the first success, then do that many trials again, we can expect to average one success in the second series of trials. ∎

EXERCISES FOR SECTION 9.1

1. Consider a Bernoulli process with success probability p. Let Y be the number of successes in the first X trials, where X is a random variable having the Poisson distribution with parameter λ. Find the following.
   a. EY b. VY c. The mass function of Y

2. We do n Bernoulli trials with success probability p, and X is the number of successes. Then we do n more Bernoulli trials, but the success probability is X/n. Find the expected number of successes in the second series.

3. **The sum of a random number of i.i.d. random variables.** Let $Y_1, Y_2, Y_3, \ldots$ be i.i.d. random variables, with $EY_j = \mu$ and $VY_j = \sigma^2$. Let $Y = Y_1 + Y_2 + \cdots + Y_X$, where X is a positive integer-valued random variable. Show that

$$EY = \mu EX \quad \text{and} \quad VY = \mu^2VX + \sigma^2EX$$

(provided, of course, that EX and VX exist). [*Hint:* Given $X = x$, Y has conditionally the same distribution as $Y_1 + Y_2 + \cdots + Y_x$. Consequently, you can write down $E(Y|X = x)$ and $V(Y|X = x)$. Then use the procedures described in (9.1.26) and (9.1.37).]

Many real-world chance phenomena involve the sum of a random number of random variables. To take a simple example, let X be the number of claims an insurance company has to pay in a given month, and let $Y_1, Y_2, \ldots$ be the dollar amounts of these claims. Then $Y_1 + Y_2 + \cdots + Y_X$ is the total amount the company must pay in claims.

4. Suppose we observe a Poisson process with arrival rate λ per minute, but that we are not able to detect all the arrivals; rather each arrival has probability p of being observed, independently of the others. Let Y be the number of arrivals observed in the first minute. Find the mass function of Y and identify the distribution by name. [*Hint:* Condition on the total number of arrivals, observed and unobserved, in the first minute. See Exercise 1 above. Also see Exercise 10 of Section 3.4, where we got the same result with a lot more work, because we did not then have the conditioning formulas.]

5. You watch a Poisson process (with arrival rate λ per minute) for a minute and count the arrivals. How much longer will you have to watch to see the same number of arrivals again? Let Y be the additional time; would you guess that EY is 1 minute? Think about it, and then find EY and VY.

6. You do n Bernoulli trials and count the successes; then you do a second series of trials with the same success probability, stopping after you have obtained the same number of successes you got in the first n. Would you guess that the expected number of additional trials is n? Think about it; then let Y be the number of *failures* in the second series.
a. Find EY and VY.
b. Find the expected number of *trials* in the second series.

7. Let X and Y be as in Example (9.1.14): the mass function of X is $p_X(n) = pq^{n-1}$ for $n = 1, 2, 3, \ldots$, and given that $X = n$, Y has the Erl(n, λ) density. In Example (9.1.24) we found EY. Find the following.
a. VY **b.** The distribution of Y [Find a density.]

8. Let Y be the number of successes in X Bernoulli trials (the success probability is p), where X has the Pois(λ) distribution.
a. Write down the joint mass function of X and Y.
b. Find the marginal mass function of Y.
c. Find the posterior mass function of X, given $Y = y$.

Parts a and b of this exercise are a repetition, in the language of Bayesian inference, of Exercise 4. A special case of this exercise is in Example (9.1.35).

9. **The average of a sample of random size.** In Exercise 3 we found the expected value and variance of $Y = Y_1 + \cdots + Y_X$, where X is a positive integer-valued random variable and the Y_j's are i.i.d. with expected value μ and variance σ^2. Now let $Z = Y/X$, the average of a random number of Y_j's. That is,

$$\text{given } X = x,\ Z \text{ has conditionally the same distribution as } \frac{Y_1 + \cdots + Y_x}{x}.$$

a. Write down $E(Z|X = x)$ and $V(Z|X = x)$.
b. Now write down the two random variables $E(Z|X)$ and $V(Z|X)$.
c. Show that $EZ = \mu$ and $VZ = \sigma^2 E(1/X)$.

Notice that whereas in Exercise 3 the results depend on the existence of EX and VX, in this exercise EZ and VZ exist whether EX and VX exist or not. This is true because $E(Z|X)$ is a constant random variable, whose expected value and variance both exist, and $V(Z|X)$ is a constant times $1/X$, a random variable that is between 0 and 1 and therefore has finite moments.

10. In a certain Bernoulli process, with success probability p, the successes are independently subjected to further inspection, and are called "unqualified successes" if they pass. The probability that a success is unqualified is α. Let Y be the number of unqualified successes in n trials. Would you guess that Y has the bin$(n, \alpha p)$ distribution? Think about it, and then find the distribution of Y by conditioning on X, the number of successes. Y has (conditionally) the bin(X, α) distribution.

11. Consider a Poisson process with arrival rate λ. Let a be a fixed time point, and let Y be the time from the last arrival before a, to a itself. By conditioning on the number of arrivals in the interval $(0, a)$, show that Y has (unconditionally) the exp(λ) distribution. [*Hint:* Use the following fact, noted in Exercise 13 of Section 8.6, and earlier in Exercise 11 of Section 2.4. In a Poisson process, given that there were n arrivals in interval I, they have conditionally the same distribution as the order statistics of a sample of size n from the uniform distribution on I. Consequently, the conditional probability that Y exceeds y is the same as the probability that all of n uniformly chosen points in $(0, a)$ happen to fall in $(0, a - y)$.]

Note 1: Conditioning on the number of arrivals in a given interval is a useful technique in many calculations involving Poisson processes, because of the fact that the arrival times behave conditionally like uniformly chosen points.

Note 2: The result of this exercise points to an interesting "paradox" of Poisson processes. Let Y be as in the exercise, and let Z be the time from a to the next arrival after a. Then Y and Z are independent, each with the exp(λ) distribution, and so $Y + Z$ has the Erl$(2, \lambda)$ distribution. The expected value of $Y + Z$ is $2/\lambda$. On the other hand, $Y + Z$ seems to be just the time between two arrivals, and so it should have the exp(λ) distribution. Its expected value is $1/\lambda$. Which is correct?

This is sometimes called the "bus stop" paradox. If buses are coming to a bus stop according to a Poisson process and you get there at a random time, then your arrival somehow doubles the expected time between the previous bus and the next one.

The truth is that $Y + Z$ is not just the time between two arbitrary arrivals. It is the length of the time interval between *the two arrivals on either side of a previously chosen time point*, which tends to be longer on average than the interval between two arbitrary arrivals. Its expected value is $2/\lambda$.

12. **The proportion of people that are firstborns.** In Exercises 20 and 21 of Section 4.2 we considered Lotka's model for family sizes. In this model, if X is the number of children in a randomly chosen family, then the mass function of X is given by $p(0) = 1 - \beta q$ and $p(k) = \beta p q^k$ for $k \geq 1$. (According to Lotka's data, $p = .2642$, $q = .7358$, and $\beta = .8785$.) We found the probability that a random person is a firstborn by dividing the expected number of firstborns in a large group of families (which is the expected number of families with one or more children) by the expected number of children in the group. The answer is $p = .2642$.

a. Show that this is the reciprocal of the conditional expected value of X, given that $X \geq 1$.

b. By contrast, find the conditional expected value of $1/X$, given that $X \geq 1$. [See (A.5.12) in Appendix A for the formula needed.]

[a]**c.** Compare the answers to parts a and b for Lotka's values of p and q. Can you explain why the answer to part b is not the correct probability that a randomly chosen person is a firstborn?

ANSWERS

1. a. λp **b.** λp **c.** $Y \sim \text{Pois}(\lambda p)$

2. np

4. $\text{Pois}(\lambda p)$

5. $EY = 1$; $VY = 2/\lambda$

6. a. $EY = nq$; $VY = \dfrac{nq}{p}(1 + q^2)$ **b.** n

7. a. $1/(\lambda p)^2$ **b.** Exponential with parameter λp

8. b. $p(y) = \dfrac{(\lambda p)^y e^{-\lambda p}}{y!}$ for $y = 0, 1, 2, \ldots$

c. $P(X = x \mid Y = y) = \dfrac{(\lambda q)^{x-y} e^{-\lambda q}}{(x - y)!}$ for $x = y, y + 1, y + 2, \ldots$; that is, the posterior distribution of $X - y$ is $\text{Pois}(\lambda q)$.

9. a. μ; $\dfrac{\sigma^2}{x}$

10. Yes, $Y \sim \text{bin}(n, \alpha p)$

12. a. Conditional expected value is $\sum_1^\infty k \cdot \dfrac{p(k)}{1 - p(0)} = \dfrac{EX}{P(X \geq 1)}$; in Section 4.2 we calculated $\dfrac{N \cdot P(X \geq 1)}{N \cdot EX}$.

b. $\sum_1^\infty \dfrac{1}{k} \cdot \dfrac{p(k)}{1 - p(0)} = -\dfrac{p}{q} \ln p = .4779$

c. The answer to part b is the probability that we get a firstborn if we choose a family at random, then choose a child at random from the family. This experiment is not the same as choosing a person at random from the population; in the experiment of first choosing a family, then a child at random, people from large families are less likely to be chosen.

9.2 CONDITIONING ON AN ABSOLUTELY CONTINUOUS RANDOM VARIABLE

In the last section we saw a number of instances of the "principle of conditioning." To find some quantity—say EY or $P(B)$—by conditioning on a discrete random variable X, we start with the conditional quantity—$E(Y \mid X = x)$

or $P(B|X = x)$—multiply by the mass function of X, and sum the product over all possible values of X.

In this section we will see that all those formulas are valid as well when X is absolutely continuous; we only need to modify the principle of conditioning. Instead of multiplying a conditional quantity by the mass function of X and adding, we multiply it by a density for X and integrate. The only difficulty is the question of what these conditional quantities are. If X is absolutely continuous, the event $(X = x)$ has probability 0 for any x. What can we possibly mean in this case by $E(Y|X = x)$ or $P(B|X = x)$?

We get an answer by looking at how we would like such a quantity to behave. Our goal is to get the principle of conditioning to work; so we look at the conditioning formulas, which are integrals, and decide how the integrands should be defined so that the formulas are valid.

We start with conditional densities.

(9.2.1) **Finding a density for Y by conditioning on an absolutely continuous X.** Let X and Y be absolutely continuous random variables. They have a joint density function $f(x, y)$ that determines their joint distribution, and they also have their own densities, $f_X(x)$ and $f_Y(y)$, that can be found as marginal densities by integrating the joint density. These matters are discussed in Section 3.5. The main thing to keep in mind is that

(9.2.2)
$$f_Y(y) = \int_{-\infty}^{\infty} f(x, y)dx.$$

How shall we define $f_Y(y|X = x)$? So that the principle of conditioning produces a density for Y. That is, we should define it so that the conditioning formula holds:

(9.2.3)
$$f_Y(y) = \int_{-\infty}^{\infty} f_Y(y|X = x) \cdot f_X(x)dx.$$

If we knew a function of x and y whose integral with respect to x gives $f_Y(y)$, we could set it equal to the integrand of (9.2.3), and use that to define $f_Y(y|X = x)$. But (9.2.2) shows us such a function—the joint density func $f(x, y)$.

So we ought to define $f_Y(y|X = x)$ so that $f_Y(y|X = x) \cdot f_X(x) = f(x, y)$. that is what we do; the ***conditional density of Y, given X = x***, is defined b

(9.2.4)
$$f_Y(y|X = x) = \frac{f(x, y)}{f_X(x)} \quad \text{for all } x \text{ such that } f_X(x) > 0.$$

Notice the requirement that $f_X(x)$ be positive, which is satisfied on the set of possible values of X. This replaces the impossible requirement that $P(X = x)$ be positive, which is never satisfied in the absolutely continuous case.

As we saw in the last section, in most applications of these ideas we do not get a conditional density by computing it from the definition; more often we assume its value and use it in conditioning.

(9.2.5) **Example: An exponential distribution with random parameter.** Let Y have an exponential distribution, but let the parameter λ of the distribution be a random variable—call it X rather than λ—whose density is

$$f_X(x) = \frac{x^2}{2}e^{-x} \quad \text{for } x > 0.$$

(This is the Erlang density with $n = 3$ and $\lambda = 1$, but that is immaterial here.)

What is a density for Y? What we are given is that a conditional density for Y, given $X = x$, is the exp(x) density:

$$f_Y(y|X = x) = xe^{-xy} \quad \text{for } y > 0.$$

So according to (9.2.3), a density for Y is

$$f_Y(y) = \int_{-\infty}^{\infty} xe^{-xy} \cdot \frac{x^2}{2}e^{-x}dx = \frac{1}{2}\int_{-\infty}^{\infty} x^3e^{-x(y+1)}dx,$$

and we can get this using the gamma formula (8.3.4) (first seen in Exercise 10 of Section 4.3):

$$f_Y(y) = \frac{3}{(y+1)^4} \quad \text{for } y > 0.$$

∎

(9.2.6) **Theorem (Bayes's theorem for two absolutely continuous random variables).** Suppose X and Y are absolutely continuous, and that we are given $f_X(x)$ and $f_Y(y|X = x)$. Then Bayes's theorem states that we can find the "reverse" conditional density $f_X(x|Y = y)$ using

$$f_X(x|Y = y) = \frac{f_Y(y|X = x)f_X(x)}{\int_{-\infty}^{\infty} f_Y(y|X = z)\,f_X(z)dz}.$$

This is clearly similar to the corresponding Theorem (9.1.34) for two discrete random variables. The difference is that densities replace mass functions, and instead of summing over possible values of X, we integrate.

Notice that the denominator in the preceding formula is $f_Y(y)$, because of (9.2.3). Accordingly, it might be better to remember the formula as

(9.2.7)
$$f_X(x|Y = y) = \frac{f_Y(y|X = x)f_X(x)}{f_Y(y)},$$

and to remember that the denominator can be found by the conditioning formula (9.2.3).

Proof. The proof of (9.2.7) is very simple. By (9.2.4), the numerator equals the joint density $f(x, y)$. So by (9.2.4) again, but with X and Y interchanged, the fraction equals $f_X(x|Y = y)$. □

(9.2.8) **Example.** In Example (9.2.5) Y has the exponential distribution, but the parameter λ is a random variable X whose density is $\frac{1}{2}x^2e^{-x}$ for $x > 0$. We found that an unconditional density for Y is $3/(y + 1)^4$ for $y > 0$. Now we ask: given $Y = y$, what is a conditional density for the parameter X?

Bayes's formula gives

$$f_X(x|Y = y) = \frac{xe^{-xy} \cdot (x^2/2)e^{-x}}{3/(y + 1)^4} = \frac{(y + 1)^4}{3!}x^3e^{-(y+1)x} \quad \text{for } x > 0.$$

That is, the conditional density of X, given $Y = y$, is the Erlang density with parameters $n = 4$ and $\lambda = y + 1$. ▮

This example is typical of the kind of calculation done in Bayesian inference, as we discussed in (9.1.36). We can think of the density $\frac{1}{2}x^2e^{-x}$ as being a prior density for the parameter X; then the conditional density found in the example is a posterior density for X, given that Y is observed to equal y.

(9.2.9) **Finding a mass function for discrete Y by conditioning on absolutely continuous X.** We proceed as we did in (9.2.1). We want to find a way to define a conditional mass function, $p_Y(y|X = x) = P(Y = y|X = x)$, even though $P(X = x)$ is 0. Our goal is to make the conditioning formula hold; that is,

(9.2.10)
$$p_Y(y) = \int_{-\infty}^{\infty} p_Y(y|X = x) \cdot f_X(x)dx.$$

So we need to find some function $g(x, y)$ such that

(9.2.11)
$$\int_{-\infty}^{\infty} g(x, y)dx = p_Y(y);$$

then we can define $p_Y(y|X = x) = g(x, y)/p_Y(y)$, and (9.2.10) will hold. We have not seen such a function before. It is the ***joint density-mass function*** for X and Y, which takes the place of the joint density $f(x, y)$ of two absolutely continuous random variables, or the joint mass function $p(x, y)$ of two discrete random variables. Because X is absolutely continuous and Y is discrete, the two variables in $g(x, y)$ play different roles.

The essential facts about a joint density-mass function are, first, that it satisfies (9.2.11); and second, that

(9.2.12)

$$P(X \leq x \text{ and } Y = y) = \int_{-\infty}^{x} g(t, y)dt.$$

Summing this over all y gives

$$\begin{aligned} F_X(x) &= P(X \leq x) \\ &= \int_{-\infty}^{x} \sum_y g(t, y)dt; \end{aligned}$$

as usual we justify summing inside the integral sign by appealing to the uniform convergence of the sum as a function of t. Finally, differentiating this integral (assuming continuity of the integrand) gives

(9.2.13)

$$f_X(x) = \sum_y g(x, y).$$

That is, we can get a marginal density for X by summing, and the marginal mass function of Y by integrating, the joint density mass function.

Putting this together, we see that we should define the ***conditional mass function of Y, given X = x***, by

(9.2.14)

$$p_Y(y|X = x) = \frac{g(x, y)}{f_X(x)}, \quad \text{for all } x \text{ such that } f_X(x) \neq 0,$$

where $g(x, y)$ is a function satisfying (9.2.11), (9.2.12), and (9.2.13)—that is, a joint density-mass function for X and Y.

Once again, we rarely encounter a joint density-mass function as given; nearly always it is the conditional mass function that we write down first, as in the following example.

(9.2.15) **Example: Bernoulli trials with random success probability.** Let Y be the number of successes in n Bernoulli trials, but let the success probability p be a random variable—call it X instead of p—that is uniformly distributed on (0, 1). What is the mass function of Y?

We know that $X \sim \text{unif}(0, 1)$, and that, given $X = x$, $Y \sim \text{bin}(n, x)$. So

$$p_Y(k|X = x) = \binom{n}{k} x^k(1 - x)^{n-k} \quad \text{for } k = 0, 1, 2, \ldots, n$$

and

$$f_X(x) = 1 \quad \text{for } 0 < x < 1.$$

Therefore, by the conditioning formula (9.2.10),

$$p_Y(k) = \int_0^1 \binom{n}{k} x^k(1 - x)^{n-k} \cdot 1\, dx \quad \text{for } k = 0, 1, 2, \ldots, n.$$

Now the beta formula (8.6.1) implies that

$$\int_0^1 x^i(1-x)^j dx = \frac{i!j!}{(i+j+1)!},$$

and so we have

$$p_Y(k) = \binom{n}{k}\frac{k!(n-k)!}{(n+1)!};$$

that is,

$$p_Y(k) = \frac{1}{n+1} \quad \text{for } k = 0, 1, 2, \ldots, n.$$

Thus, Y does not have a binomial distribution at all; its $n + 1$ possible values are equally likely. This is not surprising when we consider that all success probabilities are equally likely, in the sense that the success probability has the unif(0, 1) distribution. ∎

(9.2.16) **Example.** Let Y have a Poisson distribution in which the parameter λ is a random variable X that has some given distribution on $(0, \infty)$. Suppose X has density $f_X(x)$ for $x > 0$, and that, conditional on $X = x$, $Y \sim$ Pois(x). Then unconditionally, Y does not have a Poisson distribution; its mass function is

$$p_Y(k) = \int_0^\infty \frac{x^k e^{-x}}{k!} \cdot f_X(x)dx.$$

Suppose, for example, that X has the exponential distribution with parameter β. Then

$$p_Y(k) = \int_0^\infty \frac{x^k e^{-x}}{k!} \cdot \beta e^{-\beta x}dx,$$

and you can check using the gamma formula that this equals pq^k, where $p = \beta/(\beta + 1)$ and $q = 1/(\beta + 1)$. ∎

(9.2.17) **Theorem (Bayes's theorem for absolutely continuous *X* and discrete *Y*).** The previous conditioning formula (9.2.10) starts with the conditional mass function of Y, $p_Y(y|X = x)$, and an unconditional density for X, $f_X(x)$, and gives the unconditional mass function for Y. We can now continue and get the "reverse" conditional quantity, which now is a conditional density, $f_X(x|Y = y)$. The formula is

$$f_X(x|Y = y) = \frac{p_Y(y|X = x) \cdot f_X(x)}{\int_{-\infty}^{\infty} p_Y(y|X = z) \cdot f_X(z)dz}.$$

Notice that the denominator is just the unconditional mass function $p_Y(y)$, by (9.2.10), and so we might prefer to remember this in the form

(9.2.18)
$$f_X(x \mid Y = y) = \frac{p_Y(y \mid X = x) \cdot f_X(x)}{p_Y(y)},$$

using (9.2.10) for the denominator. The proof of this goes as follows.

Proof. By (9.2.14), the numerator in (9.2.18) equals $g(x, y)$, and so we need to prove that

(9.2.19)
$$g(x, y) = f_X(x \mid Y = y) \cdot p_Y(y).$$

Now $f_X(x \mid Y = y)$ was defined in (9.1.8) to be the derivative of $F_X(x \mid Y = y) = P(X \leq x \mid Y = y)$, and so the right side of (9.2.19) is the derivative of $P(X \leq x$ and $Y = y)$. By (9.2.12), this is precisely $g(x, y)$. ▯

(9.2.20) **Example.** In Example (9.2.15), $Y \sim \text{bin}(n, X)$, where the random success probability X has the unif(0, 1) distribution. Think of this uniform distribution as the prior distribution on the unknown success probability. Given that we observe $Y = k$, where k is one of the possible values $0, 1, 2, \ldots, n$ of Y, what is the posterior distribution of X? That is, what is a conditional density for X, given $Y = k$?

We apply Bayes's theorem in the form (9.2.18). Looking back at Example (9.2.15), we see that we know all three of the quantities in (9.2.18); the result is

$$f_X(x \mid Y = k) = \frac{\binom{n}{k} x^k (1 - x)^{n-k} \cdot 1}{1/(n + 1)};$$

that is,

$$f_X(x \mid Y = k) = \frac{(n + 1)!}{k!(n - k)!} x^k (1 - x)^{n-k} \qquad \text{for } 0 < x < 1 \quad \text{and } k = 0, 1, 2, \ldots, n.$$

This is the beta density with parameters $k + 1$ and $n - k + 1$.

We can summarize the results of this example and Example (9.2.15) as follows: If $Y \sim \text{bin}(n, X)$ where $X \sim \text{unif}(0, 1)$, then unconditionally the possible values $0, 1, 2, \ldots, n$ of Y are equally likely; and given $Y = k$, the unknown X has conditionally the beta$(k + 1, n - k + 1)$ distribution.

This is related to the "converse problem" proposed and solved by Thomas Bayes in his 1763 paper, which led to the use of his name for the subject. See the note at the end of Section 2.3 for Bayes's statement of the problem. The above is not the complete solution, nor is it the solution Bayes gave; we have simply assumed a certain prior distribution [uniform on (0, 1)] for the success probability, and found the resulting posterior distribution. ▮

As we discussed in (9.1.36), this leads to a Bayesian estimator of the unknown success probability. The reasoning goes like this: at the outset, we have no reason to believe that the success probability is more likely to be in one part of the unit interval than another, so we assume that it is a random variable with the unif(0, 1) distribution. This is the prior distribution for the parameter. After observing $Y = k$ successes in n trials, the posterior distribution is beta($k + 1$, $n - k + 1$). The expected value of this distribution is $(k + 1)/(n + 2)$; the mode (x at which the density is maximized) is k/n. The latter is the "classical" estimator of the success probability when k successes are observed. The former is the Bayesian estimator. The question of the relative merits of these two estimators is discussed in statistics courses.

(9.2.21) Theorem (Bayes's theorem for discrete *X* and absolutely continuous *Y*). As we will point out in (9.2.23), this is the last of a series of eight formulas—four conditioning formulas for finding densities and mass functions and four versions of Bayes's formula. This one involves a discrete X, but we could not cover it in Section 9.1 because it also involves $p_X(x \mid Y = y)$ for an absolutely continuous Y. The formula is

$$p_X(x \mid Y = y) = \frac{f_Y(y \mid X = x)p_X(x)}{\sum_z f_Y(y \mid X = z)p_X(z)}.$$

But because by (9.1.20) the denominator is $f_Y(y)$, we can remember this in the form

$$p_X(x \mid Y = y) = \frac{f_Y(y \mid X = x)p_X(x)}{f_Y(y)}. \tag{9.2.22}$$

Proof. The proof of (9.2.22) is similar to the proofs of the other three versions, once we realize that the numerator is a function $h(x, y)$ that is a *joint mass-density function* for X and Y. It behaves exactly like the joint density-mass function defined earlier at (9.2.11)–(9.2.13), except that it is a mass function of x and a density function of y instead of the other way around. ▯

(9.2.23) Summary: Conditioning and Bayes's theorem for various combinations of discrete and absolutely continuous random variables. There are four possible combinations for X and Y: both discrete, both absolutely continuous, and two mixed cases. There is a conditioning formula and a version of Bayes's theorem in each case—eight formulas in all—and we have now seen all eight.

But of course no one would try to memorize all eight of them separately; they are all versions of two generic formulas, which we can summarize as follows.

Conditioning formula

$$\begin{pmatrix}\text{mass function}\\ \text{or density of } Y\end{pmatrix} = \left(\sum \text{ or } \int\right)\begin{pmatrix}\text{conditional mass function or}\\ \text{density of } Y\text{, given } X = x\end{pmatrix} \cdot \begin{pmatrix}\text{unconditional mass}\\ \text{function or density of } X\end{pmatrix}$$

Bayes's theorem

$$\begin{pmatrix}\text{conditional mass function or}\\ \text{density of } X\text{, given } Y = y\end{pmatrix} = \frac{\begin{pmatrix}\text{conditional mass function or}\\ \text{density of } Y\text{, given } X = x\end{pmatrix} \cdot \begin{pmatrix}\text{unconditional mass}\\ \text{function or density of } X\end{pmatrix}}{\begin{pmatrix}\text{mass function}\\ \text{or density of } Y\end{pmatrix}}$$

For the record, here is where to find the eight formulas in this chapter.

X	Y	Conditioning formula	Bayes's theorem
Discrete	Discrete	(9.1.17)	(9.1.34)
Discrete	Absolutely continuous	(9.1.20)	(9.2.22)
Absolutely continuous	Discrete	(9.2.10)	(9.2.18)
Absolutely continuous	Absolutely continuous	(9.2.3)	(9.2.7)

The generic conditioning formula is of course the "principle of conditioning," which we described in the first paragraph of this section. It works for finding not only densities and mass functions, but also probabilities and expected values. Variances are the only common type of quantity for which it fails and a different formula is needed. We saw all this for discrete X in the previous section, and we will see it next for absolutely continuous X.

(9.2.24) Conditional expected values and variances, given an absolutely continuous X. These definitions are just what you would expect.

If Y is discrete, the conditional expected value and variance of Y, given $X = x$, is

(9.2.25)
$$E(Y|X = x) = \sum_y y p_Y(y|X = x)$$

and

(9.2.26)
$$V(Y|X = x) = \sum_y (y - E(Y|X = x))^2 p_Y(y|X = x).$$

If Y is absolutely continuous, then its conditional expected value and variance are

(9.2.27)
$$E(Y|X = x) = \int_{-\infty}^{\infty} y f_Y(y|X = x)dy$$

and

(9.2.28)
$$V(Y|X = x) = \int_{-\infty}^{\infty} (y - E(Y|X = x))^2 f_Y(y|X = x)dx.$$

Just as when X is discrete, these are defined only provided that the sums or integrals converge absolutely. As with discrete random variables, it can be proved that if the sums or integrals for the unconditional quantities EY and VY converge absolutely, then so do the sums or integrals above, for any random variable X.

It can also be shown with a little algebra that in both cases,

(9.2.29)
$$V(Y|X = x) = E(Y^2|X = x) - [E(Y|X = x)]^2.$$

(9.2.30) **Finding EY by conditioning on an absolutely continuous X.** Here as in (9.1.22) there is only one formula, but two proofs [as in (9.1.25)]. The formula is exactly what you would expect; it is another version of the ***law of total expectation***:

$$EY = \int_{-\infty}^{\infty} E(Y|X = x) f_X(x)dx.$$

Proof for discrete Y. By (9.2.25), the right side equals

$$\int_{-\infty}^{\infty} \sum_y y p_Y(y|X = x) f_X(x)dx,$$

and, as in other cases, we can reverse the order of summing and integrating. We get

$$\sum_y y \int_{-\infty}^{\infty} p_Y(y|X = x) f_X(x)dx,$$

and by the conditioning formula (9.2.10) this equals $\sum_y y p_Y(y)$, which is EY.

Proof for absolutely continuous Y. This is similar to the proof for discrete Y, except that we first use (9.2.27) on the right side, then reverse the order of the double integral, and then use (9.2.3) to see that the result is EY. ∎

(9.2.31) **The law of total expectation as an expected value.** As in Section 9.1, we notice here that if we define

$$h(x) = E(Y|X = x),$$

then the law of total expectation (9.2.30) reads $EY = Eh(X)$. Furthermore, if we denote $h(X)$ by $E(Y|X)$, then it reads

$$EY = E(E(Y|X)).$$

We can make sense of this by remembering that

> $E(Y|X)$ is a random variable that is a function of X; its value when $X = x$ is $E(Y|X = x)$. Its expected value is EY.

Alternatively, we can remember the procedure:

> To find EY by conditioning on X, first find $E(Y|X = x)$ as a function of x. Then make it a function of X, and find its expected value.

We will see some examples of this after giving the variance formula.

(9.2.32) **Finding VY by conditioning on an absolutely continuous X.** We get the same remarkable formula that we got in Section 9.1:

$$VY = E(V(Y|X)) + V(E(Y|X)).$$

This appeared as (9.1.40), and the derivation began at (9.1.37). Although we were dealing with discrete X there, the derivation did not depend on that assumption, and it works perfectly well in general. We recapitulate it here.

Let Z denote the random variable $E(Y|X)$. Then by the law of total expectation

$$EZ = EY,$$

and by (9.2.29),

$$E(V(Y|X)) = E(E(Y^2|X)) - E(Z^2),$$

which says that

$$E(Y^2) = E(V(Y|X)) + E(Z^2).$$

Therefore,

$$VY = E(Y^2) - [EY]^2 = E(V(Y|X)) + E(Z^2) - [EZ]^2;$$

that is,

$$VY = E(V(Y|X)) + VZ = E(V(Y|X)) + V(E(Y|X)).$$

As we mentioned in the previous section, the way to make sense of this formula is to remember that $E(Y|X)$ and $V(Y|X)$ are random variables that are functions of X; their values when $X = x$ are $E(Y|X = x)$ and $V(Y|X = x)$. Another way to understand it is to think about the procedure for using it:

> Find $E(Y|X = x)$ and $V(Y|X = x)$ as functions of x. Make them functions of X, take the variance of one and the expected value of the other, and add them.

(9.2.33) **Example.** In Example (9.2.15), we found the mass function of a random variable Y which, given $X = x$, is conditionally bin(n, x), where X has the uniform distribution on $(0, 1)$. We can find EY and VY very quickly without bothering with the mass function, if we just remember that $EX = \frac{1}{2}$, $E(X^2) = \frac{1}{3}$, and $VX = \frac{1}{12}$:

$$E(Y|X = x) = nx \quad \text{and} \quad V(Y|X = x) = nx(1 - x);$$

that is,

$$E(Y|X) = nX \quad \text{and} \quad V(Y|X) = nX(1 - X).$$

Therefore,

$$EY = E(nX) = nEX = n \cdot \tfrac{1}{2},$$

and

$$VY = V(nX) + E(nX(1 - X)) = n^2VX + nEX - nE(X^2)$$
$$= \frac{n^2}{12} + \frac{n}{2} - \frac{n}{3} = \frac{n(n + 2)}{12}.$$

You can check that these are correct by finding EY, EY^2, and VY directly from the mass function that we found in Example (9.2.15). ▮

(9.2.34) **Example.** Let X have some unspecified distribution on $(0, \infty)$ with $EX = \mu$ and $VX = \sigma^2$. Conditional on $X = x$, let Y have the Pois(x) distribution. We discussed this in Example (9.2.16). Here we simply give EY and VY; they are found using the law of total expectation and the variance formula. Exercise 7 at the end of the section asks you to check them.

$$EY = \mu \quad \text{and} \quad VY = \mu + \sigma^2$$

The only fact you need is that the expected value and variance of the Pois(λ) distribution are both equal to λ. ▮

(9.2.35) Finding generating functions by conditioning. If we know the conditional distribution of Y, given $X = x$, and the unconditional distribution of X, we can find the distribution of Y using one of the four conditioning formulas discussed in (9.2.23). But sometimes we can make it simpler than that, by using the law of total expectation to find the MGF or the PGF of Y. This is possible because the MGF or PGF of Y is the expected value of a certain function of Y. Consequently, we can use the following.

The PGF of Y is $\pi_Y(t) = E(t^Y) = E(E(t^Y|X))$.

The MGF of Y is $m_Y(s) = E(e^{sY}) = E(E(e^{sY}|X))$.

The quantities $E(t^Y|X)$ and $E(e^{sY}|X)$ are the ***conditional*** **PGF** and the ***conditional*** **MGF** of Y, given X; they are random variables that are functions of X, just as other conditional expected values are. They are denoted by $\pi_Y(t|X)$ and $m_Y(s|X)$.

An example or two should make the method clear.

(9.2.36) **Example.** In Example (9.2.16) we saw the following situation: $X \sim \exp(\beta)$ and, given $X = x$, Y has conditionally the Pois(x) distribution. We found there that the unconditional distribution of Y is a geometric distribution. Here we rederive this result by finding the PGF of Y as described in (9.2.35).

The conditional distribution of Y, given $X = x$, is Pois(x); so the conditional PGF of Y is the Pois(x) PGF, which is

$$\pi_Y(t|X = x) = e^{x(t-1)} \quad (t \in \mathbb{R}).$$

We can write this as

$$\pi_Y(t|X) = e^{X(t-1)} \quad (t \in \mathbb{R}).$$

So the unconditional PGF of Y is the expected value of this:

$$\pi_Y(t) = E(e^{X(t-1)}).$$

How do we find this? Simple—remember that the MGF of X is defined as $m_X(s) = E(e^{sX})$. So

$$\pi_Y(t) = m_X(t - 1).$$

Now $X \sim \exp(\beta)$, and so

$$m_X(s) = \frac{\beta}{\beta - s}.$$

Therefore,

$$\pi_Y(t) = \frac{\beta}{\beta + 1 - t}.$$

Is this really a geometric PGF? Yes, it equals $p/(1 - qt)$ where $p = \beta/(\beta + 1)$ and $q = 1 - p$. ∎

(9.2.37) **Example: A normal distribution with a random parameter.** Let Y have the $N(X, \tau^2)$ distribution, where X is a random variable having the $N(\lambda, \sigma^2)$ distribution. (The symbol τ is a lowercase Greek tau.) We will do four things here:

a. We will find EY and VY using the law of total expectation and the variance formula; these calculations are very quick.
b. Then we will indicate what would be involved in finding an unconditional density for Y using the conditioning formula (9.2.3). The calculation uses quite a bit of algebra, as do many calculations for normal distributions. It is unnecessary because the next step does it more simply, but we include it to show what it involves.
c. We will condition on X to find the MGF of Y, and will recognize it as the MGF of a certain normal distribution. This is considerably simpler than the preceding calculation of the unconditional distribution of Y.
d. Finally, we will use Bayes's theorem to find the conditional (posterior) distribution of X, given $Y = y$.

Solutions.

a. **Finding EY and VY.** We are given that

$$EX = \mu \quad \text{and} \quad VX = \sigma^2,$$

and also that

$$E(Y|X) = X \text{ and } V(Y|X) = \tau^2.$$

Therefore,

$$EY = EX = \mu,$$

and

$$VY = V(X) + E(\tau^2) = \sigma^2 + \tau^2.$$

b. **Conditioning on X to find a density for Y.** A conditional density for Y, given $X = x$, is

$$f_Y(y|X = x) = \frac{1}{\sqrt{2\pi\tau^2}} e^{-(y-x)^2/2\tau^2} \quad \text{for } y \in \mathbb{R},$$

and so we get an unconditional density for Y by multiplying by the density of X and integrating:

$$f_Y(y) = \int_{-\infty}^{\infty} \frac{1}{\sqrt{2\pi\tau^2}} e^{-(y-x)^2/2\tau^2} \cdot \frac{1}{\sqrt{2\pi\sigma^2}} e^{-(x-\mu)^2/2\sigma^2} dx.$$

It takes some work, but if we combine the two exponents, do some algebra, and factor the right constants out of the integral, the rest becomes the integral of a normal density for X, which equals 1. What is left outside the integral is a density for Y, which turns out to be a normal density. Knowing that, we know which normal density, because we found EY and VY above.

c. **Conditioning on X to find the MGF of Y.** We know that, given $X = x$, Y has conditionally the $N(x, \tau^2)$ distribution, and we know the MGF of that distribution. It is

$$m_Y(s|X = x) = e^{xs+\tau^2 s^2/2} \quad (s \in \mathbb{R}).$$

That is,

$$m_Y(s|X) = e^{Xs+\tau^2 s^2/2} \quad (s \in \mathbb{R}),$$

and so the unconditional MGF of Y is the expected value of this function of X:

$$m_Y(s) = e^{\tau^2 s^2/2} E(e^{sX}).$$

But $E(e^{sX})$ is exactly the MGF of X, which is $e^{\mu s+\sigma^2 s^2/2}$, because $X \sim N(\mu, \sigma^2)$. Therefore,

$$\begin{aligned} m_Y(s) &= e^{\tau^2 s^2/2}\, e^{\mu s+\sigma^2 s^2/2} \\ &= e^{\mu s+(\sigma^2+\tau^2)s^2/2}. \end{aligned}$$

This is the MGF of the $N(\mu, \sigma^2 + \tau^2)$ distribution, and therefore that is the distribution of Y.

d. **The conditional distribution of X, given $Y = y$.** We use (9.2.7), which is Bayes's theorem for two absolutely continuous random variables:

$$f_X(x|Y = y) = \frac{f_Y(y|X = x) f_X(x)}{f_Y(y)}.$$

All three quantities on the right are normal densities:

$$f_Y(y|X = x) = \frac{1}{\sqrt{2\pi\tau^2}} e^{-(y-x)^2/2\tau^2}, \quad \text{the } N(x, \tau^2) \text{ density;}$$

$$f_X(x) = \frac{1}{\sqrt{2\pi\sigma^2}} e^{-(x-\mu)^2/2\sigma^2}, \quad \text{the } N(\mu, \sigma^2) \text{ density;}$$

and, as we have just learned using MGFs,

$$f_Y(y) = \frac{1}{\sqrt{2\pi(\sigma^2 + \tau^2)}} e^{-(y-\mu)^2/2(\sigma^2+\tau^2)}, \quad \text{the } N(\mu, \sigma^2 + \tau^2) \text{ density.}$$

Putting these together involves some unavoidable algebra. The result, as you can check if you wish, is the $N((\mu\tau^2 + y\sigma^2)/(\sigma^2 + \tau^2), \sigma^2\tau^2/(\sigma^2 + \tau^2))$ density.

∎

(9.2.38) **Conjugate distributions in Bayesian inference.** The typical setup in Bayesian inference is as follows.

1. Start with a random variable Y whose distribution involves an unknown parameter X. The goal is to infer something about X by observing Y.
2. View X as a random variable. Because the distribution of Y involves X, it is now a conditional distribution.
3. If X is to be a random variable, it needs a distribution. Choose one that reflects knowledge about X. This is called a *prior* distribution for X, because it is decided on before any observations of Y are made.
4. Now use the principle of conditioning to get the unconditional distribution of Y.
5. Finally, do the experiment and observe Y; let the observed value be y. Use Bayes's theorem to find the conditional distribution of X, given $Y = y$. This is the *posterior* distribution for X. The expected value of this distribution is a *Bayesian estimator* of X, and the variance gives information about how confident we can be of the estimator.

We saw an example of this kind of setup in (9.1.36), and in a sense all the examples of Bayes's theorem in this chapter illustrate it.

The choice of a prior distribution for X is of course a key step. It turns out that in many cases, the computations can be greatly simplified if the prior distribution is chosen carefully, so that it fits well, in a certain sense, with the conditional distribution of Y, given X.

For example: Suppose Y has the Poisson distribution with unknown parameter X. If the prior distribution for X is an Erlang distribution, then as is shown in Exercise 5 at the end of this section, the posterior distribution is another Erlang distribution. On the other hand, if we use a distribution of some other family as a prior—say, for example, the density $f_X(x) = r/(x + 1)^{r+1}$—the computations become much more difficult, and the posterior distribution has no convenient connection with the prior.

There is considerable advantage, even apart from computational simplicity, in having the posterior distribution in the same family as the prior. For one thing, we can compare the parameters of the prior and the posterior and see quickly how the observation of Y has changed our knowledge of X. For another, we can take the posterior distribution as a new prior, take another independent observation of Y, and compute a new posterior that reflects both observations.

However, a warning is in order: it is unwise to base the choice of a prior distribution solely on computational convenience. In any real situation the prior should actually reflect one's belief about the unknown parameter. In addition, Bayesian inferences are often done more than once, with different priors, to assess the effects of the choice.

We say that the Erlang family is ***conjugate*** to the Poisson family because of the phenomenon described above: if the unknown parameter of a Poisson distribution has an Erlang prior distribution, then it has an Erlang posterior distribution.

We list here without proof the familiar families of distributions and their conjugate families. Some important families of distributions for Y—geometric and exponential, in particular, as well as $\mathrm{Erl}(n, X)$—appear to be missing, but in fact they are special cases of families included here.

Distribution of Y, given X	Conjugate prior distribution on X
$\mathrm{Pois}(X)$	Erlang
$\mathrm{Erl}(X, \lambda)$	Poisson
$\mathrm{Bin}(n, X)$	Beta
$\mathrm{N}(X, \tau^2)$	Normal
$\mathrm{Gamma}(\alpha, X)$	Gamma
Neg $\mathrm{bin}(n, X)$	Beta

EXERCISES FOR **SECTION 9.2**

1. Let X have the $\exp(\lambda)$ distribution. Given that $X = x$, let Y conditionally have the normal distribution with expected value μX and variance $\sigma^2 X^2$. Find EY and VY.

2. Let X have the mass function $p_X(n) = pq^{n-1}$ for $n = 1, 2, 3, \ldots$. Given that $X = n$, let Y have conditionally the $\mathrm{Erl}(n, \lambda)$ distribution, whose density is

$$\frac{\lambda^n}{(n-1)!} y^{n-1} e^{-\lambda y} \quad \text{for } y > 0.$$

Find an unconditional density function for Y.

This has an interpretation that we have seen before. In a Poisson process with arrival rate λ, suppose an arrival is a success with probability p and a failure with probability q. Then X is the number of arrivals we must count to get the first success, and Y is the time of the first successful arrival. The distribution of Y is in keeping with intuition.

3. Let $Y \sim \mathrm{bin}(n, X)$ where X has the density $3x^2$ on $(0, 1)$. Find the following.
a. The unconditional mass function of Y
b. A conditional density for X, given $Y = y$
[*Hint:* Use the beta formula in the form given in Example (9.2.15).]

4. If $EX = \mu$ and $VX = \sigma^2$ and, conditional on $X = x$, $Y \sim \exp(1/X)$, find EY and VY.

5. **The Erlang family is conjugate to the Poisson family.** Let Y have a Poisson distribution, whose parameter is a random variable X having the $\mathrm{Erlang}(n, \beta)$ density,

$$f(x) = \frac{\beta^n}{(n-1)!} x^{n-1} e^{-\beta x} \quad \text{for } x > 0.$$

a. Show, using (9.2.10), that the unconditional distribution of Y is the neg bin(n, $\beta/(1 + \beta)$) distribution.
b. Show that the conditional distribution of X, given $Y = k$, is the Erl($n + k$, $\beta + 1$) distribution.

6. **The beta family is conjugate to the binomial family.** Let Y have the bin(n, X) distribution, where X has the beta(i, j) density,

$$f(x) = \frac{(i + j - 1)!}{(i - 1)!(j - 1)!} x^{i-1}(1 - x)^{j-1} \quad \text{for } 0 < x < 1.$$

Show that the conditional distribution of X, given $Y = k$, is beta($i + k$, $n + j - k$).

7. Confirm the result of Example (9.2.34): If $EX = \mu$ and $VX = \sigma^2$, and if, given $X = x$, Y has conditionally the Pois(x) distribution, then $EY = \mu$ and $VY = \mu + \sigma^2$.

8. Let X have density $f_X(x)$ on $(0, \infty)$, and MGF $m(s)$. Let Y have the Poisson distribution with parameter X; that is, conditional on $X = x$, $Y \sim$ Pois(x).
a. Show that the PGF of Y is $\pi(t) = m(t - 1)$.
b. Use the result of part a to confirm the result of Exercise 5a above.
[This exercise is similar to Example (9.2.36).]

9. Show that if Y has (conditional on X) the exponential distribution with parameter X, and if X has the Erl(n, λ) prior distribution, then the posterior distribution of X (that is, the conditional distribution of X, given $Y = y$) is also an Erlang distribution.

10. **An improper prior distribution.** Given $X = x$, let Y have conditionally the Pois(X) distribution. Let the prior "density" on X be $f(x) = 1$ for $x > 0$. Notice that this is not a density at all; its integral does not exist. Nevertheless, you can use the conditioning formula (9.2.10) to get a "density" for Y, and then use Bayes's formula (9.2.18) to get a conditional density for X, given $Y = y$.
a. Show that the unconditional "density" for Y produced by (9.2.10) is also an improper density.
b. Find and identify by name the conditional (posterior) density for X, given $Y = y$.

Notice that the posterior density for X turns out to be a member of the family conjugate to the Poisson. This method—finding the posterior for a constant prior density, even though it is improper—can often be used to find conjugate prior families. Exercises 11, 12, and 14 below also involve improper prior densities.

11. Let Y have the exp(X) distribution, and let X have the "density" $f(x) = 1/x$ for $x > 0$. [This is an improper prior density; see Exercise 10 above.]
a. Show that the unconditional "density" for Y produced by the conditioning formula (9.2.3) is also an improper density.
b. Find the posterior distribution for X, given $Y = y$.

12. Let Y have the normal distribution with expected value X and variance 1. Let X have the improper prior density $f(x) = 1$ for $x \in \mathbb{R}$. Show that the unconditional density of Y is also 1, and that the posterior density of X, given Y, is normal with expected value Y and variance 1.

13. The arrival rate of a Poisson process is a positive random variable X having some distribution which is unknown except for its expected value and variance. Let Y be the number of arrivals in a time interval of length 1. Find EY and VY in terms of EX and VX.

14. Repeat Exercise 11 with the improper prior density $f(x) = 1$ instead of $1/x$.

15. If $Z(x)$ denotes the position at time x of a particle moving on a line according to a process called a one-dimensional *Brownian motion*, then $Z(x)$ is a random variable with the $N(\mu x, \sigma^2 x)$ distribution, whose MGF we know to be

$$m_{Z(x)}(s) = e^{\lambda xs + \sigma^2 xs^2/2}.$$

Suppose we choose a random time X to observe this particle. Let X be independent of $Z(x)$ for every x, and let X have the $\exp(\lambda)$ distribution. Let Y be the position $Z(X)$ of the particle at the chosen time. Then the MGF is really the conditional MGF of Y, given $X = x$.

a. Write down $m_Y(s|X)$. [See (9.2.35).]
b. Find the unconditional MGF of Y, $m_Y(s)$.
c. Find EY and VY. Find them in two ways—using the MGF and using the conditioning formulas given in (9.2.31) and (9.2.32).

16. It is instructive to use conditioning on X to rederive the well-known formulas for the expected value and variance of a sum, $E(X + Y) = EX + EY$ and (when X and Y are uncorrelated) $V(X + Y) = VX + VY$.

a. Write down $E(X + Y|X = x)$ and $V(X + Y|X = x)$.
b. Write down the two random variables $E(X + Y|X)$ and $V(X + Y|X)$.
c. Verify the two formulas using (9.2.31) and (9.2.32).
[a]**d.** But now remember that $V(X + Y) = VX + VY$ is not true in general, but only if X and Y are uncorrelated. We appear to have proved it even when they are not, so there must be a mistake. Can you find a step that is not justified without the assumption that X and Y are uncorrelated?

17. The previous exercise reveals a trap that one must be careful to avoid, but it also shows a useful technique. The key is part a, which rests on the fact that the conditional distribution of $X + Y$, given $X = x$, is the same as the conditional distribution of $x + Y$, given $X = x$. More generally, if $h(X, Y)$ is any function of X and Y, then:

> The conditional distribution of $h(X, Y)$, given $X = x$, is the same as the conditional distribution of $h(x, Y)$, given $X = x$.

If X and Y are independent, then this conditional distribution is the same as the unconditional distribution of $h(x, Y)$.

a. Use this fact with the function $h(X, Y) = XY$ to rederive the formula $E(XY) = EX \cdot EY$ for independent random variables.
b. Find a formula for $V(XY)$ (for independent random variables) by conditioning on X.
c. Find another formula for $V(XY)$ by conditioning on Y instead of X.
d. Show that these two formulas are really the same, and that both are equal to the formula $E(X^2Y^2) - (EX)^2(EY)^2$, which we found in Exercise 18 of Section 4.3.

The point of Exercises 16 and 17 is not to show a better way to find expected values and variances of sums and products; these are all found more simply using the ordinary algebra of expected values and variances, as in Chapter 4. The point is to illustrate the use of conditioning on X to find something about the distribution of a function of X and Y.

18. **$EF(X)$ is never less than $\frac{1}{2}$.** Let X and Y be independent and identically distributed, and let $F(x)$ be their common distribution function. By conditioning on X show that $P(Y \le X)$ is the expected value of the random variable $F(X)$. Then, using the result of Exercise 11 in Section 3.5, conclude that if F is the CDF of X, then $EF(X) \ge \frac{1}{2}$.

$EF(X)$ is exactly $\frac{1}{2}$ when F is continuous, because in that case it can be shown that the random variable $F(X)$ has the uniform distribution on $(0, 1)$. [See (8.5.6).] This exercise shows the somewhat surprising result, pointed out to the author by E. Carlstein, that $EF(X) \ge \frac{1}{2}$ for any CDF whatever. The difference between $EF(X)$ and $\frac{1}{2}$ is just the probability $P(X = Y)$, which is 0 if the distribution is continuous.

ANSWERS

1. $EY = \mu\lambda$; $VY = (\mu^2 + 2\sigma^2)/\lambda^2$
2. $\text{Exp}(\lambda p)$
3. a. $p_Y(y) = \dfrac{3(y + 2)(y + 1)}{(n + 3)(n + 2)(n + 1)}$ for $y = 0, 1, \ldots, n$
   b. $\text{beta}(y + 3, n - y + 1)$
4. $EY = \mu$; $VY = 2\sigma^2 + \mu^2$
9. It is the $\text{Erl}(n + 1, \lambda + y)$ distribution.
10. a. It is $p_Y(y) = 1$ for $y = 0, 1, 2, \ldots$.
    b. $\text{Erl}(y + 1, 1)$.
11. a. It is $1/y$ for $y > 0$. b. Exponential with parameter y
13. $EY = EX$; $VY = EX + VX$
14. Posterior distribution of X, given $Y = y$, is $\text{Erl}(2, y)$
15. a. $e^{\mu s X + \sigma^2 s^2 X/2}$ b. $\dfrac{\lambda}{\lambda - \mu s - \frac{1}{2}\sigma^2 s^2}$
    c. $\dfrac{\mu}{\lambda}$; $\dfrac{\sigma^2}{\lambda} + \dfrac{\mu^2}{\lambda^2}$
16. a. $E(X + Y|X = x) = E(x + Y) = x + EY$; $V(X + Y|X = x) = V(x + Y|X = x) = V(Y|X = x) = VY$
    b. $E(X + Y|X) = X + EY$ (the random variable X plus a constant); $V(X + Y|X) = VY$ (a constant random variable)
    c. The step $V(Y|X = x) = V(Y)$ in part a is correct when X and Y are uncorrelated, but not in general.
17. b. $VX \cdot (EY)^2 + VY \cdot E(X^2)$

SUPPLEMENTARY EXERCISES FOR **CHAPTER 9**

1. Let X be the number of heads in three tosses of a fair coin, and given $X = x$, let Y have the normal distribution with expected value x and variance 1. Find EY and also the probability that Y is positive. [Finding EY can be done instantly using the conditioning formulas of Section 9.2; for $P(Y > 0)$, go back to Section 2.3.]

2. Let X have the $\exp(\lambda)$ distribution and let Y, conditional on $X = x$, have the exponential distribution with parameter $1/x$. Find EY and VY.

3. Let X have the $\text{bin}(n, p)$ distribution. Let Y be the number of tosses of a fair coin needed to produce $X + 1$ heads. Find EY and VY.

4. Let X be the number of spots that show when a fair die is rolled. Then let Y be the number of spots that show when X fair dice are rolled.
   a. Find EY and VY.
   b. Find the probability that X equals 1, given that Y equals 3.

5. Let X have the uniform distribution on (0, 1), and let Y have (conditionally) the uniform distribution on $(0, X)$.
   a. Find an unconditional density for Y.
   b. Find a conditional density for X, given $Y = y$.

6. Let $X \sim \text{unif}(0, 1)$ and let Z be a random variable independent of X, with a distribution that is unknown except that $EZ = \mu$ and $VZ = \sigma^2$. Find the expected value and variance of $Y = ZX$ by conditioning on Z.

   Notice that the distribution of Y in this example is the same as in the following situation: Z has some distribution and Y has the uniform distribution on the interval between 0 and Z. This situation produces a most interesting result (unrelated to this exercise). It can be shown that in this case, no matter what the distribution of Z is, the distribution of Y is *unimodal*—that is, has a density with a single maximum. Conversely, every unimodal distribution is of this form. See Section V.9 in Volume 2 of William Feller's *An Introduction to Probability Theory and Its Applications*.

7. **Random mixtures.**
   a. Let X, U, and W be independent random variables, each with the uniform distribution on (0, 1), and let $Y = XU + (1 - X)W$. [This is a "random mixture" of U and W. The percentage of U is X, and the percentage of W is $1 - X$.] Find EY and VY by conditioning on X. Conclude that Y does not have the uniform distribution on (0, 1).
   b. More generally, let U and W have some arbitrary distribution with expected value μ and variance σ^2. As in part a, $X \sim \text{unif}(0, 1)$ and $Y = XU + (1 - X)W$, and X, U, and W are still independent. Show that $EY = \mu$ and $VY = \frac{2}{3}\sigma^2$. Conclude that Y cannot have the same distribution as U and W unless they are constant.

[a]8. In Exercise 8 of Section 3.4, X and Y had the joint mass function

$$p(x, y) = \frac{x^y e^{-x}}{N \cdot y!} \quad \text{for } x = 1, 2, 3, \ldots, N \quad \text{and} \quad y = 0, 1, 2, 3, \ldots.$$

Identify the distribution of X and the conditional distribution of Y, given $X = x$.

ANSWERS

1. 1.5; .8693
2. $1/\lambda$; $3/\lambda^2$
3. $2(np + 1)$; $4npq + 2np + 2$
4. **a.** $\frac{49}{4}$; $\frac{2,205}{48}$ **b.** $\frac{36}{49}$
5. **a.** $f_Y(y) = -\ln y$ for $0 < y < 1$
   b. $f_X(x|Y = y) = -1/(x \cdot \ln y)$ for $y < x < 1$
6. $EY = \frac{\mu}{2}$; $VY = \frac{\mu^2}{12} + \frac{\sigma^2}{3}$
7. **a.** $EY = \frac{1}{2}$; $VY = \frac{1}{18}$: VY would be $\frac{1}{12}$ if Y had the unif(0, 1) distribution.
8. X is a random choice from $\{1, 2, \ldots, N\}$; given $X = x$, Y is conditionally Pois(x).

Suggested Reading

Of all the books listed in the following bibliography, **Feller's** two volumes have had the greatest influence on this text and its author. Readers who want to pursue probability theory further, other than by taking courses in measure theory, stochastic processes, or statistical inference, would do well to look into them. Much of Volume 1 is at the same level as this text; Volume 2 introduces measure theory along with probability in a somewhat unorthodox way. Both volumes include material on stochastic processes that is not in this text.

Some other texts in probability at approximately the same level as this one, which readers could use as alternate sources, are **Ross**'s *A First Course in Probability*; **Breiman**'s *Probability and Stochastic Processes*; the probability volume by **Hoel, Port, and Stone;** the book by **Ash;** and the one by **Grimmett and Welsh. Neuts**'s book is at a slightly higher level, but still does not require measure theory. Readers interested in pursuing more of the computer simulation that is hinted at in this text should look at **Ross**'s *A Course in Simulation* or the book by **Snell.** The book by **Romano and Siegel** is an interesting and informative supplement.

Stochastic processes constitute one of the main areas of application of the material in this text. **Ross**'s *Stochastic Processes* is accessible to readers of this text, as is **Parzen**'s book. The book on the subject by **Hoel, Port, and Stone** is a sequel to their probability text. Material on stochastic processes can be found also in **Breiman**'s *Probability and Stochastic Processes*, in **Feller**'s two volumes, or in **Ross**'s *Introduction to Probability Models.*

Another main application of probability—statistical inference—is covered in a large number of texts. Again there is a sequel by **Hoel, Port, and Stone.** Most other texts on statistical inference at this level begin with coverage of the essential probability theory; the books by **Hogg and Tanis** and by **Larson** are of this type.

Readers who want to pursue probability theory at a more rigorous level will need to begin with measure theory. A good way to do this and learn rigorous probability at the same time is with **Billingsley**'s book. Another option is to use **Feller**'s two volumes.

A more common, though not necessarily better, approach is to study measure theory first, then rigorous probability. On measure theory, **Royden**'s book on real analysis contains a good introduction, and one of the classic texts is the one by **Halmos.** Following measure theory, a number of probability texts are available; among them are **Breiman**'s *Probability* (recently reprinted after long unavailability), and the book by **Chung.**

Finally, two classic, encyclopedic books at the research level are the one by **Loève** on probability, which includes essentially a full course in measure theory, and the one by **Doob** on stochastic processes.

BIBLIOGRAPHY

Andrews, D. F., and A. M. Herzberg. *Data.* New York: Springer–Verlag, 1985.

Ash, Robert B. *Basic Probability Theory.* New York: John Wiley & Sons, 1970.

Billingsley, Patrick. *Probability and Measure*, 2nd ed. New York: John Wiley & Sons, 1986.

Breiman, Leo. *Probability and Stochastic Processes.* Palo Alto: The Scientific Press, 1986.

Breiman, Leo. *Probability.* Philadelphia: Society for Industrial and Applied Mathematics, 1992.

Carroll, Lewis. *Through the Looking Glass.* In: Martin Gardner, *The Annotated Alice.* New York: Clarkson N. Potter, Inc., 1960.

Chung, Kai Lai. *A Course in Probability Theory*, 2nd ed. New York: Academic Press, 1974.

Doob, J. L. *Stochastic Processes.* New York: John Wiley & Sons, 1953.

Feller, William. *An Introduction to Probability Theory and Its Applications*, Vol. 1, 3rd ed. New York: John Wiley & Sons, 1968.

Feller, William. *An Introduction to Probability Theory and Its Applications*, Vol. 2, 2nd ed. New York: John Wiley & Sons, 1971.

Grimmett, Geoffrey, and Dominic Welsh. *Probability, an Introduction.* Oxford: Oxford University Press, 1986.

Halmos, Paul R. *Measure Theory.* New York: Springer–Verlag, 1974.

Hoel, Paul G., Sidney C. Port, and Charles J. Stone. *Introduction to Probability Theory.* Boston: Houghton Mifflin Company, 1971.

Hoel, Paul G., Sidney C. Port, and Charles J. Stone. *Introduction to Statistical Theory.* Boston: Houghton Mifflin Company, 1971.

Hoel, Paul G., Sidney C. Port, and Charles J. Stone. *Introduction to Stochastic Processes.* Boston: Houghton Mifflin Company, 1972.

Hogg, Robert V., and Elliot A. Tanis. *Probability and Statistical Inference*, 4th ed. New York: Macmillan Publishing Company, 1993.

Kolmogorov, A. N. *Foundations of the Theory of Probability.* New York: Chelsea Publishing Company, 1950.

Kotz, Samuel, and Norman L. Johnson, eds. *Encyclopedia of the Statistical Sciences.* New York: John Wiley & Sons, 1982–1988.

Larson, Harold J. *Introduction to Probability and Statistical Inference,* 3rd ed. New York: John Wiley & Sons, 1982.

Loève, Michel. *Probability Theory I*, 4th ed. New York: Springer–Verlag, 1977.

Lotka, A. J. *"Théorie Analytique des associations biologiques II" (Actualités Scientifiques et industrielles* No. 780). Paris: Hermann et Cie., 1939.

Malkiel, Burton. *A Random Walk down Wall Street*, 5th ed. New York: W. W. Norton & Company, 1990.

Marbe, K. *Die Gleichförmigkeit in der Welt (Uniformity in the World).* Munich: C. H. Beck, 1916.

Mosteller, Frederick. *Fifty Challenging Problems in Probability, with Solutions.* Reading, Mass.: Addison–Wesley, 1965.

Neuts, Marcel F. *Probability.* Boston: Allyn and Bacon, Inc., 1973.

Parzen, Emanuel. *Stochastic Processes.* San Francisco: Holden–Day, Inc., 1962.

Romano, Joseph P., and Andrew F. Siegel. *Counterexamples in Probability and Statistics.* Belmont, California: Wadsworth, Inc., 1986.

Ross, Sheldon M. *Introduction to Probability Models.* New York: Academic Press, 1972.

Ross, Sheldon M. *Stochastic Processes*. New York: John Wiley & Sons, 1983.

Ross, Sheldon M. *A First Course in Probability*, 3rd ed. New York: Macmillan Publishing Company, 1988.

Ross, Sheldon M. *A Course in Simulation*. New York: Macmillan Publishing Company, 1990.

Royden, H. L. *Real Analysis*, 3rd ed. New York: Macmillan Publishing Company, 1988.

Rucker, Rudy. *Infinity and the Mind*. New York: Bantam Books, 1983.

Snell, J. Laurie. *Introduction to Probability*. New York: Random House/Birkhäuser, 1988.

APPENDIX

Review of Some Topics from Calculus and Set Theory

A.1 Sets of Numbers

A.2 Set Operations

A.3 Finite, Countably Infinite, and Uncountable Sets

A.4 Basic Combinatorics

A.5 Sums and Series

A.6 L'Hôpital's Rule

A.7 Integration by Substitution and Integration by Parts

A.8 Improper Integrals

A.9 Integrals over Subsets of the Plane

A.1 SETS OF NUMBERS

Some of the standard sets of numbers:

$\mathbb{R}$, the set of all real numbers

$\mathbb{Z}$, the set of all integers (positive and negative)

$\mathbb{N}$, the set $\{1, 2, 3, \ldots\}$ of positive integers

$\mathbb{Q}$, the set of all rational numbers (i.e., ratios of integers, which are the same as repeating decimals)

$[a, b]$, a bounded closed interval (containing its endpoints)

(a, b), a bounded open interval (not containing its endpoints)

$[a, b)$ and $(a, b]$, bounded half-open intervals (containing one endpoint but not the other)

$[a, \infty)$, (a, ∞), $(-\infty, b]$, $(-\infty, b)$, $(-\infty, \infty)$, unbounded intervals (which have one or no endpoints)

Notice that ∞ and $-\infty$ never refer to numbers. In this context they are symbols indicating that an interval is unbounded, and has no endpoint, in one direction or the other.

Some sets of vectors:

$\mathbb{R}^2$, the set of all pairs (x, y) of real numbers, i.e., the set of points in the plane

$\mathbb{R}^3$, the set of all triples (x, y, z) of real numbers, i.e., the set of points in 3-space

$\mathbb{R}^n$, the set of all n-tuples $(x_1, x_2, \ldots, x_n)$ of real numbers, i.e., the set of points in n-space

Examples of "set-builder" notation:

In $\mathbb{R}$, $\{x: 3 \leq x < 6\}$ is the set of all x such that $3 \leq x < 6$, i.e., the interval $[3, 6)$.

In $\mathbb{R}^2$, $\{(x, y): (x - 1)^2 + (y + 3)^2 < 4\}$ is the set of all points (x, y) whose distance from $(1, -3)$ is less than 2, i.e., the interior of the circle of radius 2 centered at $(1, -3)$ in the plane.

A.2 SET OPERATIONS

The ***union*** of two sets A and B, denoted by $A \cup B$, is the set of all objects that are members of one or the other or both of A and B. That is, $A \cup B = \{x: x \in A \text{ or } x \in B\}$, where (as always in mathematical writing) "or" is interpreted as "and/or." The union of two sets is shaded in the Venn diagram of Figure A.2a.

The union of any collection of sets is the set of all objects that are members of one or more of the sets in the collection.

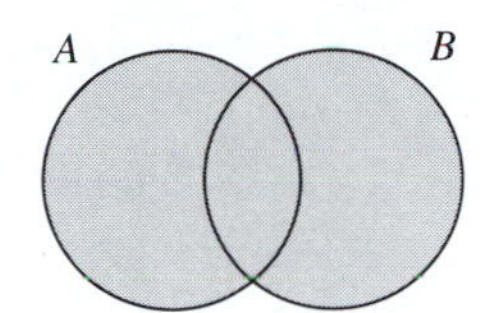

FIGURE A.2a The union $A \cup B$ is shaded.

The ***intersection*** of two sets A and B, denoted by $A \cap B$, is the set of all objects that are members of both A and B. That is, $A \cap B = \{x: x \in A \text{ and } x \in B\}$. The intersection of two sets is shaded in Figure A.2b.

The intersection of any collection of sets is the set of all objects that are members of all the sets in the collection.

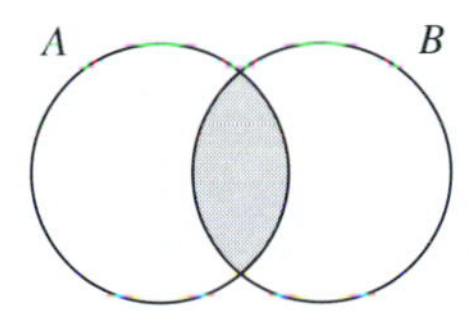

FIGURE A.2b The intersection $A \cap B$ is shaded.

Sets A and B are ***disjoint*** if they have no members in common—that is, if their intersection is the ***empty set***, which is denoted by $\varnothing$.

The ***complement*** of a set A, denoted by A^c, makes sense only with respect to some "universal" set in which the discussion is taking place. In probability theory, the universal set is usually the outcome set Ω. It is defined as the set of all members of the universal set that are not members of A. That is, $A^c = \{x: x \notin A\}$. The complement of a set is shaded in Figure A.2c (page 568).

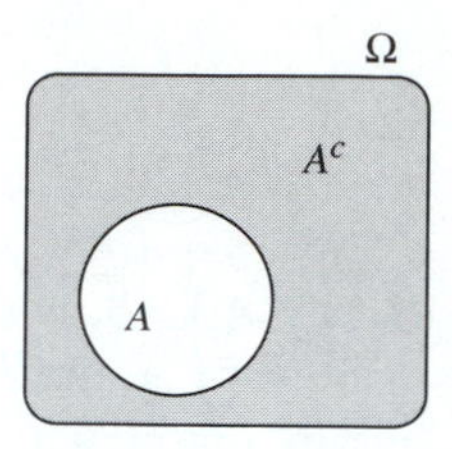

FIGURE A.2c The complement of A is shaded.

Unions and intersections are commutative,

$$A \cup B = B \cup A \quad \text{and} \quad A \cap B = B \cap A,$$

and associative,

$$A \cup (B \cup C) = (A \cup B) \cup C \quad \text{and} \quad A \cap (B \cap C) = (A \cap B) \cap C,$$

and each is distributive with respect to the other,

$$A \cup (B \cap C) = (A \cup B) \cap (A \cup C) \quad \text{and} \quad A \cap (B \cup C) = (A \cap B) \cup (A \cap C).$$

We also have ***De Morgan's Laws***, which say that the complement of a union is the intersection of the complements,

$$(A \cup B)^c = A^c \cap B^c,$$

and that the complement of an intersection is the union of the complements,

$$(A \cap B)^c = A^c \cup B^c.$$

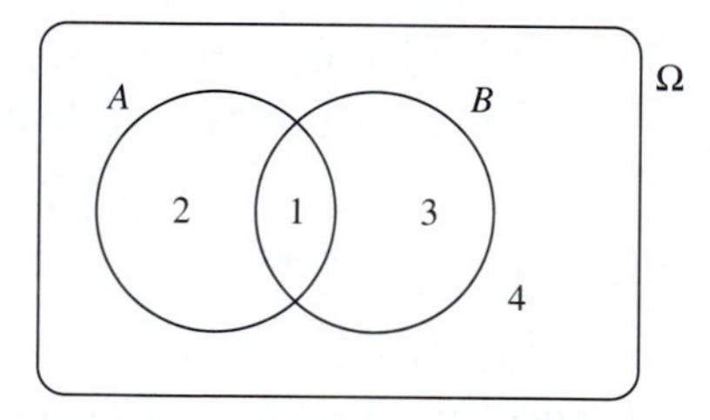

FIGURE A.2d Venn diagram for confirming some set identities

As in Figure A.2d, we can identify by number the regions in a Venn diagram corresponding to various sets, and thus confirm set identities like the commutative and associative laws and De Morgan's Laws. For example, in Figure A.2d, we can confirm the first of De Morgan's Laws by noting that

$A \cup B \leftrightarrow$ regions 1, 2, and 3; so $(A \cup B)^c \leftrightarrow$ region 4;

$A^c \leftrightarrow$ regions 3 and 4, and $B^c \leftrightarrow$ regions 2 and 4; so $A^c \cap B^c \leftrightarrow$ region 4.

For three sets there are eight regions in the Venn diagram. It is possible to draw a Venn diagram for four sets (there are 16 regions), but Venn diagrams are rarely used for more than three sets.

The laws above involve unions and intersections of pairs of sets, but they apply as well to arbitrary collections of sets. For example, if $A_1, A_2, A_3, \ldots$ are sets in any universal set Ω, then

$$\bigcup_{n=1}^{\infty} A_n = \{x: x \in A_n \text{ for one or more } A_n\text{'s}\} \quad \text{and} \quad \bigcap_{n=1}^{\infty} A_n = \{x: x \in A_n \text{ for all } A_n\text{'s}\}$$

define the same sets regardless of how the A_n's are ordered or grouped. The distributive laws,

$$B \cap \bigcup_{n=1}^{\infty} A_n = \bigcup_{n=1}^{\infty} (B \cap A_n) \quad \text{and} \quad B \cup \bigcap_{n=1}^{\infty} A_n = \bigcap_{n=1}^{\infty} (B \cup A_n),$$

and De Morgan's Laws,

$$\left(\bigcup_{n=1}^{\infty} A_n\right)^c = \bigcap_{n=1}^{\infty} (A_n)^c \quad \text{and} \quad \left(\bigcap_{n=1}^{\infty} A_n\right)^c = \bigcup_{n=1}^{\infty} (A_n)^c,$$

also hold.

A.3 FINITE, COUNTABLY INFINITE, AND UNCOUNTABLE SETS

A set A is ***countably infinite*** if it can be put into one-to-one correspondence with the set $\mathbb{N}$ of positive integers. Such a correspondence is the assignment of a member of A to each positive integer so that no member of A is unassigned. It is, in effect, an infinitely long "listing" of the members of A: the positive integer assigned to a given member of A represents its place on the list.

For example, the set $\mathbb{Z}$ of all positive and negative integers is countably infinite—even though it seems twice as big as $\mathbb{N}$—because we can make the listing

Member of $\mathbb{N}$:	1	2	3	4	5	6	...
Member of $\mathbb{Z}$:	0	1	−1	2	−2	3	...

in which it is clear that every member of $\mathbb{Z}$ appears exactly once on the list. If challenged, we can produce the function $f(z)$ that gives the member of $\mathbb{N}$ to which an arbitrary integer z is assigned: $f(z) = 2z$ if z *is* positive, and $f(z) = 1 - 2z$ if z is negative or 0. But such an exercise is not important; the important thing is to recognize the pattern of the listing and see that it includes every member of $\mathbb{Z}$ just once.

Another famous countably infinite set, which seems bigger than $\mathbb{N}$ but is not, is the set $\mathbb{Q}$ of rational numbers. Rather than exhibit a listing of all of $\mathbb{Q}$, we will be content with a listing of the members of $\mathbb{Q}$ in $(0, 1]$. (Since the union of countably many countably infinite sets can also be shown to be countably infinite, it follows that $\mathbb{Q}$ is countably infinite.)

Member of $\mathbb{N}$:	1	2	3	4	5	6	7	8	9	10	11	12	. . .
Member of $\mathbb{Q}$ in $(0, 1]$:	$\frac{1}{1}$	$\frac{1}{2}$	$\frac{1}{3}$	$\frac{2}{3}$	$\frac{1}{4}$	$\frac{3}{4}$	$\frac{1}{5}$	$\frac{2}{5}$	$\frac{3}{5}$	$\frac{4}{5}$	$\frac{1}{6}$	$\frac{5}{6}$	. . .

The principle of this listing is as follows. We start with the only fraction in $(0, 1]$ whose denominator is 1, follow with the one whose denominator is 2 (there are actually two, $\frac{1}{2}$ and $\frac{2}{2}$, but the latter equals $\frac{1}{1}$, which was already listed). Then come the ones whose denominators are 3 (excluding $\frac{3}{3}$, as before); then the ones whose denominators are 4 (there are four of these, but $\frac{2}{4}$ and $\frac{4}{4}$ are omitted because they equal fractions already listed), then the four new ones whose denominators are 5, then the ones whose denominators are 6 (only two new ones here), and so forth.

It would be difficult, or at least tedious, to work out a formula which tells exactly where any given rational number in $(0, 1]$, say $\frac{241}{256}$, appears on this list. But it should also be clear that the listing does include every such number exactly once. This listing was discovered by Georg Cantor, whose pioneering work in the 19th century laid much of the the foundation for modern rigorous thinking about the infinite. Cantor was also responsible for the famous "diagonal process" argument which shows that the whole interval $(0, 1]$—the set of *all* real numbers between 0 and 1—is *not* countably infinite. This argument shows in a simple way that any listing of real numbers in $(0, 1]$ must be missing at least one real number, by starting with a presumed listing and constructing a number that is not on the list. Readers interested in such matters will find much more in R. Rucker's book *Infinity and the Mind*.

A ***finite*** set can be defined as one which can be put into one-to-one correspondence with one of the sets $\{1, 2, 3, \ldots, n\}$ (for some integer n). A set that is neither finite nor countably infinite is called ***uncountable***. [Thus Cantor's diagonal process, mentioned above, shows that $(0, 1]$ is uncountable.] Sometimes the word ***countable*** is used to mean "finite or countably infinite."

Sometimes in probability theory we need to consider the union or intersection of a countably infinite collection of sets.

Example. For $n = 1, 2, 3, \ldots$, let A_n be the interval $[0, 1 + 1/n)$. Then the intersection of these intervals is

$$A_1 \cap A_2 \cap A_3 \cap \cdots = [0, 2) \cap \left[0, \tfrac{3}{2}\right) \cap \left[0, \tfrac{4}{3}\right) \cap \left[0, \tfrac{5}{4}\right) \cap \cdots = [0, 1].$$

That is,

$$\bigcap_{n=1}^{\infty} [0, 1 + 1/n) = [0, 1].$$

Notice that the A's do not contain their right endpoints; but because they all contain 1, their intersection contains 1. However, the intersection contains no numbers larger than 1, because for any such number there is a number of the form $1 + 1/n$ between it and 1. ∎

Example. For $n = 1, 2, 3, \ldots$, let A_n be the interval $[0, 1 - 1/n]$. Then the union of these intervals is

$$A_1 \cup A_2 \cup A_3 \cup \cdots = [0, 0] \cup \left[0, \tfrac{1}{2}\right] \cup \left[0, \tfrac{2}{3}\right] \cup \left[0, \tfrac{3}{4}\right] \cup \cdots = [0, 1).$$

That is,

$$\bigcup_{n=1}^{\infty} [0, 1 - 1/n] = [0, 1).$$

Here all the A's contain their right endpoints, but none of them contains 1; so their union does not contain 1. ∎

A.4 BASIC COMBINATORICS

The first principle is sometimes called the ***multiplication principle.*** Suppose we are to make a series of decisions. Suppose there are c_1 choices for decision 1; that for each of these choices there are c_2 choices for decision 2; that for each combination of choices for decisions 1 and 2 there are c_3 choices for decision 3, and so on. Then the number of ways the series of decisions can be made is $c_1 \cdot c_2 \cdot c_3 \cdot \cdots$. The most important instances of this principle are Theorems (A.4.1) and (A.4.3). More trivial instances are familiar, as follow.

Example. If you have three hats, four coats, and two pairs of gloves, then you have $3 \cdot 4 \cdot 2 = 24$ outerwear ensembles. If the hats are red, brown, and gray, and the coats are red, brown, gray, and green, and the hat and coat colors must not match, then you have $3 \cdot 3 \cdot 2 = 18$ ensembles. If the hat and coat colors must match, then you have $3 \cdot 1 \cdot 2 = 6$ ensembles.

The example in which the colors must *not* match involves two points worth noting. First, after you choose a hat, you have three coats to choose from. The coats you may choose depend on the hat you chose; there will be a different set of three coats for each choice of hat. The important thing is that the *number* of choices in decision 2 be the same for all choices in decision 1, not that the *same set* of choices be available.

Second, what if you wanted to choose the coat first? Then the number of choices for a hat depends on which coat you chose, and the multiplication principle does not apply so easily. But it is still possible to do it, by breaking it up into two problems. You can choose the green coat, any hat, and a pair of gloves in $1 \cdot 3 \cdot 2 = 6$ ways, or you can choose one of the other coats, a nonmatching hat, and a pair of gloves in $3 \cdot 2 \cdot 2 = 12$ ways, for a total of 18 ensembles. ∎

Example. If set A has 6 members and set B has 9, then their Cartesian product (the set of all pairs (a, b) where $a \in A$ and $b \in B$) has $6 \cdot 9 = 54$ members.

(A.4.1) Theorem. The number of k-letter words from an n-letter alphabet is n^k. □

Here, "words" are simply strings of letters, with repetitions allowed, and without regard for meaning. The reason is of course that there are n choices for the first letter in the word, then n choices for the second, and so on.

Example. The number of 2-letter words from the 3-letter alphabet $\{a, b, c\}$ is $3^2 = 9$. The 9 words are *aa*, *ab*, *ac*, *ba*, *bb*, *bc*, *ca*, *cb*, *cc*. ∎

Example. The number of 3-letter words from the 10-letter alphabet {0, 1, 2, 3, 4, 5, 6, 7, 8, 9} is $10^3 = 1{,}000$. This is not surprising when we realize that the words are 000, 001, 002, . . . , 999. ∎

(A.4.2) Corollary. The number of subsets of a set of size k is 2^k. □

This follows because there are 2^k k-letter words from the 2-letter alphabet {0, 1} (called *binary words* or *bit strings*), and these correspond in a natural way to the subsets of a set $\{p_1, p_2, \ldots, p_k\}$ if we associate the elements of the set with places in a word. For example, if $k = 3$, there are $2^3 = 8$ bit strings, and also 8 subsets of the set $\{p_1, p_2, p_3\}$. The correspondence is:

Bit string:	000	001	010	011	100	101	110	111
Subset:	$\varnothing$	$\{p_3\}$	$\{p_2\}$	$\{p_2, p_3\}$	$\{p_1\}$	$\{p_1, p_3\}$	$\{p_1, p_2\}$	$\{p_1, p_2, p_3\}$

(A.4.3) Theorem. The number of k-letter words with no repeated letters from an n-letter alphabet is $n \cdot (n-1) \cdot (n-2) \cdots (n-k+1)$. This number is denoted by $(n)_k$. Notice that it has k factors which are descending integers beginning with n. □

Example. The number of 2-letter words without repetition from the 4-letter alphabet $\{a, b, c, d\}$ is $4 \cdot 3 = 12$. The 12 words are *ab*, *ac*, *ad*, *ba*, *bc*, *bd*, *ca*, *cb*, *cd*, *da*, *db*, *dc*. ∎

(A.4.4) The case $n = k$; factorials. The number of n-letter words without repetition from an n-letter alphabet is $(n)_n = n \cdot (n-1) \cdot (n-2) \cdots 2 \cdot 1$, which is denoted by $n!$ and called *n factorial*. These words are rearrangements of the n letters of the alphabet, and are called *permutations* of the letters. For example, there are $3! = 6$ permutations of $\{a, b, c\}$; they are *abc*, *acb*, *bac*, *bca*, *cab*, *cba*.

By convention we agree that there is one "word with no letters" from any alphabet. That word plays the role of the empty set in set theory. Consequently, we define $(n)_0 = 1$. This convention also entails $0! = (0)_0 = 1$.

The formula $(n)_0 = 1$ is not provable from the definition of $(n)_k$; neither is the assumption that there is one word with no letters. These are conventions, which we adopt only because they are the most convenient interpretations of otherwise meaningless expressions. This is similar to the algebraic convention that x^0 equals 1. We cannot justify $x^0 = 1$ by thinking about the product of no factors; we adopt it as a convention because it makes the power formula $x^{a+b} = x^a \cdot x^b$ valid when b equals 0.

(A.4.5) **Theorem.** The number of subsets of size k in a set of size n is

$$\frac{(n)_k}{k!} = \frac{n \cdot (n-1) \cdot \cdot \cdot (n-k+1)}{k!}.$$

This number is denoted by $\binom{n}{k}$ and called a *binomial coefficient* for reasons that appear in (A.4.7). □

The reason that there are $(n)_k/k!$ subsets of size k in a set of size n is that among all the k-letter words without repetition from the set of size n, each subset of size k appears $k!$ different times, in all its different permutations.

Example. The number of subsets of size 3 in the set $\{a, b, c, d, e\}$ of size 5 is $\binom{5}{3} = 5 \cdot 4 \cdot 3/3! = 10$. The 10 subsets are $\{a, b, c\}$, $\{a, b, d\}$, $\{a, b, e\}$, $\{a, c, d\}$, $\{a, c, e\}$, $\{a, d, e\}$, $\{b, c, d\}$, $\{b, c, e\}$, $\{b, d, e\}$, $\{c, d, e\}$. ∎

Example. The number of subsets of size 1—"singleton" subsets—in a set of size n is of course n, because there is one singleton for every member of the set. And indeed, the formula above gives $\binom{n}{1} = n$. ∎

Example. The formula also gives $\binom{n}{0} = 1$ [because we have adopted the convention $(n)_0 = 1$], and indeed there is one subset of size 0, the empty set, in any set. ∎

Notice that $(n)_k$ equals $n!/(n-k)!$ and, consequently, $\binom{n}{k} = n!/k!(n-k)!$. But $(n)_k/k!$ is a more efficient formula for computing $\binom{n}{k}$.

Example. The formula $\binom{n}{k} = n!/k!(n-k)!$ also implies that $\binom{n}{k} = \binom{n}{n-k}$, and indeed there are just as many subsets of size k as there are subsets of size $n-k$ in any set of size n, because every subset has its unique complement. ∎

Example. To compute $\binom{100}{96}$, we use

$$\binom{100}{96} = \binom{100}{4} = \frac{100 \cdot 99 \cdot 98 \cdot 97}{4 \cdot 3 \cdot 2 \cdot 1}$$
$$= 25 \cdot 33 \cdot 49 \cdot 97 = 3{,}921{,}225. \quad \blacksquare$$

(A.4.6) Corollary. The number of n-letter words from the 2-letter alphabet $\{0, 1\}$ (that is, binary words of length n) which have k 1's and $n - k$ 0's is $\binom{n}{k}$.

This is true because the n-letter words with k 1's correspond to the subsets of size k in the set of places in an n-letter word [see Corollary (A.4.2)]. $\square$

Example. The number of 5-letter words from $\{0, 1\}$ with three 1's and two 0's is $\binom{5}{3} = 10$; the words are 11100, 11010, 11001, 10110, 10101, 10011, 01110, 01101, 01011, 00111. [Notice the correspondence between these words and the subsets in the first example under (A.4.5).] $\blacksquare$

(A.4.7) Corollary. The binomial theorem. For any numbers a and b and any positive integer n,

$$(a + b)^n = \binom{n}{0}a^0b^n + \binom{n}{1}a^1b^{n-1} + \binom{n}{2}a^2b^{n-2} + \cdots + \binom{n}{n}a^nb^0,$$

which can also be written

$$(a + b)^n = \sum_{k=0}^{n} \binom{n}{k}a^kb^{n-k}.$$

This is true because if we expand $(a + b)(a + b) \cdots (a + b)$ (n factors) but do not collect terms, we get 2^n terms, and each term is one of the n-letter words from the alphabet $\{a, b\}$. [*Example*: For $n = 3$, $(a + b)(a + b)(a + b) = aaa + aab + aba + abb + baa + bab + bba + bbb$.] When we do collect terms, the words with k a's and $n - k$ b's [there are $\binom{n}{k}$ of them] combine to form the term $\binom{n}{k}a^kb^{n-k}$. $\square$

Some special cases of the binomial theorem are the following:

$$(a + b)^2 = a^2 + 2ab + b^2;$$
$$(a + b)^3 = a^3 + 3a^2b + 3ab^2 + b^3;$$
$$(a + b)^4 = a^4 + 4a^3b + 6a^2b^2 + 4ab^3 + b^4;$$
$$(a + b)^5 = a^5 + 5a^4b + 10a^3b^2 + 10a^2b^3 + 5ab^4 + b^5;$$

etc.

A.5 SUMS AND SERIES

Some finite sums

(A.5.1)
$$1 + 2 + 3 + \cdots + n = \frac{n \cdot (n+1)}{2} \quad \text{for any } n \in \mathbb{N}.$$

(A.5.2)
$$1^2 + 2^2 + 3^2 + \cdots + n^2 = \frac{n \cdot (n+1) \cdot (2n+1)}{6} \quad \text{for any } n \in \mathbb{N}.$$

(A.5.3)
$$1 + x + x^2 + x^3 + \cdots + x^n = \begin{cases} \dfrac{1 - x^{n+1}}{1 - x} & \text{for any } x \in \mathbb{R} \textit{ except } x = 1, \\ n + 1 & \text{if } x = 1. \end{cases}$$

(A.5.4)
$$\frac{1}{1 \cdot 2} + \frac{1}{2 \cdot 3} + \frac{1}{3 \cdot 4} + \cdots + \frac{1}{n \cdot (n+1)} = 1 - \frac{1}{n+1} \quad \text{for any } n \in \mathbb{N}.$$

This holds because the left side of (A.5.4) equals

$$\left(\frac{1}{1} - \frac{1}{2}\right) + \left(\frac{1}{2} - \frac{1}{3}\right) + \left(\frac{1}{3} - \frac{1}{4}\right) + \cdots + \left(\frac{1}{n} - \frac{1}{n+1}\right),$$

a "telescoping" sum in which only the $\frac{1}{1}$ and the $1/(n+1)$ are not canceled. See (A.5.17).

Consequences of the binomial theorem

If we take $a = 1$ and $b = 1$ in the binomial theorem (A.4.7), we get

(A.5.5)
$$2^n = \binom{n}{0} + \binom{n}{1} + \binom{n}{2} + \cdots + \binom{n}{n} = \sum_{k=0}^{n} \binom{n}{k}$$

(which expresses the fact that the number of subsets in a set of size n is the sum, over all k, of the number of subsets of size k).

Similarly, putting in 1 for a and -1 for b gives

(A.5.6)
$$0 = \binom{n}{0} - \binom{n}{1} + \binom{n}{2} - \binom{n}{3} + - \cdots \pm \binom{n}{n} = \sum_{k=0}^{n} (-1)^k \binom{n}{k}$$

(which expresses the fact that in any set, there are as many subsets of even sizes as there are subsets of odd sizes).

Another useful trick is to take 1 for a and x for b in (A.4.7), getting

$$(1+x)^n = \sum_{k=0}^{n} \binom{n}{k} x^k = \binom{n}{0} + \binom{n}{1}x + \binom{n}{2}x^2 + \binom{n}{3}x^3 + \cdots + \binom{n}{n}x^n,$$

and then to differentiate with respect to x. The result is

$$\begin{aligned} n \cdot (1+x)^{n-1} &= \sum_{k=1}^{n} k \cdot \binom{n}{k} x^{k-1} \\ &= 1 \cdot \binom{n}{1} + 2 \cdot \binom{n}{2}x + 3 \cdot \binom{n}{3}x^2 + \cdots + n \cdot \binom{n}{n}x^{n-1} \end{aligned} \tag{A.5.7}$$

for any $x \in \mathbb{R}$ and $n \in \mathbb{N}$.

Again we can put in special values for x, such as $x = \pm 1$, to discover various identities.

Infinite series in general

Remember that the sum of a series $a_0 + a_1 + a_2 + a_3 + \cdots$ is defined as the limit of its sequence of partial sums, $\lim_{n\to\infty} s_n$, where $s_n = a_0 + a_1 + a_2 + \cdots + a_n$, provided this limit exists. That is,

$$\sum_{n=0}^{\infty} a_n \quad \text{is defined as} \quad \lim_{n\to\infty} \sum_{j=0}^{n} a_j.$$

Geometric series

$$1 + x + x^2 + x^3 + \cdots = \frac{1}{1-x} \quad \text{for any } x \in (-1, 1). \tag{A.5.8}$$

This is the Maclaurin series (Taylor series about 0) for $1/(1-x)$. To prove that it converges to $1/(1-x)$ for $|x| < 1$, we look at the sequence of partial sums, which are given by (A.5.3):

$$s_n = 1 + x + x^2 + x^3 + \cdots + x^n = \frac{1-x^{n+1}}{1-x}.$$

If $-1 < x < 1$, then as $n \to \infty$ this approaches $1/(1-x)$.

$$1 + 2x + 3x^2 + 4x^3 + \cdots = \frac{1}{(1-x)^2} \quad \text{for any } x \in (-1, 1). \tag{A.5.9}$$

This is proved by differentiating (A.5.8). (Differentiating a power series term by term within its radius of convergence always produces the derivative of the sum.) Continuing to differentiate we get, for $-1 < x < 1$,

$$2 \cdot 1 + 3 \cdot 2 \cdot x + 4 \cdot 3 \cdot x^2 + 5 \cdot 4 \cdot x^3 + \cdots = \frac{2 \cdot 1}{(1-x)^3},$$
$$3 \cdot 2 \cdot 1 + 4 \cdot 3 \cdot 2 \cdot x + 5 \cdot 4 \cdot 3 \cdot x^2 + 6 \cdot 5 \cdot 4 \cdot x^3 + \cdots = \frac{3 \cdot 2 \cdot 1}{(1-x)^4},$$

and, in general, for $k = 0, 1, 2, 3, \ldots$ and $-1 < x < 1$,

$$\frac{k!}{(1-x)^{k+1}} = (k)_k + (k+1)_k x + (k+2)_k x^2 + (k+3)_k x^3 + \cdots.$$

This series can be written using sigma notation in two ways:

$$\frac{k!}{(1-x)^{k+1}} = \sum_{n=k}^{\infty} (n)_k x^{n-k} = \sum_{m=0}^{\infty} (k+m)_k x^m.$$

Dividing this by $k!$ and remembering that $\frac{(n)_k}{k!} = \binom{n}{k}$ and $\frac{(k+m)_k}{k!} = \binom{k+m}{k} = \binom{k+m}{m}$, we get

(A.5.10)
$$(1-x)^{-k-1} = \sum_{n=k}^{\infty} \binom{n}{k} x^{n-k} = \sum_{m=0}^{\infty} \binom{k+m}{m} x^m \quad \text{for } k = 0, 1, 2, 3, \ldots \text{ and } -1 < x < 1.$$

Exponential and logarithmic series

The following two series are the Maclaurin series (Taylor series about 0) for the functions indicated.

(A.5.11)
$$e^x = 1 + x + \frac{x^2}{2!} + \frac{x^3}{3!} + \frac{x^4}{4!} + \cdots = \sum_{n=0}^{\infty} \frac{x^n}{n!} \quad \text{for any } x \in \mathbb{R}$$

(A.5.12)
$$-\ln(1-x) = x + \frac{x^2}{2} + \frac{x^3}{3} + \frac{x^4}{4} + \cdots = \sum_{n=1}^{\infty} \frac{x^n}{n} \quad \text{for any } x \in [-1, 1)$$

Another expression for e^x is

(A.5.13)
$$e^x = \lim_{n\to\infty} \left(1 + \frac{x}{n}\right)^n.$$

Like the Maclaurin series it is valid for all $x \in \mathbb{R}$, but it converges much more slowly.

Some other series

(A.5.14) ***p*-series.** For any real number p, the series

$$1 + \frac{1}{2^p} + \frac{1}{3^p} + \frac{1}{4^p} + \cdots = \sum_{n=1}^{\infty} \frac{1}{n^p}$$

converges to a finite sum if $p > 1$ and diverges if $p \leq 1$.

(A.5.15) **Harmonic series.** The p-series for $p = 1$,

$$1 + \frac{1}{2} + \frac{1}{3} + \frac{1}{4} + \cdots,$$

which diverges, is called the ***harmonic series.*** The *alternating harmonic series,* on the other hand, is convergent:

(A.5.16)
$$1 - \frac{1}{2} + \frac{1}{3} - \frac{1}{4} + - \cdots = \sum_{n=1}^{\infty} (-1)^{n+1} \cdot \frac{1}{n} = \ln(2).$$

However, because it is only *conditionally convergent* and not *absolutely convergent*—that is, it converges but the series of absolute values of its terms is divergent—different rearrangements of the terms of this series converge to different limits.

Some telescoping series

(A.5.17)
$$\frac{1}{1 \cdot 2} + \frac{1}{2 \cdot 3} + \frac{1}{3 \cdot 4} + \cdots = \sum_{n=1}^{\infty} \frac{1}{n \cdot (n+1)} = \sum_{n=1}^{\infty} \left(\frac{1}{n} - \frac{1}{n+1}\right) = 1$$

This can be seen by letting $n \to \infty$ in (A.5.4). The series is called "telescoping" for the same reason that the sum (A.5.4) is. This trick can be used to discover other sums; for example,

$$\frac{1}{1 \cdot 2} = \left(\frac{1}{1 \cdot 2} - \frac{1}{2 \cdot 3}\right) + \left(\frac{1}{2 \cdot 3} - \frac{1}{3 \cdot 4}\right) + \left(\frac{1}{3 \cdot 4} - \frac{1}{4 \cdot 5}\right) + \cdots$$
$$= \frac{2}{1 \cdot 2 \cdot 3} + \frac{2}{2 \cdot 3 \cdot 4} + \frac{2}{3 \cdot 4 \cdot 5} + \cdots,$$

and so

$$\frac{1}{1 \cdot 2 \cdot 3} + \frac{1}{2 \cdot 3 \cdot 4} + \frac{1}{3 \cdot 4 \cdot 5} + \cdots = \frac{1}{4}.$$

Similarly (check this and generalize it as an exercise),

$$\frac{1}{1 \cdot 2 \cdot 3 \cdot 4} + \frac{1}{2 \cdot 3 \cdot 4 \cdot 5} + \frac{1}{3 \cdot 4 \cdot 5 \cdot 6} + \cdots = \frac{1}{3 \cdot 3!}.$$

A.6 L'HÔPITAL'S RULE

To evaluate the limit of $f(x)/g(x)$ as $x \to a$ or $x \to \infty$ when the limits of $f(x)$ and $g(x)$ are both 0 or both infinite, differentiate both $f(x)$ and $g(x)$. If the limit of $f'(x)/g'(x)$ is a finite number, then that number is the limit of $f(x)/g(x)$ as well. If the limits of $f'(x)$ and $g'(x)$ are both 0 or both infinite, differentiate them both again. Continue if necessary; if at any stage the fraction has a finite limit, then that number is the limit of $f(x)/g(x)$.

Example. To find $\lim_{x\to\infty} xe^{-x}$, write xe^{-x} as x/e^x and notice that numerator and denominator both approach ∞. But differentiating numerator and denominator gives the fraction $1/e^x$, which approaches 0 as $x \to \infty$. Therefore, $\lim_{x\to\infty} xe^{-x} = 0$. ∎

Example. To find $\lim_{x\to\infty} x^3e^{-ax}$ (where a is some positive constant), write x^3e^{-ax} as x^3/e^{ax}. Numerator and denominator both approach ∞. Differentiate them both; we get $3x^2/ae^{ax}$. Numerator and denominator still approach ∞, so differentiate again; we get $6x/a^2e^{ax}$, and still they both approach ∞. But one more differentiation gives $6/a^3e^{ax}$, and this fraction approaches 0 as $x \to \infty$. Therefore, $\lim_{x\to\infty} x^3e^{-ax} = 0$. ∎

This example shows the principle by which we can see that for any positive integer k and any positive number a, $\lim_{x\to\infty} x^ke^{-ax} = 0$. A similar argument shows that it is true even when k is not an integer, and of course it is true also when k is negative (L'Hôpital's rule is not needed in that case). Thus we have

$$\lim_{x\to\infty} x^ke^{-ax} = 0 \quad \text{for any } a \in (0, \infty) \text{ and any } k \in \mathbb{R}.$$

Example. If we try to use L'Hôpital's rule to find $\lim_{x\to 0} 3/(2 + x)$, we get $\lim_{x\to 0} \frac{0}{1} = 0$. But of course we know that the limit equals $\frac{3}{2}$, not 0. L'Hôpital's rule gives the wrong answer because the limits of the numerator and denominator of the original fraction are not both 0 or both infinite. ∎

The limit of a fraction is called an "indeterminate form" when the limits of the numerator and denominator are both 0 or both infinite. L'Hôpital's rule applies only to indeterminate forms.

A.7 INTEGRATION BY SUBSTITUTION AND INTEGRATION BY PARTS

Substitution is easier to see by an example than it is to describe.

Example. To find $\int xe^{-x^2}dx$, make the substitution $u = x^2$. Then $du = 2x\,dx$, so $x\,dx = \frac{1}{2}du$. Therefore, $\int xe^{-x^2}dx = \frac{1}{2}\int e^{-u}du = -\frac{1}{2}e^{-u} + C = -\frac{1}{2}e^{-x^2} + C$. (Of course this can and should be checked by differentiating.) ▮

The general principle is as follows.

(A.7.1) **Indefinite integrals by substitution.** We can evaluate $\int f(g(x)) \cdot h(x)dx$ by substitution if $g'(x)$ is a constant times $h(x)$, say $g'(x) = ah(x)$. The substitution is $u = g(x)$, $du = g'(x)dx = ah(x)dx$, so $h(x)dx = (1/a)du$. Then we get

$$\int f(g(x)) \cdot h(x)dx = \frac{1}{a}\int f(u)du.$$

If this indefinite integral can be found, then the original indefinite integral can be found by putting in $g(x)$ for u.

In the example above, $f(g(x))$ is e^{-x^2} [$f(u)$ is e^{-u} and $g(x)$ is x^2], $h(x)$ is x, and a is 2.

It is sometimes useful to use substitution even if $g'(x)$ is not a constant times $h(x)$, to convert an integral involving one kind of function to one involving another. In such a case the point may not be to evaluate the integral, but simply to recognize it as one that arises in some other context, in order to make a connection between two apparently different problems.

(A.7.2) **Definite integrals by substitution.** One way is simply to find the indefinite integral as above, and after putting in $g(x)$ for u, then evaluate the definite integral by plugging in the two limits of integration and subtracting.

Example. To find $\int_1^3 xe^{-x^2}dx$, first find $\int xe^{-x^2}dx = -\frac{1}{2}e^{-x^2} + C$ as in the previous example; then

$$\int_1^3 xe^{-x^2}dx = \left[-\tfrac{1}{2}e^{-x^2} + C\right]_{x=1}^{x=3} = -\tfrac{1}{2}(e^{-9} - e^{-1}).$$ ▮

But sometimes it is simpler, rather than putting in $g(x)$ for u, to convert the limits of integration from limits on x to limits on u.

Example. To find $\int_1^3 xe^{-x^2}dx$, make the substitution $u = x^2$, $du = 2x\,dx$, so $x\,dx = \frac{1}{2}du$, as in the first example [of (A.7)]. Then change the limits: $x = 1 \to u = 1$ and $x = 3 \to u = 9$. So

$$\int_1^3 xe^{-x^2}dx = \frac{1}{2}\int_1^9 e^{-u}du = \left[-\tfrac{1}{2}e^{-u}\right]_{u=1}^{u=9} = -\tfrac{1}{2}(e^{-9} - e^{-1}).$$ ▮

(A.7.3) Integration by parts. The formula is

$$\int u\,dv = uv - \int v\,du,$$

or, for definite integrals,

$$\int_a^b u\,dv = \left[uv\right]_a^b - \int_a^b v\,du.$$

Example. To find the indefinite integral $\int xe^{-ax}dx$, let $u = x$ and $dv = e^{-ax}dx$. Then $du = dx$ and $v = \int e^{-ax}dx = -(1/a)e^{-ax}$. So

$$\int xe^{-ax}dx = -\frac{1}{a}xe^{-ax} - \int -\frac{1}{a}e^{-ax}dx = -\frac{1}{a}xe^{-ax} - \frac{1}{a^2}e^{-ax} + C.$$ ▮

EXERCISES FOR **SECTION A.7**

1. Find $\int x^2 e^{-ax}dx$ using integration by parts with $u = x^2$ and $dv = e^{-ax}dx$. You will need to use a second integration by parts to find $\int v\,du$. Be sure to check your answer by differentiation.
2. Find $\int x^3 e^{-ax}dx$ and $\int x^4 e^{-ax}dx$.
3. Find $\int xe^{-x^2/2}dx$. But notice that you cannot find $\int e^{-x^2/2}dx$.

ANSWERS

1. $-\dfrac{x^2}{a}e^{-ax} - \dfrac{2x}{a^2}e^{-ax} - \dfrac{2}{a^3}e^{-ax} + C$

2. $k = 3$: $-\dfrac{x^3}{a}e^{-ax} + \dfrac{3}{a}\cdot$ (answer to Exercise 1); $k = 4$: $-\dfrac{x^4}{a}e^{-ax} + \dfrac{4}{a}\cdot$ (ans. for $k = 3$)

3. $-e^{-x^2/2} + C$

A.8 IMPROPER INTEGRALS

An ***improper integral*** is a definite integral in which the integrand is defined on the open interval of integration but fails to exist at one or both of the endpoints, or in which the interval of integration is unbounded at either end. Examples include the following:

$$\int_0^\infty e^{-x}dx \quad \text{(improper at } \infty\text{)};$$

$$\int_0^\infty \frac{1}{\sqrt{x}}dx \quad \text{(improper at both ends)};$$

$$\int_0^1 \frac{1}{x(1-x)}\,dx \quad \text{(improper at both ends);}$$

$$\int_{-\infty}^{\infty} e^{-x^2/2}dx \quad \text{(improper at both ends).}$$

The value of an improper integral is defined as a limit of values of proper integrals (and thus it may fail to exist).

(A.8.1) **Integrals improper at one end.** If an integral $\int_a^b f(x)dx$ is improper at only one end, say the right end, its value is defined as

$$\int_a^b f(x)dx = \lim_{r\uparrow b}\int_a^r f(x)dx.$$

Here $\lim_{r\uparrow b}$ denotes the left-hand limit, approached by values of r less than b. If the integral were improper at a the limit would have been $\lim_{r\downarrow a}\int_r^b f(x)dx$.

Example. $\int_0^\infty e^{-x}dx = \lim_{r\to\infty}\int_0^r e^{-x}dx = \lim_{r\to\infty}(1 - e^{-r}) = 1.$ ▮

Example. To find $\int_0^\infty xe^{-ax}dx$ we evaluate $\lim_{r\to\infty}\int_0^r xe^{-ax}dx$. Integrating by parts, we see that this is

$$\begin{aligned}\lim_{r\to\infty}\left[-\frac{1}{a}xe^{-ax} - \frac{1}{a^2}e^{-ax}\right]_{x=0}^{x=r} &= \lim_{r\to\infty}\left[-\frac{1}{a}re^{-ar} + \frac{1}{a^2}e^{-ar} + 0 + \frac{1}{a^2}\right]\\ &= -0 - 0 + 0 + \frac{1}{a^2} = \frac{1}{a^2}.\end{aligned}$$

There is another, simpler, way to set out this same calculation. We integrate by parts and plug in the limits of integration, remembering that to "plug in ∞" is really to plug in r and then let $r\to\infty$:

$$\int_0^\infty xe^{-ax}dx = \left[-\frac{1}{a}xe^{-ax} - \frac{1}{a^2}e^{-ax}\right]_{x=0}^{x=\infty} = -0 - 0 - \left(-0 - \frac{1}{a^2}\right) = \frac{1}{a^2}.$$

The first 0 in the above is the result of "plugging in ∞" to $(1/a)xe^{-ax}$, that is, of taking the limit as $x\to\infty$; it is 0 by L'Hôpital's rule. The second 0 is the limit as $x\to\infty$ of $(1/a^2)e^{-ax}$. ▮

Example. We find $\int_0^\infty x^2e^{-ax}dx$ by integrating by parts as in the second calculation of the previous example. We let $u = x^2$ and $v = e^{-ax}dx$, so $du = 2x\,dx$ and $v = -(1/a)e^{-ax}$. Thus

$$\int_0^\infty x^2e^{-ax}dx = \left[-x^2\cdot\frac{1}{a}e^{-ax}\right]_0^\infty + \int_0^\infty \frac{1}{a}e^{-ax}\cdot 2x\,dx = -0 + 0 + \int_0^\infty \frac{1}{a}e^{-ax}\cdot 2x\,dx.$$

Now we integrate by parts again. We omit the details; the result is $2/a^3$. ▮

Example. To evaluate $\int_0^\infty x/(1 + x^2)dx$ we proceed as follows:

$$\int_0^\infty \frac{x}{1 + x^2}dx = \lim_{r\to\infty}\int_0^r \frac{x}{1 + x^2}dx;$$

the substitution $u = x^2$ then gives

$$\lim_{r\to\infty}\left[\tfrac{1}{2}\ln(1 + x^2)\right]_{x=0}^{x=r} = \lim_{r\to\infty}\tfrac{1}{2}\ln(1 + r^2) = \infty.$$

That is, this improper integral diverges. ∎

(A.8.2) **Integrals improper at both ends.** If an integral is improper at both ends, its value is a double limit. For example,

$$\int_{-\infty}^\infty f(x)dx = \lim_{r\to\infty}\lim_{s\to\infty}\int_{-r}^s f(x)dx.$$

But usually the simplest way to evaluate such an improper integral is to break it into two integrals that are improper at one end only:

$$\int_{-\infty}^\infty f(x)dx = \int_{-\infty}^0 f(x)dx + \int_0^\infty f(x)dx = \lim_{r\to\infty}\int_{-r}^0 f(x)dx + \lim_{s\to\infty}\int_0^s f(x)dx.$$

The important thing to remember is that it is *not* correct to use the single limit $\lim_{r\to\infty}\int_{-r}^r f(x)dx$. As the following example shows, this single limit (called the *Cauchy limit* of the improper integral) may exist even though the two single limits $\lim_{r\to\infty}\int_{-r}^0 f(x)dx$ and $\lim_{s\to\infty}\int_0^s f(x)dx$ do not.

Example. The improper integral $\int_{-\infty}^\infty x/(1 + x^2)dx$ does not exist. The reason is that, as we saw in the previous example, $\int_0^\infty x/(1 + x^2)dx$ diverges, and a similar argument shows that $\int_{-\infty}^0 x/(1 + x^2)dx$ diverges also. However, the Cauchy limit of this improper integral is

$$\lim_{r\to\infty}\int_{-r}^r \frac{x}{1 + x^2}dx,$$

and it can be checked that $\int_{-r}^r x/(1 + x^2)dx = 0$ for any r. Thus the Cauchy limit is 0 even though the correct evaluation of the improper integral is that it does not exist. ∎

EXERCISE FOR **SECTION A.8**

1. Find $\int_0^\infty x^3e^{-ax}dx$ and $\int_0^\infty x^4e^{-ax}dx$; generalize.

ANSWER

1. $\int_0^\infty x^k e^{-ax} dx = k!/a^k$ for $k = 0, 1, 2, \ldots$.
[This is related to the "gamma formula" that is used often in the text.]

A.9 INTEGRALS OVER SUBSETS OF THE PLANE

(A.9.1) Iterated integrals. To integrate a function $f(x, y)$ over a subset of the plane—that is, to find a *double integral*—the usual method (assuming the double integral is finite) is to make it into an *iterated integral*, as in the following example.

Example. To find the integral of $f(x, y) = x^2 + xy$ over the triangle T in the plane whose vertices are at (0, 0), (1, 0), and (1, 1) (see Figure A.9a), we observe that in T, x ranges from 0 to 1, and for any x, y ranges from 0 to x. (See Figure A.9b.)

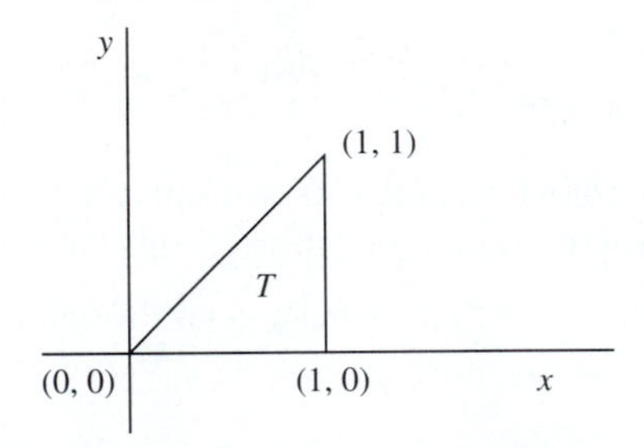

FIGURE A.9a

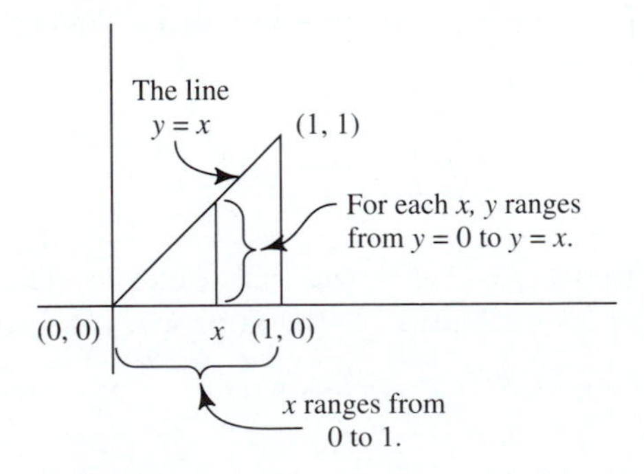

FIGURE A.9b Sketch illustrating one way to set up the integral over the triangle of Figure A.9a

Thus,

$$\iint_T (x^2 + xy)d(x, y) = \int_0^1 \left(\int_0^x (x^2 + xy)dy \right) dx = \int_0^1 \left[x^2y + \tfrac{1}{2}xy^2 \right]_{y=0}^{y=x} dx$$
$$= \int_0^1 \left(x^2 \cdot x + \tfrac{1}{2}x \cdot x^2 \right) dx = \tfrac{3}{8}.$$

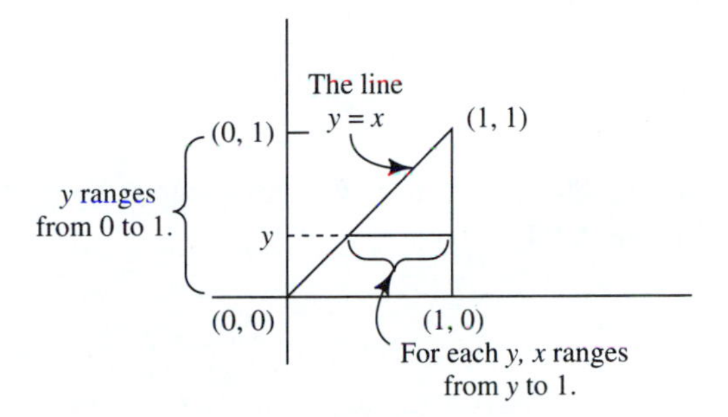

FIGURE A.9c Another way to set up the integral over the triangle of Figure A.9a

Alternatively, we could do this by observing that in T, y ranges from 0 to 1, and for any y, x ranges from y to 1. (see Figure A.9c.) Thus,

$$\iint_T (x^2 + xy)d(x, y) = \int_0^1 \left(\int_y^1 (x^2 + xy)dx \right) dy = \int_0^1 \left[\tfrac{1}{3}x^3 + \tfrac{1}{2}x^2y \right]_{x=y}^{x=1} dy$$
$$= \int_0^1 \left(\tfrac{1}{3} + \tfrac{1}{2}y - \tfrac{1}{3}y^3 - \tfrac{1}{2}y^3 \right) dy = \tfrac{3}{8}.$$ ▮

Example. To integrate the function $f(x, y) = xe^{-x+2y}$ over the set T of points (x, y) in the positive quadrant for which $y > 2x$ (an "infinite triangle"; see Figure A.9d), we observe that in T, x ranges from 0 to ∞, and that for each x, y ranges from $2x$ to ∞. (See Figure A.9e on page 586.)

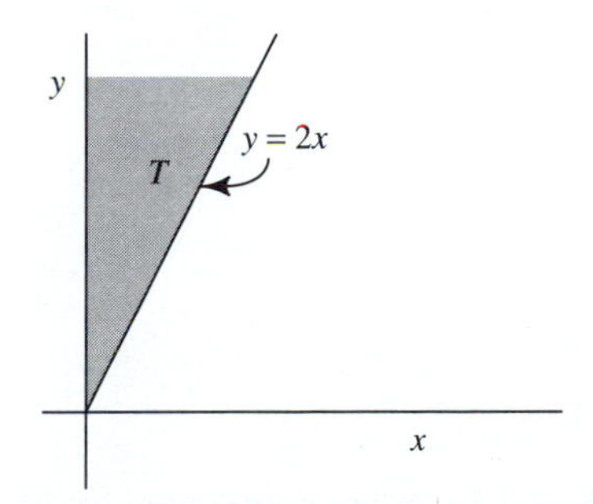

FIGURE A.9d The set of (x, y) in the positive quadrant for which $y > 2x$

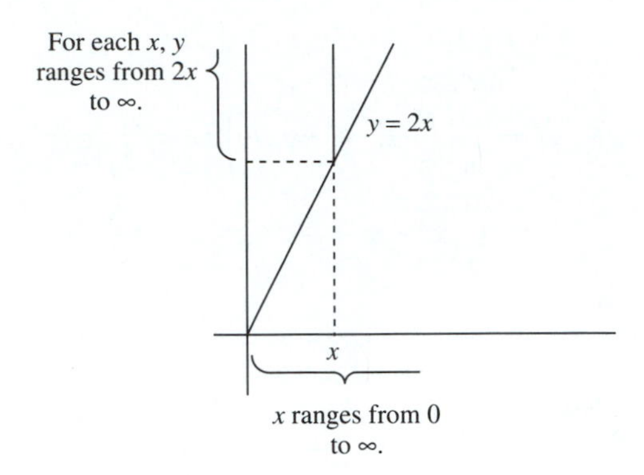

FIGURE A.9e Sketch illustrating one way to integrate over the "triangle" of Figure A.9d

Thus,

$$\iint_T xe^{-x-2y}d(x, y) = \int_0^\infty xe^{-x}\left(\int_{2x}^\infty e^{-2y}dy\right)dx$$

$$= \int_0^\infty xe^{-x} \cdot \tfrac{1}{2}e^{-4x}dx = \tfrac{1}{2}\int_0^\infty xe^{-5x}dx = \tfrac{1}{50}.$$

∎

(A.9.2) Polar coordinates. Sometimes, especially when the integrand contains $x^2 + y^2$, it is convenient to change from Cartesian coordinates (x, y) to polar coordinates (r, θ) by means of

$$x = r\cos\theta, \quad y = r\sin\theta, \quad dx\,dy = r\,dr\,d\theta.$$

(See Figure A.9f.)

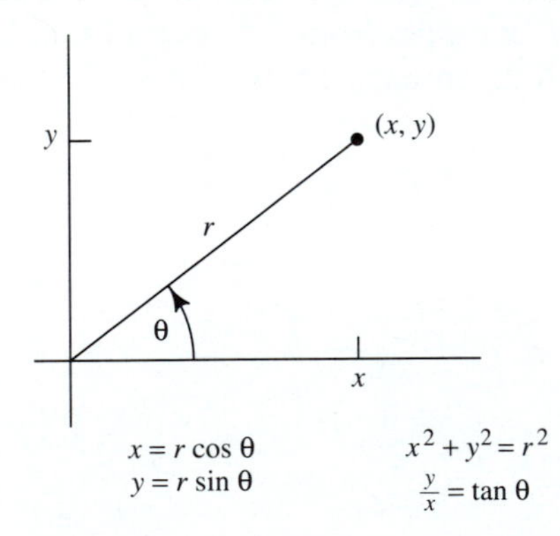

FIGURE A.9f A point (x, y) and its polar coordinates (r, θ)

With this change, $x^2 + y^2$ becomes r^2, and the r that is introduced in the differential often helps in finding the integral.

Example. To integrate $e^{-x^2-y^2}$ over the positive quadrant in the plane—namely, to find $\int_0^\infty \int_0^\infty e^{-x^2-y^2} dx\, dy$, the iterated integral is no help, because it is simply $\int_0^\infty e^{-y^2} \int_0^\infty e^{-x^2} dx\, dy$, and there is no formula for an antiderivative of e^{-x^2} in terms of standard functions.

But if we change to polar coordinates, $e^{-x^2-y^2}\, dx\, dy$ becomes $e^{-r^2} \cdot r\, dr\, d\theta$, and we can integrate this. We need only to determine the limits of integration on r and θ. For points in the positive quadrant, θ ranges from 0 to $\pi/2$, and (independently of θ) r ranges from 0 to ∞. (See Figure A.9g.)

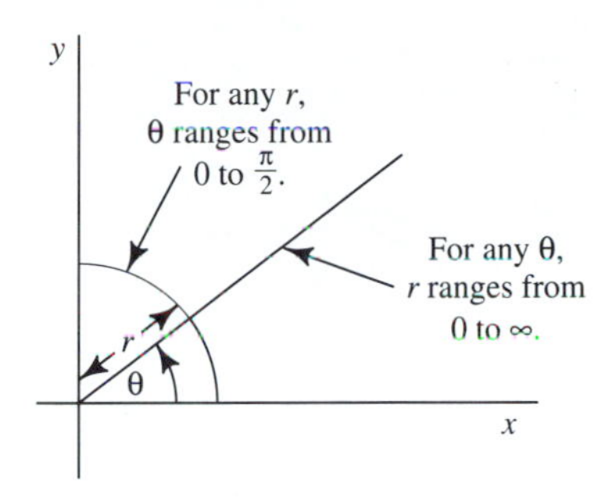

FIGURE A.9g Sketch illustrating how to integrate over the positive quadrant using polar coordinates

Thus,

$$\int_0^\infty \int_0^\infty e^{-x^2-y^2}\, dx\, dy = \int_0^{\pi/2} \int_0^\infty re^{-r^2} dr\, d\theta = \int_0^{\pi/2} \left[-\tfrac{1}{2}e^{-r^2}\right]_0^\infty d\theta = \int_0^{\pi/2} \tfrac{1}{2} d\theta = \frac{\pi}{4}. \quad \blacksquare$$

Example. A very important integral that can be evaluated in this way (with an additional trick) is

$$\int_{-\infty}^{\infty} e^{-x^2/2} dx = \sqrt{2\pi}. \tag{A.9.3}$$

Proof. If we first *square* this integral and then work backwards along the lines of (A.9.1), we get an iterated integral over the plane:

$$\int_{-\infty}^{\infty} e^{-x^2/2} dx \cdot \int_{-\infty}^{\infty} e^{-x^2/2} dx = \int_{-\infty}^{\infty} e^{-x^2/2} dx \cdot \int_{-\infty}^{\infty} e^{-y^2/2} dy$$

$$= \int_{-\infty}^{\infty} \int_{-\infty}^{\infty} e^{-(x^2+y^2)/2} dx\, dy.$$

Now we change to polar coordinates: $e^{-(x^2+y^2)/2}dx\,dy$ becomes $e^{-r^2/2} \cdot r\,dr\,d\theta$; and as (x, y) ranges over the whole plane, θ ranges from 0 to 2π and (independently of θ) r ranges from 0 to ∞. So the *square* of our desired integral is

$$\int_0^{2\pi}\int_0^{\infty} re^{-r^2/2}dr\,d\theta = \int_0^{2\pi}\left[-e^{-r^2/2}\right]_0^{\infty} d\theta = \int_0^{2\pi} 1d\theta = 2\pi.$$

This proves (A.9.3). ∎

EXERCISES FOR **SECTION A.9**

1. Find the integral of x^2y over the triangle whose vertices are at (0, 0), (0, 2), and (1, 2). (See Figure A.9h.)

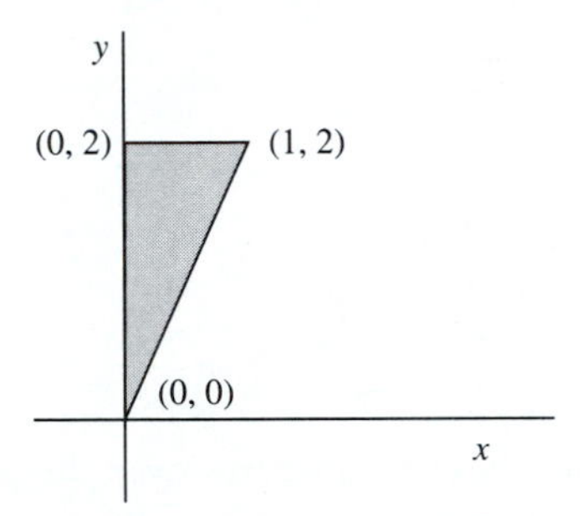

FIGURE A.9h

2. Find the integral of $e^{-x} + e^{-2y}$ over the following.
   a. The triangle T of points (x, y) for which $0 \le x \le y \le 2$ (See Figure A.9i.)
   b. The square whose vertices are at (0, 0), (2, 0), (0, 2), and (2, 2) (also shown in Figure A.9i).

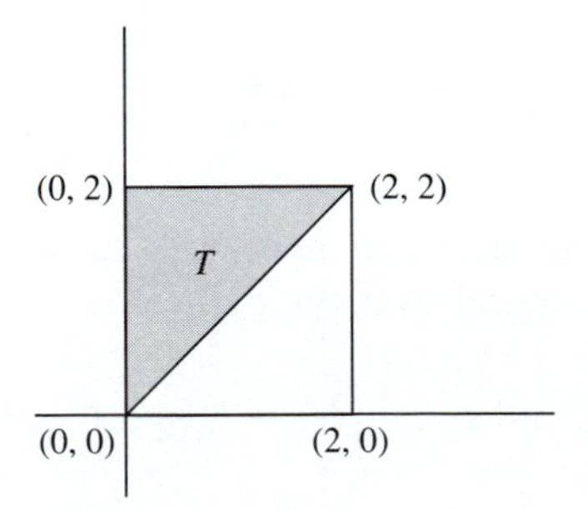

FIGURE A.9i

3. Find the integral of $\sqrt{x^2 + y^2}$ over the disk of radius 1 centered at the origin.

ANSWERS

1. $\frac{4}{15}$

2. **a.** $e^{-2} + \frac{5}{4}(1 - e^{-4})$ **b.** $3 - 2e^{-2} - e^{-4}$

3. $2\pi/3$

APPENDIX B

Summaries of Some Important Distributions

B.1 Bernoulli Distributions

B.2 Binomial Distributions

B.3 Poisson Distributions

B.4 Geometric Distributions

B.5 Negative Binomial Distributions

B.6 Sampling from a Finite Two-Type Population

B.7 Uniform Distributions

B.8 Normal Distributions

B.9 Exponential Distributions

B.10 Erlang Distributions

B.11 Gamma Distributions

B.12 Beta Distributions

B.1 BERNOULLI DISTRIBUTIONS

Parameter: p $(0 < p < 1)$ *Abbreviation:* Bern(p)
Possible values: 0 and 1

Mass function: $p(k) = \begin{cases} p & \text{if } k = 1, \\ q & \text{if } k = 0, \end{cases}$ where q denotes $1 - p$

Probability generating function: $\pi(t) = pt + q$ (for all t)
Moments: $EX = p$, $VX = pq$
In addition, for any $k = 1, 2, 3, \ldots$, $EX^k = p$.

Use: This is the distribution of an *indicator random variable.* If A is any event in a probability space, the indicator of A is the random variable X (sometimes denoted by I_A) defined to equal 1 if A occurs and 0 if not; that is,

$$X(\omega) = \begin{cases} 1 & \text{if } \omega \in A, \\ 0 & \text{if } \omega \notin A. \end{cases}$$

Then the distribution of X is Bern(p) where $p = P(A)$.

Sums of independent random variables: If $X_1, X_2, \ldots, X_n$ are independent Bernoulli random variables *with the same* p, then their sum has the binomial distribution with parameters n and p.

B.2 BINOMIAL DISTRIBUTIONS

Parameters: n (positive integer) and p $(0 < p < 1)$ *Abbreviation:* bin(n, p)
Possible values: $0, 1, 2, \ldots, n$

Mass function: $p(k) = \binom{n}{k} p^k q^{n-k}$, $k = 0, 1, 2, \ldots, n$, where q denotes $1 - p$

Probability generating function: $\pi(t) = (pt + q)^n$ (for all t)
Moments: $EX = np$, $VX = npq$

Use: $X \sim \text{bin}(n, p)$ if X is the number of successes in n Bernoulli trials (independent two-outcome trials in which one of the outcomes, called "success," has the same probability on each trial) in which the success probability is p.

Special case: If $n = 1$, then the bin(1, p) distribution is the Bern(p) distribution. Furthermore, if $X \sim \text{bin}(n, p)$, then X has the same distribution as the sum $X_1 + X_2 + \cdots + X_n$ of n independent random variables having the Bern(p) distribution. [X_j is the indicator of the event "success on the jth trial"—that is, the number of successes on the jth trial.]

Sums of independent random variables: If X and Y are independent binomially distributed random variables *with the same* p, then $X + Y$ is binomially distributed. Specifically, if $X \sim \text{bin}(n, p)$ and $Y \sim \text{bin}(m, p)$ are independent, then $X + Y \sim \text{bin}(n + m, p)$.

This result generalizes to sums of more than two random variables: the sum of n independent binomial random variables *with the same* p is a binomial random variable with that same p; its parameter n is the sum of the n's of the individual random variables.

B.3 POISSON DISTRIBUTIONS

Parameter: λ (any positive number) *Abbreviation:* Pois(λ) or $P(\lambda)$
Possible values: the nonnegative integers 0, 1, 2, 3, . . .
Mass function: $p(k) = \lambda^k e^{-\lambda}/k!$, $k = 0, 1, 2, 3, \ldots$
Probability generating function: $\pi(t) = e^{\lambda t - \lambda}$ (for all t)
Moments: $EX = \lambda$, $VX = \lambda$
Sums of independent random variables: If X and Y are independent and $X \sim$ Pois(λ) and $Y \sim$ Pois(μ), and $Z = X + Y$, then $Z \sim$ Pois($\lambda + \mu$).

This result generalizes to sums of more than two independent Poisson random variables: the sum of n independent Poisson random variables is a Poisson random variable; its parameter (expected value) is the sum of the parameters of the individual random variables.

Uses:

1. In a Poisson process with an arrival rate of λ arrivals per unit time, if X is the number of arrivals in a time interval I of length s, then $X \sim$ Pois(λs). In addition, if I and J are nonoverlapping time intervals of lengths s and r, and X and Y are the numbers of arrivals in I and J, then X and Y are independent Poisson random variables, with parameters λs and λr.
2. **The Poisson approximation to the binomial distribution.**

$$\binom{n}{k} p^k q^{n-k} \text{ is approximately equal to } \frac{\lambda^k e^{-\lambda}}{k!} \text{ where } \lambda = np,$$

if n is large, p is small, and $\lambda = np$ is moderate in value.

B.4 GEOMETRIC DISTRIBUTIONS

Parameter: p $(0 < p < 1)$ *Abbreviation:* geom(p)
Possible values: the nonnegative integers 0, 1, 2, 3, . . .
Mass function: $p(k) = pq^k$, $k = 0, 1, 2, 3, \ldots$, where q denotes $1 - p$

Probability generating function: $\pi(t) = \dfrac{p}{1 - qt} \left(-\dfrac{1}{q} \leq t < \dfrac{1}{q} \right)$

Moments: $EX = \dfrac{q}{p} = \dfrac{1}{p} - 1$, $VX = \dfrac{q}{p^2}$

Use: This is the distribution of the number of failures preceding the first success in a Bernoulli process with success probability p. It is also the distribution of the number of failures between the jth and $(j + 1)$th successes, for any j.

Sums of independent random variables: If $X_1, X_2, \ldots, X_n$ are independent random variables, each having the geometric distribution with the same parameter p, then their sum has the negative binomial distribution with parameters n and p.

B.5 NEGATIVE BINOMIAL DISTRIBUTIONS

Parameters: n (positive integer) and p $(0 < p < 1)$
Abbreviation: neg bin(n, p)
Possible values: the nonnegative integers 0, 1, 2, 3, . . .
Mass function: $p(k) = \binom{n+k-1}{n-1} p^n q^k$, $k = 0, 1, 2, 3, \ldots$

Probability generating function: $\pi(t) = \left(\frac{p}{1-qt}\right)^n \quad \left(-\frac{1}{q} < t < \frac{1}{q}\right)$

Moments: $EX = \frac{nq}{p}$, $VX = \frac{nq}{p^2}$

Use: This is the distribution of the number of failures that occur prior to the nth success in a Bernoulli process with success probability p.
Special case: When $n = 1$ we have the distribution of the number of failures preceding the first success; this is the geom(p) distribution.

Furthermore, if $X \sim$ neg bin(n, p), then X has the same distribution as the sum $X_1 + X_2 + \cdots + X_n$ of n independent random variables having the geom(p) distribution.

Sums of independent random variables: If X and Y are independent random variables having negative binomial distributions *with the same* p, then $X + Y$ has a negative binomial distribution as well. Specifically, if X and Y are independent and $X \sim$ neg bin(n, p) and $Y \sim$ neg bin(m, p), then $X + Y \sim$ neg bin$(n + m, p)$.

This result generalizes to more than two random variables. The sum of any finite collection of independent negative binomial random variables *with the same* p is also a negative binomial random variable with that same p; its parameter n is the sum of the n's of the individual random variables.

B.6 SAMPLING FROM A FINITE TWO-TYPE POPULATION

Let X be the number of S's in a sample of size n from a population consisting of σ items of type S and ϕ items of type F; population size is $\pi = \sigma + \phi$.

Let $p = \sigma/\pi$, the proportion of S's in the population, and $q = \phi/\pi$, the proportion of F's.

Sampling with replacement

Distribution of X: bin(n, p), i.e., bin$\left(n, \frac{\sigma}{\pi}\right)$

Mass function: $P(X = k) = \binom{n}{k} p^k q^{n-k} = \binom{n}{k} \left(\frac{\sigma}{\pi}\right)^k \left(\frac{\phi}{\pi}\right)^{n-k}$ $(k = 0, 1, 2, \ldots, n)$

Moments: $EX = np = n \cdot \frac{\sigma}{\pi}$ and $VX = npq = n \cdot \frac{\sigma}{\pi} \cdot \frac{\phi}{\pi}$

Sampling without replacement

Distribution of X: hypergeometric with parameters n, σ, and π
Abbreviation: hyper(n, σ, π)

Mass function: $P(X = k) = \frac{\binom{\sigma}{k}\binom{\phi}{n-k}}{\binom{\pi}{n}}$ $(k = 0, 1, 2, \ldots, n)$

Moments: $EX = np = n \cdot \frac{\sigma}{\pi}$ and $VX = npq \cdot \frac{\pi - n}{\pi - 1} = n \cdot \frac{\sigma}{\pi} \cdot \frac{\phi}{\pi} \cdot \frac{\pi - n}{\pi - 1}$

B.7 UNIFORM DISTRIBUTIONS

Parameters: a and b (real numbers with $a < b$) *Abbreviation:* unif(a, b)
Set of possible values: the interval (a, b)

Density function: $f(x) = \begin{cases} \frac{1}{b-a} & \text{if } a < x < b \\ 0 & \text{otherwise} \end{cases}$

Cumulative distribution function: $F(x) = \begin{cases} 0 & \text{if } x \le 0 \\ \frac{x-a}{b-a} & \text{if } 0 < x < 1 \\ 1 & \text{if } x \ge 1 \end{cases}$

Moment generating function: $m(s) = \frac{e^{bs} - e^{as}}{s(b-a)}$ for $s \neq 0$ (and $m(0) = 1$)

Moments: $EX = \frac{b+a}{2}$, $VX = \frac{(b-a)^2}{12}$. The kth moment is

$$EX^k = \frac{1}{k+1} \cdot \frac{b^{k+1} - a^{k+1}}{(b-a)}.$$

Special case: For $a = 0$ and $b = 1$ we have the unif(0, 1) distribution; the density is $f(x) = 1$ on $(0, 1)$ (and 0 elsewhere), and the CDF is $F(x) = x$ on $(0, 1)$ (and 0 for $x \le 0$ and 1 for $x \ge 1$).

Distribution of linear functions: If $X \sim$ unif(a, b) and $Y = cX + d$, then $Y \sim$ unif$(ca + d, cb + d)$. In particular:

If $X \sim \text{unif}(0, 1)$ and $Y = a + (b - a)X$, then $Y \sim \text{unif}(a, b)$.

If $X \sim \text{unif}(a, b)$ and $Y = (X - a)/(b - a)$, then $Y \sim \text{unif}(0, 1)$.

Uses:

1. The uniform distribution on (a, b) models a number chosen at random from the interval (a, b) in such a way that no part of the interval is favored over any other part of the same size.
2. Other distributions can be found as the distributions of functions of uniform random variables, and vice versa. To be specific:

   If a continuous function F is the CDF of X, and if $Y = F(X)$, then $Y \sim \text{unif}(0, 1)$.

   If F is an invertible function that is a CDF, and if $X \sim \text{unif}(0, 1)$ and if $Y = F^{-1}(X)$, then Y has the distribution whose CDF is F.

B.8 NORMAL DISTRIBUTIONS

Parameters: μ (a real number) and σ^2 (a positive number)
Abbreviation: $N(\mu, \sigma^2)$
Set of possible values: $\mathbb{R}$
Density function: $f(x) = \dfrac{1}{\sigma\sqrt{2\pi}} e^{-(x-\mu)^2/2\sigma^2}$ for all real x

Here σ denotes the positive square root of the variance σ^2, called the *standard deviation.*

Cumulative distribution function: $F(x) = \int_{-\infty}^{x} f(t)dt = \int_{-\infty}^{z} \varphi(u)du$, where $z = (x - \mu)/\sigma$
Moment generating function: $m(s) = e^{\mu s+\sigma^2 s^2/2}$ for all real s
Moments: $EX = \mu$, $VX = \sigma^2$
Special case: For $\mu = 0$ and $\sigma^2 = 1$, the $N(0, 1)$ distribution is called the *standard normal* distribution. Its density, CDF, and MGF are

$$\varphi(z) = \frac{1}{\sqrt{2\pi}} e^{-z^2/2} \text{ (all } z), \quad \Phi(z) = \int_{-\infty}^{z} \varphi(t)dt \text{ (all } z), \quad \text{and} \quad m(s) = e^{s^2/2} \text{ (all } s).$$

The CDF, $\Phi(z)$, is the function given in normal probability tables.

Distribution of linear functions: If $X \sim N(\mu, \sigma^2)$ and $Y = aX + b$, then $Y \sim N(a\mu + b, a^2\sigma^2)$. In particular:

If $Z \sim N(0, 1)$, then $X = \sigma Z + \mu \sim N(\mu, \sigma^2)$.

If $X \sim N(\mu, \sigma^2)$, then $Z = \dfrac{X - \mu}{\sigma} \sim N(0, 1)$.

Sums of independent random variables: If $X_j \sim N(\mu_j, \sigma_j^2)$ for $j = 1, 2, \ldots, n$ and the X_j's are independent, then $X = X_1 + X_2 + \cdots + X_n$ has the $N(\mu, \sigma^2)$ distribution, where $\mu = \mu_1 + \mu_2 + \cdots + \mu_n$ and $\sigma^2 = \sigma_1^2 + \sigma_2^2 + \cdots + \sigma_n^2$.

Uses:

1. In many situations, a measurement made on a randomly chosen member of some large population can be modeled as a normally distributed random variable.
2. According to the ***Central Limit Theorem***, if a random variable X can be regarded as the sum of a large number of independent random variables X_1, $X_2, \ldots, X_n$ having a common distribution with a finite variance, then X has approximately a normal distribution.
3. A special case of the Central Limit Theorem is the ***De Moivre–Laplace limit theorem***, or the ***normal approximation to the binomial distribution:*** If $X \sim \text{bin}(n, p)$ then the distribution of X is approximately $N(\mu, \sigma^2)$, where $\mu = np$ and $\sigma^2 = npq$. The approximation is usually good if both np and nq are at least 5.
   The "continuity correction": $P(X = k)$ is well approximated by using the normal distribution to find $P\left(k - \frac{1}{2}, k + \frac{1}{2}\right)$. $P(k \le X \le l)$ is approximated by $P\left(k - \frac{1}{2} \le X \le l + \frac{1}{2}\right)$.

B.9 EXPONENTIAL DISTRIBUTIONS

Parameter: λ (a positive real number) *Abbreviation:* $\exp(\lambda)$

Set of possible values: the interval $(0, \infty)$

Density function: $f(x) = \begin{cases} \lambda e^{-\lambda x} & \text{if } x > 0 \\ 0 & \text{otherwise} \end{cases}$

Cumulative distribution function: $F(x) = \begin{cases} 0 & \text{if } x \le 0 \\ 1 - e^{-\lambda x} & \text{if } x > 0 \end{cases}$

Moment generating function: $m(s) = \dfrac{\lambda}{\lambda - s}$ for $s < \lambda$

Moments: $EX = \dfrac{1}{\lambda}$, $VX = \dfrac{1}{\lambda^2}$. The kth moment is $EX^k = \dfrac{k!}{\lambda^k}$.

Distribution of linear functions: If $X \sim \exp(\lambda)$ and $Y = aX$ for a positive number a, then $Y \sim \exp(\lambda/a)$. This is easily seen using moment generating functions [the MGF of aX is $m_X(as)$].

If $Y = aX + b$ for some nonzero b, then Y has a "translated exponential" distribution; the density is $f(y) = (\lambda/a)e^{-(y-b)}$ for $y > b$.

Sums of independent random variables: If $X_1, X_2, \ldots, X_n$ are independent, each having the exponential distribution with the same parameter λ, then $X_1 + X_2 + \cdots + X_n$ has the gamma (Erlang) distribution with parameters n and λ.

Uses: In a Poisson process with an arrival rate of λ arrivals per unit time:

1. If X is the time between the jth and $(j + 1)$th arrivals, then $X \sim \exp(\lambda)$.
2. If X is the time from a fixed time point t_0 to the next arrival after $t = t_0$, then $X \sim \exp(\lambda)$.

B.10 ERLANG DISTRIBUTIONS

(Gamma distributions with integer parameter)

Parameters: n (a positive integer) and λ (a positive number)
Abbreviation: gamma(n, λ) or Erl(n, λ)
Set of possible values: the interval $(0, \infty)$

Density function: $$f(x) = \begin{cases} \dfrac{\lambda^n}{(n-1)!} x^{n-1} e^{-\lambda x} & \text{if } x > 0 \\ 0 & \text{otherwise} \end{cases}$$

Cumulative distribution function: $$F(x) = 1 - \sum_{k=0}^{n-1} \frac{(\lambda x)^k e^{-\lambda x}}{k!}$$

Moment generating function: $m(s) = \left(\dfrac{\lambda}{\lambda - s}\right)^n$ for $s < \lambda$

Moments: $EX = \dfrac{n}{\lambda}$, $VX = \dfrac{n}{\lambda^2}$. The kth moment is

$$EX^k = \frac{n(n+1)(n+2)\cdots(n+k-1)}{\lambda^k}.$$

Special case: When $n = 1$, the Erl$(1, \lambda)$ distribution is the exp(λ) distribution. Furthermore, if $X \sim$ Erl(n, λ), then X has the same distribution as the sum of n independent random variables having the exp(λ) distribution.

Distribution of linear functions: If $X \sim$ Erl(n, λ) and $Y = aX$ for a positive number a, then $Y \sim$ Erl$(n, \lambda/a)$.

Sums of independent random variables: If X and Y are independent random variables having Erlang distributions *with the same* λ, then $X + Y$ has an Erlang distribution as well. Specifically, if X and Y are independent and $X \sim$ Erl(n, λ) and $Y \sim$ Erl(m, λ), then $X + Y \sim$ Erl$(n + m, \lambda)$.

This result generalizes to more than two random variables: the sum of any finite collection of Erlang random variables with the same λ is also an Erlang random variable with that same λ; its parameter n is the sum of the n's of the individual random variables.

Uses: In a Poisson process with an arrival rate of λ arrivals per unit time:

1. If X is the time of the nth arrival, then $X \sim$ Erl(n, λ).
2. If X is the time between the jth and $(j + n)$th arrivals, then $X \sim$ Erl(n, λ).

B.11 GAMMA DISTRIBUTIONS

Parameters: α and λ (positive numbers) *Abbreviation:* gamma(α, λ)
Set of possible values: the interval $(0, \infty)$

Density function: $$f(x) = \begin{cases} \dfrac{\lambda^\alpha}{\Gamma(\alpha)} x^{\alpha-1} e^{-\lambda x} & \text{if } x > 0 \\ 0 & \text{otherwise} \end{cases}$$

Cumulative distribution function: There is no convenient formula unless α is a positive integer.

Moment generating function: $m(s) = \left(\frac{\lambda}{\lambda - s}\right)^{\alpha}$ for $s < \lambda$

Moments: $EX = \frac{\alpha}{\lambda}$, $VX = \frac{\alpha}{\lambda^2}$. The kth moment is

$$EX^k = \frac{\alpha(\alpha+1)(\alpha+2)\cdots(\alpha+k-1)}{\lambda^k}.$$

Distribution of linear functions: If $X \sim$ gamma(α, λ) and $Y = cX$ for a positive number c, then $Y \sim$ gamma$(\alpha, \lambda/c)$.

Sums of independent random variables: If X and Y are independent random variables having gamma distributions *with the same* λ, then $X + Y$ has a gamma distribution as well. Specifically, if X and Y are independent and $X \sim$ gamma(α, λ) and $Y \sim$ gamma(β, λ), then $X + Y \sim$ gamma$(\alpha + \beta, \lambda)$. This result generalizes to more than two random variables: the sum of any finite collection of gamma random variables with the same λ is also a gamma random variable; its parameter α is the sum of the α's of the individual random variables.

Special cases:

1. When $\alpha = 1$, the gamma$(1, \lambda)$ distribution is the exp(λ) distribution.
2. When $\alpha = n/2$ for a positive integer n (odd or even) and $\lambda = \frac{1}{2}$, the gamma$\left(\frac{n}{2}, \frac{1}{2}\right)$ distribution is the chi-squared distribution with parameter n (n "degrees of freedom"), denoted by $\chi^2(n)$. The expected value, variance, and MGF are

$$EX = n, \quad VX = 2n, \quad \text{and} \quad m(s) = \left(\frac{1}{1 - 2s}\right)^{n/2} \quad \text{for } s < \tfrac{1}{2}.$$

Uses:

1. The $\chi^2(n)$ distribution is the distribution of the sum of the squares of n independent random variables, each with the standard normal distribution.
2. Gamma distributions with arbitrary α and λ are useful for modeling many kinds of real populations which have positive scores that tend to cluster near some value but to have some very large values—a "long right tail."

B.12 BETA DISTRIBUTIONS

Parameters: α and β (positive numbers) *Abbreviation:* beta(α, β)

Set of possible values: the interval $(0, 1)$

Density function: $f(x) = \begin{cases} \dfrac{\Gamma(\alpha+\beta)}{\Gamma(\alpha)\Gamma(\beta)} x^{\alpha-1}(1-x)^{\beta-1} & \text{if } 0 < x < 1 \\ 0 & \text{otherwise} \end{cases}$

Cumulative distribution function: There is no convenient formula except in special cases.

Moment generating function: There is no convenient formula except in special cases.

Moments: $EX = \dfrac{\alpha}{\alpha + \beta}$, $VX = \dfrac{\alpha\beta}{(\alpha + \beta)^2(\alpha + \beta + 1)}$

The kth moment is $EX^k = \dfrac{\alpha(\alpha + 1)(\alpha + 2) \cdots (\alpha + k - 1)}{(\alpha + \beta)(\alpha + \beta + 1) \cdots (\alpha + \beta + k - 1)}$. (See note below.)

Special cases: When $\alpha = \beta = 1$, the beta(1, 1) distribution is the uniform distribution on (0, 1).

Uses:

1. Beta distributions arise in Bayesian analysis of the unknown success parameter p of a Bernoulli process. If Y has the beta(α, β) distribution and, given that $Y = p$, X has conditionally the bin(n, p) distribution; then, given $X = k$, Y has conditionally the beta($\alpha + k$, $\beta + n - k$) distribution.
2. They also appear in statistics as the distributions of certain ratios of sums of squares of normally distributed random variables.

Note: The name "beta" comes from the fact that the *beta function* is the function $B(\alpha, \beta)$ of two positive variables defined by

$$B(\alpha, \beta) = \int_0^1 x^{\alpha-1}(1 - x)^{\beta-1}dx,$$

which can be shown to equal

$$\frac{\Gamma(\alpha)\Gamma(\beta)}{\Gamma(\alpha + \beta)}.$$

This formula is used to derive the moments of the distribution.

APPENDIX C

Standard Normal Cumulative Distribution Function

$\Phi(z) = P(-\infty, z] = \int_{-\infty}^{z} \varphi(x)dx$

$y = \varphi(x)$

Example. $\Phi(1.62) = P(-\infty, 1.62] = .9474$

z	.00	.01	.02	.03	.04	.05	.06	.07	.08	.09
.00	.5000	.5040	.5080	.5120	.5160	.5199	.5239	.5279	.5319	.5359
.10	.5398	.5438	.5478	.5517	.5557	.5596	.5636	.5675	.5714	.5753
.20	.5793	.5832	.5871	.5910	.5948	.5987	.6026	.6064	.6103	.6141
.30	.6179	.6217	.6255	.6293	.6331	.6368	.6406	.6443	.6480	.6517
.40	.6554	.6591	.6628	.6664	.6700	.6736	.6772	.6808	.6844	.6879
.50	.6915	.6950	.6985	.7019	.7054	.7088	.7123	.7157	.7190	.7224
.60	.7257	.7291	.7324	.7357	.7389	.7422	.7454	.7486	.7517	.7549
.70	.7580	.7611	.7642	.7673	.7704	.7734	.7764	.7794	.7823	.7852
.80	.7881	.7910	.7939	.7967	.7995	.8023	.8051	.8078	.8106	.8133
.90	.8159	.8186	.8212	.8238	.8264	.8289	.8315	.8340	.8365	.8389
1.00	.8413	.8438	.8461	.8485	.8508	.8531	.8554	.8577	.8599	.8621
1.10	.8643	.8665	.8686	.8708	.8729	.8749	.8770	.8790	.8810	.8830
1.20	.8849	.8869	.8888	.8907	.8925	.8944	.8962	.8980	.8997	.9015
1.30	.9032	.9049	.9066	.9082	.9099	.9115	.9131	.9147	.9162	.9177
1.40	.9192	.9207	.9222	.9236	.9251	.9265	.9279	.9292	.9306	.9319
1.50	.9332	.9345	.9357	.9370	.9382	.9394	.9406	.9418	.9429	.9441
1.60	.9452	.9463	.9474	.9484	.9495	.9505	.9515	.9525	.9535	.9545
1.70	.9554	.9564	.9573	.9582	.9591	.9599	.9608	.9616	.9625	.9633
1.80	.9641	.9649	.9656	.9664	.9671	.9678	.9686	.9693	.9699	.9706
1.90	.9713	.9719	.9726	.9732	.9738	.9744	.9750	.9756	.9761	.9767
2.00	.9772	.9778	.9783	.9788	.9793	.9798	.9803	.9808	.9812	.9817
2.10	.9821	.9826	.9830	.9834	.9838	.9842	.9846	.9850	.9854	.9857
2.20	.9861	.9864	.9868	.9871	.9875	.9878	.9881	.9884	.9887	.9890
2.30	.9893	.9896	.9898	.9901	.9904	.9906	.9909	.9911	.9913	.9916
2.40	.9918	.9920	.9922	.9925	.9927	.9929	.9931	.9932	.9934	.9936
2.50	.9938	.9940	.9941	.9943	.9945	.9946	.9948	.9949	.9951	.9952
2.60	.9953	.9955	.9956	.9957	.9959	.9960	.9961	.9962	.9963	.9964
2.70	.9965	.9966	.9967	.9968	.9969	.9970	.9971	.9972	.9973	.9974
2.80	.9974	.9975	.9976	.9977	.9977	.9978	.9979	.9979	.9980	.9981
2.90	.9981	.9982	.9982	.9983	.9984	.9984	.9985	.9985	.9986	.9986
3.00	.9987	.9987	.9987	.9988	.9988	.9989	.9989	.9989	.9990	.9990
3.10	.9990	.9991	.9991	.9991	.9992	.9992	.9992	.9992	.9993	.9993
3.20	.9993	.9993	.9994	.9994	.9994	.9994	.9994	.9995	.9995	.9995
3.30	.9995	.9995	.9995	.9996	.9996	.9996	.9996	.9996	.9996	.9997
3.40	.9997	.9997	.9997	.9997	.9997	.9997	.9997	.9997	.9997	.9998
3.50	.9998	.9998	.9998	.9998	.9998	.9998	.9998	.9998	.9998	.9998
3.60	.9998	.9998	.9999	.9999	.9999	.9999	.9999	.9999	.9999	.9999

The Greek Alphabet

Name	Lowercase	Uppercase
alpha	α	A
beta	β	B
gamma	γ	Γ
delta	δ	Δ
epsilon	ϵ	E
zeta	ζ	Z
eta	η	H
theta	θ	Θ
iota	ι	I
kappa	κ	K
lambda	λ	Λ
mu	μ	M
nu	ν	N
xi	ξ	Ξ
omicron	o	O
pi	π	Π
rho	ρ	P
sigma	σ	Σ
tau	τ	T
upsilon	υ	Υ
phi	ϕ or φ	Φ
chi	χ	X
psi	ψ	Ψ
omega	ω	Ω

Index

A page number marked with an asterisk indicates a reference that is more definitive or complete, or sometimes more fundamental, than other references to the same item.

Absolute convergence, required for expected value, 218, 220, 229, 232
Absolutely continuous, 46, 160
AIDS testing, *see* False positives
Air in room, all moves to one half of room, 137
All subsets of an interval, impossibility of assigning probabilities to, 37
Andrews, D. F., 70, 562
Arrival rate in Poisson process, 63
Ask Marilyn, *see* Paradox, Monty Hall
Asteroid deflection technology, 427
Automobile accidents, 20
Average of sample, *see* Sample mean
Axioms for Probability, *see* Probability

Bad luck, *see* Waiting line, choosing the wrong *or* Luck, bad and getting worse
Batting average, *see* Brett, George
Bayes, Rev. Thomas, 119, 545
Bayes's theorem, 117, 545
  for random variables, 531, 541, 544, 546–547
Bayesian inference, 533
Beating the odds, *see* Casinos
Bell-shaped curve, 42
Bernoulli, Daniel, 241, 371
Bernoulli distributions, 251, 408, 592
Bernoulli, James, 241
Bernoulli, Nicholas, 241
Bernoulli process, 57
  computer simulation of, *see* Computer simulation
  connection to Poisson process, 67
  connection to uniform distribution, 68
  lack of memory in, 144, 415, 425
  rules for probabilities in, 59
Bernoulli trials, 12, 20*
  simulation of, *see* Computer simulation
Beta distributions, 47, 507, 599
Beta formula, 506
Betting a fixed proportion of current holdings, 297
Bienaymé, I.-J., 293
Billingsley, P., 40, 69, 561–562
Binomial coefficients, 573
Binomial distributions, 171, 406, 592
Binomial probabilities, 20*, 403, 406
  approximating, 372
  efficient computation of, 380
Binomial theorem, 574–575
  Newton's, 423
Birthday problem, *see* Paradox
Bivariate normal distributions, *see* Normal distributions
Bold play vs. cautious in gambling, 123
Bombs hitting boxcars, 412
Borel, Emile, 39
Borel sets, 37
Bourbaki, Nicolas, 49
Box–Muller transformation, *see* Computer generation of normally distributed random variable
Brett, George, 383
Bridge hands, perfect, 137
Brownian motion, 557
Bus stop paradox, *see* Paradox

Cantor, Georg, 570
Carlstein, E., 558
Carroll, Lewis, 154, 562
Cartesian product, required for independence, 198
Casinos, chances of making money in, 72, 123, 449*
Cauchy distributions, 230, 457
Cauchy–Schwarz inequality, 283
Causes, probability of, *see* Bayes's Theorem
Cautious play vs. bold, *see* Bold play
CDF, *see* Cumulative distribution function
CDF method, 308
Central limit theorem, history, 369, 475
  informal statement and use, 361
  rigorous statement and proof, 471–472
Central moments, *see* Moments
Certainty of eventual success, *see* Success
Chain rule for probabilities, 102

Change-of-variable theorem, 326, 333, 335
Characteristic function, 457
Chebyshev, P. L., 293
Chevbyshev's inequality, 290
  cannot be improved on in general, 294
  leads to law of large numbers, 290–291
Chevbyshev's law of large numbers, *see* Laws of large numbers
Chessboard, random, 413
Chi-squared distributions, 494, 599
Chuck-a-luck, 239
Coal mine disasters, 70
Coin tossing, 2
Comet Swift-Tuttle, 428, 451*
Common errors, *see* Pitfalls
Computer simulation *or* generation
  of arbitrary absolutely continuous random variable, 332, 503
  of Bernoulli process, 23, 29
  of Cauchy random variable, 320
  of exponentially distributed random variable, 315, 503
  of normally distributed random variable, 344
  of Poisson process, 68, 315
  rejection method, 322
  of uniformly distributed random variable, 31, 36, 56
Conditional CDF, 520
Conditional density, 202, 520, 540
Conditional expected value, 519, 547
Conditional mass function, 185, 519
Conditional MGF *or* PGF, 550
Conditional probabilities, 97, 98*
  in joint distributions, 185, 518
  obey axioms for probabilities, 105
Conditional variance, 520, 548
Conditioning, 113, 519
  on an absolutely continuous random variable, 539
  on a discrete random variable, 519, 529
  on the events in a partition, 113
  formulas summarized, 546
Conjugate prior distribution, 554–556
Constant random variables, 237
Continuity correction, 376, 378*
Convolutions, 318
Correlated, positively *or* negatively, 272
Correlation coefficient, 279
Corrupt senators, 442
Countably infinite *or* countable set, 569
Covariance, 272
Cramer, Gabriel, 260
Cumulative distribution function, 139, 156*
  characteristic properties, 164

Degrees of freedom, 495–497
De Méré, Chevalier, 71, 412
De Moivre, Abraham, 366
De Moivre–Laplace Limit Theorem, *see* Normal approximation for binomial probabilities
DeMorgan's laws, 568
Density function, 46, 151
Derangement, 88, 95
Dice, rolling two fair, 9–10
Disjoint events, 74, 84, 567*
Distribution, 149. *Also see specific distributions by name:* Bernoulli, Beta, Binomial, Cauchy, Chi-squared, Double exponential, Erlang, Exponential, *F*, Gamma, Geometric, Hypergeometric, Lognormal, Multinomial, Negative binomial, Normal, Pareto, Poisson, *t*, Truncated Poisson, Uniform
Distribution of a random variable, 168
Double exponential distribution, 240
Double integral, *see* Integration over subsets of the plane
Doubling the bet after each loss, *see* Martingale gambling system

E, See Expected value
Eggs, insect, 432
Equal numbers of heads and tails, unlikelihood of, 27, 413
Erlang distributions, 270, 480*, 598
Erlang probabilities, converting to Poisson, 483
Event, 4, 37, 73, 74*
  in absolutely continuous probability space, 31, 37, 46
Expected long-run average value, *see* Expected value
Expected long-run relative frequency, *see* Relative frequency
Expected value, 218
  of an absolutely continuous random variable, 220
  of a discrete random variable, 218
  as expected long-run average value, 218, 223–224, 287
Expected value (*continued*)
  found by conditioning, *see* Law of total expectation
Exponential distributions, 40, 478, 597

F distributions, 496, 512
Factorial, 572
Factorization of a joint density, 340
Fair entry fee, *see* Paradox, St. Petersburg
False positives, *see* Paradox
Family sizes, 259, 538
Feller, William, 475, 506, 561–562
  encounter with parapsychologists, 447
Fermat, Pierre de, 71
First digits, not equally likely, 505
Firstborn children, 259, 538
Fisher, Sir Ronald, 432, 497
Football players, 370

Gambler's ruin problem, 122, 241
Gambling house, *see* Casinos
Gambling strategies, 72, 297
Gamma distributions, 492, 598
Gamma formula, 269, 479, 491, 584
Gamma function, 489
Gauss, Karl Friedrich, 369
Gauss–Legendre duplication formula, 498
Generating random variables with a computer, *see* Computer simulation
Geometric distributions and probabilities, 26, 59, 415, 593
Geometric series, 576
Gosset, W. S., 497
Guinness brewery, 497

Hall, Monty, *see* Paradox
Harmonic series, 228, 578
Hats, checking 242
Hazard sign , 49
Herzberg, A. M., 70, 562
HIV testing, *see* False positives
Hoeffding, W., 405
Human life, monetary value of, 427
Hypergeometric distributions *or* probabilities, 23, 433, 595
  approximations for, 439

I.I.D., 210*, 287, 289–290, 348, 361
Ideenkreis, 403, 409, 422
Improper integral, 581

Some Important Families of Discrete Distributions

(See Appendix B for more complete summaries of these and other distributions.)

Distribution		
Binomial bin(n, p)	parameter values	n is a positive integer; $0 < p < 1$
	possible values	$0, 1, 2, \ldots, n$
	mass function	$p(k) = \binom{n}{k} p^k q^{n-k}$ for $k = 0, 1, 2, \ldots, n$ $(q = 1 - p)$
	PGF	$\pi(t) = (pt + q)^n$ for all $t \in \mathbb{R}$
	expected value	np
	variance	npq
Bernoulli	Bern(p) is the same as bin $(1, p)$	
Poisson Pois(λ)	parameter values	$\lambda > 0$
	possible values	$0, 1, 2, 3, \ldots$
	mass function	$p(k) = \frac{\lambda^k e^{-\lambda}}{k!}$ for $k = 0, 1, 2, \ldots$
	PGF	$\pi(t) = e^{\lambda t - \lambda}$ for all $t \in \mathbb{R}$
	expected value	λ
	variance	λ
Geometric geom(p) [same as neg bin$(1, p)$]	parameter values	$0 < p < 1$
	possible values	$0, 1, 2, 3, \ldots$
	mass function	$p(k) = pq^k$ for $k = 0, 1, 2, \ldots$ $(q = 1 - p)$
	PGF	$\pi(t) = \frac{p}{1 - qt}$ for $-1/q < t < 1/q$
	expected value	q/p
	variance	q/p^2
Negative binomial neg bin(n, p)	parameter values	n is a positive integer; $0 < p < 1$
	possible values	$0, 1, 2, 3, \ldots$
	mass function	$p(k) = \binom{n+k-1}{n-1} p^n q^k$ for $k = 0, 1, 2, \ldots$ $(q = 1 - p)$
	PGF	$\pi(t) = \left(\frac{p}{1 - qt}\right)^n$ for $-1/q < t < 1/q$
	expected value	nq/p
	variance	nq/p^2

Standard Normal Cumulative Distribution Function

$\Phi(z) = P(-\infty, z] = \int_{-\infty}^{z} \varphi(x)dx$

$y = \varphi(x)$

Example. $\Phi(1.62) = P(-\infty, 1.62] = .9474$

z	.00	.01	.02	.03	.04	.05	.06	.07	.08	.09
.00	.5000	.5040	.5080	.5120	.5160	.5199	.5239	.5279	.5319	.5359
.10	.5398	.5438	.5478	.5517	.5557	.5596	.5636	.5675	.5714	.5753
.20	.5793	.5832	.5871	.5910	.5948	.5987	.6026	.6064	.6103	.6141
.30	.6179	.6217	.6255	.6293	.6331	.6368	.6406	.6443	.6480	.6517
.40	.6554	.6591	.6628	.6664	.6700	.6736	.6772	.6808	.6844	.6879
.50	.6915	.6950	.6985	.7019	.7054	.7088	.7123	.7157	.7190	.7224
.60	.7257	.7291	.7324	.7357	.7389	.7422	.7454	.7486	.7517	.7549
.70	.7580	.7611	.7642	.7673	.7704	.7734	.7764	.7794	.7823	.7852
.80	.7881	.7910	.7939	.7967	.7995	.8023	.8051	.8078	.8106	.8133
.90	.8159	.8186	.8212	.8238	.8264	.8289	.8315	.8340	.8365	.8389
1.00	.8413	.8438	.8461	.8485	.8508	.8531	.8554	.8577	.8599	.8621
1.10	.8643	.8665	.8686	.8708	.8729	.8749	.8770	.8790	.8810	.8830
1.20	.8849	.8869	.8888	.8907	.8925	.8944	.8962	.8980	.8997	.9015
1.30	.9032	.9049	.9066	.9082	.9099	.9115	.9131	.9147	.9162	.9177
1.40	.9192	.9207	.9222	.9236	.9251	.9265	.9279	.9292	.9306	.9319
1.50	.9332	.9345	.9357	.9370	.9382	.9394	.9406	.9418	.9429	.9441
1.60	.9452	.9463	.9474	.9484	.9495	.9505	.9515	.9525	.9535	.9545
1.70	.9554	.9564	.9573	.9582	.9591	.9599	.9608	.9616	.9625	.9633
1.80	.9641	.9649	.9656	.9664	.9671	.9678	.9686	.9693	.9699	.9706
1.90	.9713	.9719	.9726	.9732	.9738	.9744	.9750	.9756	.9761	.9767
2.00	.9772	.9778	.9783	.9788	.9793	.9798	.9803	.9808	.9812	.9817
2.10	.9821	.9826	.9830	.9834	.9838	.9842	.9846	.9850	.9854	.9857
2.20	.9861	.9864	.9868	.9871	.9875	.9878	.9881	.9884	.9887	.9890
2.30	.9893	.9896	.9898	.9901	.9904	.9906	.9909	.9911	.9913	.9916
2.40	.9918	.9920	.9922	.9925	.9927	.9929	.9931	.9932	.9934	.9936
2.50	.9938	.9940	.9941	.9943	.9945	.9946	.9948	.9949	.9951	.9952
2.60	.9953	.9955	.9956	.9957	.9959	.9960	.9961	.9962	.9963	.9964
2.70	.9965	.9966	.9967	.9968	.9969	.9970	.9971	.9972	.9973	.9974
2.80	.9974	.9975	.9976	.9977	.9977	.9978	.9979	.9979	.9980	.9981
2.90	.9981	.9982	.9982	.9983	.9984	.9984	.9985	.9985	.9986	.9986
3.00	.9987	.9987	.9987	.9988	.9988	.9989	.9989	.9989	.9990	.9990
3.10	.9990	.9991	.9991	.9991	.9992	.9992	.9992	.9992	.9993	.9993
3.20	.9993	.9993	.9994	.9994	.9994	.9994	.9994	.9995	.9995	.9995
3.30	.9995	.9995	.9995	.9996	.9996	.9996	.9996	.9996	.9996	.9997
3.40	.9997	.9997	.9997	.9997	.9997	.9997	.9997	.9997	.9997	.9998
3.50	.9998	.9998	.9998	.9998	.9998	.9998	.9998	.9998	.9998	.9998
3.60	.9998	.9998	.9999	.9999	.9999	.9999	.9999	.9999	.9999	.9999

Improper prior distribution, 556–557
Inclusion and exclusion, principle of, 86, 411
Income distributions, 363, 486, 512
Independence, product formulas for, 181, 196, 203, 211, 340
Independent events, 125, 127, 130
Independent random variables, 130, 181, 196, 203
Indicator random variables, 250, 251*, 409
Integration, techniques of, 580
Integration over subsets of the plane, 584

Johnson, N. L., 241, 562
Joint distribution, 177, 190, 209
Joint density function, 196
Joint density-mass *or* mass-density function, 542, 546
Joint mass function, 181

Kolmogorov, A. N., 289, 562
Kolmogorov's law of large numbers, *see* Laws of large numbers
Kotz, S., 241, 562

Lack of memory, *see* Poisson process *or* Bernoulli process
Lapidary style of mathematical writing, 39
Laplace, Pierre Simon, 366
Law of total expectation, 525, 527, 548
Law of total probability, 113, 523
Laws of large numbers, 287
  Chevbyshev's weak law, 290
  Kolmogorov's strong law, 288
Let's Make a Deal, *see* Paradox, Monty Hall
L'Hôpital's rule, 579
Lightbulbs, 176
Lindeberg, J. W., 475
Line, waiting, *see* Waiting line
Lognormal distribution, 320
Long-run average value, expected, *see* Expected value
Long-run relative frequency, expected, *see* Relative frequency
Lotka's model, *see* Family sizes
Luck, bad and getting worse, 228
Lyapounov, A., 475

Male child, restricting families to one, 428
Malkiel, B., 383, 562
Marbe, K., 109, 562
Marginal density functions, 196
Marginal distributions, 177
Marginal mass functions, 181
Martingale gambling system, 72, 241
Mass function, 18*, 150
Mean of random variable, *see* Expected value
Mean of sample, *see* Sample mean
Median, 239, 363, 486
Memorable proofs, calculations, formulas, 236, 250, 277, 283, 535, 549, 587
Memory, lack of, *see* Poisson process *or* Bernoulli process
Method of indicators, *see* Indicator random variables
MGF, *see* Moment generating function
Mode, 432, 486
Models vs. real world, 16, 34, 36
Moment generating function, 257, 454*
  compared with probability generating function, 456, 460*
  existence of, 457
Moments, 266
  central, 266
  existence of, 267
  using MGF to find, 458
  using PGF to find, 391–392
Money, personal taste for, 371
Monte Carlo integration, 260, 296
Montmort, P. R., 241
Monty Hall paradox, *see* Paradox
Mosteller, F., 28, 562
Multinomial distributions, 443

ℕ (set of all positive integers), 566
Negative binomial distributions, 417, 594
Nested sets theorem, 136
Neuron, modeling behavior by Poisson process, 65
Newton, Isaac, only work in probability by, 28
Newton's binomial theorem, *see* Binomial theorem
Normal approximation for binomial probabilities, 366*, 374, 408
Normal distributions, 42, 320, 328, 351*, 468, 469*, 596
  bivariate (joint), 201, 281, 341*
  frequent appearance in nature, 365
  generating with computer, *see* Computer simulation
  using table to find probabilities, 44, 353*, 601

Older child paradox, *see* Paradox
Opinion polling, *see* Polling
Order statistics of uniform samples, 509
Origin of probability theory, *see* Points, problem of
Outcome set, 4, 73–74

Paradox (apparent):
  birthday, 95
  bus stop, 66, 538*
  De Méré's, 412
  of false positives, 120
  Monty Hall, 110*, 123, 238
  older child, 100
  prisoner, 110*, 123
  St. Petersburg, 241, 260
Parapsychology, William Feller's encounter with, *see* Feller
Pareto distributions, 512
Partition, 91*
Pascal, Blaise, 71
Pepys, Samuel, 28
Permutation, random, 8, 88, 107, 133
  number of items in proper places, 95, 112, 142, 175, 225, 242, 258, 402, 405
PGF, *see* Probability generating function
Pinkham, R. S., 505
Pitfalls, 17, 49, 61, 133, 199, 244, 309, 444, 557
Points, problem of, 71
Poisson approximation to binomial, *see* Binomial probabilities
Poisson distributions, 148, 429, 593
  mode of, 27–28, 432*
Poisson probabilities, 18
  efficient computation of, 431
Poisson process, 41, 62*
  conditional uniformity of arrival times in an interval, 134, 513*
  connection to Bernoulli process, 67

Poisson process (*continued*)
  lack of memory in, 66*, 108
  rules for probabilities in, 64, 173
  simulation of, *see* Computer simulation
Poisson, Siméon de, 20
Polar coordinates, 586
  of independent normal random variables, 337
Polling, 12, 29, 378, 440
Pólya, G., 105
Pólya's urn scheme, 104*, 121
Posterior and prior distributions, 533
Prisoner paradox, *see* Paradox
Probability as expected long-run relative frequency, *see* Relative frequency
Probability, axioms for, 74*, 84
Probability density function, *see* Density function
Probability distribution, *see* Distribution
Probability generating function, 256, 390*
  compared with moment generating function, 456, 460*
  existence of, 391
Probability mass function, *see* Mass function
Probability measure, 74
Probability space, 4, 74
  absolutely continuous, 46
  discrete, 18
  general definition, 74*, 84
Problem of points, *see* Points
Product, expected value of, 252
Product of random variables, distribution of, 344
Prognosticators, skepticism about successful, 383

$\mathbb{Q}$ (set of all rational numbers), 566
Queue, *see* Waiting line

$\mathbb{R}$ (set of all real numbers), 566
Random mixtures, 559
Random number generation, *see* Computer simulation
Random sample, 211, 348
  of random size, 536–537
Random variable, 5
  on absolutely continuous probability space, 48
Random variable (*continued*)
  on discrete probability space, 18
  on general probability space, 76
Ratio of random variables, distribution of, 333, 336
Rejection method, *see* Generating random variables
Relative frequency, 3, 288, 295, 382, 386
Revised outcome space *and* Revised record of outcomes, 98
ρ, *see* Correlation coefficient
Romano, J., 561–562
Ross, S. 297, 323, 504, 561–563
Roulette, 239
  winning because of well-meaning reformer, 259
Rucker, R., 40, 563, 570
Run of heads: Does it make tails more likely? 109

St. Petersburg paradox, *see* Paradox
Sample, *see* Random sample
Sample mean, 349
Sample mean *and* Sample sum, 349
Sample space, *see* Outcome set
Sampling from a finite population, 594
Sampling with replacement vs. without, 29, 435
Seligman, D., 383, 427, 442
Selvin, S., 111
Series expansions for exponential and logarithmic functions, 577
Shared birthdays, *see* Paradox, birthday
Siegel, A. F., 561–562
σ, σ^2, *see* Standard deviation *or* Variance
σ-field, 37, 74
Singleton, 573
  probability zero in absolutely continuous model, 34
Snedecor, G., 497
Standard deviation, 270
Standard normal distribution, 42
Standardization of a random variable, 270, 352
Stick, breaking at random places, 505
Stirling numbers, 393
Stirling's formula, 414, 432
Stock market predictions, *see* Prognosticators
Strong law of large numbers, Kolmogorov's, *see* Laws of large numbers
Student, *see* Gosset, W. S.
Student's *t* distribution, *see* *t* distributions
Success, certainty of eventual, 137, 142
Swift-Tuttle, Comet, *see* Comet
Symmetric distribution *or* random variable, 255

t distributions, 497
Table, normal, *see* Normal distributions
Tails, believed more likely after a run of heads, *see* Marbe, K.
Traps, *see* Pitfalls
Truncated Poisson distribution, 432

Uncountable set, 570
Uniform distributions on intervals, 31, 79, 499, 595
Uniform distributions on subsets of the plane, 50, 55, 200, 206–207
Uniformity in the World, *see* Marbe, K.
Unimodal distributions, 559

V, *see* Variance
Value of human life, monetary, *see* Human life
Value of money, *see* Money, personal taste for
Variance, 262*
  found by conditioning, 535, 549
  of a sum of random variables, 272, 276*
Vos Savant, Marilyn, *see* Paradox, Monty Hall

Waiting in Bernoulli trials:
  for first success, 26, 59, 415, 593
  for second success, 14, 28
  for *n*th success, 417, 594
Waiting line, choosing the wrong, 207
Waiting line, single vs. multiple in banks, 207, 487
Waiting times for arrivals in Poisson process, 64
Weak law of large numbers, *see* Laws of large numbers
Wheel, reinventing, *see* *Ideenkreis*

$\mathbb{Z}$ (set of all integers), 566